INTRODUCTION TO RADAR SYSTEMS

THIRD EDITION

Merrill I. Skolnik

Boston • Burr Ridge, IL • Dubuque, IA • Madison, WI • New York • San Francisco
St. Louis • Bangkok • Bogotá • Caracas • Lisbon • London
Madrid • Mexico City • Milan • New Delhi • Seoul • Singapore
Sydney • Taipei • Toronto

McGraw-Hill Higher Education

*A Division of The **McGraw-Hill** Companies*

INTRODUCTION TO RADAR SYSTEMS

5 6 7 8 9 0 FGR/FGR 0 9 8 7 6

ISBN-13: 978-0-07-288138-7
ISBN-10: 0-07-288138-0

Vice president/Editor-in-chief: *Kevin Kane*
Publisher: *Thomas Casson*
Executive Editor: *Elizabeth A. Jones*
Sponsoring editor: *Catherine Fields Shultz*
Developmental editor: *Michelle Flomenhoft*
Marketing manager: *John Wannemacher*
Project manager: *Rebecca Nordbrock*
Senior production supervisor: *Lori Koetters*
Coordinator freelance design: *Mary Christianson*
Supplement coordinator: *Matthew Perry*
Freelance cover designer: *Kay Fulton*
Cover photograph: RAT 31 DL *L*-band 3D air-surveillance radar. Courtesy of Alenia Marconi Systems
Compositor: *The GTS Companies/York, PA Campus*
Typeface: *10/12 Times Roman*
Printer: *Quebecor Printing Book Group/Fairfield*

Library of Congress Cataloging-in-Publication Data

Skolnik, Merrill I. (Merrill Ivan), 1927-
    Introduction to radar systems / Merrill I. Skolnik.—3rd ed.
      p. cm
    Includes index.
    ISBN 0-07-288138-0 (alk. paper)
    1. Radar. I. Title.

TK6575 .S477 2001
621.3848—dc21

               00-034879

www.mhhe.com

BRIEF CONTENTS

CONTENTS

Chapter   4

Tracking Radar   210

Chapter   5

Detection of Signals in Noise   276

Chapter   6

Information from Radar Signals   313

Chapter　10

Radar Transmitters　690

Chapter　11

Radar Receiver　727

PREFACE

The technology and applications of radar have continued to grow since the previous edition of this book was published in 1980. Among the many new radar systems that have appeared since then are:

- Enhanced meteorological radars (Nexrad, Terminal Doppler Weather Radar, Wind Profiler, TRMM satellite weather radar, and airborne wind-shear detection radar)

- Planetary exploration (Magellan for Venus; Cassini for Titan, a moon of Saturn)

- Interferometric SAR, for three-dimensional images of a scene and for detection of slowly moving surface targets

- Inverse synthetic aperture radar, ISAR (APS-137 for the recognition of ships)

- Ground penetrating radar

- Serial production of phased array radars (Patriot, Aegis, Pave Paws, and B-1B bomber)

- Active-aperture phased arrays

- Ballistic missile defense radars (GBR and Arrow)

- HF over-the-horizon radars (ROTHR and Jindalee)

- Battlefield surveillance (JSTARS)

- Radars for the remote sensing of the environment

- Improved air-traffic control radars

- New multifunction airborne military fighter/attack radars with sophisticated doppler processing

In addition, there have been extensive advances in the use of digital technology for signal processing, data processing, and radar control; significant improvements in the use of doppler for detecting small moving targets in heavy clutter; better understanding of the characteristics of radar echoes from the ground and the sea; automation of the detection decision and information extraction; target recognition; advancement of solid-state transmitters as well as electron power tubes; and development of unattended, highly reliable radar systems. Furthermore, the need for significant improvements in military radar is driven by advances in stealth (low cross-section targets), high-speed attackers at low as well as high altitude, and the increased capability of electronic warfare techniques.

The third edition of *Introduction to Radar Systems*, like its prior two editions, is based on a one-year graduate course designed to introduce the fundamentals of radar and the systems aspects of radar. The book also can be used for self-study, and it is a reference tool suitable for engineers and managers working in the various areas that contribute to the development, procurement, manufacture, and application of radar and related systems.

An examination of the catalogs of most engineering schools will show there are very few courses concerned with the *systems* that are the reason electrical engineering exists at all. The heart of engineering is the system designed to perform some useful function.

Radar is a classic example of an electronic engineering system that utilizes many of the specialized elements of technology practiced by electrical engineers, including signal processing, data processing, waveform design, electromagnetic scattering, detection, parameter estimation, information extraction, antennas, propagation, transmitters, and receivers. These topics, as they affect radar systems, are part of the discussion of radar in this text. Some are touched on lightly, others are discussed in more detail.

The book also includes such specific radar topics as the range equation needed for the basic understanding of radar as well as serve as a tool for system design; the use of the doppler frequency shift to separate moving target echoes from echoes received from the stationary (clutter) environment, as in moving target indication (MTI) radar and pulse doppler radar; the tracking of targets with monopulse and conical scan radars; target tracking with surveillance radars; detection of radar signals; the matched filter which maximizes the signal-to-noise ratio; accuracy of radar measurements; and the characteristics of echoes from the natural environment that affect the performance of radar.

This third edition, just as was the second edition, has been extensively revised. Almost every paragraph has had changes or has been replaced in its entirety. The order of the chapters has been rearranged to reflect how the author covers the material in his own graduate course. The first term (Chaps. 1 through 4) introduces many of the basic concepts of radar through the development of the very important and widely used radar range equation, the use of the doppler frequency shift to extract weak moving-target echo signals from much stronger land and sea clutter echoes, and the use of radar for tracking moving targets. Most of Chap. 8, on the propagation of radar waves, usually is included in the first term, and is discussed as part of the radar equation of Chap. 2. The second term (Chaps. 5 to 7 and 9) covers the detection of signals in noise, extraction of information from radar signals, waveforms, detection of targets when clutter rather than noise is the dominant factor that limits radar performance, and the many variations of the radar antenna. When time permits, the radar transmitter and receiver are discussed, as well as a review of several major radar applications (not part of the book). Sometimes, the subject of transmitters or receivers is introduced in the first term by asking the student to provide a short paper on some aspect of either of these radar subsystems.

A number of the topics in the second edition have been expanded in this third edition because of the availability of new information and because of their increased importance. Because of the publisher's need to keep the book to a reasonable size, several topics had to be removed to accommodate the expansion of others. Some of the topics left out are those that have become obsolete; for example, several methods for phase shifting in phased-array antennas and various types of dispersive delay lines for FM pulse compression. Chapter 3 of the second edition, on CW and frequency modulated radar, has been omitted because of the decreasing utilization of this type of radar. Low-power CW and frequency modulated CW radars will still be used for some special applications, and are briefly included at the end of the current Chap. 3; but long-range, high-power CW radars, with their need to employ separate and isolated antennas for transmitting and receiving, have been largely replaced by the pulse doppler radar that uses a single antenna. Another deletion is the entire last chapter of the second edition. That chapter consisted of short summaries of various system topics such as synthetic aperture radar, HF over-the-horizon radar, air-surveillance radar, height finder and 3D radar, electronic counter-

countermeasures, bistatic radar, and millimeter wave radar. Although these are still important radar topics, they could not be part of this edition if advances in other radar topics were to be included. There are other radar system applications that would be worth discussing, but it is difficult to cover adequately all the important aspects of radar in one volume.

In the first edition there was a single chapter on radar antennas. In the second edition, antennas were covered in two chapters: one on reflector antennas and the other on phased arrays. In this third edition, the subject is again covered in one chapter, which is the longest chapter in the book. By covering antennas in one chapter instead of two, it was easier to treat subjects common to both reflectors and arrays. The antenna chapter is also large because of the distinctive and significant role that the antenna plays in radar.

There are more topics included in this text than can be covered in the usual two-term graduate course. Many of the topics are included for the benefit of the practicing engineer or manager who uses this text as a reference. Thus, the instructor should select which topics to omit, depending on the objectives of the particular course.

It is recognized that the mks system of units is commonly used in university courses and in most of the world; but in this edition the use of mixed units is continued since it seems to be the current practice found in engineering, especially in the United States. (International air-traffic control systems, for example, still give the range in nautical miles and the altitude in feet.) Also, in this book when a value of some quantity is quoted from a published paper, the units found in that paper are used rather than converted to other units. Someday all units will likely be mks, but until that time the engineer should be acquainted with mixed units since they are still found in practice.

The decibel, or dB, is used throughout radar engineering, and it is used widely in this text. Some students do not seem to be too practiced in its use. As a reminder, dB is defined as 10 times the log of a *power* ratio or a parameter related to a power unit (such as antenna gain). If a ratio of two powers is always taken when considering dB, there never need be confusion as to whether the multiplying constant is 10 or is 20; it will always be 10. Also one has to be careful about using equations when some parameters are given in dB but in the equation the parameter is a numeric. The dB has to be converted to a numeric rather than substitute the dB value directly into the equation.

This edition includes problems and questions at the end of each chapter for the benefit of readers (as well as instructors) using this book as a graduate text for a course in radar. They should also be an aid in self-study for the working engineer. Being a systems-oriented course, it is difficult to provide the type of problems usually found in the problem-solving courses often taught in engineering schools. In some of the problems or questions, the methods or equations leading to the answers can be found in the text and help reinforce what has been covered. Other problems, however, attempt to extend the material in the text so the reader will have to stretch his or her own thinking in seeking an answer. A solutions manual for these problems and questions should be available from the publisher, McGraw-Hill.

In my own radar course, I have found the comprehensive term paper to be an important learning tool. I usually try to provide a relatively straightforward concept-design task that the student is not likely to find in the published literature. (This has gotten harder to do as radar progresses over the years.) A term paper has always been assigned in the second term. Sometimes I have included a paper in the first term on a simpler topic than used in the second term. Term paper topics I have used or considered in the past are also listed in the solutions manual.

As with the other two editions, I have tried to include adequate references to acknowledge the sources of my information and to indicate where the reader interested in digging further might obtain more information on a particular topic. The *Radar Handbook*, also published by McGraw-Hill, is a good source for more advanced information about many of the topics included in this book. Each chapter of the *Radar Handbook* was written by one or more accomplished experts in the particular field covered by the chapter.

In the first edition (1962) of *Introduction to Radar Systems*, almost all of the references were "recent" since radar was relatively new. The early publications on radar began to appear in the literature in the middle to late 1940s, about 15 or 16 years prior to the publication of the first edition. Therefore, the first edition contained many "up to date" references. In this third edition, I have attempted to provide recent references, when they exist, along with early ones; but this was not always possible since some subjects, still of importance, have matured and have not been extended in recent years. Thus there are a number of references in this edition that are 30 to 40 years old. Some were included in the first or second editions, but a few of the older references were not in either of the previous editions since the subjects they cover lay dormant for many years until technology and/or the need for them caught up.

A book of this nature, which covers many diverse topics on many aspects of radar, depends on technical publications in the open literature written by radar engineers. I have relied on such publications in preparing this book, and it is with gratitude that I acknowledge the significant help I have received from the vast literature on radar that now exists.

Merrill I. Skolnik

An Introduction to Radar

1.1 BASIC RADAR

Radar is an electromagnetic system for the detection and location of reflecting objects such as aircraft, ships, spacecraft, vehicles, people, and the natural environment. It operates by radiating energy into space and detecting the echo signal reflected from an object, or target. The reflected energy that is returned to the radar not only indicates the presence of a target, but by comparing the received echo signal with the signal that was transmitted, its location can be determined along with other target-related information. Radar can perform its function at long or short distances and under conditions impervious to optical and infrared sensors. It can operate in darkness, haze, fog, rain, and snow. Its ability to measure distance with high accuracy and in all weather is one of its most important attributes.

The basic principle of radar is illustrated in Fig. 1.1. A transmitter (in the upper left portion of the figure) generates an electromagnetic signal (such as a short pulse of sinewave) that is radiated into space by an antenna. A portion of the transmitted energy is intercepted by the target and reradiated in many directions. The reradiation directed back towards the radar is collected by the radar antenna, which delivers it to a receiver. There it is processed to detect the presence of the target and determine its location. A single antenna is usually used on a time-shared basis for both transmitting and receiving when the radar waveform is a repetitive series of pulses. The range, or distance, to a target is found by measuring the time it takes for the radar signal to travel to the target and return back to the radar. (Radar engineers use the term *range* to mean *distance,* which is

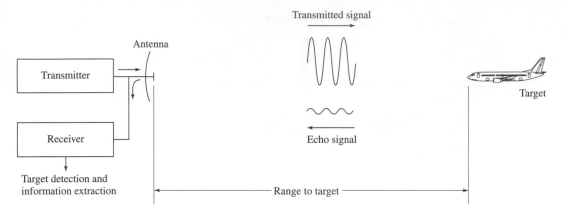

Figure 1.1 Basic principle of radar.

not the definition of range found in some dictionaries.*) The target's location in angle can be found from the direction the narrow-beamwidth radar antenna points when the received echo signal is of maximum amplitude. If the target is in motion, there is a shift in the frequency of the echo signal due to the doppler effect. This frequency shift is proportional to the velocity of the target relative to the radar (also called the radial velocity). The doppler frequency shift is widely used in radar as the basis for separating desired moving targets from fixed (unwanted) "clutter" echoes reflected from the natural environment such as land, sea, or rain. Radar can also provide information about the nature of the target being observed.

The term *radar* is a contraction of the words *ra*dio *d*etection *a*nd *r*anging. The name reflects the importance placed by the early workers in this field on the need for a device to detect the presence of a target and to measure its range. Although modern radar can extract more information from a target's echo signal than its range, the measurement of range is still one of its most important functions. There are no competitive techniques that can accurately measure long ranges in both clear and adverse weather as well as can radar.

Range to a Target   The most common radar signal, or waveform, is a series of short-duration, somewhat rectangular-shaped pulses modulating a sinewave carrier. (This is sometimes called a *pulse train.*) The range to a target is determined by the time T_R it takes the radar signal to travel to the target and back. Electromagnetic energy in free space travels with the speed of light, which is $c = 3 \times 10^8$ m/s. Thus the time for the signal to travel to a target located at a range R and return back to the radar is $2R/c$. The range to a target is then

$$R = \frac{cT_R}{2} \qquad \text{[1.1]}$$

---

*Webster's New Collegiate Dictionary defines *range* as "the horizontal distance to which a projectile can be propelled" or "the horizontal distance between a weapon and target." This is not how the term is used in radar. On the other hand, the dictionary defines *range finder* as "an instrument . . . to determine the distance to a target," which is its meaning in radar.

With the range in kilometers or in nautical miles, and T in microseconds, Eq. (1.1) becomes

$$R(\text{km}) = 0.15 \ T_R \ (\mu s) \qquad \text{or} \qquad R(\text{nmi}) = 0.081 \ T_R \ (\mu s)$$

Each microsecond of round-trip travel time corresponds to a distance of 150 meters, 164 yards, 492 feet, 0.081 nautical mile, or 0.093 statute mile. It takes 12.35 μs for a radar signal to travel a nautical mile and back.

Maximum Unambiguous Range   Once a signal is radiated into space by a radar, sufficient time must elapse to allow all echo signals to return to the radar before the next pulse is transmitted. The rate at which pulses may be transmitted, therefore, is determined by the longest range at which targets are expected. If the time between pulses T_p is too short, an echo signal from a long-range target might arrive *after* the transmission of the next pulse and be mistakenly associated with that pulse rather than the actual pulse transmitted earlier. This can result in an incorrect or ambiguous measurement of the range. Echoes that arrive after the transmission of the next pulse are called *second-time-around echoes* (or *multiple-time-around echoes* if from even earlier pulses). Such an echo would appear to be at a closer range than actual and its range measurement could be misleading if it were not known to be a second-time-around echo. The range beyond which targets appear as second-time-around echoes is the *maximum unambiguous range, R_{un}*, and is given by

$$R_{\text{un}} = \frac{cT_p}{2} = \frac{c}{2f_p} \qquad\qquad \textbf{[1.2]}$$

where T_p = pulse repetition period = $1/f_p$, and f_p = pulse repetition frequency (prf), usually given in hertz or pulses per second (pps). A plot of the maximum unambiguous range as a function of the pulse repetition frequency is shown in Fig. 1.2. The term *pulse repetition rate* is sometimes used interchangeably with *pulse repetition frequency*.

Radar Waveforms   The typical radar utilizes a pulse waveform, an example of which is shown in Fig. 1.3. The peak power in this example is $P_t = 1$ MW, pulse width $\tau = 1$ μs, and pulse repetition period $T_p = 1$ ms = 1000 μs. (The numbers shown were chosen for illustration and do not correspond to any particular radar, but they are similar to what might be expected for a medium-range air-surveillance radar.) The pulse repetition frequency f_p is 1000 Hz, which provides a maximum unambiguous range of 150 km, or 81 nmi. The average power (P_{av}) of a repetitive pulse-train waveform is equal to $P_t\tau/T_p = P_t\tau f_p$, so the average power in this case is $10^6 \times 10^{-6}/10^{-3} = 1$ kW. The *duty cycle* of a radar waveform is defined as the ratio of the total time the radar is radiating to the total time it could have radiated, which is $\tau/T_p = \tau f_p$, or its equivalent P_{av}/P_t. In this case the duty cycle is 0.001. The energy of the pulse is equal to $P_t\tau$, which is 1 J (joule). If the radar could detect a signal of 10^{-12} W, the echo would be 180 dB below the level of the signal that was transmitted. A short-duration pulse waveform is attractive since the strong transmitter signal is not radiating when the weak echo signal is being received.

With a pulse width τ of 1 μs, the waveform extends in space over a distance $c\tau = 300$ m. Two equal targets can be recognized as being resolved in range when they

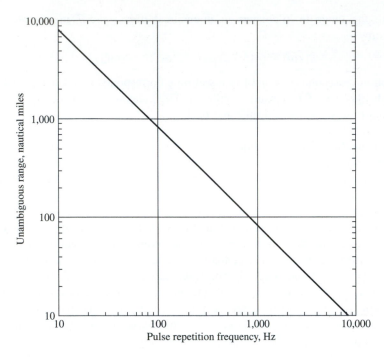

Figure 1.2 Plot of Eq. (1.2), the maximum unambiguous range R_{un} as a function of the pulse repetition frequency f_p.

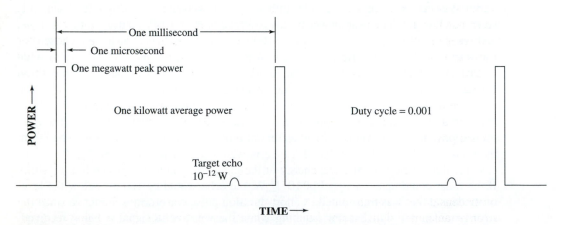

Figure 1.3 Example of a pulse waveform, with "typical" values for a medium-range air-surveillance radar. The rectangular pulses represent pulse-modulated sinewaves.

are separated a distance half this value, or $c\tau/2$. The factor of one-half results from the two-way travel of the radar wave. For example, when $\tau = 1$ μs, two equal size targets can be resolved if they are separated by 150 m.

A very long pulse is needed for some long-range radars to achieve sufficient energy to detect small targets at long range. A long pulse, however, has poor resolution in the range dimension. Frequency or phase modulation can be used to increase the spectral width of a long pulse to obtain the resolution of a short pulse. This is called *pulse compression,* and is described in Sec. 6.5. Continuous wave (CW) waveforms have also been used in radar. Since they have to receive while transmitting, CW radars depend on the doppler frequency shift of the echo signal, caused by a moving target, to separate in the frequency domain the weak echo signal from the large transmitted signal and the echoes from fixed clutter (land, sea, or weather), as well as to measure the radial velocity of the target (Sec. 3.1). A simple CW radar does not measure range. It can obtain range, however, by modulating the carrier with frequency or phase modulation. An example is the frequency modulation (FM-CW) waveform used in the radar altimeter that measures the height (altitude) of an aircraft above the earth.

Pulse radars that extract the doppler frequency shift are called either *moving target indication* (MTI) or *pulse doppler* radars, depending on their particular values of pulse repetition frequency and duty cycle. An MTI radar has a low prf and a low duty cycle. A pulse doppler radar, on the other hand, has a high prf and a high duty cycle. Both types of doppler radars are discussed in Chap. 3. Almost all radars designed to detect aircraft use the doppler frequency shift to reject the large unwanted echoes from stationary clutter.

1.2   THE SIMPLE FORM OF THE RADAR EQUATION

The radar equation relates the range of a radar to the characteristics of the transmitter, receiver, antenna, target, and the environment. It is useful not only for determining the maximum range at which a particular radar can detect a target, but it can serve as a means for understanding the factors affecting radar performance. It is also an important tool to aid in radar system design. In this section, the simple form of the radar range equation is derived.

If the transmitter power P_t is radiated by an isotropic antenna (one that radiates uniformly in all directions), the *power density* at a distance R from the radar is equal to the radiated power divided by the surface area $4\pi R^2$ of an imaginary sphere of radius R, or

$$\text{Power density at range } R \text{ from an isotropic antenna} = \frac{P_t}{4\pi R^2} \qquad \text{[1.3]}$$

Power density is measured in units of watts per square meter. Radars, however, employ *directive* antennas (with narrow beamwidths) to concentrate the radiated power P_t in a particular direction. The *gain* of an antenna is a measure of the increased power density radiated in some direction as compared to the power density that would appear in that direction from an isotropic antenna. The maximum gain G of an antenna may be defined as

$$G = \frac{\text{maximum power density radiated by a directive antenna}}{\text{power density radiated by a lossless isotropic antenna with the same power input}}$$

The power density at the target from a directive antenna with a transmitting gain G is then

$$\text{Power density at range } R \text{ from a directive antenna} = \frac{P_t G}{4\pi R^2} \qquad \text{[1.4]}$$

The target intercepts a portion of the incident energy and reradiates it in various directions. It is only the power density reradiated in the direction of the radar (the echo signal) that is of interest. The *radar cross section of the target* determines the power density returned to the radar for a particular power density incident on the target. It is denoted by σ and is often called, for short, *target cross section, radar cross section,* or simply *cross section*. The radar cross section is defined by the following equation:

$$\text{Reradiated power density back at the radar} = \frac{P_t G}{4\pi R^2} \cdot \frac{\sigma}{4\pi R^2} \qquad \text{[1.5]}$$

The radar cross section has units of area, but it can be misleading to associate the radar cross section directly with the target's physical size. Radar cross section is more dependent on the target's shape than on its physical size, as discussed in Sec. 2.7.

The radar antenna captures a portion of the echo energy incident on it. The power received by the radar is given as the product of the incident power density [Eq. (1.5)] times the effective area A_e of the receiving antenna. The effective area is related to the physical area A by the relationship $A_e = \rho_a A$, where ρ_a = antenna aperture efficiency. The received signal power P_r (watts) is then

$$P_r = \frac{P_t G}{4\pi R^2} \cdot \frac{\sigma}{4\pi R^2} \cdot A_e = \frac{P_t G A_e \sigma}{(4\pi)^2 R^4} \qquad \text{[1.6]}$$

The maximum range of a radar R_{max} is the distance beyond which the target cannot be detected. It occurs when the received signal power P_r just equals the minimum detectable signal S_{min}. Substituting $S_{min} = P_r$ in Eq. (1.6) and rearranging terms gives

$$R_{max} = \left[\frac{P_t G A_e \sigma}{(4\pi)^2 S_{min}} \right]^{1/4} \qquad \text{[1.7]}$$

This is the fundamental form of the *radar range equation*. (It is also called, for simplicity, the *radar equation* or *range equation*.) The important antenna parameters are the transmitting gain and the receiving effective area. The transmitter power P_t has not been specified as either the average or the peak power. It depends on how S_{min} is defined. In this text, however, P_t denotes the peak power.

If the same antenna is used for both transmitting and receiving, as it usually is in radar, antenna theory gives the relationship between the transmit gain G and the receive effective area A_e as[1,2]

$$G = \frac{4\pi A_e}{\lambda^2} = \frac{4\pi \rho_a A}{\lambda^2} \qquad \text{[1.8]}$$

where λ = wavelength. (Wavelength $\lambda = c/f$, where c = velocity of propagation and f = frequency.) Equation (1.8) can be substituted in Eq. (1.7), first for A_e and then for G, to give two other forms of the radar equation

$$R_{max} = \left[\frac{P_t G^2 \lambda^2 \sigma}{(4\pi)^3 S_{min}} \right]^{1/4} \qquad \textbf{[1.9]}$$

$$R_{max} = \left[\frac{P_t A_e^2 \sigma}{4\pi \lambda^2 S_{min}} \right]^{1/4} \qquad \textbf{[1.10]}$$

These three forms of the radar equation [Eqs. (1.7), (1.9), and (1.10)] are basically the same; but there are differences in interpretation. For example, from Eq. (1.9) it might be concluded that the maximum range varies as $\lambda^{1/2}$, but Eq. (1.10) indicates the variation with range as $\lambda^{-1/2}$, which is just the opposite. On the other hand, Eq. (1.7) gives no explicit wavelength dependence for the range. The correct interpretation depends on whether the antenna gain is held constant with change in wavelength, or frequency, as implied by Eq. (1.9); or the effective area is held constant, as implied by Eq. (1.10). For Eq. (1.7) to be independent of frequency, two antennas have to be used. The transmitting antenna has to have a gain independent of wavelength and the receiving antenna has to have an effective aperture independent of wavelength. (This is seldom done, however.)

These simplified versions of the radar equation do not adequately describe the performance of actual radars. Many important factors are not explicitly included. The simple form of the radar range equation predicts too high a value of range, sometimes by a factor of two or more. In Chap. 2 the simple form of the radar equation is expanded to include other factors that allow the equation to be in better agreement with the observed range performance of actual radars.

1.3 RADAR BLOCK DIAGRAM

The operation of a pulse radar may be described with the aid of the simple block diagram of Fig. 1.4. The transmitter may be a *power amplifier,* such as the klystron, traveling wave tube, or transistor amplifier. It might also be a power oscillator, such as the magnetron. The magnetron oscillator has been widely used for pulse radars of modest capability; but the amplifier is preferred when high average power is necessary, when other than simple pulse waveforms are required (as in pulse compression), or when good performance is needed in detecting moving targets in the midst of much larger clutter echoes based on the doppler frequency shift (the subject of Chap. 3). A power amplifier is indicated in Fig. 1.4. The radar signal is produced at low power by a *waveform generator,* which is then the input to the power amplifier. In most power amplifiers, except for solid-state power sources, a modulator (Sec. 10.7) turns the transmitter on and off in synchronism with the input pulses. When a power oscillator is used, it is also turned on and off by a *pulse modulator* to generate a pulse waveform.

The output of the transmitter is delivered to the *antenna* by a waveguide or other form of transmission line, where it is radiated into space. Antennas can be mechanically steered parabolic reflectors, mechanically steered planar arrays, or electronically steered phased arrays (Chap. 9). On transmit the parabolic reflector focuses the energy into a narrow

Figure 1.4
Block diagram
of a conven-
tional pulse
radar with a
superheterodyne
receiver.

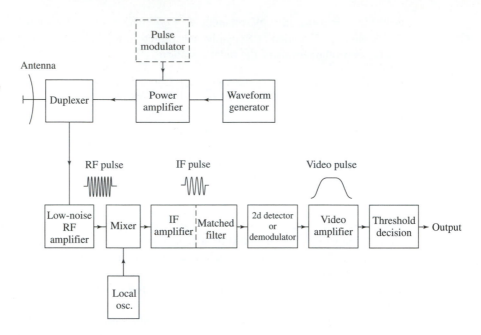

beam, just as does an automobile headlight or a searchlight. A phased array antenna is a collection of numerous small radiating elements whose signals combine in space to produce a radiating plane wave. Using phase shifters at each of the radiating elements, an electronically steered phased array can rapidly change the direction of the antenna beam in space without mechanically moving the antenna. When no other information is available about the antenna, the beamwidth (degrees) of a "typical" parabolic reflector is often approximated by the expression 65 λ/D, where D is the dimension of the antenna in the same plane as the beamwidth is measured, and λ is the radar wavelength. For example, an antenna with a horizontal dimension $D = 32.5$ wavelengths has an azimuth beamwidth of 2°. At a frequency of 3 GHz ($\lambda = 10$ cm), the antenna would be 3.25 m, or 10.7 ft, in extent. The rotation of a surveillance radar antenna through 360° in azimuth is called an antenna *scan*. A typical scan rate (or rotation rate) for a long-range civil air-traffic control air-surveillance radar might be 6 rpm. Military air-surveillance radars generally require a higher rotation rate.

The *duplexer* allows a single antenna to be used on a time-shared basis for both transmitting and receiving. The duplexer is generally a gaseous device that produces a short circuit (an arc discharge) at the input to the receiver when the transmitter is operating, so that high power flows to the antenna and not to the receiver. On reception, the duplexer directs the echo signal to the receiver and not to the transmitter. Solid-state ferrite circulators and receiver protector devices, usually solid-state diodes, can also be part of the duplexer.

The receiver is almost always a *superheterodyne*. The input, or RF,* stage can be a low-noise transistor amplifier. The mixer and local oscillator (LO) convert the RF signal

---

*In electrical engineering, RF is an abbreviation for *radio frequency*; but in radar practice it is understood to mean *radar frequency*. RF also is used to identify that portion of the radar that operates at RF frequencies, even though the inclusion of "frequencies" in this expression might seem redundant.

to an intermediate frequency (IF) where it is amplified by the IF amplifier. The signal bandwidth of a superheterodyne receiver is determined by the bandwidth of its IF stage. The IF frequency, for example, might be 30 or 60 MHz when the pulse width is of the order of 1 μs. (With a 1-μs pulse width, the IF bandwidth would be about 1 MHz.) The IF amplifier is designed as a *matched filter* (Sec. 5.2); that is, one which maximizes the output peak-signal-to-mean-noise ratio. Thus the matched filter maximizes the detectability of weak echo signals and attenuates unwanted signals. With the approximately rectangular pulse shapes commonly used in many radars, conventional radar receiver filters are close to that of a matched filter when the receiver bandwidth B is the inverse of the pulse width τ, or $B\tau \approx 1$.

Sometimes the low-noise input stage is omitted and the mixer becomes the first stage of the receiver. A receiver with a mixer as the input stage will be less sensitive because of the mixer's higher noise figure; but it will have greater dynamic range, less susceptibility to overload, and less vulnerability to electronic interference than a receiver with a low-noise first stage (Sec. 11.3). These attributes of a mixer input stage might be of interest for military radars subject to the noisy environment of hostile electronic countermeasures (ECM).

The IF amplifier is followed by a crystal diode, which is traditionally called the *second detector,* or *demodulator.* Its purpose is to assist in extracting the signal modulation from the carrier. The combination of IF amplifier, second detector, and video amplifier act as an *envelope detector* to pass the pulse modulation (envelope) and reject the carrier frequency. In radars that detect the doppler shift of the echo signal, the envelope detector is replaced by a *phase detector* (Sec. 3.1), which is different from the envelope detector shown here. The combination of IF amplifier and video amplifier is designed to provide sufficient amplification, or gain, to raise the level of the input signal to a magnitude where it can be seen on a display, such as a cathode-ray tube (CRT), or be the input to a digital computer for further processing.

At the output of the receiver a decision is made whether or not a target is present. The decision is based on the magnitude of the receiver output. If the output is large enough to exceed a predetermined threshold, the decision is that a target is present. If it does not cross the threshold, only noise is assumed to be present. The threshold level is set so that the rate at which false alarms occur due to noise crossing the threshold (in the absence of signal) is below some specified, tolerable value. This is fine if the noise remains constant, as when receiver noise dominates. If, on the other hand, the noise is external to the radar (as from unintentional interference or from deliberate noise jamming) or if clutter echoes (from the natural environment) are larger than the receiver noise, the threshold has to be varied adaptively in order to maintain the false alarm rate at a constant value. This is accomplished by a *constant false alarm rate* (CFAR) receiver (Sec. 5.7).

A radar usually receives many echo pulses from a target. The process of adding these pulses together to obtain a greater signal-to-noise ratio before the detection decision is made is called *integration.* The integrator is often found in the video portion of the receiver.

The *signal processor* is that part of the radar whose function is to pass the desired echo signal and reject unwanted signals, noise, or clutter. The signal processor is found in the receiver before the detection decision is made. The matched filter, mentioned previously, is an example of a signal processor. Another example is the doppler filter that separates desired moving targets (whose echoes are shifted in frequency due to the doppler effect) from undesired stationary clutter echoes.

Some radars process the detected target signal further, in the *data processor,* before displaying the information to an operator. An example is an *automatic tracker,* which uses the locations of the target measured over a period of time to establish the track (or path) of the target. Most modern air-surveillance radars and some surface-surveillance radars generate target tracks as their output rather than simply display detections. Following the data processor, or the decision function if there is no data processor, the radar output is displayed to an operator or used in a computer or other automatic device to provide some further action.

The signal processor and data processor are usually implemented with digital technology rather than with analog circuitry. The analog-to-digital (A/D) converter and digital memory are therefore important in modern radar systems. In some sophisticated radars in the past, the signal and data processors were larger and consumed more power than the transmitter and were a major factor in determining the overall radar system reliability; but this should not be taken as true in all cases.

A typical radar display for a surveillance radar is the PPI, or *plan position indicator* (the full term is seldom used). An example is shown in Fig. 1.5. The PPI is a presentation

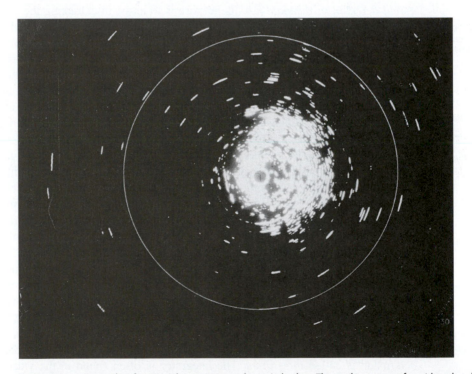

Figure 1.5   Example of a PPI (plan position indicator) display. This is the output of an L-band radar with an antenna beam-width of 3°, without MTI processing. The range ring has a radius of 50 nmi. Clutter is seen in the near vicinity of the radar. Aircraft approaching an airport are seen in the northwest direction.
| (Courtesy of George Linde of the Naval Research Laboratory.)

that maps in polar coordinates the location of the target in azimuth and range. The PPI in the past has been implemented with an intensity-modulated CRT. The amplitude of the receiver output modulates (turns on or off) the electron-beam intensity (called the z-axis of the CRT) as the electron beam is made to sweep outward (the range coordinate) from the center of the tube. The sweep of the electron beam rotates in angle in synchronism with the pointing of the antenna beam. A B-scope display is similar to a PPI except that it utilizes a rectangular format, rather than the polar format, to display range versus angle. Both the PPI and the B-scope CRT displays have limited dynamic range since they are intensity modulated. An A-scope is sometimes used for special purposes. It is an amplitude-modulated rectangular display that presents the receiver output on the y-axis and the range (or time delay) on the x-axis. (An example is shown in Fig. 7.21.) It is more suited for tracking radar or continuous staring applications than as a display for surveillance radar.

The early radars displayed to an operator *raw video,* which is the output of the radar receiver without further processing (with the exception of the matched filter). Modern radars usually present *processed video,* which is the output of the radar after signal processing and threshold detection or after automatic tracking. Only processed target detections or target tracks are presented. This relieves the burden on the operator, but processed video can also eliminate information about the environment and unusual operational situations that a trained and alert operator might be able to recognize and interpret.

Radars can operate in various modes by radiating different frequencies, with different polarizations. (The polarization of the radar wave is defined by the direction of the electric field vector.) The radar can also employ various waveforms with different pulse widths, pulse repetition frequencies, or other modulations; and different forms of processing for suppressing different types of clutter, interference, and jamming. The various waveforms and processing need to be selected wisely. A trained operator can fulfill this function, but an operator can become overloaded. When there are many available system options, the radar can be designed to automatically determine the proper mode of operation and execute what is required to implement it. The mode of radar operation is often changed as a function of the antenna look-direction and/or range, according to the nature of the environment.

1.4   RADAR FREQUENCIES

Conventional radars generally operate in what is called the *microwave* region (a term not rigidly defined). Operational radars in the past have been at frequencies ranging from about 100 MHz to 36 GHz, which covers more than eight octaves. These are not necessarily the limits. Operational HF over-the-horizon radars operate at frequencies as low as a few megahertz. At the other end of the spectrum, experimental millimeter wave radars have been at frequencies higher than 240 GHz.

During World War II, letter codes such as *S, X,* and *L* were used to designate the distinct frequency bands at which microwave radar was being developed. The original purpose was to maintain military secrecy; but the letter designations were continued after the war as a convenient shorthand means to readily denote the region of the spectrum at which a radar operated. Their usage is the accepted practice of radar engineers. Table 1.1 lists the radar-frequency letter-band designations approved as an IEEE Standard.[3] These are related to the specific frequency allocations assigned by the International Telecommunications Union (ITU) for radiolocation, or radar. For example, *L* band officially extends from 1000 MHz to 2000 MHz, but *L*-band radar is only allowed to operate within the region from 1215 to 1400 MHz since that is the band assigned by the ITU.

There have been other letter-band designations, but Table 1.1 is the only set of designations approved by the IEEE for radar. It has also been recognized by being listed in the U. S. Department of Defense Index of Specifications and Standards.[4] A different set of letter bands has been used by those working in electronic warfare. It was originally formulated by the U.S. Department of Defense for use only in conducting electronic countermeasure exercises.[5] Sometimes it is incorrectly extended to describe radar frequencies, but the use of the electronic warfare letter bands for radar can be confusing and it is not appropriate. (There may be *J*-band jammers, but according to the IEEE Letter-Band Standards there are no *J*-band radars.) Usually the context in which the nomenclature is employed can aid in distinguishing whether the letters refer to radar or to EW.

Table 1.1   IEEE standard radar-frequency letter-band nomenclature*

Band Designation	Nominal Frequency Range	Specific Frequency Ranges for Radar based on ITU Assignments in Region 2
HF	3–30 MHz	
VHF	30–300 MHz	138–144 MHz 216–225 MHz
UHF	300–1000 MHz	420–450 MHz 850–942 MHz
L	1–2 GHz	1215–1400 MHz
S	2–4 GHz	2300–2500 MHz 2700–3700 MHz
C	4–8 GHz	5250–5925 MHz
X	8–12 GHz	8500–10,680 MHz
K_u	12–18 GHz	13.4–14.0 GHz 15.7–17.7 GHz
K	18–27 GHz	24.05–24.25 GHz
K_a	27–40 GHz	33.4–36 GHz
V	40–75 GHz	59–64 GHz
W	75–110 GHz	76–81 GHz 92–100 GHz
mm	110–300 GHz	126–142 GHz 144–149 GHz 231–235 GHz 238–248 GHz

*From "IEEE Standard Letter Designations for Radar-Frequency Bands," IEEE Std 521–1984.

Letter-band nomenclature is not a substitute for the actual numerical frequencies at which a radar operates. The specific numerical frequencies of a radar should be used whenever appropriate, but the letter designations of Table 1.1 should be used whenever a short notation is desired.

1.5 APPLICATIONS OF RADAR

Radar has been employed to detect targets on the ground, on the sea, in the air, in space, and even below ground. The major areas of radar application are briefly described below.

Military   Radar is an important part of air-defense systems as well as the operation of offensive missiles and other weapons. In air defense it performs the functions of surveillance and weapon control. Surveillance includes target detection, target recognition, target tracking, and designation to a weapon system. Weapon-control radars track targets, direct the weapon to an intercept, and assess the effectiveness of the engagement (called *battle damage assessment*). A missile system might employ radar methods for guidance and fuzing of the weapon. High-resolution imaging radars, such as synthetic aperture radar, have been used for reconnaissance purposes and for detecting fixed and moving targets on the battlefield. Many of the civilian applications of radar are also used by the military. The military has been the major user of radar and the major means by which new radar technology has been developed.

Remote Sensing   All radars are remote sensors; however, this term is used to imply the sensing of the environment. Four important examples of radar remote sensing are (1) weather observation, which is a regular part of TV weather reporting as well as a major input to national weather prediction; (2) planetary observation, such as the mapping of Venus beneath its visually opaque clouds; (3) short-range below-ground probing; and (4) mapping of sea ice to route shipping in an efficient manner.

Air Traffic Control (ATC)   Radars have been employed around the world to safely control air traffic in the vicinity of airports (Air Surveillance Radar, or ASR), and en route from one airport to another (Air Route Surveillance Radar, or ARSR) as well as ground-vehicular traffic and taxiing aircraft on the ground (Airport Surface Detection Equipment, or ASDE). The ASR also maps regions of rain so that aircraft can be directed around them. There are also radars specifically dedicated to observing weather (including the hazardous downburst) in the vicinity of airports, which are called Terminal Doppler Weather Radar, or TDWR. The Air Traffic Control Radar Beacon System (ATCRBS and Mode-S) widely used for the control of air traffic, although not a radar, originated from military IFF (Identification Friend or Foe) and uses radar-like technology.

Law Enforcement and Highway Safety   The radar speed meter, familiar to many, is used by police for enforcing speed limits. (A variation is used in sports to measure the speed of a pitched baseball.) Radar has been considered for making vehicles safer by warning of pending collision, actuating the air bag, or warning of obstructions or people behind a vehicle or in the side blind zone. It is also employed for the detection of intruders.

Aircraft Safety and Navigation The airborne weather-avoidance radar outlines regions of precipitation and dangerous wind shear to allow the pilot to avoid hazardous conditions. Low-flying military aircraft rely on terrain avoidance and terrain following radars to avoid colliding with obstructions or high terrain. Military aircraft employ ground-mapping radars to image a scene. The radio altimeter is also a radar used to indicate the height of an aircraft above the terrain and as a part of self-contained guidance systems over land.

Ship Safety Radar is found on ships and boats for collision avoidance and to observe navigation buoys, especially when the visibility is poor. Similar shore-based radars are used for surveillance of harbors and river traffic.

Space Space vehicles have used radar for rendezvous and docking, and for landing on the moon. As mentioned, they have been employed for planetary exploration, especially the planet Earth. Large ground-based radars are used for the detection and tracking of satellites and other space objects. The field of radar astronomy using Earth-based systems helped in understanding the nature of meteors, establishing an accurate measurement of the Astronomical Unit (the basic yardstick for measuring distances in the solar system), and observing the moon and nearby planets before adequate space vehicles were available to explore them at close distances.

Other Radar has also found application in industry for the noncontact measurement of speed and distance. It has been used for oil and gas exploration. Entomologists and ornithologists have applied radar to study the movements of insects and birds, which cannot be easily achieved by other means.

Some radar systems are small enough to be held in one's hand. Others are so large that they could occupy several football fields. They have been used at ranges close enough to almost touch the target and at ranges that reach to the planets.

1.6 THE ORIGINS OF RADAR[6,7,8]

The basic concept of radar was first demonstrated by the classical experiments conducted by the German physicist Heinrich Hertz from 1885 to 1888.[9] Hertz experimentally verified the predictions of James Clerk Maxwell's theory of the electromagnetic field published in 1864. Hertz used an apparatus that was similar in principle to a pulse radar at frequencies in the vicinity of 455 MHz. He demonstrated that radio waves behaved the same as light except for the considerable difference in frequency between the two. He showed that radio waves could be reflected from metallic objects and refracted by a dielectric prism.

Hertz received quick and widespread recognition for his work, but he did not pursue its practical applications. This was left to others. The potential of Hertz's work for the detection and location of reflecting objects—which is what radar does—was advanced by another German, Christian Hulsmeyer. In the early 1900s he assembled an instrument that would today be known as a monostatic (single site) pulse radar. It was much improved over the apparatus used by Hertz. In 1904 he obtained a patent in England[10] and other countries. Hulsmeyer's radar detected ships, and he extensively marketed it for preventing collisions at sea. He demonstrated his apparatus to shipping companies and to the German Navy. Although it was a success and much publicized, there apparently was no interest for a ship collision-avoidance device. His invention and his demonstrations faded from memory and were all but forgotten. Radar would have to be rediscovered a few more times before it eventually became an operational reality.

During the 1920s other evidence of the radar method appeared. S. G. Marconi, the well-known pioneer of wireless radio, observed the radio detection of targets in his experiments and strongly urged its use in a speech delivered in 1922 before the Institute of Radio Engineers (now the IEEE).[11] Apparently unaware of Marconi's speech, A. Hoyt Taylor and Leo C. Young of the U.S. Naval Research Laboratory in Washington, D.C. accidentally observed, in the autumn of 1922, a fluctuating signal at their receiver when a ship passed between the receiver and transmitter located on opposite sides of a river. This was called a *CW wave-interference system,* but today it is known as *bistatic CW radar.* (*Bistatic* means the radar requires two widely separated sites for the transmitter and receiver.) In 1925, the pulse radar technique was used by Breit and Tuve of the Carnegie Institution in Washington, D.C. to measure the height of the ionosphere. The Breit and Tuve apparatus was indeed a radar, but it was not recognized at the time that the same principle might be applied for the detection of ships and aircraft.[12] There were additional reported detections of aircraft and other targets by the CW wave-interference (bistatic radar) method in several countries of the world; but this type of radar did not have, and still doesn't have, significant utility for most applications.

It was the appearance of the heavy military bomber aircraft in the late 1920s and early 1930s that eventually gave rise to operational military radar. After World War I, the bomber was transformed from a fabric-coated biplane with open cockpit to an all-metal, single-wing aircraft with enclosed cockpit, which flew at high altitude over long distance with a heavy bomb load. Long-range warning of the approach of the heavy bomber became an important military need. In most of the countries that responded to this threat, the possible detection methods examined were similar even though the developments were covered by secrecy. Sound locators were the first of the sensors to be examined. They were deployed in many armies up to the start of World War II even though they were recognized much earlier to be inadequate for the task. Attempts were made to detect the spark-plug ignition noise radiated at radio frequencies by the aircraft engine; but they were abandoned once it was realized that the radiated noise could be suppressed by proper shielding. Infrared was examined but it did not have adequate range capability; it was not all-weather; and it did not determine the range of the target. The bistatic CW radar then followed from the accidental detection of aircraft, ships, or other targets as they passed between the transmitter and receiver of a radio system. This two-site configuration was cumbersome and merely acted as a trip wire to detect the passage of a target as it crossed the line connecting transmitter and receiver. The radar method did not become truly useful until the transmitter and receiver were colocated at a single site and pulse waveforms were used.

In the 1930s, radar was rediscovered and developed almost simultaneously and essentially independently in the United States, United Kingdom, Germany, Soviet Union, France, Italy, Japan, and Netherlands. These radars operated at frequencies much lower than those generally used in modern radar. Most early radars employed frequencies in the vicinity of 100 to 200 MHz; but the British Chain Home radars operated at 30 MHz, the low end of the pre-war radar spectrum, while the German 600-MHz Wurzburg radars represented the highest frequencies used operationally early in the war. The technologies employed in these early radars were mostly bold extensions of then current leading-edge technology from the field of radio communications. Compared to microwave radars, they had some limitations, but they did their intended job well.

United States The U.S. Naval Research Laboratory tried to initiate bistatic CW radar development in 1922 and later in 1931; but it was in 1934 that serious effort started when A. Hoyt Taylor and Leo C. Young realized that a single site, pulse radar was necessary for success. By the time of the Japanese attack on the United States at Pearl Harbor on December 7, 1941, 132 radar sets were delivered to the U.S. Navy and 79 were installed on various ships.[13] Twenty of these were the 200 MHz CXAM placed on battleships, aircraft carriers, and cruisers. The Navy radars in the Pacific during the Battle of Midway in 1942 were instrumental in allowing the United States to use its limited forces effectively against the Japanese Navy, which did not have a similar radar capability.

The U.S. Army initiated pulse-radar development in the spring of 1936. At the time of Pearl Harbor, it had developed and deployed overseas a number of 200-MHz SCR-268 antiaircraft searchlight-control radars. The Army also received 112 production units of the 100-MHz SCR-270 long-range air-search radar, one of which detected the Japanese attack on Pearl Harbor. (Radar did its job at Pearl Harbor, but the command and control system to utilize the information was lacking.)

United Kingdom In the mid-1930s, the British felt the urgency of the approaching war far more than did the United States. Although the United Kingdom started later that the United States, they turned on their first operational radar system, the 30-MHz Chain Home radar, by the end of the summer of 1938 (a year before the start of World War II on September 3, 1939). The Chain Home system of radars remained in operation for the entire duration of the war. These early radars have been given much credit for helping turn back the major German air attacks on the British Isles during what was called the Battle of Britain in the late summer of 1940. By 1939, the British also developed a 200-MHz airborne-intercept radar for the detection and intercept of hostile aircraft, especially at night and during conditions of poor visibility. This radar was later modified to detect surface ships and submarines. A highly significant advance in radar technology was made with the British invention of the high-power microwave magnetron in 1940, which opened the higher frequencies to radar. Being overextended by the everyday needs of fighting a major war, the British disclosed the magnetron to the United States in the fall of 1940 for its further development. The cavity magnetron provided the basis for the extensive and rapid development of microwave radar at the MIT Radiation Laboratory and the Bell Telephone Laboratories during World War II, as well as the development of microwave radar in the United Kingdom.

Germany   The country of Hertz and Hulsmeyer also had to rediscover radar. By the end of 1940, Germany had three major operational radars:

- The 125-MHz Freya air-search radar was originally developed for the Navy, but the German Air Force also became interested in it as a transportable ground-based early warning radar. Orders were placed for its production by the German Air Force in 1939 and it was employed as a Ground Control of Intercept radar for the control of night fighters. By the start of the war, however, only eight Freyas were available.

- The Wurzburg fire-control radar was used in conjunction with the Freya and similar air-search radars. The 565-MHz radar was introduced in the spring of 1940, and by the end of the war over 4000 Wurzburg radars were procured.

- The 500-MHz Seetakt shipboard radar for ranging on ships to control gunfire was installed by the German Navy on four major ships during 1937–1938. It was used as early as 1937 in support of the rebel side during the Spanish Civil War. The Seetakt was the first military radar deployed operationally by any country. (It took a long time for the British to recognize the existence of the Seetakt radars. One of the German ships carrying this radar was sunk by the British Navy in shallow water off Montevideo in South America in December 1939. Its radar antenna remained visible above the water. Although a British radar expert inspected the antenna and reported it to be part of a radar, the British Admiralty did not pay attention.[14]) Over 100 Seetakt sets were in production at the beginning of the war.

In 1940, Germany probably was ahead of all other countries in radar technology, something the United States and the United Kingdom did not realize until after the end of the war in 1945. Neither the United States nor the United Kingdom had at that time an operational shipboard radar like the Seetakt, and the Wurzburg was considerably in advance of any Allied equivalent.[15] The German military, however, did not exploit their early technical advantage. They realized their mistake when the British commenced large-scale bomber raids on their country. When the Germans finally acted, it was too late to catch up with the fast-moving British and Americans.

USSR   The Soviet Union started the pursuit of radar early in the 1930s, and had production radars available by the time of the German invasion of their country in June 1941. Both production radars and development radars were employed for the air defense of Leningrad and Moscow. The first production radar, the RUS-1, was a bistatic CW system that operated at 75 MHz with a 35-km separation between the transmitter and receiver. As mentioned previously, the bistatic radar was not all that might be desired of a sensor system; so at the start of hostilities, the Soviets moved these radars "to the east" and deployed the monostatic truck-mounted RUS-2 pulse-radar system, also at 75 MHz, with a range of 150 km. The German invasion forced the relocation of the major Soviet radar development institutions, which seriously reduced further development.

Italy   The Italians did not believe in the importance of radar until their decisive naval defeat in March 1941 by the British Navy in the night Battle of Cape Matapan, where British radar found and fired upon the surprised Italian ships during darkness. This defeat led to the Italian production of a series of 200-MHz shipboard radars, called the Owl. A significant number were installed on Italian ships. Italian work on radar essentially ceased in 1943 when the country was invaded by the Allies (the United Kingdom and United States).

France   As did other countries, the French investigated CW wave-interference (bistatic) radar in the 1930s. They carried the bistatic method further than other countries and installed a system early in the war. It was deployed as a triple fence laid out in a Z-pattern so as to obtain the direction of travel, speed, and altitude of an aircraft target. In the mid-1930s, the French received much publicity for the civilian application of a CW obstacle-detection radar installed on a luxury ocean liner. In 1939, development finally began on a 50-MHz pulse radar, but the occupation of France by the Germans in 1940 virtually closed down their radar development.

Japan   The Japanese discovered and developed the microwave magnetron before the British. Although they had a number of different microwave magnetron configurations, they never were able to convert this capability to microwave radars that were comparable to those of the Allies. The Japanese explored bistatic radar in 1936 and deployed it for the defense of their homeland during the war. They initiated pulse radar in 1941, much later than other countries, after the Germans disclosed to them the British use of VHF pulse radar. Japan increased their efforts in radar after being surprised by its successful use at night by the U.S. Navy during the Battle of Guadalcanal in October 1942. Although the Japanese possessed the microwave magnetron and had a fine technical capability, they failed to employ radar as effectively as did other countries because of the dominance of the military over the civilian engineers and scientists in technical matters, the excessive secrecy imposed, and a shortage of engineers and materials.[16,17]

Netherlands   Even a small country such as the Netherlands produced radar in time for World War II. A naval UHF pulse radar for air defense was configured, and production of 10 prototypes was started. A demonstration was planned for the military on May 10, 1940, but this turned out to be the day of the German invasion of their country. Nevertheless, the Netherlands managed to put one of their radars, known as the Type 289, on a Dutch ship in combat during the war (operating out of the United Kingdom).

Microwave Magnetron   As mentioned, a major advance in radar was made with the invention of the high-power microwave cavity magnetron at the University of Birmingham in England early in World War II. The magnetron dramatically changed the nature of radar as it existed up to that time by allowing the development of radars with small antennas that could be carried on ships and aircraft, and by land-mobile systems. Most countries involved in early radar research recognized the importance of obtaining high power at microwave frequencies and tried to push the conventional magnetron upwards in power, but it was the British who succeeded and insured its use in operational radar. The MIT Radiation Laboratory was organized by the United States in the fall of 1940 to pursue development of microwave radar based on the use of the British magnetron. They were highly successful in applying the new microwave technology to military radar for air, land, and sea.[6] The Radiation Laboratory developed more than 100 different radar systems during the war years for such purposes as early warning of air attack, antiaircraft fire control, air intercept, blind bombing, and ship detection.

After World War II   During the war, radar technology and systems grew rapidly, but there was still much left to do. In the years that immediately followed the war, radar development was mainly concentrated on the things that were not completed during the war. Since that time, radar capability has continued to advance. The following is a list of some of the many major accomplishments of radar:[8]

- The use of the doppler effect in the *MTI* (moving target indication) pulse radar was perfected to separate desired aircraft targets from undesired large ground echoes.

- *High-power stable amplifiers* such as the klystron, traveling wave tube, and solid-state transistor allowed better application of the doppler effect, use of sophisticated waveforms, and much higher power than could be obtained with the magnetron.

- Highly accurate angle tracking of targets became practical with *monopulse radar.*

- *Pulse compression* allowed the use of long waveforms to obtain high energy and simultaneously achieve the resolution of a short pulse by internal modulation of the long pulse.

- The airborne *synthetic aperture radar* (SAR) provided high resolution map-like imaging of ground scenes.

- Airborne radars using doppler processing methods gave rise to *airborne MTI and pulse doppler radars,* which were able to detect aircraft in the midst of heavy ground clutter.

- The electronically steered *phased array antenna* offered rapid beam steering without mechanical movement of the antenna.

- *HF over-the-horizon radar* extended the detection range for aircraft by a factor of ten, to almost 2000 nmi.

- Radar became more than a "blob" detector by extracting information from the echo signal to provide *target recognition.*

- Radar has become an important tool for the meteorologist and as an aid for safe and efficient air travel by observing and measuring precipitation, warning of dangerous wind shear and other hazardous weather conditions, and for providing timely measurements of the vertical profile of wind speed and direction.

- The rapid advances in digital technology made many theoretical capabilities practical with *digital signal processing* and *digital data processing.*

New technology, new radar techniques, and new radar applications have fueled the continuous growth of radar since its inception in the 1930s. At the time this is written, growth in radar is continuing and is expected to continue. Illustrations of a small sample of various types of radars are shown in Figs. 1.6 to 1.15. Some of these, such as the airport surveillance radar and the airport surface detection equipment, might be familiar to the reader since they can be seen at major airports. Others shown here are not so usual and are included to indicate the diversity found in radar systems. Additional examples of radar systems will be found in Chapters 3, 4, and 9.

Figure 1.6 The airport surveillance radar, ASR-9, which operates at *S* band (2.7 to 2.9 GHz) with a klystron transmitter having a peak power of 1.3 MW. Pulse width is 1.0 μs. The antenna has a beamwidth of 1.4°, a gain of 34 dB, and rotates at 12.5 rpm. There are two vertical feeds for this antenna that produce two beams that overlap in elevation. The array antenna on top of the ASR-9 reflector antenna is for the FAA Air Traffic Control Radar Beacon System (ATCRBS). Airport surveillance radars have a range of about 50 to 60 nmi and provide coverage of the air traffic in the vicinity of airports. This radar is similar to the ASR-12 radar, except that the ASR-12 employs a solid-state transmitter and requires different waveforms to accommodate the high duty factor required when using a solid state transmitter.
| (Courtesy of Northrop Grumman Corporation.)

Figure 1.7   This ASDE-3 airport surface detection equipment is shown located on the control tower at the Pittsburgh International Airport. It produces a high-resolution ground map of the airport so as to monitor and expedite the movement of airport vehicle traffic and taxiing aircraft. This K_u-band radar operates from 15.7 to 17.7 GHz with a peak power of 3 kW, 40 ns pulse width, and 16 kHz pulse repetition frequency. The antenna and its rotodome rotate together at 60 rpm. The antenna gain is 44 dB, the azimuth beamwidth is 0.25° (a very narrow beamwidth for a radar), and its elevation beamwidth is 1.6°. It radiates circular polarization and has pulse-to-pulse frequency agility over 16 pulses, both of which help to increase the detectability of aircraft and vehicles in the rain. The display range is from 600 to 7300 m. The hole seen at the top of the rotodome in this figure is an access hatch that is not usually open. (It was open when this picture was made to allow a surveying instrument to be used for precisely locating the antenna.) Four lightning rods are shown around the rotodome.
I (Courtesy of Northrop Grumman Corporation Norden Systems.)

Figure 1.8 The TPS-117 is a transportable 3D military air-surveillance radar with an antenna that is 5.70 m (18.7 ft) high and 4.75 m (15.6 ft) wide and has a gain of 36 dB. It operates at *L* band over a 14 percent bandwidth centered at 1.3 GHz. Its planar phased array antenna rotates in azimuth at 6 rpm while electronically scanning its pencil beam over an elevation angle from 0 to 20°. The azimuth beamwidth is 3.4° and elevation beamwidth is 2.7°. For long range the radar employs a pulse width of 410 μs, compressed to 0.8 μs by means of linear frequency modulation pulse compression. Its solid-state transmitter is distributed over the 34 rows of the antenna, and produces 3.4 kW average power with 19 kW peak power. It can be assembled within three-quarters of an hour by a crew of eight.
| (Courtesy of Lockheed Martin. Copyright 1999.)

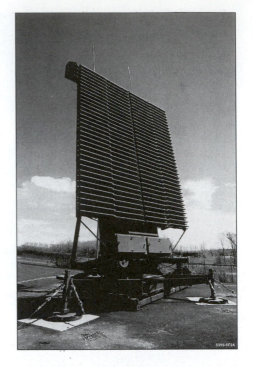

Figure 1.9 This is the 55G6U VHF (180 to 220 MHz) 3D radar, a Russian long-range air-surveillance radar developed by the Nizhny Novgorod Scientific-Research Radiotechnical Institute. The horizontal antenna structure of this radar, which is 27 m wide by 6 m high (89 by 20 ft), obtains the range and the azimuth angle of detected targets. The vertical antenna structure is 30 m high by 6 m wide (98 by 20 ft) and is electronically steered to provide the elevation (and the height) of targets. The antenna assembly extends about 42 m (138 ft) above the surface of the ground. It rotates at 6 rpm. Its accuracy in range is 100 m, in azimuth it is 0.2°, and in elevation the accuracy is 400 m. This large radar is transportable and can be deployed or redeployed in 22 hours.
| (Courtesy of A. A. Zachepitsky, Designer General, Scientific Research & Manufacturing, Nizhny Novgorod, Russia.)

Figure 1.10   A UHF (404 MHz) vertically looking *wind-profiler radar* that provides the speed and direction of the wind as a function of altitude. The phased array antenna is 12 m by 12 m and has a beamwidth of 4°. Three beams are generated. One looks vertically, a second is tilted 15° to the north, and the third is tilted 15° to the east. The orientation of these three beams allows the horizontal and vertical components of the wind direction and speed to be determined. This radar obtains an echo from the clear air, which is due to the random inhomogeneities of the atmosphere caused by turbulence. The radar has an average power of 2.2 kW and a peak power of 16 kW. It can measure winds up to an altitude of 16 km. Wind profilers are important not only for weather observation and forecasting, but also for efficient (greater fuel savings) and safe routing of aircraft. This radar requires no on-site personnel for its operation or maintenance. (Courtesy of D. W. van de Kamp and the NOAA Forecast Systems Laboratory, with the help of M. J. Post.)

Figure 1.11   The Path Finder Utility Mapping System is a ground-penetrating radar shown here being pushed by an operator to locate underground cable, pipes, power lines, and other buried objects. Operator is seen using a wearable computer with a heads-up display that provides 3D data. Depth of penetration varies with soil conditions but is about 2 to 3 m in many types of soil. This is an example of an ultrawideband radar operating in the VHF and lower UHF region of the spectrum. (Courtesy of Geophysical Survey Systems, Inc., North Salem, NH.)

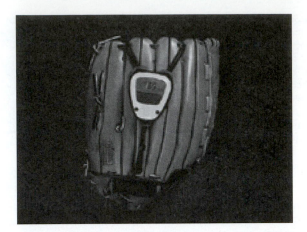

Figure 1.12 The Glove Radar is perhaps the smallest and cheapest radar one can buy. It is shown here attached to the back of a baseball glove to measure the speed of an approaching ball just before it is caught. It is 2.5 in. wide, 3.5 in. long, and 1.25 in. thick. Weight is 3 oz. An LCD displays speed in miles per hour. The frequency is 5.6 GHz and the CW transmitter power is 1 mW. I (Courtesy of Al Dilz and Sport Sensors, Inc., Cincinnati, OH.)

Figure 1.13 The Longbow K_a band (35 GHz) fire control radar for the U.S. Army's Apache helicopter shown mounted in a radome on top of the helicopter mast that rotates the blades. I (Courtesy of Northrop Grumman Corporation.)

Figure 1.14   *L*-band electronically scanned 360° Wedgetail airborne early warning and control (AEW&C) radar mounted on a 737–700 aircraft. It was developed for the Royal Australian Air Force. The "top hat" on top the aircraft provides a practical solution for fore and aft coverage while maintaining the low drag profile of the dorsal array system.
| (Photo: The Boeing Co.)

Figure 1.15   A long-range VHF Soviet space-surveillance radar that was located at Krasnoyarsk in central Siberia. The 30-story high receiving phased array antenna is shown on the right. It is approximately the size of a football field. The 11-story transmitting phased array antenna is barely visible in the background. This radar was later dismantled, but it is similar to other large phased array radars developed and operated by the former Soviet Union.
| (Courtesy of William J. Broad of the
| *New York Times*.)

REFERENCES

1. Kraus, J. D. *Antennas.* 2nd ed. New York: McGraw-Hill, 1988, Sec. 2.22.

2. Silver, S. *Microwave Antenna Theory and Design.* MIT Radiation Laboratory Series, vol. 12. New York: McGraw-Hill, 1949, Sec. 6.4.

3. IEEE Standard Letter Designations for Radar-Frequency Bands, IEEE Std 521-1984.

4. DoD Index of Specifications and Standards, Supplement Part II, 1 March 1984, p. 101.

5. Performing Electronic Countermeasures in the United States and Canada, *U.S. Navy OPNAVINST* 3430.9B, 27 Oct. 1969. Similar versions issued as *U.S. Air Force* AFR 55–44, *U.S. Army* AR 105–86, and *U.S. Marine Corps* MCO 3430.1.

6. Guerlac, H. *Radar in World War II.* American Institute of Physics, Tomash Publishers, 1987.

7. Swords, S. S. *Technical History of the Beginnings of Radar.* London: Peter Peregrinus, 1986.

8. Skolnik, M. I. "Fifty Years of Radar." *Proc. IEEE,* vol. 73 (February 1985), pp. 182–197.

9. Hertz, H. *Electric Waves.* New York: Dover Publications, 1962. (Republication of the work first published in 1893 by Macmillan and Company.)

10. British Patent 13,170 issued to Christian Hulsmeyer, Sept. 22, 1904, entitled "Hertzian-Wave Projecting and Receiving Apparatus Adapted to Indicate or Give Warning of the Presence of a Metallic Body, Such as a Ship or a Train, in the Line of Projection of Such Waves."

11. Marconi, S. G. "Radio Telegraphy." *Proc. IRE,* vol. 10, no. 4 (1992), p. 237.

12. Oral history of Leo C. Young, recorded October 15, 1953, regarding the origin of radar at the Naval Research Laboratory, from the *Rear Admiral Stanford C. Hopper collection of the Library of Congress, History of Naval Radio,* LWO 4934 R23B, Reels 150 and 151.

13. Allison, D. K. "New Eye for Navy Radar: The Origin of Radar at the Naval Research Laboratory." *Naval Research Laboratory Report 8466,* Washington, D.C., Sept. 28, 1981.

14. House, D. *Radar at Sea.* Annapolis, MD: Naval Institute Press, 1993, pp. 45–49.

15. Price, A. *Instruments of Darkness.* New York: Charles Scribner's, 1978, p. 61.

16. Nakajima, S. "The History of Japanese Radar Development to 1945." In *Radar Development to 1945,* Russel Burns, Ed. London: Peter Peregrinus, 1988, pp. 243–258.

17. Nakagawa, Y. *Japanese Radar and Related Weapons of World War II.* Laguna Hills, CA: Aegean Park Press, 1997.

PROBLEMS

1.1 *a.* What should be the pulse repetition frequency of a radar in order to achieve a maximum unambiguous range of 60 nmi?

    b. How long does it take for the radar signal to travel out and back when the target is at the maximum unambiguous range?

    c. If the radar has a pulse width of 1.5 μs, what is the extent (in meters) of the pulse energy in space in the range coordinate?

    d. How far apart in range (meters) must two equal-size targets be separated in order to be certain they are completely resolved by a pulse width of 1.5 μs?

    e. If the radar has a peak power of 800 kW, what is its average power?

    f. What is the duty cycle of this radar?

1.2 A ground-based air-surveillance radar operates at a frequency of 1300 MHz (*L* band). Its maximum range is 200 nmi for the detection of a target with a radar cross section of one square meter ($\sigma = 1$ m^2). Its antenna is 12 m wide by 4 m high, and the antenna aperture efficiency is $\rho_a = 0.65$. The receiver minimum detectable signal is $S_{min} = 10^{-13}$ W. Determine the following:

    a. Antenna effective aperture A_e (square meters) and antenna gain G [numerically and in dB, where G (in dB) $= 10 \log_{10} G$ (as a numeric)].

    b. Peak transmitter power.

    c. Pulse repetition frequency to achieve a maximum unambiguous range of 200 nmi.

    d. Average transmitter power, if the pulse width is 2 μs.

    e. Duty cycle.

    f. Horizontal beamwidth (degrees).

1.3 *a.* What is the peak power of a radar whose average transmitter power is 200 W, pulse width of 1 μs, and a pulse repetition frequency of 1000 Hz?

    b. What is the range (nmi) of this ground-based air-surveillance radar if it has to detect a target with a radar cross section of 2 m^2 when it operates at a frequency of 2.9 GHz (*S* band), with a rectangular-shaped antenna that is 5 m wide, 2.7 m high, antenna aperture efficiency ρ_a of 0.6, and minimum detectable signal S_{min} equal to 10^{-12} W (based on P_t in the radar equation being the peak power)?

    c. Sketch the received echo signal power as a function of range from 10 to 80 nmi.

1.4 The moon as a radar target may be described as follows: average distance to the moon is 3.844×10^8 m (about 208,000 nmi); experimentally measured radar cross section is 6.64×10^{11} m^2 (mean value over a range of radar frequencies); and its radius is 1.738×10^6 m.

    a. What is the round-trip time (seconds) of a radar pulse to the moon and back?

    b. What should the pulse repetition frequency (prf) be in order to have no range ambiguities?

    c. For the purpose of probing the nature of the moon's surface, a much higher prf could be used than that found in (b). How high could the prf be if the purpose is to observe the echoes from the moon's front half?

   d. If an antenna with a diameter of 60 ft and aperture efficiency of 0.6 were used at a frequency of 430 MHz with a receiver having a minimum detectable signal of 1.5×10^{-16} W, what peak power is required? Does your answer surprise you; and if so, why?

   e. The radar cross section of a smooth perfectly conducting sphere of radius a is πa^2. What would be the radar cross section of the moon if it were a sphere with a perfectly smooth, conducting surface? Why might the measured cross section of the moon (given above) be different from this value?

1.5   A radar mounted on an automobile is to be used to determine the distance to a vehicle traveling directly in front of it. The radar operates at a frequency of 9375 MHz (X band) with a pulse width of 10 ns (10^{-8} s). The maximum range is to be 500 ft.

   a. What is the pulse repetition frequency that corresponds to a range of 500 ft?

   b. What is the range resolution (meters)?

   c. If the antenna beamwidth were 6°, what would be the cross-range resolution (meters) at a range of 500 ft? Do you think this value of cross-range resolution is sufficient?

   d. If the antenna dimensions were 1 ft by 1 ft and the antenna efficiency were 0.6, what would be the antenna gain (dB)?

   e. Find the average power required to detect a 10 m^2 radar cross section vehicle at a range of 500 ft, if the minimum detectable signal is 5×10^{-13} W.

1.6   Determine (a) the peak power (watts) and (b) the antenna physical area (m^2) which make the cost of the following radar a minimum:

   Frequency: 1230 MHz (L band)

   Antenna aperture efficiency: 0.6

   Receiver minimum detectable signal: 3×10^{-13} W

   Unit cost of transmitter: $2.20 per watt of peak power

   Unit cost of antenna: $1400 per square meter of physical area

   Cost of receiver and other items: $1,000,000

   The radar must detect a target of 2 m^2 cross section at a range of 200 nmi.

   (You will have to use one of the simple forms of the radar range equation.)

   c. What is the cost of the antenna and the cost of the transmitter?

   d. In a new radar design, how would you try, as a first attempt, to allocate the costs between the antenna and the transmitter (based only on the answers to the above problem)?

1.7   Who invented radar? (Please explain your answer.)

1.8   Three variants of the simple form of the radar equation have been given. In Eq. (1.9) the wavelength is in the numerator, in Eq. (1.10) the wavelength is in the denominator, and in Eq. (1.7), there is no explicit indication of wavelength. How would you respond to the question: "How does the radar range vary with the radar wavelength, everything else being the same?"

1.9   If the weight of a transmitter is proportional to the transmitter power (i.e., $W_T = k_T P_t$) and if the weight of an antenna is proportional to its volume (so that we can say its weight is proportional to the 3/2 power of the antenna aperture area A, or $W_A = k_A A^{3/2}$), what is the relationship between the weight of the antenna and the weight of the transmitter that makes the total weight $W = W_T + W_A$ a minimum, assuming a fixed range? (You will need the simple form of the radar equation to obtain a relationship between P_t and A.)

chapter
2

The Radar Equation

2.1   INTRODUCTION

The simple form of the radar equation, derived in Sec. 1.2, expressed the maximum radar range $R_{\max}$ in terms of the key radar parameters and the target's radar cross section when the radar sensitivity was limited by receiver noise. It was written:*

$$R_{\max} = \left[\frac{P_t G A_e \sigma}{(4\pi)^2 S_{\min}} \right]^{1/4}$$

[2.1]

where

$$P_t = \text{transmitted power, W}$$
$$G = \text{Antenna gain}$$
$$A_e = \text{Antenna effective aperture, m}^2$$
$$\sigma = \text{Radar cross section of the target, m}^2$$
$$S_{\min} = \text{Minimum detectable signal, W}$$

Except for the target's radar cross section, the parameters of this simple form of the radar equation are under the control of the radar designer. It states that if long ranges are desired, the transmitted power should be large, the radiated energy should be concentrated into a narrow beam (large transmitting gain), the echo energy should be received by a large antenna aperture (also synonymous with large gain), and the receiver should be sensitive to weak signals.

| *This can also be written in terms of gain or effective aperture by using the relationship $G = 4\pi A_e/\lambda^2$.

In practice, however, this simple form of the radar equation does not adequately predict the range performance of actual radars. It is not unusual to find that when Eq. (2.1) is used, the actual range might be only half that predicted.[1] The failure of the simple form of the radar equation is due to (1) the statistical nature of the minimum detectable signal (usually determined by receiver noise), (2) fluctuations and uncertainties in the target's radar cross section, (3) the losses experienced throughout a radar system, and (4) propagation effects caused by the earth's surface and atmosphere. The statistical nature of receiver noise and the target cross section requires that the maximum radar range be described probabilistically rather than by a single number. Thus the specification of range must include the probability that the radar will detect a specified target at a particular range, and with a specified probability of making a false detection when no target echo is present. The range of a radar, therefore, will be a function of the *probability of detection, P_d,* and the *probability of false alarm, P_{fa}.*

The prediction of the radar range cannot be performed with arbitrarily high accuracy because of uncertainties in many of the parameters that determine the range. Even if the factors affecting the range could be predicted with high accuracy, the statistical nature of radar detection and the variability of the target's radar cross section and other effects make it difficult to accurately verify the predicted range. In spite of it not being as precise as one might wish, the radar equation is an important tool for (1) assessing the performance of a radar, (2) determining the system trade-offs that must be considered when designing a new radar system, and (3) aiding in generating the technical requirements for a new radar procurement.

In this chapter, the simple radar equation will be extended to include many of the important factors that influence the range of a radar when its performance is limited by receiver noise. A pulse waveform will be assumed, unless otherwise noted. In addition to providing a more complete representation of the radar range, this chapter introduces a number of basic radar concepts.

A thorough discussion of all the factors that influence the prediction of radar range is beyond the scope of a single chapter. For this reason, many subjects may appear to be treated lightly. More detailed information can be found in some of the subsequent chapters and in the references listed, especially those by Lamont Blake.[2,3]

2.2 DETECTION OF SIGNALS IN NOISE

The ability of a radar receiver to detect a weak echo signal is limited by the ever-present noise that occupies the same part of the frequency spectrum as the signal. The weakest signal that can just be detected by a receiver is the *minimum detectable signal.* In the radar equation of Eq. (2.1) it was denoted as S_{min}. Use of the minimum detectable signal, however, is not common in radar and is not the preferred method for describing the ability of a radar receiver to detect echo signals from targets, as shall be seen in Sec. 2.3.

Detection of a radar signal is based on establishing a threshold at the output of the receiver. If the receiver output is large enough to exceed the threshold, a target is said to be present. If the receiver output is not of sufficient amplitude to cross the threshold, only noise is said to be present. This is called *threshold detection*. Figure 2.1 represents the output of a radar receiver as a function of time. It can be thought of as the video output displayed on an A-scope (amplitude versus time, or range). The fluctuating appearance of the output is due to the random nature of receiver noise.

When a large echo signal from a target is present, as at *A* in Fig. 2.1, it can be recognized on the basis of its amplitude relative to the rms noise level. If the threshold level is set properly, the receiver output should not normally exceed the threshold if noise alone were present, but the output would exceed the threshold if a strong target echo signal were present along with the noise. If the threshold level were set too low, noise might exceed it and be mistaken for a target. This is called a *false alarm*. If the threshold were set too high, noise might not be large enough to cause false alarms, but weak target echoes might not exceed the threshold and would not be detected. When this occurs, it is called a *missed detection*. In early radars, the threshold level was set based on the judgment of the radar operator viewing the radar output on a cathode-ray tube display. In radars with automatic detection (electronic decision making), the threshold is set according to classical detection theory described later in this chapter.

The output that is shown in Fig. 2.1 is assumed to be from a *matched-filter* receiver. A matched filter, as was mentioned in Sec. 1.3, is one that maximizes the output signal-to-noise ratio. (This is discussed in detail in Sec. 5.2). Almost all radars employ a matched filter or a close approximation. A matched filter does not preserve the shape of the input waveform. For example, a rectangular-like pulse will be somewhat triangular in shape at the output of the matched filter. For this reason the receiver output drawn in this figure is more a series of triangular-like pulses rather than rectangular. The fact that the matched filter changes the shape of the received signal is of little consequence. The filter is not designed to preserve the signal shape, but to maximize detectability.

A threshold level is shown in Fig. 2.1 by the long dash line. If the signal is large enough, as at *A*, it is not difficult to decide that a target echo signal is present. But consider the two weaker signals at *B* and *C*, representing two target echoes of equal amplitude. The noise accompanying the signal at *B* is assumed to be of positive amplitude and adds to the target signal so that the combination of signal plus noise crosses the threshold and is declared a target. At *C* the noise is assumed to subtract from the target signal, so that the resultant of signal and noise does not cross the threshold and is a missed

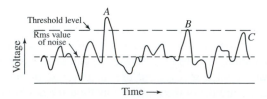

Figure 2.1   Envelope of the radar receiver output as a function of time (or range). A, B, and C represent signal plus noise. A and B would be valid detections, but C is a missed detection.

detection. The ever-present noise, therefore, will sometimes enhance the detection of marginal signals, but it may also cause loss of detection.

The signal at C would have been detected if the threshold were lower. But too low a threshold increases the likelihood that noise alone will exceed the threshold and be improperly called a detection. The selection of the proper threshold is therefore a compromise that depends upon how important it is to avoid the mistake of (1) failing to recognize a target signal that is present (missed detection) or (2) falsely indicating the presence of a target signal when none exists (false alarm).

The signal-to-noise ratio, as has been mentioned, is a better measure of a radar's detection performance than is the minimum detectable signal. The relationship between the two is developed next.

2.3   RECEIVER NOISE AND THE SIGNAL-TO-NOISE RATIO

At microwave frequencies, the noise with which the target echo signal competes is usually generated within the receiver itself. If the radar were to operate in a perfectly noise-free environment so that no external sources of noise accompany the target signal, and if the receiver itself were so perfect that it did not generate any excess noise, there would still be noise generated by the thermal agitation of the conduction electrons in the ohmic portion of the receiver input stages. This is called *thermal noise* or *Johnson noise*. Its magnitude is directly proportional to the bandwidth and the absolute temperature of the ohmic portions of the input circuit. The available thermal-noise power (watts) generated at the input of a receiver of bandwidth B_n (hertz) at a temperature T (degrees Kelvin) is[4]

$$\text{available thermal-noise power} = kTB_n \qquad \text{[2.2]}$$

where k = Boltzmann's constant = 1.38×10^{-23} J/deg. (The term *available* means that the device is operated with a matched input and a matched load.) The bandwidth of a superheterodyne receiver (and almost all radar receivers are of this type) is taken to be that of the IF amplifier (or matched filter).

In Eq. (2.2) the bandwidth B_n is called the *noise bandwidth,* defined as

$$B_n = \frac{\displaystyle\int_0^\infty |H(f)|^2 \, df}{|H(f_0)|^2} \qquad \text{[2.3]}$$

where $H(f)$ = frequency-response function of the IF amplifier (filter) and f_0 = frequency of the maximum response (usually occurs at midband). Noise bandwidth is not the same as the more familiar half-power, or 3-dB, bandwidth. Equation (2.3) states that the noise bandwidth is the bandwidth of the equivalent rectangular filter whose noise-power output is the same as the filter with frequency response function $H(f)$. The *half-power bandwidth,* a term widely used in electronic engineering, is defined by the separation between the points of the frequency response function $H(f)$ where the response is reduced 0.707 (3 dB in power) from its maximum value. Although it is not the same as the noise bandwidth, the half-power bandwidth is a reasonable approximation for many practical radar receivers.[5,6] Thus the half-power bandwidth B is usually used to approximate the noise bandwidth B_n, which will be the practice in the remainder of the chapter.

The noise power in practical receivers is greater than that from thermal noise alone. The measure of the noise out of a real receiver (or network) to that from the ideal receiver with only thermal noise is called the *noise figure* and is defined as

$$F_n = \frac{\text{noise out of practical receiver}}{\text{noise out of ideal receiver at std temp } T_0} = \frac{N_{\text{out}}}{kT_0BG_a} \qquad \text{[2.4]}$$

where $N_{\text{out}} = $ noise out of the receiver, and $G_a = $ available gain. The noise figure is defined in terms of a standard temperature T_0, which the IEEE defines as 290 K (62°F). This is close to room temperature. (A standard temperature assures uniformity in measurements that might be made at different temperatures.) With this definition, the factor kT_0 in the definition of noise figure is 4×10^{-21} W/Hz, a quantity easier to remember than Boltzmann's constant. The available gain G_a is the ratio of the signal out, S_{out}, to the signal in, S_{in}, with both the output and input matched to deliver maximum output power. The input noise, N_{in}, in an ideal receiver is equal to kT_0B_n. The definition of noise figure given by Eq. (2.4) therefore can be rewritten as

$$F_n = \frac{S_{\text{in}}/N_{\text{in}}}{S_{\text{out}}/N_{\text{out}}} \qquad \text{[2.5]}$$

This equation shows that the noise figure may be interpreted as a measure of the degradation of the signal-to-noise ratio as the signal passes through the receiver.

Rearranging Eq. (2.5), the input signal is

$$S_{\text{in}} = \frac{kT_0BF_nS_{\text{out}}}{N_{\text{out}}} \qquad \text{[2.6]}$$

If the minimum detectable signal S_{min} is that value of S_{in} which corresponds to the minimum detectable signal-to-noise ratio at the output of the IF, $(S_{\text{out}}/N_{\text{out}})_{\text{min}}$, then

$$S_{\text{min}} = kT_0BF_n\left(\frac{S_{\text{out}}}{N_{\text{out}}}\right)_{\text{min}} \qquad \text{[2.7]}$$

Substituting the above into Eq. (2.1), and omitting the subscripts on S and N, results in the following form of the radar equation:

$$R_{\text{max}}^4 = \frac{P_tGA_e\sigma}{(4\pi)^2kT_0BF_n(S/N)_{\text{min}}} \qquad \text{[2.8]}$$

For convenience, R_{max} on the left-hand side is usually written as the fourth power rather than take the fourth root of the right-hand side of the equation.

The minimum detectable signal is replaced in the radar equation by the *minimum detectable signal-to-noise ratio* $(S/N)_{\text{min}}$. The advantage is that $(S/N)_{\text{min}}$ is independent of the receiver bandwidth and noise figure; and, as we shall see in Sec. 2.5, it can be expressed in terms of the probability of detection and the probability of false alarm, two parameters that can be related to the radar user's needs.

The signal-to-noise ratio in the above is that at the output of the IF amplifier, since maximizing the signal-to-noise ratio at the output of the IF is equivalent to maximizing the video output where the threshold decision is made.[7]

Before continuing the development of the radar equation, it is necessary to digress and briefly review the concept of the *probability density function* in order to describe the signal-to-noise ratio in statistical terms. Those familiar with this subject can omit the next section.

2.4 PROBABILITY DENSITY FUNCTIONS

In this section, we introduce the concept of the probability density function and give some examples that are important in the detection of radar signals.

Noise is a random phenomenon; hence, the detection of signals in the presence of noise is also a random phenomenon and should be described in probabilistic terms. *Probability* is a measure of the likelihood of the occurrence of an event. The scale of probability ranges from 0 to 1. (Sometimes probabilities are expressed in percent—from 0 to 100 percent—rather than 0 to 1.) An event that is certain has a probability of 1. An impossible event has a probability 0. The intermediate probabilities are assigned so that the more likely an event, the greater is its probability. Probabilities represent discrete events. Continuous functions, such as random noise, are represented by the *probability density function*, abbreviated *pdf*.

Consider the variable x as representing the value of a random process such as a noise voltage or current. Imagine each x to define a point on a straight vertical line corresponding to the distance from a fixed reference point. The distance x from the reference point might represent the value of the noise voltage or noise current. Divide the line into small segments of length Δx and count the number of times that x falls within each interval. The probability density function is then defined as

$$p(x) = \lim_{\substack{\Delta x \to 0 \\ N \to \infty}} \frac{(\text{number of values within } \Delta x \text{ at } x)/\Delta x}{\text{total number of values} = N} \qquad \textbf{[2.9]}$$

Thus $p(x)$ expresses probability as a density rather than discrete values, and is more appropriate for continuous functions of time as is noise in a radar receiver.

The probability that a particular value of x lies within the infinitesimal interval dx centered at x is simply $p(x)dx$. The probability that the value of x lies within the finite range from x_1 to x_2 is found by integrating $p(x)$ over the range of interest, or

$$\text{probability } (x_1 < x < x_2) = \int_{x_1}^{x_2} p(x)\,dx \qquad \textbf{[2.10]}$$

The probability density function, by definition, is always positive. Since every measurement must yield some value, the integral of the probability density function over all values of x must equal unity; that is,

$$\int_{-\infty}^{\infty} p(x)\,dx = 1 \qquad \textbf{[2.11]}$$

This condition is used to normalize the pdf. The average value of a variable function $\phi(x)$ that is described by the probability density function $p(x)$ is

$$\langle \phi(x) \rangle_{av} = \int_{-\infty}^{\infty} \phi(x)p(x) \, dx \qquad [2.12]$$

This follows from the definitions of an average value and the probability density function. From the above, the mean, or average, value of x is

$$\langle x \rangle_{av} = m_1 = \int_{-\infty}^{\infty} xp(x) \, dx \qquad [2.13]$$

and the mean square value of x is

$$\langle x^2 \rangle_{av} = m_2 = \int_{-\infty}^{\infty} x^2 p(x) \, dx \qquad [2.14]$$

The quantities m_1 and m_2 are called the *first* and *second moments* of the random variable x. If x represents an electric voltage or current, m_1 is the d-c component. It is the value read by a direct-current voltmeter or ammeter. The mean square value m_2 of the current, when multiplied by the resistance, gives the average power. (In detection theory, it is customary to take the resistance as 1 ohm, so that m_2 is often stated to be the average power.) The *variance* σ^2 is the mean square deviation of x about its mean m_1 and can be expressed as

$$\sigma^2 = \langle (x - m_1)^2 \rangle_{av} = \int_{-\infty}^{\infty} (x - m_1)^2 p(x) \, dx = m_2 - m_1^2 \qquad [2.15]$$

It is sometimes called the *second central moment*. If the random variable x is a noise current, the product of the variance and resistance is the mean power of the a-c component. The square root of the variance is the *standard deviation* and is the root mean square (rms) of the a-c component. It is usually designated as σ. We next consider four examples of probability density functions.

Uniform pdf This is shown in Fig. 2.2a and is defined as

$$p(x) = k \qquad \text{for } a < x < a + b$$
$$= 0 \qquad \text{for } x < a \text{ and } x > a + b$$

where k is a constant. It describes the phase of a random sinewave relative to a particular origin of time, where the phase of the sinewave is, with equal probability, anywhere from 0 to 2π radians. The uniform pdf also describes the distribution of the round-off (quantizing) error in numerical computations and in analog-to-digital converters.

The constant k is found to be equal to $1/b$ by requiring that the integral of the probability density function over all values of x equal unity [Eq. (2.11)]. From Eq. (2.13) the average value of the uniform distribution is $a + (b/2)$, which could have been found from inspection in this simple example. The variance from Eq. (2.15) equals $b^2/12$.

Gaussian pdf The gaussian pdf is important in detection theory since it describes many sources of noise, including receiver thermal noise. Also, it is more convenient to

Figure 2.2 Examples of probability density functions. (a) uniform; (b) gaussian; (c) Rayleigh (voltage); (d) Rayleigh (power), or exponential.

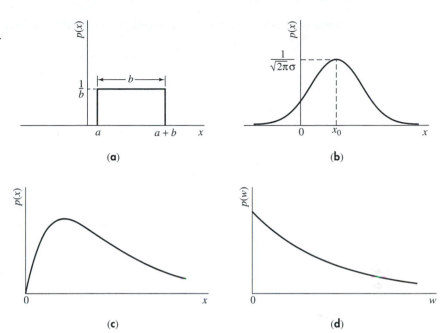

manipulate mathematically than many other pdfs. The gaussian probability density function has a bell-shaped appearance, Fig. 2.2b, and is given by

$$p(x) = \frac{1}{\sqrt{2\pi\sigma^2}} \exp\left[-\frac{(x - x_0)^2}{2\sigma^2} \right] \qquad [2.16]$$

where exp [·] is the exponential function, σ^2 is the variance of x, and x_0 is the mean value of x.

The ability of the gaussian pdf to represent many statistical phenomena is a consequence of the *central limit theorem*. It states that, under very general assumptions, the probability density function of the sum of a large number of independently distributed quantities approaches the gaussian pdf no matter what the individual pdfs might be.

Rayleigh pdf   This is of interest for radar since the envelope of a narrowband filter (such as the IF filter of a radar receiver) is described by the Rayleigh pdf when the input noise voltage is gaussian. The statistical behavior of the radar cross section of some types of targets and some types of clutter also fit this pdf. It is given as

$$p(x) = \frac{2x}{m_2} \exp\left(-\frac{x^2}{m_2} \right) \qquad x \geq 0 \qquad [2.17]$$

where $m_2 = \langle x^2 \rangle_{av}$ is the mean square value of x. The Rayleigh pdf is shown in Fig. 2.2c. It is a one-parameter pdf (the mean square value), and has the property that its standard deviation is equal to $\sqrt{(4/\pi) - 1} = 0.523$ times the mean value.

Exponential pdf   When x^2 in the Rayleigh pdf is replaced by w, the pdf becomes

$$p(w) = \frac{1}{w_0} \exp\left(-\frac{w}{w_0}\right) \qquad w \geq 0 \qquad \text{[2.18]}$$

where w_0 is the mean value of w. This is the *exponential* pdf, but is sometimes called the *Rayleigh-power* pdf, Fig. 2.2d. If the parameter x in the Rayleigh pdf is the voltage, then w represents the power and w_0 is the average power. The standard deviation of the exponential pdf is equal to the mean.

Other pdfs   Later in this chapter the Rice, log normal, and chi-square pdfs will be mentioned as statistical models describing the fluctuations of the target's radar cross section. Section 7.5 provides further discussion of probability density functions as applied to the statistics of clutter. (Clutter is the echo from land, sea, or weather that interferes with the detection of desired targets.)

Statistical phenomena can also be represented by the probability distribution function $P(x)$, which is related to the probability density function $p(x)$ by

$$P(x) = \int_{-\infty}^{x} p(x)\, dx \quad \text{or} \quad p(x) = \frac{d}{dx} P(x) \qquad \text{[2.19]}$$

Example of the Use of Probability Density Functions   The following is a simple example of the use of probability density functions. It involves the calculation of the mean value (d-c component) of the voltage output of a half-wave linear rectifier when the input is gaussian noise voltage of zero mean (thermal noise).[8] The answer itself is of little consequence for our interest in radar detection, but the method used is similar to the more complicated procedures for finding the statistical outputs of a radar receiver mentioned in the next section.

The probability density function (pdf) of the zero-mean gaussian noise voltage x at the input is

$$p(x)\, dx = \frac{1}{\sqrt{2\pi}\sigma} \exp\left(-\frac{x^2}{2\sigma^2}\right) dx \qquad -\infty < x < \infty$$

The output y of a half-wave rectifier for an input x is given as

$$y = ax \qquad x \geq 0,$$

and

$$y = 0 \qquad x < 0$$

where $a = $ constant. To find the mean value of the output, we have to find the pdf at the output. There are three components to the output pdf. The first component is the probability that the rectifier output, $y > 0$, will lie between y and $y + dy$. It is the same as the probability that x lies between x and $x + dx$ when $x > 0$. Thus,

$$p(y)\, dy = p(x)\, dx = \frac{1}{\sqrt{2\pi}a\sigma} e^{-\frac{y^2}{2a^2\sigma^2}} dy \qquad y > 0$$

The second component is the probability that $y = 0$, which is the same as the probability that $x < 0$, which is 1/2 (since half the time the noise voltage is negative and is not passed by the half-wave rectifier). This is represented by $(1/2)\delta(y)$, where $\delta(y)$ is the delta function which has the value 1 when $y = 0$, and is 0 otherwise. The third component is when $y < 0$. There is no output from a rectifier when $y < 0$, thus the probability is 0 that $y < 0$. Combining the three components gives

$$p(y)\,dy = \frac{1}{\sqrt{2\pi}a\sigma}\,e^{-\frac{y^2}{2a^2\sigma^2}}\,dy + \frac{1}{2}\delta(y)\,dy + 0 \qquad y \geq 0$$

The d-c component is $m_1 = \displaystyle\int_{-\infty}^{\infty} yp(y)\,dy$, or

$$m_1 = \frac{1}{\sqrt{2\pi}a\sigma}\int_0^{\infty} ye^{-\frac{y^2}{2a^2\sigma^2}}\,dy + \frac{1}{2}\int_{-\infty}^{\infty} y\delta(y)\,dy$$

The second integral containing the delta function is zero, since $\delta(y)$ has a value only when $y = 0$. The first integral is easily evaluated. The result is the d-c component, which is $a\sigma/\sqrt{2\pi}$.

In this example we started with the pdf describing the input and found the pdf describing the output. In the next section we will follow a similar procedure to find the probabilities of detection and false alarm, but will only provide the answers rather than go through the more elaborate mathematical derivation.

2.5 PROBABILITIES OF DETECTION AND FALSE ALARM

Next it is shown how to find the minimum signal-to-noise ratio required to achieve a specified probability of detection and probability of false alarm. The signal-to-noise ratio is needed in order to calculate the maximum range of a radar using the radar range equation as was given by Eq. (2.8). The basic concepts for the detection of signals in noise may be found in a classical review paper by Rice[9] or one of several texts on detection theory.[10]

Envelope Detector   Figure 2.3 shows a portion of a superheterodyne radar receiver with IF amplifier of bandwidth B_{IF}, second detector,* video amplifier with bandwidth B_v, and a threshold where the detection decision is made. The IF filter, second detector, and video filter form an *envelope detector* in that the output of the video amplifier is the envelope, or modulation, of the IF signal. (An envelope detector requires that the video bandwidth $B_v \geq B_{IF}/2$ and the IF center frequency $f_{IF} >> B_{IF}$. These conditions are usually met in

---

*The diode stage in the envelope detector of a superheterodyne receiver has traditionally been called the *second detector* since the mixer stage, which also employs a diode, was originally called the first detector. The mixer stage is no longer known as the first detector, but the name second detector has been retained in radar practice to distinguish it from other forms of detectors used in radar receivers (such as phase detectors and phase-sensitive detectors).

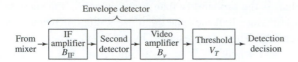

Figure 2.3   Portion of the radar receiver where the echo signal is detected and the detection decision is made.

radar.) The envelope detector passes the modulation and rejects the carrier. The second detector is a nonlinear device (such as a diode). Either a linear or a square-law detector characteristic may be assumed since the effect on the detection probability is relatively insensitive to the choice. (A square-law characteristic is usually easier to handle mathematically, but a linear law is preferred in practice since it allows a larger dynamic range than the square law.) The bandwidth of the radar receiver is the bandwidth of the IF amplifier. The envelope of the IF amplifier output is the signal applied to the threshold detector. When the receiver output crosses the threshold, a signal is declared to be present.

Probability of False Alarm   The receiver noise at the input to the IF filter (the terms *filter* and *amplifier* are used interchangeably here) is described by the gaussian probability density function of Eq. (2.16) with mean value of zero, or

$$p(v) = \frac{1}{\sqrt{2\pi\Psi_0}} \exp\left(-\frac{v^2}{2\Psi_0}\right)$$

[2.20]

where $p(v)\, dv$ is the probability of finding the noise voltage v between the values of v and $v + dv$ and Ψ_0 is the mean square value of the noise voltage (mean noise power). S. O. Rice has shown in a *Bell System Technical Journal* paper[9] that when gaussian noise is passed through the IF filter, the probability density function of the envelope R is given by a form of the Rayleigh pdf:

$$p(R) = \frac{R}{\Psi_0} \exp\left(-\frac{R^2}{2\Psi_0}\right)$$

[2.21]

The probability that the envelope of the noise voltage will exceed the voltage threshold V_T is the integral of $p(R)$ evaluated from V_T to ∞, or

$$\text{Probability } (V_T < R < \infty) = \int_{V_T}^{\infty} \frac{R}{\Psi_0} \exp\left(-\frac{R^2}{2\Psi_0}\right) dR = \exp\left(\frac{-V_T^2}{2\Psi_0}\right)$$

[2.22]

This is the *probability of a false alarm* since it represents the probability that noise will cross the threshold and be called a target when only noise is present. Thus, the probability of a false alarm, denoted P_{fa}, is

$$P_{\text{fa}} = \exp\left(-\frac{V_T^2}{2\Psi_0}\right)$$

[2.23]

By itself, the probability of false alarm as given by Eq. (2.23) does not indicate whether or not a radar will be troubled by excessive false indications of targets. The time between false alarms is a better measure of the effect of noise on radar performance.

Figure 2.4 illustrates the occurrence of false alarms. The average time between crossings of the decision threshold when noise alone is present is called the *false-alarm time*, T_{fa}, and is given by

$$T_{\text{fa}} = \lim_{N \to \infty} \frac{1}{N} \sum_{k=1}^{N} T_k \qquad \text{[2.24]}$$

where T_k is the time between crossings of the threshold V_T by the noise envelope. The false-alarm time is something a radar customer or operator can better relate to than the probability of false alarm. The false-alarm probability can be expressed in terms of false-alarm time by noting that the false-alarm probability P_{fa} is the ratio of the time the envelope is actually above the threshold to the total time it could have been above the threshold, or

$$P_{\text{fa}} = \frac{\sum_{k=1}^{N} t_k}{\sum_{k=1}^{N} T_k} = \frac{\langle t_k \rangle_{\text{av}}}{\langle T_k \rangle_{\text{av}}} = \frac{1}{T_{\text{fa}} B} \qquad \text{[2.25]}$$

where t_k and T_k are shown in Fig. 2.4, and B is the bandwidth of the IF amplifier of the radar receiver. The average duration of a threshold crossing by noise $\langle t_k \rangle_{\text{av}}$ is approximately the reciprocal of the IF bandwidth B. The average of T_k is the *false-alarm time*, T_{fa}. Equating Eqs. (2.23) and (2.25) yields

$$T_{\text{fa}} = \frac{1}{B} \exp\left(\frac{V_T^2}{2\psi_0}\right) \qquad \text{[2.26]}$$

A plot of T_{fa} as a function of $V_T^2/2\Psi_0$ is shown in Fig. 2.5. If, for example, the bandwidth of the IF amplifier were 1 MHz and the average time between false alarms were specified to be 15 min, the probability of a false alarm is 1.11×10^{-9}. The threshold voltage,

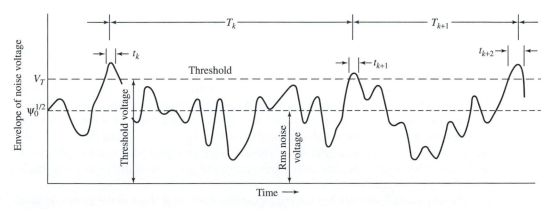

Figure 2.4   Envelope of the receiver output with noise alone, illustrating the duration of false alarms and the time between false alarms.

Figure 2.5 Average time between false alarms as a function of the threshold level V_T and the receiver bandwidth B; Ψ_0 is the mean square noise voltage.

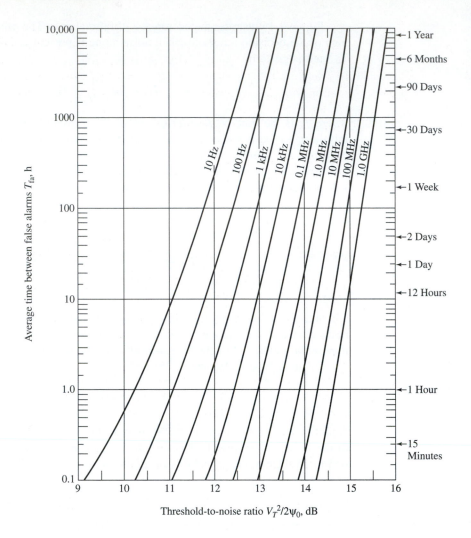

from Eq. (2.26), is 6.42 times the rms value of the noise voltage, or the power ratio V_T^2/Ψ_0 is 16.2 dB.

The false-alarm probabilities of radars are generally quite small since a decision as to whether a target is present or not is made every $1/B$ second. The bandwidth B is usually large, so there are many opportunities during one second for a false alarm to occur. For example, when the bandwidth is 1 MHz (as with a 1-μs pulse width) there are 1 million decisions made every second as to whether noise or signal plus noise is present. If there were to be, on average, one false alarm per second, the false-alarm probability would be 10^{-6} in this specific example.

The exponential relationship between the false-alarm time T_{fa} and the threshold level V_T [Eq. (2.26)] results in the false-alarm time being sensitive to small variations in the threshold. For example, if the bandwidth were 1 MHz, a value of $10 \log(V_T^2/2\Psi_0) = 13.2$

dB results in a false-alarm time of about 20 min. A 0.5-dB decrease in the threshold to 12.7 dB decreases the false-alarm time by an order of magnitude, to about 2 min.

If the threshold is set slightly higher than required and maintained stable, there is little likelihood of false alarms due to thermal noise. In practice, false alarms are more likely to occur from clutter echoes (ground, sea, weather, birds, and insects) that enter the radar and are large enough to cross the threshold. In the specification of the radar's false-alarm time, however, clutter is almost never included, only receiver noise.

Although the crossing of the threshold by noise is called a false alarm, it is not necessarily a *false-target report*. Declaration of a target generally requires more than one detection made on multiple observations by the radar (Sec. 2.13). In many cases, establishing the track of a target is required before a target is declared as being present. Such criteria can allow a higher probability of false alarm for each detection; hence, the threshold can be lowered to improve detection without obtaining excessive false-target reports. In the present chapter, however, most of the discussion relating to the radar equation concerns a detection decision based on a single crossing of the threshold.

If the receiver were turned off (or gated) for a small fraction of the time, as it normally would be during the transmission of the radiated pulse, the false-alarm probability will be *increased* by the fraction of time the receiver is not operative. This assumes the false-alarm time remains constant. The effect of gating the receiver off for a short time seldom needs to be taken into account since the resulting change in the probability of false alarm and the change in the threshold level are small.

Probability of Detection   So far, we have discussed only the noise input at the radar receiver. Next, consider an echo signal represented as a sinewave of amplitude A along with gaussian noise at the input of the envelope detector. The probability density function of the envelope R at the video output is given by[9]

$$p_s(R) = \frac{R}{\Psi_0} \exp\left(-\frac{R^2 + A^2}{2\Psi_0}\right) I_0\left(\frac{RA}{\Psi_0}\right) \qquad \text{[2.27]}$$

where $I_0(Z)$ is the modified Bessel function of zero order and argument Z. For large Z, an asymptotic expansion for $I_0(Z)$ is

$$I_0(Z) = \frac{e^Z}{\sqrt{2\pi Z}}\left(1 + \frac{1}{8Z} + \cdots\right) \qquad \text{[2.28]}$$

When the signal is absent, $A = 0$ and Eq. (2.27) reduces to Eq. (2.21), the pdf for noise alone. Equation (2.27) is called the *Rice probability density function.*

The probability of detecting the signal is the probability that the envelope R will exceed the threshold V_T (set by the need to achieve some specified false-alarm time). Thus the probability of detection is

$$P_d = \int_{V_T}^{\infty} p_s(R)\, dR \qquad \text{[2.29]}$$

When the probability density function $p_s(R)$ of Eq. (2.27) is substituted in the above, the probability of detection P_d cannot be evaluated by simple means. Rice[9] used a series approximation to solve for P_d. Numerical and empirical methods have also been used.

The expression for P_d, Eq. (2.29), along with Eq. (2.27), is a function of the signal amplitude A, threshold V_T, and mean noise power Ψ_0. In radar systems analysis it is more convenient to use signal-to-noise power ratio S/N than $A^2/2\Psi_0$. These are related by

$$\frac{A}{\Psi_0^{1/2}} = \frac{\text{signal amplitude}}{\text{rms noise voltage}} = \frac{\sqrt{2} \ (\text{rms signal voltage})}{\text{rms noise voltage}}$$

$$= \left(2\frac{\text{signal power}}{\text{noise power}}\right)^{1/2} = \left(\frac{2S}{N}\right)^{1/2}$$

The probability of detection P_d can then be expressed in terms of S/N and the ratio of the threshold-to-noise ratio $V_T^2/2\Psi_0$. The probability of false alarm, Eq. (2.23) is also a function of $V_T^2/2\Psi_0$. The two expressions for P_d and P_{fa} can be combined, by eliminating the threshold-to-noise ratio that is common to each, so as to provide a single expression relating the probability of detection P_d, probability of false alarm P_{fa}, and the signal-to-noise ratio S/N. The result is plotted in Fig. 2.6.

Albersheim[11,12] developed a simple empirical formula for the relationship between S/N, P_d, and P_{fa}, which is

$$S/N = A + 0.12AB + 1.7B \tag{2.30}$$

where

$$A = ln \ [0.62/P_{fa}] \quad \text{and} \quad B = ln \ [P_d/(1 - P_d)]$$

Figure 2.6 Probability of detection for a sinewave in noise as a function of the signal-to-noise (power) ratio and the probability of false alarm.

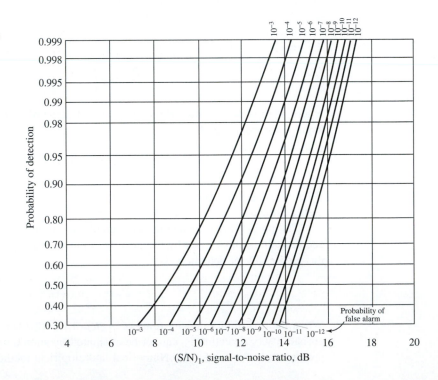

The signal-to-noise ratio in the above is a numeric, and not in dB; and ln is the natural logarithm. Equation (2.30) is said to be accurate to within 0.2 dB for P_{fa} between 10^{-3} and 10^{-7}, and P_d between 0.1 and 0.9. (It is probably suitable for rough calculations for even greater values of P_d and lower values of P_{fa}.) From such an expression or from a graph such as Fig. 2.6, the minimum signal-to-noise ratio required for a particular probability of detection and a specified probability of false alarm can be found and entered into the radar range equation. The above applies for a single pulse. The case for multiple pulses is given later.

The detection probability and the false-alarm probability are specified by the system requirements as derived from the customer's needs. From the specified detection and false-alarm probabilities, the minimum signal-to-noise ratio is found. For example, suppose that the average time between false alarms is required to be 15 min. If the bandwidth were 1 MHz, Eq. (2.25) gives a false-alarm probability of 1.11×10^{-9}. Figure 2.6 indicates that a signal-to-noise ratio of 13.05 dB is required for a probability of detection of 0.50, 14.7 dB for $P_d = 0.90$, and 15.75 dB for $P_d = 0.99$. Thus, a change of less than 3 dB can mean the difference between highly reliable detection (0.99) and marginal detection (0.50).

At first glance one might be inclined to think that the signal-to-noise ratios required for reliable detection in the above example are relatively high. They are of this magnitude because in the above example it is required that, on average, there be no more than one false alarm every 900 million possible detection decisions. Compared to telecommunications services, however, the signal-to-noise ratios for radar are relatively low. For TV the signal-to-noise ratio is said[13] to be 40 dB; there is some snow with 35 dB; objectionable interference at 30 dB; and the picture is all snow with 25 dB. Telephone service is said to require a signal-to-noise ratio of 50 dB. By comparison, radar detection is highly efficient.

The material in this section assumed only a single pulse was being used for detection. Most radars, however, utilize more than one pulse to make the detection decision, as will be discussed next.

2.6   INTEGRATION OF RADAR PULSES

The number of pulses returned from a point target by a scanning radar with a pulse repetition rate of f_p Hz, an antenna beamwidth θ_B degrees, and which scans at a rate of $\dot{\theta}_s$ degrees per second is

$$n = \frac{\theta_B f_p}{\dot{\theta}_s} = \frac{\theta_B f_p}{6\omega_r} \qquad [2.31]$$

where ω_r = revolutions per minute (rpm) if a 360° rotating antenna. The number of pulses received n is usually called *hits per scan* or *pulses per scan*. It is the number of pulses within the one-way beamwidth θ_B. Example values for a long-range ground-based air-surveillance radar might be 340-Hz pulse repetition rate, 1.5° beamwidth, and an antenna rotation rate of 5 rpm (30°/s). These numbers, when substituted into Eq. (2.31), yield $n = 17$ pulses per scan. (If n is not a whole number it can be either rounded off or the number can be used as is. It will make little difference in the calculation of radar range whichever you choose to do, unless n is small.)

The process of summing all the radar echoes available from a target is called *integration* (even though an *addition* is actually performed). Many techniques have been considered in the past to provide integration of pulses. A common integration method in early radars was to take advantage of the persistence of the phosphor of the cathode-ray-tube display combined with the integrating properties of the eye and brain of the radar operator. Analog storage devices, such as narrowband filters, can act as integrators; but they have been replaced with digital methods.

Integration that is performed in the radar receiver before the second detector (in the IF) is called *predetection integration* or *coherent integration.* Predetection integration is theoretically lossless, but it requires the phase of the echo signal pulses to be known and preserved in order to combine the sinewave pulses in phase without loss. Integration after the second detector is known as *postdetection integration* or *noncoherent integration.* It is easier to accomplish than predetection integration since the phases of the echoes are not preserved and only the envelopes of the pulses need be aligned to perform addition. There is a theoretical integration loss, however, with the use of postdetection integration.

If *n* pulses, all of the same signal-to-noise ratio, were perfectly integrated by an ideal lossless predetection integrator, the integrated signal-to-noise (power) ratio would be exactly *n* times that of a single pulse. Therefore, in this case, we can replace the single-pulse signal-to-noise ratio $(S/N)_1$ in the radar equation with $(S/N)_n = (S/N)_1/n$, where $(S/N)_n$ is the required signal-to-noise ratio per pulse when there are *n* pulses integrated predetection without loss. If the same *n* pulses were integrated by an ideal postdetection device, the resultant signal-to-noise ratio would be less than *n* times that of a single pulse. This loss in integration efficiency is caused by the nonlinear action of the second detector, which converts some of the signal energy to noise energy in the rectification processes. An integration efficiency for postdetection integration may be defined as

$$E_i(n) = \frac{(S/N)_1}{n(S/N)_n} \qquad \text{[2.32]}$$

where the symbols have been defined in the above. The improvement in signal-to-noise ratio when *n* pulses are integrated is called the *integration improvement factor* $I_i(n) = nE_i(n)$. It can also be thought of as the *equivalent number of pulses integrated* $n_{eq} = nE_i(n)$. For postdetection integration n_{eq} is less than *n*; for ideal predetection integration $n_{eq} = n$. Thus for the same integrated signal-to-noise ratio, postdetection integration requires more pulses than predetection, assuming the signal-to-noise ratio per pulse in the two cases is the same.

The postdetection integration efficiency and the required signal-to-noise ratio per pulse $(S/N)_n$ may be found by use of statistical detection theory, similar to that outlined in the previous section. This was originally undertaken for radar application in the classic work of J. I. Marcum.[14] (His work originally appeared in 1954 as a highly regarded, widely distributed, but not generally available, Rand Corporation report.) Marcum defined an integration loss in dB as $L_i(n) = 10 \log [1/E_i(n)]$. The integration loss and the integration improvement factor are plotted in Fig. 2.7. They vary only slightly with probability of detection or probability of false alarm.

Marcum used the *false-alarm number* n_f in his calculations rather the probability of false alarm. His false-alarm number is the reciprocal of our false-alarm probability

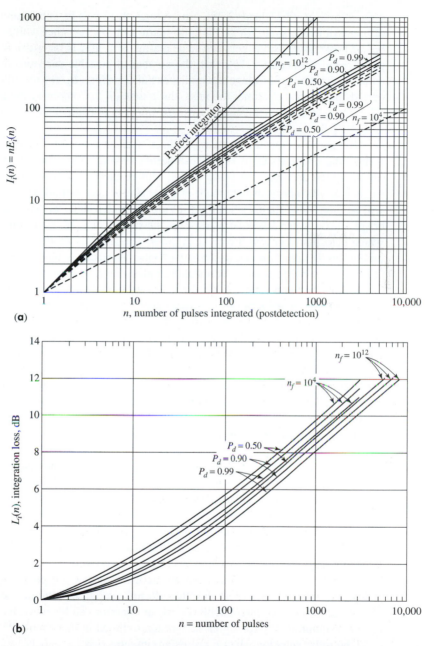

Figure 2.7   (a) Integration-improvement factor (or equivalent number of pulses integrated) for a square-law detector, where P_d = probability of detection, n_f = false-alarm number = $T_{fa}B$, T_{fa} = average time between false alarms, B = receiver bandwidth; (b) integration loss as a function of n, the number of pulses integrated, P_d, and n_f.

| (After Marcum,[14] courtesy *IRE Trans.*)

defined by Eqs. (2.23) and (2.25). On average, there will be one false-alarm decision out of n_f possible decisions within the false-alarm time T_{fa}. In other words, the average number of possible decisions between false alarms is n_f. If τ is the pulse width and T_p is the pulse repetition period $= 1/f_p$, then the number of possible decisions n_f in the time T_{fa} is equal to the number of range intervals per pulse period (T_p/τ) times the number of pulse periods per second (f_p) times the false-alarm time (T_{fa}). Combining the above, we get $n_f = (T_p/\tau) \times f_p \times T_{\text{fa}} = T_{\text{fa}}/\tau$. Since $\tau \approx 1/B$, where $B =$ bandwidth, the false-alarm number is $n_f = T_{\text{fa}}B = 1/P_{\text{fa}}$.

The above assumed that the radar made decisions at a rate equal to the bandwidth B. If the radar integrates n pulses per scan before making a target detection decision, then the rate at which decisions are made is B/n. This results in a false-alarm probability n times as great as when decisions are made at a rate B times per second. This does not mean there will be more false alarms when n pulses are integrated since we have assumed that the average time between false alarms remains the same when pulses are integrated. The rate at which detection decisions are made is lower. The above is another reason why false-alarm probability is not as good a descriptor of false alarms as is the average false-alarm time. A probability by itself has little meaning unless the rate at which events occur is known.

Following the practice of Marcum, P_{fa} will be taken in this text as the reciprocal of $T_{\text{fa}}B = n_f$, even when n pulses are integrated. Some authors, on the other hand, prefer to define a false-alarm number $n'_f = n_f/n$ that accounts for the number of pulses integrated. Therefore, caution should be exercised when using different authors' computations or different computer programs for finding the signal-to-noise ratio as a function of probability of detection and probability of false alarm (or false-alarm number). There is no standardization of definitions. Correct values of signal-to-noise ratio for use in the radar equation can be obtained from most sources provided the particular assumptions used by the sources are understood.

The solid straight line in Fig. 2.7a represents a perfect lossless predetection integrator. When only a few pulses are integrated (implying large signal-to-noise ratio per pulse), Fig. 2.7a shows that the performance of the postdetection integrator is not much different from the predetection integrator. When a large number of pulses are integrated (small signal-to-noise ratio per pulse), the difference between predetection and postdetection integration is more pronounced.

The dash straight line in this figure is proportional to $n^{1/2}$. In the early days of radar it was thought that the integration improvement factor of a radar operator viewing a cathode-ray-tube display, such as a PPI, was equal to $n^{1/2}$. This is not necessarily correct. The $n^{1/2}$ relation was based on an incorrect theory and poor displays. When individual pulses are displayed so that they do not overlap or saturate the phosphor screen, the integration improvement achieved by an operator can be equivalent to that predicted by the Marcum theory for signal integration outlined in this chapter.[15]

The radar equation when n pulses are integrated is

$$R_{\max}^4 = \frac{P_t G A_e \sigma}{(4\pi)^2 k T_0 B F_n (S/N)_n} \qquad \text{[2.33]}$$

where the parameters are the same as previously used, except that $(S/N)_n$ is the signal-to-noise ratio of each of the n equal pulses that are integrated. Also, the half-power bandwidth B is used instead of the noise bandwidth B_n, as mentioned in Sec. 2.3. To employ this form of the equation it is necessary to have, for each value of n, a set of curves for $(S/N)_n$ similar to those of Fig. 2.6 for $n = 1$. Such curves are available[16] but are not necessary since only Figs. 2.6 and 2.7 are needed. Substituting Eq. (2.32) for $(S/N)_n$ into (2.33) gives

$$R_{max}^4 = \frac{P_t G A_e \sigma n E_i(n)}{(4\pi)^2 k T_0 B F_n (S/N)_1}$$

[2.34]

The value of $(S/N)_1$ is found from Fig. 2.6, and the integration improvement factor $nE_i(n)$ is found from Fig. 2.7a.

An approximation for the signal-to-noise ratio per pulse is given by an empirical formula due to Albersheim,[11,12] which is an extension of Eq. (2.30).

$$(S/N)_n = -5 \log_{10} n + \left(6.2 + \frac{4.54}{\sqrt{n + 0.44}}\right) \times \log_{10}(A + 0.12AB + 1.7B)$$ [2.35]

where the signal-to-noise ratio per pulse $(S/N)_n$ is in dB, n is the number of independent (pulse) samples integrated, and A and B are the same as defined for Eq. (2.30). This equation is said to have an error of less than 0.2 dB over the range of $n = 1$ to $n = 8096$, $P_d = 0.1$ to 0.9, and $P_{fa} = 10^{-3}$ to 10^{-7}. (As noted with Eq. (2.30), Eq. (2.35) is probably a good approximation for rough calculations when P_d is even greater and P_{fa} is even lower than the above.)

The discussion of integration loss or efficiency in this section is theoretical loss. In addition, there can be loss due to the actual method used for implementing the integration process in a radar.

2.7   RADAR CROSS SECTION OF TARGETS

The radar cross section σ is the property of a scattering object, or target, that is included in the radar equation to represent the magnitude of the echo signal returned to the radar by the target. In the derivation of the simple form of the radar equation in Sec. 1.2 the radar cross section was defined in terms of Eq. (1.5), which was

$$\text{Reradiated power density back at the radar} = \frac{P_t G}{4\pi R^2} \cdot \frac{\sigma}{4\pi R^2}$$

[1.5]

A definition of the radar cross section found in some texts on electromagnetic scattering is

$$\sigma = \frac{\text{power reflected toward source/unit solid angle}}{\text{incident power density}/4\pi} = 4\pi R^2 \frac{|E_r|^2}{|E_i|^2}$$

[2.36]

where R is the range to the target, E_r is the electric field strength of the echo signal back at the radar, and E_i is the electric field strength incident on the target. It is assumed in the above that the target is far enough from the radar that the incident wave can be considered to be planar rather than spherical. Equation (2.36) is equivalent to the simple form of the radar equation derived in Sec. 1.2. Sometimes the radar cross section σ is said to be a (fictional) area that intercepts a part of the power incident at the target which, if scattered uniformly in all directions, produces an echo power at the radar equal to that produced at the radar by the real target. Real targets, of course, do not scatter the incident energy uniformly in all directions.

The power scattered from a target in the direction of the radar receiver, and hence the radar cross section, can be calculated by solving Maxwell's equations with the proper boundary conditions applied or by computer modeling. The radar cross section can also be measured, based on the radar equation, using either full-size or scale models of targets.

Radar cross section depends on the characteristic dimensions of the object compared to the radar wavelength. When the wavelength is large compared to the object's dimensions, scattering is said to be in the *Rayleigh region*. It is named after Lord Rayleigh who first observed this type of scattering in 1871, long before the existence of radar, when investigating the scattering of light by microscopic particles. The radar cross section in the Rayleigh region is proportional to the fourth power of the frequency, and is determined more by the volume of the scatterer than by its shape. At radar frequencies, the echo from rain is usually described by Rayleigh scattering.

At the other extreme, where the wavelength is small compared to the object's dimensions, is the *optical region*. Here radar scattering from a complex object such as an aircraft is characterized by significant changes in the cross section when there is a change in frequency or aspect angle at which the object is viewed. Scattering from aircraft or ships at microwave frequencies generally is in the optical region. In the optical region, the radar cross section is affected more by the shape of the object than by its projected area.

In between the Rayleigh and the optical regions is the *resonance region* where the radar wavelength is comparable to the object's dimensions. For many objects, the radar cross section is larger in the resonance region than in the other two regions. These three distinct scattering regions are illustrated by scattering from the sphere described next.

Simple Targets   The sphere, cylinder, flat plate, rod, ogive, and cone are examples of simple targets. Analytical expressions exist for the radar cross sections of some of these objects. Sometimes the radar cross sections of complex targets can be calculated by describing the target as a collection of simple shapes whose cross sections are known. The total cross section is obtained by summing vectorially the contributions from the individual simple shapes. A few examples will be presented to illustrate the nature of radar cross section behavior.

Sphere   The sphere is the simplest object for illustrating radar scattering since it has the same shape no matter from what aspect it is viewed. Its calculated radar cross section is shown in Fig. 2.8 as a function of $2\pi a/\lambda$, the circumference measured in wavelengths, where a is the radius of the sphere and λ is the radar wavelength. The cross section in this figure is normalized by the projected physical area of the sphere, πa^2. The three different scattering regions that characterize the sphere are labeled in the figure. In the Rayleigh region where $2\pi a/\lambda \ll 1$, the radar cross section is proportional to f^4, where f = frequency = c/λ, and c = velocity of propagation.

Figure 2.8 Normalized radar cross section of a sphere as a function of its circumference $(2\pi a)$ measured in wavelengths. a = radius; λ = wavelength.

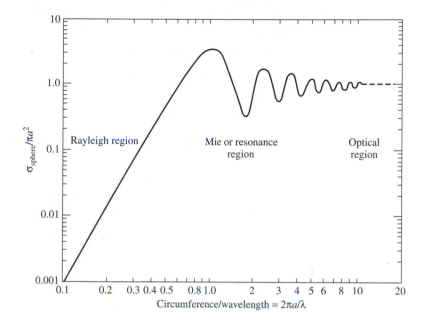

In the optical region, $2\pi a/\lambda \gg 1$, the radar cross section approaches the physical area of the sphere as the frequency is increased. This unique circumstance can mislead one into thinking that the geometrical area of a target is a measure of its radar cross section. It applies to the sphere, but not to other targets. In the optical region, scattering does not take place over the entire hemisphere that faces the radar, but only from a small bright spot at the tip of the smooth sphere. It is more like what would be seen if a polished metallic sphere, such as a large ball bearing, were photographed with a camera equipped with a flash. The only illumination is at the tip, rather than from the entire hemispherical surface. A diffuse sphere or rough-surface sphere, such as a white billiard ball, would reflect from its entire surface, as does the full moon when viewed visually.

The radar cross section of the sphere in the *resonance region* oscillates as a function of frequency, or $2\pi a/\lambda$. Its maximum occurs at $2\pi a/\lambda = 1$, and is 5.6 dB greater than its value in the optical region. The first null is 5.5 dB below the optical region value. Changes in cross section occur with changing frequency because there are two waves that interfere constructively and destructively. One is the direct reflection from the front face of the sphere. The other is the *creeping wave* that travels around the back of the sphere and returns to the radar where it interferes with the reflection from the front of the sphere. The longer the electrical path around the sphere, the greater the loss, so the smaller will be the magnitude of the fluctuation with increasing frequency.

Figure 2.9 illustrates the backscatter that would be produced by a very short pulse radar that can resolve the specular echo reflected from the forward part of the sphere from the creeping wave which travels around the back of the sphere. The incident waveform in this figure is a shaped pulse of sinewave of the form $0.5[1 + \cos(\pi t/t_0)]$, where the pulse extends from $-t_0$ to $+t_0$.[17] The radius of the sphere in this example is equal to the

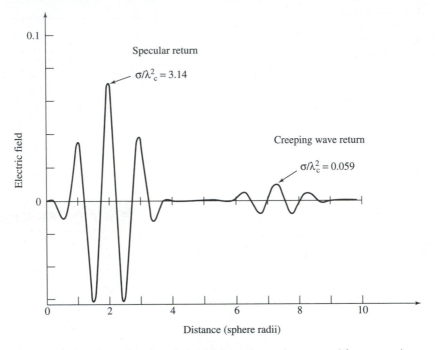

Figure 2.9 Backscattered electric field from a short pulse scattered from a conducting sphere showing the specular return from the front of the sphere and the creeping wave that travels around the back. Radius of sphere is equal to the radar wavelength λ_c.
I (From J. Reinstein,[17] © 1965 IEEE.)

carrier wavelength λ_c, and the pulse width $2t_0 = 4\lambda_c$. The creeping wave lags the specular return by the time required to travel one-half the circumference of the sphere plus the diameter.

The behavior of other simple reflecting objects as a function of frequency is similar to that of the sphere, but with some differences.[18–20] It is only the sphere, however, that has a radar cross section independent of viewing aspect and polarization. (With apologies to the avant-garde writer Gertrude Stein, a sphere is a sphere is a sphere—no matter how you look at it.)

Long Thin Wire or Rod (and the Surface Traveling Wave) The experimentally measured radar cross section of a long, thin wire is shown in Fig. 2.10.[21] The wire in this case is 16.5λ long and 0.01λ in diameter. When viewed broadside ($\theta = 90°$) the radar cross section is relatively large. As the viewing angle θ departs from $90°$, the cross section decreases rapidly, as expected from classical physical optics scattering theory. However, as the viewing angle decreases, an angle is reached where the backscatter levels off and then increases. This is due to a surface traveling wave, which is not predicted from physical optics theory.

This behavior was first demonstrated experimentally with a long thin rod by Leon Peters.[22] The incident electromagnetic wave couples onto the wire which then travels the length of the rod and reflects from the discontinuity at the far end. In addition to the wire

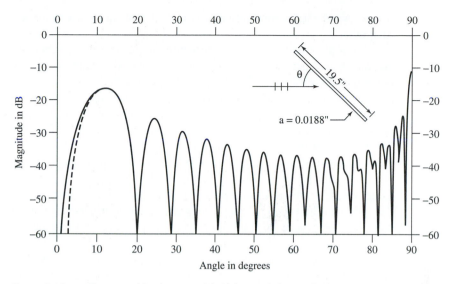

Figure 2.10 Theoretical backscattered field from a long, perfectly conducting thin wire 16.5 wavelengths long and 0.01 wavelength in diameter. Perpendicular incidence is at $\theta = 90°$. Solid curve is an approximate theory from Shamansky et al.; dashed curve is a method of moments solution. Wire dimensions shown in the figure are in inches and the frequency was 10 GHz. (From Shamansky et al.,[21] © 1989 IEEE.)

or rod, a surface traveling wave can occur with objects such as the flat plate, ogive, and the wing of an aircraft. The traveling wave is launched when the incident electric field has components perpendicular and parallel to the surface and lies in the plane of incidence defined by the surface normal and the direction of incidence. A surface traveling wave is not excited if there is no electric field component in the plane of incidence. (In the geometry of Fig. 7.2, vertical polarization, when the electric field is vertical, can excite a traveling wave, but horizontal polarization cannot.) The effect of the traveling wave is prominent when the grazing angle is small and when there is a discontinuity at the far end of the body that reflects the traveling wave back to the radar. According to Knott et al.,[23] the traveling-wave mechanism even has an effect near broadside incidence, as can be seen in Fig. 2.10. (The creeping wave, mentioned previously, is also an example of a surface traveling wave.) The traveling-wave portion of the echo is reduced if the surface is made of resistive material which causes attenuation as the wave travels down the surface and back.

Flat Plate and Corner Reflector At normal incidence (broadside) the radar cross section of a flat plate of area A in the optics region is $4\pi A^2/\lambda^2$, which can give quite large values for a modest size plate. For example, at 3-cm wavelength (*X* band), a square plate 0.3 m on a side (about one foot) has a radar cross section of 113 m^2 when viewed normal to the surface. The magnitude of the backscatter from a flat plate drops off rapidly with angle from the normal. The radar cross section of a dihedral or a trihedral corner reflector of projected area A is also given by the same expression as the flat plate, but it applies over a much larger viewing angle. (A dihedral corner reflector is a structure formed by two flat faces perpendicular to one another which returns the incident wave back to its source. A trihedral corner reflector is made up of three faces mutually perpendicular to each other.)

A flat plate may be a simple structure, but its scattering behavior is not simple. Figure 2.11 is the measured cross section of a square flat plate 5λ on a side for vertical and horizontal polarization.[24,25] (The polarization of a radar wave is defined by the orientation of the electric field. Vertical polarization, for example, is when the electric field is in the vertical direction.) Note that there is a traveling-wave backscatter component. Also shown in this figure are theoretical predictions based on two different scattering theories. The theoretical prediction of scattering from this simple object does not account for the experimental observations at angles far from normal incidence.

Cone-Sphere   This is a cone whose base is capped with a sphere. The first derivatives of the cone and the sphere contours are equal at the join between the two (that it, the generatrix of the cone is tangent to the sphere at the cone-sphere junction). Figure 2.12 is a plot of the calculated nose-on radar cross section of a cone-sphere with 30° cone angle as a function of $2\pi a/\lambda$, where a is the radius of the sphere.[26] An example of the radar cross section as a function of aspect angle for a cone-sphere with 25° cone angle is shown in Fig. 2.13.[27]

The cross section of the cone-sphere can be very low from nose-on to near normal incidence on the side of the cone. It is a shape sometimes considered for low cross-

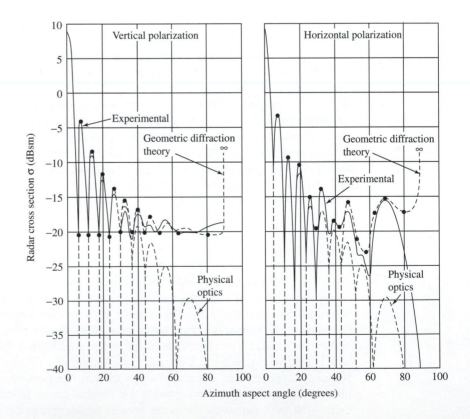

Figure 2.11   Measured radar cross section of a square flat plate, 5 wavelengths on a side compared with predictions based on physical optics and geometric diffraction theory, for vertical and horizontal polarizations. Perpendicular incidence is at 0°. The ordinate is in dBsm, which is dB relative to one square meter.
| (From Ross,[24] © 1966 IEEE.)

Figure 2.12 Theoretical normalized nose-on radar cross section of a cone-sphere based on an approximate impulse analysis; 15° half cone-angle (30° included cone-angle), a = radius of the sphere, and λ = wavelength. The dashed curve represents the approximation
$\sigma \approx 0.1 \; \lambda^2$.
| (After David Moffatt,[26] Ohio State University.)

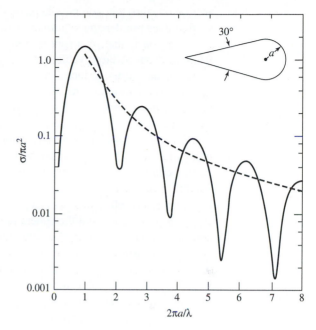

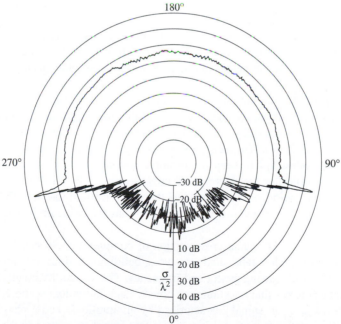

Figure 2.13 Measured radar cross section (σ/λ^2 given in dB) of a large cone-sphere with 25° cone angle and radius of spherical base = 10.4λ, for perpendicular polarization (electric field is perpendicular to the plane containing the direction of propagation and the cone axis; sometimes called horizontal polarization). For large cones the differences between perpendicular and parallel polarization are small.
| (From Pannell et al.[27] MIT Lincoln Laboratory.)

section (stealth) ballistic missile reentry vehicles. From the nose-on aspect, the radar cross section does not depend significantly on the cone angle or the volume. From the rear, the cross section is that of a sphere; hence, it is much larger than the cross section viewed from the front. A large value of cross section also occurs when the radar views the cone perpendicular to its surface. These features can be seen in Fig. 2.13. An approximate expression for estimating the radar cross section from the nose-on direction, when convenience is more important than accuracy, is $0.1\lambda^2$. This is shown by the dashed curve of Fig. 2.12.

The cone-sphere is a good example to illustrate some of the fundamental characteristics of radar scattering. Radar backscattering from metallic objects is due to discontinuities comparable to the radar wavelength and from discontinuities in the surface curvature. In the cone-sphere, echoes come from the tip and the join between the cone and the sphere, as well as from a creeping wave that travels around the sphere. The first derivatives of the cone and the sphere at the join are usually equal, but the curvature is not continuous. There is no backscatter from the sides of the cone when viewed nose-on, assuming the surface of the cone is smooth compared to the wavelength. In practice, the tip will not be of zero radius, but will be rounded. With very low cross-section targets, this rounding, even if small, can contribute to the total cross-section of the cone-sphere so that the cross section no longer varies as λ^2 for small wavelengths. With sufficiently small wavelength where the surface roughness is a significant fraction of the wavelength, scattering will be affected by the roughness and the ability to mechanically maintain the shape of the object with high precision (relative to a wavelength). The result of the rounded tip, the surface roughness, and the imprecision of the target surface contour is that the cone-sphere cross section will no longer decrease with increasing frequency, but will level off and even increase at the higher frequencies where these other factors dominate.

Effect of Target Shape   In the optical region where the wavelength is small compared to the object's dimensions, the shape of a scattering object has far greater effect on its radar cross section than does its physical size. For example, at a frequency of 3 GHz (S band) a corner reflector (or a flat plate) of physical area 1 m^2 has a radar cross section of about 1000 m^2 (rounding off to keep the example simple). On the other hand, a cone-sphere with a 1 m^2 projected area has a radar cross section at 3 GHz equal to 0.001 m^2 (based on the approximation $\sigma \approx 0.1\lambda^2$). Thus there is a million-to-one difference in the radar cross section of two objects, even though each has the same projected area.

Complex Targets   The radar cross section of complex targets such as aircraft, missiles, ships, ground vehicles, fabricated structures, buildings, and terrain can vary considerably depending on the viewing aspect and frequency. The variability results from the multiple individual scatterers that constitute the object. Each individual scatterer of a complex target produces an echo signal characterized by an amplitude and a phase. These echo signals combine at the radar to produce a resultant signal. A change in the relative phases of the echo signals from the individual scatterers will occur if the relative positions of the scatterers change with viewing aspect or there is a change in radar frequency. An example of the variation of radar cross section as a function of aspect angle from a "simple" complex target is shown in Fig. 2.14. The target consists of two equal scattering objects, such as small spheres, separated in (a) by one wavelength and in (b) by four wavelengths. If the cross section of each of the two equal scatterers is σ_0, the resultant cross section σ_s of the two scatterers considered as one target is

$$\frac{\sigma_r}{\sigma_0} = 2\left[1 + \cos\left(\frac{4\pi l}{\lambda}\sin\theta\right)\right] \qquad [2.37]$$

where l is the separation and θ is the viewing angle with respect to the normal of the line joining the two scatterers. The value of σ_r in this example varies from a minimum of zero to a maximum of four times the cross section σ_0 of an individual scatterer. As the separation in wavelengths between the two scatterers becomes larger, the scattering lobes become narrower and more numerous.

Although this is a rather simple example of a complex target, its behavior is indicative of more complicated targets. Practical targets are composed of many individual scatterers, each with different scattering properties. Also, there may be interactions between the individual scatterers, such as multiple scattering in a corner reflector, which further complicate cross-section characteristics.

Aircraft An example of the variation of the backscatter from a propeller-driven aircraft, the two-engine B-26 medium bomber which saw much service in World War II, is shown in Fig. 2.15.[28] The aircraft was mounted on a turntable in surroundings free from other reflecting objects and was illuminated by a radar operating at 3 GHz (*S* band). The aircraft's propellers were running during the measurement and produced a modulation of the order of 1 to 2 kHz. Changes in the radar cross section by as much as 15 dB can occur for a change in aspect of only 1/3 degree. The maximum echo signal occurred in the vicinity of broadside, where the projected area of the aircraft was largest and there were large, relatively flat surfaces that produced a large return.

The following description of the major scattering features of a jet aircraft is based on information found in Knott, Shaeffer, and Tuley[29] and in Crispin and Siegel.[30] The radar cross section of a jet aircraft from the nose-on aspect is determined to a large extent by reflections from the jet engines and their intake ducts. The compressor blades on the rotating jet engines can also modulate the echo, just as do the propellers of a prop aircraft.

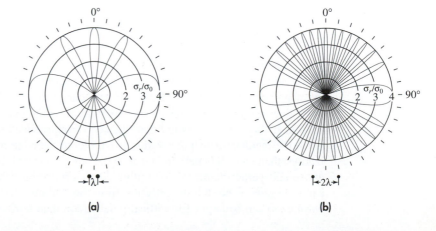

Figure 2.14 Polar plot of the radar cross section σ_r of a two-scatterer "complex" target when each scatterer has a cross section σ_0 [plot of Eq. (2.37)]. (a) separation $l = \lambda$; (b) $l = 4\lambda$.

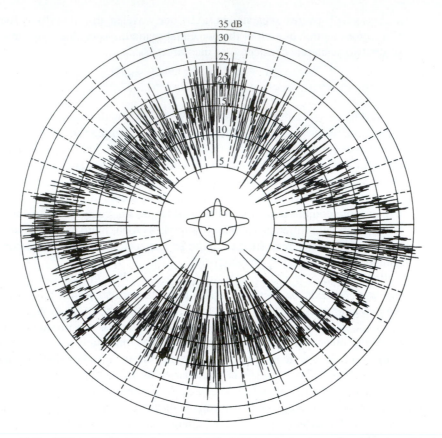

Figure 2.15 Backscatter from a full-scale B-26 two-engine (propeller driven) medium-bomber aircraft at 10-cm wavelength as a function of azimuth angle. This figure has appeared in many books on radar and radar scattering since it is one of the few examples readily available in the literature of the measurement of backscatter from a full-size aircraft without averaging over a range of angles.
| (From Ridenour,[28] courtesy McGraw-Hill Book Company, Inc.)

If there is a radar antenna in the nose, the cross section can be large when the aircraft's radar antenna is pointing in the direction of the viewing radar. At angles a few degrees off the nose, the cross section generally decreases until echoes from the leading edge of the wing become important. If the radius of curvature of the leading edge of the wing is more than a few wavelengths, the echo from the wing can be significant for any polarization. If this condition is not met, the echo from the leading edge of the wing is larger when the incident radar polarization is parallel to the edge of the wing (horizontal polarization) than when it is perpendicular. If the wing is viewed from slightly below or slightly above and perpendicular to its trailing edge, a traveling wave can be induced and there can be a radar echo due to reflection from the trailing edge of the wing. This traveling-wave echo can be larger for vertical polarization than horizontal.

In the vicinity of broadside, the fuselage and the engine nacelles are the main sources of backscatter. At broadside the vertical stabilizer also can contribute to the echo. If the aircraft is viewed from considerably above the plane of the wing, there can be significant backscatter from the corner reflector formed by the wing and the fuselage. There can also be a strong echo from the corner reflector formed by the vertical fin and the horizontal stabilizer. From the vicinity of the tail-on aspect, the trailing edge of the wings contribute to the echo. Echoes are also obtained near tail-on from the engine exhaust ducts. External weapons and fuel tanks of military aircraft also contribute to the radar cross section.

An example of the radar cross section of a one-fifteenth scale model Boeing 737 commercial jet aircraft is shown in Fig. 2.16.[31] The cross section in the ordinate is in dBsm, which is decibels relative to one square meter. The values shown are for the scale model at the measured frequency of 10 GHz. At the full-scale frequency of 667 MHz, the values of cross section given by the ordinate in this figure should be increased by 23.5 dB.

The most realistic method for obtaining the radar cross section of aircraft is to measure the actual aircraft in flight. The dynamic radar cross-section measurement facility of the U.S. Naval Research Laboratory[32] was able to measure the backscatter signals from aircraft at *L, S, C,* and *X* bands with horizontal or vertical linear polarization, circular polarization, as well as the cross (orthogonal) polarizations on receive. Pulse-to-pulse radar measurements were recorded, but for convenience in presenting the data, the cross-section values plotted usually were an average of a large number of values taken within a 10 by 10° aspect angle. Examples are given in Fig. 2.17.

An unusual method for obtaining the scattering characteristics of aircraft in flight is the airborne synthetic aperture radar utilized by Metratek, Inc. of Reston, Virginia.[33] This is a high-resolution multifrequency imaging radar mounted on the tail of an A-3 aircraft to provide 300° coverage. The aircraft carrying the radar is maneuvered from side to side to obtain good resolution in the cross-range dimension by means of synthetic aperture processing. It has been used to image large (200-ft) aircraft in flight with high

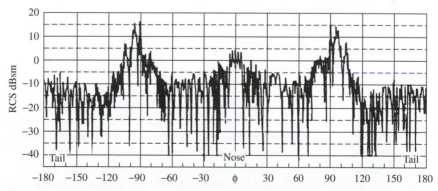

Figure 2.16   Measured radar cross section of a one-fifteenth scale model Boeing 737 commercial jetliner at the model frequency of 10 GHz with vertical polarization. The full-scale measurement would correspond to a frequency 15 times lower, which is 667 MHz. The cross section at the full-scale frequency is 23.5 dB greater than the ordinate values shown in the figure. Model was measured on a pulse-gated compact range.
(From N. A. Howell,[31] © 1970 IEEE.)

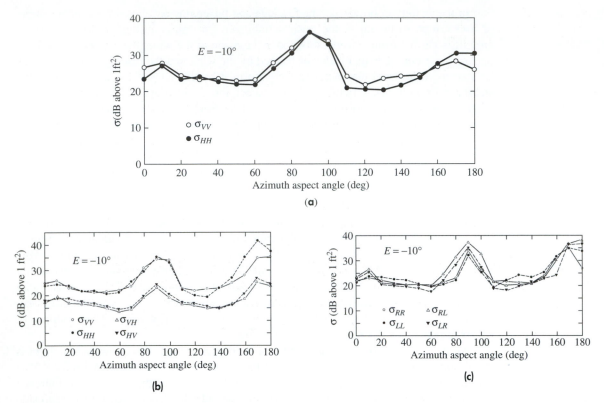

Figure 2.17 Measured radar cross section of the C-54 aircraft at a constant elevation angle of −10° measured from the aircraft. (The C-54 was a military version of the four-piston-engine DC-4 commercial aircraft with a wing span of 36 m and a length of 29 m.) Values are averaged over a 10 by 10° aspect angle. (a) 1300 MHz, linear polarization; (b) 9225 MHz, linear polarization; (c) 9225 MHz, circular polarization. The first subscript is the polarization transmitted, the second subscript is the polarization received; V = vertical, H = horizontal, R = right-hand (clockwise rotation) circular, L = left-hand (counterclockwise) circular polarization. Note that the ordinate is the cross section in dB above 1 ft². Subtract 10 dB for dB above 1 m².
| (Courtesy of I. D. Olin and F. D. Queen, Naval Research Laboratory.)

resolution from VHF to at X band. An example of an aircraft image is shown in Fig. 2.18. A similar radar system mounted in the nose of a TA-3B has been reported by Hughes Aircraft Co.[34] The nose mounting permits the rear of the target aircraft to be imaged better than can a rear-mounted radar. The advantage of the high-resolution imaging method is that the individual scatterers which contribute to the backscatter can be readily recognized and their contribution to the total cross section determined.

Although the cross section of aircraft can fluctuate over a large range of values (perhaps as much as 60 dB around the entire target[29]), its average value in the microwave region usually does not vary significantly with frequency. Aircraft cross sections, however, are often larger at lower frequencies (such as VHF) than at microwave frequencies. A military propeller aircraft, such as the old AD-4B,* had a measured cross section

| *The AD-4B was a 1950s propeller-engine naval attack bomber with a 50-ft wing span and 39-ft length.

Figure 2.18 Air-to-air radar image of a KC-135, a military version of the 707 aircraft showing its major scattering centers at VHF.[33] The outline of the aircraft is shown for comparison. Resolution is about 4 ft. Aspect is nose-on and elevation angle is zero.
I (Courtesy of Ray Harris, Metratek, Inc.)

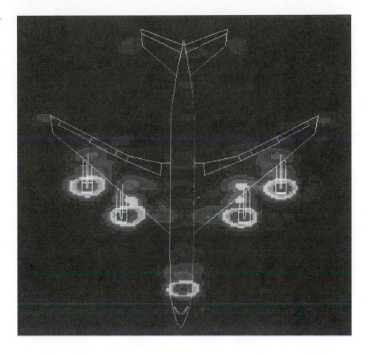

approximately five times as great at VHF than at *L* band. At HF, the cross section of large aircraft might be more than an order of magnitude greater than at microwave frequencies.

Ships   A "folk theorem" for the radar cross section of a ship is that the cross section in square meters can be approximated by the ship's displacement in tons. (The origin of this is uncertain, but the author believes he first learned of it many years ago from the General Electric Co. of Valley Forge, Pennyslvania.) Thus a 10,000-ton ship might be said to have a radar cross section of the order of 10,000 m^2. This empirical relationship is a convenient measure when no better information is available, when a single number is acceptable to describe the cross section, and when the grazing angle is not near zero degrees.

At low grazing angles near zero degrees, the median (50th percentile) value of the radar cross section can be expressed by the empirical relation[35]

$$\sigma = 52\, f^{1/2} D^{3/2} \qquad\qquad \textbf{[2.38]}$$

where σ = radar cross section in square meters, f = radar frequency in megahertz, and D is the ship's (full load) displacement in kilotons. This value of cross section was the average taken about the port and starboard bow and quarter aspects of a number of ships (omitting the peak at broadside). Measurements on which this expression is based were at *X, S,* and *L* bands, with conventional naval ship displacements from 2000 to 17,000 tons. A plot of this equation is shown in Fig. 2.19 for the three frequencies for which measurements were taken. The values of ship cross sections at zero grazing angle are much

Figure 2.19 Estimate of the median value of the radar cross section of ships at grazing incidence, averaged over the bow and quarter aspects. Based on the empirical relation of Eq. (2.38).

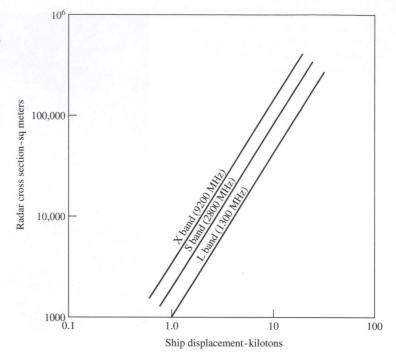

higher than the cross sections at higher grazing angles. At zero degrees a ship's super-structure presents many vertical surfaces and corner reflectors, both of which have large backscatter echoes. An example of the measured radar cross section at grazing incidence of a 16,000-ton naval auxiliary ship is shown in Fig. 2.20.

Other Targets   Missile cross sections can vary over large values depending on the type of missile. A ballistic missile with its tanks attached can be large, but the ballistic missile reentry vehicle, a small cruise missile, or antiship missile can be many orders of magnitude smaller. The range profile of a small missile viewed along its longitudinal axis is shown in Fig. 2.21.[36] The radar resolution is 0.2 m. The various scattering centers and their relative cross sections are indicated.

Automobiles generally can have a surprisingly large radar cross section. From the front the cross section might vary from 10 to 200 m^2 at X band, with 100 m^2 being a typical value.[37] This has been attributed to the flat surfaces, such as the engine's radiator, that the radar can view. Not all automobiles should be expected to have these large values, however.

The average radar cross section of small pleasure boats 20 to 30 ft in length might be in the vicinity of a few square meters at X band.[38] Boats from 40 to 50 ft in length might have a cross section of the order of 10 m^2. The measured radar cross section of a man[39] over frequencies from UHF to X band has been reported to vary from 0.03 to 1.9 m^2, with a value of one square meter usually stated as being "typical." Values of the radar cross sections of rain, snow, birds, and insects are given in Chap. 7.

Figure 2.20   Azimuth variation of the radar cross section of a large naval auxiliary ship at *X* band, horizontal polarization, showing 20th, 50th, and 80th percentiles.

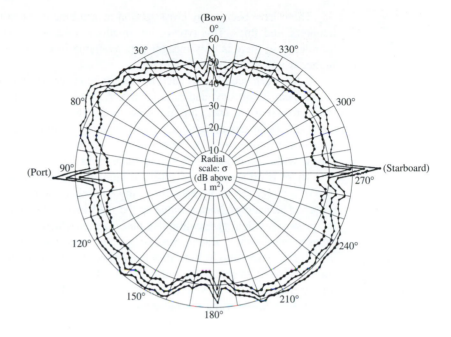

Figure 2.21   High-resolution range profile (backscatter) of a small missile, showing the individual scattering centers.
(From D. L. Mensa.[36] Reprinted with permission from Artech House, Inc., Norwood, MA, USA.)

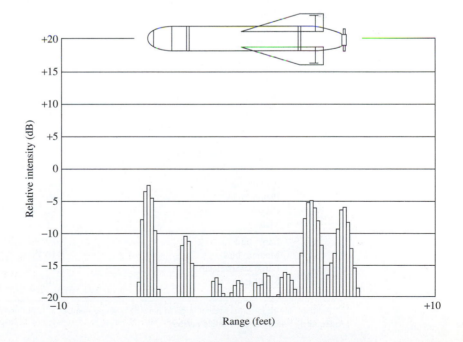

There have been many cross section measurements of military aircraft, missiles, helicopters, and ships. This type of information, however, is usually not available for public release so there is only meager data available in the published literature on the radar scatter from current operational targets.

It has been seen in this section that the radar cross section of radar targets can vary with aspect, frequency, and polarization. A single number is not a complete measure of radar cross section. There are times, however, when a single number is desired to describe a class of radar target. Table 2.1 lists "example" single values of radar cross sections at microwave frequencies for various classes of targets. Within each class, of course, there can be wide variations in the single-value cross section given here.

There is no standard agreed-upon method for specifying the single-value cross section of a target (probably because a single number by itself is seldom used to describe a target). The average (mean) value or the median value might be taken. These depend on the probability density function that describes the fluctuations of the cross section. Sometimes a "minimum" value is used, such as the value exceeded 99, 95, or 90 percent of the time. (The reader might note that the author did not have the courage to state how the single-value cross sections listed in Table 2.1 were defined.) In any case, the variable nature of the radar cross section of real targets must be taken into account when predicting radar performance using the radar range equation.

Table 2.1   Examples of radar cross sections at microwave frequencies*

	Square meters
Conventional winged missile	0.1
Small, single engine aircraft	1
Small fighter, or four-passenger jet	2
Large fighter	6
Medium bomber or medium jet airliner	20
Large bomber or large jet airliner	40
Jumbo jet	100
Helicopter	3
Small open boat	0.02
Small pleasure boat (20–30 ft)	2
Cabin cruiser (40–50 ft)	10
Ship at zero grazing angle	See Eq. (2.38)
Ship at higher grazing angles	Displacement tonnage in m^2
Automobile	100
Pickup truck	200
Bicycle	2
Man	1
Large bird	10^{-2}
Medium bird	10^{-3}
Large insect (locust)	10^{-4}
Small insect (fly)	10^{-5}

* Although the radar cross section is given here by a single number, it is not usual that the target echo can be adequately described by a single number.

2.8   RADAR CROSS-SECTION FLUCTUATIONS

A small change in viewing aspect of a radar target such as an aircraft or ship can result in major changes in the radar cross section, as was illustrated by the cross section of an aircraft shown in Fig. 2.15. Complex targets are made up of a number of individual scattering centers, or *scatterers*. The scatterers for an aircraft might be the engines, cockpit, nose, wings, tail, and external stores. The echo from each scattering center has an amplitude and phase that usually is independent of the amplitude and phase of the echoes from other scattering centers. The phase of each scatterer is determined primarily by the distance of the individual scattering center from the radar. At the radar, the echoes from all the individual scatterers add vectorially to form a resultant amplitude and phase. If the target can be represented as a collection of independent point scatterers, the form of the echo signal $s_r(t)$ can be written as

$$s_r(t) = \sum_{i=1}^{N} a_i \sin(2\pi ft + \phi_i) = A \sin(2\pi ft + \Phi) \tag{2.39}$$

where

$$A = \left[\left(\sum_i a_i \sin \phi_i\right)^2 + \left(\sum_i a_i \cos \phi_i\right)^2\right]^{1/2} \quad \text{and} \quad \Phi = \arctan \frac{\sum_i a_i \sin \phi_i}{\sum_i a_i \cos \phi_i}$$

In the above a_i = amplitude of the ith point scatterer, $\phi_i = 2\pi fT_i$, T_i is the round-trip time to the ith scatterer, and f is the radar frequency. A point scatterer is assumed to be one whose a_i and ϕ_i are independent of the viewing aspect. Equation (2.39) is a relatively simple model for scattering from a complex target, and it has limitations. It does not take account of multiple reflections, the shadowing of one scatterer by another, or that the target might not always be able to be represented as a collection of point scatterers. Also, the distance to the radar is assumed to be very large compared to the extent of the target so that the R^{-4} variation of the echo signal power with range R need not be taken into account for the individual scatterers. Nevertheless, it is an adequate target representation for many purposes.

If the target aspect changes relative to the radar, there will be changes in the distances to the scattering centers and the times T_i. These can cause a change in the relative phases of the echo signals from the various scatterers that make up the target. A relative phase shift greater than 2π radians can yield a significant change in the resultant phase and amplitude of the composite echo signal, which results in target cross-section fluctuations. (Sometimes the term *fading* is used in the literature for what here is called *fluctuations*.)

One straightforward method to account for a fluctuating radar cross section in the radar equation is to select a small value of cross section that has a high probability of being exceeded almost all of the time. This procedure has the advantage of being simple. It is not precise, but neither are the more widely used analytical methods, mentioned later in this section, that require knowledge of the actual statistics of the target fluctuations.

The method often employed for finding the minimum detectable signal-to-noise ratio when the target cross section is not constant is based on the *probability density function* that describes the cross-section fluctuations. The probability density function, or pdf, gives the probability of finding a particular value of the target cross section between the values of σ and $\sigma + d\sigma$. In addition to the pdf, the variation or *correlation* of the cross section with time, or pulse to pulse, must also be known. The time variation of cross-section fluctuations differs from that of receiver noise since receiver noise is statistically independent, or uncorrelated, from pulse to pulse.

Swerling Target Models　A popular method for representing the fluctuations of targets are the four statistical models described by Peter Swerling.[40] For each of these he calculated the signal-to-noise ratio required as a function of the probability of detection, probability of false alarm, and the number of pulses integrated. The four Swerling fluctuating target models are:

Case 1.　The echo pulses received from a target on any one scan are of constant amplitude throughout the entire scan, but are independent (uncorrelated) from scan to scan. A target echo fluctuation of this type is called *scan-to-scan fluctuation.* It is also known as *slow fluctuations.* The probability density function for the cross section σ is

$$p(\sigma) = \frac{1}{\sigma_{av}} \exp\left(-\frac{\sigma}{\sigma_{av}}\right) \qquad \sigma \geq 0 \qquad \text{[2.40]}$$

where σ_{av} is the average over all values of target cross section. This probability density function, or pdf, applies to a target consisting of many independent scatterers of comparable echo areas; that is, no scatterer is large compared to the others. Although the pdf assumes a large number of scatterers, it has been said that as few as four or five produce a reasonably close approximation for many purposes.[40,41] The target model in Case 1 is sometimes called a *Rayleigh scatterer.* The pdf of Eq. (2.40), however, is that of the exponential [Eq. (2.18)] rather than the Rayleigh [Eq. (2.17)]. It might be recalled that the exponential pdf represents the statistics of the square of a voltage that is described by a Rayleigh pdf.

Case 2.　The probability-density function is the same as that of Case 1, but the fluctuations are independent from pulse to pulse rather than from scan to scan. This is sometimes called *fast fluctuations.*

Case 3.　As in Case 1, the radar cross section is assumed to be constant within a scan and independent from scan to scan; but the probability density function is given by

$$p(\sigma) = \frac{4\sigma}{\sigma_{av}^2} \exp\left(-\frac{2\sigma}{\sigma_{av}}\right) \qquad \sigma \geq 0 \qquad \text{[2.41]}$$

Swerling states that this probability density function is representative of targets that can be modeled as one large scatterer together with a number of small scatterers.

Case 4.　The fluctuation is pulse to pulse, but with the same pdf as Case 3. The cross section to be substituted in the radar equation for these four cases is the average value σ_{av}.

A comparison of the four Swerling target fluctuation models and the nonfluctuating case, here called Case 0, is illustrated in Fig. 2.22 for $n = 10$ hits integrated

Figure 2.22 Comparison of the detection probabilities for the four Swerling models and the nonfluctuating model, for $n = 10$ pulses integrated and a false-alarm number $n_f = 10^8$.
| (Adapted from Swerling.[40])

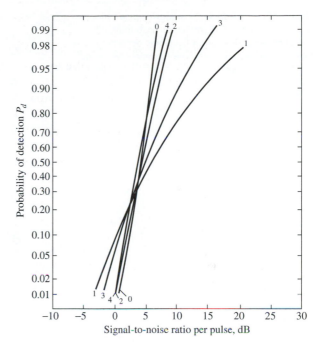

postdetection and false-alarm number of 10^8 (inverse of the false-alarm probability). In general, fluctuating targets require a larger signal-to-noise ratio than nonfluctuating targets. For example, Fig. 2.22 shows that if the probability of detection is required to be 0.95, a signal-to-noise ratio of 6.2 dB per pulse is necessary when the radar cross section is constant (Case 0); but a 16.8 dB per pulse signal-to-noise ratio is required if the target fluctuates according to Case 1. Thus if the radar had been designed on the basis of a constant cross section equal to σ_{av}, but in reality the cross section fluctuated according to the model of Case 1, there would be a reduction in range by a factor of 1.84 because of the 10.6 dB difference between the two cases.

Case 1 puts more demand on the radar than do the other cases. Figure 2.22 also indicates that for probabilities of detection greater than about 0.3 (which is always true in practical situations), a greater signal-to-noise ratio is required when the fluctuations are independent scan to scan (Cases 1 and 3) than when the fluctuations are independent pulse to pulse (Cases 2 and 4). The pulse-to-pulse fluctuations of Cases 2 and 4 tend to average to that of the constant cross-section case (Case 0) as the number of independent pulses integrated increases.

The statistical theory of detection can be applied to find for each Swerling case the signal-to-noise per pulse required for a given probability of detection, probability of false alarm (or false-alarm number), and number of pulses integrated. Curves similar to those of Fig. 2.6 for the constant cross section are available for the four fluctuating models.[40,42,43] They are also included in commercially available computer programs for the radar equation.[68]

The required signal-to-noise ratio $(S/N)_1$ for Swerling Cases 1 and 3 when $n = 1$ (single-pulse detection) can be obtained by adding the value found from Fig 2.23 (dB) to

Figure 2.23 Additional signal-to-noise ratio required to achieve a particular probability of detection when the target fluctuates according to a Swerling model. The ordinate is sometimes called the *fluctuation loss, L_f.*

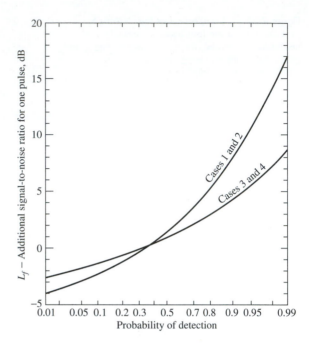

the signal-to-noise ratio (dB) for the nonfluctuating single-pulse signal-to-noise ratio of Fig. 2.6.

Barton[44,45] calls the ordinate of Fig. 2.23 the *fluctuation loss,* which is denoted as L_f. With scan-to-scan target fluctuations (Cases 1 and 3), the integration improvement factor $nE_i(n)$ is the same as that given by Eq. (2.7) for a constant (nonfluctuating) cross section.

To account for a fluctuating target in the radar range equation when the target is Swerling Cases 1 and 3, $(S/N)_1$ is replaced with $(S/N)_1 L_f$ and the same integration improvement factor is used as for a constant cross section. In practice, Swerling Case 1 is used more than the other three cases to describe target cross-section statistics since it requires a larger (more conservative) value of signal-to-noise ratio than the others. When the number of pulses integrated is greater than about 10 to 20, Case 2 approaches Case 0, and Case 4 approaches Case 0. Therefore, they do not have to be treated separately. The procedures for using the Swerling Cases in the radar equation are summarized later, after the discussion of partial correlation.

Partial Correlation of Target Cross Section   The Swerling Cases assume that the pulses received as the antenna scans by the target are either completely correlated pulse to pulse (correlation coefficient $\rho = 1$) or completely uncorrelated pulse to pulse ($\rho = 0$). When the correlation of the pulses is some value of ρ other than 0 or 1, Barton[46] provides an empirical relationship for the required signal-to-noise ratio which compares favorably with an exact analysis by Kanter.[47] Barton states that the fluctuation loss when n_e independent samples are integrated is approximately

$$L_f(n_e) = (L_f)^{1/n_e} \qquad \text{[2.42]}$$

where L_f is the fluctuation loss as was given by Fig. 2.23 for detection with a single pulse for either Case 1 or 3. (Note that L_f in Fig. 2.23 is in dB; so that $L_f(n_e)_{dB}$ is equal to $L_f(1)_{dB}/n_e$.) The number of independent samples received from a target and integrated during the observation time t_0 is the smaller of n and n_e, where n_e is given as

$$n_e = 1 + \frac{t_0}{t_c} \leq n \qquad\qquad [2.43]$$

where t_c is the correlation time of the target echo signal and n is the total number of pulses available. Barton states that the correlation time is the inverse of the effective noise bandwidth of the two-sided fluctuation spectrum of the echo signal. It is the time required for the target cross section to change to a new value. It is not easy to specify and is highly dependent on the particular scenario.

Kanter[47] provides exact curves for the signal-to-noise ratio when expressed in terms of the correlation coefficient ρ rather than n_e. He gives the relationship between the two as

$$n_e = 1 + (n-1) \, ln\left(\frac{1}{\rho}\right) \leq n \qquad\qquad [2.44]$$

The values of n_e in the above cannot be greater than the number of pulses n available for integration.

Radar Equation for Swerling Cases   The radar equation for partially correlated Swerling Cases can be written

$$R^4_{max} = \frac{P_t GA_e \sigma n E_i(n)}{(4\pi)^2 kT_0 BF_n (S/N)_1 (L_f)^{1/n_e}} \qquad\qquad [2.45]$$

where L_f is the fluctuation loss, or the additional signal-to-noise ratio for one pulse for a fluctuating target (given by Fig. 2.23); and n_e is the equivalent number of independent samples integrated [given by Eq. (2.43) or (2.44)]. The curve in Fig. 2.23 labeled "Cases 1 and 2" is used when the radar cross section fluctuates according to the Rayleigh pdf of Eq. (2.40). The curve labeled "Cases 3 and 4" is used when the target cross-section pdf is described by Eq. (2.41). Equation (2.45) applies to the partially correlated Swerling Cases 1 and 3 when the partial correlation is represented by the effective number of pulses integrated $n_e < n$. It also applies to the Swerling cases when the target echo is not partially correlated; i.e., $\rho = 0$ or $\rho = 1$, as summarized below:

Case 1.   Completely correlated: $\rho = 1$, $n_e = 1$, L_f is that labeled for Cases 1 and 2 in Fig. 2.23.

Case 2.   Completely uncorrelated: $\rho = 0$, $n_e = n$, and L_f is that labeled for Cases 1 and 2 in Fig. 2.23. When n is large ($n > 20$, $P_d \leq 0.9$), Case 2 can be replaced by Case 0 (nonfluctuating target).

Case 3.   Completely correlated: $\rho = 1$, $n_e = 1$, and L_f is that labeled for Cases 3 and 4 in Fig. 2.23.

Case 4. Completely uncorrelated: $\rho = 0$, $n_e = n$, and L_f is that labeled Cases 3 and 4 in Fig. 2.23. When n is large ($n > 10$, $P_d \leq 0.9$), Case 4 can be replaced by Case 0.

Case 0. Set $L_f = 1$.

In all cases, including partial correlation, $(S/N)_1$ is given by Fig. 2.6 and the integration improvement factor $nE_i(n)$ is given by Fig. 2.7. The above procedures give results that are suitable for many engineering purposes. When more exact values are required, one should use the procedures given in the previously cited references or a suitable radar equation computer software package.

Decorrelation by Frequency Diversity or Frequency Agility A change in the radar cross section of a complex target can be obtained by a change in radar frequency, even when the target aspect is fixed. In other words, the target echo can be decorrelated by a sufficient change in frequency. If a target can be considered as made up of only two major scatterers separated in the range (radial) coordinate by a distance D, the phase difference of the echo signals from the two is $4\pi(f/c)D$. A change in frequency Δf produces a change in phase $\Delta\phi = 4\pi(\Delta f/c)D$. If the phase change is greater than 2π radians, the resultant signal is considered independent of the previous signal. Thus a change in frequency $\Delta f \geq c/2D$ decorrelates the echo. (The separation D in an actual target may be some characteristic target dimension projected along the radial coordinate. It might be the projected separation between two major scatterers that are near the extremes of the target. It is not necessarily the projected physical length of the target since the echoes from the target extremes might be small compared to those from other scatterers that make up the target.)

Decorrelation of the echo signal can be obtained by either frequency diversity or frequency agility. *Frequency diversity* means that more than one transmitter, each at a different frequency, is utilized in parallel with each transmitter channel operating as a separate radar. (The separate frequencies can also be obtained from one wideband transmitter, but that is not what is envisioned in this particular discussion.) Most air-traffic-control radars usually employ two transmitters in order to achieve the redundancy necessary for reliable civil air-traffic control. Rather than have each transmitter operate at the same frequency, they generally operate at two different frequencies in order to obtain two independent echo signals. If the echo is very small on one frequency, it will likely be larger on the other if the second frequency is widely separated from the first. The improvement that might be gained in using frequency diversity can be found from Fig. 2.23 for the fluctuation loss. For example, if the target cross section is described by Case 1 (scan-to-scan independence), the fluctuation loss for a 0.8 probability of detection (the value usually specified by the U.S. Federal Aviation Agency for its air-traffic-control radars) is 5.4 dB. When two independent frequencies are used, the fluctuation loss (in dB) is one-half this value; so that there is a gain due to frequency diversity of 2.7 dB, in addition to the 3 dB provided by the second transmitter.

Although there is a gain in using frequency diversity, it is more difficult to accurately quantify than indicated by the simple example above. It depends on the probability of detection. If the probability of detection were $P_d = 0.9$, the improvement due to frequency diversity is 4 dB when two independent frequencies are used, rather than the 2.7 dB when $P_d = 0.8$. With $P_d = 0.5$, the diversity improvement is only 0.8 dB, so that use of two frequencies in this case offers little value. On the other hand, when the P_d is required to be 0.99, two independent frequencies provide a diversity improvement of 8.6 dB. Thus frequency diversity can be of value when a high probability of detection is required and the pulses are very highly correlated.

There is not much to gain, however, in continuing to add transmitters beyond two or three. If the fluctuation loss is 5.4 dB (as for $P_d = 0.8$ in the above example) two transmitters give a diversity improvement of 2.7 dB, three give a diversity improvement of 3.6 dB, and four give an improvement of 4.1 dB. Most applications of frequency diversity radar are satisfied with the gain offered by just two transmitters.

Pulse-to-pulse change in frequency is called *frequency agility*. The target cross section is decorrelated if the change in radar frequency is greater than $\Delta f = c/2D$, where D is the effective radial size of the target. The total number of independent samples available within a total bandwidth B is

$$n_e = 1 + \frac{B}{\Delta f} \qquad [2.46]$$

Frequency agility can be accomplished with a single wideband transmitter. It is mentioned here as a method for increasing the detectability of a target. It is also a tactic available to combat narrowband hostile jamming. Pulse-to-pulse frequency agility, however, is not suitable for radars that require doppler processing to detect moving targets in clutter (as in the MTI and pulse doppler radars discussed in Chap. 3) since these radars must utilize multiple pulses at the same frequency in order to extract the doppler frequency shift. Doppler processing, on the other hand, is compatible with frequency diversity.

Chi-Square Target Model   The two probability density functions of Eqs. (2.40) and (2.41) that are used in the four Swerling models are special cases of the chi-square probability density function of degree $2m$.[41,48] The chi-square pdf is

$$p(\sigma) = \frac{m}{(m-1)! \, \sigma_{av}} \left(\frac{m\sigma}{\sigma_{av}} \right)^{m-1} \exp\left(-\frac{m\sigma}{\sigma_{av}} \right) \qquad \sigma > 0 \qquad [2.47]$$

where σ_{av} is the average, or mean, value of the cross section σ. The chi-square pdf is also known as the *gamma probability density function*. In statistics texts, $2m$ is an integer equal to the number of degrees of freedom. When applied to target cross-section models, however, $2m$ is not restricted to integer values. It can be any positive, real number. When $m = 1$, the chi-square reduces to the exponential, or Rayleigh-power, pdf that applies to Swerling Cases 1 and 2. Cases 3 and 4 are equivalent to the chi-square with $m = 2$. The ratio of the standard deviation to the mean value of the chi-square pdf is equal to $m^{-1/2}$. The larger the value of m, the less will be the fluctuations; that is, the fluctuations are more constrained. The limit as $m \to \infty$ corresponds to the nonfluctuating target.

The chi-square pdf with parameter m between 0.3 and 2 has been found to approximate the statistics of certain simple shapes, such as cylinders or cylinders with fins.[49]

Other Target Models   The Rice probability density function that was encountered in Sec. 2.5 has also been suggested as a target cross-section model when the target can be represented as one dominant scatterer together with a collection of small independent scatterers.[50] The Rice pdf is

$$p(\sigma) = \frac{1+s}{\sigma_{av}} \exp\left[-s - \frac{\sigma}{\sigma_{av}}\left(1 + s^2\right)\right] I_0\left(2\sqrt{\frac{\sigma}{\sigma_{av}}s(1 + s)}\right) \qquad \sigma > 0 \quad \text{[2.48]}$$

where s is the ratio of the radar cross section of the single dominant scatterer to the total cross section of the small scatterers, and $I_0(\cdot)$ is the modified Bessel function of zero order. Its description appears similar to the description of the chi-square with $m = 2$ (Swerling Cases 3 and 4), but it is not the same.

The log-normal pdf has also been considered for representing some types of target echo fluctuations. It is given by

$$p(\sigma) = \frac{1}{\sqrt{2\pi}s_d\sigma} \exp\left\{-\frac{1}{2s_d^2}\left[ln\left(\frac{\sigma}{\sigma_m}\right)\right]^2\right\} \qquad \sigma > 0 \qquad \text{[2.49]}$$

where s_d = standard deviation of $ln\,(\sigma/\sigma_m)$, and σ_m = median value of σ. The ratio of the mean to the median value of σ is exp $(s_d^2/2)$. There is no theoretical target model that leads to the log-normal pdf, but it has been suggested that it can approximate cross-section statistics from some satellite bodies, ships, cylinders, plates, and arrays.[51,52] The log-normal pdf usually has higher values of mean-to-median than other pdfs. It also has a higher probability of obtaining abnormally high values than other pdfs (usually expressed as its distribution having "high tails").

There are a number of other statistical models that might be used. Shlyakin[53] provides a comprehensive review of many of the statistical representations that have been considered for modeling radar signals. Although most of these are two-parameter pdfs, he includes three- and four-parameter statistical models as well. The more parameters, the closer it might fit experimental data. Only a few of these statistical pdfs, however, can be derived from some basic physical model. It seems that the only justification for using most of the available analytical statistical models is that they can be made to "curve fit" some particular set of experimental data.

Which Target Model to Use   A valid concern is how to model real target behavior so that radar performance can be predicted from the radar range equation. If the statistics of the target's cross section can be determined or assumed, application of the classical theory of detection of signals in noise can provide the required signal-to-noise ratio as a function of the probabilities of detection and false alarm, and the number of pulses integrated. As has been indicated, however, this is not practical except in special cases. It is made difficult because the cross-section statistics can vary with aspect, duration of observation, and frequency.

Although few real targets uniquely fit a mathematical model with any precision, it has been said that measurements[54] of aircraft targets indicate that in many cases the Rayleigh model (chi-square with $m = 1$) provides a closer fit than other models. On the other hand, experimentally measured values of the parameter m in the chi-square pdf were found to change with aspect and could vary from about 0.5 to almost 20. There were even examples when no value of m could fit the data. Thus it is difficult to reliably fit a target with a statistical model that is applicable over a variety of conditions.

Another concern when using statistical target models, such as the chi-square, is what average value of target cross section σ_{av} to insert in the radar equation. If the cross section is averaged over the entire 4π steradians of viewing aspect, the average will be dominated by the very high values that occur at and near the broadside aspect as well as the top and bottom aspects. One can argue that these should be eliminated from the average since targets such as aircraft, missiles, or ships are seldom viewed from these aspects. If the average is to be taken only over the aspects that occur in a given scenario, then the scenario (the nature of the trajectories that determine the range of viewing aspects) needs to be specified (something not always easy to do). To make the situation worse, the average value of cross section is rarely reported. (Note that the average cross section is not the average of its values given in decibels, but the average of its numerical values.)

It can be said that there may be no analytical statistical target model or models that can reliably represent the statistics of complex target cross sections, except in special situations. This may be true, but the radar engineer has to make predictions of radar performance using the radar equation whether or not there is perfection. Engineers often have to work with compromise solutions that give usable results, if not absolute accuracy. There are at least three approaches that have been used for dealing with the radar cross section of practical targets:

1. Use the best guess for a statistical target model based as much as possible on (successful) past experience.

2. Use a constant cross-section model, but select a value of cross section that is exceeded a large fraction of the time, say 90 or 95 percent or higher. This is the simplest method.

3. Use the Swerling Case 1 model (chi-square with $m = 1$ and scan-to-scan fluctuations). It produces a conservative estimate (large signal-to-noise ratio). This model is often used by radar procurement agencies when specifying the contractual performance required of a new radar. Its large values of signal-to-noise ratios when the probability of detection is high provide a bit of a cushion to the radar design.

Because of its simplicity, the author prefers using the constant cross section (no. 2 above). The Swerling Case 1, however, is probably the most widely employed. It is a good approach if a large probability of detection is used, such as $P_d = 0.9$ or higher. Sometimes a lower P_d has been used with Case 1, but this defeats its advantage as a conservative estimate. There is little difference among the Swerling models and the constant cross-section model with a P_d of 0.5. There is no point, therefore, in specifying a sophisticated target model for low values of P_d.

Sellers of radars like to quote the range performance of their products with P_d of 0.5; buyers of radars want to see the performance based on a P_d of 0.9 or higher. The author's preference is that radar performance be based on a high P_d, such as 0.9.

2.9 TRANSMITTER POWER

The power P_t in the simple radar equation derived in Sec. 1.2 was not actually specified but is usually *peak power* of the pulse. (It is not the *instantaneous* peak power of a pulse of sinewave, but one-half the instantaneous peak value.) The *average power* P_{av} of a radar is also of interest since it is a more important measure of radar performance than the peak power. It is defined as the average transmitter power over the duration of the total transmission. If the transmitter waveform is a train of rectangular pulses of width τ and constant pulse-repetition period $T_p = 1/f_p$ (f_p = pulse repetition frequency), the average power is related to the peak power by

$$P_{av} = \frac{P_t \tau}{T_p} = P_t \tau f_p \qquad [2.50]$$

The radar *duty cycle* (sometimes called *duty factor*) can be expressed as P_{av}/P_t, or τ/T_p, or τf_p. Pulse radars might typically have duty cycles of from 0.001 to 0.5, more or less. A CW radar has a duty cycle of unity. The duty cycle depends on the type of waveform, the pulse width, whether or not pulse compression is used, the problems associated with range and doppler ambiguities, and the type of transmitter employed.

Writing the range equation of Eq. (2.45) in terms of average power by substituting Eq. (2.50) for P_t gives

$$R^4_{max} = \frac{P_{av} G A_e \sigma n E_i(n)}{(4\pi)^2 k T_0 F_n (B\tau)(S/N)_1 f_p} \qquad [2.51]$$

For simplicity, the fluctuation loss L_f, as included in Eq. (2.45), has been set equal to unity. It should be reinserted, of course, when needed to account for Swerling target models. The symbols in this equation have been defined previously, but the reader can find them listed under Eq. (2.61) in Sec. 2.13. The bandwidth and the pulse width are grouped together since the product of the two is approximately unity in a well-designed radar.

From the definition of duty cycle given above, the energy per pulse, $E_p = P_t \tau$, is also equal to P_{av}/f_p. Substituting the latter into Eq. (2.51) gives the radar equation in terms of energy, or

$$R^4_{max} = \frac{E_p G A_e \sigma n E_i(n)}{(4\pi)^2 k T_0 F_n (B\tau)(S/N)_1} = \frac{E_T G A_e \sigma E_i(n)}{(4\pi)^2 k T_0 F_n (B\tau)(S/N)_1} \qquad [2.52]$$

where E_T is the total energy of the n pulses, which equals nE_p.

2.10 PULSE REPETITION FREQUENCY

The pulse repetition frequency (prf) is often determined by the maximum unambiguous range beyond which targets are not expected. As in Sec. 1.1, the prf corresponding to a maximum unambiguous range, R_{un}, is given by $f_p = c/2R_{un}$, where c is the velocity of propagation. There are times, however, when echoes might appear from beyond the maximum unambiguous range, especially for some unusually large target or clutter source (such as a mountain), or when anomalous propagation conditions (Sec. 8.5) occur to extend the normal range of the radar beyond the horizon. Echo signals that arrive at a time later than the pulse-repetition period are called *second-time-around echoes*. They are also called *multiple-time-around echoes,* particularly when they arrive from ranges greater than $2R_{un}$. The apparent range of these ambiguous echoes can result in error and confusion.

Another problem with multiple-time-around echoes is that clutter echoes from ranges greater than R_{un} can mask unambiguous target echoes at the shorter ranges.

Some types of radars, such as pulse doppler radars, always operate with a prf that can result in range ambiguities. As described in Sec. 3.9, range ambiguities are tolerated in a pulse doppler radar in order to achieve the benefits of a high prf when detecting moving targets in the midst of clutter. Resolving the range ambiguities is an important part of the operation of pulse doppler radars.

The existence of multiple-time-around echoes cannot be readily recognized with a constant prf waveform. Consider the three targets labeled *A*, *B*, and *C* in Fig. 2.24a. Target *A* is within the unambiguous range interval R_{un}, target *B* is at a distance greater than R_{un} but less than $2R_{un}$, while target *C* is greater than $2R_{un}$ but less than $3R_{un}$. Target *B* is a second-time-around echo; target *C* is a multiple-time-around echo. When these three pulse repetition intervals, or sweeps, are superimposed on a radar display such as the A-scope of Fig. 2.24b or a PPI, the ambiguous echoes (*B* and *C*) look no different from the unambiguous-range echo of *A*. Only the range of *A* is correct, but it cannot be determined from this display that the other two are not at their apparent range.

Ambiguous-range echoes can be recognized by changing the prf of the radar. When the prf is changed, the unambiguous echo (at a range less than R_{un}) remains at its true range. Ambiguous-range echoes, however, appear at different apparent ranges for each

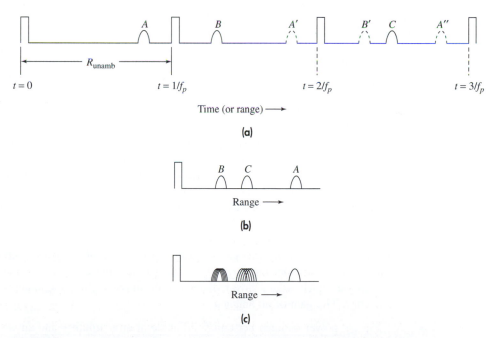

Figure 2.24   Multiple-time-around radar echoes that give rise to ambiguities in range. (a) Three targets A, B, and C, where A is within the unambiguous range R_{un}, B is a second-time-around echo, and C is a multiple-time-around echo; (b) the appearance of the three echoes on an A-scope; (c) appearance of the three echoes on the A-scope with a changing prf.

prf. An example of how these three echoes might appear on an A-scope is shown in Fig. 2.24c. A similar effect would be seen on the PPI. Thus the ambiguous target ranges can be readily identified.

If the first pulse repetition frequency f_1 has an unambiguous range R_{un1}, and if the apparent range measured with prf f_1 is denoted R_1, then the true range is one of the following

$$R_{true} = R_1, \text{ or } (R_1 + R_{un1}), \text{ or } (R_1 + 2R_{un1}), \text{ or } \dots$$

Anyone of these might be the true range. To find which is correct, the prf is changed to f_2 with an unambiguous range R_{un2}, and if the apparent measured range is R_2, the true range is one of the following

$$R_{true} = R_2, \text{ or } (R_2 + R_{un2}), \text{ or } (R_2 + 2R_{un2}), \text{ or } \dots$$

The correct range is that value which is the same with the two prfs. In theory, two prfs can resolve the range ambiguity; but in practice, three prfs are often used for increased accuracy and avoiding false values.

The pulse repetition frequency may be changed pulse to pulse, every half beamwidth (with a scanning antenna), or on every rotation of the antenna. Each method has been used, and there are benefits as well as limitations of each.

2.11 ANTENNA PARAMETERS

Almost all radars use *directive antennas* with relatively narrow beamwidths that direct the energy in a particular direction. The antenna is an important part of a radar. As was found from the derivation of the radar equation in Sec. 1.2, it serves to place energy on target during transmission, collect the received echo energy reflected from the target, and determine the angular location of the target. There is always a trade between antenna size and transmitter size when long-range performance is required. If one is small the other must be large to make up for it. This is one reason why large antennas are generally desirable in most radar applications when practical considerations do not limit their physical size. Thus far, the antenna has been thought of as a mechanically steered reflector. Radar antennas can also be electronically steered or mechanically steered phased arrays, as described in Chap. 9.

Antenna Gain   The antenna gain $G(\theta,\phi)$ is a measure of the power per unit solid angle radiated in a particular direction by a directive antenna compared to the power per unit solid angle which would have radiated by an omnidirectional antenna with 100 percent efficiency. The gain of an antenna is

$$G(\theta,\phi) = \frac{\text{power radiated per unit solid angle at an azimuth } \theta \text{ and an elevation } \phi}{(\text{power accepted by the antenna from the transmitter})/4\pi} \quad \text{[2.53]}$$

This is the *power gain* and is a function of direction. If it is greater than unity in some directions, it must be less than unity in other directions. There is also the *directive gain*, which has a similar definition except that the denominator is the power *radiated* by the antenna per 4π steradians rather than the power *accepted* from the transmitter. The difference between the two is that the power gain accounts for losses within the antenna. The power gain is more appropriate for the radar equation than the directive gain, although there is usually little difference between the two in practical radar antennas, except for the phased array. The power gain and the directive gain of a radar antenna are usually considered to be the same in this text. When they are significantly different, then the distinction between the two must be made. In the radar equation, it is the maximum power gain that is meant by the parameter G.

Effective Area and Beamwidth   It was mentioned in Sec 1.2 that the directive gain G and the effective area A_e (sometimes called *effective aperture*) of a lossless antenna are related by[55]

$$G = \frac{4\pi A_e}{\lambda^2} = \frac{4\pi \rho_a A}{\lambda^2} \qquad\qquad \text{[2.54]}$$

where λ = wavelength, ρ_a = antenna aperture efficiency, and A = physical area of the antenna. The gain of an antenna is approximately equal to

$$G \approx \frac{26{,}000}{\theta_B \phi_B} \qquad\qquad \text{[2.55]}$$

where θ_B and ϕ_B are the azimuth and elevation half-power beam widths, respectively, in degrees. This results in a gain of 44 dB for a one-degree pencil beam. [The justification for this equation is given in Sec. 9.2 after Eq. (9.5c).]

The half-power beamwidth of an antenna also depends on the nature of the aperture illumination and, therefore, the sidelobe level. When no specific information is available regarding the nature of the antenna, the following relation between beamwidth and antenna dimension D is sometimes used

$$\theta_B = 65\, \lambda/D \qquad \text{degrees} \qquad\qquad \text{[2.56]}$$

where the wavelength λ has the same units as the aperture dimension D. When D is the horizontal dimension of the antenna, the beamwidth θ_B is the azimuth beamwidth; when D is the vertical dimension, θ_B is the elevation beamwidth. Equation (2.56) might apply for an antenna with 25 to 28 dB peak sidelobe level.

The half-power beamwidth of an antenna can be measured somewhat accurately, but the antenna gain "is probably one of the least accurate measurements made on an antenna system."[56] It has been questioned whether "many gain estimates are more accurate than $\pm$ 0.5 dB.[57] Therefore, it is not necessary to give the gain of the usual radar antenna (in dB) to more than one decimal place.

Revisit Time   The *revisit time* is the time that an antenna takes to return to view the same region of space. It usually represents a compromise between (1) the need to collect sufficient energy (a sufficient number of pulses) for the detection of weak targets and (2) the need to have a rapid re-measurement of the location of a moving target so as to quickly determine its trajectory. The revisit time is also called the *scan time;* and both are inversely related to the *rotation rate* (rpm) of a scanning antenna. The revisit times of long-range civil air-traffic-control radars are generally in the vicinity of 10 to 12 s, corresponding to an antenna rotation rate of 6 to 5 rpm. Military air-surveillance radars, unlike civil radars, have to detect and track high-speed maneuvering targets. A revisit time of 10 to 12 seconds is too long. Revisit times for long-range military radars are more like 4 seconds (15 rpm). Short-range military radars that must detect and quickly respond to low-flying high-speed targets that pop up over the near horizon generally require revisit times of 1 or 2 seconds (60 or 30 rpm), depending on the radar type and design. A small civil marine radar commonly found on boats and ships might have a rotation rate of about 20 rpm (3-s revisit time). High-resolution radars which monitor the ground traffic at major airports, as the ASDE shown in Fig. 1.6, generally have rotation rates of 60 rpm.

Beam Shape   Radars employ either *fan beams* or *pencil beams.* The beam width of the pencil beam, Fig. 2.25a, in the horizontal plane is equal or almost equal to the beamwidth in the vertical plane. Its beamwidth is generally less than a few degrees; one degree might be typical. It is found in radars that must have accurate location measurements and resolution in both azimuth and elevation. The pencil beam is popular for tracking radars, 3-D radars (rotating air-surveillance radars that obtain elevation angle measurement as well as azimuth and range), and many phased array radars.

The fan-beam antenna, Fig. 2.25b, has one angle small compared to the other. In air-surveillance radars that use fan beams, the azimuth beamwidth might typically be one or a few degrees, while the elevation beamwidth might be from perhaps four to ten times the azimuth beamwidth. Fan beams are found with 2-D (range and azimuth) air-surveillance radars that have to search out a large volume of space. The narrow beamwidth is in the horizontal coordinate so as to obtain a good azimuth angle measurement. The elevation beamwidth is broad in order to obtain good elevation coverage, but at the sacrifice of an elevation angle measurement.

Figure 2.25   Typical radar antenna pattern types: (a) pencil beam; (b) fan beam; (c) stacked beams, or 3-D; and (d) shaped beam, such as cosecant-squared shaping.

Pencil beam

(a)

Fan beam

(b)

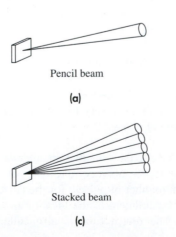

Stacked beam

(c)

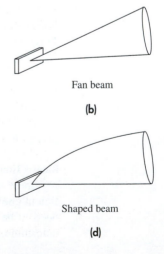

Shaped beam

(d)

A single pencil beam has difficulty searching out a large angular volume. Employing a number of scanning pencil beams (3 to 9 have been used) can solve this problem, as is found in some 3-D radars.[58] Sometimes in a 3-D radar a *stacked-beam* coverage is used in the vertical dimension. This consists of a number of contiguous fixed pencil beams, as in Fig. 2.25c. Six to sixteen contiguous beams have been typical in the past.

Usually the shape of a fan beam has to be modified to obtain more complete coverage. An example is the cosecant-squared shaped beam, indicated by Fig. 2.25d.

Cosecant-Squared Antenna Pattern   The coverage of a simple fan beam is not adequate for the detection of aircraft targets at high altitudes close to the radar. The detection of such targets requires a very broad fan beam and therefore an undesirable low gain. In addition, a fan beam radiates more energy at the higher elevation angles than is needed since aircraft are not at long ranges when viewed at high elevation angles. To obtain better illumination of close-in targets at high elevation angles, the fan-beam pattern is modified so that its gain in the elevation-angle coordinate is proportional to the square of the cosecant of the elevation angle, Fig. 2.25d. In other words, a *cosecant-squared antenna* is one whose elevation gain as a function of elevation angle ϕ is

$$G(\phi) = G(\phi_0) \, \frac{\csc^2 \phi}{\csc^2 \phi_0} \qquad \text{for } \phi_0 \leq \phi \leq \phi_m \qquad\qquad [2.57]$$

where $G(\phi)$ is the gain at the elevation angle ϕ, and ϕ_0 and ϕ_m are the angular limits between which the beam follows a cosecant-squared shape. Section 9.7 discusses this type of antenna and gives an approximate expression for the reduction in gain and effective area that occurs because of the cosecant-squared beam shape. The loss in gain or effective area can be as much as 3 dB. Also discussed in Sec. 9.7 are other forms of beam shaping that can have an effect on the gain and radar coverage.

In designing a cosecant-squared antenna for a surface-based air-surveillance radar, an elevation beamwidth ϕ_B is selected, based on the elevation coverage required at the longest range. The elevation beamwidth should be as narrow as conditions will allow (so as to obtain as high a gain as practical). One method for selecting the basic beamwidth ϕ_B is to place the maximum of the antenna beam at an elevation $\phi_B/2$, so that the direction corresponding to the lower half-power point of the radiation pattern is directed towards the horizon ($\phi = 0$ degrees). The upper half-power direction is set so that it passes through the point defined by the maximum range and the maximum target altitude. This elevation angle is $\phi_0 = \phi_B$. Then the antenna gain is made to vary as cosecant-squared from ϕ_0 to the maximum elevation angle ϕ_m. This assumes the earth is flat. With long-range radars, the round earth must be considered and the above characterization of cosecant-squared coverage must be further modified. (It will not be as simple.)

Most surface-based 2-D air-surveillance radars use some form of shaped beam, such as the cosecant-squared, to provide the desired coverage. Some airborne ground-surveillance radars also use the cosecant-squared beam shape, but inverted compared to the surface-based air-surveillance radar.

The cosecant-squared antenna for an air-surveillance radar has the property that the echo power P_r received from a target of constant cross section flying at constant altitude h is independent of the target's range R from the radar. This assumes a flat earth and that the surface-reflection multipath which causes lobing is ignored. (The effect of lobing due to multipath is discussed in Sec. 8.2). Substituting the gain of the cosecant-squared antenna [Eq. (2.57)] into a variation of the simple form of the radar equation [Eq. (1.6) with $A_e = G\lambda^2/4\pi$] gives

$$P_r = \frac{P_t G^2 \lambda^2 \sigma}{(4\pi)^3 R^4} = \frac{P_t G^2(\phi_0)\, \csc^4\phi\, \lambda^2\sigma}{(4\pi)^3\, \csc^4\phi_0\, R^4} = K_1\, \frac{\csc^4\phi}{R^4} \qquad\qquad \textbf{[2.58]}$$

where K_1 is a constant. The height h of the target is assumed constant, and since $\csc\phi = R/h$, the received power becomes $P_r = K_1/h^4 =$ constant. The echo signal is therefore independent of range when the antenna elevation pattern is proportional to the cosecant-squared of the elevation angle ϕ, with the assumptions stated.

2.12 SYSTEM LOSSES

At the beginning of this chapter it was said that one of the important factors omitted from the simple radar equation was the loss that occurs throughout the radar system. The loss due to the integration of pulses and the loss due to a target with a fluctuating cross section have been already encountered in this chapter. Propagation losses in the atmosphere are considered later, in Chap. 8. This section considers the various *system losses,* denoted L_s, not included elsewhere in the radar equation. Some system losses can be predicted beforehand (such as losses in the transmission line); but others cannot (such as degradation when operating in the field). The latter must be estimated based on experience and experimental observations. They are subject to considerable variation and uncertainty. Although the loss associated with any one factor may be small, there can be many small effects that add up and result in significant total loss. The radar designer, of course, should reduce known losses as much as possible in the design and development of the radar. Even with diligent efforts to reduce losses, it is not unusual for the system loss to vary from perhaps 10 dB to more than 20 dB. (A 12-dB loss reduces the range by one-half.)

All numerical values of loss mentioned in this section, including the above values of system loss, are meant to be illustrative. They can vary considerably depending on the radar design and how the radar is maintained.

System loss, L_s (a number greater than one), is inserted in the denominator of the radar equation. It is the reciprocal of efficiency (number less than one). The two terms (loss and efficiency) are sometimes used interchangeably.

Microwave Plumbing Losses   There is always loss in the transmission line that connects the antenna to the transmitter and receiver. In addition, there can be loss in the various microwave components, such as the duplexer, receiver protector, rotary joint, directional couplers, transmission line connectors, bends in the transmission lines, and mismatch at the antenna.

Transmission Line Loss   The theoretical one-way loss in decibels per 100 ft for standard waveguide transmission lines is shown in Table 2.2.[59] Since the same transmission line

Table 2.2 Attenuation of Rectangular Waveguides*

Frequency Band	EIA Waveguide Designation†	Frequency Range (GHz) for Dominant TE_{10} Mode	Outer Dimensions and Wall Thickness, inches	Theoretical Attenuation, Lowest to Highest Frequency, dB/100 ft (one-way)
UHF	WR-2100	0.35–0.53	$21.25 \times 10.75 \times 0.125$	0.054–0.034
L band	WR-770	0.96–1.45	$7.95 \times 4.1 \times 0.125$	0.201–0.136
S band	WR-284	2.6–3.95	$3.0 \times 1.5 \times 0.08$	1.102–0.752
C band	WR-187	3.95–5.85	$2.0 \times 1.0 \times 0.064$	2.08–1.44
X band	WR-90	8.2–12.40	$1.0 \times 0.5 \times 0.05$	6.45–4.48
K_u band	WR-62	12.4–18.0	$0.702 \times 0.391 \times 0.04$	9.51–8.31
Ka band	WR-28	26.5–40.0	$0.36 \times 0.22 \times 0.04$	21.9–15.0

*After "Reference Data for Engineers," 8th ed., M. E. Van Valkenburg, editor-in-chief, Chap. 30, *Waveguides and Resonators*, by T. Itoh, SAMS Prentice Hall Computer Publishing, Carmel, Indiana, 1993.

†UHF and L-band guides are made of aluminum, K_a band is silver, the rest are copper-zinc alloy.

generally is used for both transmission and reception, the loss to be inserted in the radar equation is twice the one-way loss. Flexible waveguides and coaxial lines can have higher losses than conventional waveguides. At the lower radar frequencies, the transmission line introduces little loss unless its length is exceptionally long. At the higher frequencies, attenuation may not always be small and may have to be taken into account. When practical, the transmitter and receiver should be placed close to the antenna to keep the transmission-line loss small. Additional loss can occur at each connection or bend in the line. Connector losses are normally negligible, but if the connection is poorly made, it can contribute measurable attenuation.*

Duplexer Loss   The loss due to a gas duplexer that protects the receiver from the high power of the transmitter is generally different on transmission and reception. It also depends, of course, on the type of duplexer used. Manufacturers' catalogs give values for a duplexer's *insertion loss* and (for a gas duplexer) the *arc loss* when in the fired condition. The radar might also have a waveguide shutter, with some insertion loss, that closes when the radar is shut down so as to protect the receiver from extraneous high-power signals when its duplexer is not activated. A solid-state receiver protector is often used as well as a solid-state attenuator in the receiver transmission line for applying sensitivity time control (STC). The duplexer and other related devices that might be used could, in some cases, contribute more than 2 dB of two-way loss.

*At a particular radar laboratory many years ago, an old *L*-band air-surveillance radar was used as an experimental test-bed system. Its range was poor, and the engineers attributed this to its "age"—whatever that meant. Its poor performance was tolerated for many years. One day, a technician working near the transmission line to the antenna happened to find, accidentally, that one of the transmission-line connectors had not been properly secured. He tightened a few bolts and the radar "miraculously" achieved a significant increase in performance. Sometimes, it's the little things that count!

Example   Although each radar can have different losses, an *S*-band (3-GHz) radar might have, by way of illustration, two-way microwave plumbing losses as follows:

100 ft of RG-113/U aluminum waveguide line	1.0 dB
Duplexer and related devices	2.0 dB
Rotary joint	0.8 dB
Connectors and bends (estimate)	0.3 dB
Other RF devices	0.4 dB
Total "example" microwave plumbing loss	4.5 dB

Antenna Losses   The antenna efficiency, discussed in Chap. 9, is not included as a system loss. It should be accounted for in the antenna gain. Shaping of the antenna pattern, for example, to provide a csc^2 pattern (Sec. 2.11), results in a loss that is included as an additional lowering of the antenna gain (Sec. 9.11) rather than as a system loss. The beam-shape loss of a surveillance radar, however, is usually included as part of the system losses.

Beam-Shape Loss   The antenna gain that appears in the radar equation is assumed to be a constant equal to the maximum value. But in reality the train of pulses returned from a target by a scanning antenna is modulated in amplitude by the shape of the antenna beam, Fig. 2.26. Only one out of *n* pulses has the maximum antenna gain *G,* that which occurs when the peak of the antenna beam is in the direction of the target. Thus the computations of probability of detection (as given earlier in this chapter) have to take account of an amplitude-modulated train of pulses rather than constant-amplitude pulses. Some published probability of detection computations and computer programs for the radar equation account for the beam-shape loss. Others do not. When using published values of detection probability one needs to determine whether the beam-shape effect has been included or whether it must be accounted for separately. In this text, the approach is to

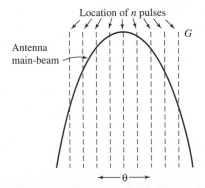

Figure 2.26   Nature of the beam-shape loss. The simple radar equation assumes *n* pulses are integrated, all with maximum antenna gain *G*. The dashed lines represent *n* pulses with maximum gain, and the solid curve is the antenna main-beam pattern $G(\theta)$. Except for the pulse at the center of the beam, the actual pulses illuminate the target with a gain less than the maximum.

assume a constant-amplitude pulse train as determined by the maximum antenna gain, and then add a beam-shape loss to the total system losses in the radar equation. This is a simpler, albeit less accurate, method. It is based on calculating the reduction in total signal energy received from a modulated train of pulses compared to what would have been received from a constant-amplitude pulse train. As defined, it does not depend on the probability of detection.

To obtain the beam-shape loss, the one-way-power antenna pattern is approximated by a gaussian shape given by $\exp[-2.78\,\theta^2/\theta_B^2]$, where θ is the angle measured from the center of the beam and θ_B is the half-power beamwidth. If n_B is the number of pulses received within the one-way half-power beamwidth θ_B, and n the total number of pulses integrated (n does not necessarily equal n_B), the beam-shape loss is

$$\text{Beam-shape loss} = \frac{n}{1 + 2\sum_{k=1}^{(n-1)/2} \exp[-5.55k^2/(n_B-1)^2]} \qquad [2.59]$$

This expression applies for an odd number of pulses with the middle pulse appearing at the beam maximum. For example, if $n = 11$ pulses are integrated, all lying uniformly between the 3-dB beamwidth ($n = n_B$), the beam-shape loss is about 2 dB.

The above applies to a fan beam. It also applies to a pencil beam if the target passes directly through the center of the beam. If the target passes through any other part of the pencil beam, the maximum signal will be reduced. The beam-shape loss is increased, therefore, by the square of the maximum antenna gain seen (if the antenna were to pass through the beam center) to the square of the maximum gain actually seen (when the antenna passes through other than the beam maximum). The ratio is the square because of the two-way radar propagation.

When a large number of pulses are integrated, the scanning loss was found by Blake[60] to be 1.6 dB for a fan beam scanning in one coordinate and 3.2 dB for a pencil beam scanning in two coordinates. Blake's values are commonly used as the beam-shape loss in the radar equation, unless the number of pulses integrated is small.

A similar loss must be taken into account when searching a volume with a step-scanning pencil beam antenna (as with a phased array) since not all regions of space are illuminated with the same value of antenna gain. (In *step scanning,* the antenna beam is stationary and dwells in a fixed direction until all n pulses are collected, and then rapidly switches to dwell in a new direction.) Some tracking radars, such as conical scan, also have a loss due to the antenna beam not illuminating the target with maximum gain.

Scanning Loss   When the antenna scans rapidly enough, relative to the round-trip time of the echo signal, the antenna gain in the direction of the target on transmit might not be the same as that on receive. This results in an additional loss called the *scanning loss.* It can be important for some long-range scanning radars such as those designed for space surveillance or ballistic missile defense, rather than for most air-surveillance radars.

Radome   Loss introduced by a radome (Sec. 9.17) will depend on the type and the radar frequency. A "typical" ground-based metal space-frame radome might have a two-way transmission loss of 1.2 dB at frequencies ranging from L to X band.[61] Air-supported radomes can have lower loss; the loss with dielectric space-frame radomes can be higher.

Phased Array Losses Some phased array radars have additional transmission-line losses due to the distribution network that connects the receiver and transmitter to each of the many elements of the array. These losses are more correctly included as a reduction in the antenna power gain, but they seldom are. When not included as a loss in antenna gain, they should be included under the system losses.

Signal Processing Losses Sophisticated signal processing is prevalent in modern radars and is very important for detecting targets in clutter and in extracting information from radar echo signals. Unfortunately, signal processing can introduce loss that has to be tolerated. The factors described below can introduce significant loss that has to be accounted for; doppler processing radars might have even greater loss.

Nonmatched Filter There can be from 0.5 to 1.0 dB of loss due to a practical, rather than ideal, matched filter (Sec. 5.2) A similar loss can occur with a pulse-compression filter (which is an example of a matched filter).

Constant False-Alarm Rate (CFAR) Receiver As mentioned in Sec. 5.7, this loss can be more than 2.0 dB depending on the type of CFAR.

Automatic Integrators The binary moving-window detector, for example, can have a theoretical loss of 1.5 to 2.0 dB (Sec. 5.6). Other automatic integrators might have more or less loss.

Threshold Level A threshold is established at the output of the radar receiver to achieve some specified probability of false alarm or average false-alarm time (Sec. 2.5). Because of the exponential relationship between false-alarm time and the threshold level, the threshold might be set at a slightly higher level as a safety factor to prevent excessive false alarms. Depending on how accurately the threshold can be set and maintained, the loss might be only a small fraction of a dB.

Limiting Loss Some radars might use a limiter in the radar receiver. An example is pulse-compression processing to remove amplitude fluctuations in the signal. The so-called Dicke-fix, an electronic counter-countermeasure to reduce the effects of impulsive noise, employs a hard limiter. Early MTI radars used hard limiters, but this is now considered poor practice and is almost never used since they reduce the clutter attenuation that can be obtained (Sec. 3.7). Analysis of an ideal bandpass hard limiter shows that, for small signal-to-noise ratio, the reduction in the signal-to-noise ratio of a sinewave imbedded in narrowband gaussian noise is theoretically $\pi/4$, which is about 1 dB.[62] Hard limiting with some forms of pulse compression can introduce greater loss (Table 6.6).

Straddling Loss A loss, called the *range straddling loss,* occurs when range gates are not centered on the pulse or when, for practical reasons, they are wider than optimum.

Likewise in a doppler filter bank (Sec. 3.4) there can be a *filter straddling loss* when the signal spectral line is not centered on the filter. These occur in both analog and digital processing.

Sampling Loss   When digital processing is employed, a related loss to the straddling loss can occur when the video signal after the matched filter is sampled prior to digitizing by the analog-to-digital (A/D) converter. If there is only one sample per pulse width, sampling might not occur at the maximum amplitude of the pulse. The difference between the sampled value and the maximum pulse amplitude represents a *sampling loss*. The loss is about 2 dB when the sampling is at a rate of once per pulse width (applies for a probability of detection of 0.90 and probability of false alarm of 10^{-6}).[63] The larger values occur with higher probability of detection. Decreasing the sampling interval rapidly decreases the loss. When two samples per pulse are taken, the loss is approximately 0.5 dB; and with three samples per pulse, it is under 0.2 dB.

Losses in Doppler-Processing Radar*   When range and/or doppler ambiguities exist (as discussed in Sec. 2.10 and Chap. 3 for MTI and pulse doppler radars), multiple redundant waveforms may be used to resolve the ambiguities or to prevent "blind speeds." Compared to a radar that has no ambiguities the redundant waveforms can represent a significant loss. The use of redundant waveforms is seldom considered as a system loss; but it can certainly affect the range of a radar. For example, in some medium prf pulse doppler radars (Sec. 3.9), eight different waveforms, each with a different prf, might need to be transmitted so as to obtain at least three dwells (with no missed targets due to blind speeds) in order to resolve the range ambiguities. The represents a loss of signal of 8/3, or 4.3 dB.

There can be an eclipsing loss in pulse doppler radars when echoes from (ambiguous) multiple-time-around targets arrive back at the radar at the same time that a pulse is being transmitted. MTI doppler processing also introduces loss due to the shape of the doppler (velocity) filters if the target velocity does not correspond to the maximum response of the doppler filter. Fill pulses in MTI and pulse doppler radar sometimes are necessary, but they represent wasted pulses from the point of view of detection of signals in noise.

Losses due to doppler processing are not always included as part of the system losses. This is justified in an MTI radar by noting that the clutter that the doppler processing is designed to remove generally does not occur at the maximum range of a radar, which is where the range performance of a radar usually is determined. Nevertheless, it should be recognized that MTI radars that employ the doppler frequency shift to detect moving targets in the presence of large clutter echoes (as discussed in Chap. 3) can seriously reduce the ability of a radar to detect a target when no clutter is present. This is why MTI processing is disconnected at ranges beyond where clutter is not expected.

Fill pulses are sometimes used in an MTI radar when the pulses are processed in batches, as in the MTD radar of Sec. 3.6. They are also sometimes used with high prf

*Doppler processing is described in Chap. 3. The reader not familiar with MTI and pulse doppler radar can skip this subsection.

doppler radar to avoid the effects of multiple-time-around clutter echoes. They are necessary for performing some types of doppler processing, but they are wasted pulses when signal-to-noise, and not signal-to-clutter, is important. They have not usually been considered as introducing a system loss, but they could if the fill pulses affect the detection of signals in noise.

Collapsing Loss   If the radar were to integrate additional noise samples along with signal-plus-noise pulses, the added noise would result in a degradation called the *collapsing loss*. An example is in a 3-D radar that has a "stack" of multiple independent pencil beams in elevation. If the outputs from the N beams are superimposed on a single PPI display, at a given range resolution cell that contains the target echo the display will add $N-1$ noise samples along with the single target echo. A collapsing loss can also occur when the output of a high-resolution radar is shown on a display whose resolution is coarser than that inherent in the radar. If the radar receiver output is automatically processed and thresholded rather than rely on an operator viewing a display to make the detection decision, there need not be a collapsing loss in the above two examples.

The mathematical derivation of the collapsing loss, assuming a square-law detector, may be carried out as suggested by Marcum.[64] He has shown that the integration of m noise pulses along with n signal-to-noise pulses with signal-to-noise ratio per pulse $(S/N)_n$, is equivalent to the integration of $m+n$ signal-to-noise pulses each with signal-to-noise ratio $n(S/N)_n/(m+n)$. Thus the collapsing loss, $L_c(m,n)$, is equal to the ratio of the integration loss L_i (Sec. 2.6) for $m+n$ pulses to the integration loss for n pulses, or

$$L_c(m,n) = \frac{L_i(m+n)}{L_i(n)} \qquad [2.60]$$

For example, assume there are 10 signal-plus-noise pulses integrated along with 30 noise-only pulses, and that $P_d = 0.90$ and $P_{fa} = 1/n_f = 10^{-8}$. From Fig. 2.8a, $L_i(40) = 3.5$ dB and $L_i(10) = 1.7$ dB, so that the collapsing loss according to Eq. (2.60) is 1.8 dB.

The above applies for a square-law detector. Trunk[65] has shown that the collapsing loss for a linear detector can be much greater than that for the square-law detector when the number of pulses integrated is small and the collapsing ratio is large, where collapsing ratio is defined as $(m+n)/n$. As the number of pulses becomes large, the difference between the two detectors becomes less, especially for low values of collapsing loss.

Operator Loss   Most modern high-performance radars provide the detection decision automatically without intervention of a human operator. Processed information is presented directly to an operator or to a computer for some other action. In the early days of radar, operators were depended upon to find targets on a display. Sometimes, when the radar range performance was less than predicted, the degradation of performance was attributed to an operator loss. As engineers began to learn more about radar and the performance of the operator, it was found that an alert, motivated, well-trained operator can perform as well as indicated by theory for electronic detection. For this reason, an operator loss factor is seldom included even if the operator makes the detection decision. (When an operator is used to make detection decisions from the output of a radar display, he or she should be replaced with a rested, alert operator every 20 to 30 min, or else performance can seriously degrade.)

Equipment Degradation   It is not uncommon for radars operated under field conditions to have lower performance than when they left the factory. This loss of performance can be recognized and corrected by regularly testing the radar, especially with built-in test equipment that automatically indicates when equipment deviates from specifications. It is not possible to be precise about the amount of loss to be assigned to field degradation. From one to three dB might be used when no other information is available.

Propagation Effects   The effect of the environment on the propagation of radar waves can be significant and can make the actual range considerably different from that predicted as if the radar were operated in free space. Propagation effects can increase the free-space range as well as decrease it. The major effects of propagation on radar performance are: (1) reflections from the earth's surface, which cause the breakup of the antenna elevation pattern into lobes; (2) refraction, or bending, of the propagating wave by the variation of the atmosphere's index of refraction as a function of altitude, which usually increases the radar's range; (3) propagation in atmospheric ducts, which can significantly increase the range at low altitudes; and (4) attenuation in the clear atmosphere or in precipitation, which usually is negligible at most radar frequencies.* Propagation effects are not considered part of the system losses. They are accounted for separately by a *propagation factor,* usually denoted as F^4 and, when appropriate, by an *attenuation factor* $\exp[-2\alpha R]$, where α is the attenuation coefficient (nepers per unit distance[†]), and R is the range. (This assumes the attenuation coefficient is independent of range.) The factor F^4 mainly includes the effects of lobing of the elevation antenna pattern due to reflection from the earth's surface (Sec. 8.2), but it can include all other propagation effects except attenuation. Both the propagation and the attenuation factors, as written, are in the numerator of the radar equation.

The effect of the environment on radar propagation and performance is the subject of Chap. 8.

Radar System Losses—the Seller and the Buyer   There is no universally agreed upon procedure for determining system losses or what losses should be considered when predicting radar performance. It is natural for a person selling a radar to be optimistic about the total system loss and claim a lower loss than might a potential buyer of the radar or an independent evaluator of a radar's performance. The advertised performance predicted by a radar manufacturer cannot be adequately verified or compared to the advertised predictions for similar radars by other manufacturers without complete knowledge of the losses that each radar designer has included.

---

*Although attenuation in rain is usually not a factor in radar performance, the reflection from rain that competes with the target echo can seriously degrade the performance of a radar at the higher microwave frequencies, as discussed in Sec. 7.6.

[†]Neper is a dimensionless unit for expressing the ratio of two values of amplitude, and thus is used to express attenuation. The number of nepers is the natural logarithm of the amplitude ratio $\ln(A_2/A_1)$. Radar engineers more usually describe attenuation in decibels. One decibel equals 8.686 nepers.

2.13 OTHER RADAR EQUATION CONSIDERATIONS

Prediction of Radar Range   This chapter discussed many, but not all, of the factors that might enter into the radar equation for the prediction of range, when limited by receiver noise. The simple form of the radar equation we started with as Eq. (2.1), with the modifications indicated in this chapter, now becomes

$$R_{max}^4 = \frac{P_{av}GA\rho_a\sigma nE_i(n)F^4 e^{-2\alpha R_{max}}}{(4\pi)^2 kT_0 F_n(B\tau)f_p(S/N)_1 L_f L_s}$$

[2.61]

where

R_{max} = Maximum radar range, m

P_{av} = Average transmitter power, W

G = Antenna gain

A = Antenna area, m^2

ρ_a = Antenna aperture efficiency

σ = Radar cross section of the target, m^2

n = Number of pulses integrated

$E_i(n)$ = Integration efficiency

F^4 = Propagation factor

α = Attenuation coefficient, nepers per unit distance

k = Boltzmann's constant = 1.38×10^{-23} J/deg

T_0 = Standard temperature = 290 K

F_n = Receiver noise figure

B = Receiver bandwidth, Hz

τ = Pulse width, s

f_p = Pulse repetition frequency, Hz

$(S/N)_1$ = Signal-to-noise ratio required as if detection were based on only a single pulse

L_f = Fluctuation loss (for a Swerling target model)

L_s = System loss

The product $kT_0 = 4 \times 10^{-21}$ w/Hz. In most radar designs the product $B\tau \approx 1$. The average power can be expressed as $P_{av} = P_t\tau f_p = E_p f_p$, where E_p is the energy in a transmitted pulse. The total energy transmitted in n pulses is $E_T = nE_p$. The signal-to-noise ratio for a rectangular pulse can be expressed as an energy ratio since $S/N = (E/\tau)/N_0 B = E/(N_0 B\tau)$, where E is the energy of the received pulse, and N_0 is the receiver noise power per unit bandwidth. When $B\tau = 1$, then $(S/N)_1 = (E/N_0)_1$. Omitting the propagation factors, atmospheric attenuation, and the fluctuation loss, the radar equation can be written

$$R_{max}^4 = \frac{E_T GA\rho_a\sigma E_i(n)}{(4\pi)^2 kT_0 F_n(E/N_0)_1}$$

[2.62]

This radar equation can be applied to any waveform, not just a rectangular pulse, so long as a matched filter is used on reception and the energy parameters are properly defined.

The radar equation, Eq. (2.61), developed in this chapter for a pulse waveform can be modified for other radars, such as CW, FM-CW, pulse doppler, and MTI. It also can be adapted to specialized radar applications such as the surveillance-radar equation, derived next. Tracking radars, synthetic aperture radars, HF over-the-horizon radars, and other specialized radars require modification of the classical radar equation to account for the special attributes of different radar systems.

When radar performance is limited by clutter echoes rather than receiver noise, the radar equation takes on a completely different form from the equations presented here, as discussed in Chap. 7. When assessing radar performance when hostile ECM noise jamming dominates, receiver noise in the denominator of the radar equation is replaced by the jamming noise that enters the radar receiver.

Surveillance-Radar Range Equation   The radar equation described so far applies to a radar that dwells on the target for n pulses. The radar equation for a surveillance radar, however, is slightly different since it must account for the defining characteristic of a surveillance radar which is that it search a specified angular region in a given time. The *scan time,* or *revisit time,* is t_s, in seconds. The angular region to be searched is denoted by Ω, in steradians. (A steradian is the area subtended by a solid angle Ω on the surface of a sphere of unit radius. The total solid angle about a point is therefore 4π steradians. If the region Ω, for instance, represents 360° in azimuth and 30° in elevation, the solid angle in steradians is $2\pi \sin 30° = \pi$ steradians.)

The scan time, t_s, is equal to $t_0\Omega/\Omega_0$, where $t_0 = n/f_p$ is the time that the radar beam dwells on the target, n is the number of pulses received as the antenna scans past the target, $f_p = $ prf, and Ω_0 is the solid angular beamwidth that is approximately equal, for small beamwidths, to the product of the azimuth beamwidth θ_a times the elevation beamwidth θ_e in radians. (This assumes θ_A/θ_a and θ_E/θ_e are integers, where θ_A is the total azimuth coverage and θ_E the total elevation coverage.) The antenna gain is approximately $G = 4\pi/\Omega_0$. With the above substitutions into a slightly simplified Eq. (2.61), the surveillance-radar equation becomes

$$R^4_{\max} = \frac{P_{av}A_e\sigma E_i(n)}{4\pi k T_0 F_n (S/N)_1 L_s} \frac{t_s}{\Omega} \qquad [2.63]$$

This equation shows that the important parameters of a surveillance radar under the control of the radar designer are the *average power* and the *effective aperture.* The *power-aperture product,* therefore, is an important measure of the capability of a radar to perform long-range surveillance. The frequency does not appear explicitly. In practice, however, it is easier to achieve high power and large antennas at lower rather than higher frequencies. Furthermore, weather effects are less of a bother at the lower frequencies, which is something not indicated by this form of the surveillance-radar equation.

Although Eq. (2.63) illustrates the basic radar characteristics that affect the range of a surveillance radar, it is not a good equation on which to base a radar design. Too many factors are not explicitly stated. It is better to use Eq. (2.61) and the several auxiliary equations that relate to the surveillance application, such as the number of pulses received per scan [Eq. (2.31)], and the relationship between the scan time and the coverage volume.

The surveillance radar equation does not explicitly contain the number of pulses per dwell. There should, of course, be at least one pulse; but in most cases there need to be more than one pulse. If only one or two pulses are obtained from a target, the beam-shape loss is large. In an MTI or pulse doppler radar for the detection of moving targets in clutter, the greater the time on target, the larger the number of pulses processed, and the greater will be the reduction in clutter (as discussed in Sec. 3.7 on antenna scanning modulation in MTI radar.) The surveillance radar equation given above, therefore, might need to be modified when doppler processing requires a fixed dwell time or minimum number of pulses.

M-out-of-N Criterion   What has been discussed thus far is the probability of detection based on a single scan, or single observation, as the radar antenna scans by the target. A surveillance radar, however, seldom makes a detection decision that a target is present based on only a single observation. One criterion for announcing that a target is present is based on requiring M detections on N scans, where $1 < M \leq N$. For instance, the criterion for detection might be to require a detection (threshold crossing) on each of 2 successive scans, or 2 detections over 3 scans, or 3 out of 4, or 3 out of 5. Denoting the probability of detection on a single trial (scan) by P, the probability of detecting a target on M out of N trials, is given by the classical expression

$$\text{prob } [M \text{ out of } N] = \sum_{k=M}^{N} \frac{N!}{k!(N-k)!} \, P^k(1-P)^{N-k} \qquad \textbf{[2.64]}$$

From this expression the probability of detecting a target on $M = 2$ out of $N = 3$ scans is $3P^2 - 2P^3$. With a 2-out-of-3 criterion, the detection probability is larger than that of a single scan when the single-scan probability is greater than 0.5.

The probability of false alarm with the M-out-of-N criterion also can be found from Eq. (2.64). It will be much lower than the single-scan probability of false alarm. This means that a higher false-alarm probability per scan can be tolerated in order to achieve a specified overall false-alarm probability. For example, if the single-scan false-alarm probability were 10^{-8}, Eq. (2.64) shows that the probability of obtaining a false report of a target, when the detection criterion is 2 out of 3, would be 3×10^{-16}, which is a very low number. If the false-report probability is to be 10^{-8} when the detection criterion is 2 out of 3, the single-scan false-alarm probability can be set equal to 0.6×10^{-4}, which results in a lowering of the required detection threshold with a concurrent savings in transmitter power (or its equivalent).

Track Establishment as a Detection Criterion   Many modern air-surveillance radars declare that a target is present when a track is established rather than when a single detection decision is made. As discussed in Sec. 4.9, establishment of a track requires multiple observations of a target. Since the likelihood is very small that noise alone can establish a logical track, the single-observation probability of false alarm can be relaxed. Thus the false-alarm probability of a threshold-crossing with noise alone might be as high as 10^{-3} without excessive false track-reports. One criterion used to establish a track is that there be target echoes detected on at least 3 out of 5 scans. This is similar to the *M*-out-of-*N* criterion, except for the added constraint that the track should be within an expected speed range and not exhibit unusual changes in its trajectory. When the establishment of a valid track is taken as the criterion for the report of a target's presence, a false alarm can be made an exceedingly rare event with a well-designed radar and well-designed tracking algorithms.

Cumulative Probability of Detection If a target is observed over multiple scans, the cumulative probability of detection can be large even though the single-scan probability is small. Cumulative probability is the probability that the target is detected *at least* once on *N* scans. Consider a radar that can observe a target on *N* successive scans of the rotating antenna. It is assumed, for convenience of discussion, that the range does not change significantly over the *N* scans so that the change in received signal power with range need not be taken into account. The probability of detecting a target at least once during the *N* scans is called the *cumulative probability of detection, P_c,* and is written

$$P_c = 1 - (1 - P_d)^N \qquad\qquad [2.65]$$

where P_d = single-scan probability of detection. (The maximum radar range based on the cumulative probability of detection has been said to vary as the third power rather than the more usual fourth power variation based on the single-scan probability.[66,67]) The cumulative probability of detection, however, is *not* a good measure of radar performance since a target-detection decision can seldom be made on the basis of a single threshold crossing. Generally, more than one observation of a detection is needed before a reliable report of the presence of a target can be announced.

Verification of Predicted Range This chapter has shown that there are many factors affecting the range of a radar, and they are not always known accurately. The prediction of the maximum range is not something that can always be done as well as might be desired. Even if one could make an accurate prediction of range, there is the problem of trying to verify the prediction experimentally. Suppose, for example, an air-surveillance radar is required to have a 0.90 probability of detecting a one square meter target at a range of 200 nmi, with a false-alarm probability of 10^{-8}. A large number of observations are needed to insure that the probability of detection is actually 0.90 and not 0.80 or 0.95. It is often difficult to account for the varying target cross section not being exactly one square meter (or whatever other value the radar is designed to detect), especially when the cross section varies with viewing aspect. The effect of atmospheric refraction and multipath lobing of the elevation pattern must be known. One can, and should, experimentally determine the range performance of a radar, but one should not expect highly precise measurements.

Radar range performance of a ground-based air-surveillance radar is sometimes determined experimentally by measuring the *blip-scan ratio* as a function of range. The blip-scan ratio is an experimental approximation to the single-scan probability of detection. It is typically found by having an aircraft fly back and forth at constant altitude on a radial course relative to the radar, and on each scan of the antenna it is recorded whether or not a target blip is detected. This process is repeated many times until sufficient data are obtained to compute, as a function of range, the number of scans (blips) on which the target was detected, divided by the number of times it could have been detected (scans). It provides a measure of performance for a particular aircraft flying at a particular altitude when viewed from the head-on and tail-on aspects. Ducting and other anomalous propagation effects that might occur during testing can make for difficulty, and ought to be avoided if possible.

Just as in tossing a coin many times to determine the fraction of events that it is heads, the blip-scan ratio is a statistical quantity whose accuracy depends on how many times the measurement is attempted. This is a classical statistical problem in Bernoulli trials. It is related to Eq. (2.64), which describes the probability of obtaining at least M successes out of N trials. In the experimental verification of a radar's performance, however, the problem is: given M successes out of N trials, what is the underlying probability of detection. When the number of trials is large, the probability of detection approaches M/N. Figure 2.27, provided to the author by Fred Staudaher, shows pairs of curves that statistically bound the correct values, with a confidence level of 90 percent, when the number of trials N equals 10, 100, and 1000. The abscissa is the experimentally measured blip-scan ratio of M/N. The ordinate is the range of values within which the true value of blip-scan ratio might be, with a specified probability of confidence. For example, assume that

Figure 2.27 The ordinate of each pair of similarly labeled curves gives the range of confidence that the abscissa (measured blip-scan ratio) represents the true value of the blip-scan ratio (or single-scan probability of detection) with a confidence coefficient of 90 percent. N is the number of trials. See text for example.
| (Courtesy of Fred Staudaher.)

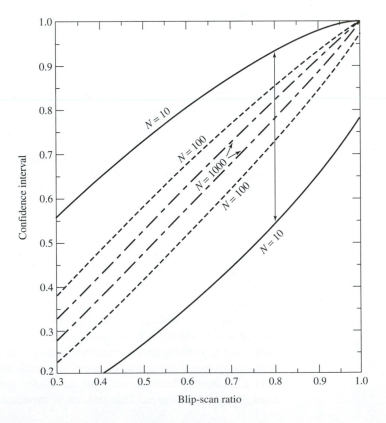

the measured blip-scan ratio *M/N* is 0.80 after 10 trials. Figure 2.27 states that there is 90 percent confidence that the true value lies between 0.54 and 0.93 (see the vertical line in Fig. 2.27). If there were only 10 trials, one would not have a good idea of the true value of the blip-scan ratio. If the measured blip-scan ratio were again 0.80 after 100 trials, its true value has 90 percent confidence of lying between 0.73 and 0.86. With 1000 trials, the true value lies within 0.78 and 0.82 with 90 percent confidence. Thus a relatively large number of trials might be required to be sure the radar meets its performance specifications.

The prediction of the range of a radar is not as exact as might be desired, and the accurate experimental measurement of its range performance is not easy. For this reason, the acceptance of a new radar system by a buyer is not usually based on the costly experimental measurement of performance. One might make a limited number of trials to insure that radar performance is not far out of line. However, for contracting purposes when buying a radar, the performance of the individual subsystems is specified (such as transmitter power, antenna gain, receiver noise figure, receiver dynamic range, and so forth) since these can be measured and used in calculations to predict what the actual performance might be when the radar is operated as a system. If each of the subsystem specifications are met, and if the specifications are properly devised, the buyer can be confident that the radar will perform as predicted.

Accuracy of the Radar Range Computation   There are those who believe that each parameter that enters into the computation of range should be determined with the highest accuracy possible. The author, however, is of the opinion that the limited accuracy of many of the values that enter the radar equation, as well as the difficulty in experimentally verifying the predicted radar performance, do not justify a precision approach to radar prediction. One can't be sloppy, of course, but in engineering one cannot always be overly precise. In spite of difficulties, the engineer has to be able to guarantee that the radar can perform as required.

Calculation of the Radar Equation   The range of a radar can be obtained with nothing more complicated than a simple calculator and a set of tables or graphs similar to what has been presented in this chapter. There exist, however, computer programs on the market for calculation of the radar equation.[68] They make the calculation of range easy; but they are not necessary except when it is required to plot coverage diagrams that take into account the effect of the earth's surface and other propagation factors mentioned in Chap. 8.

This chapter has been concerned with radar detection when receiver noise is the dominant factor hindering detection. The prediction of radar range when clutter echoes from the land, sea, or rain are larger than receiver noise requires a different formulation of the radar range equation, as well as a different design of the radar. Detection of targets in clutter is the subject of Chap. 7. It has been suggested[69] that computer simulation on a computer can be employed when clutter echoes, jamming noise, and receiver noise have to be considered and cannot be faithfully represented by gaussian statistics. The success of simulation depends on how well the clutter and other interfering effects can be modeled. They need to be accurate, which is not always easy to achieve in the "real world."

Radar Equation in Design   In this chapter the radar equation has been discussed mainly as a means for predicting the range of a radar. It also serves the important role of being the basis for radar system design. Some parameters that enter into the radar equation are given by the customer based on the nature of the task the radar is to perform. Examples are the required range, coverage, and target characteristics. Generally there might be trade-offs and compromises required in selecting values of the other parameters that are under the control of the radar designer. It is the radar equation that is used to examine the effect of the trade-offs among the various parameters, such as the trade between a large antenna or a large transmitter power. The decision as to the radar frequency is also something that should come from examining the radar equation. Thus almost all radar design starts with the radar equation.

Conservative Design   Because of the lack of precision in knowing many of the parameters that enter into the radar equation, it is advisable to design a radar as conservatively as possible. This means taking full account of all the factors affecting performance that are knowable—and then adding a safety factor to increase the signal-to-noise ratio. (One method for doing this is to specify the Swerling Case 1 target model with a probability of detection of 0.9 or greater.) Such practice has produced excellent radars that do the job required. In a few cases the radar signal-to-noise ratio was 20 dB greater than actually needed. This may be high, however, by today's standards. Only a few radars have been built with a large safety factor, but they have verified the validity of this approach. Unfortunately, the procurement practices of most agencies that buy radars, as well as the competitive nature of the marketplace, usually do not permit this degree of conservative design to occur very often.

REFERENCES

1. Ridenour, L. N. *Radar System Engineering,* MIT Radiation Laboratory Series, vol. 1, p. 592. New York: McGraw-Hill, 1947.

2. Blake, L. V. "Prediction of Radar Range." In *Radar Handbook,* 2nd ed., ed. by M. Skolnik. New York: McGraw-Hill, 1990, Chap. 2.

3. Blake, L. V. *Radar Range-Performance Analysis.* 2nd ed. Norwood, MA: Artech House, 1986.

4. Johnson, J. B. "Thermal Agitation of Electricity in Conductors." *Phys. Rev.* 32 (July 1928), pp. 97–109.

5. Lawson, J. L., and G. E. Uhlenbeck, eds. *Threshold Signals,* MIT Radiation Laboratory Series, vol. 24, p. 17. New York: McGraw-Hill, 1950.

6. Dixon, R. C. *Radio Receiver Design.* Sec. 5.5. New York: Marcel Dekker, 1998.

7. Van Vleck, J. H., and D. Middleton. "A Theoretical Comparison of the Visual, Aural, and Meter Reception of Pulsed Signals in the Presence of Noise," *J. Appl. Phys.* (November 1946), pp. 940–971.

8. Bennett, W. R. "Methods of Solving Noise Problems," *Proc. IRE* 44, (May 1956), pp. 609–638.

9. Rice, S. O. "Mathematical Analysis of Random Noise." *Bell System Tech. J.,* vol. 23, pp. 282–332, 1944; and vol. 24, pp. 46–156, 1945.

10. Davenport, W. B., and W. L. Root. *Introduction to the Theory of Random Signals and Noise.* New York: IEEE Press, 1987.

11. Albersheim, W. J. "A Closed-Form Approximation to Robertson's Detection Characteristics," *Proc. IEEE* 69 (July 1981), p. 839.

12. Tufts, D. W., and A. J. Cann. "On Albersheim's Detection Equation," *IEEE Trans.* AES-19, (July 1983), pp. 643–646.

13. Scott, A. W. *Understanding Microwaves.* New York: John Wiley, 1993.

14. Marcum, J. I. "A Statistical Theory of Target Detection by Pulse Radar, Mathematical Appendix," *IRE Trans.* IT-6 (April 1960) pp. 145–267.

15. Skolnik, M. I., and D. G. Tucker. "Discussion on 'Detection of Pulse Signals in Noise: Trace-to-Trace Correlation in Visual Displays.'" *J. Brit. IRE* 17 (December 1957), pp. 705–706.

16. Meyer, D. P., and H. A. Mayer. *Radar Target Detection.* New York: Academic, 1973.

17. Rheinstein, J. "Scattering of Short Pulses of Electromagnetic Waves." *Proc. IEEE* 53, (August 1965), pp. 1069–1070.

18. Crispin, J. W., and K. M. Siegel. *Methods of Radar Cross-Section Analysis.* New York: Academic, 1968.

19. Ruck, G. T., D. E. Barrick, W. D. Stuart, and C. K. Krichbaum. *Radar Cross Section Handbook* (2 vols.). New York: Plenum, 1970.

20. Knott, E. F., J. F. Shaeffer, and M. T. Tuley. *Radar Cross Section.* 2d ed. Norwood, MA: Artech House, 1993.

21. Shamansky, H. T., A. K. Dominek, and L. Peters, Jr. "Electromagnetic Scattering by a Straight Thin Wire," *IEEE Trans.* AP-37 (August 1989), pp. 1019–1025.

22. Peters, L., Jr. "End-fire Echo Area of Long, Thin Bodies." *IRE Trans.* AP-6 (January 1958), pp. 133–139.

23. Knott, E. F. et al. Ref. 20, Sec. 1.4, p. 228, and Sec. 6.3.

24. Ross, R. A. "Radar Cross Section of Rectangular Flat Plates." *IEEE Trans.* AP-14 (May 1966), pp. 329–335. Also discussed in Ref. 20, Sec. 6.3.

25. Knott, E. F. "Radar Cross Section." In *Radar Handbook* (ed. M. Skolnik), New York: McGraw-Hill, 1990, Chap. 11.

26. Kennaugh, E. M., and D. L. Moffatt. "On the Axial Echo Area of the Cone-Sphere Shape," *IRE Trans.* AP-10 (February 1962), pp. 199–200. See also Moffatt. D. L. "Low Radar Cross Sections, The Cone-Sphere," Ohio State University Dept. of Elect. *Eng. Rept.* 1223-5, Contract No. AF 33(616)-8039, Columbus, Ohio, May 15, 1962.

27. Pannell, J. H., J. Rheinstein, and A. F. Smith. "Radar Scattering from a Conducting Cone-Sphere," MIT Lincoln Laborataory Tech. Rept. no. 349, Mar. 2, 1964.

28. Ridenour, L. N. *Radar System Engineering.* MIT Radiation Laboratory Series, vol. 1. New York: McGraw-Hill, 1947, Fig. 3.8.

29. Knott, E. F., et al. Ref. 20, Sec. 6.5.

30. J. W. Crispin, Jr., and K. M. Siegel. *Methods of Radar Cross Section Analysis.* New York: Academic, 1962, Chap. 10.

31. Howell, N. A. "Design of Pulse Gated Compact Radar Cross Section Range," *1970 IEEE G-AP Int. Prog & Dig.,* IEEE Publ. 70c 36-AP, pp. 187–195, Sept. 1970. (Also available in refs. 20 and 25.)

32. Olin, I. D., and F. D. Queen. "Dynamic Measurement of Radar Cross Sections." *Proc. IEEE* 53 (August 1965), pp. 954–961.

33. Harris, R., R. Redman, B. Freburger, and J. Hollis. "Dynamic Air-to-Air Imaging Measurement System." *Conf. Proc. of the 14th Annual Meeting of the Antenna Measurements Techniques Assoc.* October 19–23, 1992, pp. 6-11–6-16.

34. Jain, A., and I. Patel. "Dynamic Imaging and RCS Measurements of Aircraft." *IEEE Trans.* AES-31 (January 1995), pp. 211–226.

35. Skolnik, M. I. "An Empirical Formula for the Radar Cross Section of Ships at Grazing Incidence." *IEEE Trans.* AES-10 (March 1972), p. 292.

36. Mensa, D. L. *High Resolution Radar Cross-Section Imaging.* Norwood, MA: Artech House, 1991, Sec. 4.1.

37. Chandler, R. A., and L. E. Wood. "System Considerations for the Design of Radar Braking Sensors." *IEEE Trans.* VT-26 (May 1977), pp. 151–160.

38. Williams, P. D. L., H. D. Cramp, and K. Curtis. "Experimental Study of the Radar Cross-Sections of Maritime Targets." *IEE (London) J. Electronic Circuits and Systems* 2, no. 4 (July 1978), pp. 121–136.

39. Schultz, F. V., R. C. Burgener, and S. King. "Measurement of the Radar Cross Section of a Man." *Proc. IRE* 46 (February 1958), pp. 476–481.

40. Swerling, P. "Probability of Detection for Fluctuating Targets." *IRE Trans.* IT-6 (April 1960), pp. 269–308.

41. Nathanson, F. E. *Radar Design Principles,* 2nd ed. New York: McGraw-Hill, 1991, Sec. 5.3.

42. Meyer, D. P., and H. A. Mayer. *Radar Target Detection.* New York: Academic, 1973.

43. DiFranco, J. V., and W. L. Rubin. *Radar Detection.* Englewood Cliffs, NJ: Prentice-Hall, 1968, Chap. 11.

44. Barton, D. K. *Modern Radar System Analysis.* Norwood, MA: Artech House, 1988, Sec. 2.4.

45. Barton, D. K., C. E. Cook, and P. Hamilton. *Radar Evaluation Handbook.* Norwood, MA: Artech House, 1991, Sec. 4.5.

46. Barton, D. K. "Simple Procedures for Radar Detection Calculations." *IEEE Trans.* AES-5 (September 1969), pp. 837–846.

47. Kanter, I. "Exact Detection Probability for Partially Correlated Rayleigh Targets." *IEEE Trans.* AES-22 (March 1986), pp. 184–196.

48. Swerling. P. "Radar Probability of Detection for Some Additional Fluctuating Target Cases." *IEEE Trans* AES-33 (April 1997), pp. 698–709.

49. Weinstock, W. "Target Cross Section Models for Radar Systems Analysis," doctoral dissertation, Univ. of Pennsylvania, Philadelphia, 1964.

50. Jao, J. K., and M. Elbaum. "First-Order Statistics of a Non-Rayleigh Fading Signal and Its Detection." *Proc. IEEE* 66 (July 1978), pp. 781–789.

51. Heidbreder, G. R., and R. L. Mitchell. "Detection Probabilities for Log-Normally Distributed Signals." *IEEE Trans.* AES-3 (January 1967), pp. 5–13.

52. Pollon, G. E. "Statistical Parameters for Scattering from Randomly Oriented Arrays, Cylinders, and Plates." *IEEE Trans.* AP-18 (January 1970), pp. 68–75.

53. Shlyakhin, V. M. "Probability Models of Non-Rayleigh Fluctuations of Radar Signals (A Review)." *Soviet J. Communications Technology and Electronics* 33, no. 1 (January 1988), pp. 1–16.

54. Nathanson, F. E. Ref. 41, Sec. 5.4.

55. Kraus, J. D. *Antennas,* 2nd ed. New York: McGraw-Hill, 1988, Sec. 2–22.

56. Stegen, R. J. "The Gain-Beamwidth Product of an Antenna." *IEEE Trans.* AP-12 (July 1964), pp. 505–506.

57. Evans, G. E. *Antenna Measurement Techniques.* Norwood, MA: Artech House, 1990.

58. Murrow, D. J. "Height Finding and 3D Radar." In *Radar Handbook,* 2nd ed., ed. M. Skolnik. New York: McGraw-Hill, 1990, Chap. 20.

59. Van Valkenburg, M. E., ed. *Reference Data for Engineers.* 8th ed. SAMS, Carmel, IN: Prentice-Hall Computer Publishing, 1993, Chap. 30.

60. Blake, L. V. "Prediction of Radar Range." In *Radar Handbook,* ed. M. Skolnik. New York: McGraw-Hill, 1990, Sec. 2.7.

61. Information obtained from Electronic Space Systems Corporation (Essco), Concord, MA. See also Dicaudo, V. J. "Radomes." In *Radar Handbook,* 1st ed., ed. M. Skolnik. New York: McGraw-Hill, 1970, Chap. 14.

62. Davenport, W. B., Jr. "Signal-to-Noise Ratios in Band-Pass Limiters." *J. Appl. Phys.* 24 (June 1953), pp. 720–727.

63. D'Aloisi, D., A. DiVito, and G. Galati. "Sampling Losses in Radar Signal Detection." *J. IERE* 56 (June/July 1986), pp. 237–242.

64. Marcum, J. I. Ref. 14, pp. 213–215.

65. Trunk. G. V. "Comparison of the Collapsing Losses in Linear and Square-Law Detectors." *Proc. IEEE* 80 (June 1972), pp. 743–744.

66. Mallett, J. D., and L. E. Brennan. "Cumulative Probability of Detection for Targets Approaching a Uniformly Scanning Search Radar." *Proc. IEEE* 51 (April 1963), pp. 596–601, and 52 (June 1964), pp. 708–709.

67. Brookner, E. *Radar Technology.* Dedham, MA: Artech House, 1977, Chap. 3.

68. Barton, D. K., and W. F. Barton. *Modern Radar System Analysis Software and User's Manual.* Boston, MA: Artech House, 1993.

69. Schleher, D. C. "Solving Radar Detection Problems using Simulation." *IEEE AESS Systems Magazine* 10 (April 1995), pp. 36–39.

PROBLEMS

2.1 If the noise figure of a receiver is 2.5 dB, what reduction (measured in dB) occurs in the signal-to-noise ratio at the output compared to the signal-to-noise ratio at the input?

2.2 What is the noise bandwidth B_n of a low-pass RC filter whose frequency-response function is $H(f) = \dfrac{1}{1 + j\,(f/B_v)}$, where B_v is the half-power bandwidth? That is, find the ratio B_n/B_v.

2.3 The random variable x has an exponential probability density function:

$$p(x) = a \exp\,[-bx] \qquad x > 0$$

where a and b are constants.
a. Determine the relation between a and b required for normalization.
b. Determine the mean m_1 and the variance σ^2 for the normalized $p(x)$.
c. Sketch $p(x)$ for $a = 1$.
d. Find the probability distribution function $P(x)$, and sketch the result for a = 1.

2.4 Show that the standard deviation of the Rayleigh probability density function [Eq. (2.17)] is proportional to the mean value. You should use integral tables when integration cannot be performed in a simple manner. (This result is used in deriving the form of the log-FTC clutter suppression circuit described in Sec. 7.8.)

2.5 The average time between false alarms is specified as 30 min and the receiver bandwidth is 0.4 MHz.
a. What is the probability of false alarm?
b. What is the threshold-to-noise power ratio (V_T^2/Ψ_0)?
c. Repeat (a) and (b) for an average false alarm time of one year (8760 h).
d. Assume the threshold-to-noise power ratio is to be set to achieve a 30-min false-alarm time [value as in part (b)]; but, for some reason, the threshold is actually set lower by 0.3 dB than the value found in part (b). What is the resulting average time between false alarms with the lower threshold?
e. What would be the average time between false alarms if the threshold were to increase by 0.3 dB?
f. Examine the two values of threshold-to-noise ratio you have calculated in (d) and (e) and comment on the practicability of precisely achieving a specified value of false-alarm time.

2.6 A radar has a bandwidth $B = 50$ kHz and an average time between false alarms of 10 min.
a. What is the probability of false alarm?
b. If the pulse repetition frequency (prf) were 1000 Hz and if the first 15 nmi of range were gated out (receiver is turned off) because of the use of a long pulse, what would be the new probability of false alarm? (Assume the false-alarm time has to remain constant.
c. Is the difference between (a) and (b) significant?
d. What is the pulse width that results in a minimum range of 15 nmi?

2.7 A transmission line with loss L is connected to the input of a receiver whose noise figure is F_r. What is the overall noise figure of the combination?

2.8 A radar at a frequency of 1.35 GHz has an antenna of width $D = 32$ ft, a maximum unambiguous range of 220 nmi, and an antenna scan time (time to make one rotation of the antenna) of 10 s.

  a. What is the number of echo pulses per scan received by the radar from a point target? [Use the relationship that the antenna half-power beamwidth in radians is $\theta_B = 1.2\lambda/D$ (λ = wavelength).]

  b. What is the integration loss and the integration-improvement factor when the probability of detection is 0.9 and the probability of false alarm is 10^{-4}?

2.9 Show that the far right-hand side of Eq. (2.36), a definition of the radar cross section, is the same as the simple radar equation, Eq. (1.6). [It easier to start with Eq. (2.36) and obtain Eq. (1.6), than vice versa.]

2.10 *a.* What frequency will result in the maximum radar cross section of a metallic sphere whose diameter is 1 m?

  b. At what frequency will the radar cross section of a ball bearing one millimeter in diameter be maximum?

2.11 *a.* What is the maximum radar cross section (square meters) of an automobile license plate that is 12 inches wide by 6 inches high, at a frequency of 10.525 GHz (the frequency of an X-band speed radar)?

  b. How many degrees in the vertical plane should the plate be tilted in order to reduce its cross section by 10 dB? For purposes of this problem you may assume the license plate is perfectly flat. The radar cross section of a flat plate as a function of the incidence angle ϕ may be written, for ϕ not too large, as:

$$\sigma(\phi) \approx \sigma_{max} \frac{\sin^2 [2\pi(H/\lambda) \sin \phi]}{[2\pi(H/\lambda) \sin \phi]^2}$$

  where σ_{max} = maximum radar cross section of a flat plate = $4\pi A^2/\lambda^2$, A = area of plate, λ = radar wavelength, and H = height of the plate. (Be careful of units. You will have to sketch a portion of the cross section pattern as a function of ϕ to find the value of ϕ corresponding to -10 dB.)

  c. What other parts of an automobile might contribute to its radar cross section when viewed directly from the front?

2.12 Describe briefly the behavior of the radar cross section (in the microwave region) of a raindrop and a large aircraft with respect to its dependence on (*a*) frequency and (*b*) viewing aspect.

2.13 Describe the chief characteristic of the radar echo from a target when its radar cross section is in the (*a*) Rayleigh region, (*b*) resonance region, and (*c*) optical region.

2.14 A typical value of an individual "sea spike" echo at X band (wavelength = 3.2 cm), as discussed in Sec. 7.4, might be 1 m^2. What is the dimension of the side of a square flat plate that produces the same radar echo when the plate is viewed at normal incidence?

2.15 A radar noncoherently integrates 18 pulses, each of uniform amplitude (the nonfluctuating case). The IF bandwidth is 100 kHz.

    a. If the average time between false alarms is 20 min, what must be the signal-to-noise ratio per pulse $(S/N)_n$ in order to achieve a probability of detection of 0.80? (Suggest the use of Albersheim's equation.)

    b. What is the corresponding value of $(S/N)_1$?

    c. What would $(S/N)_1$ be if the target cross section fluctuated according to a Swerling Case 1 model?

2.16 Why does the cross section of a complex target, such as that in Fig. 2.15, fluctuate so rapidly with a small change in aspect angle when the radar wavelength is small compared to the target's dimensions?

2.17 Show that the probability density function for the Swerling Case 1 model is the same as the chi-square of degree 2 [Eq. (2.47)].

2.18 *a.* What signal-to-noise ratio is required for a radar that makes a detection on the basis of a single pulse, when the probability of detection is 0.50 and the probability of false alarm is 10^{-6}? Assume a nonfluctuating target echo.

    b. Repeat for a 0.99 probability of detection and the same probability of false alarm.

    c. Repeat parts (*a*) and (*b*), but for a Swerling Case 1 fluctuating target.

    d. Compare your results in a table. What conclusions can you obtain from this?

2.19 A radar measures an apparent range of 7 nmi when the prf is 4000 Hz, but it measures an apparent range of about 18.6 nmi when the prf is 3500 Hz. What is the true range (nmi)?

2.20 *a.* Show that the echo signal power P_r received from an aircraft flying at a constant height *h* over a perfectly conducting flat earth is independent of the range *R*, when the antenna elevation gain varies as the cosecant-squared of the elevation angle ϕ (that is; $G = G_0 \csc^2 \phi$).

    b. In addition to having a received signal that is independent of the range (requiring less dynamic range in the receiver), what is another reason for employing an antenna with a $\csc^2 \phi$ elevation pattern for an air-surveillance radar when compared to a conventional unshaped fan-beam elevation pattern?

    c. What are the limitations in applying the simple result of (*a*) to a radar in the real world?

2.21 A surface-based air-surveillance radar with a fan-beam antenna that rotates 360° in azimuth has a maximum range of 150 nmi and height coverage to 60,000 ft. Its maximum elevation-angle coverage is 30°. What percentage of the total available volume coverage is lost because of the overhead "hole" compared to a radar with complete angular coverage up to 90° (no hole in the coverage)? Assume, for simplicity, a flat earth.

2.22 A radar receives five pulses within its half-power (3 dB) beamwidth as the antenna beam scans past a point target. The middle of the five pulses is transmitted when the maximum of the antenna pattern points in the direction of the target. The first and the fifth pulses are transmitted when the leading and trailing half-power points are, respectively, directed at the target. What is the two-way beam-shape loss (dB) in this case?

2.23 Five identical radars, each with a receiver having a square-law detector, have partial overlap in their radar coverages so that not all radars are guaranteed to see each target. The outputs of all five radars are combined before a detection decision is made. If a target is seen on only one of the five radars and the other four radars see only receiver noise, what is the collapsing loss when the detection probability is 0.5 and the false-alarm probability is 10^{-4}?

2.24 A civil marine radar is employed on boats and ships for observing navigation buoys, detecting land-sea boundaries, piloting, and avoiding collisions. Consider the following civil-marine radar:

> frequency: 9400 MHz (X band)
> antenna: horizontal beamwidth = 0.8°
>   vertical beamwidth = 15°
>   gain = 33 dB
>   azimuth rotation rate = 20 rpm
> peak power: 25 kW
> pulse width: 0.15 μs
> pulse repetition rate: 4000 Hz
> receiver noise figure: 5 dB
> receiver bandwidth: 15 MHz
> system losses: 12 dB
> average time between false alarms: 4 hours

a. Plot the single-scan probability of detection as function of range (nmi), assuming a constant cross-section target of 10 m² (a navigation buoy) and free-space propagation. [You will find it easier to select the probability of detection and find the corresponding signal-to-noise ratio, rather than the reverse. You need only consider probabilities of detection from 0.30 to 0.99. You may, for purposes of this problem, select a single (average) value of the integration improvement factor rather than try to find it as a function of P_d (since the curve in the text does not permit otherwise).]
b. Repeat (a) for a Swerling Case 1 target fluctuation model with average cross section of 10 m². Plot on the same diagram as (a).
c. Comment on whether the average power of this radar is too low, just right, or too high for the job it has to perform here.
d. Why do you think this ship-mounted radar antenna has a 15° elevation beamwidth when all the targets are located on the surface of the sea?

2.25 Consider the following air-surveillance radar:

> frequency: 2.8 GHz (S band)
> peak power: 1.4 MW
> pulse width: 0.6 μs
> pulse repetition frequency: 1040 Hz
> receiver noise figure: 4 dB
> antenna rotation rate: 12.8 rpm

antenna gain: 33 dB

antenna azimuth beamwidth: 1.35 deg

system losses : 12 dB

average false-alarm time: 20 min

target cross section: 2 m^2

Plot each of the following on the same coordinates (with range as the abscissa):

a. The free-space single-scan probability of detection as a function of range (in nautical miles) for a constant cross-section target. [You will find it easier to select the probability of detection and find the corresponding signal-to-noise ratio, rather than the reverse. You need only consider probabilities of detection from 0.30 to 0.99. You may, for purposes of this problem, select a single (average) value of the integration improvement factor rather than try to find it as a function of P_d (since the curve in the text does not permit otherwise).]

b. The probability of detection as a function of range for the same situation as part (a) but with the detection criterion that the target must be found on at least 2 out of 3 scans of the rotating antenna. [You may assume that the range and the received signal power do not change appreciably over the three scans. For convenience of this calculation, you may assume that the single-scan false-alarm probability is the same as used in part (a).]

c. Repeat (a) for a Swerling Case 1 with average target cross section of 2 m^2.

d. Repeat (b) for a Swerling Case 1 with average target cross section of 2 m^2.

e. Is the prf adequate for avoiding range ambiguities?

(The radar in this problem is similar to the airport surveillance radar known as the ASR.)

2.26 Starting with Eq. (2.51) derive the surveillance radar equation [Eq. (2.63)]. You may omit the propagation factor, attenuation, and fluctuation loss.

2.27 What is the effect of receiver bandwidth on the maximum range of a well-designed radar, assuming the average power remains constant? Explain your answer.

2.28 a. What is the probability of detecting a target on at least 2 out of 4 scans when the single-scan probability of detection is 0.8?

b. What is the corresponding probability of false alarm in this case when the single-scan false-alarm probability is 10^{-8}?

c. What should be the single-scan false-alarm probability if the overall false-alarm probability with a detection criterion of 2 out of 4 scans is 10^{-8}?

d. When the higher single-scan probability of false alarm of part (c) is employed rather than a 10^{-8} single-scan probability of false alarm, what reduction in the signal-to-noise ratio can be obtained?

2.29 In this problem, it is assumed that the targets for an air-surveillance radar are characterized by a Swerling Case 1 model. There are n pulses received from a target. Half of the n pulses are at one frequency and the other half are at a second frequency that is far enough removed from the first to completely decorrelate the second set of $n/2$ pulses relative to the first set of $n/2$ pulses. What is the improvement in signal-to-noise ratio obtained from this use of frequency diversity when (*a*) $P_d = 0.95$ and (*b*) $P_d = 0.6$? (*c*) If the radial extent (in range) of a target is 30 m, what must the difference in the two frequencies be to decorrelate the target echo?

2.30   *a*. Make a list of the system losses that might occur in a long-range air-surveillance radar, and estimate an approximate value for the loss due to each factor. You need not include losses due to doppler processing. (There is, of course, no unique answer for this question.)

   b. Using the total system loss you have estimated, what is the reduction in radar range that occurs due to the system losses if the radar range without losses is 200 nmi?

2.31   Question 1.8 of Chapter 1 asked "How does radar range depend on the wavelength?" Based on Chapter 2, how would you now answer this question for an air-surveillance radar? (Please justify your answer.)

2.32   An experimental measurement of the blip-scan ratio (single-scan probability of detection) of a particular target at a particular range gives a value of 0.5 after 10 trails (antenna scans).

   a. What is the confidence interval of this measurement if the confidence level has to be 90 percent?

   b. What is the confidence interval (with the same confidence level of 90 percent) after 100 scans, assuming the measured blip-scan ratio is still 0.5?

chapter

3

MTI and Pulse Doppler Radar

3.1 INTRODUCTION TO DOPPLER AND MTI RADAR

The radars discussed in the previous chapter were required to detect targets in the presence of noise. In the real world, radars have to deal with more than receiver noise when detecting targets since they can also receive echoes from the natural environment such as land, sea, and weather. These echoes are called *clutter* since they can "clutter" the radar display. Clutter echoes can be many orders of magnitude larger than aircraft echoes. When an aircraft echo and a clutter echo appear in the same radar resolution cell, the aircraft might not be detectable. Chapter 7 describes the characteristics of clutter and discusses methods for reducing these unwanted echoes in order to detect the desired target echoes. However, the most powerful method for detecting moving targets in the midst of large clutter is by taking advantage of the doppler effect, which is the change of frequency of the radar echo signal due to the relative velocity between the radar and the moving target. The use of the doppler frequency shift with a pulse radar for the detection of moving targets in clutter is the subject of this chapter.

Radar deserves much credit for enabling the Allies (chiefly the United Kingdom and the United States) in the first half of World War II to prevail in the crucial air battles and night naval engagements against the Axis powers. Almost all of the radars used in World War II, however, were by today's standards relatively simple pulse systems that did not employ the doppler effect. Fortunately, these pulse radars were able to accomplish their mission without the use of doppler. This would not be possible today. All high-performance military air-defense radars and all civil air-traffic control radars for the detection and tracking of aircraft depend on the doppler frequency shift to separate the large clutter echoes from the much smaller echoes from moving targets. Clutter echoes can be greater than the desired target echoes by as much as 60 or 70 dB, or more, depending on the type of radar and the environment.

MTI Radar and Pulse Doppler Radar   A pulse radar that employs the doppler shift for detecting moving targets is either an MTI (moving target indication) radar[1] or a pulse doppler radar.[2] The MTI radar has a pulse repetition frequency (prf) low enough to not have any range ambiguities as defined by Eq. (1.2), $R_{un} = c/2f_p$. It does, however, have many ambiguities in the doppler domain. The pulse doppler radar, on the other hand, is just the opposite. As we shall see later in this chapter, it has a prf large enough to avoid doppler ambiguities, but it can have numerous range ambiguities. There is also a medium-prf pulse doppler that accepts both range and doppler ambiguities, as discussed in Sec. 3.9.

   In addition to detecting moving targets in the midst of large clutter echoes, the doppler frequency shift has other important applications in radar; such as allowing CW (continuous wave) radar to detect moving targets and to measure radial velocity, synthetic aperture radar and inverse synthetic aperture radar for producing images of targets, and meteorological radars concerned with measuring wind shear. These other uses of the doppler frequency shift are not discussed in this chapter.

Doppler Frequency Shift   The doppler effect used in radar is the same phenomenon that was introduced in high school physics courses to describe the changing pitch of an audible siren from an emergency vehicle as it travels toward or away from the listener. In this chapter we are interested in the doppler effect that changes the frequency of the electromagnetic signal that propagates from the radar to a moving target and back to the radar. If the range to the target is R, then the total number of wavelengths λ in the two-way path from radar to target and return is $2R/\lambda$. Each wavelength corresponds to a phase change of 2π radians. The total phase change in the two-way propagation path is then

$$\phi = 2\pi \times \frac{2R}{\lambda} = 4\pi R/\lambda \qquad [3.1]$$

If the target is in motion relative to the radar, R is changing and so will the phase. Differentiating Eq. (3.1) with respect to time gives the rate of change of phase, which is the angular frequency

$$\omega_d = \frac{d\phi}{dt} = \frac{4\pi}{\lambda}\frac{dR}{dt} = \frac{4\pi v_r}{\lambda} = 2\pi f_d \qquad [3.2]$$

where $v_r = dR/dt$ is the radial velocity (meters/second), or rate of change of range with time. If, as in Fig. 3.1, the angle between the target's velocity vector and the radar line of sight to the target is θ, then $v_r = v \cos \theta$, where v is the speed, or magnitude of the vector velocity. The rate of change of ϕ with time is the angular frequency $\omega_d = 2\pi f_d$, where f_d is the *doppler frequency shift*. Thus from Eq. (3.2),

$$f_d = \frac{2v_r}{\lambda} = \frac{2f_t v_r}{c} \qquad [3.3]$$

Figure 3.1 Geometry of radar and target in deriving the doppler frequency shift. Radar, target, and direction of target travel all lie in the same plane in this illustration.

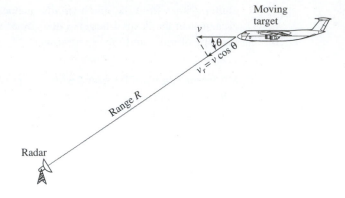

The radar frequency is $f_t = c/\lambda$, and the velocity of propagation $c = 3 \times 10^8$ m/s. If the doppler frequency is in hertz, the radial velocity in knots (abbreviated kt), and the radar wavelength in meters, we can write

$$f_d \text{ (Hz)} = \frac{1.03 v_r \text{(kt)}}{\lambda \text{ (m)}} \approx \frac{v_r \text{(kt)}}{\lambda \text{ (m)}} \qquad \textbf{[3.4]}$$

The doppler frequency in hertz can also be approximately expressed as $3.43 v_r f_t$, where f_t is the radar frequency in GHz and v_r is in knots. A plot of the doppler frequency shift is shown in Fig. 3.2 as a function of the radial velocity and the various radar frequency bands.

Figure 3.2 Doppler frequency shift from a moving target as a function of the target's radial velocity and the radar frequency band.

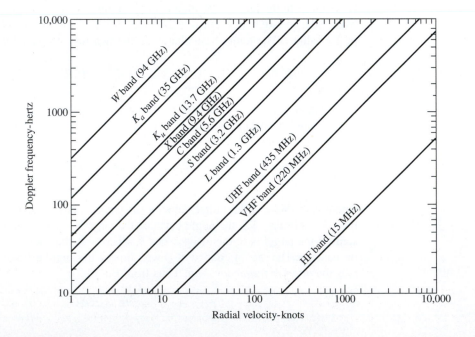

Simple CW Doppler Radar   Before discussing the use of doppler in pulse radar, it is instructive to begin by considering the doppler frequency shift experienced with a CW radar. The block diagram of a very simple CW radar that utilizes the doppler frequency shift to detect moving targets is shown in Fig. 3.3a. Unlike a pulse radar, a CW radar transmits while it receives. Without the doppler shift produced by the movement of the target, the weak CW echo signal would not be detected in the presence of the much stronger signal from the transmitter. Filtering in the frequency domain is used to separate the weak doppler-shifted echo signal from the strong transmitter signal in a CW radar.

The transmitter generates a continuous (unmodulated) sinusoidal oscillation at frequency f_t, which is then radiated by the antenna. On reflection by a moving target, the transmitted signal is shifted by the doppler effect by an amount $\pm f_d$, as was given by Eq. (3.3). The plus sign applies when the distance between radar and target is decreasing (a closing target); thus, the echo signal from a closing target has a larger frequency than that which was transmitted. The minus sign applies when the distance is increasing (a receding target). To utilize the doppler frequency shift a radar must be able to recognize that the received echo signal has a frequency different from that which was transmitted. This is the function of that portion of the transmitter signal that finds its way (or leaks) into the receiver, as indicated in Fig. 3.3a. The transmitter leakage signal acts as a reference to determine that a frequency change has taken place. The detector, or mixer, multiplies the echo signal at a frequency $f_t \pm f_d$ with the transmitter leakage signal f_t. The doppler filter allows the difference frequency from the detector to pass and rejects the higher frequencies. The filter characteristic is shown in Fig. 3.3a just below the doppler-filter block. It has a lower frequency cutoff to remove from the receiver output the transmitter leakage signal and clutter echoes. The upper frequency cutoff is determined by the maximum

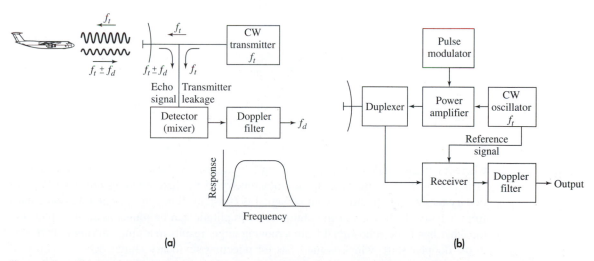

Figure 3.3   (a) Simple CW radar block diagram that extracts the doppler frequency shift from a moving target and rejects stationary clutter echoes. The frequency response of the doppler filter is shown at the lower right. (b) Block diagram of a simple pulse radar that extracts the doppler frequency shift of the echo signal from a moving target.

radial velocity expected of moving targets. The doppler filter passes signals with a doppler frequency f_d located within its pass band, but the sign of the doppler is lost along with the direction of the target motion. CW radars can be much more complicated than this simple example, but it is adequate as an introduction to a pulse radar that utilizes the doppler to detect moving targets in clutter.

Pulse Radar That Extracts the Doppler Frequency-Shifted Echo Signal   One cannot simply convert the CW radar of Fig. 3.3a to a pulse radar by turning the CW oscillator on and off to generate pluses. Generating pulses in this manner also removes the reference signal at the receiver, which is needed to recognize that a doppler frequency shift has occurred. One way to introduce the reference signal is illustrated in Fig. 3.3b. The output of a stable CW oscillator is amplified by a high-power amplifier. The amplifier is turned on and off (modulated) to generate a series of high-power pulses. The received echo signal is mixed with the output of the CW oscillator which acts as a *coherent reference* to allow recognition of any change in the received echo-signal frequency. By *coherent* is meant that the phase of the transmitted pulse is preserved in the reference signal. The change in frequency is detected (recognized) by the doppler filter.

The doppler frequency shift is derived next in a slightly different manner than was done earlier in this section. If the transmitted signal of frequency f_t is represented as $A_t \sin(2\pi f_t t)$, the received signal is $A_r \sin[2\pi f_t(t - T_R)]$, where A_t = amplitude of transmitted signal and A_r = amplitude of the received echo signal. The round-trip time T_R is equal to $2R/c$, where R = range and c = velocity of propagation. If the target is moving toward the radar, the range is changing and is represented as $R = R_0 - v_r t$, where v_r = radial velocity (assumed constant). The geometry is the same as was shown in Fig. 3.1. With the above substitutions, the received signal is

$$V_{\text{rec}} = A_r \sin\left[2\pi f_t\left(1 + \frac{2v_r}{c}\right)t - \frac{4\pi f_t R_0}{c}\right] \qquad [3.5]$$

The received frequency changes by the factor $2f_t v_r/c = 2v_r/\lambda$, which is the doppler frequency shift f_d.* If the target had been moving away from the radar, the sign of the doppler frequency would be minus, and the received frequency would be less than that transmitted.

The received signal is heterodyned (mixed) with the reference signal $A_{\text{ref}} \sin 2\pi f_t t$ and the difference frequency is extracted, which is given as

$$V_d = A_d \cos (2\pi f_d t - 4\pi R_0/\lambda) \qquad [3.6]$$

where A_d = amplitude, $f_d = 2v_r/\lambda$ = doppler frequency, and the relation $f_t\lambda = c$ was used. (The cosine replaces the sine in the trigonometry of the heterodyning process.) For stationary targets $f_d = 0$ and the output signal is constant. Since the cosine takes on values from $+1$ to -1, the sign of the clutter echo amplitude can be minus as well as plus. On the other hand, the echo signal from a moving target results in a time-varying output (due to the doppler shift) which is the basis for rejecting stationary clutter echoes (with zero doppler frequency) but allowing moving-target echoes to pass.

| *The terms *doppler frequency shift*, *doppler frequency*, and *doppler shift* are used interchangeably in this chapter.

If the radar pulse width is long enough and if the target's doppler frequency is large enough, it may be possible to detect the doppler frequency shift on the basis of the frequency change within a single pulse. If Fig. 3.4a represents the RF (or IF) echo pulse train, Fig. 3.4b is the pulse train when there is a recognizable doppler frequency shift. To detect a doppler shift on the basis of a single pulse of width τ generally requires that there be at least one cycle of the doppler frequency f_d within the pulse; or that $f_d\tau > 1$. This condition, however, is not usually met when detecting aircraft since the doppler frequency f_d is generally much smaller than $1/\tau$. Thus the doppler effect cannot be utilized with a *single* short pulse in this case. Figure 3.4c is more representative of the doppler frequency for aircraft-detection radars. The doppler is shown sampled at the pulse repetition frequency (prf). More than one pulse is needed to recognize a change in the echo frequency due to the doppler effect. (Figure 3.4c is exaggerated in that the pulse width is usually small compared to the pulse repetition period. For example, τ might be of the order of 1 μs, and the pulse repetition period might be of the order of 1 ms.)

Sweep-to-Sweep Subtraction and the Delay-Line Canceler   Figures 3.5a and b represent (in a very approximate manner) the bipolar video (both positive and negative amplitudes) from two successive sweeps* of an MTI (moving target indication) radar defined at the beginning of this chapter. The fixed clutter echoes in this figure remain the same from sweep to sweep. The output of the MTI radar is called *bipolar video,* since the signal has negative as well as positive values. (*Unipolar video* is rectified bipolar video with

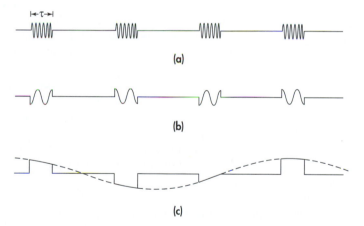

(a)

(b)

(c)

Figure 3.4   (a) Representation of the echo pulse train at either the RF or IF portion of the receiver; (b) video pulse train after the phase detector when the doppler frequency $f_d > 1/\tau$; (c) video pulse train for the doppler frequency $f_d < 1/\tau$, which is usually the case for aircraft-surveillance radar. The doppler frequency signal is shown dashed in (c), as if it were CW. Note that the pulses in (c) have an exaggerated width compared to the period of the doppler frequency.

---

*Sweep as used here is what occurs in the time between two transmitted pulses, or the pulse repetition interval. It is a more convenient term to use than is *pulse repetition period,* but the latter is more descriptive. The term *sweep* originally signified the action of moving the electron beam of a cathode ray tube display across the face of the tube during the time of a pulse repetition period.

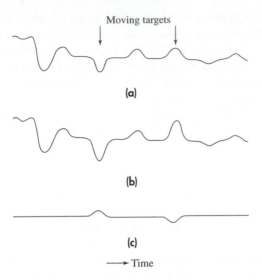

(a)

(b)

(c)

→ Time

Figure 3.5 Two successive sweeps, (a) and (b), of an MTI radar A-scope display (amplitude as a function of time, or range). Arrows indicate the positions of moving targets. When (b) is subtracted from (a), the result is (c) and echoes from stationary targets are canceled, leaving only moving targets.

only positive values.) If one sweep is subtracted from the previous sweep, fixed clutter echoes will cancel and will not be detected or displayed. On the other hand, moving targets change in amplitude from sweep to sweep because of their doppler frequency shift. If one sweep is subtracted from the other, the result will be an uncancelled residue, as shown in Fig. 3.5c.

Subtraction of the echoes from two successive sweeps is accomplished in a *delay-line canceler,* as indicated by the diagram of Fig. 3.6. The output of the MTI receiver is digitized and is the input to the delay-line canceler (which performs the role of a doppler filter). The delay T is achieved by storing the radar output from one pulse transmission, or sweep, in a digital memory for a time equal to the pulse repetition period so that $T = T_p = 1/f_p$. The output obtained after subtraction of two successive sweeps is *bipolar (digital) video* since the clutter echoes in the output contain both positive and negative amplitudes [as can be seen from Eq. (3.6) when $f_d = 0$]. It is usually called *video,* even though it is a series of digital words rather than an analog video signal. The absolute value

Figure 3.6
Block diagram of a single delay-line canceler.

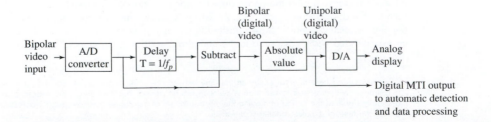

of the bipolar video is taken, which is then *unipolar video.* Unipolar video is needed if an analog display is used that requires positive signals only. The unipolar digital video is then converted to an analog signal by the digital-to-analog (D/A) converter if the processed signal is to be displayed on a PPI (plan position indicator). Alternatively, the digital signals may be used for automatically making the detection decision and for further data processing, such as automatic tracking and/or target recognition. The name *delay-line canceler* was originally applied when analog delay lines (usually acoustic) were used in the early MTI radars. Even though analog delay lines have been replaced by digital memories, the name delay-line canceler is still used to describe the operation of Fig. 3.6.

MTI Radar Block Diagram The block diagram of Fig. 3.3 illustrated the reference signal necessary for an MTI radar, but it is oversimplified. A slightly more elaborate block diagram of an MTI radar employing a power amplifier as the transmitter is shown in Fig. 3.7. The local oscillator of an MTI radar's superheterodyne receiver must be more stable than the local oscillator for a radar that does not employ doppler. If the phase of the local oscillator were to change significantly between pulses, an uncancelled clutter residue can result at the output of the delay-line canceler which might be mistaken for a moving target even though only clutter were present. To recognize the need for high stability, the local oscillator of an MTI receiver is called the *stalo,* which stands for *stable local oscillator.* The IF stage is designed as a matched filter, as is usually the case in radar. Instead of an amplitude detector, there is a *phase detector* following the IF stage. This is a mixer-like device that combines the received signal (at IF) and the reference signal from the *coho* so as to produce the difference between the received signal and the reference signal frequencies.[3] This difference is the doppler frequency. The name *coho* stands for

Figure 3.7
Block diagram of an MTI radar that uses a power amplifier as the transmitter.

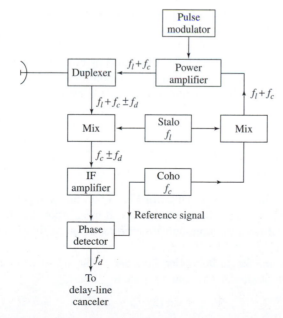

coherent oscillator to signify that it is the reference signal that has the phase of the transmitter signal. Coherency with the transmitted signal is obtained by using the sum of the coho and the stalo signals as the input signal to the power amplifier. Thus the transmitter frequency is the sum of the stalo frequency f_l and the coho frequency f_c. This is accomplished in the mixer shown on the upper right side of Fig. 3.7. The combination of the stalo and coho sometimes is called the *receiver-exciter* portion of the MTI radar. Using the receiver stalo and coho to also generate the transmitter signal insures better stability than if the functions were performed with two different sets of oscillators. The output of the phase detector is the input to the delay-line canceler, as in Fig. 3.6. The delay-line canceler acts as a high-pass filter to separate the doppler-shifted echo signals of moving targets from the unwanted echoes of stationary clutter. The doppler filter might be a single delay-line canceler as in Fig. 3.6; but it is more likely to be one of several other more elaborate filters with greater capability, as described later in this chapter.

The power amplifier is a good transmitter for MTI radar since it can have high stability and is capable of high power. The pulse modulator turns the amplifier on and off to generate the radar pulses. The klystron and the traveling wave tube have usually been the preferred type of vacuum-tube amplifier for MTI radar. The crossed-field amplifier has also been used, but it is generally noisier (less stable) than other devices; hence, it might not be capable of canceling large clutter echoes. Triode and tetrode vacuum tubes have also been used with success for radars that operate at UHF and lower frequencies, but they have been largely replaced at these lower radar frequencies by the solid-state transistor. The transistor has the advantage of good stability and it does not need a pulse modulator.

Before the development of the high-power klystron amplifier for radar application in the 1950s, the only suitable RF power generator at microwave frequencies was the magnetron oscillator. In an oscillator, the phase at the start of each pulse is random so that the receiver-exciter concept of Fig. 3.7 cannot be used. To obtain a coherent reference in this case, a sample of each transmitter pulse is used to lock the phase of the coho to that of the transmitted pulse until the next pulse is generated. The phase-locking procedure is repeated with each pulse. The RF locking-pulse is converted to IF in a mixer that also uses the stalo as the local oscillator. This method of establishing coherence at the receiver sometimes is called *coherent on receive*. Further information on the MTI using an oscillator may be found in previous editions of this text or in the chapter on MTI in the *Radar Handbook*.[1]

3.2   DELAY-LINE CANCELERS

The simple MTI delay-line canceler (DLC) of Fig. 3.6 is an example of a *time-domain filter* that rejects stationary clutter at zero frequency. It has a frequency response function $H(f)$ that can be derived from the time-domain representation of the signals.

Frequency Response of the Single Delay-Line Canceler   The signal from a target at range R_0 at the output of the phase detector can be written

$$V_1 = k \sin (2\pi f_d t - \phi_0) \tag{3.7}$$

where f_d = doppler frequency shift, ϕ_0 = a constant phase equal to $4\pi R_0/\lambda$, R_0 = range at time equal to zero, λ = wavelength, and k = amplitude of the signal. [For convenience, the cosine of Eq. (3.6) has been replaced by the sine.] The signal from the previous radar transmission is similar, except it is delayed by a time T_p = pulse repetition interval, and is

$$V_2 = k \sin [2\pi f_d(t - T_p) - \phi_0]$$ [3.8]

The amplitude k is assumed to be the same for both pulses. The delay-line canceler subtracts these two signals. Using the trigonometric identity $\sin A - \sin B = 2 \sin[(A - B)/2] \cos [(A + B)/2]$, we get

$$V = V_1 - V_2 = 2k \sin (\pi f_d T_p) \cos \left[2\pi f_d\left(t - \frac{T_p}{2}\right) - \phi_0\right]$$ [3.9]

The output from the delay-line canceler is seen to consist of a cosine wave with the same frequency f_d as the input, but with an amplitude $2k \sin (\pi f_d T_p)$. Thus the amplitude of the canceled video output depends on the doppler frequency shift and the pulse repetition period. The frequency response function of the single delay-line canceler (output amplitude divided by the input amplitude k) is then

$$H(f) = 2 \sin (\pi f_d T_p)$$ [3.10]

Its magnitude $|H(f)|$ is sketched in Fig. 3.8.

The single delay-line canceler is a filter that does the job asked of it: it eliminates fixed clutter that is of zero doppler frequency. Unfortunately, it has two other properties that can seriously limit the utility of this simple doppler filter: (1) the frequency response function also has zero response when moving targets have doppler frequencies at the prf and its harmonics, and (2) the clutter spectrum at zero frequency is not a delta function of zero width, but has a finite width so that clutter will appear in the pass band of the delay-line canceler. The result is there will be target speeds, called *blind speeds,* where the target will not be detected and there will be an uncanceled clutter residue that can interfere with the detection of moving targets. These limitations will be discussed next.

Blind Speeds   The response of the single delay-line canceler will be zero whenever the magnitude of $\sin (\pi f_d T_p)$ in Eq. (3.10) is zero, which occurs when $\pi f_d T_p = 0, \pm\pi, \pm2\pi, \pm3\pi, \dots$. Therefore,

$$f_d = \frac{2v_r}{\lambda} = \frac{n}{T_p} = nf_p \qquad n = 0,1,2,\dots$$ [3.11]

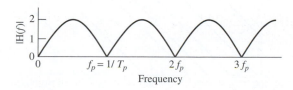

Figure 3.8   Magnitude of the frequency response $|H(f)|$ of a single delay-line canceler as given by Eq. (3.10), where f_p = pulse repetition frequency and $T_p = 1/f_p$.

This states that in addition to the zero response at zero frequency, there will also be zero response of the delay-line canceler whenever the doppler frequency $f_d = 2v_r/\lambda$ is a multiple of the pulse repetition frequency f_p. (The doppler shift can be negative or positive depending on whether the target is receding or approaching. When considering the blind speed and its effects, the sign of the doppler can be ignored—which is what is done here.) The radial velocities that produce blind speeds are found by equating Eqs. (3.11) and (3.3), and solving for the radial velocity, which gives

$$v_n = \frac{n\lambda}{2T_p} = \frac{n\lambda f_p}{2} \qquad n = 1,2,3\ldots \tag{3.12}$$

where v_r has been replaced by v_n, the nth blind speed. Usually only the first blind speed v_1 is considered, since the others are integer multiples of v_1. If λ is measured in meters, f_p in hertz, and the radial velocity in knots, the first blind speed can be written

$$v_1 \ (\text{kt}) = 0.97 \ \lambda \ (\text{m}) \ f_p \ (\text{Hz}) \approx \lambda \ (\text{m}) \ f_p \ (\text{Hz}) \tag{3.13}$$

A plot of the first blind speed as a function of the pulse repetition frequency and the various radar frequency bands is shown in Fig. 3.9.

Blind speeds can be a serious limitation in MTI radar since they cause some desired moving targets to be canceled along with the undesired clutter at zero frequency. Based on Eq. (3.13) there are four methods for reducing the detrimental effects of blind speeds:

Figure 3.9 Plot of the first blind speed, Eq. (3.13), as a function of the pulse repetition frequency for the various radar frequency bands.

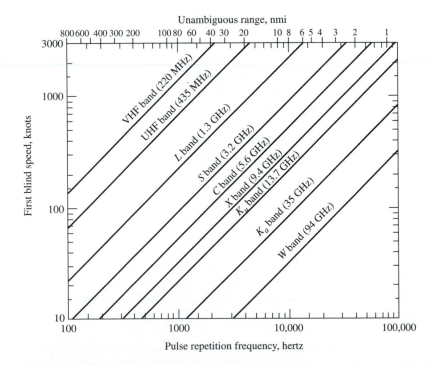

1. Operate the radar at long wavelengths (low frequencies).
2. Operate with a high pulse repetition frequency.
3. Operate with more than one pulse repetition frequency.
4. Operate with more than one RF frequency (wavelength).

Combinations of two or more of the above are also possible to further alleviate the effect of blind speeds. Each of these four methods has particular advantages as well as limitations, so there is not always a clear choice as to which to use in any particular application.

Consider the case where a low RF frequency is chosen to avoid blind speeds. If, for example, the first blind speed is to be no lower than 640 kt (approximately Mach 1) and the prf is selected as 330 Hz (which gives an unambiguous range of 245 nmi), then the radar wavelength from Eq. (3.13) is 2 m. This corresponds to a frequency of 150 MHz (the VHF region of the spectrum). Although there were many radars at VHF built in the 1930s and early 1940s and there are still advantages to operating a radar at these frequencies, VHF is not usually considered a desirable frequency choice for a long-range air-surveillance radar for a number of reasons: (1) resolution in range and angle are poor due to narrow bandwidths and large beamwidths, (2) this portion of the electromagnetic spectrum is crowded with other than radar services (such as broadcast FM and TV), and (3) low-altitude coverage generally is poor. Thus attempting to use low frequencies to avoid the blind speed problem is not usually a desirable option for many radar applications.

On the other hand, if we choose to operate at a high RF frequency and increase the prf to avoid blind speeds, we would then have to tolerate the many range ambiguities that result. For example, if the first blind speed was again chosen to be 640 kt and the wavelength were 0.1 m (an S-band frequency of 3000 MHz), the prf would have to be 6600 Hz. This results in a maximum unambiguous range of 12.3 nmi, which is small for many radar applications. (Such an approach, however, is used successfully in pulse doppler radars, as discussed later in this chapter.)

When two or more prfs are used in a radar, the blind speeds at one prf generally are different from the blind speeds at the other prfs. Thus targets that are highly attenuated with one prf might be readily seen with another prf. This technique is widely used with air-surveillance radars, especially those for civil air-traffic control. A disadvantage of a multiple-prf waveform is that multiple-time-around clutter echoes (from regions beyond the maximum unambiguous range) are not canceled.

A radar that can operate at two or more RF frequencies can also unmask blind speeds, but the required frequency change is often larger than might be possible within the usual frequency bands allocated for radar use. A limitation of multiple frequencies is the need for greater system bandwidth.

In some circumstances, it might be desirable to tolerate the blind speeds rather than accept the limitations of the above methods. As in many aspects of engineering, there is no one single solution best for all cases. The engineer has to decide which of the above limitations can be accepted in any particular application.

Blind speeds occur because of the sampled nature of the pulse radar waveform. Thus it is sampling that is the cause of ambiguities, or aliasing, in the measurement of the doppler frequency, just as sampling in a pulse radar (at the prf) can give rise to ambiguities in the range measurement.

Clutter Attenuation   The other limitation of the single delay-line canceler is insufficient attenuation of clutter that results from the finite width of the clutter spectrum. The single delay-line canceler whose frequency response was shown in Fig. 3.8 does what it is supposed to do, which is to cancel stationary clutter with zero doppler shift. In the "real world," however, the clutter spectrum has a finite width due to such things as the internal motions of the clutter, instabilities of the stalo and coho oscillators, other imperfections of the radar and its signal processor, and the finite signal duration. (The factors that widen the clutter spectrum will be discussed later in Sec. 3.7.) For present purposes we will assume the clutter power spectral density is represented by a gaussian function, and is written as

$$W(f) = W_0 \exp\left(-\frac{f^2}{2\sigma_c^2}\right) = W_0 \exp\left(-\frac{f^2\lambda^2}{8\sigma_v^2}\right) \qquad f \geq 0 \qquad \textbf{[3.14]}$$

where W_0 = peak value of the clutter power spectral density, at $f = 0$, σ_c = standard deviation of the clutter spectrum in hertz, and σ_v = standard deviation of the clutter spectrum in meters/second. The relation between the two forms of the clutter-spectrum standard deviation is based on applying the doppler frequency expression of Eq. (3.3) such that $\sigma_c = 2\sigma_v/\lambda$. The advantage of using the standard deviation σ_v is that it is often independent of the frequency; whereas σ_c is in hertz and depends on the radar frequency. Nevertheless, we will generally use σ_c in this chapter for the clutter spectrum.

The consequences of a finite-width clutter spectrum can be seen from Fig. 3.10. The frequency response of the single delay-line canceler shown by the solid curve encompasses a portion of the clutter spectrum; therefore, clutter will appear in the output. The greater the standard deviation σ_c, the greater the amount of clutter that will be passed by the filter to interfere with moving target detection. The clutter attenuation (CA) produced by a single delay-line canceler is

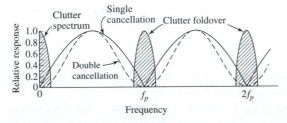

Figure 3.10   Relative frequency response of a single delay-line canceler (solid curve) and the double delay-line canceler (dashed curve), along with the frequency spectrum of the clutter (shaded area). Note the clutter spectrum is folded over at the prf and its harmonics because of the sampled nature of a pulse radar waveform.

$$CA = \frac{\int_0^\infty W(f)\, df}{\int_0^\infty W(f)|H(f)|^2\, df} \qquad [3.15]$$

where $H(f)$ is the frequency response of the delay-line canceler. Substituting for $H(f)$ [Eq. (3.10)], the clutter attenuation becomes

$$CA = \frac{\int_0^\infty W_0 \exp\left[-f^2/2\sigma_c^2\right]\, df}{\int_0^\infty W_0 \exp[-f^2/2\sigma_c^2]\, 4\sin^2(\pi f T_p)\, df} = \frac{0.5}{1 - \exp\left(-2\pi^2 T_p^2 \sigma_c^2\right)} \qquad [3.16]$$

If the exponent in the denominator at the right-hand part of this equation is small, the exponential term $\exp[-x]$ can be replaced by $1 - x$, so that

$$CA \approx \frac{f_p^2}{4\pi^2\sigma_c^2} = \frac{f_p^2 \lambda^2}{16\pi^2\sigma_v^2} \qquad [3.17]$$

In this equation the pulse repetition period T_p has been replaced by $1/f_p$. The clutter attenuation provided by a single delay-line canceler is not sufficient for most MTI radar applications.

If a second delay-line canceler is placed in cascade, the frequency response of the two filters is the square of the single delay-line canceler, or

$$H(f) = 4\sin^2(\pi f T_p) \qquad [3.18]$$

This is indicated in Fig. 3. 10 by the dashed curve (except that we have plotted the relative response rather than the absolute response). Less of the clutter spectrum is included within the frequency response of the double delay-line canceler; hence, it attenuates more of the clutter. The clutter attenuation for the double delay-line canceler is

$$CA \approx \frac{f_p^4}{48\pi^4\sigma_c^4} = \frac{f_p^4 \lambda^4}{768\pi^4\sigma_v^4} \qquad [3.19]$$

Additional delay-line cancelers can be cascaded to obtain a frequency response $H(f)$ which is the nth power of the single delay-line canceler given by Eq. (3.10), where n is the number of delay-line cancelers.

MTI Improvement Factor   The clutter attenuation is a useful measure of the performance of an MTI radar in canceling clutter, but it has an inherent weakness if one is not careful. The clutter attenuation can be made infinite by turning off the radar receiver! This, of course, would not be done knowingly since it also eliminates the desired moving-target echo signals. To avoid the problem of increasing clutter attenuation at the expense of desired signals, the IEEE defined[4] a measure of performance known as the *MTI improvement factor* which includes the signal gain as well as the clutter attenuation. It is defined as "The signal-to-clutter ratio at the output of the clutter filter divided by the signal-to-clutter ratio at the input of the clutter filter, averaged uniformly over all target radial velocities of interest." It is expressed as

$$\text{improvement factor} = I_f = \frac{(\text{signal/clutter})_{\text{out}}}{(\text{signal/clutter})_{\text{in}}}\bigg|_{f_d} = \frac{C_{\text{in}}}{C_{\text{out}}} \times \frac{S_{\text{out}}}{S_{\text{in}}}\bigg|_{f_d}$$

$$= CA \times \text{average gain} \qquad [3.20]$$

The vertical line on the right in the above equation indicates that the average is taken with respect to doppler frequency f_d. The improvement factor can be expressed as the clutter attenuation $CA = (C_{\text{in}}/C_{\text{out}})$ times the average filter gain. The average gain is determined from the filter response $H(f)$ and is usually small compared to the clutter attenuation. The average gain for a single delay-line canceler is 2 and for a double delay-line canceler is 6. The improvement factors for single and double delay-line cancelers are

$$I_f(\text{single DLC}) \approx \frac{1}{2\pi^2(\sigma_c/f_p)^2} = \frac{\lambda^2}{8\pi^2(\sigma_v/f_p)^2} \qquad [3.21]$$

$$I_f(\text{double DLC}) \approx \frac{1}{8\pi^4(\sigma_c/f_p)^4} = \frac{\lambda^4}{128\pi^4(\sigma_v/f_p)^4} \qquad [3.22]$$

The general expression for the improvement factor for a canceler with n delay-line cancelers in cascade is[5]

$$I_f(n \text{ cascaded DLCs}) \approx \frac{2^n}{n!}\left(\frac{1}{2\pi(\sigma_c/f_p)}\right)^{2n} \qquad [3.23]$$

As with the previous expressions, this applies when σ_c/f_p is small. A plot of the improvement factor as a function of σ_c/f_p is provided later in Fig. 3.13. The ratio σ_c/f_p is a measure of the amount of "doppler space" occupied by clutter. Equation (3.23) also applies for the so-called *N-pulse canceler* with $N = n + 1$, to be discussed next.

N-Pulse Canceler   A double delay-line canceler is shown in Fig. 3.11a. A canceler with two delay lines that has the same frequency response as the double delay-line canceler, but which is arranged differently, is shown in Fig. 3.11b. This is known as a *three-pulse canceler* since three pulses are added, with appropriate weights as shown. To obtain a $\sin^2(\pi f T_p)$ response, the weights of the three pulses are $+1, -2, +1$. When the input is $s(t)$, the output of the three-pulse canceler is then

$$s(t) - 2s(t + T_p) + s(t + 2T_p)$$

which is the same as the output from the double delay-line canceler

$$s(t) - s(t + T_p) - [s(t + T_p) - s(t + 2T_p)].$$

Thus the double delay-line canceler and the three-pulse canceler have the same frequency response function.

A four-pulse canceler with weights $+1, -3, +3, -1$ has a frequency response proportional to $\sin^3(\pi f T_p)$. A five-pulse canceler has weights $+1, -4, +6, -4, +1$. If n is

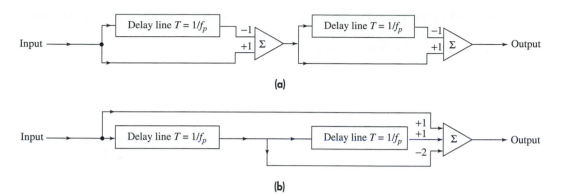

Figure 3.11 (a) Double delay-line canceler; (b) three-pulse canceler. The two configurations have the same frequency response. The three-pulse canceler of (b) is an example of a transversal filter.

the number of delay lines, there are $n + 1 = N$ pulses available to produce a frequency response function proportional to $\sin^n(\pi f T_p)$ when the weights are the coefficients of the expansion of the binomial series $(1 - x)^n$ with alternating signs. The binomial weights with alternating sign are given by

$$w_i = (-1)^{i-1} \frac{n!}{(n - i + 1)!(i - 1)!} \qquad i = 1, 2, \cdots, n + 1 \qquad \textbf{[3.24]}$$

The N-pulse canceler has the same frequency response as n single delay-line cancelers in cascade, where $n = N - 1$. The greater the value of N, the greater will be the clutter attenuation.

Transversal (Nonrecursive) Filter The three-pulse canceler of Fig. 3.11b is an example of a *transversal filter*. Its general form with n delay lines is shown in Fig. 3.12. The weights w_i are applied to the $N = n + 1$ pulses and then combined in the summer, or adder. The transversal filter is a time-domain filter with feed-forward lines and taps with weights w_i.

Figure 3.12
Transversal, or nonrecursive, filter for MTI signal processing.

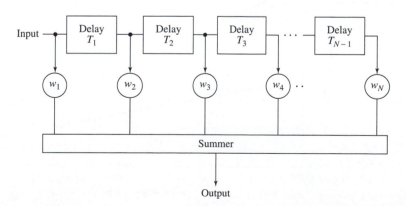

The delays T_i are usually equal, but they need not be (Sec. 3.3). The configuration of Fig. 3.12 is also known as a *nonrecursive filter, feed-forward filter, finite impulse-response (FIR) filter,* or *finite memory filter.*

"Optimum" MTI Filter   Transversal filters can have other than a $\sin^n(\pi f T_p)$ frequency response by choosing other than binomial weights. It is logical to ask what might be an optimum MTI filter. If one speaks of an optimum, there must be some criterion on which it is based. In the case of the so-called *optimum MTI filter*[6] the criterion is that it maximize the *improvement factor,* Eq. (3.20). It can also be considered as one which maximizes the clutter attenuation. It happens that a close approximation to the "optimum" MTI filter is a transversal filter like that in Fig 3.12 with binomial weights of alternating sign that has a frequency response function proportional to $\sin^n(\pi f T_p)$, where n is the number of delay lines, as was discussed above. The optimum weights $(1, -1)$ of a transversal filter with a single delay line are the same as that of the single delay-line canceler, if the clutter spectrum can be modeled as a gaussian function.[7] The difference between the three-pulse canceler and the "optimum" three-pulse MTI filter is less than 2 dB.[8] The difference is also small for higher values of n. Figure 3.13 is a plot of the improvement factor as a function of σ_c/f_p, where σ_c = the standard deviation of the clutter spectrum assuming it is of gaussian shape, and f_p = the pulse repetition frequency.[9] The solid curves apply for optimum weights and the dashed curves apply for binomial weights. There are two things to be noted from this plot. First, the differences between the two sets of curves are small so that we will take the optimum MTI filter to be adequately approximated by the filter with binomial weights whose response is proportional to $\sin^n(\pi f T_p)$. Second, sufficient improvement factor for many applications might be obtained with no more than two

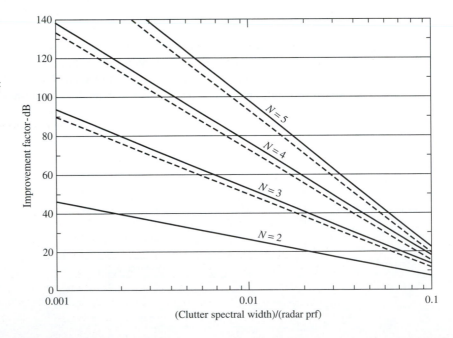

Figure 3.13   MTI improvement factor for an *N*-pulse delay-line canceler with binomial weights (dashed curves) and optimum weights (solid curves) as a function of the clutter spectral width σ_c. | (After Andrews.[9])

or three delay lines, if the improvement factor is the major criterion for the design of the MTI filter.

The term *optimum* is sometimes mistaken for the *best* that can be achieved. Optimum, however, is defined as the best under some implied or specified conditions. A so-called optimum solution might not be desired if the conditions for which it is defined are not suitable. This happens to be the case for the "optimum" MTI filter mentioned above. It is optimum if one wishes to have a filter that maximizes the improvement factor or the clutter attenuation. This may seem to be a suitable criterion, but it is not necessarily what one wants to achieve in a MTI filter. As the number n of delay lines increases in a filter with $H(f) \sim \sin^n (\pi f T_p)$, the response of $H(f)$ narrows and more and more clutter is rejected. The narrower bandwidth of the filter also means that fewer moving targets will be detected. If, for example, the -10 dB width of $H(f)$ is taken as the threshold for detection, and if all targets are uniformly distributed across the doppler frequency band, the following reductions in performance occur:

- 20 percent of all targets will be rejected by a two-pulse canceler

- 38 percent of all targets will be rejected by a three-pulse canceler

- 48 percent of all targets will be rejected by a four-pulse canceler

Thus if the "optimum" clutter filter is used, the loss of desired target detections is another reason it should not employ a large value of n.

"Rectangular-like" Transversal Filter Response   If one examines the clutter spectrum, such as the shaded region in Fig. 3.10, it can be seen that a desirable filter should approximate a rectangular passband that attenuates the clutter but has uniform response over as much of the doppler space as practical. It would not have as much clutter attenuation as an "optimum" filter of the same number of delay lines; but as we have seen from Fig. 3.13, the clutter attenuation of the "optimum" generally is far greater than can be used in practice when the number of delay lines n is large. A transversal filter, as in Fig. 3.12, can be designed to approximate a rectangular passband if it contains a sufficient number of delay lines and if the weights w_i are chosen appropriately.

Some examples of transversal, or nonrecursive, filters for MTI applications that have appeared in the literature have been summarized by Y. H. Mao.[10] Procedures for nonrecursive filter design can be found in classical text books on digital filters. An early example due to Houts and Burlage,[11] based on a Chebyshev filter response that employs 15 pulses, is in Fig. 3.14. Also shown for comparison is the response of a three-pulse canceler with binomial weights and the response of a five-pulse canceler with "optimum" weights. Generally, the goal in such filter design is to achieve the necessary improvement factor by choosing the attenuation in the stopband of a bandpass filter, the extent of the stopband, the extent of the passband, and the ripple that can be tolerated in the passband.

The large improvement factor that results with the "optimum" MTI filter when n is large can be traded for increased doppler-frequency passband. For example, when $\sigma_c/f_p = 0.001$, Eq. (3.22) indicates that the three-pulse-canceler, or double delay-line canceler, (which is close to the "optimum") has a theoretical improvement factor of 91 dB. This is a large improvement factor and is usually more than is required for

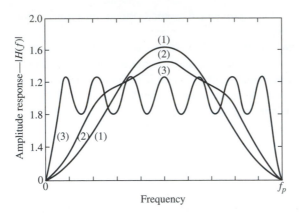

Figure 3.14 Amplitude response for three MTI delay-line cancelers. (1) Classical three-pulse canceler, (2) five-pulse delay-line canceler with optimum weights, and (3) 15-pulse Chebyshev design.
| (After Houts and Burlage.[11])

routine MTI radar applications. Furthermore, it might be difficult to achieve such high values in practice considering the problems of equipment instability and other factors that can limit the improvement factor. The five-pulse "optimum" of Fig. 3.14 is indicated by Houts and Barlage[11] as having an improvement factor of 85 dB for this clutter spread and the 15-pulse Chebyshev design has an improvement factor of 52 dB. If there are many pulses available for MTI processing, an approximately rectangular filter response may be preferred over the "optimum" since increased doppler-frequency passband is more important than extremely large theoretical values of improvement factor which are not needed or cannot be achieved in practice. It has been said[11] that even with only five pulses available, a five-pulse Chebyshev design provides significantly wider doppler space than the five-pulse "optimum" design.

When only a few pulses are available for processing, there is probably little that can be done to control the shape of the nonrecursive filter characteristic, and there might not be much gained by using other than a filter with binomial weights that has a $\sin^n (\pi f T_p)$ response.

Recursive Filters The N-pulse nonrecursive canceler discussed above allows the designer N zeros for synthesizing the frequency response using the classical z-plane procedure for filter design. Each feedforward line and its weight w_i correspond to a zero in classical filter design on the z-plane. Filter design using only zeros does not have the flexibility of filter design based on poles as well as zeros. Poles can be obtained with delay-line cancelers by employing feedback. With both feedback and feedforward lines providing both poles and zeros, arbitrary filter frequency-response functions can be synthesized from cascaded delay lines, within the limits of realizability.[12] These are known as *recursive filters* or *infinite impulse response (IIR) filters*. Significantly fewer delay lines (and fewer pulses) are needed to achieve desirable frequency-response functions than with nonrecursive filters that only have zeros available for design in the z-plane.

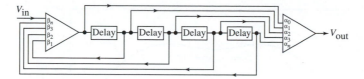

Figure 3.15   Canonical configuration of a recursive delay-line filter with both feedforward and feedback.
| (After White and Ruvin, *IRE Natl. Conv. Rec.*, vol. 5, pt. 2, 1957.)

The canonical configuration of a time-domain filter with both feedback and feedforward is shown in Fig. 3.15. More usually, the canonical configuration is broken into sections with feedback and feedforward around individual delay lines. An example is the three-pole Chebyshev filter of Fig. 3.16a. The frequency response of this recursive filter is shown in Fig. 3.16b, with 0.5 dB ripple in the passband.[13] The width of the passband can be changed with different sets of weights. Figure 3.17, due to J. S. Shreve,[14] compares the response of a nonrecursive and recursive filter. It is seen that the recursive filter provides a frequency response that better resembles the rectangular shape than the nonrecursive, and does so with only two delay lines rather than the four of the nonrecursive filter.

The recursive filter provides more flexibility in shaping the frequency response to produce better response than the nonrecursive filter, and it does it with fewer delay lines. Unfortunately, its utility is limited by poor transient response because of the feedback. Large discrete clutter echoes, interference pulses from a nearby radar, deliberate pulse jamming, or the beginning of a dwell by a step-scan radar (a phased array radar, for example) can all produce transient ringing that can mask the target signal until the transient dies out. A frequency response with steep sides might allow 15 to 30 or more pulses to be generated at the filter output from a single input pulse because of the feedback.[15] It

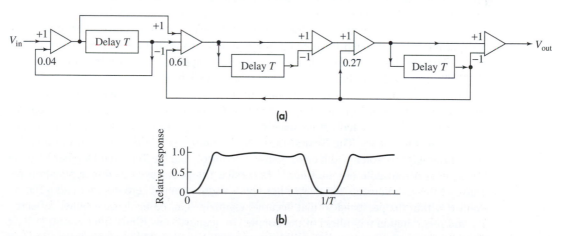

Figure 3.16   Shown in (a) is a recursive delay-line filter whose frequency response in (b) is based on a three-pole Chebyshev design with 0.5-dB ripple in the passband.
| (After W. D. White.[13])

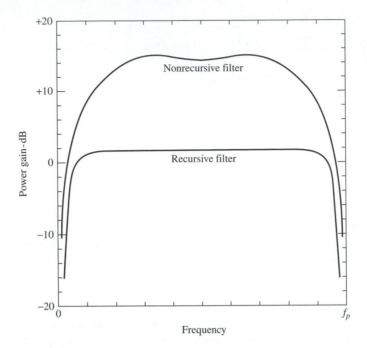

Figure 3.17 Example of the frequency response of a recursive filter with two delay lines and a nonrecursive filter with four delay lines. For the recursive filter the feedforward weights are 1, −2, and 1 and its feedback weights are 1, 1.561, and −0.641. The feedforward weights for the nonrecursive filter are 1, 1.25, −4.75, 1.75, and 0.75.
I (After J. S. Shreve.[14])

has been suggested[16] that the undesirable transient in a stepped-scan radar can be mitigated by using the first return from each new beam position to apply initial conditions to the MTI recursive filter to cancel or reduce the transient effects. The clutter returns can be approximated by a step-input equal in magnitude to the first return for each beam position. The steady-state values that would normally appear in the (digital) filter are calculated and loaded into the filter to suppress the transient response.

Because of the poor transient response of recursive filters (infinite impulse response), they are usually avoided in military radars that might be subject to deliberate electronic countermeasures. They have been widely used, however, to eliminate clutter in civilian doppler weather radars. The Nexrad radar, for example, uses a five-pole (fifth-order) elliptic filter with its notch width under software control, and the Terminal Doppler Weather Radar uses a four-pole elliptic filter.[17] Recursive filters are used in this application because of the requirements for sharp filter notch widths around zero doppler and a flat response within the passband so that accurate estimates can be made of weather reflectivity and precipitation regardless of the doppler frequency. The elliptic filter is said to have the advantage of narrower notch widths than any other filter of the same order.

Although the recursive filter seems attractive for MTI, its transient response makes it of limited application. There are several other approaches to achieving an MTI filter.

The use of multiple staggered prfs will be discussed next, which increases the first blind speed as well as provide limited filter shaping. This will be followed by the doppler filter bank that provides capabilities not available with the filters discussed previously.

The multiple delay-line canceler, the N-pulse canceler, the transversal (nonrecursive) filter, and the recursive filter are all examples of time-domain filters rather than frequency-domain filters. The doppler filter bank in Sec. 3.4 is also considered as a time-domain filter.

3.3   STAGGERED PULSE REPETITION FREQUENCIES

The use of multiple waveforms with different pulse repetition frequencies allows the detection of moving targets that would otherwise be eliminated with a constant-prf waveform if their radial velocities were at, or in the vicinity of, a blind speed [as defined by Eq. (3.12)]. A simple illustration is shown in Fig. 3.18 which graphs the frequency response of a single delay-line canceler with two different pulse repetition frequencies (prfs). At prf f_1, blind speeds (nulls) occur when the doppler frequency is f_1 or $2f_1$ (and other integer multiples, which are not shown). With prf $f_2 = 2f_1/3$, blind speeds occur when the doppler frequency equals f_2, $2f_2$, or $3f_2$. It can be seen in Fig. 3.18 that targets not detectable because of a blind speed in the frequency response of one prf will be detectable with the other prf. A target is lost on both prfs, however, when the blind speeds occur simultaneously, as when $2f_1 = 3f_2$. Thus the first blind speed at prf f_1 has been doubled in this simple example. The above illustrates the benefit of using more than one prf to reduce the effects of blind speeds; but it might be cautioned that it is not usual to use prfs with the relatively large ratio of 3/2.

There are several methods for employing multiple prfs to avoid losing target echoes due to blind speeds. The prfs can be changed (1) scan to scan, (2) dwell to dwell, or (3) pulse to pulse (usually called a staggered prf). A *dwell* is the *time on target,* and is

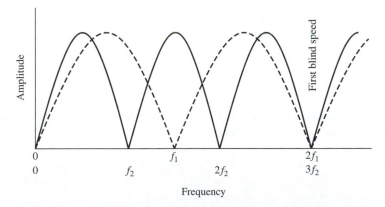

Figure 3.18   Frequency response of two prfs f_1 (dash curve) and f_2 (solid curve), where $f_1 = 3f_2/2$. The blind speeds of the first prf are shown at f_1 and $2f_1$, and the blind speeds of the second prf are shown at f_2, $2f_2$, and $3f_2$. A target at the blind speeds of one prf are "covered" by the response of the other prf, except when $2f_1 = 3f_2$, where both prfs have the same blind speed.

usually the time it takes to scan the antenna beam over a beamwidth or some fraction of a beamwidth. A dwell can be the time to scan one half-beamwidth, if there are only two prfs (as in the MTD discussed in Sec. 3.6). Staggered prfs have been popular for air-traffic control radars. An example of the four intervals of a staggered prf waveform is given in Fig. 3.19. The four-interval sequence is then repeated.

Each of the three strategies for applying the prfs for relieving the problem of blind speeds has its particular advantages as well as limitations, and each has been used in operational radars systems.

Staggered prfs   In pulse-to-pulse staggered prfs, as in Fig. 3.19, the time between pulses is an interval or a period. The term *interval* is more appropriate, but the term *period* arose historically and its use remains even though it might not always be used precisely. In this discussion, the term *period* will be used for the interval between pulses of a staggered waveform. Multiple staggered prfs can be processed with a transversal filter, as was shown in Fig. 3.12. The filter samples the doppler frequency at nonuniform times rather than the uniformly spaced time-samples when the prf is constant. The frequency response of this filter is

$$H(f) = w_0 + w_1 e^{j2\pi f T_1} + w_2 e^{j2\pi f(T_1+T_2)} + \cdots + w_n e^{j2\pi f(T_1+T_2+\cdots+T_n)} \qquad \text{[3.25]}$$

The selection of the $n + 1$ weights w_i and the n pulse repetition periods T_i is generally constrained by several design factors:

1. The minimum period should not result in range ambiguities.
2. The sequence of the periods should be selected so as not to stress the transmitter by employing a widely varying duty cycle or a duty cycle for which the transmitter is not designed.
3. The maximum period should not be too long since any range beyond the maximum unambiguous range represents "dead time" to a radar.
4. The response in the filter stopband should produce the required MTI improvement factor for detection of targets in clutter.
5. The deepest null in the passband should not be excessive. Usually the deepest null occurs at a frequency equal to the inverse of the average period.
6. The variation (or ripple) of the response over the passband should be minimized and relatively uniform.

Not all of these conditions can be satisfied simultaneously. Design of a staggered prf and its processing is often a compromise.

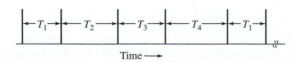

Time →

Figure 3.19   Staggered pulse-train with four different pulse periods, or intervals.

An example of the frequency response with a four-period (five-pulse) stagger with dual-canceler weighting is shown in Fig. 3.20. This is due to W. W. Shrader[18] and might be typical for a long-range en-route civil air-traffic control radar. The ratios of the pulse repetition periods are 25:30:27:31. These were found by computer search. Shrader states that with four periods, the stagger ratios can be obtained by adding the numbers -3, 2, -1, and 3 to the desired first blind speed expressed as v_1/v_B, where v_1 is the first blind speed and v_B is the blind speed that would have been obtained if a waveform with a constant prf had been used whose pulse repetition period was the average of the four staggered periods. This ratio is given by Eq. (3.26). If v_1/v_B were 28, adding the four numbers given above by Shrader results in the particular stagger ratios used in Fig. 3.20. [Substituting these four period-ratios into Eq. (3.26), however, gives $v_1/v_B = 28.25$ (instead of 28).]

Although the staggered prfs remove the blind speeds that would have been obtained with a constant prf, there will eventually be a new blind speed that occurs when the n prfs have the following relationship $\eta_1 f_1 = \eta_2 f_2 = \cdots = \eta_n f_n$, where $\eta_1, \eta_2, \ldots, \eta_n$ are relatively prime integers (with no common divisor other than 1). The ratio of the first blind speed v_1 with a staggered prf waveform to the first blind speed v_B of a waveform with constant prf equal to the average period is

$$\frac{v_1}{v_B} = \frac{\eta_1 + \eta_2 + \cdots + \eta_n}{n} \qquad \textbf{[3.26]}$$

In deriving the above, the average period is $T_{\text{av}} = (T_1 + T_2 + \cdots + T_n)/n$ where $T_i = 1/f_i$, and the new first blind speed is given by $v_1 = \lambda \eta_1 f_1/2 = \lambda \eta_1/2T_1$. A slightly different expression is obtained if the average prf were used instead of the average period.

Another example of the frequency response is that for a medium-range civil air-traffic control airport surveillance radar shown in Fig. 3.21. Note that there is a change in scale of the abscissa. The amplitude weighting is based on a dual canceler. The solid curve of Fig. 3.22 shows the response of a four-period canceler with weightings of $\frac{7}{8}$, 1, $-3\frac{3}{4}$, 1, $\frac{7}{8}$, and four interpulse periods of -15 percent, -5 percent, $+5$ percent, and $+15$ percent of the fixed period equal to the average period.[19] The deepest null is 6.6 dB. The dashed curve is the frequency response for the same amplitude weights, but with a fixed period.

Figure 3.20   Frequency response of a five-pulse (four-period) stagger.
(From Shrader and Gregers-Hansen,[1] courtesy McGraw-Hill Book Co.)

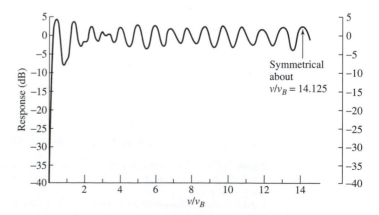

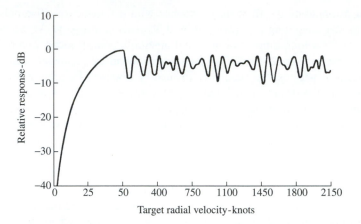

Figure 3.21 Example of the frequency response of a staggered MTI with four periods for an S-band ASR air-traffic control radar as a function of the target's radial velocity. The four periods are 876, 961, 830, and 1177 μs, with amplitude weighting based on a dual-canceler response. (Note change of scale of the abscissa.)

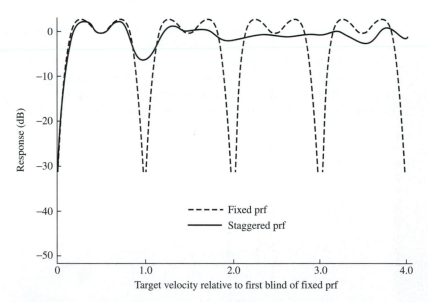

Figure 3.22 Frequency response of a weighted four-period staggered waveform compared with the frequency response using a constant prf waveform. Dashed curve, constant prf; solid curve, staggered prf.
| (From Zverev,[19] Courtesy IEEE.)

Null Depth and Improvement Factor   There have been two methods proposed for finding the effect of the stagger periods on the depth of the null and the MTI improvement factor. One is based on computer search, the other on an analytic formulation.

Computer Search   The depth of the deepest null within the MTI filter passband when using a staggered waveform is shown in Fig. 3.23, as found by Shrader based on computer search for the best sets of pulse periods.[18] He found that the null was relatively independent of the type of canceler (whether single, double, or triple) or the number of hits per beamwidth. The null depth depended mainly on the ratio of the maximum-to-minimum period, which is the abscissa of Fig. 3.23.

There have been other methods proposed for finding in a logical manner the values of the weights w_i and the periods T_i.[20-23] Some of these are concerned mainly with minimizing the extent of the deepest null and do not examine the effect on the improvement factor. Computer search on a trial-and-error basis, such as that of Shrader, has probably produced as good a result as any.

Analytic Formulation   Using the rms value σ_n of the periods of a staggered-prf waveform, Cleetus[24] extended earlier work of McAulay[25] and Wardrop[26] to arrive at an estimate of the maximum improvement factor as

$$I_{\max} = \frac{1}{4\pi^2 \sigma_c^2 \sigma_n^2} \tag{3.27}$$

where σ_c^2 = total variance of a gaussian clutter spectrum as defined by Eq. (3.14), σ_n^2 = mean square value of $\epsilon_n = T_i - T_{\mathrm{av}}$, T_i = period of the ith stagger, and T_{av} = average value of the n periods. This is called the maximum improvement factor since it assumes an ideal rectangular MTI filter. It depends only on the spread of the T_i's and not

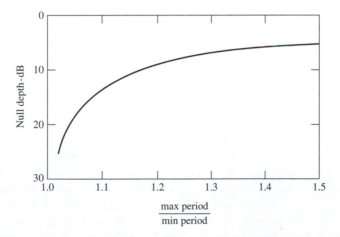

Figure 3.23   Approximate depth of the first null in the frequency response of a pulse-to-pulse staggered prf MTI as a function of the ratio of the maximum period to the minimum period. (After W. W. Shrader and V. Gregers-Hansen.[1] Courtesy McGraw-Hill Publishing Co.)

on the number of stagger pulses. The depth of the deepest null was also explored by Clee-tus. From the graph presented in his paper (his Fig. 3) the following expression can be obtained for the depth of the first null

$$\text{null depth} = 40 \left(\frac{\sigma_n}{T_{\text{av}}} \right)^2 \qquad [3.28]$$

This applies for values of σ_n/T_{av} less than approximately 0.09 (which produces a null depth of 5 dB). At higher values of σ_n/T_{av} the null depth rises slowly and approaches zero at $\sigma_n/T \approx 0.3$. Comparing Eqs. (3.27) and (3.28) shows that the null depth and improve-ment factor cannot be selected independently since they both depend on the variance σ_n^2 of the stagger periods. The computer-derived null depth given in Fig. 3.23 depends only on the ratio $T_{\text{max}}/T_{\text{min}}$, whereas Eq. (3.28) depends on the variance of the periods.

Shrader also presents a computer-derived expression for the improvement factor, which is

$$I = \left[\frac{2.5 n_B}{(T_{\text{max}}/T_{\text{min}}) - 1} \right]^2 \qquad [3.29]$$

where n_B = number of pulses per beamwidth. This is based on the width of the clutter spectrum being determined mainly by the antenna scan modulation. It may seem quite different from the improvement factor of Eq. (3.27), but the two can be shown to be some-what similar. If we make the following very gross assumptions, this equation can be put in almost the same form as Eq. (3.27): (1) the number of pulses $n_B \approx t_0/T_{\text{av}}$, where t_0 = time on target and T_{av} = average stagger period; (2) the value of t_0 is replaced by $(\sigma_e = 1/3.77 t_0)$ [Eq. (3.45)], the standard deviation for antenna scan modulation; (3) the antenna scan modulation σ_e is replaced by the more general σ_c, the symbol for the stan-dard deviation of the clutter spectrum, whatever its cause; and (4) the distribution of stag-ger periods is assumed uniform between T_{max} and T_{min} so that its standard deviation σ_n can be written (Sec. 2.4) as $\sigma_c = (T_{\text{max}} - T_{\text{min}})/2\sqrt{3}$. With these assumptions,

$$I = \left(\frac{1}{5.22 \sigma_c \sigma_n} \frac{T_{\text{min}}}{T_{\text{av}}} \right)^2 \qquad [3.30]$$

This resembles Eq. (3.27), and would be identical if $(T_{\text{av}}/T_{\text{min}}) = 2\pi/5.22 = 1.2$.

Comparison of Multiple-prf Methods   It was said earlier that there are three methods for applying multiple prfs to reduce the effects of blind speeds. These were (1) staggering a number of periods pulse to pulse, (2) alternating between two periods every half-beamwidth, and (3) changing periods (prfs) from one antenna scan to the other. Each has advantages and limitations.[27] Pulse-to-pulse stagger with several pulse periods can sig-nificantly increase the blind speed compared to a constant prf waveform and it doesn't have large regions of doppler space where moving targets are not detected as, for exam-ple, when a double or triple delay-line canceler is used. Also, compared to recursive filters, staggered-prf filter design is simplified since it does not require the feedback found in recursive filters nor does it experience poor transient response. Staggering the prfs degrades slightly the improvement factor by transferring clutter energy to the doppler-frequency space and it is more difficult to stabilize a transmitter that doesn't have a constant prf. Also, it does not cancel multiple-time-around clutter (clutter that originates beyond the maximum unambiguous range of the pulse period). In many radar systems that employ pulse-to-pulse stagger there is usually a constant prf waveform available that can be selected when the radar looks in those directions where multiple-time-around clutter is encountered.

The radar that changes its prf from scan to scan is easier to implement and it can cancel multiple-time-around clutter. It takes a longer time (more than one scan time) to unmask doppler blind speeds compared to pulse-to-pulse stagger. Seldom would more than two different prfs be used to unmask blind speeds in a scan-to-scan staggered system. With only two prfs, the reduction in blind speeds is not as effective as when many periods are employed, as in a staggered waveform. Dwell-to-dwell stagger has similar advantages as scan-to-scan changes, and unmasking of blind speeds is done quicker. Half-beamwidth dwells have been used effectively with filter-bank processing, as in the MTD radar discussed later in Sec. 3.6.

3.4   DOPPLER FILTER BANKS

A doppler filter bank is a set of contiguous filters for detecting targets as shown in Fig. 3.24. A filter bank has several advantages over the single filters that have been considered previously in this chapter:

1. Multiple moving targets can be separated from one another in a filter bank. This can be particularly important when one of the echo signals is from undesired moving clutter; such as a rain storm or birds with a nonzero doppler shift. When the clutter and target echo signal appear in different doppler filters, the clutter echo need not interfere with the detection of the desired moving target.

2. A measure of the target's radial velocity can be obtained. It might be ambiguous, but a change in the prf can resolve the ambiguity in the radial velocity,[28] just as changing the prf can resolve range ambiguities.

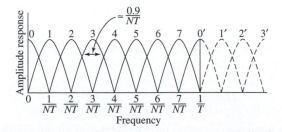

Figure 3.24   MTI doppler filter bank resulting from the processing of $N = 8$ pulses with the phase weights of Eq. (3.31), yielding the response of Eq. (3.34). N is the number of pulses processed and the number of filters generated; T is the pulse repetition period. The sidelobe structure of the filters is not shown.

3. The narrowband doppler filters exclude more noise than do the MTI delay-line cancelers described previously, and provide coherent integration. In general, this improvement in signal-to-noise ratio by means of coherent integration is seldom the major reason for using filter banks in MTI radar. The gain due to the coherent integration of n pulses is not that much greater than the gain due to the noncoherent integration of n pulses when n is not too large.

The price paid for these advantages is greater complexity, difficulty in achieving filters with low enough sidelobes to reduce clutter, and the need for a significant number of pulses to produce desirable filter characteristics. The basic method for achieving a doppler filter bank is to employ the transversal filter with complex weights rather than real (amplitude) weights as in the transversal filters discussed before. Complex weights mean that phase shifts as well as amplitude weights are employed.

Consider the transversal filter of Fig. 3.12, with N pulses (N taps) and $N - 1$ delay lines. With proper phase weights this will form N contiguous filters covering the frequency range from 0 to f_p (or from $-f_p/2$ to $+f_p/2$), where f_p is the pulse repetition frequency. The time delay between each tap of the transversal filter is $T = 1/f_p$. Not shown in this simple depiction of a transversal filter are the N parallel outputs at each of the N taps, one for each filter. The weights $w_{i,k}$ for each of N taps, with k outputs at each tap can be expressed as

$$w_{i,k} = e^{j[2\pi(i-1)k/N]}$$ [3.31]

where $i = 1, 2, \ldots, N$ represents the N taps and k is an index from 0 to $N - 1$ that corresponds to a different set of N weights, each for a different filter. In this example, the amplitude is the same at each tap, only the phases are different. The N filters generated by the index k constitute the filter bank. If there were eight pulses available to generate eight filters, the phase weights $w_{i,0}$ for the filter $k = 0$ [found from Eq. (3.31)] are all of zero phase. For the next filter, $k = 1$, Eq. (3.31) gives the phase weights $w_{i,1}$ as 0, 45, 90, 135, 180, 225, 270, and 315 degrees, respectively. The phase weights for the higher filters (k from 2 to 7) are the same as those for $k = 1$, but multiplied by k, modulo 360.

We next obtain the frequency response of the N filters of the filter bank. The impulse response of the transversal filter of Fig. 3.12 with the weights given by Eq. (3.31) can be formulated (almost) by inspection as

$$h_k(t) = \sum_{i=1}^{N} \delta[t - (i-1)T]e^{j2\pi(i-1)k/N}$$ [3.32]

where $\delta(t)$ is the delta function. The Fourier transform of the impulse response is the frequency response function, so that

$$H_k(f) = \sum_{i=1}^{N} e^{-j2\pi(i-1)[fT - k/N]}$$ [3.33]

The magnitude of the frequency response function is the amplitude passband characteristic of the filter, which is

$$|H_k(f)| = \left| \sum_{i=1}^{N} e^{-j2\pi(i-1)[fT - k/N]} \right| = \left| \frac{\sin[\pi N(fT - k/N)]}{\sin[\pi(fT - k/N)]} \right|$$ [3.34]

This is sketched in Fig. 3.24. The peak response occurs whenever the denominator is zero, or when $\pi(fT - k/N) = 0, \pi, 2\pi, \ldots$. The numerator also will be zero when the denominator is zero. The value of $0/0$ is indeterminate; but by using L'Hopitals rule, the peak value of Eq. (3.34) is found to be N when both numerator and denominator are zero. Nulls in the frequency response function occur when the numerator is zero and the denominator is not zero. The width of the main response as defined by the spacing between the first pair of zeros is $2/NT$. The half-power width is approximately $0.9/NT$. When the doppler filter bank is shown, as in Fig. 3.24, the sidelobes are not usually included. This is done for clarity; but they are there and can limit the amount of clutter attenuation obtained. The shape of an individual filter is as sketched in Fig. 3.25.

When $k = 0$, the peak response occurs at $f = 0, 1/T, 2/T, \ldots$. This defines a filter with peak response at zero frequency, at the prf ($f_p = 1/T$), and at harmonics of the prf. Thus the $k = 0$ filter corresponds to the clutter spectrum and does not reject clutter as do the other $N - 1$ filters. When $k = 1$, the peak response occurs at $f = 1/NT$, as well as $1/T + 1/NT, 2/T + 1/NT$, and so forth. For $k = 2$, the peaks occur at $f = 2/NT, 1/N + 2/NT, 2/N + 2/NT$, etc. Thus each value of k corresponds to one of N separate doppler filters, as in Fig. 3.24. Together the N filters cover the frequency region from 0 to $f_p = 1/T$ (or, equivalently, from $-f_p/2$ to $+f_p/2$). In this particular case where the amplitude weights at each tap are all the same, the first nulls of each filter are at the peaks of the adjacent filters. Being a periodic signal, the remainder of the frequency domain is covered by similar filters, but with ambiguity and aliasing.

The generation of N filters from the output of N taps of a transversal filter requires a total of $(N - 1)^2$ digital multiplications. The process is equivalent to that of a discrete Fourier transform. In many cases, the fast Fourier transform (FFT) can be used to speed computations if the value of N is some power of 2; which is why the above example was for $N = 8 = 2^3$.

The improvement factor for each of the eight filters of an eight-pulse filter bank with uniform amplitude weights is shown in Fig. 3.26 as a function of σ_c/f_p, where σ_c is the standard deviation of the clutter spectrum, which is assumed to be gaussian.[29] The average improvement factor for all filters is indicated by the dotted line. This might be compared to the improvement factor for an N-pulse canceler with response $\sin^N \pi f T_p$, which was shown in Fig. 3.13. Note that the improvement factor for a two-pulse canceler is about the same as the average of the eight-pulse doppler filter bank. The three-pulse canceler is even better. Thus lower sidelobes are needed if large values of the improvement factor are to be obtained with a doppler filter bank.

Figure 3.25 Sketch of an individual filter of the doppler filter bank of Fig. 3.24, including the sidelobes.

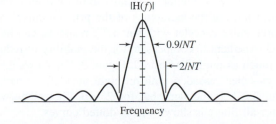

Figure 3.26 Improvement factor for each filter of an eight-pulse doppler filter bank, with uniform amplitude weighting, as a function of the clutter spectral width (standard deviation). The average improvement factor is indicated by the dotted curve. | (From Andrews.[29])

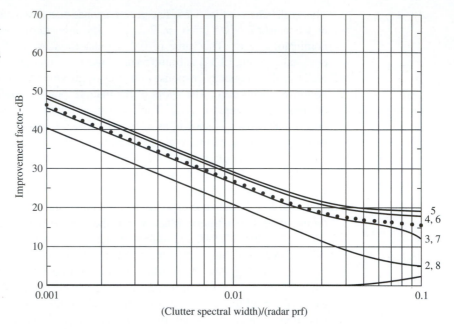

Reducing the Filter Sidelobes The frequency-response function of each filter of the filter bank derived above is of the form $(\sin NX)/\sin X$; thus the highest sidelobes are 13.2 dB below the peak response. These are relatively high and are not good at rejecting clutter, especially for those filters near zero frequency or near a blind speed. The sidelobes can be reduced by employing amplitude weights in addition to phase weights. Shrader and Gregers-Hansen[30] indicate that when only a few pulses are available (about six or so) in a coherent processing interval (CPI), empirical design procedures can be used since there is little flexibility for more sophisticated design. With a larger number of pulses available a more systematic approach can be employed based on Chebyshev filter design.[31] The frequency response of the Chebyshev filter has all its sidelobes equal and the width between the nulls of the main response is a minimum for a given sidelobe level.[32] (This is similar to the Dolph-Chebyshev array antenna pattern synthesis technique known to antenna engineers.) When the sidelobes are lowered, however, the main response is widened, the peak gain is reduced, but the straddling loss at filter crossover is less. For example, with 68-dB sidelobe level the null width is 3.5 times the half-power bandwidth, and the peak is 2.1 dB lower than that of a filter with uniform weighting.[30]

To further reduce the clutter, especially in those filters close to zero frequency or the clutter spectral lines at the harmonics of the prf, a relatively simple delay-line canceler such as a three-pulse canceler with $\sin^2 \pi f T_p$ response can be placed ahead of the filter bank. It is an expedient to compensate for the inability to reduce the sidelobes of the filter bank as much as might be desired. Figure 3.27a shows the improvement factor for a three-pulse canceler cascaded with an eight-pulse filter bank with uniform amplitude weights; Fig. 3.27b is the same but for 25-dB Chebyshev weights.[29] The average improvement for all filters is shown by the dotted curves.

Figure 3.27 Improvement factor for a three-pulse canceler followed by an eight-pulse doppler filter bank. (a) Uniform amplitude weights and (b) 25-dB Chebyshev weights. The average improvement factor for all filters is indicated by the dotted curve. (From Andrews.[29])

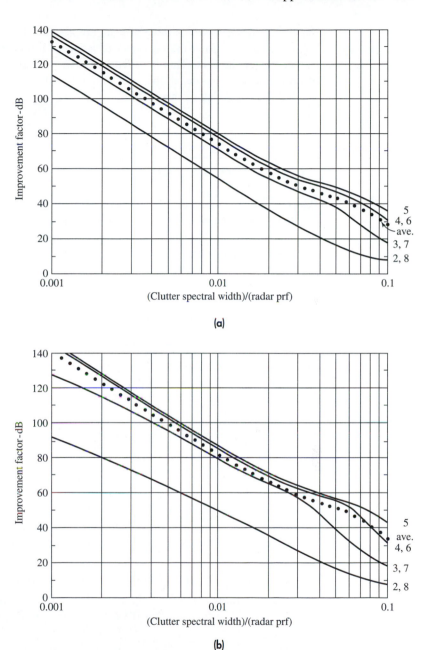

(a)

(b)

Other Limitations In addition to the difficulty in obtaining low filter sidelobes and large improvement factors, the doppler filter bank is more complex than the simpler delay-line cancelers discussed previously in this chapter, it generally requires more pulses for good performance, and it requires a larger signal-to-noise ratio if the true radial velocity is to be extracted when two or more prfs are employed. There will also be a reduction in the signal-to-noise ratio at the crossover of adjacent filters, relative to the peak response at the center of the filter This is called a *straddling loss*. The doppler filter bank is used, however, since it has capabilities not available with delay-line cancelers, as was enumerated at the beginning of this section.

3.5 DIGITAL MTI PROCESSING

Most of the basic theoretical aspects of MTI filter design were formulated when the delay lines were analog acoustic devices. Sophisticated MTI doppler filters were difficult to implement with analog methods, so it was rare for an MTI radar to employ more than two analog delay lines in a delay-line canceler. The rapid development of digital technology beginning in the early 1970s, however, allowed the delays to be obtained by storing digital words in a memory for whatever length of time was required. This greatly increased the options open to the radar signal-processing engineer. Digital doppler filters with many delay lines are now practical so that sophisticated filters can be readily obtained when a large number of pulses are available for processing. Thus the theoretical aspects of MTI doppler filters, which were only of academic curiosity when analog delay lines were all that was available, now can be implemented using digital methods. In addition to making practical the design of more sophisticated filters, the advantages offered by digital MTI processing include:

- Compensation for "blind phases," which cause a loss due to the difference in phase between the echo signal and the MTI reference signal. This is achieved by use of I and Q processing (in-phase and quadrature), something that was always known to be of value for MTI processing, but which was not convenient to implement with analog methods.

- Greater dynamic range can be obtained than was possible with acoustic delay lines.

- Unwanted changes in the delay times of acoustic delay lines due to temperature changes are eliminated by the accurate timing of digital methods.

- There is no problem in making the delay time in the digital memory synchronous with the radar's prf, something difficult to do with acoustic delay lines.

- The flexibility offered by digital methods allows signal processors to be readily obtained with many different filter characteristics. Digital processors can be made reprogrammable.

- Digital MTI is more stable and reliable than analog MTI, and requires less adjustments during operation in the field.

Although digital processing does not have the serious weaknesses associated with analog delay lines, it has other characteristics that have to be understood if full advantage is to be taken of its capabilities for significantly improving doppler processing in an MTI radar.

Blind Phases, *I* and *Q* Channels   The block diagram of an MTI radar that was shown in Fig. 3.7 had a single phase detector and filter channel. With a single phase detector and single processing channel, there is a loss when the doppler-shifted signal is not sampled at the peak positive and negative values of the sinewave. When the phase between the doppler signal and the sampling at the prf results in a loss, it is called a *blind phase*. A blind phase is different from the blind speed discussed previously. It will be recalled that a blind speed occurs when the sampling pulse appears at the same point in the doppler cycle at each sampling time, as in Fig. 3.28a. Figure 3.28b illustrates the loss due to a

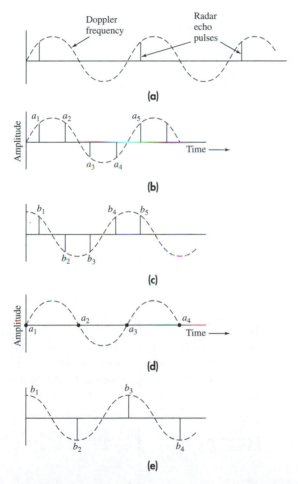

Figure 3.28   (a) Example of a blind speed in an MTI radar. The target's doppler frequency is equal to the prf. (b) Example of the effect of a blind phase in the *I* channel, and (c) in the *Q* channel. (d) The *I* channel of another special example of a blind speed. The prf is twice the doppler frequency and the phase of the sampling is such that there is no response at all since the sampling is at the zeros of the doppler frequency. Nothing is detected. (e) The *Q* channel for the example of (d) in which the sampling is at the positive and negative peaks of the doppler frequency so that there is complete recovery of the signal.

blind phase. The sampled signals in this particular example are all of the same amplitude and with a spacing such that when pulse a_2 is subtracted from pulse a_1, the result is zero. When pulse a_3 is subtracted from a_2, however, there is a finite output; but when a_4 is subtracted from a_3 the result is zero, and so on. Thus in this particular case, half of the signal energy is lost. The other half of the signal energy can be recovered if a second identical processing channel is used and there is a 90° phase change of the coho (reference) signal that is applied to its phase detector. This second channel is called the Q, or *quadrature,* channel. The original channel is called the I or in-phase channel. If the coho signal in the I channel is $\sin 2\pi f_{if}t$, the coho in the Q channel is $\cos 2\pi f_{if}t$. The result of the 90° phase change in the Q channel is shown in Fig. 3.28c. Those pulse pairs that had zero output in the I channel now have a finite residue in the Q channel. Likewise those pulse pairs which had a finite residue in the I channel now have zero output in the Q channel. What was lost in the I channel is recovered in the Q channel, and vice versa. The combination of the I and Q channels results in a uniform output with no loss.

The example of Fig. 3.28b and c is a special case. Another very special case of a blind phase is when the prf is twice the doppler frequency (this is not a blind speed) and the phase of the prf is such that the samples occur whenever the doppler signal in the I channel passes through a zero crossing, Fig. 3.28d. There is no output in this case. In the Q channel, Fig. 3.28e, where the phase of the prf is shifted 90°, the doppler signal is sampled whenever it is at a positive or a negative peak. The maximum signal is obtained. Again, what was lost in one channel is recovered in the other channel.

Block Diagram   The block diagram of a digital MTI signal processor with I and Q channels is shown in Fig. 3.29. The signal from the IF amplifier is split into two channels. The phase detectors in each channel extract the doppler-shifted signal. In the I channel the doppler signal is represented as $A_d \cos(2\pi f_d t + \phi_0)$ and in the Q channel it is the same except that sine replaces the cosine. The signals are then digitized by the analog-to-digital (A/D) converter. A sample-and-hold circuit usually is needed ahead of the A/D converter for more effective digitizing. Sample and hold is often on the same chip as the A/D converter. (Some A/D converters, such as the flash type, do not require a sample and hold.[33]) The digital words are stored in a digital memory for the required delay time(s) and are processed with a suitable algorithm to provide the desired doppler filtering. The magnitude of the doppler signal is obtained by taking the square root of $I^2 + Q^2$. Sometimes, for simplicity, the sum of the magnitudes of the two channels, $|I| + |Q|$, is taken;

Figure 3.29   Block diagram of a digital MTI doppler signal processor.

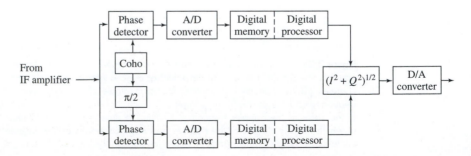

or the "greater of" of the two channels might be used instead.[34] The I and Q processor of Fig. 3.29 has a square-law detector characteristic. A linear-law detector can be approximated by the greater of $|I| + |Q|/2$ or $|Q| + |I|/2$.

If required, the combined unipolar output can be converted to an analog signal by a digital-to-analog (D/A) converter for display. Otherwise the digital output might be subject to further processing.

There are several methods for implementing an A/D converter, depending on the speed and number of bits required. Since there are two channels (the I and the Q), sampling in each can be at one-half the Nyquist rate, which generally makes the implementation of the A/D converters simpler. (The Nyquist rate is twice the signal bandwidth.) The number of quantization levels in the A/D is generally given as 2^N. D/A converters are usually easier to achieve than A/D converters of the same resolution.

The available technology of A/D converters has sometimes been a serious limitation to MTI performance, but there have been continual advances in this technology. A 16-bit A/D, for example, might have a sampling rate of several MHz; a 14-bit A/D, a 50-MHz sampling rate; a 12-bit AD, a 100-MHz sampling rate; and an 8-bit A/D, a 1-GH$_z$ sampling rate.

The output of the IF amplifier is usually made to limit at a level consistent with the MTI improvement factor (as in Sec. 3.7) and the full-scale range of the A/D converter. The IF portion of the receiver is then a linear-limiting amplifier. The limiting action means that the peak excursion of the signal that the A/D converter must handle is known, and the A/D can be designed to cover this excursion. The signal should not be allowed to exceed the full-scale range of the A/D converter, since the output would then be degraded and severe harmonics generated. [35]

Limitation on the Improvement Factor Due to the A/D Converter   All analog signals that lie within the same quantization step of the A/D converter are represented by the same digital word. Since the rms value of the noise accompanying the signal is usually greater than the quantization step of the A/D converter, the digital word can change slightly from pulse to pulse in a noiselike manner. Thus the quantization of the analog signal results in noise or uncertainty, called *quantization noise,* which can limit the MTI improvement factor.

Shrader and Gregers-Hansen[36] give the limitation to the improvement factor due to quantization noise as

$$I_q = 20 \log [(2^N - 1)\sqrt{0.75}] \qquad \text{(dB)} \qquad \textbf{(3.35)}$$

where N = number of bits. This is approximately 6 dB per bit. (Each bit represents a factor of two in amplitude resolution.) Thus a 10-bit A/D theoretically limits the improvement factor to about 60 dB. In practice, the A/D converter generally requires one or more additional bits to achieve the desired performance.

The limitation on the improvement factor based on Eq. (3.35) assumed that the quantization error was independent from pulse to pulse. This happens when the rms noise is greater than a quantization step. It was suggested by Glen Preston that there will be a quieting effect that reduces the quantization noise if the pulse-to-pulse clutter samples are correlated rather than independent. Brennan and Reed[37] derived an expression for the quieting that is a function of the correlation, and they found quieting to be small. When it occurs, the quantization step is large compared to the rms noise, a condition that is not preferred for best performance in a practical radar system.

Dynamic Range The dynamic range is the maximum signal-to-noise ratio that can be handled by an A/D converter without saturation. The noise level relative to the quantization step affects the dynamic range. The available dynamic range (power ratio) is given as[35]

$$\text{dynamic range} = 2^{2N-3}/k^2 \qquad \text{[3.36]}$$

where N = number of bits in the A/D converter (the sign bit is included), and k = rms noise level divided by the quantization level. The larger k is, the less the dynamic range. A value of k less than one means that the receiver noise at the input to the A/D converter is less than the quantization noise, which results in a loss of detectability. Generally k equals 1 or 2. With $k = 2$ (a value recommended by Shrader and Gregers-Hansen), the dynamic range for a 10 bit A/D according to Eq. (3.36) is 45.2 dB, which is considerably less than the 60 dB improvement factor based on a value of 6 dB per bit. If the limitation on the improvement factor is taken as 2^{2N} (6 dB per bit), the dynamic range given by Eq. (3.36) when $k = 2$ is seen to be about 15 dB less than the improvement factor. Thus the limitation on the improvement factor might be determined by the dynamic range of Eq. (3.36) rather than by the quantization noise limitation usually given as 6 dB per bit. (This could be the reason why, in practice, one or more bits are added to the A/D converter when the improvement factor limitation is taken as 6 dB per bit.)

Other Limitations The above has, for the most part, assumed ideal A/D converters and ideal I and Q detection. There are several practical conditions that need to be considered when good MTI performance is required. Errors and reduced performance can be due to:[35,38] (1) other than a 90° phase difference between the I and Q reference signals,[39] (2) gain and phase imbalance in the two channels, (3) timing jitter in the sample-and-hold circuits, (4) nonlinearity in the A/D,[40] and (5) range straddling loss due to the sampling not being at the peak of the output of the matched filter.

The degradation in performance caused by the inability to precisely match the I and Q channels in phase, amplitude, and frequency spectrum can be significantly reduced by employing a single channel whose signal is down converted (from IF or RF, by mixing) to obtain the in-phase and quadrature channels. There have been several methods proposed in the past to perform this form of *bandpass sampling*, also sometimes called *digital down conversion*, A technique described by Rader[41] for a band-limited signal of bandwidth B down-converts (mixes) the IF or RF signal to a center frequency of B. The signal is sampled at a rate of $4B$. Two filters are realized as a pair of 90° phase splitting networks with several symmetries which are used to reduce the computations. The in-phase and quadrature samples are obtained at an output rate of B. It has been said[42] that the oversampling by a factor of 2 from the Nyquist rate ($2B$) results in a simple and efficient implementation of the filters. Only one mixer and one A/D converter are needed, but the sampling is four times faster.

In another method for baseband sampling,[43] IF samples taken near the Nyquist rate are interpolated, based on a number of stored samples, to provide the in-phase and quadrature channels. Experimentally it was found that the phase errors could be reduced by an order of magnitude compared with conventional baseband I,Q processing.

Fast Fourier Transform (FFT)   Digital filtering involves the use of the Fourier transform. The fast Fourier transform requires less computational effort, and it has been popular for many applications. It has some limitations, however, compared to the classic Fourier transform. The number of samples used has to be expressed as 2^N. If a filter bank is being generated, all filters have identical responses, they will be uniformly spaced in frequency, and the weighting coefficients are not optimum since they cannot be chosen independently for each filter. The filters possible with a non-FFT filter bank also can achieve greater attenuation of moving clutter (such as rain or chaff) because of the greater flexibility available in their design. There are times, therefore, when the classical Fourier transform may be more advantageous than the FFT even though the FFT might be quicker and require less complexity.[44]

Loss Due to Blind Phases in Using a Single Channel[45]   With only a single channel, the effect of a blind phase can be anything from complete loss of the signal to no loss at all, as was indicated by the special cases mentioned earlier in this section in the discussion of the I,Q processor. With a nonfluctuating target and only a single channel there is a loss of 2.8 dB and 13.7 dB for a probability of detection of 0.5 and 0.9, respectively, and a probability of false alarm of 10^{-6}.[34] When the target fluctuates pulse to pulse with a Rayleigh probability density function (Swerling Case 2) and with postdetection integration of from 2 to 8 pulses, the loss in using only a single channel is about 2.4 dB when $0.5 < P_d < 0.9$ and $10^{-6} < P_{fa} < 10^{-10}$. For a very large number of pulses integrated (greater than 1000), the loss approaches 1.5 dB. When the target fluctuations are described by Swerling Case 1, there is a loss of 3.2 dB in using a single channel with postdetection integration of only two pulses.[46] The loss is about 2 dB when more than 20 pulses are integrated.

It might be concluded that these losses are moderate, so that a single channel might be acceptable when it is important to reduce hardware complexity. This was the conclusion reached in the early days of MTI radar (the 1950s and 1960s) when the delay lines were analog. Hardware complexity, however, is seldom a consideration with digital MTI processing, so that both the I and Q channels are always employed. A block diagram of a digital processor, therefore, almost always shows I and Q processing.

3.6   MOVING TARGET DETECTOR

The Moving Target Detector (MTD) is an example of an MTI processing system that takes advantage of the various capabilities offered by digital techniques to produce improved detection of moving targets in clutter. It was originally developed by the MIT Lincoln Laboratory for the airport-surveillance radar (ASR), a 60-nmi radar found at major airports for control of local air traffic. The introduction of the MTD represented an innovative and significant advance in radar detection of aircraft in the presence of clutter.

Original MTD[47–49]   The original MTD concept was designed for a radar similar to the FAA's ASR-8, which operated at S band (2.7–2.9 GHz) with a pulse width of 0.6 μs, 1.35° azimuth beamwidth, antenna rotation rate of 12.8 rpm, average prf of 1040 Hz, and an average power of 875 W. The ASR-8 employed four staggered prfs; but staggering of the prfs was not used with the MTD. The original MTD included the following:

- An *eight-pulse FFT digital filter bank* with eight filters, preceded by a three-pulse delay-line canceler. The three-pulse canceler reduced the dynamic range of the signals which the doppler filter bank had to handle, and it compensated for the lack of adequate cancellation of stationary clutter in the doppler filters. The doppler filter bank separated moving targets from moving weather clutter if they appeared in different doppler filters.

- *Frequency-domain weighting* to reduce the doppler-filter sidelobes for better clutter attenuation.

- *Alternate prfs* to eliminate blind speeds and to unmask aircraft echoes from weather clutter.

- *Adaptive thresholds* to take advantage of the nonuniform nature of clutter.

- *Clutter map* to detect crossing targets with zero radial velocity that would otherwise be canceled by an ordinary MTI.

- *Centroiding* of multiple reports from the same target for more accurate location measurements.

The range coverage of this processor totaled 47.5 nmi.

The MTI processor was preceded by a large dynamic range receiver to avoid the reduction in improvement factor caused by limiting (Sec. 3.7). The output of the receiver IF amplifier was fed to I and Q phase detectors. From there the A/D converters changed the analog signals to 10-bit digital words. Figure 3.30 is a block diagram of the MTD, but further explanation is probably required before the block diagram becomes clear.

Coherent Processing Interval (CPI)   The range was quantized into $\frac{1}{16}$-nmi intervals, which were approximately equal to the range resolution of the pulse. The azimuth angle

Figure 3.30
Block diagram of the original Moving Target Detector (MTD) signal processor.

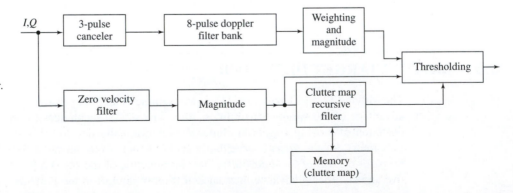

was quantized into $\frac{3}{4}$-degree intervals, which were about one-half-beamwidth. Thus there were a total of 365,000 range-azimuth resolution cells within the 47.5-nmi range coverage. In each $\frac{3}{4}$-degree azimuth cell there were 10 pulses transmitted at a constant prf. On receive, these ten pulses were called a *coherent processing interval (CPI)*. The ten pulses of the CPI were processed by an eight-pulse filter bank that was preceded by a three-pulse delay-line canceler. The filter bank produced eight contiguous filters. There were then 2,920,000 range-azimuth-doppler resolution cells. Each of these cells had its own adaptive threshold (described later) which was controlled by the amount of clutter echo seen in the vicinity of the target. In the next $\frac{3}{4}$-degree azimuth cell, the prf of the 10-pulse CPI was changed to eliminate blind speeds and to unmask moving targets hidden by moving weather clutter (as will be described later). Changing the prf every 10 pulses (every half-beamwidth) eliminated second-time-around clutter echoes that would normally degrade a conventional MTI using pulse-to-pulse stagger of the prf.

Filter Bank The filter bank was implemented by a FFT. Although there were 10 pulses in the CPI, there are only eight doppler filters. Since the three-pulse canceler requires all three pulses before it can cancel clutter, the first two output pulses were discarded. The discarded pulses are called *fill pulses* since they are needed to "fill" the canceler before cancellation of the clutter can begin.

The frequency response of each filter of the FFT filter bank had a $(\sin x)/x$ shape with the first sidelobes 13.2 dB down from the peak. These sidelobes were high and generally did not have sufficient rejection of nearby targets or clutter. The sidelobes were reduced in this MTD by subtracting from the output of each filter $\frac{1}{4}$ the sum of the two adjacent filters.[50] [If A, B, and C represent the unweighted outputs of three contiguous filters, the weighted output applied to filter B is $B - (A/4 + C/4)$.]

Ideally the three-pulse canceler is not needed if the filter bank can be designed to obtain the necessary clutter attenuation for both stationary and moving clutter. It was not possible, however, to eliminate the three-pulse canceler in the original MTD. In the block diagram of Fig. 3.30, the *magnitude* is the operation $(I^2 + Q^2)^{1/2}$, or an approximation thereto.

Clutter Map A conventional MTI processor eliminates stationary clutter, but it also eliminates aircraft moving on a crossing trajectory (one perpendicular to the radar line of sight) which causes the aircraft's radial velocity to be zero. This is unfortunate since the radar cross section of an aircraft is relatively large when viewed at the broadside aspect presented by a crossing trajectory. The MTD took advantage of this large cross section to detect targets that normally would be lost to a simple MTI radar. It did this with the aid of a clutter map that stored the magnitude of the clutter echoes in a digital memory. The clutter map established the thresholds used for detecting those aircraft targets which produced zero radial velocity.

Since the three-pulse canceler removed all echoes with zero velocity, the zero-velocity filter had to be re-established in order to produce the clutter map, as indicated in Fig. 3.30. Each of the 365,000 range-azimuth resolution cells of the clutter map stored the average value of the output of the zero-velocity filter received during the last eight scans (32 s). On each scan, one-eighth of the output of the zero-velocity filter was added to seven-eighths of the value stored in the map. The map, therefore, was built up in a recursive manner. About 10 to 20 scans were required to establish steady-state values. This number of scans was necessary so that aircraft echoes would not affect the threshold. (They didn't stay in one cell too long.) As rain moved into the area or as propagation conditions changed, the clutter map changed accordingly. The values in the clutter map were multiplied by an appropriate constant to establish the threshold which allowed the detection of moving targets with zero radial velocity. The MTD proved highly successful in using this technique to detect crossing targets that would have been rejected by a simpler MTI processor.

Adaptive Thresholds   As mentioned, each of the 2,920,000 range-azimuth-doppler resolution cells had an adaptive threshold to allow detection of moving targets in either stationary or moving clutter. Consider the eight doppler filters diagrammed at the top of Fig. 3.31 (indicated as prf-1). (For the moment concentrate only on this one prf and its set of filters and ignore the rest of the figure which will be discussed in the next subsection.) There are three different criteria for selecting the thresholds, depending on the location of the filter with respect to zero velocity.

The adaptive threshold setting for filter no. 1 (which was the filter centered at zero radial velocity) was determined by what was stored in the clutter map. The adaptive thresholds for filters nos. 3 through 7 were set by a clutter CFAR similar to what is described in Sec. 5.7 and shown in Fig. 5.7. Briefly, the clutter CFAR in the MTD established a threshold based on the mean level of the signals received in the 16 range cells centered around the range cell of interest. The 16 range cells correspond to a distance one-half mile behind and one-half mile ahead of the range cell of interest. The mean value of the signals in these 16 cells, when multiplied by an appropriate constant, established a threshold at each of the nonzero-velocity filters 3 through 7. Thus the threshold continually adapted to the local environment as the radar pulse traveled in time through space. For the remaining two filters, 2 and 8, which lie adjacent to the zero-velocity filter, the threshold was selected as the larger of that given by the clutter map and the clutter CFAR. Large echoes from moving clutter, which might be confused as real target echoes by a conventional MTI with a single filter covering the doppler space, are not detected in the MTD.

Unmasking Moving Targets in Moving Clutter   An advantage of a filter bank is that it allows moving targets to be separated from moving clutter (such as rain) so that detection can take place. Since the doppler-shifted radar echo from a moving target usually (but not always) has a radial velocity greater than the maximum unambiguous velocity (or first blind speed), its doppler frequency can fold over into one of the eight filters of the filter bank. This is shown happening with prf-2 of Fig. 3.31. The true target velocity is represented by the solid vertical arrow lying to the right of the maximum unambiguous velocity. In this case it folds over into filter no. 8, as indicated by the dashed arrow. The rain echo is also in filter no. 8; hence, it masks the echo of the aircraft on the CPI with prf-2. On prf-1, however, the aircraft echo folds into a different filter, in this case filter no. 7. The rain echo remains in filter no. 8 since its radial velocity is less than the maximum unambiguous velocity. In this manner the change in prf and the use of a doppler filter bank allow moving aircraft echoes to appear in at least one CPI that is free from

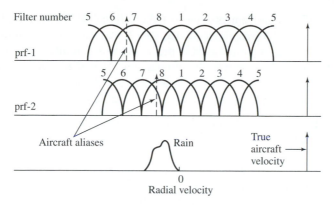

Figure 3.31   Detection of an aircraft in rain by use of two prfs and a doppler filter bank, illustrating the unmasking of the aircraft's echo on prf-1 due to doppler foldover when it is masked by rain on prf-2.
| (From Muehe,[52] Copyright 1974 IEEE.)

rain echoes. A change in prf of about 20 percent was used in the MTD to allow this unmasking of the aircraft echo. (The echo at S band from precipitation might typically have a spectral width of about 25 to 30 kt; and might be centered anywhere from -60 to $+60$ kt, depending on the wind conditions and the antenna pointing direction.[51])

Interference and Saturation   Pulses from a nearby radar might be received by the radar and appear as a large signal among the 10 pulses normally processed as a CPI. In the MTD an *interference eliminator* compares the magnitude of each of the 10 pulses against the average of the 10. If any pulse is greater than five times the average, all information from that range-azimuth cell is discarded. A *saturation detector* detects whether any of the 10 pulses within a processing interval saturates the A/D converter and, if it does, the entire 10 pulses of that particular CPI are discarded.

Post-processor Centroiding   The output of the MTD processor was a *hit report* which contained the azimuth, range, and amplitude of the target echo as well as the filter number and prf. On a particular scan, a large aircraft echo might be reported from more than one doppler filter, from several coherent processing intervals, and from adjacent range cells. As many as 20 hit reports might be generated by a single large target.[52] A *post-processor* grouped together all reports that appear to originate from the same target and interpolated to estimate the target's azimuth, range, amplitude, and radial velocity. The target amplitude and doppler were used as the basis to eliminate small cross-section and low-speed echoes (such as birds and insects) before the target reports were delivered to the automatic tracking computer. The automatic tracker also eliminated false reports that did not form logical tracks. The output of the automatic tracker is what is displayed to the operator. Since the MTD processor eliminated a large amount of the clutter and had a low false-detection rate, its output could be reliably transmitted to a different location, if desired, by narrow-bandwidth telephone lines.

Performance The improvement factor of the original MTD measured on an airport-surveillance radar (ASR) was about 45 dB, which was said to be a 20-dB increase over the ASR's conventional three-pulse MTI processor with a limiting IF amplifier. In addition, the MTD achieved a narrower clutter rejection notch at zero velocity and at the blind speeds. The MTD has proven to be an important model for the design of radars for detecting moving targets in stationary as well as moving clutter.

Second Generation MTD[53] The original MTD, as does any electronic development, used the technology that was available when it was first developed. Advances in digital hardware occurred rapidly, however, and an improved version soon appeared. The original hard-wired processor was replaced by a parallel microprogrammed processor (PMP) which operated out to the full instrumented 60 nmi range of the radar. The characteristics of the second generation MTD are summarized as follows:

- The filter bank was implemented as a *generalized transversal (FIR) filter* rather than as an FFT so as to provide more flexibility in design and reduce the doppler-filter sidelobes.

- A *two-pulse delay-line canceler* (rather than the three-pulse canceler of the original) preceded the filter bank.

- The *zero radial-velocity filter* was also designed as a transversal filter. It utilized the Chebyshev criterion to produce relatively uniform filter gain across the portion of the doppler space not covered by the nonzero-doppler filters.

- The *clutter map* had one cell for each range-azimuth cell rather than one cell for each range-CPI.

- The *nonzero-doppler filters* were based on the method of DeLong and Hoffstetter (mentioned in ref. 53). They provided sidelobe levels 10 dB lower than the original MTD, and thus gave better performance in rain. When the two-pulse delay line preceded the filter bank, the number of bits required in the transversal filter decreased, so that the filter weights only had to be 3 or 4 bits plus sign.

- The *CPI* consisted of eight pulses. One CPI had a 900-μs period, and the other had a period of 1100 μs. The prfs were 1111 Hz and 909 Hz.

- *Correlation and interpolation algorithms* clustered the threshold crossings in range, azimuth, and doppler to provide in a single report the best values of the radar observables of a single target, whether it be aircraft, automobile, or bird. Target signal strength was one of the observables as well as range, angle, and doppler. The doppler interpolation was one part in 64 across the band of eight doppler cells.

- An *area CFAR* was used after the MTD thresholding to eliminate single CPI false alarms due to birds, interference, and weather clutter that were not removed by the signal processing algorithms.

- The *scan-to-scan correlator* was an automatic tracker that deleted those targets whose scan-to-scan behavior was not what would be expected of an aircraft. This removed low-speed targets as well as echoes that lacked spatial correlation from scan to scan. The scan-to-scan correlator determined when a track was of a quality sufficient to allow the data associated with it to be displayed.

Experimental demonstration of the second-generation MTD installed on an ASR-7 at MIT Lincoln Laboratory showed that the PMP processor produced approximately 500 to 600 range-azimuth-doppler threshold crossings per scan out of a total of about 3,000,000 cells within the radar coverage. There were typically 5 to 15 threshold crossings per aircraft target.

The FFT of the original MTD constrained the type of filter characteristic that could be obtained. Performance was sacrificed for reduced cost. By employing FIR filters in the second generation MTD, the frequency response of each filter of the filter bank could be shaped as desired. Less power was wasted by using the eight-pulse FIR filters since only one fill pulse was required rather than the two fill pulses of the 10-pulse FFT. The reduced number of pulses allowed better performance with a shorter observation time.

Both the first and second generation MTDs were developed by the MIT Lincoln Laboratory. The MTD concept further evolved into a third configuration used in the ASR-9 airport-surveillance radar developed for the FAA by Westinghouse (later known as Northrop-Grumman) at Baltimore, MD.

Third Generation MTD for the ASR-9[54,55]   The first operational air-traffic control implementation of the MTD concept was for the ASR-9, a radar employed at the major airports of the United States and elsewhere in the world. It utilized the basic philosophy of the second generation Lincoln Laboratory MTD and significant extensions by Westinghouse.

The pulse periods of the two CPIs in the ASR-9 were not equal, but were in the ratio of 9/7. The longer-pulse-period CPI processed 8 pulses and the shorter-period CPI processed 10 pulses, for a total of 18 pulses. The different number of pulses allowed the doppler space covered by the two CPIs to be more equal than the unequal coverage of the two CPIs of the original MTD (whose filters were sketched in Fig. 3.31). There were approximately 21 pulses received over the radar's half-power beamwidth, but only 18 of these were used for establishing the clutter map that helped to control the false-alarm rate. Twelve-bit filter weights were used rather than the five-bit weights of the Lincoln MTD. This enabled the filter to be designed to achieve 40-dB doppler sidelobes for rain rejection and 44-dB ground-clutter rejection for the heavy clutter filters. In mountain clutter, an alternate filter design was used that provided 52 dB of attenuation at the expense of less rain-clutter rejection. Sensitivity time control (STC) was employed to reduce the large clutter echoes that exceeded the available dynamic range of the system.

The range and azimuth resolutions of this radar were improved compared to the original MTD. In a simple CFAR, the threshold increases when there are two nearby targets, thus reducing the probability of resolving them in range. The ASR-9 CFAR algorithm, however, omitted the strongest cluster of samples from the estimate of interference (that determined the threshold) so that a neighboring target of comparable echo strength didn't cause the threshold to be raised. Without this cluster editing, the two targets would not be resolved and only the larger target would be detected.

When detections were obtained that extended for more than two beamwidths in azimuth, the shape of the response in azimuth was compared to what would be expected from a single target. A significant deviation from the expected response indicated that the extended azimuth signature was caused by two targets, thus increasing the probability of resolution.

Accuracy with the MTD[56]   The accuracy of the original MTD was not as good as might be obtained with a conventional scanning radar. The angle in the original MTD was determined on the basis of the two CPIs rather than as is usually done in a scanning radar that uses all the pulses received from a target to estimate its center by beam splitting.

In the Westinghouse, or third generation, version a typical aircraft might give rise to 35 individual threshold crossings (which were called *primitive reports*) in range, azimuth, doppler, and amplitude. A large aircraft might produce 100 such reports. These have to be centroided to produce a single report of range, azimuth, doppler, and amplitude. The radar must also provide an accurate measurement of angle and be able to resolve two nearby targets. Algorithms were developed to resolve unequal-size targets separated in azimuth by less than two beamwidths and separated in range by less than two range cells, which was $\frac{1}{8}$ nmi.* Since a large target might affect up to five range cells, range resolution in the MTD had to be obtained by using a simplified form of pulse-shape matching. For azimuth resolution and estimate of angle, one of four algorithms was used, depending on the extent of the data in angle. The most elaborate of the four employed beamshape matching when target primitive reports were obtained on three or more CPIs with the same prf. The typical number of primitive reports with this algorithm was greater than 30. The rms angle accuracy was said to be about 0.04 beamwidth.

When the received signal was so strong that it saturated the A/D converter, the accuracy of centroiding and the ability to resolve two targets were degraded. In the first two versions of the MTD, any CPI that experienced saturation by a large target or interference signal was eliminated. It was found, however, that eliminating the CPI caused splitting of the target due to premature termination of a report. The Westinghouse MTD avoided this problem by providing information about its *presence* (rather than nothing) when the data saturated.

Hit, or Primitive, Reports   In the above discussion of the various evolutions of the MTD, there were various values given for the number of hit reports, or primitive reports, generated by a single aircraft target. According to the references cited in this section, these varied from 15 to 100. Apparently the number depends on the type of MTD, the radar with which it was used, and the nature of the trials that were conducted.

Extension to Other Radars   The three generations of MTD systems described above were well suited for use in an *S*-band medium-range airport-surveillance radar. The MTD has also been adapted for use with a solid-state transmitter, as in the ASR-12 air-traffic control radar.[57] Although the individual techniques that constitute the MTD have applicability to other types of air-surveillance radars, it is more difficult to apply in its entirely the MTD systems as described above to long-range radars or those at other frequencies. Some features of the MTD concept described here might not be suitable for military air-surveillance radar applications since they might make the radar more vulnerable to countermeasures.

---

*It is generally accepted that a good radar can resolve two targets of equal size if the targets are separated in angle by at least 0.8 beamwidth. Similarly, two targets can be separated in range, in the absence of CFAR, if they are at least 0.8 pulse width apart.

3.7   LIMITATIONS TO MTI PERFORMANCE

In this section we consider the degradation in performance of MTI radars caused by (1) antenna scanning modulation, (2) internal fluctuations of clutter, (3) equipment instabilities, and (4) limiting. The adverse effects of the A/D converter and other aspects of digital processing on MTI performance were considered in Sec. 3.5.

The limitations to MTI performance to be discussed cause the clutter spectrum to widen. More clutter energy is then passed by the doppler filter, which lowers the improvement factor. If the clutter power spectral density can be expressed as a gaussian function with a standard deviation σ_c in Hz, it can be represented as

$$W(f) = W_0 \exp\left(-\frac{f^2}{2\sigma_c^2}\right) \qquad \text{[3.37]}$$

where W_0 is the peak value of the clutter power spectral density at $f = 0$. This equation was given previously as Eq. (3.14). The clutter standard deviation is sometimes written in terms of the radial velocity, in units of meters/second, and denoted σ_v. The two may be related by $\sigma_c = 2\,\sigma_v/\lambda$, where λ = wavelength in meters. (The standard deviation expressed in velocity, σ_v, is sometimes, but not always, independent of frequency for windblown clutter.) If the MTI filter has a frequency response $H(f)$, the clutter attenuation is [a repeat of Eq. (3.15)]

$$CA = \frac{\displaystyle\int_0^\infty W(f)\,df}{\displaystyle\int_0^\infty W(f)\,|H(f)|^2\,df} \qquad \text{[3.38]}$$

The improvement factor, Eq. (3.20), is found by multiplying the clutter attenuation by the average gain of the filter.

The improvement factors for a single delay-line canceler and a double delay-line canceler were derived in Sec. 3.2 when the clutter spectrum is represented by a gaussian spectrum. The general expression for the improvement factor with n delay-line cancelers in cascade was given by Eq. (3.23) as

$$I_f \approx \frac{2^n}{n!}\left(\frac{f_p}{2\pi\sigma_c}\right)^{2n} \qquad \text{[3.39]}$$

where f_p is the pulse repetition frequency, or prf. We shall use this expression to determine the improvement factor when the clutter spectrum is gaussian.

If there are N different effects that contribute to the widening of the clutter spectrum, and if each is gaussian and independent of one another, the overall standard deviation is

$$\sigma_c = (\sigma_1^2 + \sigma_2^2 + \sigma_3^2 + \cdots + \sigma_N^2)^{1/2} \qquad \text{(3.40)}$$

where the subscripts indicate the various effects, such as oscillator stability, quantization of the A/D converter, antenna scanning modulation, and so forth. The overall improvement factor is given by the following expression when multiple effects contribute to the total (whether or not their clutter spectra are gaussian)

$$\frac{1}{I_f} = \frac{1}{I_1} + \frac{1}{I_2} + \frac{1}{I_3} + \cdots + \frac{1}{I_N} \qquad [3.41]$$

where I_i represents the limit on the improvement factors due to each effect.

Antenna Scanning Modulation   The duration of the echo signal received from a target or a clutter scatterer as the antenna of a pulse radar scans past is given by $t_0 = n_B/f_p = \theta_B/\dot\theta_s$, where n_B = number of pulses received, f_p = pulse repetition frequency, θ_B = antenna beamwidth, degrees, and $\dot\theta_s$ = antenna scanning rate in degrees per second. The frequency spectrum has a bandwidth inversely proportional to the time duration t_0. Consequently, even if the clutter scatterers were perfectly stationary and there were no instabilities in the radar equipment, there would still be a finite spectral spread due to the finite duration of the echo signal. This limitation has been called *antenna scanning modulation,* but it is basically due to the finite time on target. The longer the time on target, the less will be the spread in the clutter spectrum.

The pulse-to-pulse difference in echo amplitude due to the antenna pattern shape also contributes to the clutter residue that is part of the antenna scanning modulation. In many MTI radars, the finite duration of the signal is what sets the limit on the clutter attenuation that can be obtained. It is also the reason why a long signal duration is needed for good clutter rejection. Short duration signals, therefore, usually are not consistent with good MTI performance.

The two-way voltage waveform of the received echo signal from clutter will be modified in amplitude by the square of the one-way antenna electric-field-strength pattern. The two-way voltage is equal to the one-way antenna power pattern, which can often be approximated by a gaussian function such as

$$G(\theta) = G_0 \exp\left(-\frac{2.776\theta^2}{\theta_B^2}\right) \qquad [3.42]$$

where G_0 = maximum antenna gain, θ = angle coordinate, and θ_B = beamwidth. The scanning antenna beam modulates the amplitude of the received pulse-train. If the antenna scans at a rate of $\dot\theta_s$ deg/s, the modulation with time of the echo pulse-train is found by dividing both the numerator and denominator of the exponent of Eq. (3.42) by $\dot\theta_s$. Then we let $\theta/\dot\theta_s = t$, which is the time variable, and we let $\theta_B/\dot\theta_s = t_0$, the time duration of the signal (or time on target). The modulation of the received echo signal from an individual clutter scatterer due to the antenna pattern is then

$$s_a(t) = k \exp\left(-\frac{2.776t^2}{t_0^2}\right) \qquad [3.43]$$

where k = constant. The power spectrum of $s_a(t)$ is given by the square of the Fourier transform of the above equation, which is

$$|S(f)|^2 = K \exp\left(-\frac{\pi^2 f^2 t_0^2}{1.388}\right) \qquad [3.44]$$

where K = constant. Since this is a gaussian function with exponent of the form $f^2/2\sigma_a^2$ the standard deviation σ_a of the clutter spectrum caused by antenna scanning modulation is

$$\sigma_a = \frac{1}{3.77t_0} = \frac{0.265f_p}{n_B} \qquad \text{[3.45]}$$

When this is substituted into Eq. (3.39), the limitation on the improvement factor due to antenna scanning, assuming n delay-line cancelers in cascade, is

$$I_f = \frac{2^n}{n!}(0.6n_B)^{2n} \qquad \text{[3.46]}$$

where $n_B = f_p t_0$ is the total number of pulses received from a scatterer during the time the antenna scans past it. For a single delay-line canceler, $I_f = 0.72\,n_B^2$, and for a double delay-line canceler, $I_f = 0.26\,n_B^4$. A plot of the limitation to the improvement factor due to antenna scanning modulation is shown in Fig. 3.32.

A broadening of the clutter spectrum can be obtained even with a nonscanning antenna beam, such as happens with a phased array radar whose antenna beam dwells in one direction for a finite number of pulses before it jumps to another direction to again dwell for a fixed number of pulses. (This is called *step scan*.) The batch of N_T pulses obtained in this manner will have a frequency spectrum with a bandwidth of approximately the inverse of the time on target or $(N_T T_p)^{-1}$. If processed by a conventional filter, there will be incomplete attenuation of the clutter and there will be a limitation on the improvement factor. This limitation does not occur with digital processing if sufficient fill pulses are used to initialize the filter before canceled outputs are obtained.[58] For

Figure 3.32 Limitation to the improvement factor due to antenna scanning or a finite time on target, as a function of the number of pulses received over the half-power beamwidth. The antenna pattern is assumed to have a gaussian shape.

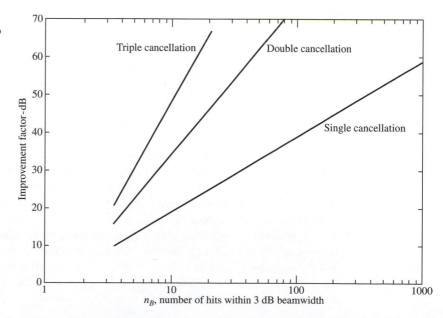

example, with a three-pulse binomial canceler (as was used in the original MTD), there are two fill pulses and a useful output doesn't occur until the third pulse is processed. The energy associated with these two fill pulses is not utilized for detection in noise.

Internal Modulation of Clutter   Previously in this section, we have assumed that clutter was perfectly stationary. Echoes from mountains, rocks, buildings, water towers, fences, thick tree trunks, and bare hills usually can be considered stationary. Many other sources of clutter echoes, however, can be in motion. These include echoes from the sea, rain, and chaff, as well as trees, large vegetation, and structures blowing in the wind. The amplitude and phase fluctuations of windblown trees and vegetation can result in a widened frequency spectrum of the clutter echo that can be a limitation on the achievable MTI improvement factor, which is this subject of this section.

The first model of clutter spectrum suggested in World War II was based on the *gaussian* function.[59] A gaussian spectrum was, and still is, popular since it fit the early experimental clutter measurements and it is relatively easier to manipulate mathematically. Later, when more sensitive radars were employed, it was found that experimentally measured clutter spectra did not decrease with increasing frequency as rapidly as predicted by the gaussian function. Overly optimistic predictions of radar performance were therefore obtained. This began to occur for clutter spectral values in the vicinity of 15 to 20 dB below the peak of the clutter spectrum at zero doppler frequency. A *power-law* spectrum, which did not fall off as fast as the gaussian, was then proposed to account for the available measured spectral data. It improved on the gaussian, but when even more sensitive modern radars became available with capability to measure clutter spectra 60 to 80 dB below the peak at zero doppler, it was found that the power-law spectrum did not adequately describe the measurements much beyond 35 to 40 dB below the peak. Experimental windblown clutter measurements were found to decrease faster with frequency than predicted by the power law. Thus the power law resulted in overly pessimistic predictions of radar performance and could be excessively demanding when designing high-performance MTI radars. It was then found that the *exponential* model could adequately represent measured clutter spectra over a wide dynamic range down to 80 dB below the peak. Although no single model fits all data over a wide dynamic range, the exponential seems to be the best of those proposed to account for experimental measurements of windblown clutter over a wide variety of conditions and a large dynamic range of values. These three clutter spectral models are described in turn.

Gaussian Spectrum   The gaussian model was widely used in the past to describe the frequency spectrum of clutter, especially windblown land clutter. It is a convenient representation and easy to work with. It fits reasonably well to measured clutter data so long as the required MTI improvement factor is not large. The gaussian spectrum of Eq. (3.37) is described by its variance or standard deviation. It is generally assumed that the standard deviation of windblown clutter echo fluctuations when measured in meters/second is independent of frequency. Here the standard deviation in meters/second is denoted σ_v, and the standard deviation in hertz is denoted σ_c. Substituting the standard deviation in Eq. (3.39) gives the limitation on the MTI improvement factor due to windblown clutter for n delay-line cancelers in cascade.

The often-quoted "representative values" of the gaussian model standard deviation of clutter spectra, based on experimental measurements as first reported by Barlow[60] at a frequency of 1 GHz, were:

- Heavily wooded hills, 20-mph wind: 1.5 Hz (0.22 m/s)
- Sparsely wooded hills, calm day: 0.11 Hz (0.02 m/s)
- Sea echo, windy day: 6.0 Hz (0.9 m/s)
- Rain clouds: 13.4 Hz (2.0 m/s)
- Chaff: 7.1 Hz (1.1 m/s)

The nature of the clutter examined by Barlow was not precisely defined. Barton[61] gives the following range of values of σ_v for clutter spectra:

- Wooded hills: 0.01–0.32 m/s
- Sea echo: $v_w/8$, where v_w is the wind speed in meters/second
- Rain and chaff: 1 to 2 m/s

Based on a composite of many sources of data, Nathanson[62] provides for wooded terrain a plot of the standard deviation of a gaussian clutter spectrum as a function of wind speed. The mean and 90th percentile values of the standard deviation, both in meters/second, can be approximated by the following expressions

$$\sigma_v(\text{mean}) = 0.0115w^{1.12} \qquad \text{[3.47a]}$$

$$\sigma_v(90\%) = 0.021w^{1.10} \qquad \text{[3.47b]}$$

where w = wind speed in meters/second. This was said to apply over a frequency range from about 3 GHz to 24 GHz and wind speeds from 1 to 25 m/s. Polarization did not seem to be a significant factor. Values of the ratio of the d-c to the a-c components of land clutter spectra were also given by Nathanson.

The use of a gaussian spectrum may be convenient, but as mentioned above, it falls off more rapidly with frequency than experimental measurements indicate. Its application is limited, therefore, to situations when the clutter attenuation or the improvement factor are not large.

Power-Law Model   The normalized power-law spectrum is given by

$$P(f) = \frac{1}{1 + (f/f_c)^n} \qquad \text{[3.48]}$$

This was first suggested by Fishbein et al.[63] as being a better description of measured clutter spectra than the gaussian. At X band, they found the value of the exponent n to be equal to 3. The clutter spectrum *characteristic frequency,* denoted f_c, is the value of the spectral density when it is reduced by one-half its zero-frequency value. It was said to equal $k_1 \exp[k_2w]$ (in hertz), where $k_1 = 1.33$ Hz, $k_2 = 0.1356$ kt^{-1}, and w = wind speed in knots. Their work was extended by Li Neng-jing[64] who stated that experimental measurements at X, S, and L bands suggested that the characteristic frequency f_c increased with increasing wind and the exponent n varied with radar frequency and wind speed.

Based on measurements at L band by the Chinese Airforce Radar Institute, Li states that as the wind speed increased from low values (0 to 3 m/s) to high values (13 to 15 m/s), the exponent n decreased from 3.3 to 2.2, and f_c increased from 0.8 to 1.9. Experiments by Georgia Institute of Technology indicate that the power-law expression also can be applied for frequencies from 35 to 95 GHz with appropriate selection of the two parameters f_c and n.[65]

As the clutter spectrum decreases below about 40 dB from its peak zero doppler-frequency value, it is found that the power-law spectrum does not decrease as fast as indicated by experimental measurements. (Note that with $n = 3$, the falloff is 30 dB per decade of frequency.) Consequently, the power-law model will give erroneous results by predicting less clutter attenuation than actual. A modified exponential law provides a better representation of the windblown land clutter faced by high-performance doppler radars, as discussed next.

Exponential Law The information in this subsection is taken from an excellent examination of low grazing angle land-clutter spectra as reported by J. B. Billingsley of the MIT Lincoln Laboratory.[66] His work was based on extensive, well-calibrated clutter measurements made over a long period of time at many different sites at frequencies from VHF to X band, with equipment sensitive enough to measure clutter spectral levels 60 to 80 dB below the zero-doppler clutter level. (These clutter measurements are discussed further in Sec. 7.3.) The measurements were accurately made to the very low spectral power-density levels important for the design of high-performance radars that must see small moving targets in clutter. The Lincoln Laboratory clutter spectral model from windblown vegetation and trees (forest) is given by the next three equations, the first of which is

$$\text{total power spectral density: } P_{\text{tot}}(v) = \frac{r}{r+1}\delta(v) + \frac{1}{r+1}P_{\text{ac}}(v) \qquad \text{[3.49]}$$

where v = doppler velocity in meters/second, $-\infty < v < \infty$, r = ratio of the d-c power in the spectrum to the a-c power, $\delta(v)$ = delta function that represents the shape of the d-c component of the spectrum, and $P_{\text{ac}}(v)$ is the shape of the a-c component of the spectrum. Note that this is written in terms of velocity v rather than frequency f. (Billingsley actually takes the near zero doppler, or quasi d-c power region, to be $0 < |v| < 0.25$ m/s.) $P_{\text{ac}}(v)$ is normalized such that its integral over all frequencies equals unity, and is the quantity usually represented by analytical expressions of the spectral shape. P_{tot} represents the measured clutter spectra.

The a-c portion of the spectrum is given by the two-sided exponential

$$P_{\text{ac}}(v) = \frac{\beta}{2}\exp\left[-\beta|v|\right] \qquad -\infty < v < \infty \qquad \text{[3.50]}$$

where β is the *exponential shape parameter*, and is given in Table 3.1. Billingsley[66] also gives an empirical relation for β as

$$\beta^{-1} = 0.105\left[\log_{10} w + 0.476\right]$$

Table 3.1   Exponential a-c shape-parameter β versus wind speed*

Wind Conditions	Wind Speed (knots)	β typical $(m/s)^{-1}$	β worst case $(m/s)^{-1}$
Light air	1–6	12	—
Breezy	6–12	8	—
Windy	12–25	5.7	5.2
Gale force (est.)	25–45	4.3	3.8

*After J. B. Billingsley, ref. no. 66

with w the wind speed in knots. Measurements made from VHF to X band indicate that the values in this table are largely independent of frequency. It follows that the doppler velocity is relatively independent of frequency for windblown vegetation and trees. Equations (3.49) and (3.50) can be expressed in hertz rather than velocity by use of the equation $f = 2v/\lambda$.

The value of r, the d-c/a-c ratio, is dependent on both wind speed and radar frequency. It is given by

$$r = 394 \, w^{-1.55} f_0^{-1.21} \qquad [3.51]$$

where the wind speed is in knots and f_0 is the carrier frequency in gigahertz.

Equations (3.49), (3.50), and (3.51) and Table 3.1 as given by Billingsley describe the exponential clutter spectral model for windblown vegetation and trees over spectral dynamic ranges down to 60 to 80 dB below the zero-doppler peak. At these low levels, the observed clutter doppler velocities are limited to 3 to 4 m/s.

The shape of the a-c component is invariant with frequency from VHF to X band. It is independent of polarization, but strongly depends on the wind speed. The d-c component, however, is much larger at VHF than at X band. Within the limits of the experimental situations covered by these measurements, there is little noticeable dependence on (1) the type of trees (species, density, growth stage), (2) the season of the year (whether the leaves are on or off), (3) resolution cell size, (4) polarization, (5) grazing angle, (6) wind direction, and (7) angle of illumination.*

The exponential clutter model was derived to describe the spectrum of windblown trees, but Billingsley states that it seems to apply as well to the clutter spectra from scrub desert, range land, and cropland vegetation by suitably adjusting the d-c/a-c term. He also estimates that the correlation time for windblown trees might be about 4 s for light wind and about 1 s for windy days.

Figure 3.33 is an example of a windblown X-band clutter spectrum for a forested area showing the effect of wind speed.[67] The radar cell contained mixed deciduous and

*This, and the related land-clutter measurements mentioned in Sec 7.3, tend to question the utility of radar for the remote sensing of the land-surface environment. It might explain why there has been so little success for radar remote sensing in such applications. It also seems to indicate that explanations for radar echoes from trees might be in doubt if the scattering models assume that significant scattering is obtained from the leaves.

Figure 3.33 Power spectra of X-band radar echoes showing the effect of the wind.
(From J. B. Billingsley,[67] reprinted with permission of MIT Lincoln Laboratory, Lexington, MA.)

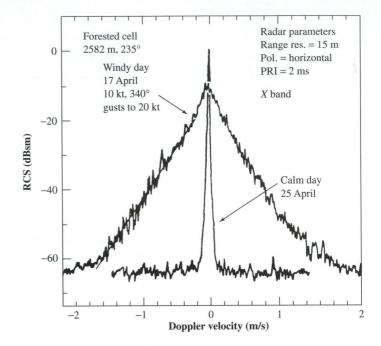

evergreen trees to an approximate height of 60 to 70 ft. On the windy day, the wind speed was recorded as 10 kt with gusts to 20 kt. An example of the frequency dependence of clutter spectra is shown in Fig. 3.34. This is cropland (wheat) in North Dakota. The delta-function-like component at d-c is clearly indicated.

The three different models of clutter spectra that have been discussed (gaussian, power law, and exponential) are compared in Fig. 3.35. The three are normalized so that they each have $\int_{-\infty}^{\infty} P_{ac}(v) \, dv = 1$, which allows the comparison to be made on equivalent total a-c spectral power. (This type of normalization is seldom found in the past literature of clutter spectra, but Billingsley recommends it should be done. The value at zero frequency, when normalized in this manner, is not necessarily unity.) The gaussian spectrum may be appropriate for low clutter doppler velocities, the power law for moderate velocities, but the exponential is best for covering the largest clutter velocities as well as representing the entire range of clutter spectral values.

Improvement Factor for Exponential Clutter   Billingsley gives the MTI improvement factor for an exponential clutter spectrum with a single delay-line canceler as[66]

$$I_\beta = (r + 1)\left(\frac{\lambda f_p \beta}{4\pi}\right)^2 = (r + 1)\left(\frac{v_1}{\pi}\right)^2\left(\frac{1}{2\sigma_\beta^2}\right) \qquad \textbf{[3.52]}$$

where the standard deviation of the exponential spectrum is $\sigma_\beta = \sqrt{2}/\beta$ (σ_β is in units of meters/second) and $v_1 = \lambda f_p/2$ is the first blind speed as was given by Eq. (3.12). The other symbols have been defined previously. If in the right hand portion of this equation

Figure 3.34 Power spectra of North Dakota wheatland measured on four different days under various windy conditions. (From J. B. Billingsley,[66] reprinted with permission of MIT Lincoln Laboratory, Lexington, MA.)

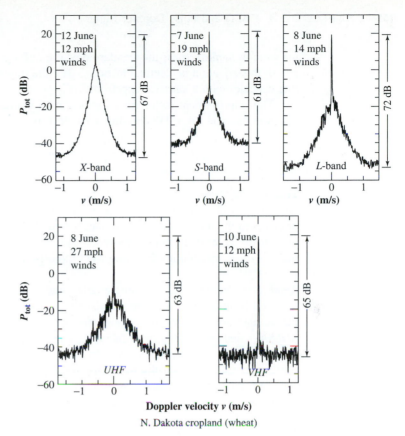

N. Dakota cropland (wheat)

Figure 3.35 Three analytic spectral shapes, each normalized to unit spectral power. The parameter g in the gaussian spectrum is called the gaussian shape parameter and is equal to $1/(2\sigma_v^2)$, where σ_v is the standard deviation in meters/second. (From J. B. Billingsley,[66] reprinted with permission of MIT Lincoln Laboratory, Lexington, MA.)

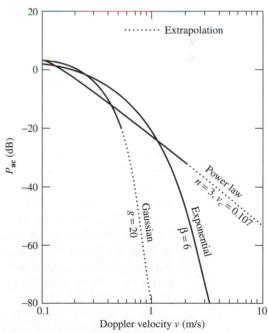

we were to identify σ_β as the standard deviation of a gaussian distribution and if we take $r = 0$, this equation is identical to Eq. (3.21), the improvement factor for a single delay-line canceler with a gaussian spectrum.

The improvement factor for an exponential clutter spectrum with n delay-line cancelers in cascade [$(n + 1)$ delay-line canceler], is

$$I_\beta(n) = (r + 1) \left(\frac{\lambda f_p \beta}{2\pi} \right)^{2n} \frac{1 \times 3 \times 5 \times \cdots \times (2n - 1)}{2^n n! \, (2n)!} \qquad [3.53]$$

System Instabilities   Changes in amplitude, frequency, or phase of the stalo or coho oscillators as well as changes in the pulse-to-pulse characteristics of the transmitted signal or errors in the timing can result in uncanceled clutter echoes and cause a limit to the improvement factor that can be achieved.[68] In this subsection we briefly review some of the major system instabilities that limit MTI radar performance.

Amplitude Changes   If in a single delay-line canceler, the amplitude of the first pulse received from a stationary clutter scatterer is A and that of the second pulse is $A + \Delta A$, the voltage output of the delay-line canceler is ΔA. The clutter attenuation is therefore $A^2/(\Delta A)^2$, and the improvement factor is twice this.

Nathanson[69] presents a slightly different way of looking at the limitation on the improvement factor that results from amplitude pulse-to-pulse changes. Let ΔV_1 and ΔV_2 represent the voltage variation of each of the two pulses about the mean value V_m. The clutter attenuation for a single delay-line canceler is then

$$\text{CA} = \frac{V_m^2}{(\Delta V_1 - \Delta V_2)^2} = \frac{V_m^2}{\Delta V_1^2 + \Delta V_2^2} = \frac{V_m^2}{2\sigma_v^2} \qquad [3.54]$$

where σ_v^2 is the variance of ΔV. The clutter attenuation for a double delay-line canceler is $V_m^2/6\sigma_v^2$. When the average gain is included, the limitation to the improvement factor for both the single and the double delay-line canceler due to amplitude changes is

$$I_1 = I_2 = V_m^2/\sigma_v^2 \qquad [3.55]$$

Phase Changes   If the echo received from the first pulse from stationary clutter is represented as $A \sin(\omega t + \phi)$, and if the echo from the second pulse is $A \sin(\omega t + \phi + \Delta\phi)$, there will be an uncanceled residue from a single delay-line canceler equal to the difference, $2A \sin(\Delta\phi/2)$, where $\Delta\phi$ is the phase change between pulses. For small phase changes, the output voltage is $A\Delta\phi$. The clutter attenuation is then $(1/\Delta\phi)^2$ and the improvement factor is twice this.

For the double delay-line canceler, Nathanson describes two different cases. In the first case, the phase errors from one pulse to the next are assumed equal ($\Delta\phi_1 = \Delta\phi_2$), as in an oscillator with a constant drift in phase. In the second case, the phase errors from pulse to pulse are assumed to be statistically independent. The limitations in improvement factor for a double delay-line canceler in these two cases are given by Nathanson as

$$I_2 = \frac{6}{(\Delta\phi)^4} \qquad \text{for } \Delta\phi_1 = \Delta\phi_2 \qquad [3.56]$$

$$I_2 = \frac{3}{(\Delta\phi)^2} \qquad \text{for } \Delta\phi \text{ statistically independent} \qquad \textbf{[3.57]}$$

Phase Noise   Noise due to phase fluctuations associated with the stalo and coho oscillators can be a major limitation to the improvement factor of high-performance MTI radars. Generally, phase noise has a much larger effect than noise caused by amplitude instabilities. The phase noise from oscillators in the exciter of a power amplifier affect the transmitted signal as well as the signal in the receiver. The spectrum of a CW oscillator is not a classical delta function; but has a finite spectrum, as in Fig. 3.36a.[70] The ordinate is dBc/Hz, which is the noise power within a one-hertz bandwidth relative to the noise at the carrier frequency (at $f = 0$) given in decibels. Phase noise decreases as the frequency increases from the carrier and then levels off to a constant value when thermal noise dominates. Significant spectral energy can appear in the doppler space of the echo signal. This can limit the minimum velocity with which a moving target can be detected. The nature

Figure 3.36   (a) An example of an oscillator phase noise spectrum. (b) Example of large spurious signals (spurs) resulting from the power supply.
(From C. L. Everett,[70] Courtesy of Bell Helicopter TEXTRON and the *Microwave Journal*.)

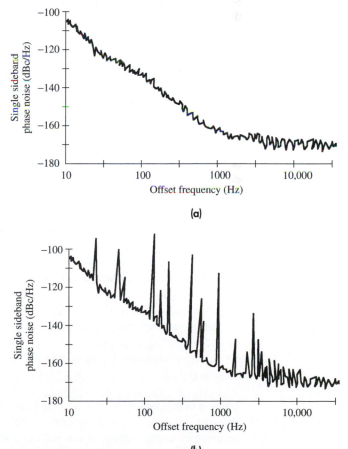

of the noise spectrum depends on the type of oscillator, whether it is a transistor, a crystal oscillator in a temperature-controlled oven, SAW oscillator, dielectric resonator oscillator, or whatever.[71] The shape of the noise spectrum is influenced by the Q of the resonator or resonant circuit associated with the oscillator. Sometimes there are discrete frequency spikes appearing in the spectrum, as in Fig. 3.36b. These are called *spurious,* or *spurs,* and are often associated with the power supply or mechanical vibrations. Phase noise can be introduced by mixers, by the high-power amplifier, and it can ride on the echo signals from clutter.

Stable, low-noise oscillators are usually easier to achieve at the lower frequencies (10 MHz, for instance). They can be multiplied to a higher frequency by using a diode or varactor to generate harmonics. A filter selects the desired harmonic. When oscillators are multiplied in frequency, their phase noise increases as the square of the frequency multiplication, or 20 dB per decade of multiplication. (For example, a 10-MHz oscillator multiplied in frequency to 10 GHz would have its sideband noise levels increased by 60 dB.)

It has been said[72] that if a 1-GHz doppler radar, with a doppler bandwidth of 10 kHz, is to detect a 1-m^2 target at a range of 100 km with a minimum detectable radial velocity of at least 40 m/s (about 80 kt) in clutter with a cross section per unit area $\sigma^0 = 0.02$, the subclutter visibility has to be 80 dB and the transmitter noise sidebands must be better than -120 dB/Hz at sideband frequencies greater than 200 Hz from the carrier.

The phase noise of an oscillator can be reduced by utilizing superconductive resonators to increase its Q.[72] This technique becomes more attractive with advances in high-temperature superconductors that can operate with liquid nitrogen cooling at 77 K rather than the 4 K of liquid helium.[73] The use of dielectric resonator oscillators and high-temperature superconducting films is said to provide at 10 GHz a phase noise lower than -140 dBc/Hz offset 1 kHz from the carrier.[74]

Phase noise can significantly increase due to mechanical vibrations and acceleration as experienced in airborne platforms. Quartz crystal resonators are particularly sensitive to vibrations. Filler and Vig[75] state that mechanical vibrations might increase the phase noise of a crystal oscillator almost 50 dB at a frequency 100 Hz from the carrier under a 0.1 g^2 random vibration (assuming 10^{-9} per g acceleration sensitivity). The dielectric resonator oscillator is less sensitive to shock and vibration than others since it is small and has a high degree of rigidity.[71] Reduced sensitivity to acceleration and vibration (3×10^{-10} per g) has been claimed for oscillators based on "surface transverse waves" (STW), which is a surface-skimming bulk wave with most of the energy confined near the surface but not as tightly confined as in the case of SAW propagation.[76]

The calculation of the limitation on the MTI improvement factor due to the oscillator phase-noise spectrum is described by Shrader and Gregers-Hansen (pp. 15.48 to 15.50 of Ref. 1) and by R. Kerr.[77]

Other limitations due to system instabilities can be found in the discussion of Table 15.4 in Ref. 1.

Limiting in MTI Radar[78,79] Clutter echoes often can be large enough to saturate the radar receiver, obscure target echoes on a display, and cause false alarms. Saturation of the receiver by clutter echoes also results in a spreading of the clutter spectrum that reduces the improvement factor. If the receiver is of large enough dynamic range, and there are sufficient bits in the A/D converter, and if the improvement factor is large enough to make the uncanceled clutter residue smaller than receiver noise, there will be no problem since there will be no limiting. Large dynamic range and cancellation of all the large clutter, however, are not the usual situation.

A limiter in the MTI receiver has sometimes been used to reduce the clutter to the level of receiver noise. The radar literature of World War II recommended that a hard limiter be used for an MTI radar,[80] but we now know this can cause quite serious degradation of the MTI performance. Instead, the limiter should be set above receiver noise by an amount equal to the improvement factor. If it is less than this amount, the improvement factor will not be as predicted when it is assumed that the receiver is linear.

Shrader and Gregers-Hansen[79] provide the following example for a particular air-surveillance radar which receives 16.4 pulses per scan from a target. When the receiver is designed so that a point clutter echo doesn't exceed the limit level, the improvement factor for a two-pulse, three-pulse, and four-pulse canceler is 22.4, 42.2, and 59.9 dB, respectively. If, on the other hand, the limit level is lowered so that the point clutter echo exceeds the limit level by 20 dB, these numbers reduce to 20.4, 29.4, and 33.3 dB, respectively. In this situation with the 20 dB limit level, a four-pulse canceler has very little advantage over a three-pulse canceler. Thus it is important to have the limiter set correctly if the expected MTI performance is to be achieved.

Clutter Map[81]   Sensitivity time control, small radar resolution cells, reduced antenna gain near the horizon, and CFAR are also used to keep clutter from saturating the receiver. In spite of efforts to reduce the amount of clutter viewed by a radar, uncancelled clutter residue might exist which must be surpressed to avoid saturation of the display and/or excessive false alarms in an automatic detection and tracking system. With improvements in digital memories, it became practical to employ digital clutter maps to establish thresholds based on the clutter echo in each radar resolution cell (or combinations of cells). This acts as a type of CFAR, and is superior to conventional cell-averaging CFAR (Sec. 5.7) especially when the clutter is not spatially homogeneous, as with typical land clutter. The use of a clutter map avoids the problems of a limiter in an MTI radar, but introduces some limitations of its own. As with other forms of CFAR, there is a loss in sensitivity with the clutter map, which could be two dB, more or less. The loss increases as fewer past scans are used to establish the clutter map.[82]

3.8 MTI FROM A MOVING PLATFORM (AMTI)

When the radar is on a moving platform such as a ship, aircraft, or spacecraft, the doppler frequency shift from clutter is no longer at zero frequency. The doppler frequency of the clutter echo depends on the relative velocity with respect to the moving radar platform so it will vary with the speed of the radar platform as well as the azimuth and elevation of the clutter cell with respect to the radar. The doppler shift from the clutter can seriously degrade the MTI improvement factor if it is not taken into account.

If the radar were on a moving ship and the antenna is not scanning too fast, it might be practical to compensate for the change in clutter doppler frequency by changing the frequency of the coherent reference oscillator (coho) in an open-loop manner if the speed of the platform and the direction of antenna pointing are known. Compensation for the clutter doppler frequency can be achieved by mixing the output of the coho with a signal from a tunable oscillator whose frequency is made equal to the clutter doppler. It might also be possible in some cases to widen the doppler filter rejection notch so as to exclude the nonzero doppler from clutter if it doesn't remove too much of the available doppler space. With airborne radar, however, the clutter doppler frequency shift can be too large and too rapidly varying with time to compensate open loop. A form of closed-loop compensation must be used.

There is another problem associated with a moving radar platform that can degrade the performance of a moving MTI radar. Not only does the center frequency of the clutter spectrum vary, but so does the clutter spectral width. The widening of the clutter spectrum is due to the finite beamwidth of the antenna which makes the doppler frequency shift from the clutter scatterers within the radar resolution cell differ depending on their location within the antenna beam. For example, the doppler from a scatterer at the center of the beam will differ from that of a scatterer located a half-beamwidth away. These two effects, a nonzero clutter doppler shift and a widening of the clutter spectrum, can seriously degrade the performance of a radar on a moving platform. They are compensated by two different techniques: the change in center frequency of the clutter spectrum by what is known as TACCAR, and the widening of the clutter spectrum by DPCA. These are discussed in this section.

An MTI radar on a moving platform that uses these two methods (TACCAR and DPCA) for compensating for *platform motion* is known as an *AMTI radar*.[83] Although the *A* originally stood for airborne, the term is now almost universally applied to MTI radar on any moving platform that uses these two methods of platform motion compensation.

Compensation for Clutter Doppler Shift (TACCAR)[84]   The doppler frequency shift from clutter can be compensated by using the clutter echo signal itself to set the frequency of the reference oscillator, or coho, so that the rejection notch of the MTI doppler filter attenuates the clutter echo. This has been called *clutter-lock MTI* in the past, but it is more usually known as *TACCAR*, which stands for *time-averaged-clutter coherent airborne radar*. TACCAR was originally developed by MIT Lincoln Laboratory for a particular airborne MTI system, but it now refers to the clutter-lock technique that was the special feature of that system. The chief feature of the original TACCAR was the use of a voltage-controlled oscillator arranged in a phase-lock loop that averaged the measurement of the clutter doppler frequency within a sampled range interval. It also compensates for drift in various system components. Further information on TACCAR can be found in Ref. 84, by Fred Staudaher.

Moving Clutter   A similar problem occurs in a stationary radar when the clutter has a velocity of its own, as might occur for ocean currents, weather, and windblown chaff. TACCAR can also be applied for reducing the effects of moving clutter as seen by a stationary radar. Generally, such techniques do not adequately eliminate both stationary and moving clutter simultaneously when they appear within the same radar resolution cell. A TACCAR, for example, might be designed to reject ground clutter close-in or to reject moving weather clutter at a different doppler at ranges beyond the ground clutter; but not to cancel both simultaneously.[85] A doppler filter bank in the MTD, however, can detect moving targets in the presence of both moving clutter as well as stationary clutter if their doppler frequency shifts are different. When the clutter spectrum consists of both stationary and moving components (a bimodal clutter spectrum), the doppler filter of an MTI radar might be made adaptive to reject the moving clutter as well as have a rejection notch at zero frequency to remove stationary clutter.[86]

Compensation for Clutter Doppler Spread (DPCA)[87]   The doppler frequency shift of the echo from a stationary clutter scatterer when viewed by a moving radar is given as $f_c = 2(v/\lambda) \cos \theta$, where v = velocity of the platform carrying the radar, λ = wavelength, and θ = angle between the velocity vector of the vehicle carrying the radar and the line of sight to the clutter scatterer. The angle θ includes both a horizontal (azimuth) and a vertical (elevation) component; but for simplicity in the discussion here, we will consider only the azimuth component. The doppler frequency from those scatterers located along the center of the antenna beam, Fig. 3.37, will be compensated by TACCAR so that it appears at zero frequency. The echoes from all other scatterers that do not lie along the center of the main beam will not be compensated by TACCAR. Thus there will be a spread in the clutter doppler frequency spectrum due to the finite antenna beamwidth of Fig. 3.37. The clutter spectral spread due to the many clutter scatterers within the finite beamwidth can be found by taking the differential of the doppler frequency, which is

$$\Delta f_c = \frac{2v}{\lambda} \sin \theta \, \Delta\theta = \frac{2v}{\lambda} \theta_B \sin \theta \qquad [3.58]$$

Figure 3.37
Geometry of scattering from a single clutter scatterer located at an angle α from the beam center. Antenna beamwidth = θ_B.

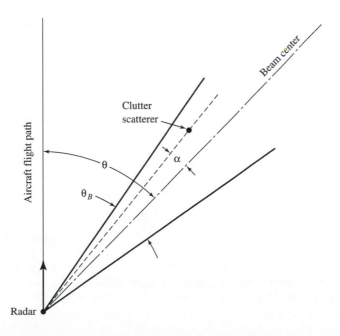

where the antenna beamwidth θ_B is taken to be the same as $\Delta\theta$. (The negative sign introduced by differentiating the cosine is ignored.) The spectral spread Δf_c varies with the angle θ. When the antenna points in the direction of the platform velocity vector ($\theta = 0$), the doppler shift of the clutter echo is maximum, but the width Δf_c of the spectrum is theoretically zero. On the other hand, when the antenna is directed perpendicular to the direction of the platform velocity ($\theta = 90°$, or broadside), the clutter doppler center frequency is zero, but the spread is maximum. The widening of the clutter spectral spread as a function of the azimuth angle can set a limit on the improvement factor.

Figure 3.38 is an example of the spread of the clutter spectrum caused by motion of the radar platform. It assumes a UHF airborne radar at a speed of 400 kt, with an antenna beamwidth of 7 degrees. The width of the clutter spectrum and its center frequency both depend on the angle θ that the radar beam makes with the platform vector velocity. In Fig. 3.38a, the clutter spectra are shown as a function of angle θ. The prf is assumed to be high enough to not include any foldover of the clutter. (At $\theta = 0$ degrees, the clutter spectrum is shown as an infinitesimal line, but in reality there will be effects other than platform motion that contribute to its widening.) In Fig. 3.38b, the foldover of the clutter spectrum at $\theta = 90°$ is shown when the prf is 360 Hz, which corresponds to a maximum unambiguous range of 225 nmi. This figure illustrates the relatively large portion of the frequency domain (doppler space) occupied by the clutter spectrum because of the travel of the aircraft carrying the radar, or platform motion. The widening of the clutter spectrum needs to be reduced in order for AMTI radar to be effective.

There would be no spreading of the clutter spectrum ($\Delta f_c = 0$) if the antenna could be made to think that it is stationary. It is possible to do this with two separate antennas. Consider the case where the two antennas are separated in line with the platform's velocity vector, and with their beams pointing broadside at $\theta = 90°$. The spacing between the phase centers of the two antennas is made equal to vT_p, where v is the velocity of the aircraft and T_p is the pulse repetition period. One is the forward antenna, the other is the trailing antenna. The first pulse is transmitted and received by the forward antenna. When the trailing antenna arrives at the spot where the forward antenna was when it transmitted its pulse, the trailing antenna transmits its pulse. Thus the two pulses from the two

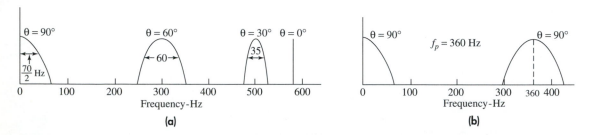

Figure 3.38 Change in shape and location of the clutter spectrum due to platform motion as a function of azimuth angle. (a) Case for a very high prf or CW. (b) $\theta = 90°$, with spectrum folded about the prf at 360 Hz.

antennas are transmitted (and received) at the same point in space. To the radar, it is as if there were a single stationary antenna. The two pulses are then processed in a delay-line canceler and there is no effect of platform motion.

There are some problems, however, with this solution. The prf must be synchronized with the aircraft speed and the spacing between the two antennas has to vary with the cosine of the angle θ if the antenna scans in azimuth. The antennas are coincident when looking in the direction of the aircraft's velocity vector at $\theta = 0$ and furthest apart at $\theta = 90°$. In fact, the antennas always overlap, even at 90°. A single sidelooking phased array can be used with a fraction of its columns operating on the first pulse and an overlapping fraction of its columns operating on the second pulse, so that the phase center is shifted in the horizontal plane. On the first pulse, the phase center might be slightly ahead of the physical center of the array antenna and on the second pulse the phase center might be slightly behind the physical center, with the locations of the phase center in space being the same on each of the two pulses. This is an example of a *displaced phase center antenna* (DPCA). Phased arrays for this purpose have been built and flown, but there is another, usually simpler, method to obtain the same effect using a single antenna with two squinted beams.

Consider a mechanically rotating antenna that generates two overlapping (squinted) beams, Fig. 3.39. This can act as a DPCA when the outputs of the two squinted beams are properly combined. The sum (Σ) and the difference (Δ) of the two squinted beams are taken (similar to the sum and difference beams of a monopulse tracking radar as described in Sec. 4.2). The sum beam is used for transmit, and both the sum and difference beams are used on receive. The received signal from the first pulse is processed to form $\Sigma_r + jK\Delta_r$, where $+j$ represents a 90° phase advance applied to the difference signal, K is a constant that determines the amount of the phase center shift, and the subscript r denotes the received signals in the sum and difference channels (as will be described later).

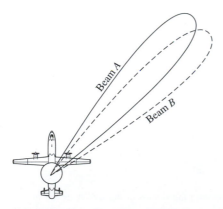

Figure 3.39   Mechanically rotating airborne antenna (within a rotodome) with two squinted beams. The sum of these two beams is used for transmitting, and the echo signal is received on both the sum and the difference of the two beams.

The constant K depends on the aircraft velocity, the prf, and the angle θ. This process applies an apparent forward displacement to the phase center on the first pulse. From the second pulse is formed $\Sigma_r - jk\Delta_r$, where $-j$ is a 90° phase lag. This applies an apparent rearward displacement to the phase center. The two signals are then subtracted from one another, resulting in the cancellation of the doppler spread in the clutter. It is not obvious from the above explanation why this works. A bit of mathematical analysis, however, will show why the doppler spread does not occur.

In Fig. 3.40a, the phasor E_1 represents the clutter amplitude and phase of the echo from the first pulse. Phasor E_2 represents the clutter amplitude and phase from the second pulse. The single clutter scatterer is shown in Fig. 3.37 at an angle α with respect to the beam center at θ. If both the radar and the clutter were stationary, the phase difference 2η between the two echoes would be zero and the two echoes would cancel when subtracted in a delay-line canceler. Thus there would be no clutter residue. The movement of the platform carrying the radar, however, will result in the clutter echo having a doppler frequency shift δf_c relative to the echoes from the TACCAR-compensated beam center. In the time T_p between two pulses, the phase shift resulting from the clutter doppler frequency is

$$\Delta\phi_c = 2\eta = 2\pi(\delta f_c)T_p \qquad [3.59]$$

From Fig. 3.37, the doppler frequency shift from a clutter scatterer at an angle α is

$$\delta f_c = \frac{2v}{\lambda}\cos(\theta - \alpha) - \frac{2v}{\lambda}\cos\theta \approx \frac{2\alpha}{\lambda}v\sin\theta \qquad [3.60]$$

The far-right expression assumes α is small. Substituting the above into Eq. (3.59) gives

$$2\eta = \frac{4\pi T_p\alpha}{\lambda}(v\sin\theta) \qquad [3.61]$$

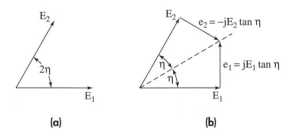

(a)   (b)

Figure 3.40   Phasor diagrams illustrating the nature of the DPCA action to compensate for platform motion using the sum and difference patterns of two squinted beams formed from a single antenna. (a) Phasors representing the clutter echo signals received on two successive pulse transmissions. The angle 2η represents the phase change due to the travel (platform motion) of the radar antenna between pulses. (b) Corrections applied to each pulse so that a cancellation of the clutter echoes occurs on two successive pulses.

This phase shift gives rise to the angular separation of the two phasors of Fig. 3.40a so that when the two are subtracted in a delay-line canceler, there will be an uncanceled, unwanted clutter residue.

If correction signals e_1 and e_2 are applied to each pulse as indicated in Fig. 3.40b, the two resultant signals cancel after passing through a delay-line canceler and (theoretically) there would be no clutter residue. Thus there is no effect of the clutter spread due to platform motion. From the geometry of Fig. 3.40b, the correction to the clutter echo from the first pulse needs to be $e_1 = +jE_1 \tan \eta$, and the correction to the second pulse is $e_2 = -jE_2 \tan \eta$. The $+j$ in e_1 means that the phase of the correction signal is advanced $90°$, and the $-j$ in e_2 means that its phase is delayed $90°$. These corrections can be obtained from the signals received in the sum and difference beams from the two squinted beams of Fig. 3.39.

The DPCA radar transmits on the sum Σ of the two squinted beams and receives on both the sum Σ and difference Δ of the two squinted beams. Because of the two-way path of radar signals, the received sum and difference signals may be represented as

$$\Sigma_r(\alpha) = k_1 \, \Sigma^2(\alpha) \qquad \text{[3.62a]}$$

$$\Delta_r(\alpha) = k_2 \, \Sigma(\alpha) \, \Delta(\alpha) \qquad \text{[3.62b]}$$

where the constants k_1 and k_2 account for the parameters of the radar equation. The amplitudes of the two pulses are assumed equal so that

$$|E_1| = |E_2| = |\Sigma_r(\alpha)| = k_1 |\Sigma^2(\alpha)| \qquad \text{[3.63]}$$

From a uniformly illuminated aperture of dimension D, the sum pattern is proportional to

$$\Sigma(\alpha) = \frac{\sin\,[\pi(D/\lambda)\,\sin\,\alpha]}{\pi(D/\lambda)\,\sin\,\alpha} \qquad \text{[3.64]}$$

and the difference pattern is

$$\Delta(\alpha) = \frac{1 - \cos\,[\pi(D/\lambda)\,\sin\,\alpha]}{\pi(D/\lambda)\,\sin\,\alpha} \qquad \text{[3.65]}$$

Using Eqs. (3.61) and (3.64), the correction signals e_1 and e_2 are

$$e_{1,2} = \pm jk_1 \frac{\sin^2\,[\pi(D/\lambda)\,\sin\,\alpha]}{[\pi(D/\lambda)\,\sin\,\alpha]^2}\,\tan\left[\frac{2\pi T_p \alpha}{\lambda}(v\,\sin\,\theta)\right] \qquad \text{[3.66]}$$

For small α and small values of the arguments

$$e_{1,2} = \pm jk_1 \frac{2\pi T_p \alpha}{\lambda}(v\,\sin\,\theta) \qquad \text{[3.67]}$$

On substituting Eqs. (3.64) and (3.65) into the received difference signal of Eq. (3.62b), and when the arguments are small, the signal in the difference channel can be approximated as

$$\Delta_r(\alpha) \approx k_2\frac{\pi}{2}\frac{D}{\lambda}\alpha \qquad \text{[3.68]}$$

Substituting the above into Eq. (3.67) gives the correction signals as

$$e_{1,2} = j\frac{k_1}{k_2}\frac{4\,v\,\sin\theta}{D}\,T_p\Delta_r(\alpha) = \pm jk'\,v\,\sin\theta\,\Delta_r(\alpha) \qquad [3.69]$$

Thus the above are applied to the received signals so that the input to the delay-line canceler is

$$\text{Pulse No. 1} = \Sigma_r(\alpha) + jk\,(v\,\sin\theta)\,\Delta_r(\alpha) \qquad [3.70a]$$

$$\text{Pulse No. 2} = \Sigma_r(\alpha) - jk\,(v\,\sin\theta)\,\Delta_r(\alpha) \qquad [3.70b]$$

The constant k accounts for the difference in gains between the sum and difference channels, as well as the factor $4T_p/D$.

Thus one takes the received difference signal, multiplies it by $k(v\sin\theta)$, shifts its phase forward or back by 90° (depending on which pulse), and adds it to the received sum signals. Note that the multiplier $k(v\sin\theta)$ is bipolar in that it must be capable of changing sign when the antenna pointing direction changes from one side of the platform to the other.

Compensation for Antenna Scan Modulation.[88–90]   In Sec. 3.7, the limitation on the MTI improvement factor due to antenna scan modulation was discussed, and Eq. (3.46) gave the limitation when the MTI processor consisted of n delay-line cancelers in cascade. Antenna scan modulation is affected by the change in amplitude from pulse to pulse caused by the shape of the antenna pattern. It is possible in an AMTI radar to compensate for this change in amplitude by adding half of the needed correction to one pulse and subtracting half from the other pulse. The mechanization for this form of scan-motion compensation is similar to the DPCA compensation described previously except that the difference signal is applied in phase with the sum signal and is amplified by an amount determined by the antenna rotation between pulses. It is also possible to combine the DPCA and antenna scan-motion compensation by properly scaling and applying the difference pattern both in phase and quadrature.

Antenna Sidelobes.[91]   If the antenna sidelobes of an AMTI radar are not sufficiently low, the clutter that enters the receiver via the sidelobes can set a limit to the improvement factor equal to

$$I_{\text{SL}} = \frac{K\displaystyle\int_{-\pi}^{+\pi} G^2(\theta)\,d\theta}{\displaystyle\int_{\text{SL}} G^2(\theta)\,d\theta} \qquad [3.71]$$

where K is the average gain of the delay-line canceler, and $G^2(\theta)$ is the one-way power gain of the antenna in the plane of the ground surface. The integral in the denominator is taken outside the main-beam region, and it assumed the sidelobes are well distributed in azimuth. This limitation to the improvement factor would be one of the many inserted into Eq. (3.41) to find the overall MTI improvement factor for the radar.

Space-Time Adaptive Processing (STAP)   When greater clutter attenuation is required of an AMTI radar than can be achieved with conventional DPCA, adaptivity in both the antenna pattern and the doppler processing can be considered. Both forms of adaptivity, spatial and temporal, are used together in what is known as *STAP*. Spatial and temporal adaptivity will be briefly discussed individually before discussing their combination as STAP.

Adaptive Antennas   The adaptive antenna is a phased array that automatically adjusts the phase and amplitude at each element to place one or more antenna-pattern nulls in the direction of external noise, jamming, or interference for the purpose of reducing or eliminating unwanted signals that enter the receiver via the antenna.[92–94] It is also possible to employ an adaptive antenna to place nulls in the direction of large clutter echoes so as to act as a *spatial* filter, something that can be of importance for AMTI.[95] On the basis of the signals received at the individual elements of the array antenna, algorithms in the adaptive processor determine the phase and amplitude weights that should be applied at each element of the receiving array to maximize the signal-to-noise ratio. *Noise* in this case can be considered as all unwanted signals that arrive at the antenna. It can include clutter echoes, jamming, and interference. Adaptive antennas can automatically compensate for mechanical or electrical errors in an antenna by sensing the errors and applying corrective signals. They can also compensate for failed elements, radome effects, and blockage of the aperture from nearby structures. Any adaptive system must be able to distinguish desired from undesired signals. Thus an adaptive antenna requires some a priori knowledge of the desired signal, such as it direction, waveform, doppler, or statistical properties.

Adaptive MTI   The adaptive antenna places its nulls so as to minimize the clutter (or other interference) that enters via the antenna sidelobes. A transversal filter, such as was shown in Fig. 3.12, is similar in some respects to an array antenna in that there are a number of equally spaced taps whose outputs can be weighted to provide the doppler-filter response to select moving targets and reject clutter. Analogous to the adaptive antenna, it is possible to have an *adaptive MTI* (an adaptive doppler filter) which selects the necessary weights for the taps of the transversal filter to adaptively maximize the signal-to-clutter ratio on the basis of its doppler frequency spectrum.[96,97] If there were only a single point-clutter echo, the adaptive MTI filter would place the nulls it has available at the frequency that corresponds to its doppler frequency. If the clutter were distributed over a range of doppler frequencies (as might be the echoes from rain in an atmosphere with wind shear), the nulls of the adaptive MTI filter would be automatically placed over the doppler region occupied by the clutter. A three-pulse adaptive canceler, for example, with three zeros to place, can adaptively locate three nulls at three different frequencies, or it can place the three nulls so as to make a single wide notch depending on the nature of the clutter spectrum, or it can place its three nulls at the same doppler frequency for greater attenuation. Also, if the clutter spectrum varies with range, the filter can automatically adapt and change its weights so as to continue to maximize the signal-to-clutter ratio. Adaptive MTI can also handle bimodal clutter, as when surface clutter and weather clutter occur in the same range resolution cell.

STAP[98–100]   The benefits of adaptive antenna (spatial) processing and adaptive doppler (temporal) processing can be combined in a synergistic manner to achieve improved performance of AMTI radar. The combination of the two is called *space-time adaptive processing,* (STAP). A highly simplified block diagram of STAP is shown in Fig. 3.41. The antenna is a linear phased array of N elements. At each element of the array there is a tapped delay line that processes K pulses. The $K - 1$ time delays between the taps are each equal to the pulse repetition interval. The K pulses constitute a coherent processing interval (CPI). At each range resolution cell there are $N \times K$ outputs that are processed by a suitable algorithm to obtain W_{nk}, the $N \times K$ adaptive weights that are necessary for maximizing the signal-to-noise ratio. At the output of each tap there is an adaptive processor (not shown) that compares the received signal, V_{nk}, with the output of the summer, V_Σ, to develop the proper weight. The signal V_{kn} is the output of the kth pulse from the nth element of the array. The system output is

$$V_\Sigma = \sum_{k=1}^{K} \sum_{n=1}^{N} W_{nk} V_{nk}$$ [3.72]

These quantities are complex in that both phase and amplitude are employed in the adaptive process. The signal processing is linear and simultaneously controls the antenna pattern and the doppler filter response. The direction of the receiving antenna beam is determined by steering signals injected into the adaptive array control circuits at each element. Being a phased array with control of the amplitude and phase weights, the antenna can be made to perform DPCA to compensate for the broadening of the clutter spectrum caused by platform motion. Adaptive processing reduces the response in the spatial or the temporal domains only when there is clutter or other interference that is to be eliminated.

When nulls are placed in the sidelobe region, the shape of the main beam can be distorted. If this occurs within the CPI, the main-beam clutter will not be fully canceled. In an adaptive antenna such as STAP, some of the available spatial degrees of freedom can be used to maintain the shape of the main beam when nulls are placed in the sidelobe region.

Figure 3.41
A highly simplified block diagram of space-time adaptive processing, consisting of an array antenna of N elements and a doppler processor with K weights.

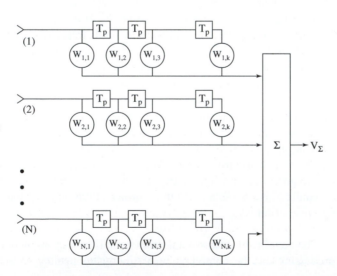

In brief, STAP can provide the following capabilities for an AMTI radar:

- Reduction in clutter by adaptively locating the nulls of the doppler filter response to attenuate the clutter spectrum.

- Reduction of clutter, jamming, and interference that enters the receiver via the antenna sidelobes by adaptively placing the antenna nulls in the direction of the unwanted clutter echoes and interfering signals. This takes some of the burden off the doppler processor, especially when the clutter is not at zero doppler (as would be the echoes from rain or chaff).

- When clutter echo that enters the antenna sidelobes has the same doppler frequency as an aircraft target echo in the main beam, doppler filtering alone might not be able to separate the clutter echo from the target echo, but the antenna pattern nulls introduced by STAP will.

- Adaptive selection of antenna weights to provide platform motion compensation.

- Ability to compensate for radar scattering from aircraft structures near the antenna, failed antenna elements, or errors in the antenna array.

The more elements in the array and the more pulses processed during each CPI, the greater will be the degrees of freedom that can be used in placing spatial and temporal nulls. That flexibility, however, comes with a price—sometimes a severe price. The greater the number of antenna elements and the greater the number of pulses to be processed, the greater the demand placed on the digital processing and the longer it might take for the adaptive weights to be computed. In practice, therefore, there will be a limit on the number of adaptive elements that can be employed and the number of pulses to be processed. For this reason, it is unusual to have a fully adaptive array (an adaptive processor at each element) when the number of elements in the array antenna is large.

3.9   PULSE DOPPLER RADAR

In an MTI radar the prf is chosen so that there are no range ambiguities, but there are usually many doppler ambiguities, or blind speeds. The MTI radar has proven to be a fine method for the detection of moving targets in clutter if the effect of the blind speeds can be tolerated. There are important situations, however, where the proliferation of blind speeds can eliminate a large portion of the available *doppler space* (the doppler region where desired moving targets can be detected). A reduction in the available doppler space causes loss in the detection of moving targets. Blind speeds become troublesome when the radar frequency is increased. An increase in frequency, when the prf remains constant, means a decrease in the first blind speed [according to Eq. (3.12)] and more of them will appear within the desired doppler space. The degradation of performance caused by blind speeds can be especially difficult in an airborne radar that has to operate at a high microwave frequency in order to have a narrow beamwidth with the relatively small antenna size that can be tolerated in an aircraft. Also, as described in the previous section, airborne radars suffer a widening of the clutter spectral width due to platform motion, further aggravating the reduction in the doppler space available for detection of targets. The result is that at the higher microwave frequencies, the MTI technique needs to be replaced since it cannot perform satisfactorily. The luxury of having no range ambiguities in a low-prf MTI radar has to be sacrificed in order to eliminate the serious consequences of the ambiguous doppler and its accompanying blind speeds. Increasing the prf increases the first blind speed [Eq. (3.12)] and reduces the number of nulls found within the doppler space. A high prf, however, can increase the problem of range ambiguities. Thus the trading of doppler ambiguities (blind speeds) for range ambiguities is something that has to be tolerated in order to obtain good detection of moving targets at the high microwave frequencies.

A radar that increases its prf high enough to avoid the problems of blind speeds is a called a *pulse doppler radar*. More precisely, a *high-prf pulse doppler radar* is one with no blind speeds within the doppler space. In some situations, however, it may be beneficial to operate at a slightly lower prf and accept both range and doppler ambiguities. Such a radar is called a *medium-prf pulse doppler radar*. Thus there are three different types of pulse radars that use doppler. They differ in their prfs and the type of ambiguities they are willing to tolerate. These are:

1. The *MTI* with no range ambiguities and many doppler ambiguities.

2. The *high-prf pulse doppler* with just the opposite: many range ambiguities and no doppler ambiguities.

3. The *medium-prf pulse doppler radar* with some of each.

At one time there was a significant difference in the type of transmitter and the signal processing techniques used for MTI and pulse doppler radar. In the early days of MTI radar, a magnetron oscillator was commonly used as the transmitter. A pulse doppler radar, on the other hand, usually used a high-power amplifier transmitter such as the klystron. Both MTI and pulse doppler radars now use a high-power amplifier of some sort. The MTI radar originally used analog delay-line cancelers and the pulse doppler radar used analog filter banks. Both now use digital processing and an MTI radar can use a filter bank (as did the MTD described in Sec. 3.6). Thus the equipment differences between the two are no longer significant enough to distinguish one from the other. The basic difference between an MTI and a pulse doppler radar is the prf and duty cycle that each employ. Another significant difference is that the pulse doppler radar generally receives much more clutter than an MTI radar (because of the foldover in range of clutter echoes when ambiguous prfs are employed), so that the pulse doppler radar requires a much greater improvement factor than does an MTI radar of comparable performance.

High-prf Pulse Doppler Radar   The pulse doppler radar will be described here in terms of an airborne radar[101,102] such as AWACS (Airborne Warning and Control System).* Figure 3.42 is a sketch of the airborne geometry showing the *main beam* of the antenna, the *antenna sidelobes* which illuminate clutter over a wide range of viewing angles, and the *altitude return* which is reflected from the ground directly below the radar. It will be recalled that a train of pulses, like that of a pulse doppler radar, produces a line spectrum,

---

| *The terms AEW, AWACS, and AEW&C have all been used to indicate an *airborne air-surveillance radar*.

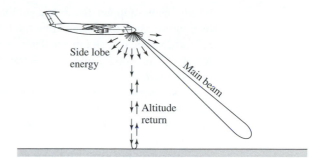

Figure 3.42   Sketch of the geometry of an airborne pulse doppler radar illustrating the scanning main beam, the antenna sidelobes illuminating ground clutter, and the strong altitude return from directly below the radar.

Fig. 3.43a, with the separation between spectral lines equal to the prf. Figure 3.43b illustrates the carrier frequency f_0 and the two adjacent spectral lines at $f_0 + f_p$ and $f_0 - f_p$, where f_p is the prf. The spectrum of the received echo signal is not a pure line spectrum because of the finite time on target and other factors such as modulations introduced by the clutter echoes. At the carrier frequency f_0 there will be a large return due to echoes received from directly below the radar. Such echoes can be relatively large. This altitude

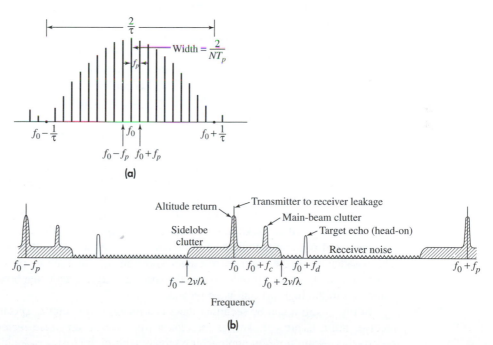

Figure 3.43   (a) Transmitted spectrum of a train of N rectangular sinewave pulses of carrier frequency f_0, width τ, pulse repetition frequency $f_p = 1/T_p$, and total duration NT_p. Null width of the spectral lines (not indicated in the figure) is $2/NT_p$. (b) Portion of the received signal spectrum in the vicinity of the RF carrier frequency f_0, for a high-prf pulse doppler radar.

return has no doppler frequency shift since its relative velocity is zero. Because of fold-over, or aliasing, the altitude line (and the rest of the spectrum) is repeated at frequencies $f_0 \pm nf_p$, where n is an integer. There can also be leakage of the transmitter signal into the receiver. The altitude line and the transmitter leakage can be removed with a notch filter centered at f_0.

The antenna sidelobes will illuminate clutter over a large range of incidence angles (from 0 to almost 90°). Thus there can be clutter echoes extending almost $\pm 2v/\lambda$ about the carrier frequency and the other spectral lines, where v is the radar's absolute velocity. For convenience, the shape of the sidelobe clutter spectral region is shown in Fig. 3.43b as uniform. In practice, however, the shape will not be uniform as shown. Sidelobe clutter echoes usually decrease in amplitude the farther their doppler-shifted echo frequency is from the carrier.

The antenna sidelobe clutter is large in a high-prf pulse doppler radar since there are many range-ambiguous pulses simultaneously illuminating the clutter. With a duty cycle of 50 percent, the antenna sidelobes are simultaneously illuminating half the total clutter within the radar's coverage, much more than would be seen by a low-prf AMTI radar. The large sidelobe clutter viewed by a pulse doppler radar is a reason why it requires a higher improvement factor than an AMTI radar of equivalent performance. The greater improvement factor required of a pulse doppler radar does not necessarily mean it can detect targets in clutter better than the AMTI. (It just means the pulse doppler radar designer has a tougher task to accomplish than does the AMTI radar designer.) One must look further to determine the relative capability of the two radars and not rely only on the improvement factor as the measure of relative merit.

The sidelobe clutter region is seen to extend over a relatively large portion of the doppler space. To detect aircraft targets within the sidelobe clutter region, a bank of narrowband doppler filters with adaptive thresholds can be used. To keep the sidelobe clutter from swamping the detection of small targets, the antenna must have exceptionally low sidelobes compared to conventional antennas. The success of the *S*-band AWACS high-prf pulse doppler radar depended on achieving, for the first time, antennas with ultralow sidelobes (Sec. 9.13).[103]

The clutter echo seen by the main beam can be relatively large in amplitude, and is shown in Fig. 3.43b. It lies somewhere within the sidelobe clutter region. As the antenna beam scans in angle, the doppler frequency of the main-beam clutter varies, so its location within the echo-signal spectrum will also vary. Main-beam clutter can be removed by a tunable filter that tracks the doppler-shifted frequency of the clutter echo. Platform motion can also affect the width of the main-beam clutter just as it does with an AMTI. The broadening of the main-beam clutter spectrum in this type of pulse doppler radar, however, usually is not a problem since the spectral width, even when broadened, is small compared to the high prf that is employed.

In Fig. 3.43b it can be seen that there is a region of the clutter spectrum where only receiver noise is present. This is the *clutter-free* or *receiver-noise region,* and it corresponds to viewing nose-on approaching targets with high closing velocities that result in high doppler shifts. The presence of a clutter-free region is an important advantage of a high-prf pulse doppler radar, especially if detecting high-speed closing targets at long range is required. On the other hand, if the relative velocity of the target is low, as when targets are viewed tail-on or when the target is on a crossing trajectory, the echo can fall within the sidelobe clutter region and its detectability is much less than a high-speed target in the clutter-free region. Such targets with low doppler frequency shifts are detected by a high-prf pulse doppler radar at significantly shorter ranges than high-speed targets that are in the clutter-free region.

The prf of an X-band high-prf pulse doppler radar designed for use with a military airborne fighter/interceptor might be from 100 to 300 kHz. With such a high prf, there is not much room for long pulses. It is not unusual for the duty cycles of high-prf pulse doppler radars to have values from 0.3 to 0.5. Such radars might operate with only one range gate. Only one set of doppler filters is needed, rather than one set for each of the many range gates of radars that employ lower duty cycles.

As mentioned, the prf is chosen high enough so that there are no doppler ambiguities and no foldover of the clutter spectrum. The prf required can be determined from an examination of the spectrum of Fig. 3.43b. It has been shown[104,105] that if the center of the doppler filter bank is always maintained at the frequency of the main-beam clutter, the minimum prf that can be used is $4v_T/\lambda$, where v_T is the maximum ground speed of the target. For this to occur, the main-beam clutter frequency must be known and a tracking device used to keep the filter bank centered at the main-beam clutter frequency. If, on the other hand, the doppler filter bank is fixed with its center always at the radar transmitted frequency, f_0, the prf might be as large as $4v_T/\lambda + 2v_R/\lambda$, depending on the maximum angle scan in azimuth, where v_R = radar platform velocity. When the doppler filter bank is centered on the radar frequency, more filters are required than when it is centered on the doppler of the main-beam clutter.

Eclipsing Loss   Since the pulse doppler radar cannot receive when it is transmitting, the high duty cycle can result in a loss if the echo signal arrives when a pulse is being radiated and the receiver is gated (turned) off. This is called *eclipsing loss,* and can be anything between zero and a large value depending on the exact location of the received echo pulse within the timing of the transmission.[106] There are some things that might be considered, however, to reduce the effects of eclipsing loss. The degree of eclipsing varies as the target range changes with time so eclipsing can cause periodic holes in the coverage. A rapidly approaching target, on the other hand, will not remain eclipsed for long, so that detections will occur at a slightly shorter range when eclipsing is present. With low relative-velocity targets, the time during which eclipsing occurs may be reduced by using more than one prf. If a single target is being tracked, a change in prf at the appropriate time can keep the target echo from arriving when the receiver is gated off. A reduction of the duty cycle and an increase in the number of range gates will reduce the effect of eclipsing, but if the prf is too low, one no longer has a high-prf pulse doppler radar.[107]

Resolution of Range Ambiguities   The high-prf pulse doppler radar results in many range ambiguities. These can be resolved by use of multiple prfs, although other methods that involve modulating the RF frequency or the pulse characteristics are possible.[108] From the measurement of the ambiguous ranges on two prfs, the true range can be determined similar to what was outlined in Sec. 2.10. Three different prfs are usually used rather than two, in order to increase the unambiguous range that can be achieved and to reduce the possibility of "ghosting." (A "ghost" can occur if there is more than one target present and if an ambiguous range of one target is the same as one of the ambiguous ranges of another target. The false coincidence of the two will result in a decision being made that this is the true range of a target, which it is not.) The ratios of the prfs are usually related by relatively prime integers (that is, they do not have a common factor among them). If, for example, the prfs were in the ratio 7, 8, and 9, the unambiguous range from these three prfs would be $7 \times 9 = 63$ times that of the middle prf. The Chinese remainder theorem has been used to calculate the true range from the unambiguous measurements,[108] but there are other algorithms available for the same purpose.[109]

Other Considerations   A high-prf pulse doppler radar that uses three different prfs to obtain the true range measurement of a target requires that a target detection and range measurement be made on each of the three prfs. This results in a power-aperture product (or whatever other appropriate figure of merit is used) three times greater than a radar that doesn't require redundant transmissions. For a given range performance, the high-prf pulse doppler radar will therefore require much larger transmitter average power than an AMTI radar, if everything else remains unchanged. The high prf and high duty cycle also result in poor resolution of multiple targets and may make it difficult to determine the number of military aircraft that might be flying in close formation. A high-prf pulse doppler radar is more complex and more costly than an AMTI. On the other hand, because of the absence of doppler ambiguities and good doppler processing, this radar produces a good measurement of the target's radial velocity. Detection of aircraft targets based only on extracting the doppler without attempting to resolve range ambiguities can produce large detection ranges. This is called the *velocity search mode* and is often employed for the initial detection of targets at long range with a multimode airborne radar.

The airborne air-surveillance radar application can be accomplished equally well with the AMTI approach (the U.S. Navy's E2 aircraft radar at UHF) or the pulse doppler approach (the U.S. Air Force E3 AWACS aircraft radar at *S* band). The performance of these two surveillance radars can be said to be comparable;[110] but in the past, the cost of the E2 AMTI system has been considerably less than that of the high-prf pulse doppler AWACS E3 system. The AMTI method, however, cannot be used at the higher microwave frequencies that are necessary for military fighter/attack aircraft.

Medium-prf Pulse Doppler Radar   A medium-prf pulse doppler radar is one whose prf is between that of the high-prf pulse doppler and the MTI radar. It therefore has both range and doppler ambiguities. The medium prf results in less clutter being seen by the antenna sidelobes than the high-prf radar since there are fewer pulses viewing ambiguous range cells. The lower clutter level in the sidelobe region of the echo signal spectrum allows better detection of moving targets with low doppler velocity (such as those viewed at tail-aspect as well as those with near-crossing trajectories), and will detect slow targets at longer ranges than can the high-prf pulse doppler. The reduced prf, however, causes the prf lines to become much closer, the sidelobe clutter regions will overlap, and there is no clutter-free region as in the high-prf pulse doppler. Although the lower prfs cause the sidelobe clutter from higher prf lines in the spectrum to fold over and increase the sidelobe clutter, it does not completely eliminate the benefit obtained by lowering the clutter when fewer pulses are received with the medium prf.

In engineering, few things are ever perfect. Compromises are often necessary. This is true when trying to select which of the two types of pulse doppler radars should be used. The high-prf pulse doppler radar has good performance against high-speed closing targets; but poorer performance against targets with low relative velocity. The opposite is true for the medium-prf pulse doppler. It does much better against slower targets than does the high prf, but it is much poorer against high-speed targets. According to Morris,[111] the high-prf pulse doppler radar can typically provide 50 percent greater detection range against high-speed, head-on targets than can a medium-prf pulse doppler radar with the same average transmitter power. On the other hand, the medium-prf pulse doppler radar is superior for receding (low relative velocity) targets at altitudes less than 10,000 ft. In the original *S*-band AWACS procurement, one major bidder proposed high-prf pulse doppler, the other major bidder proposed medium prf. The high-prf radar was selected, but either probably could have done the job.

In the *X*-band airborne radar designed for fighter/interceptor aircraft, it is not unusual to employ both high-prf and medium-prf modes interleaved so as to obtain the advantages of each. Such airborne radars operate with a number of modes, including a low-prf radar mode used when the antenna beam is looking up where it doesn't see clutter and doesn't need doppler processing. When high-prf and medium-prf pulse doppler waveforms are interleaved, more time is required to do both (if nothing else is done to maintain constant the time to make a detection). The transmitter power and antenna scanning must be increased to allow a detection decision in shorter time or there must be a sacrifice in the detection range and/or the time to make a decision.

The altitude line in the medium-prf pulse doppler can be removed by range gating or by filtering, whereas in the high-prf pulse doppler it can only be removed by filtering. Range gating permits the detection of low relative-velocity targets near the radar transmitted frequency that might be rejected in a high-prf system where the altitude return is rejected by filtering.

Because of the lower duty cycle of the medium-prf pulse doppler radar, its range accuracy and ability to resolve multiple targets in range are better than those of a high-prf system. The medium-prf pulse doppler, just as with the high-prf, cannot employ sensitivity time control (STC) to reduce the effects of clutter at short ranges as can a low-prf radar.

It is important in the medium-prf pulse doppler, just as with the high-prf pulse doppler, to have antennas with reduced sidelobe levels in order to minimize the clutter that enters the radar via the sidelobes.

Resolution of Range Ambiguities[112]   As with the high-prf pulse doppler radar, range ambiguities that occur in the medium-prf system are resolved by using transmissions on three different prfs. In the medium-prf pulse doppler radar, however, there are regions of doppler space where the main-beam clutter might be so strong that detection of targets is not practical. In order to insure that a target will be detected outside of these *blind zones* on at least three prfs, a medium-prf pulse doppler has to transmit on seven or eight different prfs. An example is shown in Fig. 3.44. The upper part of the figure depicts the

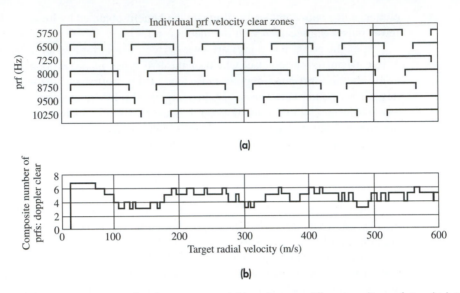

(a)

(b)

Figure 3.44   (Upper) The clear zones available with seven different medium prfs in which target detection can take place. (Lower) The number of prfs in which a target can be seen in a clear zone, as a function of the target's radial velocity.
| (This illustration was originally provided the author by the late Frederick C. Williams.)

clear zones of seven prfs. The lower part of the figure shows the number of clear zones available as a function of the target's radial velocity. The minimum number of clear zones is three in this case, as required for resolving range ambiguities. The fact that seven or eight different transmissions have to be used means that the radar's power-aperture product has to be much larger than that of a high-prf pulse doppler radar if equivalent range performance is to be had, and even higher than that of an AMTI radar with equivalent detection range. Each time the prf is changed there needs to be a settling time before processing begins, which can reduce the range of detection. If the processor requires *fill pulses* for each prf to increase the doppler filter performance, this can also reduce the range performance.

The prfs used in a medium-prf pulse doppler radar might extend over a range of values of almost two-to-one. If the pulse width remains constant, the range of the radar will vary as the prf is changed since the transmitter duty cycle and total energy of the radiated signal will depend on the prf (assuming the peak power is constant, which it usually is). This variability in the radar range with prf is undesirable. Furthermore, high-power radar transmitters generally like to operate with a constant duty cycle in order to minimize the changes in operating conditions and heating cycles. It is important, therefore, to maintain the duty-cycle constant. This can be done in the medium-prf pulse doppler radar by changing the pulse width as the prf is changed. The change in pulse width requires that the bandwidth of the receiver's matched filter be altered accordingly.

The Three prfs[113]   Throughout this section, there have been comparisons made among the high-prf, medium-prf, and low-prf radars. Table 3.2 summarizes "typical" values that characterize these three prf regimes for airborne radar. The values of prf and duty cycle

Table 3.2   Comparison of the three prfs

Radar	prf*	Duty Cycle*
X-band high-prf pulse doppler	100–300 kHz	< 0.5
X-band medium-prf pulse doppler	10–30 kHz	0.05
X-band low-prf pulse radar	1–3 kHz	0.005
UHF low-prf AMTI	300 Hz	Low

*These are illustrative values only. Actual radars can have values that extend to either side of what is shown.

listed in this table are illustrative only (especially the duty cycles). Airborne radars can be, and often are, found outside these values. The three entries for *X* band apply to radars used for fighter/interceptor applications. The low-prf *X* band airborne pulse radar does not have doppler processing to detect moving aircraft in clutter. It is mainly used in a *look-up* mode where no clutter is seen rather than in a *look-down* mode that sees clutter. For pulse doppler radars at frequencies other than *X* band, the prfs would be in proportion to the RF frequency. For comparison, a UHF AMTI wide-area surveillance radar with low prf is included in the table. Not included are the wide-area surveillance radars at *S* band with either medium or high-prf waveforms. The characteristics of airborne doppler radars at each of the three prf regimes are summarized below:

High-prf pulse doppler:

- No ambiguities in doppler frequency, no blind speeds, but many range ambiguities.
- Range ambiguities can be resolved by transmitting three redundant waveforms, each at a different prf.
- Transmitter leakage and altitude return are removed by filtering.
- Main-beam clutter is removed by a tunable filter.
- High closing-speed aircraft are detected at long range in the clutter-free region.
- There is poor detection of low radial-speed targets that are masked in the frequency domain by short-range sidelobe clutter folded over in range.
- Often only a single range gate is employed, but with a large doppler filter bank.
- For comparable performance, a much larger improvement factor is required than lower prf systems since the high prf results in more clutter being viewed by the antenna sidelobes.
- The antenna sidelobes must be quite low in order to minimize the sidelobe clutter.
- Range accuracy and the ability to resolve multiple targets in range are poorer than with other radars.

Medium-prf pulse doppler:

- Has both range and doppler ambiguities.
- There is no clutter-free region as in the high-prf system, so detection of high-speed targets will not be as good as with the high-prf system.

- Fewer range ambiguities mean that there will be less clutter seen in the antenna side-lobes so that targets with low relative speeds will be detected at longer range than with a high-prf system.
- The trade-off of detectability of high-speed targets for better detection of low-speed targets often makes the medium-prf system preferred over the high-prf system for airborne fighter/interceptor radar applications, if only a single system is used.
- The altitude return can be eliminated by range gating.
- More range gates are required than in the high-prf system, but the number of doppler filters at each range gate is less.
- Seven or eight different prfs must be used to insure that a target will have the proper doppler frequency to be detected on at least three prfs in order to resolve range ambiguities.
- For comparable range performance the large number of redundant waveforms means that the transmitter must be larger.
- Better range accuracy and range resolution are available than with high-prf systems.
- The antenna must also have low sidelobes to reduce the sidelobe clutter.

Low-prf AMTI:

- No range ambiguities, but many doppler ambiguities (blind speeds).
- Requires TACCAR and DPCA to remove effects of platform motion.
- Operates clutter free at long range where no clutter is seen due to curvature of the earth.
- Sidelobe clutter is usually not as important as it is in pulse doppler systems.
- Best employed at UHF or perhaps *L* band. Increased blind speeds and the lower effectiveness of platform motion compensation prevent its use at the higher microwave frequencies.
- The lower RF frequency (UHF) of the AMTI radar results in wider antenna beamwidths than a higher frequency (*S* band) pulse doppler radar whose mission is wide-area air-surveillance.
- Because there are no range ambiguities to be resolved, redundant waveforms with multiple prfs are not needed.
- For comparable performance the required product of average power and antenna aperture is less than for pulse doppler radars.
- Usually simpler than pulse doppler radars.
- Cost is generally much less than pulse doppler radars of comparable performance.
- AMTI cannot be used in fighter/interceptor *X*-band radars for look-down detection of targets in clutter. The low-prf mode without doppler processing, however, is used to advantage when there is no clutter present, as when the antenna is looking up or the target is at a range less than the height of the radar over the ground so that near-in ground clutter that enters via the sidelobes can be gated out.

Airborne Air-Surveillance Radar Antennas Airborne air-surveillance radars such as the U.S. Navy's E2 AEW (AMTI) radar at UHF (Fig. 3.45a) and the U.S. Air Force's E3 AWACS (high-prf pulse doppler) radar at *S* band (Fig. 3.45b) both employ a large externally mounted rotodome antenna. (The *rotodome* is a radome-enclosed antenna in which

Figure 3.45 Photos of four different types of airborne air-surveillance radars. (a) U.S. Navy aircraft-carrier-based E2C Hawkeye Airborne Early Warning (AEW) radar operating at UHF. A rotating array of Yagi end-fire antenna elements is enclosed within a 24-ft by 2.5-ft rotodome. The small cone-shaped addition on top of the rotodome is a satellite communications (SATCOM) antenna.
| (Courtesy of Northrop Grumman Corp.)

(b) U.S. Air Force E3 AWACS *S*-band pulse doppler radar with a rotating 24-ft by 5-ft ultralow-sidelobe planar array in a rotodome. This particular example is that of the Japan Air Self-Defense Force mounted in a 767 aircraft. The original AWACS was in a 707 aircraft.
| (Courtesy of Northrop Grumman Corporation.)

(c) Phalcon *L*-band Airborne Early Warning System that employs "conformally" mounted electronically scanned phased array radars mounted on the left and right side of the forward part of the fuselage. This example is on a 707 aircraft. The arrays are 12 by 2 m (39 by 6.6 ft), and extend 1.5 ft out from the side of the fuselage. Each side array contains 600 solid-state T/R (transmit/receive) modules. The 3-m (9.8-ft) diameter radome in the nose of the aircraft provides coverage of the forward sector. A total of 280° of azimuth coverage is obtained. Provision was made for installing smaller 4-m by 2-m (13-ft by 6.6-ft) phased arrays along the sides of the rear fuselage to extend the coverage to 360°.
| (Courtesy IAI ELTA Electronics Industries, Ltd., Israel.)

(d) Erieye *S*-band AEW&C pulse doppler radar with back-to-back electronically scanned phased array antennas mounted in a dorsal unit on top of a Saab 340B aircraft. There are 192 solid-state T/R modules in the 8-m-long array. Each side of the array covers an azimuth sector of 120°.
| (Courtesy of Ericsson Microwave Systems AB, Molndal, Sweden.)

the antenna and radome rotate together.) The UHF radar antenna is an array of Yagi end-fire radiators that provide the equivalent of vertical aperture even though it has a low physical profile. The antenna for the *S*-band AWACS is an ultralow-sidelobe array of slotted waveguides. Both the E2 and E3 antennas have comparable effective aperture areas, even though their antenna gains are different. Russian (formerly Soviet Union) AEW radars also have used rotodomes. Phased arrays mounted conformally around the aircraft are employed in the Israeli Phalcon *L*-band AEW radar (Fig. 3.45c). The Swedish *S*-band Eyrieye AEW radar (Fig. 3.45d) uses a phased array mounted along the top of the aircraft's fuselage to view on either side of the aircraft. The Eyrieye was designed for a small aircraft and had a shorter range for the same size target than larger systems. An electronically steerable phased array antenna, as in the Phalcon, can be mounted to conform to the aircraft's structure or externally along the fuselage. It is difficult with either fixed phased array method to achieve 360° azimuth coverage or to obtain as large an antenna as can be placed in a rotodome. The rotodome can be mounted relatively high above the aircraft's fuselage so as to minimize blockage from the wings and engines that might result in unwanted sidelobes. It can be more difficult to avoid antenna blockage from the aircraft's structure with flush-mounted phased arrays. The mechanical disadvantages of a rotodome have not been a limitation for past AEW radars. The rotodome can have aerodynamic lift to compensate in part for its weight. In spite of its limitations and the aesthetic desire to replace it with conformal antennas, the rotodome has proven to be an excellent choice for long-range AEW aircraft that must have 360° coverage. Another method for using an electronically scanned phased array is the Wedgetail system shown in Fig. 1.14.

Surface-Based Pulse Doppler Radars   Most of the discussion in this section on pulse doppler radar has been concerned with airborne applications. Pulse doppler radars have also been used with land-based and shipborne radar to detect small targets in clutter, especially high-speed hostile targets rapidly approaching a military air-defense system. Pulse doppler surface-based radars have been used at *L* band for both short and long-range detection of aircraft and missiles, and at *X* band for detection of low altitude antiship missiles as they appear at short range just as they come within the horizon. A *C*-band shipborne radar for detection of missiles at ranges of 10 km has been described[114] as using a prf of 4 kHz and a 45-rpm antenna, with unambiguous range of 37.5 km and unambiguous velocity of 110 m/s.

3.10   OTHER DOPPLER RADAR TOPICS

Noncoherent MTI   In an MTI radar there must be a reference signal to recognize that the echo signal of a moving target is shifted in frequency by the doppler effect. In the type of MTI radars discussed thus far, often called *coherent MTI* radars, the reference signal is an oscillator whose phase is referenced to the transmitter signal. The echo signal from clutter also has the characteristics of the transmitted signal and can be used as a reference to extract the doppler frequency shift of the target echo signal. Since the clutter echo and the moving target echo appear together at the input to the receiver, an internal reference signal is not needed. A radar that uses the clutter echo as the reference signal to extract the doppler-shifted target echo is known as a *noncoherent MTI* radar. It might also be called an *externally coherent MTI.*

A simple block diagram of a noncoherent MTI with a power oscillator as the transmitter is shown in Fig. 3.46. It is similar to the block diagram of a conventional pulse radar, except for the MTI filter, which is usually a delay-line canceler. The detector following the IF amplifier is a more or less conventional amplitude detector rather than a phase detector that requires a coho as reference. It is a nonlinear device which acts to multiply the clutter signal and the doppler-shifted moving target signal to produce the doppler (difference frequency) that is then extracted by the MTI filter.

The advantage of the noncoherent MTI is its relative simplicity. It was used in the past for both land-based and airborne MTI applications. A limitation is that it requires that clutter echo be present along with the target echo. This seems, at first glance, to be a reasonable requirement to achieve since an MTI radar would not be needed if there were no clutter. Clutter, however, is seldom spatially uniform. Since the target will be canceled by the delay-line canceler if there were no clutter present, the delay-line canceler must be switched out when there is no clutter within the radar resolution cell that contains the target. Some method is required to recognize the presence or absence of clutter and to rapidly switch the delay-line canceler in and out. Because of the patchy nature of clutter, the delay-line canceler might have to be switched in and out multiple times as the pulse travels out in range.

The improvement factor of a noncoherent MTI will usually be much poorer than that of a coherent MTI.[115] Since the clutter "reference signal" is not separately available, a Q channel cannot be employed. Significant loss, therefore, can occur due to blind phases (Sec. 3.3). Furthermore, the clutter echo as the reference signal is likely to be noisier than the internal reference oscillator (coho) of a coherent MTI. The combined effect of all these less than desirable features of a noncoherent MTI does not make this radar an attractive candidate for most MTI applications that require large improvement factors. Noncoherent radar has been successfully used, however, to detect moving ground vehicles with a side-looking airborne radar (SLAR) described in the next subsection.

Airborne Radar Detection of Ground Moving Targets[116,117]   AMTI radar and airborne pulse doppler radars for the detection of aircraft can also detect moving ground vehicles when

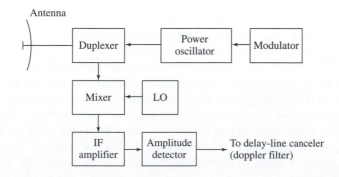

Figure 3.46
Block diagram of a noncoherent MTI radar.

operating over land. Truck and auto traffic on major highways can be present in large numbers, are of large cross section, and travel at speeds that are within the doppler pass-bands of these radars. The result can be large numbers of unwanted echoes that can over-load the radar's tracking computer.

On the other hand, ground vehicles are of interest for military purposes. The mission of airborne battlefield surveillance radars is to detect and track such targets. There are at least four different airborne radar methods for detection of ground moving targets:

- A variant of AMTI called GMTI (ground moving target indication).
- Side-looking airborne radar (SLAR) with noncoherent MTI.
- Synthetic aperture radar (SAR) with doppler filtering.
- Interferometric SAR (InSAR)

Each will be briefly discussed.

GMTI The modern airborne strike/fighter aircraft has as one of its many air-to-surface radar modes the detection of ground moving targets, usually called *ground moving target indication* (GMTI). When the targets of interest are ground vehicles, radar echoes from aircraft are the nuisance targets.

The GMTI mode uses a low or a medium prf, chosen to be unambiguous over the range and doppler frequency coverage of interest. A goal in the design of such radars is to have a low minimum detectable velocity. For a fixed minimum detectable doppler frequency, the minimum detectable velocity will be lower the higher the frequency. Thus K_u or K_a band air-to-surface radars might be more capable of detecting slow-speed targets than X band or lower frequency radars.

The AN/APG-67, a small low-cost X-band airborne radar (weighing only 270 lb) for the F-20 aircraft used a medium-prf waveform to detect ground vehicles within a $\pm 45°$ sector when the vehicle velocity was greater than 5 kt.[118] The maximum range in the GMTI mode was 50 nmi. Tank-type targets could be detected out to 20 nmi range.

A low-prf MTI mode, similar to the GMTI mode, can be used to detect moving ships. The AWACS radar employs a special "maritime mode" specifically designed to detect ships. (Since ships are relatively large targets and sea clutter is much smaller than land clutter, ships can be readily detected with a modest radar with modest resolution, usually without the need for doppler processing.)

SLAR[119, 120] A side-looking airborne radar (SLAR) employs a relatively conventional high-resolution radar in an unconventional manner to obtain a maplike image of a surface scene as well as detect moving targets using noncoherent MTI. An example is the Motorola SLAMMR which is an X-band airborne imaging radar with a 16-ft fixed side-looking antenna, 0.2-μs pulse width, and a magnetron transmitter with 60-W average power. Noncoherent MTI processing uses the clutter echo signal as the reference to extract the doppler-shifted moving target echo. An attenuated version of the SLAR image of the ground usually is superimposed on the moving target detections to aid in geographically locating targets. At each range cell, SLAMMR digitally generates a filter bank with 32 contiguous filters. A form of adaptive threshold is employed at the lower frequency filters to reduce the effect of the spread of the clutter spectrum due to platform motion. This dynamic biasing of the low-frequency filters allows the filters to adapt to the requirements imposed by the ground clutter in each range cell.[119] It has been said that such a SLAR with noncoherent MTI can extract small moving targets with speeds down to five miles per hour at a range of 50 nmi. Larger, faster targets can be detected out to 80 nmi.

SAR with MTI[121]   A synthetic aperture radar (SAR) produces a high-resolution image, or map, of a scene by synthesizing in its processor the equivalent of a large antenna to obtain good resolution in the cross-range direction. High resolution in the range direction is obtained by either a short pulse or pulse compression. A good SAR might have a resolution in range and cross-range of one meter, but it can be much less if desired. A SAR is more complex than the SLAR, but it produces much higher resolution images.

A conventional SAR is normally designed to image stationary objects. Moving targets are smeared, distorted, and misplaced when seen by a SAR. Moving targets can be detected with a SAR if they have a doppler frequency shift greater than the spectral bandwidth of the stationary ground clutter echo. (Clutter in this case *is* the desired signal for a SAR.) This technique has been limited, however, because it needs a prf high enough to avoid doppler foldover of echo signals. A high prf, on the other hand, might give rise to range ambiguities. This method for extracting moving targets with a SAR might not be able to detect moving targets that have a low radial velocity.

Interferometric SAR   The resolution cells, or pixels, of a SAR image usually are described by the amplitude of the signal within the pixel. In addition to the amplitude, there is also a phase that can be associated with each pixel. An interferometric SAR is one that takes advantage of the phase information to detect moving targets. The phase information can also be used to obtain a measurement of the height of the object within a pixel,[122,123] but this is not the subject of this subsection.

The original interferometric SAR employed two identical sidelooking antennas mounted longitudinally along the line of travel of the aircraft or space vehicle. They were separated in the horizontal dimension and looked 90° (broadside) to the line of travel. A SAR image of the same scene was formed by each antenna, but at a slightly different time because of the forward travel of the aircraft. Thus the information at a particular pixel obtained by the forward antenna was at a slightly earlier time than the information at the corresponding pixel obtained by the aft (trailing) antenna. If there were a moving target in a particular pixel, the phase would be different in the two images. The phase of a stationary object, however, would be unchanged. Moving targets could then be detected by examining corresponding pairs of pixels in the two images to determine if there were a phase change. (A change in phase with time means there was a doppler frequency shift.) The amount of phase change depended on the separation between the forward and aft antennas, the speed of the aircraft carrying the radar, as well as the speed of the target within the pixel. This technique was capable of detecting slowly moving targets and was a good method for detecting moving targets that appear on SAR images. One of the first applications for the interferometric SAR was the detection and measurement of the radial velocity of ocean currents.[124]

A more sophisticated form of interferometric SAR has been employed for the detection of ground vehicles along with imaging the ground and fixed surface targets, as in the *X*-band Joint STARS (Joint Surveillance Target Attack Radar System) AN/APY-3 developed by Northrop Grumman. This is a U.S. Air Force/Army airborne battle-management system that detects both moving and fixed targets by an interferometric SAR that uses a side-looking antenna operating in a Boeing 707 aircraft designated E-8C. According to Northrop-Grumman,[125] Joint STARS is supposed "to detect, locate, identify, classify, and track trucks, tanks, and a variety of other military targets." "In its Wide Area Surveillance/ Moving Target Indicator (MTI) mode, the radar covers a nearly 50,000-square kilometer area . . . [and] can even differentiate tracked and wheeled vehicles." "It can see vehicles at ranges in excess of 200 kilometers, moving at walking speeds. And, it can accurately locate them within meters of their actual position."

The Joint STARS slotted-array antenna is 24 ft wide by 2 ft high with 456 by 28 horizontally polarized elements.[126] The beam can be electronically scanned $\pm 60°$ in azimuth and is rotated mechanically in elevation. The antenna is divided into three independent apertures, but the entire antenna can be fed to act as a single aperture with a narrow beamwidth. This is called the *sum mode* and is for transmitting and SAR mapping. The three independent apertures are each 8 ft in width and have three times the beamwidth of the sum mode. They provide the *interferometer mode* for moving target detection and tracking. Although two separated apertures are sufficient for detecting and accurately locating targets, at least three apertures are needed when spatially correlated ground clutter appears at the same range and with the same doppler as the target.

Moving target echoes detected by the three antenna apertures are distinguished from surface clutter echoes by taking advantage of the different angles of arrival associated with targets and clutter that have the same range and doppler frequency. This is accomplished by adding the signals from adjacent antenna apertures (left and center as one pair, and center and right as the other pair) and applying a relative weighting that depends on the doppler. This is designed to generate a null in the receiving antenna pattern in the direction of clutter while retaining, as much as possible, the target signals that arrive from a slightly different direction. This interferometric clutter cancellation results in two separate clutter-canceled channels which, through the application of classical detection and angle measurement methods, allow the residual signals from the surface moving target to be readily detected and located.[127]

The left and center apertures as well as the center and right apertures of the Joint STARS antenna act as displaced phase center antenna (DPCA) to cancel main-beam clutter and produce two clutter-free channels containing moving target information.[128,129]

The interferometric SAR with three beams as used on Joint STARS can detect targets moving with smaller relative velocities than can a conventional airborne MTI with only a single beam. To be detected by a conventional airborne MTI system, a moving ground target located near the center of the illuminated clutter area would have to have a radial speed that produced a doppler frequency shift that was outside the doppler clutter spectrum. For example, a side-looking antenna with 2° beamwidth, centered at broadside, located on an airborne platform traveling at 240 m/s (about 470 kt), produces a ground-clutter velocity spectrum that subtends approximately ± 4 m/s (about 8 kt). A target located near the center of the illuminated area would have to have a radial component of velocity greater than ± 4 m/s before it could be detected. A target approaching the radar platform from a direction located near the trailing edge of the antenna beam would need a ground speed greater than 8 m/s before it was sufficiently separated from the background clutter spectrum. An interferometric SAR, on the other hand, that was able to detect and measure the angle of arrival of the target and the angle of arrival of the clutter to an accuracy of 0.1 beamwidth ($\pm 0.1°$) would theoretically be capable of detecting targets with one-tenth the radial speed, or in this case ± 0.4 m/s (about 0.8 kt).[127]

The two modes for detecting stationary targets and moving targets are time shared (interleaved). The detection of parked tanks and other vehicles is determined by the SAR mode, and slowly moving targets are found with the interferometer doppler mode.

Joint STARS is a long-range wide-area surveillance system that uses a relatively large antenna in a large aircraft. The same basic technology has been employed in a much smaller radar, the Northrup Grumman AN/APG-76 K_u-band system, which is small enough to fit in a pod or in the nose of a fighter/attack aircraft.[130] Such an interferometric SAR uses more than two antennas, the antennas do not have to be identical, and they need not be side-looking or mounted longitudinally along the line of travel of the aircraft. The AN/APG-76 uses forward-looking bulkhead-mounted mechanically gimbaled antennas with a set of fixed interferometric arrays whose mechanical boresights are physically steered toward the target. It provides a SAR image and moving target detection with an interferometer antenna. It also has apertures displaced in the vertical dimension to provide on a pixel-to-pixel basis altitude information of scattering objects, to produce a 3-D SAR image. The APG-76 has 10-ft resolution with 20 ft elevation sensitivity per pixel.[131]

Effect of MTI on Signal-to-Noise Ratio   An MTI delay-line canceler generally attempts to cancel clutter with fewer pulses than are available from the target. Therefore, coherent integration of the available pulses following the MTI processor is usually performed to increase the output signal-to-noise ratio. It has been shown, however, that there will be a degradation in detection performance because the MTI processing correlates the independent input noise when the MTI pulses are integrated.[132,133] This results in an MTI noise integration loss that is approximately 1.0, 1.8, and 2.2 dB for single, double, and triple delay-line cancelers, respectively.[134] The loss can be reduced by applying amplitude weights to the pulses to be integrated.[135]

The loss in signal-to-noise ratio, however, is generally not of major concern in many applications of MTI radar. MTI is needed at the shorter radar ranges where clutter is a serious limitation to the detectability of targets, and receiver noise is not an issue. This is why MTI processing is usually turned off after the pulse travels beyond the maximum range at which clutter appears. If MTI processing were to remain in the radar signal-processing channel at long range where there is no clutter, other losses in signal to noise could occur because of the attenuation by the doppler processor of targets with radial velocities at or in the vicinity of the blind speeds.

The lesser concern for the detectability of signals when clutter is much larger than noise is evidenced by the willingness of the radar designer (and the radar buyer) of high-performance pulse doppler radars to tolerate the use of redundant waveforms for resolving range ambiguities and the use of fill pulses for sophisticated doppler processors in order to successfully detect moving targets in heavy clutter.

Pulse-Burst Radar[136-138]   This is a waveform that attempts to obtain the benefits of a pulse doppler waveform without some of its disadvantages. It consists of a burst of pulses at a high or medium prf, which are repeated at a prf low enough for the burst to be un-ambiguous in range. The *pulse burst* has also been called a *burst pulse* or *pulse train*. Nathanson[137] states that the number of pulses within a burst might be from 8 to 32; but, of course, these are not meant to be firm limits and depend on the nature of the radar ap-plication. Sometimes amplitude and phase weightings are employed across the burst to further reduce the sidelobes of the waveform (and increase the clutter attenuation), espe-cially in clutter that extends in range further than the range extent of the burst waveform. Amplitude weighting of the radiated signal, however, can present demands on the radar transmitter that might not be easily met.

Single-Pulse Doppler Radar   Ambiguities in the doppler and range measurement occur when the radar waveform is not continuous, and such ambiguities cause problems in MTI and pulse doppler radars, as has been described previously in this chapter. There are some limited situations, however, when a single pulse might be used for detection of the doppler frequency shift. In order to detect a doppler shift on the basis of a single pulse, there must be at least one cycle of doppler frequency during the pulse duration. This requires that the pulse duration τ be greater than the period T_d of the doppler frequency $f_d = 2v_r/\lambda = 1/T_d$, which means that $f_d\tau > 1$. Thus $\tau > \lambda/2v_r$, where λ = wavelength and v_r = radial velocity. If the minimum detectable velocity is v_{min}, then $v_{min} = \lambda/2\tau$. (If the clut-ter is strong, the minimum detectable velocity might be greater than this value.) A low v_{min} requires either long pulses and/or very short wavelengths. If moving targets are to be detected in clutter, however, the expression for the minimum detectable velocity as given above is too simplistic, and more precise methods are needed to assess MTI performance. Longer pulses might be required for good doppler discrimination than is indicated by the above expression for v_{min}.

The single-pulse doppler method was used in early space surveillance radars for the detection of ballistic missiles and satellites. However, such radars did not have to detect targets in ground or weather clutter. The single-pulse doppler measurement was employed in these radars to extract the radial velocity to aid in establishing the track of the target. Such radars operated at UHF with unusually long pulses. If the pulse width were 2 ms and the wavelength were 70 cm (UHF), the minimum detectable velocity is 175 m/s, or about 350 kt. This is high for detection of aircraft, but is suitable for detection of ex-traterrestrial targets with their much higher speeds.

At the other extreme, that of short wavelength, a millimeter-wave radar operating at a frequency of 94 GHz would require a pulse duration of 32 μs in order to have a minimum detectable velocity of 100 kt, based on the above criterion for v_{min}. The pos-sible advantage of a single-pulse doppler waveform for millimeter-wave radar is that it would not experience the many blind speeds that occur with a conventional pulse doppler waveform. In order to obtain range resolution with a long single pulse, pulse compres-sion can be employed. For detection of moving targets in clutter, a single pulse wave-form, with or without pulse compression, would have to be designed with good spec-tral characteristics and doppler filtering so as to have an acceptable minimum detectable velocity.

Area MTI Originally an *area MTI* radar was one that stored the amplitude of the echoes received at each range-azimuth resolution cell over an entire scan of the radar, and subtracted these stored signals from similar data received from the next scan. This results in cancellation of echoes from stationary clutter, but echoes from moving targets that appear in a different resolution cell from scan to scan (because of their movement) are not canceled. Area MTI does not depend on the doppler frequency shift to discriminate moving targets from stationary clutter; hence, a coherent radar is not needed. It does not have the limitations of blind speeds or doppler and range ambiguities that are experienced by doppler moving target detection methods. This form of area MTI is a video detection method that requires clutter to be stationary from scan to scan. Early forms of area MTI used analog storage tubes to record the video data. The poor analog storage technology and the limitation of other early radar equipment restricted the ability of this form of area MTI to extract small targets in clutter. A digital clutter map would be better for this application than the analog storage tube.

A more suitable area MTI is one that uses very short pulses and subtracts them pulse to pulse rather than scan to scan.[139] The subtraction can be done in the video rather than coherently. It can require less memory than scan-to-scan subtraction and avoids many of the problems found with doppler forms of MTI. For detection of aircraft, however, pulse widths need to be of the order of nanoseconds. It is more attractive at the higher microwave frequencies where the available bandwidths are large and the usual doppler MTI suffers from the detrimental effects of excessive blind speeds.

Area MTI has also been examined for detection of moving targets using ultrawideband (UWB) radar. The concept of doppler is quite different with UWB radar than with the narrow relative bandwidths usually employed in radar. In conventional-bandwidth radar, the bandwidth is narrow enough that the doppler frequency shift can be considered constant over the entire spectral width of the signal. With UWB systems, this assumption is no longer true. The variation of the doppler frequency over the signal's frequency spectrum has to be taken into account. For this reason, the time domain is more useful for UWB than is the frequency domain. The potential for extremely high resolution and its independence of doppler make area MTI attractive for UWB application. A particular UWB radar concept examined was a 30 percent bandwidth radar centered on *X* band. Its purpose was to detect sea-skimmer missiles flying at high speed, low over water, at ranges where they just came over the local radar horizon.[140] The minimum detectable velocity of this conceptual radar was 100 kt and its prf was 8 kHz. With the high resolution (5 cm) of a 3-GHz bandwidth pulse, the sea echo is from "sea spikes" which are nonstationary in both time and space, as discussed in Sec. 7.4. The ultrahigh resolution and the nonstationarity of the sea spikes allows detection of targets in the region where there is no clutter. (This is sometimes called *interclutter visibility*). When the target happens to coincide with a clutter spike, the area MTI cancels the clutter, but preserves moving target echoes. A serious limitation of an UWB area MTI radar, however, is its need for large allocations of the electromagnetic frequency spectrum.

Coherent on Receive The term *coherent on receive MTI* is sometimes used to describe an MTI radar that uses a power oscillator, such as the magnetron, as the transmitter rather than a power amplifier. Each time a power oscillator is pulsed, the phase of the pulse is random from pulse to pulse. To provide a coherent reference at the receiver so that the doppler signal may be detected, the phase of the coho (coherent oscillator) is set equal to the phase of the power oscillator each time the pulse is transmitted. That is, the phase of the coho is "locked" to the phase of the transmitter on each pulse, as was mentioned at the end of Sec. 3.1. This method of operation is sometimes called *coherent on receive* to distinguish it from an MTI that uses a power amplifier.

The amplifier MTI has been called *fully coherent* by some radar engineers to distinguish it from coherent on receive, but this is unnecessary nomenclature. Almost all modern high-performance MTI radars are "fully coherent" in the sense that they employ *I* and *Q* processing and a power amplifier transmitter, so the use of the term "fully coherent" is often superfluous.

The power oscillator MTI that employs a magnetron as the transmitter was the original form of MTI used for many years until the much more capable high-power amplifier became available. The power oscillator, however, is seldom used for high-performance radars. It finds application in those radars where high clutter attenuation is not required and where low cost is more important than high performance.

Adaptive MTI[141]   In this chapter adaptive processing in MTI, AMTI, and pulse doppler radar has already been mentioned. Each of the filters of the MTD doppler filter bank, for example, had an adaptive threshold that was set by the amount of clutter seen in each filter. Similarly, the filters of a pulse doppler radar also have individual thresholds whose level is determined by the clutter. In an AMTI radar, TACCAR adaptively adjusts the frequency of the reference oscillator (coho) to compensate for clutter whose doppler frequency is other than zero. STAP is also a form of adaptive processing that employs both the spatial (antenna) and temporal (doppler) domains.

The technique usually known as *adaptive MTI,* however, is different from these. It can be considered as an *n*-pulse delay-line canceler whose filter weights, and therefore its frequency response, are determined by the clutter seen by the radar. It is analogous to what is known as coherent sidelobe cancellation which places nulls in the radiation pattern of an antenna in the direction of unwanted noise sources such as jamming.[142] For example, if a four-pulse canceler is used, three nulls can be adaptively placed at three different frequencies depending on the nature of the clutter spectrum. If the clutter were concentrated at d-c, all three nulls would be placed at d-c. If the clutter were narrowband and in motion, the nulls would be placed at the doppler frequency of the clutter. When the clutter is rain or chaff, the spectrum can be broad and the nulls would be staggered in frequency to cover the spectral width of the clutter. The wider the clutter spectral width, the wider will be the adaptive filter notch, but it would not be as deep. If the clutter is bimodal in that both stationary surface clutter and moving weather clutter were present in the same resolution cell, the adaptive filter would place its nulls to cancel both as best it can. Adaptive MTI has been said to be potentially desirable when there is bimodal clutter or when the clutter characteristics change with range, angle, or time.

Although the adaptive MTI appears to have some attractive features, it has not seen much use. This might be due to the success of other methods, such as the MTD, for dealing with bimodal clutter. It might also be due in part to the difficulty in assuring that the adaptive MTI doesn't cancel the target echo along with the clutter.

Frequency Agility and MTI   Pulse-to-pulse frequency agility has the important benefit of increasing detectability of some targets, as when the pulse-to-pulse correlated echoes of a Swerling Case 1 target are decorrelated by frequency agility to produce a Swerling Case 2 fluctuating target echo. As mentioned in Sec. 2.8, a Swerling pulse-to-pulse fluctuating echo allows a lower detection threshold than one that is not. Frequency agility can also reduce the harmful effects of glint in a tracking radar and makes for more accurate target tracking (Sec. 4.4). In military radars, pulse-to-pulse frequency agility is important in causing a hostile jammer to spread its power over a wide bandwidth and not concentrate all its energy within the narrow bandwidth of a fixed-frequency radar receiver.

Unfortunately, frequency agility is not compatible with doppler processing. If the frequency is changed by an amount Δf, the phase of the echo return from a stationary point target at a range R is changed by $4\pi\Delta fR/c$. The change in phase with a change in frequency means there will be an uncanceled residue from the clutter if each pulse of the doppler processor is not of the same frequency. The effect is even worse when there are multiple clutter scatterers within the radar resolution cell. Thus frequency agility and doppler processing are not compatible.

One possible compromise that allows different frequencies to be used is to transmit multiple frequencies in close sequence (or pulse to pulse) over a relatively wide band.[143] For MTI purposes, however, the radar has to eventually retransmit these same frequencies. The order of transmission of the multiple frequencies can be varied so as to make fixed narrowband jamming difficult. If narrowband jamming takes out one or more of the radar frequencies, those frequencies can be blanked at the receiver.

An interesting method for using multiple frequencies with MTI radar was demonstrated with the Senrad experimental long-range air-surveillance radar by the Radar Division of the Naval Research Laboratory.[144] This radar operated in two frequency bands simultaneously: the normal L band from 1215 to 1400 MHz and the band from 850 to 942 MHz, sometimes called L_u. There were various waveforms used for this radar, but the clear-sky MTI waveform was a three-pulse MTI. The three MTI pulses have to be at the same frequency in order to perform MTI, which makes them vulnerable to narrowband electronic countermeasures if only a single MTI frequency is used. In Senrad, however, the three MTI waveforms are radiated on multiple frequencies in both sub-bands as the antenna scans by the target. Thus a three-pulse MTI waveform is simultaneously transmitted in both the lower and the upper sub-bands. As the antenna scans by the target there are six sets of three MTI pulses radiated in the lower sub-band and four sets of three MTI pulses in the upper sub-band, with each set of three pulses at a different frequency. (The lower sub-band provided more opportunities for putting pulses on the target because its beamwidth was wider than that of the upper sub-band.) The same antenna was used for both sub-bands, but a separate transmitter was used for each sub-band. A jammer would have to spread (dilute) its power to completely cover both sub-bands simultaneously. This can significantly reduce the amount of noise jamming that the radar encounters.

Clutter Decorrelation with Frequency Agility   Sometimes it has been claimed that frequency agility can improve the detection of stationary or slowly moving targets in clutter by decorrelating the clutter echoes by changing the radar frequency pulse to pulse. This is supposed to make the clutter echo look more like receiver noise rather than correlated clutter. Doppler processing is not employed. Integration of the signal and the pulse-to-pulse decorrelated clutter echoes can increase the signal-to-clutter ratio compared to when clutter is correlated pulse to pulse. This may be correct, but it is usually not a good method for detecting targets in clutter. Random or pseudorandom frequency hopping does not decorrelate the clutter effectively when the clutter statistics are not described by the Rayleigh probability density function (which it often isn't). Degradation can also occur if the frequency change is large enough to decorrelate the target echo as well as the clutter echo.

If one is given a fixed wide bandwidth to use to improve detectability of stationary targets in clutter, it is usually much better to use that bandwidth in a coherent fashion to radiate a narrow pulse or the equivalent pulse compression waveform rather than employ frequency agility. High range resolution increases the signal-to-clutter ratio by excluding clutter echoes outside the vicinity of the target. In this manner, one uses the available bandwidth to reduce clutter, rather than depend on the less efficient clutter-rejection method of a random-like hopping of a narrowband transmission over the same bandwidth. If the target is moving, however, doppler processing can provide greater detectability in clutter than can high range resolution.

Two-Frequency MTI   It has been proposed in the past that an MTI radar transmit two different frequencies and on receive operate with the difference frequency rather than either of the RF frequencies. The difference frequency, being much lower than either of the two RF frequencies, was thought to have higher blind speeds than if the radar operated on only one of the RF frequencies. If, for example, the two frequencies of the MTI radar were 9.3 and 9.4 GHz, the difference frequency would be 100 MHz and would have blind speeds 93 times that of the lower RF frequency. This possibility has attracted the attention of radar engineers more than once.

It is true that when two or more RF frequencies are used, the first blind speed can be increased just as it can when two or more prfs (pulse repetition frequencies) are used, as mentioned later in this subsection. There are some serious limitations, however, that negate the supposed advantages of two-frequency MTI when the difference frequency is used. One doesn't obtain the target scattering characteristics of a low-frequency radar by operating with the difference frequency. Also, the width of the clutter spectrum seen at the difference frequency is not that of a 100 MHz radar (in the above example), but is greater than the clutter spectrum seen with either RF frequency.[145] The potential blind speed advantage of a two-frequency MTI and its claimed ability to see slower targets are largely nullified by the increased clutter spectral width caused by the mixing process. Furthermore, there can be potential problems with spurious signals generated whenever nonlinear operations of multiple signals are performed as they are in this system to extract the difference frequency.

Meyer and Muehe[146] characterize the problem nicely when they say that ideas like the two-frequency MTI (and others) are nonsolutions that "crop up again and again, and for some reason, continue to attract attention. Much labor and money is used re-examining these ideas with the result that attention is diverted from the principal job." These nonsolutions are often reinvented by those not familiar with the past.

Delta-k Radar   This was another two-frequency radar with nonlinear processing that extracted the difference frequency, but for different purposes than the two-frequency MTI. It was promoted by remote sensing scientists, rather than by radar engineers, as providing information about the environment just as would a radar that radiated only the difference frequency. If, for example, an *X*-band radar radiated at 9.3 and 9.4 GHz and the 100-MHz difference signal was extracted at the receiver (by mixing the two *X*-band signals), the result was said to be the same as obtained by a 100-MHz radar. Assuming that the target dimensions were such that scattering was in the resonance region (Fig. 2.8) at 100 MHz (and thus had a large radar cross section), it was suggested that similar behavior would be obtained by extracting the 100 MHz difference frequency when two microwave frequencies were used. This does not happen.

The delta-*k* concept has also been proposed for investigation of sea clutter based on the Bragg scatter model. (It is mentioned in Sec. 7.4 that the Bragg model for sea clutter applies at HF and VHF, but not at the higher microwave frequencies.) The limited information obtained about the sea that is provided by the nonlinear processing of a delta-*k* radar with two (or a few) frequencies is suspect, and is overshadowed by the information that can be obtained by a radar with a short pulse.

Sometimes the delta-*k* radar used more than two frequencies. The ultimate would be to have many frequencies so as to obtain a filled spectrum, as with a single short pulse that does not use nonlinear processing. Nonlinear processing, as used in the delta-*k*, can degrade the information contained in a signal and generate spurious responses, especially when multiple signals are present or when signal-to-noise ratios are modest. Furthermore, the technology of short pulse radar is such that there does not seem to be any hardware advantages in using a delta-*k* radar, even if it did perform as its proponents hope. One should be very cautious, therefore, when considering using the delta-*k* or any other radar-like method that uses similar nonlinear processing.

Blind Speeds with Multiple Frequencies   There is a potentially useful application of two or more frequencies in MTI radar, but for a different purpose than the two-frequency MTI radar or delta-*k* radar discussed above. Just as two or more pulse repetition frequencies can be used to resolve doppler ambiguities or to increase the first blind speed in an MTI radar, two or more frequencies can be used instead of two or more prfs to achieve this purpose.[144] This follows from Eq. (3.12) which gives the first blind speed as $v_1 = \lambda f_p/2$. Instead of employing different f_p's, one can employ different λ's.

CW Radar   An entire chapter was devoted to CW radar in the earlier two editions of this text. It was omitted in this edition since, over time, the role of CW radar has changed. Although the use of CW radar to detect targets in clutter has been replaced by pulse doppler radar, there are still important applications of CW radar, as will be briefly discussed in this subsection.

Simple CW Radar   The block diagram of a simple CW radar that extracts the doppler frequency shift of a moving target was shown in Fig. 3.3a. This represents such CW radar applications as the familiar police speed meter,[147] the speed gun used in baseball and other sports, the artillery proximity fuze of World War II, artillery-projectile muzzle velocity measurement radar, radars for the docking of large ships, the airborne doppler navigator, and many other CW radars whose purpose is the simple noncontact measurement of velocity.[148] The relatively short-range CW radar has also been employed for vibration measurement, intruder detection, monitoring the respiration rate of humans and animals, miss-distance indication, gunfire detector, as a sensor for vehicle braking, and for the precision measurement of the ground speed[149] for both railway and automotive applications.

All of the above are short-range applications. When a CW radar is needed for long range, as for air defense, space surveillance, or ballistic missile detection, the simple CW configuration of Fig. 3.3a has several serious limitations. These include: (1) lack of isolation between the transmitter and receiver, which can cause receiver burnout if the transmitter power is large enough and/or introduce transmitter noise in the receiver which masks the detection of wanted targets, (2) introduction of flicker-effect noise because the receiver is a homodyne (zero IF frequency), (3) lack of a matched filter in the receiver, (4) lack of knowledge as to whether the target is approaching or receding, (5) increased clutter compared to pulse radars, and (6) no measurement of the range to the target.

Each of these limitations can be mitigated or sometimes eliminated. The long-range high-performance CW radar, however, even with improvements, usually is no longer competitive with a modern pulse doppler radar.

CW, however, has been and still is used with success for semiactive missile guidance because of its ability to discriminate against clutter on the basis of the doppler frequency shift, which makes it capable of operation at low altitudes.[150]

Isolation of Transmitter and Receiver   Reducing the leakage of the transmitter signal from the receiver can be achieved by using two antennas, one for receive and the other for transmit. They are physically separated so as to minimize their mutual coupling. Absorbing material and/or baffles between the two antennas further prevent the transmitter signal from entering the receiver. Also, a small sample of the transmitter signal can be directed to the receiver to adaptively cancel the portion of the transmitter signal that leaks into the receiving system via the inevitable coupling between the two antennas. Even if there were perfect isolation between transmitter and receiver, the transmitted signal can also enter the receiver via scattering from nearby clutter or other obstructions. Since the transmitter signal cannot be prevented from completely entering the receiver, it is important that it be as clean a signal as practicable (i.e., one with little accompanying noise).

Significant isolation for some CW radar applications can be obtained using only a single antenna by modulating the CW carrier with sinewave frequency modulation.[151] On reception, the difference signal (between the received signal and the transmitted signal) may be mathematically expressed in terms of Bessel functions. A harmonic of the modulation frequency is selected by filtering. Usually this has been the third harmonic ($3f_m$), which has an amplitude given by the Bessel function of third order, $J_3(D)$, where D is a function of the range. Since the higher order Bessel functions are zero at $D = 0$ (range = 0), the leakage signal is attenuated and isolation is achieved.

Reduction of Flicker-Effect Noise   The simple CW receiver of Fig. 3.3a is called a homodyne, or a superheterodyne with zero IF frequency. Semiconductor devices, including diodes and transistors, and vacuum tubes with oxide cathodes generate noise whose power is inversely proportional to frequency. It is known as $1/f$ noise or flicker-effect noise. It can be a concern at frequencies below about 100 kHz; hence, it will limit the sensitivity of a simple CW radar. Flicker-effect noise can be avoided by replacing the homodyne receiver with a superheterodyne whose IF frequency is large enough to make the flicker-effect noise negligible.

Matched Filter   The doppler filter in Fig. 3.3a uses a wide passband and is not a matched filter (one that maximizes the output signal-to-noise ratio). If T_d is the time duration of the CW signal processing, then an approximation to the matched filter is one whose bandwidth is $1/T_d$. A single tunable filter might be used in the doppler amplifier to search the entire frequency range over which the signal might appear. Time sharing a single tunable filter in this way can lead to missed targets simply because the filter is being time shared to operate over a wide frequency range. A filter bank doesn't have such limitations and is preferred. It approximates a matched filter implementation and simultaneously covers the entire expected doppler space so it will not miss targets as would a single scanning filter.

Direction of Target Motion   The simple CW radar loses the sign of the doppler in the mixing process shown in Fig. 3.3a so that one cannot tell whether the target is approaching or receding. The sign of the doppler, which gives the direction of target motion, can be obtained with a second processing channel whose reference signal is shifted by 90°, similar to the *I,Q* processing discussed in Sec. 3.5. The sign of the doppler may be determined by noting whether the output of the second channel leads or lags the output of the first by 90°.

Clutter in a CW Radar   The clutter seen by an MTI radar with a low prf is only that clutter within the range-angle resolution cell of the radar. A CW radar, on the other hand, sees clutter everywhere. Since the CW radar encounters more clutter than most other radars, it must have wider dynamic range receivers (compared to a low-prf MTI radar) and larger MTI improvement factors.

FM-CW Radar for the Measurement of Range   The determination of range requires that the CW waveform be marked in some manner so that the transit time out to the target and back can be measured. A popular method in CW radar is to linearly frequency modulate the waveform, as in Fig. 3.47a. The modulation is triangular (since what goes up eventually has to come down). The transmitted signal is shown by the solid triangular waveform. The frequency excursion Δf corresponds to the bandwidth of a pulse radar, and the frequency modulation at a rate f_m is the equivalent of the pulse repetition frequency of a pulse radar. The dashed curve represents the frequency of the received echo signal from a stationary target. It arrives back at the radar at a time $T = 2R/c$, where R = range to the target. The time-delayed received signal and the transmitted signal are multiplied in a mixer to produce a difference frequency f_r (in Fig. 3.47b). From the geometry of Fig. 3.47a, f_r can be shown to be $4Rf_m\Delta f/c$. If there is a doppler frequency shift f_d from the target, during half the modulation period the difference frequency is $f_r + f_d$ and during the

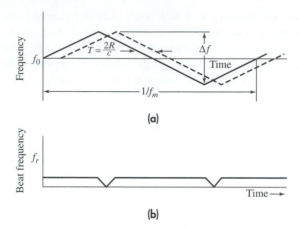

(a)

(b)

Figure 3.47   (a) Frequency-time relation in a FM-CW radar with linear triangular frequency modulation. Solid lines represent the transmitted signal, dash lines represent the receive signal delayed a time $T = 2R/c$. Δf = frequency excursion, f_m = modulation frequency. (b) Difference frequency between the transmitted and received signals.

other half of the modulation period it is $f_r - f_d$. By averaging these two difference frequencies over the period $1/f_m$ the target range can be obtained.

The FM-CW radar has been widely used in the past as an altimeter. In an altimeter there is only a single large target (the earth) and the range is not large so the radar can be relatively modest. (There is no clutter since the clutter is the target.) A similar application is the determination of distances for industrial purposes, such as the measurement of the distance to molten pig-iron in a blast furnace.

The FM-CW radar that uses the beat frequency between the transmitted and received signals to determine range does not work well when there are multiple targets. If the mixing process is not perfectly linear when multiple targets appear, there can be spurious signals generated. This may be all right for an altimeter when there is only one target, but it is not satisfactory for an air-defense radar. A better method for detecting signals and extracting range is to think of the FM-CW radar as a linear-FM pulse compression radar (discussed in Sec. 6.5) but with a unity duty cycle. The processing is with a matched filter in a superheterodyne receiver.

The U.S. Air Force, U.S. Navy (ROTHR), and the Australian (Jindalee) HF over-the-horizon radars employ FM-CW waveforms with extraction of the doppler frequency shift to separate clutter echoes from moving aircraft echoes. FM-CW radar works for this application, but with some serious penalties compared to a pulse radar. The FM-CW radar requires a greater receiver dynamic range, greater improvement factor, and two widely separated sites (to obtain isolation).

The FM-CW radar has been suggested for naval navigation radars. The chief advantage in that application is that its CW signal has lower peak power than does an equivalent pulse radar, thus making it more difficult (but not impossible) for a hostile electronic warfare receiver to intercept.[152]

Hawk Air-Defense System   The CW radar has also been used in the past for the detection of moving aircraft targets in clutter. The U.S. Army's Hawk air-defense system, whose concept dates back to the 1950s, was an example that was employed operationally by many countries of the world for many years. CW radar was the basis for the Hawk since the development of pulse doppler radar at the time of its development was not sufficiently advanced. This has changed. With the development of the high-prf pulse doppler radar, the long-range, high-power CW radar has been much less attractive. CW radar has the disadvantage that it requires two separate antennas for transmit and receive that have to be well isolated from one another. The pulse doppler radar needs only one antenna. The CW radar requires some method for measuring target range; the pulse doppler radar requires some method for resolving the ambiguities of the range it measures. The CW radar sees more clutter than a pulse doppler radar. A pulse doppler radar avoids seeing unwanted echoes from nearby clutter or reflections from other nearby objects since, unlike the CW radar, its receiver is turned off when the pulse is being radiated. The doppler processing employed in a high-prf pulse doppler radar signal after range gating is similar to the doppler processing of a CW radar. The pulse doppler radar is generally superior to the CW radar in several ways and has replaced it for the detection of aircraft in clutter.

Further information on CW radars can be found in either of the two previous editions of this text or in the *Radar Handbook*, published by McGraw-Hill.

REFERENCES

1. Shrader, W. W., and V. Gregers-Hansen. "MTI Radar." In *Radar Handbook,* 2nd ed., M. Skolnik, Ed. New York: McGraw-Hill, 1990, Chap. 15.

2. Long, W. H., D. H. Mooney, and W. A. Skillman. "Pulse Doppler Radar." In *Radar Handbook,* 2nd ed., M. Skolnik, Ed. New York: McGraw-Hill, 1990, Chap. 17.

3. Taylor, J. W., Jr. "Receivers." In *Radar Handbook,* 2nd ed., M. Skolnik, Ed. New York: McGraw-Hill, 1990, Chap. 3, Sec. 3.10.

4. IEEE Standard Radar Definitions, IEEE Std. 686-1997.

5. Andrews, G. A. "Airborne Radar Motion Compensation Techniques, Evaluation of TACCAR," Naval Research Laboratory Report 7407, Washington, D.C., Apr. 12, 1972.

6. Capon, J. "Optimum Weighting Functions for the Detection of Sampled Signals in Noise." *IEEE Trans.* IT-10 (April 1964), pp. 152–159.

7. Kretschmer, F. F. "MTI Weightings." *IEEE Trans.* AES-10 (January 1974), pp. 153–155.

8. Murakami, T., and R. S. Johnson. "Clutter Suppression by Use of Weighted Pulse Trains." *RCA Rev.* (September 1971), pp. 402–428.

9. Andrews, G. A., Jr. "Optimal Radar Doppler Processors." Naval Research Laboratory Report 7727, Washington, D.C., May 29, 1974.

10. Mao, Y. H. "MTI, MTD, and Adaptive Clutter Cancellation." In *Advanced Radar Techniques and Systems,* G. Galati, Ed. London: Peter Peregrinus, 1993, Sec. 6.6.2.

11. Houts, Y. C., and D. W. Burlage. "Maximizing the Usable Bandwidth of MTI Signal Processors." *IEEE Trans.* AES-13 (January 1977), pp. 48–55.

12. Rabiner, L. R., and C. M. Rader. *Digital Signal Processing.* New York: IEEE Press, 1972.

13. White, W. D. "Synthesis of Comb Filters." *Proc. Natl. Conf. on Aeronaut. Electronics,* 1958, pp. 279–285.

14. Shreve, J. S. "Digital Signal Processing." In *Radar Handbook,* 1st ed., M. Skolnik, Ed. New York: McGraw-Hill, 1970.

15. Ellis, J. G. "Digital MTI, A New Tool for the Radar User." *Marconi Rev.* 36 (1973), pp. 237–248, 4th qtr.

16. Fletcher, R. H., Jr., and D. W. Burlage. "An Initiation Technique for Improved MTI Performance in Phased Array Radars." *Proc. IEEE* 60 (December 1972) pp. 1551–1552. See also by the same authors: "Improved MTI Performance for Phased Array Radars in Severe Clutter Environments." *IEE Conference on Radar—Present and Future.* IEE Conference Publication No. 105, London, pp. 280–285, 1973.

17. Aalfs, D. D., E. G. Baxa, Jr., and E. M. Bracalente. "Signal Processing Aspects of Windshear Detection." *Microwave J.* 36 (September 1993), pp. 76–96.

18. Shrader, W. W., and V. Gregers-Hansen. Ref. 1, Sec. 15.9.

19. Zverev, A. I. "Digital MTI Radar Filters." *IEEE Trans.* AU-16 (September 1968), pp. 422–432.

20. Thomas, H. W., and T. M. Abram. "Stagger Period Selection for Moving Target Radars." *Proc. IEE* 123 (March 1976), pp. 195–199.

21. Hsiao, J. K., and F. F. Kretschmer, Jr. "Design of a Staggered-PRF Moving Target Indication Filter." *The Radio and Electronic Engineer* 43 (November 1973), pp. 689–693.

22. Thomas, H. W., and T. M. Abram. "Stagger Period Selection for Moving-Target Radars." *Proc. IEE* 123 (March 1976) pp. 195–199.

23. Ewell, G. W., and A. M. Bush. "Constrained Improvement MTI Radar Processors." *IEEE Trans.* AES-11 (September 1975), pp. 768–780.

24. Cleetus, G. M. "Properties of Staggered PRF Radar Spectral Components." *IEEE Trans.* AES-12 (November 1976), pp. 800–803.

25. McAulay, R. J. "The Effect of Staggered PRF's on MTI Signal Detection." *IEEE Trans.* AES-9 (July 1973), pp. 615–618.

26. Wardrop, B. "The Performance of MTI Systems When Used with P.R.F. Stagger." *The Marconi Review* 4th quarter, 1974, pp. 217–231.

27. Shrader, W. W., and V. Gregers-Hansen. Ref. 1, pp. 15–34 to 15–35.

28. Ludloff, A., and M. Minker. "Reliability of Velocity Measurement by MTD Radar." *IEEE Trans.* AES-21 (July 1985), pp. 522–528.

29. Andrews, G. A., Jr. "Performance of Cascaded MTI and Coherent Integration Filters in a Clutter Environment." *Naval Research Laboratory Report* 7533, Washington. D.C., March 27, 1973.

30. Shrader, W. W., and V. Gregers-Hansen. Ref. 1, Sec. 15.8.

31. Ward, H. R. "Properties of Dolph-Chebyshev Weighting Functions." *IEEE Trans.* AES-9 (September 1973), pp. 785–786.

32. Gregers-Hansen, V. "Optimum Pulse Doppler Search Radar Processing and Practical Approximations." *International Conference Radar—82* IEE Conference Publication No. 216 (October 1982), pp. 138–143, 18–28.

33. Mao, Y. H. "Architectures and Implementation of Radar Signal Processor." In *Advanced Radar Techniques and Systems,* G. Galati, Ed. London: Peter Peregrinus, 1993, Chap. 8.

34. Nathanson, F. E., and P. J. Luke. "Loss from Approximations to Square-law Detectors in Quadrature Systems with Postdetection Integration." *IEEE Trans.* AES-8 (January 1972), pp. 75–77.

35. Shrader, W. W., and V. Gregers-Hansen. Ref. 1, Sec. 15.12.

36. Shrader, W. W., and V. Gregers-Hansen. Ref. 1, Sec. 15.11.

37. Brennan, L. E., and I. S. Reed. "Quantization Noise in Digital Moving Target Indication Systems," *IEEE Trans.* AES-2 (November 1966), pp. 655–658.

38. Taylor, J. W., Jr. Ref. 3, Sec. 3.12.

39. Churchill, F. E., G. W. Gar, and B. J. Thompson. "The Correction of I and Q Errors in a Coherent Processor." *IEEE Trans.* AES-17 (January 1981) pp. 131–137.

40. Suresh Babu, B. N., and C. M. Sorrentino. "Analogue-to-Digital Convertor Effects on Airborne Radar Performance." *IEE Proc.-F* 139 (February 1992), pp. 73–78.

41. Rader, C. M. "A Simple Method for Sampling in-Phase and Quadrature Components," *IEEE Trans.,* vol. AES-20 (November, 1984), pp. 821–824.

42. Mitchell, R. L. "Creating Complex Signal Samples From a Band-Limited Real Signal," *IEEE Trans.,* vol. AES-25 (May, 1989), pp. 425–427.

43. Waters, W. M., and B. R. Jarrett. "Bandpass Sampling and Coherent Detection," *IEEE Trans.,* vol. AES-18 (November 1982), pp. 731–736.

44. Taylor, J. W., Jr. "Sacrifices in Radar Clutter Suppression Due to Compromises in Implementation of Digital Doppler Filters." *International Conference Radar—82,* Institution of Electrical Engineers (London), pp. 46–50, 1982.

45. Nathanson, F. E. *Radar Design Principles,* 2nd ed. New York: McGraw-Hill, 1991, Sec. 9.1.

46. Weiss, M., and I. Gertner. "Loss in Single-Channel MTI with Post-Detection Integration." *IEEE Trans.* AES-18 (March 1982), pp. 205–208.

47. Muehe, C. E. "Digital Signal Processor for Air Traffic Control Radars." *IEEE NEREM 74 Record,* Part 4: Radar Systems and Components. (October 28–31, 1974), pp. 73–82.

48. Drury, W. H. "Improved MTI Radar Signal Processor." Report no. FAA-RD-74-185, MIT Lincoln Laboratory, April 3, 1975.

49. O'Donnell, R. M., et al. "Advanced Signal Processing for Airport Surveillance Radars." *IEEE EASCON 77,* Washington, D.C. (October 1974), pp. 71A–71F.

50. Drury, W. H. "Improved MTI Radar Signal Processor." Federal Aviation Agency Report No. FAA-RD-74-185, 3 April 1975.

51. Muehe, C. E. "Digital Signal Processors for Air Traffic Control Radars." *IEEE NEREM 74 Record,* Part 4: Radar Systems and Components. IEEE Catalog no. 74 CHO 934-0 (1974), pp. 73–82.

52. Muehe, C. E. "Advances in Radar Signal Processing." *IEEE Electro '76,* Boston, MA, May 11–14, 1976.

53. O'Donnell, R. M., and C. E. Muehe. "Automated Tracking for Aircraft Surveillance Radar Systems." *IEEE Trans.* AES-15 (July 1979), pp. 508–517.

54. Taylor, J. W., Jr. "Sacrifices in Radar Clutter Suppression Due to Compromises in Implementation of Digital Doppler Filters." *International Conference Radar-82,* IEEE Conference Publication No. 216 (October 18–20, 1982), pp. 46–50.

55. Taylor, J. W., Jr., and G. Brunins. "Design of a New Airport Surveillance Radar (ASR-9)." *Proc. IEEE* 73, (February 1985), pp. 284–289.

56. Cole, E. L., M. J. Hodges, R. G. Oliver, and A. C. Sullivan. "Novel Accuracy and Resolution Algorithms for the Third Generation MTD." *IEEE 1986 National Radar Conference Proceedings,* pp. 44–47, IEEE Catalog no. 86CH2270-7.

57. Cole, E. L., et al. "ASR-12: A Next Generation Solid State Air Traffic Control Radar." *Proc. 1998 IEEE Radar Conference,* Dallas, Texas, May 11–14, 1998, pp. 9–14.

58. Shrader, W. W., and V. Gregers-Hansen. Ref. 1, Sec. 15.5.

59. Goldstein, H. "The Fluctuations of Clutter Echoes." *Propagation of Short Radio Waves,* D. E. Kerr, Ed. MIT Radiation Lab Series, vol. 13, New York: McGraw-Hill, 1951, pp. 550–587.

60. Barlow, E. J. "Doppler Radar." *Proc. IRE* 37 (April 1949), pp. 340–355.

61. Barton, D. K. *Modern Radar Analysis.* Norwood, MA: Artech House, 1988, p. 246.

62. Nathanson, F. E. Ref. 45, Fig. 7.32.

63. Fishbein, W., S. Graveline, and O. R. Rittenbach. "Clutter Attenuation Analysis." *Technical Report ECOM-2808, U. S. Army Electronics Command, Ft. Monmouth, NJ,* March, 1967. Reprinted in Schleher, D. C. *MTI Radar.* Dedham, MA: Artech House, 1978.

64. Li Neng-jing. "A Study of Land Clutter Spectrum." *Proc. Second International Symposium on Noise and Clutter Rejection in Radars and Imaging Sensors.* T. Suzuki, H. Ogura, and S. Fujimura, Eds. IEICE, Elsevier, North Holland, 1990, pp. 48–53.

65. Currie, N. C., and C. E. Brown. *Principles and Applications of Millimeter Wave Radar.* Norwood, MA: Artech House, 1987, Table 5.4.

66. Billingsley, J. B. "Exponential Decay in Windblown Radar Ground Clutter Doppler Spectra: Multifrequency Measurements and Model." *MIT Lincoln Laboratory Technical Report 997* July 29, 1996.

67. Billingsley, J. B. "Ground Clutter Measurements for Surface-Sited Radar." *MIT Lincoln Laboratory Technical Report* 786 (Revision 1), February 1, 1993.

68. Shrader, W. W., and V. Gregers-Hansen. Ref. 1, Sec. 15.11.

69. Nathanson, F. E. Ref. 45, Sec. 9.7.

70. Everett, C. L. "Phase Noise Contamination to Doppler Spectra." *Microwave J.* 39 (September 1996), pp. 105–122.

71. Ewell, G. W. "Stability and Stable Sources." In *Coherent Radar Performance Estimation.* J. A. Scheer and J. L. Kurtz, Eds. Norwood, MA: Artech House, 1993, Chap. 2.

72. Bloomfield, D. L. H. "Low-Noise Microwave Sources." *IEE International Conference on Radar—Present and Future,* London, IEE Conference Publication No. 105, pp. 178–183, October 23–25, 1973.

73. Khanna, A. P. S., M. Schmidt, and R. B. Hammond. "A Superconducting Resonator Stablized Low Phase Noise Oscillator." *Microwave J.* 34, (February 1991), pp. 127–130.

74. Mage, J. C., B. Marcilhac, P. Hartemann, and J. P. Castera. "Low Phase Noise Oscillator for Stealth Target Detection." *International Conference on Radar,* Paris, May 3–6, 1994, pp. 202–206.

75. Filler, R. L., and J. R. Vig. "Low-Noise Oscillators for Airborne Radar Applications." *Army Research Laboratory, Ft Monmouth, Research and Development Technical Report* SLCET-TR-91-26 (Rev. 1), August 1993. DTIC AD-A269 372.

76. Almar, R. C, and M. S. Cavin. "Low g-Sensitivity Fixed-Frequency Oscillators." *Microwave J.* 38 (February 1995), pp. 88–98.

77. Kerr, R. R. "MTI Systems." In *Coherent Radar Performance Estimation.* J. A. Scheer and J. L. Kurtz, Eds. Norwood, MA: Artech House, 1993, Chap. 8. Sec. 8.5.

78. Ward, H. R., and W. W. Shrader. "MTI Performance Caused by Limiting," *EASCON '68 Record.* Supplement to *IEEE Trans.* AES-4, pp. 168–194, November 1968.

79. Shrader, W. W., and V. Gregers-Hansen. Ref. 1, Sec. 15.10.

80. Ridenour, L. N. *Radar System Engineering.* MIT Radiation Laboratory Series 1, New York: McGraw-Hill, 1947, Sec. 16.8.

81. Shrader, W. W., and V. Gregers-Hansen. Ref. 1, Sec. 15.14.

82. Nitzberg, R. "Clutter Map CFAR Analysis." *IEEE Trans.* AES-22 (July 1986), pp. 419–421.

83. Dickey, F. R., Jr., M. Labitt, and F. M. Staudaher. "Development of Airborne Moving Target Radar for Long Range Surveillance." *IEEE Trans.* AES-27 (November 1991), pp. 959–972.

84. Staudaher, F. M. "Airborne MTI." In *Radar Handbook.* 2nd ed. M. I. Skolnik, Ed. New York: McGraw-Hill, 1990, Chap. 16, Sec. 16.3.

85. Shrader, W. W., "MTI Radar." In *Radar Handbook,* 1st ed., M. I. Skolnik, Ed. New York: McGraw-Hill, 1970, Chap. 17, Sec. 17.9.

86. Shrader, W. W. and V. Gregers-Hansen. Ref. 1, Sec. 15.13.

87. Staudaher, F. M. Ref. 81, Sec. 16.4.

88. Grisetti, R. S., M. M. Santa, and G. M. Kirkpatrick. "Effect of Internal Fluctuations and Scanning on Clutter Attenuation in MTI Radar." *IRE Trans.* ANE-2 (March 1955), pp. 37–42.

89. Anderson, D. B. "A Microwave Technique to Reduce Platform Motion and Scanning Noise in Airborne Moving-Target Radar." *IRE Wescon Conv. Rec.* 2 (1958), pt. 1, pp. 202–211.

90. Staudaher, F. M. Ref. 84, Secs. 16.6 and 16.6.

91. Staudaher, F. M. Ref. 84, p. 16–13.

92. Farina, A. *Antenna-Based Signal Processing Techniques for Radar Systems.* Boston, MA: Artech House, 1992.

93. Ghose, R. N. *Interference Mitigation.* New York: IEEE Press, 1996.

94. Gabriel, W. F. "Adaptive Arrays—An Introduction." *Proc. IEEE* 64 (February 1976), pp. 239–272.

95. Brennan, L. E., and I. S. Reed. "Theory of Adapative Radar." *IEEE Trans.* AES-9 (March 1973), pp. 237–252.

96. Kretschmer, F. F., Jr., B. L. Lewis, and F-L C. Lin. "Adaptive MTI and Doppler Filter Bank Clutter Processing." *Proc. 1984 IEEE National Radar Conf.* IEEE Cat. No. 84CH1963-8, pp. 69–73.

97. Shrader, W. W., and V. Gregers-Hansen. Ref. 1, Sec. 15.13.

98. Brennan, L. E., J. D. Mallett, and I. S. Reed. "Adaptive Arrays in Airborne MTI Radar." *IEEE Trans.* AP-24 (September 1976), pp. 607–615.

99. Staudaher, F. M. Ref. 84, Sec. 16.8.

100. Ward, J. "Space-Time Adaptive Processing for Airborne Radar." MIT Lincoln Laboratory, Lexington, MA, Technical Report 1015, December 13, 1994.

101. Perkins, L. C., D. H. Mooney, and H. B. Smith. "The Development of Airborne Pulse Doppler Radar." *IEEE Trans.* AES-20 (May 1984), pp. 292–303.

102. Cowdery, R. E., and W. A. Skillman. "Development of the Airborne Warning and Control System (AWACS) Radar." *IEEE Trans.* AES-31, (October 1995), pp. 1357–1356.

103. Shrank, H. E. "Low Sidelobe Phased Array Antennas." *IEEE Antennas and Propagation Society Newsletter* 25, no. 2 (April 1983), pp. 5–9.

104. Goetz, L. P., and J. D. Albright. "Airborne Pulse Doppler Radar." *IRE Trans. MIL-5* (April 1961) pp. 116–126. Reprinted in *Radars, Vol. 7, CW and Doppler Radar.* Dedham, MA: Artech House, 1978.

105. Mooney, D. H., and W. A. Skillman. "Pulse-doppler Radar." In *Radar Handbook,* 1st ed. M. Skolnik, Ed. New York: McGraw-Hill, 1970, Chap. 19 p. 19–5.

106. Long, W. H., D. H. Mooney, and W. A. Skillman. "Pulse-Doppler Radar." In *Radar Handbook,* M. Skolnik, Ed. New York: McGraw-Hill, 1990, Chap. 17, p. 17.34

107. Stimson, G. W. *Introduction to Airborne Radar,* 2nd ed., Medham, NJ: Scitech Publishing, 1998, pp. 376–378.

108. Long, W. H., et al. Ref. 106, Sec. 17.4.

109. Hovanessian, S. A. "An Algorithm for Calculation of Range in Multiple PRF Radar." *IEEE Trans.* AES-12 (March 1976), pp. 287–289.

110. Anonymous. "AWACS vs. E2C Battle a Standoff." *EW Magazine* (May/June 1976), p. 31.

111. Morris, G. V. *Airborne Pulsed Doppler Radar.* Norwood, MA: Artech House, 1988, Sec. 6.3.

112. Morris, G. V. Ref. 108, Chaps. 10 and 11.

113. Skolnik, M. I. *Radar Applications.* New York: IEEE Press, 1988, Part 4. Airborne radar.

114. Carpentier, M. H. "Pulse Doppler Radars." In *Advanced Radar Techniques and Systems,* G. Galati, Ed. London: Peter Peregrinus, 1993, Chap. 5, Sec. 5.3.

115. Bath, W. G., and F. R. Castella. "Detection Performance of a Noncoherent MTI." *Proc. 1984 IEEE National Radar Conference,* Atlanta, GA, pp. 74–78.

116. Staudaher, F. E. Ref. 84, Sec. 16.11.

117. Skolnik, M. I. Ref. 113, p. 244.

118. Nevin, R. L., and F. W. Schatz. "AN/APG-67 Multimode Radar Development." Presented at *IEEE 1985 International Radar Conf.* (but not in the convention record). May be found as paper 4.6 in M. Skolnik, *Radar Applications,* Ref. 113.

119. Kennedy, P. D. "FFT Signal Processing for Non-Coherent Airborne Radars." *Proc. 1984 IEEE National Radar Conference,* pp. 79–83.

120. Strickland, P. C. "Multiprocessor Architecture Gives SLAR New Features." *Defense Electronics* (February 1987), pp. 92–101.

121. Raney, R. K. "Synthetic Aperture Imaging Radar and Moving Targets." *IEEE Trans.* AES-7 (May 1971) pp. 499–505.

122. Zebker, H. A., T. G. Farr, R. P. Salazar, and T. H. Dixon. "Mapping the World's Topography Using Radar Interferometry. The TOPSAT Mission." *Proc. IEEE* 82 (December 1994), pp. 1774–1786.

123. Bamler, R., and P. Hartl. "Synthetic Aperture Radar Interferometry." *Inverse Problems* 14 (August 1998), pp. R1–R54.

124. Goldstein, R. M., T. P. Barnett, and H. A Zebker. "Remote Sensing of Ocean Currents." *Science* 246 (December 8, 1989), pp. 1282–1285.

125. Anonymous. Joint Stars, Advertising Supplement to *Jane's Defense Weekly* (September 5, 1992).

126. Shnitkin, H. "Joint Stars Phased Array Radar Antenna." *IEEE AES Systems Magazine* 9 (October 1994), pp. 34–41.

127. The information in this paragraph and its wording were graciously supplied by Marshall Greenspan and his colleagues at Northrop Grumman Norden Systems. It is used here with only minor modification.

128. Tang, C. H. "Engineering Analyses Associated with the Development of an Airborne Phased Array Radar Antenna." *Antenna Applications Symposium.* Allerton Park, Monticello, IL, September 25–27, 1991.

129. DiDomizio, R., et al. "Dual Cancellation Interferometric AMTI Radar," U. S. Patent 5,559,516, September 24, 1996.

130. Tobin, M. E., and M. Greenspan. "Adaptation of AN/APG-76 Multimode Radar to the Smuggling Interdiction Mission." *Proc. 1996 IEEE National Radar Conference* (May 13–16, 1996), pp. 13–18. Reprinted in *IEEE AES Systems Magazine* (November 1996).

131. Anonymous. "Norden Offering 3-D SAR System for Moving Target Detection." *Defense Electronics* (November 1993), p. 18.

132. Hall, W. M., and H. R. Ward. "Signal-to-Noise Loss in Moving Target Indicator." *Proc. IEEE* 56 (February 1968), pp. 233–234.

133. Kretschmer, F. K., Jr. "Correlation Effects of MTI Filters." *IEEE Trans.* AES-13 (May 1977), pp. 321–322.

134. Trunk, G. V. "MTI Noise Integration Loss." *Proc. IEEE* 65 (November 1977), pp. 1620–1621.

135. Muller, B. "MTI Loss with Coherent Integration of Weighted Pulses." *IEEE Trans.* AES-17 (July 1981), pp. 549–552.

136. Schleher, D. C. *MTI Radar.* Dedham, MA: Artech House, 1978, Sec. 3, Pulse-Burst MTI Radars.

137. Nathanson, F. E. Ref. 45, Chap. 11.

138. Zeoli, G. W. "Some Results on Pulse-Burst Radar Design." *IEEE Trans.* AES-7 (May 1971), pp. 486–498.

139. Cantrell, B. H. "A Short-Pulse Area MTI." *Naval Research Laboratory Report* 8162 (September 22, 1977).

140. Skolnik, M. I., G. Andrews, and J. P. Hansen. "An Ultrawideband Microwave-Radar Conceptual Design." *Record of the IEEE 1995 International Radar Conference* (May 8–11, 1995), pp. 16–21. IEEE Catalog No. 95-CH-3571-0.

141. Lewis, B. L., F. F. Kretschmer, Jr., and W. W. Shelton. *Aspects of Radar Signal Processing.* Norwood, MA: Artech House, 1986.

142. Howells, P. W. "Explorations in Fixed and Adaptive Resolution at GE and SURC." *IEEE Trans.* AP-24 (September 1976) pp. 575–584, Sec. 4.3.

143. Xiaojun, Y., L. Yongtan, and D. Fengzeng. "A Compatible Method for Frequency Agility and MTI Operation." *Proc. 1989 International Symposium on Noise and Clutter Rejection in Radars and Imaging Sensors,* IEICE, pp. 543–547, 1989.

144. Skolnik, M. "Improvements for Air-Surveillance Radar." *Proc. 1999 IEEE Radar Conference,* Waltham, MA, April 20–22, 1999, pp. 18–21.

145. Hsiao, J. K. "Analysis of a Dual Frequency Moving Target Indication System." *The Radio and Electronic Engineeer* 45 (July 1975), pp. 351–356.

146. Meyer, J. W., and C. E. Muehe. "Report of a Survey of Airborne Moving Target Indicator Radar Systems," *MIT Lincoln Laboratory Technical Note* 1970–14, 28 April 1970, AD 509740, p. 20.

147. Nichols, R. E., Jr. *Police Radar.* published by Springfield, IL: C. C. Thomas, 1982.

148. Heide, P., R. Schubert, V. Magori, and R. Schwarte. "A High Performance Multisensor System for Precise Vehicle Ground Speed Measurement." *Microwave J.* 39 (July 1996), pp. 22–34.

149. Skolnik, M. *Radar Applications.* New York: IEEE Press, 1988, Sec. 7.1.

150. Ivanov, A. "Radar Guidance of Missiles." In *Radar Handbook.* M. Skolnik, Ed. New York: McGraw-Hill, 1990, Chap. 19, Sec. 19.2.

151. Saunders, W. K. "Post-War Developments in Continuous-Wave and Frequency-Modulated Radar." *IRE Trans.* ANE-8 (March 1961), pp. 7–19.

152. Stove, A. G. "Linear FMCW Radar Techniques." *IEE Proc.-F* 139 (October 1992), pp. 343–350.

PROBLEMS

3.1 A satellite orbiting the earth in a circular orbit at an altitude of 5000 nmi has a speed of 2.7 nmi/s. (a) What is the doppler frequency shift if the satellite is observed by a ground-based UHF radar (450 MHz) lying in the plane of orbit, just as the satellite appears over the horizon? (The radius of the earth is 3440 nmi. You may ignore the effects of refraction by the earth's atmosphere and reflection from the earth's surface.) (b) What is the doppler frequency shift when the satellite is observed at the zenith?

3.2 A VHF radar at 220 MHz has a maximum unambiguous range of 180 nmi. (a) What is its first blind speed (in knots)? (b) Repeat, but for an *L*-band radar at 1250 MHz. (c) Repeat, but for an *X*-band radar at 9375 MHz. (d) What would be the unambiguous range (nmi) of the *X*-band radar of part (c) in order to give the same blind speed you found in part (a) for the VHF radar? (e) If you needed to have a radar with the first blind speed of the VHF radar of part (a), would you rather have the VHF radar of part (a) or the *X*-band radar of part (d)? Please explain your answer (There might not be a unique answer.)

3.3 An *L*-band radar (1250 MHz) has a prf of 340 Hz. It detects a rainstorm moving at a radial velocity of 12 kt. Assume that the width of the thunderstorm's doppler spectrum is very small (a narrow spectral line; which is not, of course, reality but it makes the problem easier). The radar employs a single delay-line canceler. (a) How much does the single delay-line canceler attenuate (in dB) the storm echo compared to the response if the storm were moving with a radial velocity corresponding to the velocity which gives the maximum filter response? (b) What would be the attenuation of the storm compared to the maximum response, if a double delay-line canceler were used?

3.4   (a) Show that the product of the maximum unambiguous range R_{un} and the first blind speed v_1 is equal to $c\lambda/4$, where c = velocity of propagation and λ = radar wavelength. (b) What guidance, if any, does this relation give for avoiding ambiguities?

3.5   What is the highest frequency that a radar can be operated if it is required to have a maximum unambiguous range of 200 nmi and no blind speeds less than 600 kt.

3.6   Show that a triple delay-line canceler is equivalent to a four-pulse delay-line canceler with weights equal to the coefficients of the binomial expansion with alternating sign.

3.7   (a) Derive the expression for the ratio v_1/v_B, where v_1 is the first blind speed of a staggered prf with N different prfs, and v_B is the first blind speed obtained with a constant prf waveform equal to the average of the N staggered prfs. (b) What is the ratio v_1/v_B when $N = 4$ and the prfs are related as 30:35:32:36?

3.8   (a) How can the transmission of N constant-prf radar waveforms, each at a different RF frequency, be used to avoid blind speeds? (b) Derive an expression for v_1/v_{cf}, where v_1 is the first blind speed when N different RF frequencies are transmitted, all at the same prf, and v_{cf} is the blind speed when only a single RF frequency is transmitted equal to the average of the N RF-frequencies. (c) Is there any advantage in changing both the prf and the RF frequency to avoid blind speeds?

3.9   An S-band (3.1 GHz) air-surveillance radar utilizes a staggered waveform with four different prfs, which are 1222, 1031, 1138, and 1000 Hz.
   a. What is the first blind speed (knots) if a constant prf is used which has a pulse repetition period equal to the average of the four periods of the staggered waveform?
   b. What is the first blind speed (knots) of the staggered prf waveform? Note that the n_i for these four frequencies are 27, 32, 29, 33 respectively.
   c. What is the maximum unambiguous range of the staggered prf waveform?
   d. What is the depth (dB) of the first null of the staggered prf waveform?
   e. What is the maximum MTI improvement factor for the staggered prf waveform, assuming a gaussian clutter spectrum with a standard deviation of 10 Hz?

3.10   (a) What is the first blind speed (knots) of an L-band radar (1250 MHz) when the prf has a maximum unambiguous range of 240 nmi? (b) Determine the periods of a pulse-to-pulse staggered MTI waveform with three different periods, for the purpose of increasing the radar's first blind speed found in (a) to a value no less than 1200 kt. The maximum unambiguous range of the three periods is to be no less than 240 nmi. (Note: There is no unique answer for this part. In practice the choice of the three periods also should be selected to achieve an acceptable null depth and a desired improvement factor, something beyond the scope of this particular problem.)

3.11   (a) In a digital filter bank with 16 filters, what phase increment (degrees) is required for the phase shifts at each of the 16 taps of the delay line so as to generate the filter that is adjacent to the zero doppler filter? (b) What is the null-width (the distance between the two nulls that define the main response) of the filters if the pulse repetition frequency of the radar is 2560 Hz? (c) If a four-pulse canceler precedes the 16-tap delay line of the filter bank, how many pulses are there in a coherent processing interval?

3.12 Consider an MTD processor with a bank of eight contiguous doppler filters installed in an airport surveillance radar operating at S band (2.8 GHz). The MTD uses two different pulse repetition frequencies to unmask a moving target whose echo signal might be in the same doppler filter as moving weather clutter. Let one of the two prfs be 1100 Hz. Weather clutter is assumed to have a spectrum of radial velocities extending from 0 to 25 kt. An aircraft flying at a radial velocity of 250 kt is aliased in doppler and falls into the same doppler filter (near the middle of filter no. 2—please verify this) as the weather clutter. It is masked by the clutter and not detected. (Note that filter no. 1 is designated as the filter centered at zero frequency.) (a) What should be the second prf (smaller than the first, so as not to decrease the maximum unambiguous range) in order to move the aliased aircraft velocity completely outside of the main response of filter no. 2 and into the middle of filter no. 4? (b) What percent change has been made in the first prf to arrive at the second prf? (c) Instead of changing the prf to unmask the target, how much should the RF frequency be changed in order to unmask the target (by shifting it to the middle of the fourth filter) when the prf is kept at 1100 Hz?

3.13 An MTI radar with a single delay-line canceler has the following characteristics:

       frequency = 3000 MHz

       azimuth beamwidth = 1.2 degrees

       antenna rotation rate = 10 rpm

       prf = 1000 Hz

       A/D converter with 8 bits quantization

       stalo phase stability = 0.6 degree

       clutter standard deviation, σ_v = 0.3 m/s

  a. Determine which of the following is the major limitation to the overall MTI improvement factor for this radar:

     •   stability of the stalo (due to phase changes)

     •   clutter internal motion

     •   antenna scanning modulation

     •   A/D converter noise

  b. What is the overall improvement factor (dB) for this radar?

3.14 A radar is to have a total improvement factor of 45 dB. Its frequency is 3.0 GHz, the prf is 340 Hz, and it uses a single delay-line canceler. Assume that the four factors asked for below in (a) through (d) are independent of one another and that their contributions to the overall improvement factor are allocated equally.

  a. What must be the overall phase stability (degrees)?

  b. What must be the relative amplitude stability (percent)?

  c. What is the minimum number of bits required for the A/D converter?

d. What is the maximum permissible rms velocity spread (m/s) of the clutter fluctuations?

e. What should be the limit level of the receiver?

3.15 (a) What is the limitation to the improvement factor due to antenna scanning modulation (finite time on target) with a radar that has a beamwidth of 1.5 degrees, prf of 340 Hz, and antenna scan rate of 6 rpm when a single delay-line canceler is used? (b) What two *basic* things can the radar systems designer consider to increase to at least 40 dB the limitation to the improvement factor caused by antenna scanning? (c) Which of your two methods in part (b) do you think is the better option?

3.16 In a rotating reflector antenna with dimension D, what is the relationship between the spread in doppler frequency due to the finite time on target and the doppler frequency shift from the tip (end) of the rotating antenna? (Assume that the beamwidth of the antenna is $\theta_B = \lambda/D$ radians.)

3.17 (a) Assume the MTI improvement factor is determined only by the internal motion of the clutter. How much will the improvement factor be decreased (in dB) if the frequency of a radar using a three-pulse canceler were changed from 430 MHz (UHF) to 3.3 GHz (S band)? (b) How much should the prf of the S-band radar be increased to make its improvement factor equal that of the UHF radar?

3.18 If internal motion of the clutter caused by the wind were the only factor affecting the clutter spectrum, what would be the improvement factor at a frequency of 10 GHz for a breezy wind (9 kt), prf of 1000 Hz, and a double delay-line canceler, when the clutter spectrum is modeled by an exponential clutter spectrum? (Be careful of units.)

3.19 Show that the spread in the AMTI radar clutter spectrum due to movement of the radar (platform motion), which is given by Eq. (3.58) as $\Delta f_c = \dfrac{2v}{\lambda}\, \theta_B \sin \theta$, is independent of frequency, assuming both the velocity v and the antenna dimension D are constant.

3.20 Assuming an airborne air-surveillance (AMTI) radar flying at a speed of 300 kt with a rotating fan-beam antenna whose horizontal dimension $D = 24$ ft, plot the doppler shift f_c of the ground clutter echo and the spread in doppler shift Δf_c as a function of the azimuth angle (from 0 to 180 degrees) for frequencies of 420 MHz and 3.5 GHz. (OK to assume the elevation angle is zero in this problem and that the azimuth beamwidth in degrees is $\theta_B = 65\lambda/D$, where λ is the wavelength.)

3.21 Consider an AMTI radar with a frequency of 440 MHz flying in an aircraft at a speed of 320 kt. Its azimuth beamwidth is 6 degrees and its prf is 330 Hz. (a) What is the doppler frequency of the clutter echo and the spread in the clutter doppler at azimuth angles of 0, 45, and 90 degrees, where 0 is head-on and 90 is broadside? (You may assume that the elevation angle is zero, which is unrealistic of course, but it makes the problem simpler.) (b) Assume that TACCAR is applied so that the doppler clutter frequency is fully compensated along the center of the main beam (that is, the center frequency of the clutter doppler spectrum is at zero doppler frequency). DPCA is not applied. Sketch the doppler space (the resulting clutter spectrum as a function of doppler frequency) for the case where the radar antenna is pointing broadside at 90 degrees azimuth angle. For this problem you may assume that σ_c is the same as $\Delta f_c/2$. (Draw approximately to scale along the frequency axis.) (c) What is the value of σ_c/f_p, when the antenna is pointing broadside as it is in part (b), where f_p = prf and σ_c = standard deviation of the clutter spectrum which can be approximated here by $\Delta f_c/2$? (d) How well do you think a radar of this type detects moving aircraft targets in clutter?

3.22   (a) Why does a high-prf pulse doppler radar require a much larger improvement factor than a low-prf MTI radar, assuming comparable performance in detecting moving targets in clutter? (b) Why does a high-prf pulse doppler radar (such as AWACS) generally need more average power than an AMTI radar of comparable performance? (c) Why does a high-prf pulse doppler radar not need DPCA as does an AMTI radar?

3.23   (a) What does a medium-prf pulse doppler radar do better than a high-prf pulse doppler radar? (b) What does a high-prf radar do better than a medium-prf pulse doppler radar?

3.24   Why can't the altitude line in a high-prf pulse doppler radar be eliminated by range gating rather than by filtering?

3.25   An HF over-the-horizon radar for detection of commercial aircraft targets out to a range of 2000 nmi might, for example, operate at a frequency of 15 MHz and a prf of 30 Hz. It employs doppler processing to separate moving targets from clutter. Is it an MTI radar, a pulse doppler radar, or what? Explain your answer.

4

Tracking Radar

4.1 TRACKING WITH RADAR

Types of Tracking Radar Systems   Thus far we have considered radar mainly as a surveillance sensor that detects targets over a region of space. A radar not only recognizes the presence of a target, but it determines the target's location in range and in one or two angle coordinates. As it continues to observe a target over time, the radar can provide the target's trajectory, or *track,* and predict where it will be in the future. There are at least four types of radars that can provide the tracks of targets:

- *Single-target tracker* (STT). This tracker is designed to continuously track a single target at a relatively rapid data rate. The data rate, of course, depends on the application, but 10 observations per second might be "typical" of a military guided-missile weapon-control radar. The antenna beam of a single-target tracker follows the target by obtaining an angle-error signal and employing a closed-loop servo system to keep the error signal small. (A small angle-error signal means that the radar is accurately tracking the target.) Most of this chapter will be concerned with this type of tracker. The *C*-band AN/FPQ-6, shown in Fig. 4.1a is an example of a long-range precision tracking radar that was used at missile instrumentation ranges. The major application for continuous tracking radars has been for the tracking of aircraft and/or missile targets in support of a military weapon-control system.

Figure 4.1   Two examples of missile-range precision instrumentation radars that use monopulse angle-tracking. Both operate at C band. (Top) Fixed-site AN/FPQ-6 Precision Tracking Radar with a 29-ft diameter Cassegrain reflector antenna capable of 0.1 mil tracking accuracy. (Bottom) Mobile AN/MPS-39 Multiple Object Tracking Radar (MOTR), a trainable space-fed 12-ft diameter electronically steerable lens array radar for simultaneously tracking up to 10 objects to a range accuracy of several feet and 0.2 mil angle accuracy.

(Courtesy of U.S. Army White Sands Missile Range). Both radars were manufactured by Lockheed-Martin Government Electronic Systems, Moorestown, New Jersey (formerly known as RCA Moorestown).

- *Automatic detection and track* (ADT). This performs tracking as part of an air-surveillance radar. It is found in almost all modern civil air-traffic control radars as well as military air-surveillance radars. The rate at which observations are made depends on the time for the antenna to make one rotation (which might vary from a few seconds to as much as 12 seconds). The ADT, therefore, has a lower data rate than that of the STT, but its advantage is that it can simultaneously track a large number of targets (which might be many hundreds or a few thousands of aircraft). Tracking is done open loop in that the antenna position is not controlled by the processed tracking data as it is in the STT. This type of tracking is discussed in Sec. 4.9.

- *Phased array radar tracking.* A large number of targets can be held in track with a high data rate by an electronically steered phased array radar. Multiple targets are tracked on a time-shared basis under computer control since the beam of an electronically scanned array can be rapidly switched from one angular direction to another, sometimes in a few microseconds. It combines the rapid update rate of a single-target tracker with the ability of the ADT to hold many targets in track. This is the basis for such air-defense weapon systems as Aegis and Patriot. An example of a phased array for multiple-target tracking is the *C*-band multiple-target tracking range instrumentation radar called MOTR which is shown in Fig. 4.1b.

- *Track while scan* (TWS). This radar rapidly scans a limited angular sector to maintain tracks, with a moderate data rate, on more than one target within the coverage of the antenna. It has been used in the past for air-defense radars, aircraft landing radars, and in some airborne intercept radars to hold multiple targets in track. It is briefly mentioned in Sec. 4.7. Unfortunately, the same name *track while scan* was also applied in the past to what is now usually called ADT.

A radar can track targets in range as well as angle. Sometimes tracking of the doppler frequency shift, or the radial velocity, is also performed. Most of the discussion in this chapter, however, will be on angle tracking.

Angle-Tracking   In a simple pencil-beam radar the detection of a target provides its location in angle as being somewhere within the antenna beamwidth; but more information is needed to determine the direction the antenna should be moved to maintain the target within its beam. Consider the angle measurement in a single angular coordinate. In order to determine the direction in which the antenna beam needs to be moved, a measurement has to be made at two different beam positions. Figure 4.2 shows two beam positions A and B at two different angles. The two beams are said to be *squinted,* with a squint angle $\pm\theta_q$ relative to the boresight direction. These may be two simultaneous beams, or one beam that is rapidly switched between the two angular positions. The crossover of the two beams determines the *boresight* direction. The tracking radar has to position the two beams so that the boresight is always maintained in the direction of the target; that is, the angle θ_0 is in the direction of the target angle θ_T. In this example, the relative amplitudes a_A and a_B of the echo signals received from a target measured in the two positions determine how far the target is from boresight and in what direction the two beams have to be repositioned to maintain the target on boresight. This applies for one angle coordinate. Two additional beam positions are needed in the orthogonal plane to obtain angle

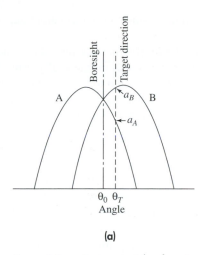
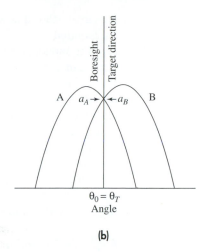

(a) (b)

Figure 4.2 Basic principle of continuous angle tracking. (a) Two overlapping antenna patterns that cross over at the boresight direction θ_0. A target is located in this example to the right of the boresight at the angle θ_T. The amplitude a_B of the target echo in beam B is larger than the amplitude a_A in beam A, which indicates that the two beams should be moved to the right to bring the target to the boresight position. (b) Boresight position θ_0 is shown located in the direction of the target θ_T when $a_A = a_B$.

tracking in the orthogonal angle coordinate. Three beam positions are the minimum needed to obtain an angle measurement in two coordinates; but, almost always, four beams have to be used.

Early tracking radars used a single time-shared beam to track in two angles. These trackers which time share a single beam are known as either *conical scan* or *sequential lobing* trackers. (Both will be discussed later in this chapter.) Modern, high-precision tracking radars, however, use the equivalent of four simultaneous beams to perform two-dimensional tracking. They are called *simultaneous lobing* trackers, of which the most popular is *monopulse,* which is described next.

4.2 MONOPULSE TRACKING[1,2]

A monopulse tracker is defined as one in which information concerning the angular location of a target is obtained by comparison of signals received in two or more simultaneous beams.[3] A measurement of angle may be made on the basis of a single pulse; hence, the name *monopulse.* In practice, however, multiple pulses are usually employed to increase the probability of detection, improve the accuracy of the angle estimate, and provide resolution in doppler when necessary. By making an angle measurement based on the signals that appear simultaneously in more than one antenna beam, the accuracy is improved compared to time-shared single-beam tracking systems (such as conical scan or sequential lobing) which suffer degradation when the echo signal amplitude changes with time. Thus the accuracy of monopulse is not affected by amplitude fluctuations of the target echo. It is the preferred tracking technique when accurate angle measurements are required.

The monopulse angle method may be used in a tracking radar to develop an angle error signal in two orthogonal angle coordinates that mechanically drive the boresight of the tracking antenna using a closed-loop servo system to keep the boresight positioned in the direction of the moving target. In radars such as the phased array, angle measurements can be obtained in an open-loop fashion by calibrating the error-signal voltage in terms of angle.

There are several methods by which a monopulse angle measurement can be made. The most popular by far has been the *amplitude-comparison monopulse* which compares the amplitudes of the signals simultaneously received in multiple squinted beams to determine the angle. When the term *monopulse* is used by itself with no other descriptors, it generally refers to the amplitude-comparison version.

Amplitude-Comparison Monopulse   For simplicity, this form of monopulse is first described for the measurement of only one angle coordinate. Two overlapping antenna patterns with their main beams pointed in slightly different directions are used, as in Fig. 4.3a. The two beams in this figure are said to be *squinted,* or *offset.* They might be generated by using two feeds slightly displaced in opposite directions from the focus of a parabolic reflector. The essence of the amplitude-comparison monopulse method is in

Figure 4.3   Monopulse antenna patterns and error signal. The left-hand sketches in (a) to (c) are in polar coordinates; right-hand sketches are in rectangular coordinates. (a) Two squinted antenna beams; (b) sum pattern of two squinted beams shown in (a); (c) difference pattern; (d) error signal as a function of the angle from boresight.

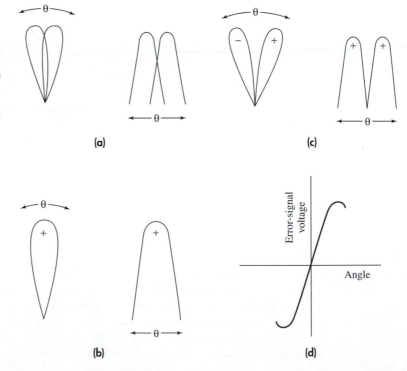

taking both the sum and the difference* of the two squinted antenna patterns, which are shown in Figs. 4.3b and 4.3c. The sum pattern is employed on transmission, while both the sum and the difference patterns are used on reception. The signal received with the difference pattern provides the magnitude of the angle error. The direction of the angle error is found by comparing the phase of the difference signal with the phase of the sum signal, as explained below. Signals received from the sum and difference patterns are amplified separately and combined in a phase-sensitive detector to produce the angle-error signal, Fig. 4.3d. The sum signal also provides target detection and range measurement, as well as act as a reference for determining the sign of the angle measurement.

Block Diagram   A simple block diagram of the amplitude-comparison monopulse tracking radar for a single angular coordinate is shown in Fig. 4.4. The two adjacent antenna feeds are connected to the two input arms of a *hybrid junction,* which is a four-port microwave device with two input and two output ports. When two signals (such as the signals from the two squinted beams) are inserted at the two input ports, the sum and difference of the two are found at the two output ports. (There are several methods for obtaining a hybrid junction, as indicated later.) On reception, the output of the sum and difference ports are each heterodyned to an intermediate frequency and amplified in the superheterodyne receiver. It is important that the sum and difference channels have the same phase and amplitude characteristics. For this reason, a single local oscillator (LO) is shared by the two channels. The transmitter is connected to the sum port of the hybrid junction. A duplexer (TR) is included in the sum channel for the protection of the sum-channel receiver. Although it might not be needed for protection of the difference-channel receiver, a duplexer is often inserted in the difference channel so as to maintain the phase and amplitude balance of the two channels. Automatic gain control, not shown, is also used to help maintain balance.

The outputs of the sum and difference channels are the inputs to the *phase-sensitive detector,* which is a nonlinear device that compares two signals of the same frequency.

Figure 4.4   Simple block diagram of the amplitude-comparison monopulse in one angle coordinate. Σ denotes the sum channel. Δ denotes the difference channel.

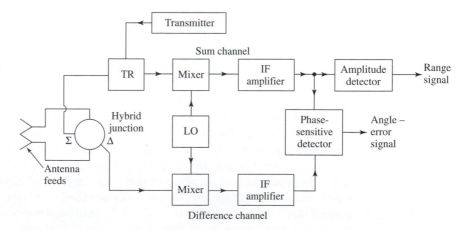

*It is sometimes said that a simultaneous-lobing tracking radar should not be called monopulse unless it employs the sum and difference patterns.

In this case, the two signals are those of the sum and difference channels. The output of a phase-sensitive detector is the angle-error signal. Its magnitude is proportional to $|\theta_T - \theta_0|$, where θ_T = target angle and θ_0 = boresight, or crossover, angle. The sign of the output of the phase-sensitive detector indicates the direction of the angle error relative to the boresight. If the sum signal in the IF portion of the receiver is $A_s \cos \omega_{IF}t$, the difference signal will be either $+A_d \cos \omega_{IF}t$ or $-A_d \cos \omega_{IF}t$, depending on which side of boresight is the target. (In the above, $A_s > 0$ and $A_d > 0$.) Since $-A_d \cos \omega_{IF}t = A_d \cos (\omega_{IF}t + \pi)$, the sign of the difference signal may be found with the phase-sensitive detector by determining whether the difference signal is in phase with the sum signal or 180° out of phase; that is, whether the output phase is 0 or π radians.

Although a phase comparison is part of the amplitude-comparison monopulse radar, the magnitude of the angle-error signal is determined by comparing the echo-signal amplitudes received with simultaneous squinted beams. The separation of the two antenna feeds is small so that the phases of the signals in the two beams are almost equal when the target angle is not far from boresight.

Hybrid Junctions[4]   As mentioned, the hybrid junction is a four-port device that provides at its two output ports the sum and difference of the signals that are at its two input ports. For monopulse radar, they are usually constructed from waveguide; but they can also be in coax or stripline. The hybrid junction known as the *magic-T,* sketched in Fig. 4.5a, consists of an E-plane T-junction (shown vertical) and an H-plane T-junction (shown horizontal) arranged as indicated. A signal, whose E-field (electric field) is indicated by the solid arrow, is shown as the input at port 1. It is divided equally in power and appears with the same phase at both ports 3 and 4. Nothing will appear at port 2. A signal whose E-field is indicated by the dashed arrow is the input at port 2. It is divided equally between ports 3 and 4, and no energy appears at port 1. The nature of the E-plane junction is such as to make the two signals at ports 2 and 3 out of phase by 180°, as is indicated by the dashed arrows being reversed in direction. Thus the output of port 4 is the *difference* of the signals at ports 1 and 2; and the output of port 3 is the *sum* of ports 1 and 2. The magic-T is inherently a broadband device. As shown, it is bulky, but its arms can be folded to make it more compact without changing its electrical characteristics. Folding means making arms 3 and 4 to be parallel to arm 2 (by folding either up or down) or they may be folded forward to be parallel to arm 1.

The *rat-race,* or *hybrid-ring junction,* is shown sketched in Fig. 4.5b. Ports 1 and 2 are the two inputs. A signal at port 1 can reach port 4 by two separate paths, one clockwise and the other counterclockwise. The two paths are of the same length (3/4 wavelength), so they reinforce and a signal will appear at this output port. The signal input at port 1 also reaches port 3 by two paths—one which travels 5/4 wavelength and the other 1/4 wavelength. They are also in phase, so a signal will appear at port 3 from port 1. At port 2, however, the two signals from port 1 are 180° out of phase (the clockwise signal travels one wavelength and the counterclockwise signal travels one-half wavelength). Thus a signal that is input at port 1 will be divided equally and appear at ports 3 and 4, but not appear at 2. Similarly, a signal input at port 2 will appear at ports 3 and 4 and not at port 1. At port 4, however, the signal from port 2 can be seen to be 180° out of phase with the signal that arrives there from port 1. Thus the output of port 4 is the *difference* of the

Figure 4.5 Examples of hybrid junctions as might be used in monopulse radar. (a) Magic-T; (b) rat-race, or hybrid-ring junction; (c) 3-dB directional coupler obtained by use of two rectangular waveguides with narrow walls touching and with quarter-wavelength spacing between the two coupling holes.

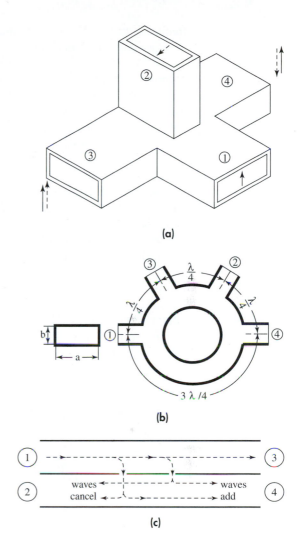

input signals at ports 1 and 2; and port 3 is the *sum* of the signals at ports 1 and 2. Since the operation of this device depends on the lengths between ports being some fraction of a wavelength, it will be frequency sensitive and not as broadband as the magic-T.

The *3-dB directional coupler* is a relatively compact form of hybrid junction that can also be used to obtain the sum and difference signals for monopulse. One method of obtaining a 3-dB directional coupler is to align two rectangular waveguides with their narrow walls touching, as in Fig, 4.5c. Microwave energy from one of the waveguides is coupled to the other by means of appropriate holes or slots between the two waveguides. Because of the quarter-wave spacing between the two coupling holes shown in the figure, this configuration is a frequency-sensitive device, but by employing more than two coupling holes or by using slots instead of holes, it can be made to operate over a useful frequency band.[5] (In the configuration of Fig. 4.5c, a 90° phase shift has to be inserted in either port 1 or 2 in order to provide the sum and difference at the other two.)

Monopulse in Two Angle Coordinates A block diagram of a monopulse radar for extracting angle-error signals in both azimuth and elevation is shown in Fig. 4.6. The cluster of four feed horns generate four partially overlapping (squinted) beams. The four feeds might be used to illuminate a parabolic reflector, Cassegrain reflector, or a space-fed phased array antenna. The arrangement of the four feeds is shown in the upper left-hand portion of the figure. All four feeds are used to generate the sum pattern on transmission and reception. The difference pattern in one plane is formed by taking the sum of two adjacent feeds and subtracting them from the sum of the other two adjacent feeds. The difference pattern in the orthogonal plane is obtained similarly. For example, based on the arrangement of the feeds shown in Fig. 4.6, the sum pattern is found from $A + B + C + D$; the azimuth difference pattern is obtained from $(A + B) - (C + D)$; and the elevation difference pattern is $(B + D) - (A + C)$. Note that the upper feeds form the lower beams when radiated by a reflector antenna. A total of four hybrid junctions are needed to obtain the sum pattern and the two difference patterns. The three mixers for the sum, elevation difference, and azimuth difference channels use a common local oscillator to better maintain the phase relationships among the three channels. Two phase-sensitive detectors extract the angle-error information; one for azimuth and the other for elevation. Range information is extracted from the output of the sum channel after envelope detection.

Since a phase comparison is made between the output of the sum channel and each of the difference channels, it is important that large relative phase differences not occur among the three channels. The phase difference between channels should be maintained to within 25° or better for reasonably proper performance.[6]

Automatic Gain Control (AGC) AGC is required in the receiving system in order to maintain a stable closed-loop servo system for angle tracking and to insure that the

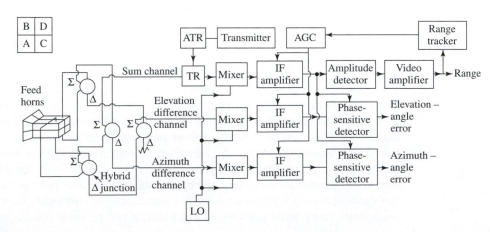

Figure 4.6
Block diagram of two-coordinate (azimuth and elevation) amplitude-comparison monopulse tracking radar. Diagram in the upper left corner represents the four-horn feed. (After Fig. 18.9, Ref. 1.)

angle-error signal is not affected by changes in the received signal amplitude. As indicated in the block diagram of Fig. 4.6, the AGC signal is obtained from the peak voltage of the sum channel and generates a negative dc voltage proportional to the peak signal voltage. The AGC signal from the sum channel is fed back to control the gain of all three channels so as to provide a constant angle sensitivity independent of changes in target cross-section fluctuations or changes in range.

Antennas for Monopulse   The assemblage of hybrid junctions, waveguides, and other microwave components needed to extract the sum and difference signals in a monopulse radar is called a *comparator.* The comparator circuitry of the early monopulse radars was quite large and bulky. If it were placed at the feed of a parabolic reflector antenna, it would cause considerable blockage of the antenna radiation and result in high sidelobes and reduced angle accuracy. For this reason, the original amplitude-comparison monopulse radar that was developed by the Naval Research Laboratory employed a metal-plate lens antenna. A lens does not cause aperture blocking, but it experiences greater loss than does a reflector antenna because there will be unwanted reflections from both the input surface and the output surface of the lens. With advances in microwave hardware technology, the size of the comparator circuitry was reduced and the AN/FPS-16 precision tracking radar (introduced in the late 1950s) was able to use a reflector antenna with the microwave circuitry at the focus of the parabolic reflector. The four waveguide transmission lines to the four feeds at the focus were made of Invar to reduce the adverse effects of temperature differences that might be experienced by the waveguides. The introduction of the Cassegrain reflector antenna (also in the late 1950s) allowed the microwave circuitry to be placed behind the parabolic reflector at its apex without aggravating the antenna blockage problem. Also, the feed system at the apex is easier to support mechanically than if it had to be placed in front of the reflector at the focus. Almost all continuous tracking precision monopulse radars employ the Cassegrain antenna. The monopulse principle can also be used with phased array antennas.

Optimum Squint Angle   The greater the squint angle, the greater will be the slope of the angle-error signal at boresight and the better will be the accuracy of the angle measurement. As the squint angle increases, however, the on-axis gain of the sum pattern decreases. Thus there will be an optimum value of the squint angle. Figure 4.7 plots the slope of the error signal as a function of the squint angle θ_q, assuming the shape of the squinted beams can be modeled by a gaussian function and that mutual coupling between the feeds can be ignored. (The basis for this curve was described in the first edition of this text.[7]) The signal received in the sum channel is proportional to the square of the sum pattern (the sum pattern on transmit times the sum pattern on receive), and the signal in the difference channel is proportional to the product of the sum and the difference patterns. The error signal is the output of the phase-sensitive detector. The optimum squint angle is found to be $\theta_q = 0.31\theta_B$, where θ_B is the half-power beamwidth of the squinted beams. This corresponds to a crossover 1.2 dB down from the peak. A different optimum squint angle, based on a different criterion, is given by both Rhodes[8] and Sherman[9] as $0.46\theta_B$, which corresponds to a crossover 2.6 dB down from the peak. Berger[10] has pointed out that the analysis of Rhodes (as well as Sherman) which gives a greater optimum squint

Figure 4.7 Slope of the angle-error signal at crossover for monopulse and conical-scan tracking radars. θ_B = half-power beamwidth, θ_q = squint angle.

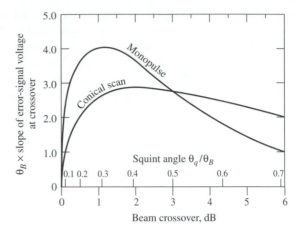

angle applies for one-way beacon tracking (instead of two-way radar tracking) since it assumes a one-way signal is being tracked, but the curve in Fig. 4.7 is based on two-way radar tracking.

Monopulse Antenna Feed Systems[11]   The question of optimum sum and difference patterns also can be examined on the basis of optimum antenna aperture illuminations rather than the concept of optimum squint angle. The desired illumination of the antenna has to be convenient to implement and produce low antenna sidelobes. The sum and difference patterns for a reflector or lens antenna are determined by the system of feeds that are used. The original four-horn feed system is the simplest to consider, but it cannot provide a difference pattern and a sum pattern that are independently optimized. The sum pattern should have maximum gain on axis, which requires a uniform aperture illumination. The difference pattern should have an aperture illumination that results in a large slope of the error signal at the beam crossover. Also, the antenna patterns should have low sidelobes and be able to maintain their favorable characteristics over a wide bandwidth. If circular polarization is needed, the demands on the feed system are further increased and often some compromise in performance must be accepted.

An improvement over the original four-horn monopulse system is shown in Fig. 4.8. This has been approximated in some precision tracking radars with a five-horn feed consisting of one horn generating the sum pattern surrounded by four horns generating the difference patterns. What makes the arrangement of Fig. 4.8 more suitable for monopulse tracking than the original four-horn feed is that analysis indicates that the size of the feed system generating the difference pattern should be about twice that of the feed generating the sum pattern.[12] This is approximately true of the feed of Fig. 4.8. Another approximation to the ideal is a 12-horn feed, but it is relatively large and complex.[13] Simpler and more compact feed systems can be obtained by using higher-order waveguide modes to obtain independent control of the sum and difference patterns. These are called *multimode feeds.*

Figure 4.8 Approximately "ideal" feed illumination for monopulse sum and difference channels.[1]

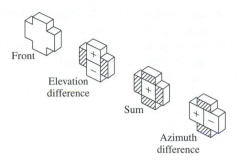

For circular polarization, a five-horn feed system can be obtained so that the antenna can be switched to operate with either horizontal, vertical, or circular polarization. This feed does not provide optimum sum and difference patterns; but it is a practical compromise between complexity and efficiency in obtaining circular polarization for monopulse trackers. In a radar with dual-polarization monopulse tracking, insufficient isolation between the polarizations can degrade the angle accuracy due to crosstalk from cross-polarized target scattering.

Phased Array Monopulse Difference Patterns If monopulse angle measurement is required with a phased array antenna, it is possible to independently control the sum and the difference patterns by means of separate beam-forming systems for the phased array. The sum pattern may be chosen for maximum gain and low sidelobes, and the difference pattern for good angle accuracy and low sidelobes. In Sec. 9.11, the synthesis of good antenna patterns based on the Taylor aperture illumination is discussed. The Taylor illumination[14] is widely used to design a pattern with predetermined peak sidelobes for radar antennas, such as the sum beam of a monopulse tracking antenna. An extension of the Taylor method, due to Bayliss,[15] is also widely used to obtain good monopulse difference patterns, which are known as Bayliss patterns.

Two-Channel and One-Channel Monopulse[16–20] Three channels, or three receivers, are required to obtain monopulse angle tracking in two orthogonal angle coordinates. Good phase and amplitude balance must be maintained among the three receivers. To simplify this problem, monopulse radars that need only two or even one receiver were considered in the past. Systems with less than three receivers that process two or even three of the monopulse channels on a shared basis were conceived when receiver hardware was large and its technology was based on vacuum tubes. Over the years, improved technology has allowed better and smaller receivers so that the need for compromising the performance of a monopulse radar by employing fewer than three receiving channels has become less important.

Conopulse[21] This is another attempt to obtain the benefits of monopulse, but with only two channels rather than three. Conopulse, also called *scan with compensation* by the Russians,[22] employs two simultaneous beams that are squinted in opposite directions from the antenna axis. The pair are mechanically rotated around the boresight axis. The two beams are similar to those of a single angle-coordinate amplitude-comparison monopulse; but their rotation allows the angle measurement in the two orthogonal coordinates to be obtained by time sharing a single channel. The sum and difference of the two squinted beams are processed similar to a conventional monopulse radar.

Since it provides a measurement of angle with simultaneous beams, the accuracy of conopulse is not degraded by amplitude fluctuations of the target as happens with conventional conical scan radars. A single-pulse measurement is not obtained as in a true monopulse, so it has a lower angular data rate than can be obtained with a three-receiver system. Although two receivers are used in conopulse rather than three, it has the disadvantage of requiring the two beams to be rotated mechanically. This can be difficult especially when the polarization has to be maintained constant on rotation. As with other one- and two-receiver monopulse systems, time and the advancement of technology have made conopulse almost obsolete.

Phase-Comparison Monopulse   In a *phase-comparison monopulse,* two antenna beams are used to obtain an angle measurement in one coordinate, just as in amplitude-comparison monopulse. The two beams, however, look in the same direction and cover the same region of space rather than be squinted to look in two slightly different directions. In order for the two beams to look in the same direction, two antennas have to be used in the phase-comparison monopulse, Fig. 4.9a, rather than using two feeds at the focus of a

Figure 4.9   Phase-comparison monopulse in one angle coordinate. (a) Two antennas radiating identical beams in the same direction; (b) geometry of the signals at the two antennas of (a) when received from a target at an angle θ, measured with respect to the perpendicular to the baseline of the two radiators.

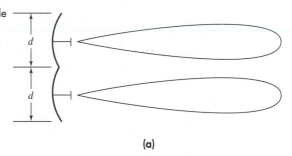

(a)

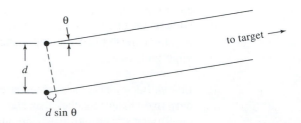

(b)

single antenna as is the case for an amplitude-comparison monopulse. The amplitudes of the signals are the same, but their phases are different. This is just the opposite of the amplitude-comparison monopulse. Consider two antennas spaced a distance d apart, as in Fig. 4.9b. If the signal arrives from a direction θ with respect to the normal to the baseline, the phase difference in the signals received in the two antennas is

$$\Delta\phi = 2\pi\frac{d}{\lambda}\sin\theta \qquad [4.1]$$

where $\lambda = w$avelength. A measurement of the phase difference of the signals received in the two antennas can provide the angle θ to the target. The phase-comparison monopulse is sometimes known as an *interferometer radar*.

The phase-comparison monopulse was invented during World War II about the same time as amplitude-comparison monopulse. Its early implementation was less efficient than that of amplitude-comparison monopulse. Four reflector antennas were used arranged in two rows of two columns. One of the antennas might be used as the transmitter (for purposes of this discussion, assume it is the aperture at the upper left of the two-by-two set of apertures). It also provides range information on reception. The other three antennas are used as receivers to obtain the azimuth and elevation angles. The upper right and the lower right antennas might obtain the phase difference in elevation which gives the elevation angle. The lower left and the lower right might obtain the azimuth angle. (The lower right antenna in this case is used for both the elevation and azimuth measurements.) A disadvantage of this method of obtaining the two angle coordinates is that only one-fourth of the available antenna area is used for transmitting and only one-half the area is used on receive to obtain each angle coordinate. Thus, the gain and effective area of a phase-comparison monopulse operating in this manner is less efficient than if the same total antenna area had been used for amplitude-comparison monopulse that generates sum and difference beams.

Angle information can also be extracted in a phase-comparison monopulse by employing sum and difference patterns and processing the signals similar to that described for the amplitude-comparison method. The full antenna aperture area can then be utilized, which is an advantage over the phase-measurement method described above. An analysis of the sum and difference patterns for phase-comparison monopulse shows that a 90° phase shift has to be introduced in the difference signal so that the output of the phase-sensitive detector is an error signal whose amplitude is a function of the sine of the angle of arrival from the target measured with respect to the perpendicular to the two antennas. (The phase-sensitive detector in this case performs a multiplication of the sum and difference signals.)

One of the limitations of phase-comparison monopulse is the effect of grating lobes due to the separation d of the two antennas each of dimension d. Grating lobes in phased array antennas are discussed in Sec. 9.5, but they apply to this situation as well. If the spacing d between the phase centers of the antenna is greater than that of the antenna diameter, high sidelobes are produced in the sum pattern and ambiguities can occur in the angle measurement. Even when the spacing is the same as the antenna diameter, a poor antenna pattern can result. In practice, the separation between the two antennas should be less than the antenna diameter d if good radiation patterns are to be obtained on transmit and angle ambiguities are to be avoided on receive. In the past when parabolic reflectors were used, a portion of the right-hand side of one antenna was sliced off (truncated), and a portion of the left-hand side of the other antenna was also sliced off so that when the two sliced-off edges were butted together, the separation between the two truncated reflector antennas could be made less than the original diameter d.

There has been little application of the phase-comparison monopulse as compared to the more popular amplitude-comparison method.

4.3   CONICAL SCAN AND SEQUENTIAL LOBING

The monopulse tracker described in the previous section utilized multiple fixed beams to obtain the angle measurement. It is also possible to time share a single antenna beam to obtain the angle measurement in a sequential manner, as was done in early tracking radars. Time sharing a single antenna beam is simpler and uses less equipment than simultaneous beams, but it is not as accurate.

Sequential Lobing   The first U.S. Army angle-tracking air-defense radar in the 1930s (SCR-268) switched a single beam between two squinted angular positions to obtain an angle measurement. This is called *lobe switching, sequential switching,* or *sequential lobing.* Figure 4.10a is a polar representation of the antenna beam in the two switched positions. The same in rectangular coordinates is in Fig. 4.10b. The error signal obtained

Figure 4.10   Lobe-switching antenna patterns and the error signal (for one angle coordinate). (a) Polar representation of the switched antenna pattern; (b) rectangular representation; (c) error signal.

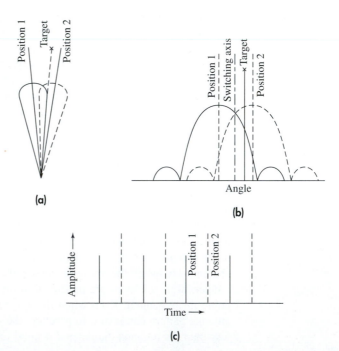

from a target not located on the switching axis (boresight) is shown in Fig. 4.10c. The difference in amplitude between the voltages obtained in the two switched positions is a measure of the angular displacement of the target from the switching axis. The direction in which to move the beam to bring the target on boresight is found by observing which beam position has the larger signal. When the echo signals in the two beam positions are equal, the target is on axis and its direction is that of the switching axis.

Two additional switching positions are needed to obtain the angle measurement in the orthogonal coordinate. Thus a two-dimensional sequentially lobing radar might consist of a cluster of four feed horns illuminating a single reflector antenna, arranged so that the right–left, up–down sectors are covered by successive antenna positions. A cluster of five feed horns might also be used, with a central feed used for transmission and four outer feeds used for reception on a sequential basis.

In a sequential lobing system, a pulse might be transmitted and received when the beam is squinted to the right, again when the beam is squinted up, when the beam is squinted to the left, and when the beam is squinted down. Thus the beam might be switched right, up, left, down, right, and so forth. After living with this type of scanning for a while, it must have become obvious that the four horns and RF switches could be replaced by a single feed that radiated a single beam squinted off axis. The squinted feed could then be continuously rotated to obtain angle measurements in two coordinates. This is a *conical-scan* radar.

Conical Scan   The basic concept of conical scan, or *con-scan*, is shown in Fig. 4.11. The angle between the axis of rotation and the axis of the antenna beam is the squint angle. Consider a target located at position *A*. Because of the rotation of the squinted beam and the target's offset from the rotation axis, the amplitude of the echo signal will be modulated at a frequency equal to the beam rotation frequency (also called the conical-scan frequency). The amplitude of the modulation depends on the angular distance between the target direction and the rotation axis. The location of the target in two angle coordinates determines the phase of the conical-scan modulation relative to the conical-scan beam rotation. The conical-scan modulation is extracted from the echo signal and applied

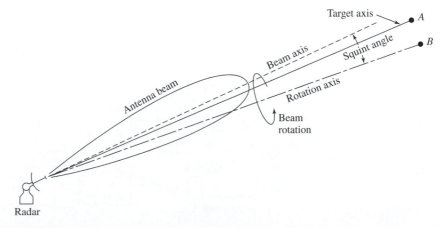

Figure 4.11 Conical-scan tracking.

to a servo control system that continually positions the antenna rotation axis in the direction of the target. It does this by moving the antenna so that the target line of sight lies along the beam rotation axis, as at position *B* in Fig. 4.11. Two servos are required, one for azimuth and the other for elevation. When the antenna is "on target," the conical-scan modulation is of zero amplitude.

Block Diagram   A block diagram of the angle-tracking portion of a conical-scan tracking radar is shown in Fig. 4.12. The antenna is mounted so that it can be mechanically positioned in both azimuth and elevation by separate motors. The antenna beam is squinted by displacing the feed slightly off the focus of the parabola.

The parabolic-antenna feed can be a rear-feed design for mechanical convenience. When the feed is designed to maintain the plane of polarization as it rotates about the axis, it is called a *nutating* feed. A *rotating* feed is one which causes the plane of polarization to rotate. The nutating feed is preferred over the rotating feed since a rotating polarization can cause the amplitude of the target echo signal to change with time even for a stationary target on-axis. A change in amplitude caused by a modulated echo signal can result in degraded angle-tracking accuracy. The nutating feed is usually more complicated, however, than the rotating feed. If the antenna is small enough (as in a missile guidance system), it might be easier to mechanically rotate the tilted reflector rather than the feed, thus avoiding the problems of either a rotary joint or a flexible RF joint for the nutating feed.

A typical conical-scan rotation speed might be in the vicinity of 30 rev/s. The same motor that provides the conical-scan rotation of the antenna beam also drives a two-phase reference generator with electrical outputs at the conical-scan frequency that are 90° apart in phase. These two outputs serve as reference signals to extract the elevation and azimuth errors as indicated in Fig. 4.12. The received echo signal is fed to the receiver from the antenna via two rotary joints (not shown in the block diagram). One rotary joint permits motion in azimuth; the other, in elevation.

Figure 4.12 Block diagram of conical-scan tracking radar.

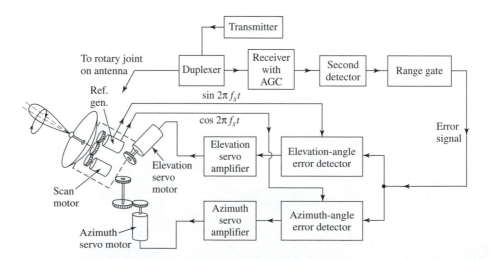

The receiver is a superheterodyne except for features related to the conical-scan tracking. The error signal is extracted in the video after the second detector. A single target is put into track by having the receiver scan a range gate to search for the target and lock on to it and then continually track it in range (as described later in the chapter). Range gating eliminates noise and excludes all targets other than the desired target. The error signal from the range gate is compared with both the elevation and azimuth reference signals in the angle-error detectors, which are phase-sensitive detectors. As described previously, the phase-sensitive detector is a nonlinear device in which the input signal is mixed with a reference signal. The magnitude of the d-c output from the angle-error detector is proportional to the angle error, and its sign (polarity) indicates the direction of the error. The angle error outputs are amplified and used to drive the antenna elevation and azimuth servo motors. The angular position of the target may be determined from the elevation and azimuth of the antenna axis.

The video signal is a pulse-train modulated by the conical-scan frequency, as in Fig. 4.13a. It is usually convenient to stretch the pulses before low-pass filtering so as to increase the energy at the conical-scan frequency and to perform analog-to-digital conversion. Pulse stretching, Fig. 4.13b, is accomplished by a *sample-and-hold* circuit; which has also been known in the past as a *boxcar* generator.

The pulse repetition frequency must be sufficiently large compared with the conical-scan frequency for proper filtering and avoiding inaccuracy of the angle measurement. There must be at least four pulses during each revolution of the conical scanning beam (so as to obtain up–down and right–left comparisons). The prf, therefore, must be at least four times that of the conical-scan frequency; but it is preferable that it be more than 10 times greater.

Automatic Gain Control As with monopulse radar, AGC is employed in the conical-scan radar. It has the purpose of maintaining constant angle-error sensitivity in spite of

Figure 4.13 (a) Pulse-train with conical-scan modulation; (b) same pulse-train after stretching by a sample-and-hold circuit.

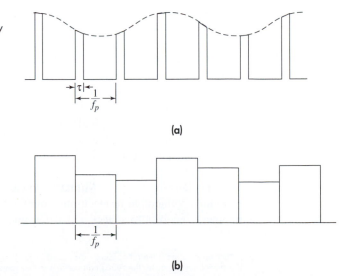

amplitude fluctuations or changes of the echo signal due to changes in range. Constant angle-error sensitivity is required to provide stable tracking. AGC is also important for avoiding saturation by large signals which could cause the loss of the scanning modulation and the accompanying error signal. It also attempts to smooth or eliminate as much of the noiselike amplitude of the target echo signal as is practical without disturbing the extraction of the desired echo signal at the conical-scan frequency. The gain of the AGC loop at the conical-scan frequency should be low so that the error signal will not be suppressed by the AGC action.

The required dynamic range* for the AGC will depend on the variation in range over which targets are tracked and the variation expected in the target cross section. If, for example, the range variation were 10 to 1, its contribution to the required dynamic range would be 40 dB. The target cross section might contribute another 40-dB variation. Another 10 dB might be allowed to account for other variations in the parameters of the radar equation. Hence, the dynamic range in this example that is required for operation of the receiver AGC might be of the order of 90 dB. In practice, a large dynamic range cannot be obtained with only one stage of AGC.[23]

Optimum Squint Angle   The greater the squint angle in the conical-scan tracker the greater will be the slope of the error signal around boresight and the more accurate will be the angle measurement. In Fig. 4.7 is shown the theoretical slope of the error signal for a conical-scan radar when the target is on boresight, or crossover, computed for an antenna pattern with a gaussian shape. The maximum slope occurs at a squint angle equal to 0.41 of the half-power beamwidth. The maximum is seen to be not too sensitive to squint angle in this case. A squint angle of $0.41\theta_B$ corresponds to a point on the antenna pattern of about 2 dB down from the peak. This means that when a conical-scan radar has a target in track, the echo signal is 4 dB less than if the target were viewed at the peak of the antenna beam. (This is sometimes called the *crossover loss.*) A monopulse radar, it will be recalled, tracks the target with the peak of the sum beam so that it does not incur such a loss. Thus the monopulse tracker will have a larger signal-to-noise ratio which provides more accurate tracking in both angle and range than that of the conical-scan tracker.

In a conical-scan tracker a compromise is often made between the range and angle accuracy by selecting a smaller squint angle than that which produces maximum angle-error-signal slope. A compromise value might be $\theta_q/\theta_B = 0.28$, which corresponds to a point on the antenna pattern about 1.0 dB below the peak. The two-way loss in antenna gain is 2.0 dB instead of 4.0 dB, which makes more accurate range tracking but lower accuracy angle tracking. If the radar is used to track a beacon on a one-way path rather than the two-way path of the radar "skin echo," the optimum squint angles are larger.[24]

Scan on Receive Only[25]   Military conical-scan and lobe-switching tracking radars are especially vulnerable to electronic countermeasures (ECM) since it is easy for a hostile intercept receiver to detect and determine the conical-scanning frequency. With such

---

*Dynamic range is the ratio, usually expressed in decibels, of the maximum to the minimum signal power over which a device (in this case, the AGC) can operate within some specified level of performance.

knowledge, a hostile ECM jammer can cause a conical-scan radar to cease tracking a target (called *break-lock*) by retransmitting the received radar signal with an amplitude modulation that is the inverse of the conical-scan frequency. This produces a return signal that is out of phase with the signal which would have been received from the skin-echo of the target, and break-lock might occur. This type of countermeasure is called *inverse gain,* and can degrade conical-scan or lobe-switching tracking systems.

To prevent the hostile ECM jammer from detecting a conical-scan frequency, a tracking radar can operate with a nonscanning transmitting beam to illuminate the target and apply conical scanning or lobe switching only on receive. This is called COSRO, which stands for *conical scan on receive only.* The analogous operation with sequential lobing is called LORO, or *lobe on receive only.*

4.4 LIMITATIONS TO TRACKING ACCURACY

In this section several of the major effects that determine the accuracy of a tracking radar will be discussed, including:

- Glint, or angle noise, which affects all tracking radars, especially at short range.
- Receiver noise, which also affects all radars, and mainly determines tracking accuracy at long range.
- Amplitude fluctuations of the target echo that bother conical-scan and sequential-lobing trackers, but not monopulse.

Other factors that influence the overall accuracy of a tracking radar include the mechanical properties of the antenna and pedestal, the servo system, the method by which the pointing of the antenna boresight is determined, the antenna beamwidth, atmospheric effects, and multipath.[26]

Glint   This has also been called *angle noise, target noise, angle fluctuations,* and *angle scintillation;* but *glint* is the term commonly used. It occurs with complex targets that have more than one scattering center within the resolution cell of the radar. A single "point" scatterer, such as a sphere, does not show the phenomenon of glint. Complex targets with multiple scattering centers, such as aircraft, can cause glint and degrade tracking. The echo from a single scatterer generally arrives at the radar antenna with a uniform planar waveform that has a tilt which depends on the angle of arrival, Fig. 4.14a. The usual method for measuring angle almost always assumes that the arriving wavefront is planar and uniform. If, however, the target consists of multiple scatterers, each at a different angle, their individual echo signals arrive at the antenna with slightly different wave tilts, as sketched in Fig. 14.14b (exaggerated to show the principle). These tilted wavefronts add vectorially across the aperture to give a composite wavefront whose amplitude and phase are not uniform across the aperture. Glint from a complex target is sometimes thought of as a distortion of the echo wavefront.

The result of having a nonuniform wavefront from a complex target, when the radar is designed to process the echo wavefront that is planar, is an error in the measurement

Figure 4.14   (a) Plane wave from one scatterer incident on the antenna. (b) Plane waves from three scatterers incident on the antenna. Resultant aperture illumination is the vector sum of the three plane waves.

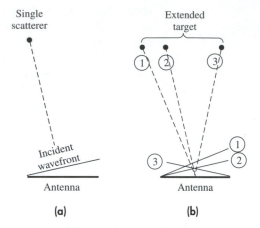

of the angle of arrival. The measured angle does not bear a simple relationship to some distinctive property of the target, such as its center, leading edge, or its largest scatterer. Furthermore, the measured angle of arrival can sometimes cause the boresight of the tracking antenna to point *outside* the angular extent of the target, which can cause the radar to break track. The greater the extent of the target in angle as seen by the radar, the worse will be the angle measurement, as we shall see. Glint, therefore, can be a major source of error when making angle measurements, especially at short range where the angular extent of the target can be relatively large. It bothers all continuous tracking radars that employ closed-loop angle tracking, whether conical-scan, sequential lobing, amplitude-comparison monopulse, or phase-comparison monopulse.

Example of Glint from a Simple Target Model   Consider a target model consisting of two independent, isotropic scatterers separated by an angular distance θ_D as measured from the radar, Fig. 4.15a. (Sometimes this is called a *dumbbell target*.) The two scatterers are assumed to be located symmetrically to each side of the perpendicular from the antenna at $\pm \theta_D/2$. Although it may be a fictitious target model chosen for reasons of simplicity, it illustrates the effects that a complex (multiple-scatterer) target has on the accuracy of a tracking radar. The relative amplitude of the echo signals from the two isotropic scatterers is taken to be a (a number less than unity) and the relative phase difference is α. Differences in the phase might be due to differences in range between the two scatterers or to differences in the reflecting characteristics of the two scatterers. The angular error $\Delta\theta$ as measured from the larger of the two isotropic scatterers is given by J. E. Meade as[27]

$$\frac{\Delta\theta}{\theta_D} = \frac{a^2 + a\cos\alpha}{1 + a^2 + 2a\cos\alpha} \qquad [4.2]$$

This is shown plotted in Fig. 4.15b. The larger of the two scatterers is at $\Delta\theta/\theta_D = 0$, and the smaller is at $\Delta\theta/\theta_D = +1$. Positive values of $\Delta\theta$ correspond to the angular region to the left of the larger of the two scatterers; negative values lie outside the target, at angles to the right of the larger target. When the echo signals from both scatterers are in phase

Figure 4.15 Plot of Eq. (4.2) of the error $\Delta\theta/\theta_D$ as a function of a and α. Insert shows two isotropic scatterers of relative amplitude a and relative phase shift α, separated by an angular extent θ_D as viewed from the radar. The angle $\Delta\theta$ is measured with respect to the larger of the two scatterers.

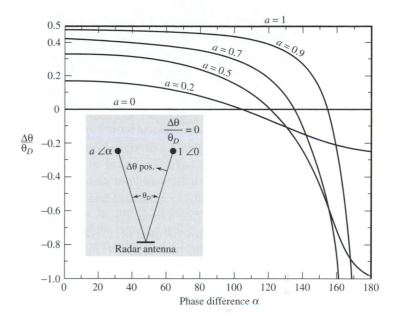

($\alpha = 0$), $\Delta\theta/\theta_D$ reduces to $a/(a + 1)$, sometimes called the "center of gravity" of the two scatterers. When the signals are of equal amplitude ($a = 1$) and the phase difference $\alpha = \pi$ radians, Eq. (4.2) indicates the antenna points to $\Delta\theta/\theta_D = -\infty$; that is, the antenna is driven well outside the bounds of the target. It should be cautioned that an assumption made in deriving Eq. (4.2) is that the voltage of the angle-error signal is directly proportional to the angle error; that is, the angle-error signal is linear. This implies that angle errors are small; but when $\Delta\theta/\theta_D \rightarrow -\infty$, the angle error is not small. As was indicated in Fig. 4.3d, as the angle error increases, the angle-error signal voltage will no longer be linear and its slope can even change sign. Equation (4.2) no longer applies when $\Delta\theta/\theta_D$ becomes large. Nevertheless, the simple model of Eq. (4.2) describes the general nature of the behavior to be expected by a radar tracker subjected to glint from a complex target.

The relative phase, α, changes if the relative ranges to the two scatterers change. This can occur, for example if the aspect angle of an aircraft changes due to its flight path or atmospheric turbulence. Thus the value of α can vary with time. The radar tracker theoretically can point outside the target region a significant fraction of the time.

Equation (4.2) indicates that the tracking error $\Delta\theta$ for the two-scatterer target is proportional to the angular extent of the target θ_D, which is why the tracking error due to glint varies inversely with range. The error becomes larger as the range becomes shorter. Although this statement is based on the simple two-scatterer target, it is a reasonable approximation to a real target provided the angular extent is not greater than the antenna beamwidth. When the angular extent of the target is greater than the antenna beamwidth, the two scatterers can be resolved and there is no glint.

A slightly more complex model than the two-scatterer target considered above is one consisting of many individual scatterers, each of the same cross section, arranged uniformly along a line of length L perpendicular to the line of sight from the radar. The resultant cross section from such a target is assumed to behave according to the Rayleigh probability density function. It was found that the probability is 13.4 percent that the apparent direction of the target measured by the radar will be outside of the target region.[28] Similar results for a two-dimensional model consisting of equal cross-section scatterers uniformly spaced over a circular area indicate that the probability the apparent direction lies outside the target is 20 percent. Howard[29] states that measurements made on actual aircraft of the rms value of angle noise, expressed in the same length units as the target extent L in cross-range, are between $0.15L$ and $0.2L$. A small single-engine aircraft viewed nose-on might have a value near $0.1L$, and larger aircraft and aircraft viewed from the side approach $0.2L$.

Although glint is a deterministic phenomenon if the target configuration and its scattering properties are known, it has sometimes been analyzed in statistical terms.[30,31] The deterministic approach, however, helps in understanding what is taking place with the target and the radar, something that statistical models cannot do as well. Glint is usually thought of as a target effect; but to some extent, the radar enters also. Glint occurs when the radar cannot resolve the individual scatterers of a complex target; so that some radars might not be affected by glint that would seriously bother others.

Methods for reducing the effects of glint on radar tracking performance are discussed later in this section. The phenomenon of glint also occurs in the range dimension, as will be discussed in Sec. 4.6. A brief survey of radar glint, along with an extensive bibliography of the early work in this field, has been given by Wright.[31]

In addition to tracking radars, it has been said in the literature that glint also occurs with scanning surveillance radars.[32] This might not be completely accurate. Surveillance radars estimate the target direction as the antenna pointing angle where the echo signal is a maximum. The estimate of angle in scanning surveillance radars is often made by *beam splitting,* or something equivalent. Closed-loop angle tracking is not employed. A complex target might cause an error in the beam-splitting angle estimate because of the nonuniform response of the target with angle, but the error does not seem to produce an angle measurement that extends beyond the angular confines of the target as does the glint that occurs in a radar with closed-loop tracking. Thus, glint does not occur with a scanning radar, only one with closed-loop tracking.

Receiver Noise The noise at the input of a radar receiver affects the accuracy of radar tracking just as it does the detection capability of a radar. In Sec. 6.3, the accuracy of radar measurements is discussed, based on a noise model described by the gaussian probability density function. All theoretical expressions for the rms value of the error of a radar measurement (such as angle) are inversely proportional to the square root of the signal-to-noise ratio. From our previous discussion of the radar range equation in Chap. 2, we know that the range of a radar is inversely proportional to the fourth root of the signal-to-noise ratio. The rms value of a radar measurement error is, therefore, directly proportional to the square of the range. Receiver noise is a major factor limiting the accuracy of a radar at long range where signal-to-noise ratios are small. Section 6.2 also indicates that the rms error in the radar measurement of angle is directly proportional to the antenna beamwidth.

The theoretical accuracy of a tracking radar, given as the rms error in the angle measurement, has been given by Barton[33] and can be found in Howard,[34] It is

$$\delta_{\text{ang}} = \frac{k\theta_B}{k_s\sqrt{B\tau(\text{S/N})(f_p/\beta_n)}} \tag{4.3}$$

where the constant $k = 1$ for a monopulse radar and 1.4 for a conical-scan radar, $\theta_B =$ half-power beamwidth, $k_s =$ slope of the angle-error signal at boresight (which is different for monopulse and conical-scan radars), $B =$ bandwidth, $\tau =$ pulse width, $(\text{S/N}) =$ signal-to-noise ratio per pulse (assumed in the derivation of this expression to be greater than 6 dB), $f_p =$ pulse repetition frequency, and $\beta_n =$ servo bandwidth. Generally, $B\tau \approx$ 1, and $f_p/2\beta_n =$ number of pulses integrated. According to Howard,[34] the value of the slope k_s for a good four-horn monopulse feed is 1.57. Its value for a conical-scan radar is 1.5 when the offset angle is chosen to optimize overall radar performance. Since the conical-scan radar does not track a target with its maximum antenna gain, the signal-to-noise ratio is lower than that of a monopulse radar. Earlier in this chapter, we have said that the conical-scan tracker might suffer a two-way loss of 2.0 dB. More elaborate expressions for the angle error for the conical-scan tracker[35] and the monopulse tracker[36] can be found in the literature, but the above expression is often suitable for many purposes.

Amplitude Fluctuations   The amplitude of the radar echo from a complex target with multiple scattering centers will fluctuate as the aspect of the target changes with respect to the radar. (Changes in aspect may be due to the motion of the target in yaw, roll, or pitch. Aspect changes also occur even if the target moves in a straight line.) Conical-scan and sequential-lobing radars interpret any change in amplitude of the target echo signal as being due to the target not being on boresight. They then direct the antenna to move in a direction to make the "error signal" zero. Thus a change in amplitude due to fluctuations in the target echo during the time interval of the sequential measurement can degrade the accuracy of the measurement. Amplitude fluctuations in the target echo signal, which are also known as *target fading,* do not affect the angle-error measurement accuracy of simultaneous lobing or monopulse systems that extract an angle-error voltage with each pulse.

Since the percentage modulation of the echo signal due to fluctuations in the target cross section is independent of range if AGC is used, the angle error as a result of amplitude fluctuations will be independent of range.

Amplitude fluctuations from aircraft targets are classified as either low frequency or high frequency. According to Howard,[29] the low-frequency pulse-to-pulse amplitude fluctuations might be concentrated mainly below 10 Hz at X band. These are due to variations in the target cross section caused by changes in the relative distances of the individual scattering centers. The amplitude spectrum does not seem to depend strongly on the size of the target. An aircraft with a large wingspan has its scattering centers (the engines, for example) spaced wider than would an aircraft with a small wingspan so that a higher fluctuation frequency would be expected if the rate of change of yaw were the same in the two cases. The large aircraft, however, has slower rates of yaw than a small aircraft so that the frequency extent of the spectra might be expected to be similar. The spectral width of the amplitude fluctuations is closely proportional to the radar frequency, since a change in the relative distances of the scatterers will result in a larger change in wavelengths if the frequency is high than if it is low.

High-frequency amplitude fluctuations can be caused by reflections from propellers and jet engines. The frequency of propeller modulation depends on the number of blades and the rotation rate. The modulation is not sinusoidal and has harmonics of the fundamental frequency. The fundamental frequency of the propeller modulation and its harmonics do not depend on the radar frequency.

The effect of amplitude fluctuations on the accuracy of conical-scan tracking can be reduced by choosing a conical-scan frequency that corresponds to a low value of the target's amplitude fluctuation spectrum. If the amplitude fluctuation noise power were large at the conical-scan or lobing frequency, it could not be readily eliminated by AGC or filtering. A typical conical-scan frequency, for example, might be 30 Hz. Generally the higher the scan frequency, the less the noise due to amplitude fluctuations. At the higher scan frequencies, however, propeller modulation might be present and needs to be avoided. A sufficiently high scanning frequency, however, will have little degradation in tracking because of amplitude fluctuations. It has been reported that experimental measurements with radars operating with pulse repetition frequencies from 1000 to 4000 Hz and a lobing or scan rate one-quarter of the prf are not limited by amplitude fluctuations of the target.[37]

Servo Noise   This is the hunting action of the tracking servomechanism which results from backlash and compliance in the gears, shafts, and structures of the antenna mount. The magnitude of the noise is independent of the target echo and will therefore be independent of range.

Summary of Errors   The contributions of glint, receiver noise, and amplitude fluctuations to the accuracy of a tracking radar as a function of range is illustrated in Fig. 4.16. The error due to glint varies inversely with range; receiver noise causes the error to vary as the square of the range; and both amplitude fluctuations and servo noise are independent of range. This is a very qualitative plot showing the general nature of each of these factors. Two resultant curves are shown. Curve A might be representative of conical-scan and sequential-lobing trackers. It assumes that the error due to servo noise is less than that due to amplitude fluctuations. Curve B might represent monopulse trackers since it does not include the effect of amplitude fluctuations. Tracking accuracy deteriorates at both long and short range with the best angular accuracy occurring at the intermediate ranges.

The best tracking radars have been able to achieve an angle accuracy of about 0.1 milliradian. Such accuracy does not come easy. It can only be achieved by full attention to the many internal and external factors affecting tracking accuracy and by timely, accurate calibration.[38]

Methods for Reducing Angle Errors Due to Target Glint   Glint can be debilitating to a military tracking radar or a radar guided missile. It is important, therefore, to reduce its adverse effects when highly accurate tracking is required. There have been a number of methods proposed to reduce glint. Some can provide significant improvement but may require operating the radar in a manner that is not always best for achieving the radar mission. Other methods might not degrade the major task of the radar, but they do not always

Figure 4.16 Relative contributions to the angle tracking error due to glint, amplitude fluctuations, receiver noise, and servo noise. Curve A represents the composite error for a conical-scan or sequential-lobing radar; curve B represents the composite error for monopulse.

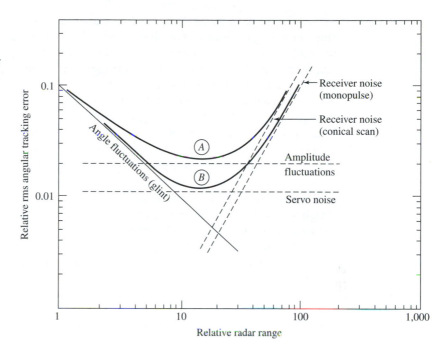

have sufficient effect on reducing glint effects. As with many things, compromises usually have to be made.

Frequency Agility   The angle error due to glint depends on the radar frequency since the relative phase α in the simple two-scatterer model depends on frequency. A change in the relative phase α can change the glint error (Fig. 4.15); it can be smaller or it can be larger. If the relative phase α is due primarily to the difference in range ΔR between the two scatterers, then the phase difference $\alpha = 4\pi(\Delta R)f/c$, where f = radar frequency, c = velocity of propagation, and $f/c = \lambda$ = radar wavelength. A change in radar frequency results in a change in the relative phase α and a change in the angle error $\Delta\theta$. There must be a sufficient change in frequency to decorrelate the phase measurement and make the measurements independent. Decorrelation occurs when the phase α changes by more than 2π radians. With this criterion, the frequency should be changed by an amount[39]

$$\Delta f \text{ (Hz)} = \frac{c \text{ (m/s)}}{2D \text{ (m)}} \quad \text{or} \quad \Delta f \text{ (MHz)} = \frac{150}{D \text{ (m)}} \quad \textbf{[4.4]}$$

where D = the depth of the target. (In the two-scatterer case, D is the same as ΔR.) The depth D of an aircraft target as seen by radar usually is not the same as the projected physical size of the target since the extremities of the target might be too small to be detected by the radar or they might be masked by other scatterers.

There are several methods that have been proposed to use the frequency dependence of glint to reduce its effect on tracking accuracy. One method is based on the observation that a large glint error is accompanied by a low echo-signal power. This inverse relationship between glint error and receiver signal power was apparently recognized as early as the mid-1950s.[40] This can be seen qualitatively with the two-scatterer model. When the amplitudes of the two signals are almost equal and their relative phase is almost 180°, Fig. 4.15 indicates that glint errors will be large. But when the two signals are almost equal in amplitude and almost 180° out of phase, they will destructively interfere at the antenna and result in a low echo-signal amplitude. Thus when the signal received by the radar is found to be low, when it is not normally expected to be low, it is likely that the angle measurement is in error. To take advantage of this effect, several different frequencies are radiated with frequency separation greater than that given by Eq. (4.4). The angle estimates of the received echo signals are weighted according to some criterion (such as proportional to their signal-to-noise ratios), and the average is taken. Several other weightings were analyzed by Loomis and Graf.[41] It was found that simply using the angle measurement associated with the largest signal amplitude provides the most accurate estimate of angle. The reduction in rms tracking error by processing only that signal whose frequency results in the largest echo amplitude is approximately

$$\delta_{\mathrm{rms}} \approx \frac{\delta_g}{N} \qquad 1 \le N \le 4 \qquad\qquad \text{[4.5]}$$

where δ_g is the single-frequency (no agility) glint error and N is the number of independent frequencies. More than four frequencies were found to offer no significant improvement in accuracy.

Although pulse-to-pulse frequency agility can reduce the effects of glint, it has some drawbacks. Detection of moving targets in clutter using MTI or pulse doppler methods is not compatible with pulse-to-pulse frequency agility. To employ doppler processing, the frequency must remain constant for a significant number of pulses. Frequency diversity is possible by transmitting either simultaneously or in sequence a number of constant-frequency pulses at each of several frequencies that are separated by more than the Δf of Eq. (4.4).

Pulse-to-pulse frequency agility reduces the glint found in conical-scan and sequential-lobing radars, but it can increase the amplitude fluctuations in such radars because of the dependence of radar cross section on frequency. It might result in a net increase in the angle error, and is not always appropriate to use with such radars.

In some crowded radar bands, it might not be easy to achieve the required range of frequencies needed to obtain a reduction in the glint error. If, for example, the effective depth D of a target were 10 m, Eq. (4.4) indicates that the change in frequency must be at least 15 MHz to decorrelate the glint error. If four different frequencies were required, there must be at least 45 MHz of bandwidth available.

Range Resolution   It was mentioned that glint occurs when there are multiple scatterers within the radar resolution cell. Range resolution can be quite good in a microwave radar, far better than the angle resolution. Therefore, if the radar has sufficiently high range-resolution to be able to resolve the multiple scatterers that constitute a target, angle glint (as well as range glint) will not occur and the tracking accuracy in angle as well as range will be improved compared to that of a low-resolution system. If, for example, the radar has to have a range resolution of one meter to resolve the various target scatterers, the required spectral width must be 150 MHz. The higher microwave frequencies are therefore more suitable for this method of glint reduction than are lower frequencies.

In some respects the use of high range-resolution for reduction of angle glint is related to the use of frequency agility for the same purpose. Both take advantage of wide bandwidth. The frequency agility method uses a finite number of discrete narrowband frequencies within a wide frequency band rather than a continuous spectrum as is required for high range-resolution. If both methods use the same extent of spectral bandwidth, it is suspected that the high-resolution method should produce more accurate tracking than the use of frequency agility. High range-resolution can be used in conjunction with MTI or pulse doppler radar if digital signal-processing technology is not limited by the wide bandwidth required for high resolution.

Servo Bandwidth and AGC Bandwidth[31]   Angular error due to glint may be reduced by keeping the servo bandwidth small. This might not be a good idea in practice since the servo bandwidth usually is determined by the requirement that the tracker be able to follow a maneuvering target. Too narrow a servo bandwidth might cause the track of the maneuvering target to be broken and the target lost.

The effects of glint also may be reduced by reducing the bandwidth (increasing the time constant) of the AGC system.[42] A narrowband AGC does not respond to rapid fluctuations in signal amplitude with the result that the echo-signal amplitude might not maintain constant signal level. This can cause a reduction in the angle-error sensitivity during large angle-noise peaks, and smaller rms tracking noise can result. This reduction in angle noise, however, is accompanied by a new component of noise due to amplitude fluctuations associated with the echo signal. Narrowing of the AGC bandwidth generates additional noise in the vicinity of zero frequency that can result in poor tracking. In spite of the potential benefits of a narrowband AGC, a wideband (fast) AGC is usually preferred especially at short and medium ranges where target maneuvers result in high angular rates and the lag in the tracking can be large if a narrowband AGC were used.

Thus narrow servo and/or AGC bandwidth as a means for reducing glint produce other undesirable effects. Bandwidths should be selected so as to be consistent with the various factors that can affect the tactical requirements that determine the acceptable tracking error and probability of breaking lock on a target.

Filtering of Angle Noise   One of the first methods for dealing with the angle error due to glint was to consider it as a noise that could be filtered if its characteristics were known.[43] This is one reason glint is sometimes called angle noise. The noise model has not been too successful, however, since the statistical description of glint is considered to be non-gaussian and nonstationary.[44] Glint can often be better modeled as deterministic or non-statistical[45] rather than statistical. Another problem in modeling glint as noise to be filtered is that glint errors are "spiky." They tend to be of large value only when the phase difference α in the two-scatterer model is near π radians. Filtering to smooth the relatively large spikes in the glint can result in a bandwidth too narrow to maintain the track of a maneuvering target, as mentioned above when discussing the servo and AGC bandwidth. It has continued to be fashionable to consider glint as a noise problem, but there has not been the success that one might desire.

Excising of Measurements Associated with Fades It has been mentioned previously that when the glint error is large, the received signal amplitude is small. If the signal level is properly monitored so as to recognize a low signal amplitude, the corresponding angle measurement can be removed (censored) and the tracking error improved. It has been suggested[46] that when a Kalman tracking filter is used in conjunction with a rank detector preprocessor to detect fades and remove the accompanying angle measurement an improvement of 15 percent in the angle tracking accuracy can be obtained.

Polarization[47] It has been suggested that polarization agility can reduce the glint error based on the expectation that the scattering centers will be different with different polarizations. The assumption is that the target echo will be produced by scatterers with widely different polarization response characteristics so that glint will be decorrelated when there is a change in polarization. In practice, however, it might not be expected that polarization agility can decorrelate the glint errors as well as can frequency agility. It has been said[44] that with polarization agility "the improvements are often modest (at best)."

Spatial and Aspect Diversity It also has been suggested that glint can be reduced with either spatial or aspect diversity.[48] By spatial diversity is meant viewing of a target from a different location. The required separation of the antennas for spatial diversity need not be large. Aspect diversity requires that the target rotate with respect to the radar so as to change its aspect with respect to the radar. Not all targets cooperate by changing their aspects sufficiently to obtain the necessaray diversity. Furthermore, a change in aspect takes time, which is usually not available with weapon control radars subject to glint. Both spatial and aspect diversity, therefore, have operational limitations that tend to restrict their practical application for glint reduction.

Avoiding Closed-Loop Tracking As mentioned previously in this section, a radar that extracts the angle of a target without performing closed-loop tracking is not susceptible to the large errors caused by glint. There have been many solutions proposed for reducing the adverse effects of glint. No one method solves all problems, each has its advantages and disadvantages, and no one solution is universally applicable. As with so many other things, compromises might have to be made in order to deal with the potential effects of glint.

4.5 LOW-ANGLE TRACKING

A radar that tracks at low elevation angles illuminates the target via two paths, as shown in Fig. 4.17. One is the direct path from radar to target. The other is the path that includes a reflection from the earth's surface. It is as though the radar were illuminating two targets, one above the surface and the other its image below the surface. This is an example of the classic two-scatterer model mentioned in the previous section on glint. An error in the measured elevation angle of the target occurs because of the effect of glint. The error can be large enough to seriously degrade the quality of the tracking. At low grazing

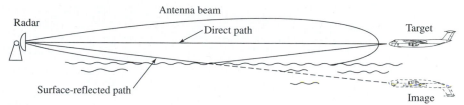

Figure 4.17   Low-angle tracking illustrating the surface-reflected path and the target's image below the surface.

angles over a perfectly smooth reflecting surface, the reflection coefficient from the surface is approximately −1 (Sec. 8.2). That is, its phase is in the vicinity of 180° and its magnitude is approximately unity so that the signal amplitude reflected from the surface is almost equal to the signal amplitude incident on the surface. This is, unfortunately, close to the worse condition for the angle error due to glint, as can be seen from the plot for the two-scatterer target model of Fig. 4.15. For this reason, the tracking of targets at low elevation angles can produce significant errors in the elevation angle and can cause loss of target track. The surface-reflected signal is sometimes called the *multipath signal* and the glint error due to the geometry of Fig. 4.17, a *multipath error.* Multipath errors can be a serious limitation to the radar guidance of missiles to targets low on the water as well as to surface or shipborne radars used for defense against low-altitude cruise missiles or sea-skimmer missile attack.

The effects of multipath depend on what part of the antenna pattern strikes the surface. Three regions can be identified,[49] according to elevation angle:

1. *Sidelobe region.* Elevation angles are such that the near-in sidelobes, rather than the main beam, illuminate the surface. The accuracy of a precision tracking radar (such as the AN/FPQ-6) begins to be degraded when the elevation angle is less than six beamwidths above the horizon.[50]

2. *Main-beam region.* The effects of multipath can begin to be severe when the elevation angle is less than about 0.8 beamwidth.

3. *Horizon region.* At grazing angles approaching zero degrees when there is specular reflection from the surface, the echo signal from the target and its image are approximately equal and out of phase so that combined direct and surface-reflected signal is very low. This reduction in signal-to-noise ratio further aggravates the accuracy problem.

Figure 4.18 shows an experimental measurement of the elevation-angle error obtained with an *S*-band radar as a function of range for an aircraft flying at low altitude.[51,52] The antenna beamwidth in this case was 2.7°. The aircraft flew out in range at a nearly constant altitude of 3300 ft. The start of the track is at about 4° elevation. At this angle the antenna sidelobes, rather than the main beam, illuminate the surface and the effect of multipath on angle accuracy is relatively small. At the center of Fig. 4.18 where the elevation angle is less than 2°, the main beam illuminates the surface and the effect of the surface-reflected wave becomes significant. Large elevation errors occur, which cause the antenna

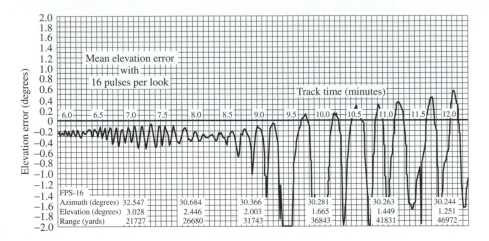

Figure 4.18 Example of the measured elevation tracking error using a phased array radar with 2.7° beamwidth. The aircraft target flew out in range at a nearly constant altitude. The numbers along the zero-error line indicate the track time in minutes.
| (From Linde.[51])

to look at the image below the ground rather than at the target. The errors are cyclic because a phase greater than 2π radians appears to the radar as if it where within 0 and 2π radians. It is seen in the example of Fig. 4.18 that there are times the antenna can point at an angle greater than 2° below the target. The tracking has been described as "wild" and great enough to cause the radar to lose track. The effect is most pronounced over a smooth water surface where the surface-reflected signal is strong. It can be so great that it becomes impossible to maintain a track at low altitudes with conventional tracking radars. According to Barton,[49] in the horizon region when the magnitude of the reflection coefficient exceeds 0.7 and the target is below 0.7 beamwidth, "there will be a strong tendency for the radar to track a centroid of reflection at or near the horizon." (This did not appear to occur for the data shown in Fig. 4.18.)

The multipath problem is similar to the glint error that results from tracking a classical two-scatterer target. Careful examination, however, of the theoretical glint error given in Fig. 4.15 and the experimental data of Fig. 4.18 show a significant difference. The two-scatterer model indicates there will be large errors when the surface reflection coefficient approaches -1. The large tracking errors appear to the side of the larger scatterer of the pair rather than the smaller scatterer. Generally, it is expected that in the multipath situation, the larger scatterer will be the target and the image will be the smaller (because the reflection coefficient is not greater than 1). The data of Fig. 4.18 show, however, that the large elevation angle errors are in the direction of the image rather than the true target. Thus the antenna is more likely to look into the ground than in the direction of the sky. Howard et al.[50] state that this difference can be explained by the failure to account for the amplitude distortion across the aperture, in addition to the phase distortion.

Although it has been easy to accept the two-scatterer target model as applied to low-angle tracking multipath case, there is a difference between the two. In the two-scatterer model there are two propagation paths: the two two-way paths from the radar to each scatterer and back. In the multipath case, there are actually four propagation paths.[53] These are: (1) the path from radar to target and return by the same path; (2) the path from radar to target via reflection from the surface and return by the same path (this is the image path); (3) the path from radar directly to target and return via reflection from the surface; and (4) the inverse of path no. 3 which is the path from radar to target via the surface reflection, or image, and return directly to the radar. Thus paths no. 3 and 4 are the same except for direction of travel. (These four paths are made use of in multipath time-delay height finding to obtain the height of a target using high-range resolution with a fan-beam antenna pattern rather than a pencil-beam pattern.) The echoes from the four paths in the multipath geometry will provide a different composite echo signal at the radar than that obtained from the two-scatterer model. It should be expected, therefore, that the nature of the glint error in the multipath situation will be different from that predicted by the two-scatterer model, as has been demonstrated by experiment.

Another complication with surface multipath not modeled by the two-scatterer target is the effect of surface roughness which results in two components of the surface reflected signal; one is specular scatter and the other is diffuse scatter.[49]

With a rough surface, the angle error due to multipath is reduced because the surface-reflected wave is of lower magnitude (the reflection coefficient is less). Note that the roughness of a surface depends on its physical variations in height relative to the radar wavelength. The higher the radar frequency, the greater will be the "electrical" roughness of the surface (in wavelengths) for a given physical roughness, so that the effect of multipath on the elevation-angle accuracy might be less at the higher frequencies. At millimeter-wave frequencies, the surface reflection is more likely to be diffuse scatter rather than specular scatter.[54]

In addition to causing errors in elevation-angle tracking, it is also possible for multipath to introduce errors in the azimuth-angle tracking channel. This can be caused by *cross talk* where a portion of the elevation-angle channel signal enters in some manner the azimuth-angle channel. It might also be caused by the target-image plane departing from the vertical as when over sloping land or when the radar is on a rolling or pitching ship.

Methods for Reducing Multipath Effects at Low Angles   There have been a number of methods demonstrated or proposed for reducing the large elevation angle errors that can occur due to the multipath experienced at low elevation angles.[49,55,56] There is not, however, a single method suitable for all applications where low-angle tracking is required. Each has its limitations. They are mainly for eliminating the large errors that occur when the main beam, rather than the sidelobes, illuminate the surface. Many of the methods mentioned below assume the use of a monopulse tracker. Only the first three seem to have had application. The others are mentioned because of some particular technical interest, and can be skipped if desired.

Narrow Beamwidth   The surest method for reducing or eliminating tracking errors due to multipath is to have a narrow antenna beam that does not illuminate the surface. The beamwidth of an antenna is approximately $\theta_B \approx \lambda/D$ radians, where λ = radar wavelength and D = antenna dimension, both in the same units. Thus a narrow beamwidth requires a large antenna, a high radar frequency, or both. Although a narrow beamwidth can eliminate the multipath problem, it is not always possible to do so in practice since there may be compelling reasons for not using a large antenna or for operating the radar at high frequency.

A method that has been successfully used operationally to obtain low-angle tracking capability is to employ two radars, one at X band (9 GHz) and the other at K_a band (35 GHz), using a single antenna system to provide operation at the two frequencies.[57,58] The lower frequency radar generally has a longer range. Target acquisition can be initiated with the lower-frequency tracker and then precision low-altitude tracking can be obtained by switching to the higher-frequency tracker. It has been said that with the dual-frequency tracker "targets at 100-ft altitude were successfully tracked over water with less than 0.4 mils multipath error to ranges in excess of 30,000 yds." Another advantage of a dual-frequency radar for military applications is that it makes hostile electronic countermeasures more difficult since both frequencies have to be jammed.

Illogical Target Trajectory   Prior knowledge of potential target behavior can be used to reduce the effects of low-angle multipath without overly complicating the radar. Since aircraft or missile targets will not likely go below the surface of the earth and are limited in their ability to accelerate upward and downward, radar tracking data indicative of unreasonable target behavior can be recognized and rejected. In some situations, the target might be flying fast enough and the inertia of the antenna might be great enough to dampen the angle-error excursions caused by multipath.

Off-Axis, or Off-Boresight, Monopulse Tracking   Advantage can be taken of the fact that large angle errors are usually limited to a region of low elevation angle predictable from the antenna pattern and the terrain. It should be possible, therefore, to determine when the target is in the low-angle region by sensing large elevation-angle errors. The antenna is then locked at a small positive elevation angle (usually about 0.7 to 0.8 beamwidth) while continuing closed-loop azimuth-angle tracking. With the beam fixed at a positive elevation angle, the target's elevation angle may be determined open-loop from the error-signal voltage. Alternatively, the elevation angle may simply be assumed to be halfway between the horizon and the antenna boresight. In the extreme, the peak-to-peak tracking error would not exceed 0.7 to 0.8 beamwidth, and the rms error would be typically about 0.3 beamwidth. The tracking accuracy is only slightly improved, but wide swings of the antenna and loss of track are avoided.

Double-Null Elevation-Difference Pattern[59,60]   A monopulse radar when tracking a single target attempts to have the null of its difference pattern pointing in the direction of the target. When this occurs at low elevation angles the echo signal that arrives in the lower beam of the difference pattern via the surface reflection can result in a glint error. By using a more complicated antenna pattern than that of the usual four-horn monopulse antenna, a second null can be independently steered to the direction of the surface-reflected echo signal (the image echo) so as to cancel it before it reaches the radar receiver. The second null can be obtained by employing a third pair of feeds in the vertical plane. The signals in elevation from the three pairs of vertical feeds are combined to produce two nulls, one in the direction of the target and the other in the direction of its image, as has been described by White in the cited references. The second null is maintained in the direction of the image, as computed by Snell's law for the measured target range.

A conventional four-element feed in a monopulse tracker has only one degree of freedom in the vertical plane (as well in the horizontal plane). It is designed to track only a single target. When multipath occurs, there are two target signals present (the actual target and its image). A conventional monopulse radar cannot cope since it is designed on the assumption that there is only a single target present. When a third pair of feed-elements is employed in the vertical, two degrees of freedom result so that there are now two nulls that can be positioned in the direction of the target and its image. This can be extended to even more feeds in the elevation plane (providing additional degrees of freedom and nulls) which enables better control of the nulling of the unwanted image.

Experiments over water have demonstrated good tracking (0.05 to 0.1 beamwidth rms) with the double-null technique to elevations as low as 0.25 beamwidth.[60]

There have been several other variants of this approach. The signal incident on the antenna aperture can be sampled at multiple points across the aperture (especially if an array antenna is used) and the maximum likelihood decision criterion employed to determine the location of the target and its image. Barton[55] states that these methods degrade in the presence of diffuse surface reflection. They also require more complicated feed structures and processing than a conventional monopulse.

High Range-Resolution   The surface-reflected signal travels a longer path than the direct signal so it may be possible in some cases to separate them by use of high range-resolution waveforms. By tracking only the direct signal, the angle errors introduced by multipath are avoided. The range resolution ΔR required to separate the direct from the ground-reflected signal is approximately

$$\Delta R = \frac{2 h_a h_t}{R} \qquad\qquad \textbf{[4.6]}$$

where h_a = antenna height, h_t = target height, and R = range. For a radar antenna height of 20 m, target height of 30 m, and a range of 4 km, the range resolution ΔR must be 0.3 m. This requires a pulse width of 2 ns and a bandwidth of 500 MHz, which is a shorter pulse and a greater bandwidth than usually found in operational radars. If the target were a sea-skimmer antiship missile at an altitude as low as 2 m above the sea, the bandwidth required to eliminate the multipath effect is too great to be practical for most applications. Thus range resolution has not usually been a satisfactory solution to the multipath problem.

Frequency Agility   It was stated in Sec. 4.4 in the discussion of the two-scatterer model that use of more than one frequency each sufficiently separated from one other could smooth the angle error due to glint and produce an average result that was less likely to have large errors. The frequencies had to be separated by the amount given by Eq. (4.4) in order to obtain independent values of the glint error. The same can occur with the elevation-angle measurement when multipath is present if D in Eq. (4.4) is taken as the difference in the path lengths; but there are two reservations. First, the average measurement of angle is more likely to be somewhere near the horizon rather than indicate the elevation angle of the target. Second, the bandwidth within which the various frequencies must occur is likely to be large. It is likely to be comparable to the bandwidth, discussed in the above, that is required of a short pulse for resolving the direct target echo from the image target echo. If a sufficiently large bandwidth were available to the radar designer, it might be better to employ that bandwidth for high range-resolution rather than for frequency agility. We have indicated previously, however, that this bandwidth is often larger than can be conveniently obtained in practice, especially when the target is at a very low altitude.

Doppler Resolution   Since the target and its image are at different elevation angles, their doppler frequency shifts are slightly different. With sufficient doppler resolution, the target can be separated from its image. In practice, however, the difference in the doppler frequencies generally is too small to be used for isolating the target from its image, unless exceptionally long observation times are employed.

Clutter Fence   A fence surrounding the radar can mask the echo from the image, especially when the near-in sidelobes are a factor in creating a multipath error. Fences can be expensive and are only of value when the radar is at a fixed site. Since the main beam illuminates the top edge of the fence, diffracted energy might illuminate the image.

Polarization   Vertical polarization, which is often used in tracking radars, reduces the surface-reflected signal from the image when the image elevation angle is in the vicinity of the Brewster angle (Sec. 8.2). It has no special advantage, however, at low grazing angles when the angle to the image is much less than the Brewster angle.

Complex Angle (CA)[49,61]   The normal monopulse receiver uses only the in-phase (or the out-of-phase) component of the difference signal. When a multipath signal is present along with the direct signal, the difference signal has a quadrature component. The in-phase and quadrature components of the error signal define a *complex angle* error signal. In the complex plane, with the in-phase and quadrature components as the two axes, the locus of the complex angle as a function of elevation angle is a spiral path. By measuring the complex angle, the target elevation can, in principle, be inferred. In using the complex-angle technique, the radar antenna is fixed at some angle above the horizon and an open-loop measurement of the complex angle is compared with a predicted set of values for the particular radar installation, antenna elevation-pointing angle, and terrain properties. A given in-phase and quadrature measurement does not give a unique value of the elevation angle since the plot of the complex angle shows multiple, overlapping turns of a spiral with increasing elevation angle of the target. The ambiguity can be resolved with frequency diversity or by continuous tracking over a long enough interval to recognize the ambiguous spirals.

This technique is limited by the need to resolve ambiguities, by the echoes from the real surface being different from theoretical when the surface is rough and the reflection is diffuse, and by random variations in the measurements which are difficult to remove by calibration because they vary rapidly with time and depend on target position.

Superresolution   In general, two equal-amplitude signals can be resolved in angle if they are separated by at least 0.8 beamwidth. Sometimes it can be better than this since resolution depends on the phase between the two signals as well as the signal-to-noise ratio. Many attempts have been made in the past to improve angular resolution beyond 0.8 beamwidth, but without the desired success. *Superresolution*[62] is an example that is based on *spectral estimation* or *spectral analysis* to provide resolution in angle. It has been of interest for low-altitude tracking because of the claim that it can produce improved resolution compared to conventional linear methods, but it does not work for radar signals. Such methods do provide enhanced resolution of uncorrelated signals (such as independent noise), as occur with multiple jamming signals or the radio astronomy observation of cosmic sources. They are not applicable, however, for the resolution of correlated signals—and radar echo signals are correlated since they originate from the same transmitter. W. D. White[63] has shown the limitation of superresolution for coherent echo signals and that it has little, if any, value in solving the radar low-angle tracking problem.

Superresolution methods produce impressive-looking plots with sharp responses indicating the target locations. Depending on the algorithm used, however, the amplitude and position of the responses might not always be related to the features of the target. Spurious responses can be obtained when the algorithm used is nonlinear. Of importance, but of lesser concern compared to its other limitations, is that the signal-to-noise ratio must be high to use these methods and many of the algorithms are computationally complex. Superresolution is a technique whose name implies more than has been delivered and whose promises have not materialized as advertised.

Maximum Likelihood Estimation   There have been many papers in the literature that describe the application of maximum likelihood estimation (MLE) to obtain the target's elevation angle in a multipath situation. Only a few are referenced here.[64–66] MLE also has been applied in conjunction with multiple frequencies[67] and with three-aperture antennas.[68,69] There have been some interesting results, but the technique has had limitations.

Electro-Optical and Infrared   Optical and IR sensors offer the advantage of far better angular resolution than can be obtained with radar. They have the important advantage of not suffering from the multipath problem. Both optical and IR have been used to supplement radar coverage at low angles. At low angles, they are of short range since they are seriously limited by high attenuation in the clear atmosphere as well as in rain. Furthermore, they do not provide range or doppler velocity measurements. If reliable all-weather capability is necessary, radar is the sensor that has to be seriously considered.

Other Comments   In addition to the above there have been several other methods proposed or investigated for mitigating the effects of glint when tracking a target at low angle. These include height diversity,[70] the use of neural nets,[71] the use of maximum likelihood estimation based on deterministic modeling that requires finding only four unknowns and which doesn't need to know the range,[72] the use of a threshold in the sum channel that must be exceeded in order to accept an elevation-angle measurement (this eliminates the large errors that can occur when the direct and the image signals are almost out of phase—the condition for a large glint error and for a small signal in the sum channel),[73] and the use of bias compensation to reduce the glint error along with a threshold on the sum signal that eliminates measurements made with small signal-to-noise ratio.[74] The large number of publications on this subject is an indication of the importance of low-angle tracking and that there has been a lack of a good all-purpose solution.

When conditions permit, the use of a narrow beamwidth is the best method to ensure accurate tracking at low angle. Off-axis tracking with the antenna at a fixed elevation, and with the elevation angle measurement made open-loop, is simple and can provide relief from the wild swings of the antenna caused by multipath glint.

The low-angle tracking problem as discussed here concentrated on the effect of multipath. Radars concerned with low-angle tracking also have to be able to detect targets in the presence of clutter echoes that can be many orders of magnitude larger than the target, so that doppler methods, as discussed in Chap. 3, need to be considered.

In this section, the radar was usually considered to be located on or near the surface and tracking a target at low altitude. The problem of low-angle multipath is also important for missile guidance at low altitudes.[75]

4.6 TRACKING IN RANGE[76,77]

In the early days of radar, tracking of a target in range was usually done manually by an operator who watched an A-scope or similar presentation and positioned a handwheel to maintain a marker on the display over the desired target pip. The setting of the handwheel was a measure of the target range and was converted to an electrical signal and supplied to a data processor. Manual tracking has many limitations and it cannot be used in systems such as missiles where there is no operator present. It was soon replaced by closed-loop automatic tracking, such as the *split-gate tracker*.

Split-Gate Tracker   The technique for automatically tracking in range is based on the split range gate. Two range gates, as indicated in Fig. 4.19, are generated. One is the *early gate* and the other is the *late gate*. The video echo pulse is shown in Fig. 4.19a, the relative position of the two gates at a particular instant is in Fig. 4.19b, and the difference signal is in Fig. 4.19c. In this example the portion of the signal in the early gate is less than that of the late gate. The signals in the two gates are integrated and subtracted to produce the difference error signal. The sign of the difference indicates the direction the two range gates have to be moved in order to have the pair straddle the echo pulse. The amplitude of the difference determines how far the pair of gates are from the "center" of the pulse, sometimes called the centroid. When the error signal is zero, the range gates are centered on the pulse and the position of the two gates gives the target's range. Deviation of the pair of gates from the center of the echo pulse increases the signal energy in one of the gates and decreases it in the other. This produces an error signal that causes the two pulses to be moved so as to reestablish equilibrium.

Range gating allows a single target to be isolated. The gate rejects unwanted signals and improves the signal-to-noise ratio by eliminating noise from other ranges. The AGC

Figure 4.19   Split-gate range tracking: (a) Echo pulse; (b) early–late range gates; (c) difference signal between early and late range gates.

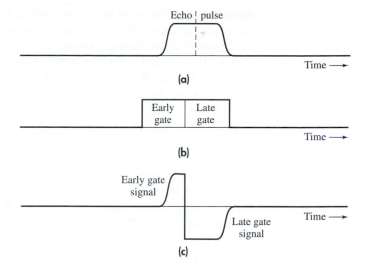

circuits respond only to the short time interval of the range gate where the echo from the target is expected. The range-gate width should be sufficiently narrow so as to minimize extraneous noise, but not so narrow that an appreciable fraction of the signal energy is excluded. Generally, the gate width is approximately equal to the pulse width. If tracking of the leading edge of the echo pulse is desired rather than its center, a bias can be inserted to move the gates forward. With leading-edge track, the gates should be "considerably narrower than with normal radar tracking."[78]

Range Glint   A target with multiple scatterers distributed in range can cause tracking errors because of glint similar to the glint error experienced in angle tracking that was discussed in Sec. 4.4.[79]

Assume a two-scatterer model, similar to that considered for angle glint, but with the scatterers separated in range rather than angle. One scatterer is at a range $R_1 = cT_1/2$ and the other is at $R_2 = cT_2/2$, where c = velocity of propagation and T_1 and T_2 are the respective two-way time delays to the two scatterers. The two scatterers are assumed to be unresolved. The error ΔT_R due to range glint in the measurement of the time delay, relative to the center of the two scatterers, is found to be[79]

$$\Delta T_R = \frac{\Delta T}{2} \cdot \frac{1 - a^2}{1 + a^2 + 2a \cos (2\pi f_0 \, \Delta T)} \qquad \text{[4.7]}$$

where $\Delta T = 2(R_2 - R_1)/c$ = time extent of the target; a = ratio of the amplitudes of the signals from the two scatterers ($a \le 1$); and f_0 is the radar carrier frequency. This expression is equivalent to the angle-glint expression of Eq. (4.2) if the error in Eq. (4.2) is referred to the center of the two scatterers as it is here rather than referred to the larger scatterer. The range measurement error can be larger than the extent of the target.

Some reservations need to be mentioned about the use of this expression for range-glint error. It is based on the time delay (or range) of a target being given as the rate of change of the echo signal phase with respect to frequency, or $\delta\phi/\delta f$. This can be seen by differentiating Eq. (3.1) as a function of frequency. Cross and Evans state[79] that the split-gate range tracker makes a measurement equivalent to finding range from the derivative of phase with respect to frequency. When the target consists of multiple scatterers, a glint error is introduced when using this criterion for extracting time delay.[80]

If, however, range is determined without a closed-loop tracker, a glint error does not occur. There can be range errors in this case due to the effect of noise, but they do not generate the type of effects that can happen when glint is possible. A glint error does not appear, for example, when the measurement is made manually by viewing the output signal on an A-scope display (amplitude versus range). A complex target with multiple scatterers within the radar resolution cell can cause inaccuracy in the measurement of the exact center of the target, but an operator would not be misled as would an automatic closed-loop tracker to position the range gate well outside of the target extent or the resolution of the pulse. Thus Eq. (4.7) does not apply to manual tracking or its electronic equivalent that doesn't used closed-loop tracking.

If range glint is a problem, some of the methods previously described for reducing angle glint might offer relief. The best approach, when sufficient bandwidth is available, is to employ high range-resolution. If the individual scatterers of a target can be resolved in range, both angle glint and range glint are not a problem. Tracking in range is generally much more accurate than tracking in angle (as measured by cross-range distance) so that any effects of range glint, if it occurs at all, is less of a concern than errors in angle, or cross-range.

4.7 OTHER TRACKING RADAR TOPICS

Target Acquisition   A tracking radar must first find and acquire (lock on to) its target before it can operate as a tracker. Most tracking radars employ a narrow pencil beam for accurate tracking in angle; but it can be difficult to search a large volume for targets when using a narrow antenna beamwidth. Some other radar, therefore, must first find the target to be tracked and then designate the target's coordinates to the tracker. These radars have been called *acquisition radars* or *designation radars* and are surveillance radars that search a large volume.

The tracker is slewed to the direction of the target based on the target coordinates supplied by the acquisition radar. These coordinates are not always accurate enough to bring the tracker directly onto the target. Some searching in both azimuth and elevation angle might have to be done by the tracker in order to find the target. There have been several different types of patterns employed to search a limited angular region, as was described in Sec. 5.7 of the 2d edition of this book; but the raster scan has been one of the most popular. The *raster,* or *TV, scan* paints a rectangular search area in a uniform manner. An airborne intercept radar, for example, might acquire its target by scanning a 3° pencil beam over a target space 60° in azimuth by 15.5° in elevation by scanning six elevation steps, or bars.[81] The search space is relatively large in this example; but it can be much smaller if the target location information provided to the tracker is more accurate.

Raster scan is a simple and convenient means for searching a limited sector. It is also known as N-bar scan, where N is the number of azimuth scans, or bars.

If a 2D air-surveillance radar (range and azimuth) is used for designating a target to a surface-based mechanical tracking radar, the tracker might acquire its target with a nodding-beam scan in elevation, which is a raster scan in the vertical rather than the horizontal. Surface-based mechanical tracking radars can be designed to slew 180° in one second and perhaps take another second to bring the target into track, so that acquisition can take place in under two seconds from the time of designation.

The target must be found in range as well as in angle. During the acquisition process, the tracking radar receiver range gate is scanned in range as the pulse propagates outwards in space. As has been mentioned, narrow range gates are important for restricting the noise the receiver must handle and to have only one target within the gate. The range gate is scanned from minimum to maximum range and is usually set to acquire the first target echo signal it detects. In some trackers there might be several contiguous range gates to shorten the acquisition time.[76]

A multifunction phased array radar for air defense usually performs as both the acquisition radar and the tracking radar. There is usually not much time that can be allocated for target acquisition in a multifunction radar that must maintain track on many targets and perform surveillance with adequate revisit times. The target-designation information obtained by a phased array needs to be accurate enough so that the tracking beam can be placed directly on the target without having to perform a search in angle. For this reason the target designation data from the surveillance portion of the multifunction phased array radar must be much more accurate than in air-defense systems that employ separate tracking and surveillance radars.

Servo System   The automatic tracking of a target in angle employs a servo system that utilizes the angle-error signals to maintain the pointing of the antenna in the direction of the target. The servo system introduces a lag in the tracking that results in an error. The lag error will depend on the nature of the target trajectory (whether it is a straight line, a gradual turn, or a rapid maneuver). The error also depends on the ability of the servo system to accommodate to changes in speed, velocity, or acceleration. A so-called *Type II servo system,* which has often been used in tracking radar, theoretically has no steady-state error when the target velocity is constant. For this reason, it is also known as a *zero velocity-error system.* A steady-state error will exist, however, for a step-acceleration input. The operation of the Type II servo is similar to the alpha-beta tracker discussed in Sec. 4.9 for automatic tracking in a scanning surveillance radar. There are other types of servo systems depending on the required tracking accuracy and the expected target characteristics.[82] The effect of velocity and acceleration on a servo system can be described by the frequency response of the tracking loop.[83]

Servo Bandwidth   The tracking bandwidth of a servo system is that of a low-pass filter. There are conflicting requirements on the choice of tracking bandwidth. On the one hand, the bandwidth should be narrow to minimize the effects of noise or jitter, reject unwanted signal components (such as the conical-scan frequency or aircraft-engine modulation), and to provide a smoothed output of the desired measurement parameters. A wide tracking bandwidth, on the other hand, is desired for following rapid changes in the target trajectory or the changes in the vehicle carrying the radar. Thus a wide bandwidth is needed so as not to lose track of a maneuvering target, but a narrow bandwidth is needed for sensitivity. The choice of servo bandwidth, therefore, usually must be a compromise.

A target at long range will have low angular rates of change and a low signal-to-noise ratio. The bandwidth can then be narrow to increase sensitivity and yet follow the target with minimum tracking lag. At short range, on the other hand, angular rates are likely to be large so that a wide bandwidth tracking filter is needed in order to follow the target without loss of track. Thus the loss in sensitivity because of the wider bandwidth is offset by the greater target signal at the shorter ranges. At shorter ranges, errors due to target glint can become a problem; hence, the bandwidth should be no wider than necessary in order to keep glint errors from becoming excessive. The tracking bandwidth can be made variable or even adaptive to conform automatically to the target conditions.

Lowest Servo Resonant Frequency   Another restriction on the tracking bandwidth of a mechanical tracker is that it should be small compared to the lowest natural resonant frequency of the antenna and its structural foundation in order to prevent the antenna system from oscillating at its resonant frequency. Figure 4.20 illustrates the bounds of measured data on the lowest resonant frequency as a function of antenna size. This figure is based on a compilation by the Aerospace Corporation of the lowest servo resonant frequencies for 190 individual radar, radio-telescope, and communications parabolic-

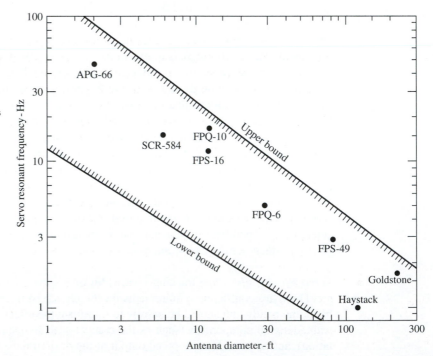

Figure 4.20   Bounds of the servo resonant frequency as a function of antenna diameter for actual tracking radars using parabolic-reflector antennas. (Based on extensive data compiled January 1994 by D. D. Pidhayny and Alan R. Lewis of the Aerospace Corporation.)

reflector antenna systems. Also shown are the resonant frequencies for several tracking radars. Howard[83] states that it is desired that the lowest resonant frequency be at least 10 times the servo bandwidth. This might be difficult to do with large antennas, and is not always practical. He mentions that the highly accurate AN/FPQ-6 radar tracker with a 29-ft diameter Cassegrain reflector antenna (Fig. 4.1a) has a resonant frequency only three times its servo bandwidth of 3.5 Hz. He also states that the smaller 12-ft antenna (such as for the AN/FPS-16) can provide a servo bandwidth up to 7 or 8 Hz.

Precision "On-Axis" Tracking[84,85]   Some of the most precise tracking radars are those associated with the instrumentation used at missile test-ranges.[86] One such class of highly precise tracking radar that achieves better than usual performance is known as *on-axis tracking*.

The output of a conventional servo system lags its input, which results in a tracking error. The on-axis tracker accounts for this lag and keeps the target being tracked in the center of the beam or on the null axis of the difference pattern. This improves tracking accuracy by reducing coupling between the azimuth and elevation angle-track channels, by minimizing the generation of cross polarization, and by reducing the effects of any system nonlinearities.

On-axis tracking includes (1) the use of adaptive tracking whose output updates a stored prediction of the target trajectory rather than control the antenna servo directly, (2) removal of the static and dynamic system biases and errors by prior calibration, and (3) use of appropriate coordinates for filtering (smoothing) the target data.

The radar's angle-error signals are smoothed and compared to a predicted measurement based on a target-trajectory model updated by the results of previous measurements. This is especially useful for situations where prior knowledge of the target trajectory is approximately known, as when tracking ballistic missiles or satellites. If the measurement is the same as the prediction, no adjustment is made and the antenna beam is pointed according to the stored prediction. If they do not agree, the target trajectory prediction is changed until they do. Thus the pointing of the antenna is performed open-loop based on the stored target-trajectory prediction which is updated by the radar measurements. The servo loop that points the antenna is made relatively wideband (a high data rate) to permit fast response when tracking targets have high acceleration. The adjustment of the predicted position based on the measured position, however, is performed with a narrow bandwidth. This error-signal bandwidth is adaptive and can be made very narrow, yet the system will continue to point open-loop based on the stored target-trajectory prediction and the wide bandwidth of the antenna-pointing servos. The radar makes its measurements of the target in the polar coordinates of range, azimuth, and elevation (r,θ,ϕ). A target on a straight-line trajectory has a curvilinear track in polar radar coordinates which can generate apparent accelerations and can complicate the processing. These can be avoided by converting the polar coordinates to rectilinear coordinates (x,y,z) for data smoothing and comparison with prediction. After updating the target prediction, the rectilinear coordinates are converted back to radar polar coordinates to drive the antenna.

Systematic tracking errors include (1) error in the zero reference of the encoders used to indicate the orientation of the antenna axis, (2) misalignment of the elevation axis with respect to the azimuth axis (nonothogonality), (3) droop or flexing of the antenna and mount caused by gravity, (4) misalignment of the antenna with respect to the elevation axis (skew), (5) noncoincidence of the azimuth plane of the mount to the local reference plane (mislevel), (6) insufficient dynamic range in the servo system, (7) finite transit time that results in the target being in a different position by the time the echo is received by the radar (which could be important for tracking of space objects), and (8) bending and additional time delay of the propagation path due to atmospheric refraction. Systematic errors are determined by prior measurement. They are then used to adjust the encoded position of the antenna to provide the correct target position.

A boresight telescope mounted on the radar antenna permits calibration of the mechanical axis of the antenna with respect to a star field. This calibration accounts for bias in azimuth and elevation, mislevel, skew, droop, and nonorthogonality. Tracking a visible satellite with the radar permits the position of the RF axis to be determined relative to the mechanical (optical) axis. The difference between the position measured by the optics and that measured by the radar is obtained after correction is made for the difference in atmospheric refraction for optical and RF propagation. This type of dynamic calibration requires that the radar be large enough to track satellites.

The calibration and compensation procedures to insure the tracking radar is of the highest accuracy might take several hours before the radar is ready for use. Such calibration might have to be repeated regularly. This is something that can be tolerated in precision instrumentation radars where time might not be important, but such times are less likely to be available for operational military radars. Also the performance of on-axis tracking can be degraded when the target performs unanticipated maneuvers.

There is nothing unique about any of the individual processes that enter into on-axis tracking. They can each be applied individually, if desired, to any tracking radar to improve the accuracy of track.[87]

Tracking in Doppler   Coherent radars can also track the doppler frequency shift of the moving target echo. This is especially important for airborne pulse doppler radars and missile-guidance radars. The tracking doppler filter not only provides a measure of relative velocity, but also improves the ability of the radar to isolate the moving target echo from much larger clutter echoes. The doppler tracking filter is sometimes called a *speed gate*. Modulations of the target echo due to moving parts such as the propellers of piston engines or the compressor stages of jet engines extend the doppler spectrum and can degrade tracking accuracy.

Track While Scan (Limited Sector Scan)   It was said at the beginning of this chapter that in the past the term *track while scan* has been applied to two different types of tracking radars. The name has been used to denote the tracking performed by a rotating-antenna air-surveillance radar which obtains target location updates each time the antenna beam rotates past the target, which might be from about 1 to 12 seconds. The name track while scan for this type of radar is now seldom used since almost all modern air-surveillance radars provide the equivalent with what is called *automatic detection and track (ADT)*, which is discussed in Sec. 4.9. Here we will briefly discuss the type of track-while-scan (TWS) radar used to rapidly scan a relatively narrow angular sector, usually in both azimuth and elevation. It combines the search function and the track function. Scanning may be performed with a single narrow-beamwidth pencil beam (or a monopulse cluster of beams) that might cover a rectangular sector in a raster fashion. Scanning can also be performed with two orthogonal fan beams, one that scans in azimuth and the other in elevation. TWS radars have been used in airport landing radars, airborne interceptors, and air-defense systems.

A difference between a continuous tracker and the TWS radar is that the angle-error signal in a continuous tracker is used in a closed-loop servo system to control the pointing of the antenna beam. In the TWS radar, however, there is no closed-loop positioning of the antenna. Its angle output is sent directly to a data processor. Another significant difference is that the TWS radar can provide simultaneous tracks on a number of targets within its sector of coverage, while the continuous tracker observes only a single target, which is why it is sometimes called a *single-target tracker* (STT). With comparable transmitters and antennas, the energy available to perform tracking is less in a TWS radar than a STT since the TWS shares its radiated energy over an angular sector rather than concentrate it in the direction of a single target. In airborne-interceptor applications, the TWS radar might be preferred when multiple targets have to be maintained in track and the tracking accuracy only has to be good enough to launch missiles which contain their own guidance systems to home on the target. On the other hand, if highly accurate tracking is needed, a single-target tracker might be preferred.[81]

Limited sector-scan TWS radars have been used in *Precision Approach Radars* (PAR) or *Ground-Controlled Approach* (GCA) systems that guide aircraft to a landing.[88] These radars allow a ground controller to direct an aircraft to a safe landing in bad weather by tracking it as it lands. The ground controller communicates to the pilot directions to change his or her heading up, down, right, or left. In the control of aircraft landing, fan beams have been used which are electromechanically scanned over a narrow sector at a rate of twice per second. The azimuth sector that is scanned might be 20° and the elevation sector 6 or 7°. Landing radars using limited-scan phased array antennas, such as the AN/TPS-19, operate differently than scanning fan beams since they electronically scan a pencil beam over a region 20° in azimuth and 15° in elevation at a rate of twice per second. The AN/TPN-19 used multiple receive beams to obtain a monopulse angle measurement. Being a phased array, the AN/TPN-19 could simultaneously track up to six aircraft at a 20-Hz data rate. In the past, radars for the control of landing aircraft have been mainly used by the military. Civilian pilots prefer to use landing systems in which the control of landing is in the aircraft's cockpit rather than from a voice originating from the ground.

Track-while-scan radars have also been used successfully for the control of weapons in surface-to-air missile systems for both land and ship-based air defense, especially by the former Soviet Union.

Generally, TWS systems using fan beams have some limitations compared to systems that operate with one or more pencil beams. The fan-beam system can see more rain and surface clutter, it is more vulnerable to electronic countermeasures, and there might be problems with associating multiple targets that appear in the two beams.

The advantage of TWS compared to a continuous tracker is that multiple targets can be tracked. Because it shares its energy over a region of space, the TWS radar needs to have a larger transmitter to obtain the same detection and tracking capabilities of a STT that dwells continuously on a single target. If the target's angle is found from the centroid of the angle measurements obtained as the TWS antenna beam scans past the target, inaccuracies can occur if the target signal fluctuates in amplitude. TWS radars are also more vulnerable to angle jamming than are continuous trackers.[89] TWS radars can use monopulse angle measurements in an open-loop manner, similar to a phased array. Since the monopulse measurement is not made with closed-loop tracking, a TWS radar should not experience the wild fluctuations in angle caused by glint when there are multiple scatterers within the resolution cell or when there is multipath at low elevation angles.

Tracking with Phased Array Radar   Tracking with a phased array radar is more like that of a track-while-scan radar or automatic tracking with a surveillance radar than the continuous single-target tracker that has been the subject of most of this chapter thus far. The advantage of a phased array is that it can have a much higher data rate than radars with mechanically scanned antennas and it can simultaneously track multiple targets by time sharing a single antenna beam. This is possible because of the rapid inertialess beam positioning that, in some systems, can switch the antenna beam from one direction to another within microseconds. There is no closed-loop feedback to control the positioning of the beam as in a STT. When it is time to update a track, the computer controlling the radar directs the beam in an open-loop fashion to the expected present position of the target to either transmit or to receive the expected echo signal. Many targets can be tracked simultaneously in this manner.

An example of a phased-array instrumentation radar used for multiple target tracking on missile ranges is the AN/MPS-39 that was shown in Fig. 4.1b. It is also known as MOTR (Multiple Object Tracking Radar). This is a transportable *C*-band radar capable of tracking up to 10 targets simultaneously with an angle accuracy of 0.2 mrad. Its space-fed lens array is mounted on an elevation-over-azimuth tracking pedestal so that its 60° cone-angle coverage can be positioned anywhere in the hemisphere. A 1° (at broadside) beamwidth is obtained with a 12-ft diameter aperture. The monopulse tracking capability is achieved with a four-horn triple-mode feed. The manufacturer's literature states it is able to can acquire a target in less than 1 s. The advantage of a phased array for range instrumentation is that a single system simultaneously can track a drone target or targets, the aircraft that launches a weapon, multiple parts of a destroyed target, and other aircraft that might be observing the test or be in the vicinity. Without the phased array, a separate single-target tracker, such as the AN/FPS-16 or equivalent, would have to be used for each target that is to be tracked.

An excellent review of tracking with phased arrays and multifunction phased array radar has been given by Barton.[90]

Signal-to-Noise Ratio   In Chapter 2, we found that reliable detection of a target required an integrated or single-pulse signal-to-noise ratio of the order of 12 to 15 dB. The tracking accuracy of the best radars is about 0.1 mrad. Such highly accurate tracking in angle requires much higher signal-to-noise ratios than required for detection. As indicated by Eq. (4.3) or the discussion of angle accuracy in Sec. 6.3, integrated or single-pulse signal-to-noise ratios need to be more than 40 dB to achieve the inherent accuracy available in the best of the precision trackers.

Distinguishing Feature of Monopulse   There are other types of simultaneous tracking methods, in addition to amplitude-comparison monopulse, that can obtain an angle measurement based on a single pulse. One might, for example determine the angle-error signal in one angular coordinate by taking the difference of the signals produced by two squinted beams at the video outputs of two separate receivers. Obtaining an accurate difference signal in this manner requires that the total gains of the two receiver channels be equal and remain that way. The high stability required to maintain balance of the two channels is not practical considering the total amount of gain there is in a receiver from the input antenna to the output difference circuit. The ineffectiveness of such a method was the reason why amplitude-comparison monopulse was invented during World War II by Robert Page of the U.S. Naval Research Laboratory.[91] By taking the sum and the difference of the two squinted beams right at the antenna, there is no need for ultrastable receivers as would be required to make an accurate measurement of angle in the video. It is the sum and difference networks at the antenna that distinguish monopulse from other simultaneous angle-measurement methods.

4.8   COMPARISON OF TRACKERS

In this chapter we have concentrated on two major tracking systems: the amplitude-comparison monopulse and the conical scan. There are several other types of trackers that were mentioned and others that could have been mentioned, but these two are the only ones that will be considered in the comparison in this section since they are quite representative of simultaneous tracking and sequential scanning.

Signal-to-Noise Ratio   When the target is being tracked, the signal-to-noise ratio from a monopulse radar is greater than that from a conical-scan radar since the monopulse antenna views the target at the peak of its sum pattern. The conical-scan radar views the target at some angle off the peak of the antenna beam. Thus the signal-to-noise ratio of monopulse might be from 2 to 4 dB greater than with conical scan.

Accuracy   The monopulse radar will have greater angle accuracy since its signal-to-noise ratio is higher (important when accuracy is limited by thermal noise). Also its angle accuracy is not affected by fluctuations in the amplitude of the echo signal as are sequential scanning systems. Both monopulse and conical-scan systems are degraded by the wandering of the apparent position of the target caused by glint. Monopulse, because of its better signal-to-noise ratio, has a better range accuracy than conical scan.

Complexity   The monopulse radar is the more complex of the two since it requires RF combining circuitry at the antenna and three receiving channels. Conical scan has only one receiving channel and uses a single feed, but it has to rotate or nutate the antenna beam at a high speed. In the early days of tracking radar, the relative complexity of monopulse was more pronounced. Receivers were big since they were based on vacuum-tube technology, and the combining circuitry was also large. Many a tracking radar development started out with monopulse, but had to switch to conical scan when its size or cost became too large. This is no longer a major consideration. Receivers are now solid state and small, and the combining circuitry has been made small by specially designed devices and the use of multimode feed systems. Thus complexity seldom need be a reason for not choosing monopulse. The Cassegrain is a popular antenna for monopulse since the combining circuitry and low-noise receiver front-ends can be placed behind the reflector where they can be better supported mechanically and not encounter the loss that can occur with long transmission lines.

A space-fed phased array radar can implement monopulse by using a multiple feed system similar to that used in a Cassegrain reflector antenna or a paraboloid reflector. With a corporate or constrained feed system in a phased array, the generation of multiple squinted beams requires a more complicated beam-forming network.

Minimum Number of Pulses   As the name implies, a monopulse radar can perform an angle measurement in two coordinates on the basis of a single pulse. A phased array radar might make such a single-pulse angle measurement if the signal-to-noise ratio received on a single pulse is large enough. Usually, a number of pulses are integrated in a monopulse single-target tracker to increase the signal-to-noise ratio and the measurement accuracy. The conical-scan tracker requires a minimum of four pulses per revolution of the beam to extract an angle measurement in two coordinates. Ten pulses per revolution is more likely than four. Generally the pulse repetition frequency (prf) is at least 10 times the conical-scan frequency. (There have been exceptions to this, however.)

The monopulse radar first makes its angle measurement and then integrates a number of measurements to obtain the required signal-to-noise ratio and to smooth (reduce) the error. The conical-scan radar, on the other hand, integrates a number of pulses first (in its narrowband filter) and then extracts the angle measurement. The two tracking radars would have to integrate the same approximate number of pulses to achieve the same signal-to-noise ratio (assuming comparable radar systems), except that the conical-scan tracker has to process more pulses than the monopulse because of its 2 to 4 dB lower signal-to-noise ratio compared to the monopulse tracker, as mentioned above.

Susceptibility to Electronic Countermeasures   The military conical-scan tracker is more vulnerable to spoofing countermeasures that take advantage of its conical-scan frequency. It can also suffer from deliberate amplitude fluctuations. A well-designed monopulse tracker is much harder to deceive.

Application   Monopulse trackers should be used when good angle accuracy is wanted and/or when susceptibility to electronic countermeasures is to be minimized. When high-performance tracking is not necessary, the conical-scan tracker might be used because of its lower cost and reduced complexity.

4.9   AUTOMATIC TRACKING WITH SURVEILLANCE RADARS

This section is concerned with tracking performed by an air-surveillance radar rather than a single target tracker, or STT. Tracking with air-surveillance radars is done at a much longer revisit time (lower data rate) between observations than STTs. STTs have revisit times of the order of a tenth of a second; air-surveillance radars have revisit times of a few to many seconds. The STT tracks only a single target; the air-surveillance radar may have many hundreds, or even thousands, of targets in track.

A long-range or a medium-range air-surveillance radar can have within its coverage a large number of aircraft targets as well as many individual clutter echoes. In the busy parts of the continental United States, a long-range radar might have more than 600 aircraft within view during the heavy air-traffic part of the day. Military radars might have to deal with many more than that number. In the early days of air-surveillance radars, when there were fewer aircraft in the sky and they didn't fly as fast as modern jets, target tracking was done manually by operators using grease pencils to mark the position of a target on each antenna scan, calculate its speed, and determine its direction. An alert, trained operator can update a target track manually at a rate of about once per two seconds.[92,93] With a civil air-traffic control radar antenna rotating at the relatively slow rate of 5 rpm, a good operator might be able to hold in track 5 or 6 aircraft. Such a rate cannot be sustained, however, for more than about 20 to 30 minutes before the operator's performance is reduced. With the higher antenna rotation rates of military radars, manual tracking of aircraft is even more limited. When there are more aircraft than can be tracked manually, automatic methods must be used for target detection, coordinate extraction, and tracking. This is called *automatic detection and track* (ADT).

ADT requires a good radar that eliminates clutter echoes and other undesired signals. This might sound like an obvious statement that can be said for many things; but when ADT was first introduced it was mistakenly applied to the then existing radars which had poor or no MTI, or any other means for reducing clutter. Its performance was a disaster when used with poor radars. A tracking system can be designed to recognize and eventually eliminate clutter echoes that do not form logical tracks; but it takes time and computer capacity which might not be available when a large number of targets must be maintained in track. Thus good tracking starts with a good radar that eliminates unwanted clutter echoes and other extraneous signals.

When clutter targets cannot be completely eliminated by doppler processing, the ADT radar has to employ CFAR to maintain a *constant false-alarm rate*. (CFAR techniques are discussed in Sec. 5.7.) A CFAR senses the local clutter and noise environment in the vicinity of the radar echo signal, and automatically adjusts the receiver decision-threshold to maintain a constant false-alarm rate. When the environment consists of range-extensive clutter such as from land, sea, or rain, the clutter signal is used to change the threshold whenever the target echo competes with clutter. The required change in threshold due to a change in the clutter environment occurs almost instantaneously as the clutter seen by the radar changes. CFAR works fine to keep the ADT from being overloaded by echoes that are not from aircraft; but it comes at a price. An increase in the detection threshold to maintain a constant false-alarm rate reduces the detection probability, thus some targets might not be detected. As discussed in Sec. 5.7, CFAR can introduce a loss of one or more dB, it can suppress desired echoes from targets in the vicinity of the target which is benefiting from the CFAR action, and it degrades the ability of the radar to resolve two closely spaced targets. CFAR may be needed in many radar systems; but one would like to have a radar that can accurately track a large number of targets without the accompanying limitations of CFAR.

Functions of an ADT[94–96]   The functions performed by an ADT system include target detection, track initiation, track association, track update, track smoothing (filtering), and track termination. Each will be briefly discussed, assuming a ground-based 2D (range and azimuth) air-surveillance radar with mechanically rotating antenna.

Automatic Detection   (This is reviewed in Sec. 5.5.) One approach is to first quantize the range and sometimes the azimuth angle (similar to what was done in the MTD discussed in Sec. 3.6). The quantization increment in range might be the pulse width and that in angle might be the azimuth beamwidth. At each range-azimuth quantization cell, the pulses received during the time the antenna scans past the target are integrated and a detection decision is made. CFAR generally is incorporated before the decision process in order to prevent excessive false alarms due to clutter echoes. Pulse integration is performed in some form of automatic detector, or integrator, such as discussed in Sec. 5.6. Another approach to automatic detection is the *moving window detector* which examines continuously the last n pulses and announces the presence of a target if at least m out of n of the pulses exceed a preset threshold.

A by-product of the automatic detection decision with a moving window detector or something similar, is an angle measurement made by *beam splitting*.[97] If n pulses are expected to be received from a target, beam splitting involves recognizing the beginning and end of the n pulses and locating their center. Angle accuracy depends on how well the beginning and end of the train of n pulses can be determined, as well as the number of pulses available and their signal-to-noise ratio. The beam splitting decision logic usually has no prior knowledge of a target's beginning. The logic must be sufficiently sensitive to quickly recognize the increased density region that signifies the start of an echo-signal pulse train, yet it must not be so sensitive that it generates false starts due to noise alone. Once a target's beginning is recognized, the device must sense the end of the increased density region. If the decision logic is too sensitive to change, it could cause a single target to split into two. A rough rule of thumb often quoted is that the accuracy of beam splitting is about one-tenth of a beamwidth when the signal-to-noise ratio is high enough to provide a good probability of detection.

Track Initiation   In principle, a track can be initiated from the target-location information obtained on two successive scans of the radar antenna. In practice, however, target information from three or more scans is usually needed to initiate a track. Two scans would be adequate when there is only one or a few aircraft within view; but when the radar has in view a large number of echoes, one or more additional scans may be needed to prevent false tracks from being initiated. Thus it is more usual to require three or more scans before establishing a track.

A *clutter map* is used to store the locations of fixed clutter echoes and prevent tracks from being initiated based on a clutter echo combined with a real target detection. Such tracks can eventually be recognized as false and can be dropped, but it takes time and computer capacity to do so when there are a large number of them. Clutter echoes for inclusion in the clutter map are those echoes that do not change their location with time or that change location too slowly to be targets of interest.

The process of initiating a track in a dense environment of targets and clutter not eliminated by the radar can be quite demanding in both computer software and hardware. Initiation of a new track may take more computer time and capability than any other aspect of ADT.

Requiring three scans for a civil air-traffic control radar to establish a track is usually not a burden. Waiting three scans for track establishment, however, may be an excessively long time for a military air-defense radar that has to direct weapon-control radars to defend against high-speed attackers that "pop up" at short range over the horizon. It is possible to quickly acquire the target on the basis of a single scan past the target if the radar can obtain a quick second look. This might be done with a *look-back beam* directed to the angle of the original detection. The quick look-back can provide confirmation of detection and an estimate of target's radial velocity. A phased array radar is well suited for this purpose, but mechanical rotating radars can also be outfitted with a fixed look-back beam. (One approach is shown in the antenna of Fig. 9.54.) Look-back might also be accomplished with a 3D radar whose electronically scanning beam in elevation is returned to the elevation angle of initial detection, before the radar beam entirely scans past the target.

Track Association   When a new detection is received that is not at the location of a clutter echo stored in the clutter map, an attempt is made to associate it with an existing track. Association with an existing track is aided by establishing for each track a small search window, or gate, within which the detection of the target on the next scan of the radar antenna is predicted to appear. The gate should be as small as possible in order to avoid having more than one echo fall within it when the traffic density is high or when two tracks are close to one another. On the other hand, a large gate region is needed if the tracker is to follow target turns or maneuvers. More than one gate size is used to overcome this dilemma. Figure 4.21 shows a small nonmaneuvering gate situated around the predicted position of the target in track. The size of the gate is determined by the estimated errors in the predicted position and the estimated errors in speed and direction of the track. The detection threshold might be lowered in the gate region to increase the probability of detection. When an echo is not found within the nonmaneuvering gate, the larger region encompassing the maneuvering gate is then searched. The size of the maneuvering gate is determined by the estimate of the maneuvering capability of the target under track.

One reason the target might not appear in the nonmaneuvering gate is that its radar cross section might decrease, or fade, so that it is not detected. When this is the case, it

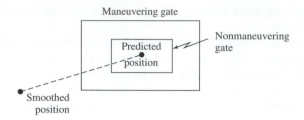

Figure 4.21   Maneuvering and nonmaneuvering gates centered at the target's predicted position. (From G. V. Trunk[110])

is possible for a false track to occur when a noise spike or an echo from another target is found in the maneuvering gate. To avoid the problem caused by a target fade and a false indication appearing in the larger maneuvering gate, the track can be divided into two tracks. (This is know as *bifurcation* of the track.) One is the original track with no new detection in the nonmaneuvering gate. The other is a new track based on the signal found in the maneuvering gate. After receiving the target position on the next scan of the radar (or sometimes after two scans), a decision is made as to which of the two tracks should be dropped.

Tracking is usually done in Cartesian coordinates, but the correlation gates are defined in polar (r, θ) coordinates.

Track Smoothing (α-β Tracker)   On the basis of a series of past target detections, the automatic tracker makes a smoothed (filtered) estimate of the target's present position and velocity, as well as its predicted position and velocity. One method for accomplishing this is with the α-β (alpha-beta) tracker that computes the present smoothed target-position $\bar{x}_n$ and smoothed velocity $\dot{\bar{x}}_n$ with the following equations

$$\text{smoothed position} = \bar{x}_n = x_{pn} + \alpha(x_n - x_{pn}) \qquad \text{[4.8a]}$$

$$\text{smoothed velocity} = \dot{\bar{x}}_n = \dot{\bar{x}}_{n-1} + \frac{\beta}{T_s}(x_n - x_{pn}) \qquad \text{[4.8b]}$$

The predicted position on the next scan (the $n + 1$st) is then

$$x_{p(n+1)} = \bar{x}_n + \dot{\bar{x}}_n T_s \qquad \text{[4.8c]}$$

where x_{pn} = predicted position of the target on the nth scan, x_n = measured position on the nth scan, α = position smoothing parameter, β = velocity smoothing parameter, and T_s = time between observations. If $\alpha = \beta = 0$, the tracker uses no current target information, only the smoothed data from prior observations. When $\alpha = \beta = 1$, no smoothing of the data is included at all. Thus the closer α and β are to zero, the more important is the smoothed track in determining the predicted track. The closer they are to 1, the more important is the currently measured data. If target acceleration is significant, a third equation can be added to describe an α-β-γ tracker, where γ = acceleration smoothing parameter.

Benedict and Bordner[98] show that if the transient response to a maneuvering target can be modeled by a ramp function, the output noise variance at steady state is minimized in an α-β tracker when $\beta = \alpha^2/(2 - \alpha)$. It was stated that the analysis does not, and cannot, specify the optimum value of α. The value of α is determined by the bandwidth and will depend on the system application. In selecting α, a compromise usually must be made between good smoothing of the random measurement errors (requiring a narrow bandwidth) and a rapid response to maneuvering targets (wide bandwidth). Trunk[99] states that an α and a β satisfying the above relation can be chosen so that the tracking filter will follow a specified g turn.

Another criterion for selecting the α-β values is based on the best linear track fitted to the radar data in a least squares sense:[100]

$$\alpha = \frac{2(2n - 1)}{n(n + 1)} \qquad \beta = \frac{6}{n(n + 1)} \qquad \text{[4.9]}$$

where n is the number of the scan or target observation ($n > 2$). The above equations for α and β are also called the Kalman gain components.[101]

The classical α-β tracker is designed to minimize the mean-square error in the smoothed position and velocity. This type of tracker is said to be relatively easy to implement, but it does not handle the maneuvering target. Some means has to be included to detect maneuvers and change the values of α and β accordingly.

The two tracking gates described in connection with Fig. 4.21 is one example of how to deal with a large maneuver. Another example is an adaptive α-β tracker which varies the smoothing parameters to achieve a variable bandwidth that allows the radar to follow target maneuvers.[102] When the target is not maneuvering the adaptive tracking algorithm provides heavy smoothing. If the target maneuvers or makes a turn, the filter bandwidth is widened so as to allow the track filter to follow. As the selection of the values of α and β become more sophisticated and requires knowledge of the statistics of the measurement errors and the prediction errors, the α-β tracker approaches the Kalman filter.

Track Smoothing (Kalman Filter)   The Kalman filter is similar to the α-β tracker except it can inherently provide for the maneuvering target.[103] A model for the measurement error has to be assumed, as well as a model for the target trajectory and the disturbance or uncertainty of the trajectory.[104] Such disturbances might be due to neglect of higher order derivatives in the model for the dynamics, random motions due to atmospheric turbulence, and deliberate target maneuvers. The Kalman filter can utilize a wide variety of models for measurement of noise and disturbance; but it is often assumed that these are described by white noise with zero mean.[105] A maneuvering target does not always fit such a model since its measurements are likely to be correlated. The proper inclusion of realistic models increases the complexity of the calculations. Furthermore, it may be difficult to describe ahead of time the precise nature of the trajectory disturbances.

When the Kalman filter is modeled with the target trajectory as a straight line, and the measurement noise and the trajectory disturbance are modeled as white, gaussian noise with zero mean, the Kalman filter equations reduce to the α-β tracker equations with α and β computed sequentially by the Kalman filter procedure.

Blackman[106] states that "Experience with airborne radars has shown the versatility of Kalman filters to be almost indispensable when dealing with problems presented by missing data, variable measurement noise statistics, and maneuvering targets with dynamic capabilities." The Kalman filter has better performance than the α-β tracker since it utilizes more information. The α-β tracker, however, might be considered when the target's maneuver statistics are not known or in a dense target environment where computational simplicity is important.

The Kalman filter[107] and the α-β tracker also can be applied to control digitally the feedback loop in the single-target tracker.

Track Termination   If the radar does not receive target information on a particular scan, the smoothing and prediction operation can be continued by properly accounting for the missing data.[108] (This is sometimes called *coasting.*) When data from a target is missing for a number of consecutive scans, the track is terminated. Although the criterion to be used for determining when to terminate a track depends on the application, it has been suggested that when three target reports are used to establish a track, five consecutive misses is a suitable criterion for termination.[109]

Tracking Performed on a Sector Basis   To avoid having to correlate all new detections with all existing tracks, the correlation and track updating process can be done on a sector basis. The 360° of azimuth coverage might be divided, for example, into 64 sectors. Figure 4.22 shows sectors 4 through 12. As given by Trunk,[110] the following actions occur:

- Radar has reported all its detections found in Sector 11 and is now obtaining detections from Sector 12.

- Detections from Sectors 9, 10, and 11 are examined to see if they correlate with clutter cells from Sector 10 in the stored clutter map. Any detections associated with clutter cells are deleted from the detection file.

- Detections from Sectors 7, 8, and 9 are examined for possible association with the firm tracks in Sector 8. At this point, all detections from clutter cells have been removed from Sector 9 and below. Detections that are found to be associated with firm tracks are deleted from the detection file and are used to update the appropriate track, as with an α-β tracker.

- Preference is given to firm tracks. Thus tentative tracks are examined two sectors behind the firm tracks.

- Remaining detections not associated with either clutter cells or tracks are used to initiate new tentative tracks. Both a tentative track and a clutter cell are established until enough information is obtained to determine which of the two can be deleted.

Maneuvering Targets   Target maneuvers and crossing trajectories can cause problems in tracking systems unprepared to cope with them. Some of the methods for handling maneuvering targets have been mentioned. Bar-Shalom[111] points out that a commercial aircraft that can turn at rates up to 3°/s completes a 90° turn in 30 s, so that a radar with a scan time of 10 s (rotation rate of 6 rpm) will obtain only three observations during the turn. He states that "very few of the existing schemes among the many in the literature (even adaptive ones with maneuver detection) can track such a target with good accuracy."

Figure 4.22   Various operations of a track-while-scan system performed on a sector basis.
| (From G. V. Trunk[110])

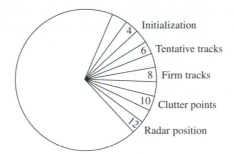

Initialization

Tentative tracks

Firm tracks

Clutter points

Radar position

He mentions algorithms and tracking methods to employ, but the radar must also be included. A 10-s scan time, or 6 rpm antenna rotation rate, is used for long-range civil air-traffic control radars where 90° turns are unlikely. Thus such radars are seldom exposed to a 90° turn in 30 s. Medium-range air-traffic control radars that observe traffic in the vicinity of airports, where aircraft are more likely to make a 90° turn, have shorter scan times than 10 s. The ASR-9, for example, has a scan time of 4.8 s. Military radars that have to deal with targets that maneuver more than do civil airliners, also have shorter scan times. In addition to the antenna scan time, the performance of an ADT system depends on the accuracy of the radar measurements, as well as the ability of the radar to resolve targets, reject clutter, eliminate discrete clutter echoes, and reject interference. Obtaining a good ADT is a total system problem, and should not be done in isolation by the control-system theorist without coordination with the radar systems designer.

Narrowband Data Communications Allowed by ADT   One of the corollary advantages of ADT it that it allows a reduction in the bandwidth of the radar output since the information content of processed video detections or processed target tracks occupy significantly less bandwidth than the raw radar data. Narrowband telephone lines can be used for transmission rather than wideband microwave data links. Thus, if circumstances permit, when the radar has to communicate its output to a remote location, it is often better to extract the information right at the radar and transmit processed information rather than raw data.

Meaning of Target Detection When Tracking is Performed   When the operators of early radars viewed the radar display, a single blip seen on the scope was taken to mean a target was detected. It was soon found that this was not sufficient, especially when clutter or noise appeared on the display. For greater reliability of detection, the operator required that echoes (blips) be seen on two consecutive scans, or two out of three (or some other combination) before the presence of a target was indicated. This is especially important when the noise or clutter has higher tails in its probability density function than the gaussian. With many modern ADT methods, the presence of a target is generally not finally confirmed until a track is formed. This criterion can employ a higher false-alarm rate per observation since multiple target observations are employed and the resulting track must be realistic. Thus the single-scan detection-threshold can be lowered and still achieve a very low probability of a false track.

Track-Before-Detect   The probability of detection can be improved for weak targets, or the detection range extended for conventional-size targets, by noncoherently integrating the radar echoes received over multiple scans of the antenna. With such long-duration integration, the target can move beyond the resolution cell of the radar and may traverse many resolution cells during the integration time. In order to perform integration, knowledge is required of the speed and direction of each target in order to properly associate the echoes from scan to scan. Target trajectory information, of course, is usually not known beforehand, so scan-to-scan integration must be performed assuming all possible trajectories. A correct trajectory is one that provides a realistic speed and direction for the type of target being observed. In other words, the target must be *tracked* before it is *detected,* which is why this method was originally called *track-before-detect.* It has also been called *retrospective detection* and *long-term integration,* as well as *scan-to-scan integration.*

Track-before-detect can provide greater sensitivity because of the large number of pulses that it integrates. Also, the single-scan probability of false alarm can be much higher because of the requirement that the pulses being integrated form a logical target track from scan to scan. It can, however, be much more demanding of computer capability than conventional ADT. Track-before-detect was first examined experimentally in the 1960s, when data processing technology was still based on vacuum tubes. What it could do was quite limited at that time. As solid-state digital computer technology became available with much better capabilities, so could the performance of track-before-detect be significantly improved.

An analytical investigation[112] of a hypothetical radar with track-before-detect for detection of sea-skimming missiles by a shipborne radar when sea clutter is a problem concluded that "long-term integration [track-before-detect] provides around 10 dB increase in detection sensitivity over conventional cumulative detection . . . in 'Beaufort' sea state 4." Track-before-detect, or retrospective detection, was experimentally investigated[113] using an AN/FPS-114 shore-based sea-surveillance radar. The conclusions were that the single-scan probability of false alarm can be of the order of 10^{-3} rather than the 10^{-5} to 10^{-8} that might be typically required for conventional automatic tracking systems.

In addition to requiring increased data processing capability, track-before-detect also requires a longer observation time, which may be something that is not be available in all tracking applications. Unusual target maneuver during the scan-to-scan integration time might also limit performance.

Weak echoes on each scan are used to form the track. They will have lower signal-to-noise ratio than a radar designed to make the detection decision based on what is received with only a single scan. Thus the conventional detection philosophy as discussed in Chap. 2 has to be modified for use with track-before-detect.

The outputs of each resolution cell, whether from weak targets or from noise, are examined over N scans to determine if they form logical tracks. Tracks may be found by an exhaustive search of all possible trajectories based on the data from all N scans. The number of possible trajectories to be examined with exhaustive search can easily become impracticably large. The processing load can be narrowed by limiting the range of target speeds and placing restrictions on the type of target trajectories. Even this is usually not sufficient. Barniv,[114] however, suggests that by applying a dynamic programming algorithm the amount of computation can be significantly reduced so as to become feasible.

It can be applied with curved as well as straight-line target trajectories. When conditions permit, dynamic programming can perform the equivalent of an exhaustive search, but in a more convenient manner and with significantly less computer capability. In his example of tracking using a set of (passive) infrared mosaic-sensor images (a nonradar example), Barniv found that the use of dynamic programming typically required five or more orders of magnitude less computations than what was needed if employing an exhaustive search.

Multiple-Radar Tracking   When multiple radars view a common volume, there can be improved tracking, the data rate can be greater than any of the radars acting alone, there is less vulnerability to electronic countermeasures, and less likelihood of having missed detections due to reduced echo-signal strength caused by nulls in one of the antenna patterns or changes in the target aspect. There are two related cases: one is when the radars are collocated, as on the same ship or at the same land site; the other, is when they are at separated locations and netted together.

Integrated Tracking from Collocated Radars at a Single Site   When more than one radar, covering approximately the same volume in space, are located in the same vicinity, their individual outputs can be combined to form a single track. The radars might operate in different frequency bands, have different antenna characteristics, and different data rates. There is more than one way to combine the outputs of multiple radars.[115] A good approach is to combine all the detections from each radar to form a single track and to update the track rather than develop separate tracks at each radar and either select the best track or combine them in some other manner.[116] The data from the various radars do not arrive at the tracker at a uniform rate. The development of a single track file by the use of the total data available from all radars produces a better track than combining the tracks developed individually at each radar. It reduces the likelihood of a loss of data as might be caused by antenna lobing, target fading, interference, and clutter since integrated processing permits the favorable weighting of the better data and lesser weighting of the poorer data. This method of combining data from multiple radars has been known as either *Automatic Detection and Integrated Tracking* (ADIT) or *Integrated Automatic Detection and Tracking* (IADT).

Integrated Tracking from Multiple Sites   When radars at multiple sites cover the same area, tracking can be combined as described above for the collocated radars; that is, the individual target data points (location and time), rather than the tracks, are transmitted to a single tracking center for processing into a single track. Farina and Studer[117] call this *centralized architecture*. They call a *distributed architecture* one in which the tracks are first formed at each radar site based only on the measurements from a single radar. These are transmitted to a tracking center which combines the tracks to establish a single track for each target. The distributed architecture allows the track information to be transmitted over telephone lines rather than wider bandwidth links, and it can employ less capable computer resources. The composite tracking accuracy it produces, however, is not as accurate as that with the centralized architecture.

To track targets over a large area as for air-traffic control or military air defense, multiple radars are located so that as a target leaves the coverage of one radar it enters the coverage of the other and a continuous track is maintained. There needs to be some overlap in the coverages so that tracks can be properly handed over from one radar to the other. One of the problems that plagued early multisite radars was being able to associate radar detections or tracks made by one radar with data or tracks made by another radar. The absolute location of a radar site, even when fixed, was not always known accurately so that the central tracking facility might generate multiple tracks from a single target. This situation is even more serious when trying to net the radars from multiple ships or aircraft. Methods were developed for merging the data from multiple sites to provide a single track when the location or orientations of the radars was inaccurate.[118] The availability of GPS (Global Positioning System), however, readily provides the accurate location of a radar and eliminates the problem of associating data from radars located far apart.

Multiradar tracking (MRT) developed for the air-traffic control system of the Rome Flight Information Region integrates five radars sited on the west side of the center and south of Italy.[119,120] Alenia ATC radars were used with ranges up to 170 nmi. They were separated approximately 135 nmi apart. The filtering algorithm used an α-β tracker. The measured tracking accuracy of this system when the target flew a straight line course, was said to be better than 0.3 nmi. The observed accuracy with an accelerating path (acceleration was not further described) was 0.8 nmi.

Combined Tracking with Radar and Passive Direction Finding   Radar provides the range and angle to a target. Passive direction finding (DF) provides angle and not range. The angle accuracy obtained by a radar is usually more accurate than the angle accuracy obtained by DF, so there is no gain in using the DF measurement in the tracking filter to produce a more accurate track. Passive DF, also called Electronic Support Measures (ESM), however, can assist in target recognition since it receives the emitted signal and can recognize the type of system that emitted it. Signals from hostile aircraft are likely to be distinctive and different from signals emitted by other aircraft, and can be used to recognize the nature of the target. This is of importance for military air-defense radars since some form of target recognition is needed to separate friendly and neutral aircraft from hostile aircraft.

Although DF systems provide target recognition and an angle measurement, they do not provide range so a track cannot be formed. If the passive angle track can be associated with a radar track, then the target being tracked by the radar can be recognized based on the information received by the passive DF system. Statistical methods for achieving this have been described.[121,122]

REFERENCES

1. Howard, D. D. "Tracking Radar." In *Radar Handbook,* M. Skolnik, Ed. New York: McGraw-Hill, 1990, Chap. 18.

2. Sherman, S. M. *Monopulse Principles and Techniques.* Norwood, MA: Artech House, 1984.

3. "IEEE Standard Radar Definitions," *IEEE Std 686–1997*, New York, 1997.

4. Sherman, S. M. Ref. 2, Sec. 4.4.

5. Sherman, S. M. Ref. 2, Sec. 4.4.2.

6. Page, R. M. "Monopulse Radar." *IRE Natl. Conv. Record* 3, pt. 8 (1955), pp. 132–134.

7. Skolnik, M. I. *Introduction to Radar Systems,* 1st ed. New York: McGraw-Hill, 1962, Sec. 5.4.

8. Rhodes, D. R. *Introduction to Monopulse.* Norwood, MA: Artech House, 1982.

9. Sherman, S. M. Ref. 2, Sec. 6.4.

10. Berger, H. "On the Optimum Squint Angles of Amplitude Monopulse Radar and Beacon Tracking Systems." *IEEE Trans.* AES-8 (July 1972), pp. 545–547.

11. Howard, D. D. Ref. 1, pp. 18.11–18.16.

12. Hannan. P. W. "Optimum Feeds for All Three Modes of a Monopulse Antenna." *IEEE Trans.* AP-9 (September 1961) pp. 444–460.

13. Barton, D. K. *Radars, Vol. 1, Monopulse Radar.* Norwood, MA: Artech House, 1974, papers nos. 11 and 12.

14. Elliott, R. S. *Antenna Theory and Design.* Englewood Cliffs, NJ: Prentice-Hall, 1981, Sec. 5.11.

15. Bayliss, E. T. "Design of Monopulse Antenna Difference Patterns with Low Side Lobes." *Bell System Tech. J.* 47 (1968), pp. 623–640.

16. Chubb, C. F., B. L. Hulland, and R. S. Noblit. Simplified Monopulse Radar Receiver, U. S. Patent 3,239,836, March 8, 1966.

17. Sherman, S. M. Ref. 2, Sec. 7.14.

18. Thomson, D. "Monopulse Design for Tactical Tracking Radar." *Microwave J.* 28 (May 1985), pp. 307–310.

19. Howard, D. D. Ref. 1, pp. 18.19–18.21.

20. Rubin, W. L., and S. K. Kamen. "SCAMP—A New Ratio Computing Technique with Application to Monopulse." *Microwave J.* 7 (December 1964) pp. 83–90.

21. Peebles, P. Z., Jr., and H. Sakamoto. "Conopulse Radar Tracking Accuracy." *IEEE Trans.* AES-16 (November 1980), pp. 870–874.

22. Bakut, P. A., and I. S. Bol'shakov. *Questions on the Statistical Theory of Radar, vol. II.* Moscow: Sovetskoye Radio, 1963, Chaps. 10 and 11. Translation available from NTIS, AD 645775, June 28, 1966.

23. Field, J. C. G. "The Design of Automatic-gain-control Systems for Auto-tracking Radar Receivers." *IEE Proc.,* Pt. C, 105 (March 1958), pp. 93–108.

24. Howard, D. D. Ref. 1, pp. 18.6–18.8.

25. Van Brunt, L. B. *Applied ECM, Vol. 2.* Dunn Loring, VA: EW Engineering, Inc., 1982, pp. 570–575.

26. Barton, D. K. *Modern Radar System Analysis.* Norwood, MA: Artech House, 1988, Chap. 11, Radar Error Analysis.

27. Meade, J. E. "Target Considerations." In *Guidance,* A. S. Locke, Ed. Princeton, NJ: D. Van Nostrand, 1955, Chap. 11, pp. 440–442.

28. Delano, R. H. "A Theory of Target Glint or Angular Scintillation in Radar Tracking." *Proc. IEEE* 41 (December 1952), pp. 1778–1784.

29. Howard, D. D. Ref. 1, Sec. 18.8.

30. Ostrovityanov, R. V., and F. A. Basalov: *Statistical Theory of Extended Radar Targets.* Norwood, MA: Artech House, 1985.

31. Wright, J. W. "Radar Glint—A Survey." *Electromagnetics* 4 (1984), pp. 205–227.

32. Howard, D. D. "Radar Target Angular Scintillation in Tracking and Guidance Systems Based on Echo Signal Phase Front Distortion." *Proc. Natl. Electronics Conf.,* vol. XV, Chicago, IL, Oct. 13–15, 1959.

33. Barton, D. K. *Radar System Analysis.* Norwood, MA: Artech House, 1977, Sec. 9.2.

34. Howard, D. D. Ref. 1, Sec. 18.10.

35. Lange, S., and A. Hammer. "Thermal Noise Analysis in Conical-Scan Radars." *IEEE Trans.* AES-14 (March 1978), pp. 400–413.

36. Peebles, P. Z., Jr., and T. K. Wang. "Noise Angle Accuracy of Several Monopulse Architectures." *IEEE Trans.* AES-18 (November 1972), pp. 712–721.

37. Dunn, J. H., D. D. Howard, and A. M. King. "Phenomena of Scintillation Noise in Radar Tracking Systems." *Proc. IRE* 47 (May 1959), pp. 855–863.

38. Howard, D. D. Ref. 1, Sec. 18.11.

39. Lind, G. "Reduction of Radar Tracking Errors with Frequency Agility." *IEEE Trans.* AES-4 (May 1968), pp. 410–416.

40. Weimer, F. C., and L. Peters, Jr. "Study of Pointing Errors in Conically Scanning and Monopulse Tracking Radars for Multipoint Targets." Antenna Lab., Ohio State University, Rept. 601-12, AD 123 476, 1956.

41. Loomis, J. M., and E. R. Graf. "Frequency Agility Processing to Reduce Radar Glint Pointing Error." *IEEE Trans.* AES-10 (November 1974), pp. 811–820.

42. Dunn, J. H., and D. D. Howard. "The Effects of Automatic Gain Control Performance on the Tracking Accuracy of Monopulse Radar Systems." *Proc. IRE* 47 (March 1959) pp. 430–435.

43. Delano, R. H. "A Theory of Target Glint or Angle Scintillation in Radar Tracking." *Proc. IRE* 41 (December 1953), pp. 1778–1784.

44. Borden, B. "What is the Radar Tracking 'Glint' Problem and can it be solved?" *Naval Air Warfare Center Weapons Division,* China Lake, CA, NAWCWPNS TP 8125, May 1993, AD-A266 509.

45. Borden, B. "Requirements for Optimal Glint Reduction by Diversity Methods." *IEEE Trans.* AES-30 (October 1994), pp. 1108–1114.

46. Guest, I. W., and C. K. Pauw. "Radar Detector Preprocessor for Glint Reduction in a Tracking Radar." *IEEE Trans.* 29 (April 1993), pp. 527–531.

47. Hatcher, J. L., and C. Cash. "Polarization Agility for Radar Glint Reduction." *IEEE Region 3 Convention,* Huntsville, Alabama, Nov. 19–21, 1969.

48. Sims, R. J., and E. R. Graf. "The Reduction of Radar Glint by Diversity Techniques." *IEEE Trans.* AES-19 (July 1971), pp. 462–468.

49. Barton, D. K. "Low-Angle Radar Tracking." *Proc. IEEE* 62 (June 1974), pp. 687–704.

50. Howard, D. D., J. T. Nessmith, and S. M. Sherman. "Monopulse Tracking Errors Due to Multipath Causes and Remedies." *IEEE EASCON '71* (1971), pp. 175–182.

51. Linde, G. J. "Improved Low-Elevation Angle Tracking with Use of Frequency Agility." Naval Research Laboratory, Washington, D.C. Report No. 7378, March 17, 1972.

52. Howard, D. D. Ref. 1, Sec. 18.9.

53. Skolnik, M. I. *Introduction to Radar Systems.* 2nd ed. New York: McGraw-Hill, 1980, p. 546.

54. Bruder, J. A., and J. A. Saffold. "Multipath Effects on Low-Angle Tracking at Millimeter-Wave Frequencies." *IEE Proc.-F* 138, No. 2 (April 1991), pp. 172–184.

55. Barton, D. K. "Low-Angle Tracking." *Microwave J.* 19 (December 1976), pp. 19–24 & 60.

56. Barton, D. K. *Modern Radar System Analysis.* Norwood, MA: Artech House, 1988, Sec. 11.2.

57. Cross, D., D. Howard, M. Lipka, A. Mays, and E. Ornstein: "TRAKX: A Dual-Frequency Tracking Radar." *Microwave J.* 19 (September 1976), pp. 39–41.

58. Klaver, L. J. "Combined X/K_a-Band Tracking Radar." *Conference Proc. Military Microwaves,* London, England, October 25–27, 1978, pp. 147–155.

59. White, W. D. "Low-Angle Radar Tracking in the Presence of Multipath." *IEEE Trans.* AES-10 (November 1974), pp. 835–852.

60. White, W. D. "Double Null Technique for Low Angle Tracking." *Microwave J.* 19 (December 1976), pp. 35–38 and 60.

61. Howard, D. D., S. M. Sherman, D. N. Thomson, and J. J. Campbell. "Experimental Results of the Complex Indicated Angle Technique for Multipath Correction." *IEEE Trans.* AES-10 (November 1974), pp. 779–787.

62. Gabriel, W. F. "Spectral Analysis and Adaptive Array Superresolution Techniques." *Proc. IEEE* 68 (June 1980), pp. 654–666.

63. White, W. D. "Angular Spectra in Radar Applications." *IEEE Trans.* AES-15 (November 1979), pp. 895–899. See also discussion in the same issue, pp. 899–904.

64. Zoltowski, M. D. "Beamspace ML Bearing Estimation Incorporating Low-Angle Geometry." *IEEE Trans.* AES-27 (May 1991), pp. 441–458.

65. Reilly, J., J. Litva, and P. Bauman. "New Angle-of-Arrival Estimator Comparative Evaluation Applied to the Low-Angle Tracking Radar Problem." *IEE Proc.* 135, Pt. F (October 1988), pp. 408–420.

66. Zoltowski, M. D. "Beamspace ML Bearing Estimation for Adaptive Phased Array Radar." In *Adaptive Radar Detection and Estimation,* S. Haykin and A. Steinhardt, Eds. New York: John Wiley, 1992, Chap. 5.

67. Bosse, E., R. M. Turner, and E. S. Riseborough. "Model-Based Multifrequency Array Signal Processing for Low-Angle Tracking." *IEEE Trans.* AES-31 (January 1995), pp. 194–209.

68. Cantrell, B. H., W. B. Gordon, and G. V. "Trunk Maximum Likelihood Elevation Angle Estimates of Radar Targets Using Subapertures." *IEEE Trans.* AES-17 (March 1981), pp. 213–221.

69. Taha, A., and J. E. Hudson. "Trigonometric High-Resolution Method to Resolve Two Close Targets." *IEE Proc.* 134, Pt. F (October 1987), pp. 597–601.

70. Giuli, D., and R. Tiberio. "A Modified Monopulse Technique for Radar Tracking with Low-Angle Multipath." *IEEE Trans.* AES-11 (September 1975), pp. 741–748.

71. Wong, T., T. Lo, H. Leung, J. Litva, and E. Bossse. "Low-Angle Radar Tracking Using Radial Basis Function Neural Network." *IEE Proc.* 140, Pt. F (October 1993), pp. 323–328.

72. Lo, T., and J. Litva. "Use of a Highly Deterministic Multipath Signal Model in Low-Angle Tracking." *IEE Proc.* 138, Pt. F (April 1991), pp. 163–171.

73. Seifer, A. D. "Monopulse-Radar Angle Tracking in Noise or Noise Jamming." *IEEE Trans.* AES-28 (July 1992), pp. 622–638.

74. Daeipour, E., W. D. Blair, and Y. Bar-Shalom. "Bias Compensation and Tracking with Monopulse Radars in the Presence of Multipath." *IEEE Trans.* AES-33 (July 1997), pp. 863–882.

75. Ivanov, A. "Radar Guidance of Missiles." In *Radar Handbook,* M. Skolnik, Ed. New York: McGraw-Hill, 1990, chap. 19, p. 19.30.

76. Howard, D. D. Ref. 1, Sec. 18.5.

77. Barton, Ref. 56, Chap. 9.

78. Van Brunt, L. B. *Applied ECM.* Dunn Loring, VA: EW Engineering, 1982, Vol. 2, pp. 309–315.

79. Cross, D. C., and J. E. Evans. Target-Generated Range Errors, Naval Research Laboratory, Washington, D.C., Memorandum Report 2719, January, 1974.

80. Skolnik, M. I. Radar Information from the Partial Derivative of the Echo Signal Phase from a Point Scatterer, Naval Research Laboratory, Washington, D.C., Memorandum Report 6148, February 17, 1988.

81. Stimson, G. E. *Introduction to Airborne Radar.* El Segundo, CA: Hughes Aircraft Co., 1983, p. 472. See also the second edition, published by Menhdam, NJ: Scitech, 1998, pp. 388–390.

82. Barton, D. K. Ref. 56, Sec. 10.2.

83. Howard, D. D. Ref. 1, Sec. 18.4.

84. Schelonka, E. P. "Adaptive Control Techniques for On-Axis Radars." *IEEE 1975 International Radar Conference,* Arlington, VA, April 21–23, 1975, pp. 396-401, IEEE Publication 75 CHO 938-1 AES.

85. Clark, B. L., and J. A. Gaston. "On-Axis Pointing and the Maneuvering Target." *IEEE NAECON '75 Record,* pp. 163–170, 1975.

86. Nessmith, J. T. "Range Instrumentation Radars. *IEEE ELECTRO '76,* Boston, MA, May 11–14, 1876. Reprinted in Skolnik, M. *Radar Applications.* New York: IEEE Press, 1988, pp. 458–468.

87. Nessmith, J. T., and Patton, W. T. "Tracking Antennas." In *Antenna Engineering Handbook,* 2d ed. R. C. Johnson and H. Jasik, Eds. New York: McGraw-Hill, 1984, Chap. 34.

88. Ward, H. R., C. A. Fowler, and H. I. Lipson. "GCA Radars: Their History and State of Development." *Proc. IEEE* 62 (June 1974), pp. 705–716.

89. Barton, D. K., and S. A. Leonov. *Radar Technology Encyclopedia.* Norwood, MA: Artech House, 1997, p. 445.

90. Barton, D. K. Ref. 56, Sec. 10.4.

91. Page, R. M. Accurate Angle Tracking by Radar, Naval Research Laboratory, Washington, D.C. Report RA 3A 222A, 28 December 1944. Reprinted in D. K. Barton, Ref. 13, paper No. 1.

92. Plowman, J. C. "Automatic Radar Data Extraction by Storage Tube and Delay Line Techniques." *J. Brit. IRE* 27 (October 1963), pp. 317–328.

93. Baker, C. H. *Man and Radar Displays.* New York: Macmillian, 1962.

94. Brookner, E. *Tracking and Kalman Filtering Made Easy.* New York: Wiley, 1998.

95. Trunk, G. V. "Automatic Detection, Tracking, and Sensor Integration." In *Radar Handbook,* 2nd ed., M. Skolnik, Ed. New York: McGraw-Hill, 1990, Chap. 8.

96. Farina, A, and F. A. Studer. *Radar Data Processing. Vol. I—Introduction and Tracking.* New York: Wiley, 1986.

97. Dinneen, G. P., and I. S. Reed. "An Analysis of Signal Detection and Location by Digital Means." *IRE Trans.* IT-2 (March 1956), pp. 29–38.

98. Benedict, T. R., and G. W. Bordner. "Synthesis of an Optimal Set of Radar Track-While-Scan Smoothing Equations." *IRE Trans.* AC-7 (July 1962) pp. 27–32.

99. Trunk. G.V. Ref. 95, Sec. 8.3.

100. Quigley, A. L. C. "Tracking and Associated Problems." *International Conf. on Radar—Present and Future,* Oct. 23–25, 1975, London, pp. 352–359, IEE Conference Publication No. 105.

101. Farina, A., and F. A. Studer. Ref. 96, Sec. 3.4.2.

102. Cantrell, B. H. "Adaptive Tracking Algorithm." Naval Research Laboratory, Washington, D.C., Memorandum Report 3037, April, 1975.

103. Kalman, R. E., and R. S. Bucy. "New Results in Linear Filtering and Prediction Theory." *J. Basic Eng.* (ASME Trans., Ser. D) 83 (March 1961), pp. 95–107.

104. Morgan, D. R. "A Target Trajectory Noise Model for Kalman Trackers." *IEEE Trans.* AES-12 (May 1986) pp. 405–408.

105. Hampton, R. L. T., and J. R. Cooke. "Unsupervised Tracking of Maneuvering Vehicles." *IEEE Trans.* AES-9 (March 1973), pp. 197–207.

106. Blackman, S. S. *Multiple-Target Tracking with Radar Applications.* Dedham, MA: Artech House, 1986, Chap. 2.

107. Biernson, G. *Optimal Radar Tracking Systems.* New York: Wiley, 1990, Chap. 8.

108. Kanyuck, A. J. "Transient Response of Tracking Filters with Randomly Interrupted Data." *IEEE Trans.* AES-6 (May 1970), pp. 313–323.

109. Leth-Espensen, L. "Evaluation of Track-While-Scan Computer Logics." In *Radar Techniques for Detection, Tracking, and Navigation,* W. T. Blackband, Ed. New York: Gordon and Breach, 1966, Chap. 29.

110. Trunk, G. V. "Survey of Radar ADT." Naval Research Laboratory, Washington, D.C., Report 8698, June 30, 1983.

111. Bar-Shalom, Y. *Multitarget-Multisensor Tracking Applications and Advances,* Vol. II, Boston, MA: Artech House, 1992, p. xiii.

112. Urkowitz, H., and M. R. Allen. Long Term Noncoherent Integration Across Resolvable Sea Clutter Areas." *Proc. 1989 IEEE National Radar Conference,* March 29–30, 1989, pp. 67–72. For more technical details, see Allen, M. R., S. L. Katz, and H. Urkowitz. "Geometric Aspects of Long-Term Noncoherent Integration." *IEEE Trans.* AES-25 (September 1989), pp. 689–700.

113. Prengaman, R. J., R. E. Thurber, and W. G. Bath. "A Retrospective Detection Algorithm for Extraction of Weak Targets in Clutter and Interference Environments." *IEE International Conference Radar-82,* October 18–20, 1982, pp. 341–345, IEE Conference Publication No. 216.

114. Barniv, Y. "Dynamic Programming Algorithm for Detecting Dim Moving Targets" In *Multitarget-Multisensor Tracking,* Vol. I. Y. Bar-Shalom, Ed. Norwood, MA: Artech House, 1990, Chap. 4.

115. Trunk, G.V. Ref. 95, Sec. 8.4.

116. Cantrell, B. H., G. V. Trunk, J. D. Wilson, and J. J. Alter. "Automatic Detection and Integrated Tracking." *IEEE 1975 International Radar Conference* pp. 391–395. Arlington, VA, Apr. 21–23, 1975.

117. Farina, A., and F. A. Studer. *Radar Data Processing. Vol. II—Advanced Topics and Applications.* New York: Wiley, 1986.

118. Bath, W. G. "Association of Multisite Radar Data in the Presence of Large Navigational and Sensor Alignment Errors." *IEE Int. Radar Conf.,* London, pp. 371–379, 1982.

119. Farina, A., and F. A. Studer. "Radar and Sensor Netting Present and Future." *Microwave J.* 29 (January 1986), pp. 97, 98, 100, 104, 106, 108, 110, 112, 114, and 124.

120. Farina, A., and F. A. Studer. Ref. No. 117, Sec. 7.2.6.

121. Trunk, G. V., and J. D. Wilson. "Association of DF Bearing Measurements with Radar Measurements." *IEEE Trans.* AES-23 (July 1987), pp. 438–447.

122. Farina, A., and B. La Scala. "Methods for the Association of Active and Passive Tracks for Airborne Sensors." *International Radar Symposium, IRS-98*, September, 15–17, 1998, Munich, Germany, pp. 735–744.

PROBLEMS

4.1   If the one-way antenna power pattern of a conical-scan tracking antenna is described by the gaussian function of Eq. (3.34), what is the loss in received signal when the target is directly at the beam crossover? The antenna half-power beamwidth is 2 degrees and the squint angle is 0.75 degree.

4.2   One reason that automatic gain control (AGC) is used in continuous-tracking radars is to prevent saturation of the receiver due to such things as the variation of the target echo signal with range and with aspect. (a) How much might a target echo vary in power (in dB) if the radar has to observe targets from a minimum range of 2 nmi to a maximum of 100 nmi? (b) How much might the echo vary because of different size aircraft targets? [See Table 2.1.] (c) How much might an aircraft echo change due to a change in aspect? [See Figs. 2.15 and 2.16.]

4.3   Compare the amplitude-comparison monopulse tracker and the conical scan tracker with respect to accuracy at long, medium, and short ranges; complexity; the number of pulses usually used for an angle measurement; and the type of application where each might be preferred.

4.4   (a) Why is the amplitude-comparison monopulse more likely to be preferred over the phase-comparison monopulse? (b) Why is the conical scan tracker more likely to be preferred over the sequential lobing, or lobe switching, tracker?

4.5   Derive the error signal in one angle coordinate for the amplitude-comparison monopulse. Show that for small angular errors, the error signal is linearly proportional to θ_T, where θ_T is the angle of the target measured from the antenna pointing direction. The angular separation between the two squinted antenna beams is $2\theta_q$. [The one-way (voltage) pattern of the two antenna beams when not squinted can be approximated by the normalized gaussian function $\exp(-a^2\theta^2/2)$; where $a^2 = 2.776/\theta_B^2$, and θ_B is the half-power beamwidth. Note that the hyperbolic cosine can be expressed as $\cosh x = (e^x + e^{-x})/2$ and the hyperbolic sine as $\sinh x = (e^x - e^{-x})/2$; and for small values of x, $\sinh x \approx x$ and $\cosh x \approx 1$. Also, $\sinh 2x = 2 \sinh x \cosh x$.].

4.6   Two echo signals from a finite size target arrive at a single-coordinate radar tracking antenna from a direction of $+\theta_D/2$ and $-\theta_D/2$, respectively, measured with respect to the antenna broadside (boresight) direction. How can one use the output of the sum channel to recognize that a serious glint error is occurring?

4.7 A tracking radar is tracking a "dumbbell" target consisting of two isotropic scatterers separated by an angular extent θ_D when seen at the location of the radar. (This is similar to the geometry shown in Fig. 4.15a in the discussion of glint.) The ratio of the echo-signal amplitudes from the two scatterers is $a = 0.5$. If the phase difference α between the two scatterers varies uniformly with time over the range of values 0 to 2π radians, what fraction of the time will the radar angle-error signal indicate an "apparent" target direction that is pointing outside the angular extent θ_D of the dumbbell target? (You may assume that the angular extent of each of the two scatterers is very small compared to θ_D.)

4.8 A monopulse radar is found to be tracking a target with an angular accuracy of 0.5 mil at a particular range. (a) What is this accuracy in degrees? (b) Assuming the accuracy is determined solely by the receiver noise, what would be the angle accuracy at this range of a similar conical scan tracker (the same frequency, beamwidth, power, noise figure, prf, number of pulses processed, and antenna effective area)? (c) If, on the other hand, the accuracy is at short range so that angle accuracy is determined solely by glint, what would be the accuracy of the conical scan tracker relative to that of the monopulse tracker?

4.9 Show that the phase of the echo from a dumbbell target (two unresolved isotropic scatterers separated by a distance D) oriented along the radial (range) direction (rather than the cross-range direction) is decorrelated if the frequency is changed by at least $c/2D$, where c is the velocity of propagation.

4.10 A target has an effective depth in the radial (range) dimension of 15 m. What must be the change in frequency in order to get a decorrelated measurement of angle glint?

4.11 This problem involves range glint. (a) A dumbbell target at a long range from the radar has its two unresolved equal cross-section isotropic scatterers located in line in the radial (range) direction and separated by 10 m. What is the phase difference between the echoes from these two scatterers when viewed by a radar at a frequency of 3 GHz? What is the range glint error in this case? (b) What change in aspect angle (such as might be caused by a rotation of the target about its center) will cause the two echoes to be 180° out of phase, resulting is a severe glint error in range? (c) What change in frequency is needed to decorrelate the echo when the target is oriented as in (b)? (d) What must be the pulse width [in (a)] in order to resolve the two scatterers (so that glint may be avoided)?

4.12 What two measures might be taken to reduce the effects of the glint error in both angle and range?

4.13 (a) Why does a tracking radar have poor accuracy at low elevation angles? (b) Summarize the two methods that may be worth considering when it is necessary to avoid poor tracking of targets at low altitudes.

4.14 One approach (mentioned in Sec. 4.5) to minimize the probability of breaking lock and the large errors that can occur in elevation angle when tracking at low-angles is to fix the antenna beam at some low elevation angle θ_e and cease closed-loop tracking until the target returns to a higher elevation angle where multipath is reduced. In such a case, all that might be said about the target's elevation angle is that it is somewhere within the half-power beamwidth θ_B of the antenna. Let $\theta_e = \theta_B/2$. (a) If it is assumed, when using this method, that the probability of the target being within the half-power elevation beamwidth

θ_B is uniform, the estimate of its location is the mean value, which is $\theta_B/2$. What is its standard deviation under these assumptions? (b) The assumption that the elevation-angle measurements are uniformly distributed within the elevation beamwidth may or may not be correct; but if it were valid, what does the result of (a) say about the use of more sophisticated methods that give rms values of the elevation that are no better than 0.1 to 0.3 beamwidths?

4.15   What might be the upper bound of the resonant frequency of the servo when the tracking antenna is 30 ft in diameter?

4.16   a. If an alert, well-trained air-surveillance radar operator can manually update an aircraft track in two seconds, how many aircraft can be held in track by a single operator when the radar antenna scans at a rate of 6 rpm?

b. If seven operators were available for performing manual tracking, how many targets do you think could be held in track with the radar of part (a) (assuming each operator had a display)? [Caution: This is a little like the question: How many engineers does it take to change a light bulb?]

4.17   (a) What is meant by beam splitting? (b) Describe briefly how is it accomplished? (c) What accuracy might it typically have?

4.18   Under what conditions does the Kalman filter perform like the α-β tracking filter?

4.19   (a) What is the chief advantage of automatic detection and tracking? (b) What are its limitations?

5

Detection of Signals in Noise

5.1 INTRODUCTION

A radar *detects the presence* of an echo signal reflected from a target and *extracts information* about the target (such as its location). One without the other has little meaning. The detection of radar signals in noise was discussed in Chap. 2, detection of moving targets in clutter was the subject of Chap. 3, and detection of stationary targets in clutter is discussed in Chap. 7. In the current chapter, additional aspects of the detection of radar signals in noise will be presented, chiefly the matched filter and related topics. The extraction of information from a target echo signal is the subject of the following chapter, Chap. 6.

Methods for the detection of desired signals and the rejection of undesired noise, clutter, and interference in radar are called *signal processing*. The matched filter, described next, is an important example of a radar signal processor.

5.2 MATCHED-FILTER RECEIVER[1,2]

Under certain conditions, usually met in practice, maximizing the output peak-signal-to-noise (power) ratio of a radar receiver maximizes the detectability of a target. A linear network that does this is called a *matched filter*. Thus a matched filter, or a close approximation to it, is the basis for the design of almost all radar receivers.

Matched Filter Frequency Response Function   The matched filter that maximizes the output peak-signal-to-mean-noise ratio when the input noise spectral density is uniform (white noise) has a frequency response function[1]

$$H(f) = G_a S^*(f) \exp(-j2\pi f t_m) \qquad\qquad [5.1]$$

where G_a is a constant, t_m is the time at which the output of the matched filter is a maximum (generally equal to the duration of the signal), and $S^*(f)$ is the complex conjugate of the spectrum of the (received) input signal $s(t)$, found from the Fourier transform of the received signal $s(t)$ such that

$$S(f) = \int_{-\infty}^{\infty} s(t) \exp(-j2\pi f t)\, dt$$

(The matched filter that maximizes the output signal-to-noise ratio should not be confused with the circuit-theory concept of impedance matching, which maximizes the power transfer between two networks.)

The received signal spectrum can be written as $S(f) = |S(f)| \exp[-j\phi_s(f)]$, where $|S(f)|$ is the amplitude spectrum and $\phi_s(f)$ is the phase spectrum. Similarly, the matched filter frequency-response function can be expressed in terms of an amplitude and phase as $H(f) = |H(f)| \exp[-j\phi_m(f)]$.

Letting the constant G_a equal unity, we can use these relations to write Eq. (5.1) as

$$|H(f)| \exp[-j\phi_m(f)] = |S(f)| \exp\{j[\phi_s(f) - 2\pi f t_m]\} \qquad [5.2]$$

Equating the amplitudes and phases in the above gives

$$|H(f)| = |S(f)| \qquad\qquad [5.3]$$

$$\phi_m(f) = -\phi_s(f) + 2\pi f t_m \qquad\qquad [5.4]$$

It is seen that the magnitude of the matched-filter frequency-response function is the same as the amplitude spectrum of the input signal, and the phase of the matched-filter frequency response is the *negative* of the phase spectrum of the signal plus a phase shift proportional to frequency. The effect of the negative sign before $\phi_s(f)$ is to cancel the phase components of the received signal so that all frequency components at the output of the filter are of the same phase and add coherently to maximize the signal.

Matched Filter Impulse Response   The matched filter may also be described by its *impulse response* $h(t)$, which is the inverse Fourier transform of the frequency response function $H(f)$ of Eq. (5.1), or

$$h(t) = \int_{-\infty}^{\infty} H(f) \exp(j2\pi f t)\, df = G_a \int_{-\infty}^{\infty} S^*(f) \exp[-j2\pi f(t_m - t)]\, df \qquad [5.5]$$

Since $S^*(f) = S(-f)$, Eq. (5.5) becomes

$$h(t) = G_a \int_{-\infty}^{\infty} S(f) \exp[j2\pi f(t_m - t)]\, df = G_a s(t_m - t) \qquad [5.6]$$

The expression on the far right comes from recognizing that the integral is an inverse Fourier transform. Equation (5.6) indicates that the impulse response of a matched filter is the time inverse of the received signal. It is the received signal reversed in time, starting from the fixed time t_m. Figure 5.1 shows an example of the impulse response $h(t)$ of the filter matched to a signal $s(t)$.

The impulse response of a filter, if it is to be realizable, must not have any output before the input signal is applied. Therefore, we must have $(t_m - t) > 0$, or $t < t_m$. This is equivalent to the condition on the frequency response function that there be a phase $\exp(-j2\pi f t_m)$, which implies a time delay of t_m. For convenience, the impulse response is often written simply as $s(-t)$ and the frequency response function as $S^*(f)$, with the realizability conditions understood.

Receiver Bandwidth   The matched filter is implemented in the IF stage of a super-heterodyne receiver since the bandwidth of a superheterodyne receiver is essentially that of the IF. (The bandwidths of the RF and the mixer stages are usually large compared to that of the IF.) Thus the maximum signal-to-noise ratio occurs at the output of the IF. The second detector and the video portion of the receiver have negligible effect on the output signal-to-noise ratio if the video bandwidth is greater than one half the IF bandwidth.

Derivation of the Matched-Filter Frequency Response   The frequency response function of the matched filter can be derived using the calculus of variations[1] or the Schwartz inequality.[3] In this section, the Schwartz inequality is used.

We wish to show that the frequency-response function of the linear, time-invariant filter that maximizes the output peak-signal-to-mean-noise ratio is

$$H(f) = G_a S^*(f) \exp(-j2\pi f t_m)$$

when the input noise is stationary and white (uniform spectral density). The ratio to be maximized is

$$R_f = \frac{|s_0(t)|^2_{max}}{N} \qquad [5.7]$$

where $|s_0(t)|_{max}$ is the maximum value of the output signal voltage and N is the mean noise power at the receiver output. (The ratio R_f is not quite the same as the

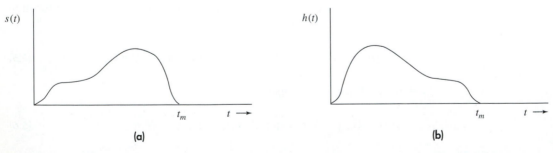

$s(t)$

t_m

$t \longrightarrow$

(a)

$h(t)$

t_m

$t \longrightarrow$

(b)

Figure 5.1   (a) Example of a received waveform $s(t)$; (b) impulse response $h(t)$ of the matched filter for the input signal $s(t)$ of (a).

signal-to-noise ratio considered previously in the radar range equation of Chap. 2. The peak power here is the peak *instantaneous* power, whereas the peak power in the discussion of the radar equation in Chap. 2 was the average value of the power over the *duration* of a pulse of sinewave.) The ratio R_f is *twice* the average signal-to-noise ratio when the input signal $s(t)$ is a rectangular sinewave pulse.) The magnitude of the output voltage of a filter with frequency-response function $H(f)$ is

$$|s_0(t)| = \left| \int_{-\infty}^{\infty} S(f)H(f) \exp(j2\pi ft) \, df \right| \qquad \text{[5.8]}$$

where $S(f)$ is the Fourier transform of the input (received) signal. The mean output noise power is

$$N = \frac{N_0}{2} \int_{-\infty}^{\infty} |H(f)|^2 \, df \qquad \text{[5.9]}$$

where N_0 is the input noise power per unit bandwidth. The factor 1/2 appears before the integral because the limits extend from $-\infty$ to $+\infty$, but N_0 is defined as the noise power per unit bandwidth *only* over positive values of f.

Substituting Eqs. (5.8) and (5.9) into (5.7), and letting t_m denote the time t at which the output $|s_0(t)|^2$ is a maximum, the ratio R_f becomes

$$R_f = \frac{\left| \int_{-\infty}^{\infty} S(f)H(f) \exp(j2\pi ft_m) \, df \right|^2}{\dfrac{N_0}{2} \int_{-\infty}^{\infty} |H(f)|^2 \, df} \qquad \text{[5.10]}$$

Schwartz's inequality states that if P and Q are two complex functions, then

$$\int P^*P \, dx \int Q^*Q \, dx \geq \left| \int P^*Q \, dx \right|^2 \qquad \text{[5.11]}$$

The equality sign applies when $P = kQ$, where k is a constant. Letting

$$P^* = S(f) \exp(j2\pi ft_m) \quad \text{and} \quad Q = H(f)$$

and recalling that $\int P^*P \, dx = \int |P|^2 \, dx$, application of the Schwartz inequality to the numerator of Eq. (5.10) gives

$$R_f \leq \frac{\int_{-\infty}^{\infty} |H(f)|^2 \, df \int_{-\infty}^{\infty} |S(f)|^2 \, df}{\dfrac{N_0}{2} \int_{-\infty}^{\infty} |H(f)|^2 \, df} = \frac{\int_{-\infty}^{\infty} |S(f)|^2 \, df}{\dfrac{N_0}{2}} \qquad \text{[5.12]}$$

Parseval's theorem, which relates the energy in the frequency domain and the energy in the time domain, states that

$$\int_{-\infty}^{\infty} |S(f)|^2 \, df = \int_{-\infty}^{\infty} |s(t)|^2 \, dt = \text{signal energy} = E \qquad \text{[5.13]}$$

Therefore,

$$R_f \leq \frac{2E}{N_0} \qquad \text{[5.14]}$$

which states that the output peak-signal-to-mean-noise ratio from a matched filter depends only on the total energy of the received signal and the noise power per unit bandwidth. It does not depend explicitly on the shape of the signal, its duration, or bandwidth; hence, these characteristics of a signal can be used to achieve radar capabilities other than signal detectability.

The frequency response function which maximizes the peak-signal-to-mean-noise ratio R_f is obtained by noting that the equality sign in Eq. (5.11) applies when $P = kQ$, or

$$H(f) = G_a S^*(f) \exp(-j2\pi f t_m) \qquad \text{[5.15]}$$

where the constant k has been set equal to $1/G_a$.

The matched filter has the interesting property that no matter what the shape, time duration, or bandwidth of the input-signal waveform, the maximum ratio of the output peak-signal-power-to-mean-noise power is simply twice the energy E contained in the received signal divided by the noise power per unit bandwidth N_0. The noise power per hertz of bandwidth is equal to kT_0F_n, where k in this case is the Boltzmann constant, T_0 is the standard temperature (290 K), and F_n is the receiver noise figure.

The concept of the matched filter assumes that the input signal is of the same form $s(t)$ as the transmitted signal (except for a difference in amplitude). This requires that the shape of the transmitted signal not change on reflection by the target or by propagation through the atmosphere. It also requires that the radial dimension of the target be small compared to the range resolution of the radar.

Output Signal from the Matched Filter   From linear filter theory the output $y_0(t)$ of a filter is the convolution of the input $y_{in}(t) = s(t) + n(t)$ and the filter's impulse response $h(t)$, where $s(t)$ is the input signal and $n(t)$ is the input noise. It may be written as

$$y_0(t) = \int_{-\infty}^{\infty} y_{in}(\lambda)h(t - \lambda)\, d\lambda \qquad \text{[5.16]}$$

It was found previously that the impulse response of a matched filter is $h(t) = s(-t)$. (Here, for convenience, $G_a = 1$ and $t_m = 0$.) Then $h(t - \lambda) = s(-t + \lambda)$ and Eq. (5.16) becomes

$$y_0(t) = \int_{-\infty}^{\infty} y_{in}(\lambda)s(\lambda - t)\, d\lambda \qquad \text{[5.17]}$$

It is seen that the output of a matched filter as given by Eq. (5.17) is the cross correlation between the received signal $y_{in}(t)$ and the signal $s(t)$ that was transmitted, since the cross-correlation function between two signals $y_1(t)$ and $y_2(t)$ is defined as

$$\Phi(t) = \int_{-\infty}^{\infty} y_1(\lambda)y_2(\lambda - t)\, d\lambda \qquad \text{[5.18]}$$

When the signal-to-noise ratio is large, $y_{in}(t) \approx s(t)$, and the output signal from the matched filter is approximated by the autocorrelation function of the transmitted signal $s(t)$.

Figure 5.2 illustrates, in a highly simplified manner, the nature of the matched filter for a perfectly rectangular pulse of sinewave when the signal-to-noise ratio is large. In Fig. 5.2 (a) is the input signal; (b) is the frequency response function of the matched filter; (c) is the output of the matched filter (in the IF); and (d) is the envelope of the output of the matched filter, which is what appears in the video portion of the receiver.

Correlation Receiver   Since the output of the matched filter is the cross-correlation function of the received signal and the transmitted signal, it is possible to implement the matched filter as a correlation process based on Eq. (5.17). In a correlation receiver the input signal $y_{in}(t)$ is multiplied by a delayed replica of the transmitted signal $s(t - T_R)$, where T_R is an estimate of the time delay of the target echo signal. The product is passed through a low-pass filter to perform the integration. If the output of the integrator (filter) exceeds a predetermined threshold at a time T_R, a target is said to be at a range $R = cT_R/2$, where c is the velocity of propagation. The cross-correlation receiver tests for the presence of a target at only a single time delay T_R. Targets at other time delays, or ranges, are found by either varying the value of T_R on successive transmissions or employing multiple channels and simultaneously performing the correlation process at all possible values of T_R. The need to search through all possible values of T_R can seriously complicate the correlation receiver.

Since the cross-correlation receiver and the matched-filter receiver are equivalent mathematically, the choice as to which to use in a particular radar application is determined by which is more practical to implement. The matched-filter receiver has almost always been preferred over the correlation receiver.

Approximation to the Matched Filter for a Rectangular-like Pulse   The early radar pioneers in the 1930s were not aware of the concept of the matched filter; yet they learned from experience how to maximize the output signal-to-noise ratio for the simple pulse waveforms that were used at that time. They found that if the receiver passband was too wide compared with the spectral bandwidth of the radar signal, extra noise was introduced (since noise power is proportional to bandwidth); and the signal-to-noise ratio was reduced. On the other hand, if the receiver bandwidth was too narrow, the noise was

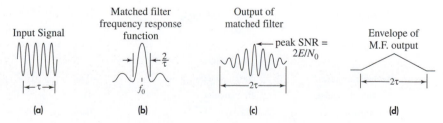

Figure 5.2   (a) Sketch of a perfectly rectangular pulse of sinewave of width τ and frequency f_0; (b) frequency-response function of the matched filter, where $H(f) = S^*(f) = S(f)$; (c) output of the matched filter; (d) envelope of matched-filter output.

reduced but so was the signal energy. Consequently, too narrow a bandwidth relative to the signal spectral width reduced the signal-to-noise ratio, and too wide a bandwidth also reduced the signal-to-noise ratio. Thus there was an optimum value of bandwidth relative to signal spectral width that maximized the signal-to-noise ratio. With rectangular-like pulses and conventional filter design, experience showed that the maximum signal-to-noise ratio occurred when the receiver bandwidth B was approximately equal to the reciprocal of the pulse width τ, or when $B\tau \approx 1$.

In practice the matched filter cannot be perfectly implemented. There will usually be some loss in signal-to-noise ratio compared to that of a theoretically perfect matched filter. The measure of efficiency is taken as the peak-signal-to-mean-noise ratio from the nonmatched filter divided by the peak-signal-to-noise ratio $(2E/N_0)$ obtained from a matched filter. Table 5.1 lists values of $B\tau$ that maximize the signal-to-noise ratio (SNR) for various combinations of hypothetical filters and pulse shapes.[4,5] Note that the rectangular pulse assumed in Table 5.1 is not a realistic waveform since it has zero rise time, which implies infinite bandwidth. Radar pulses are bandwidth limited, and the rise time is approximately $1/B$. Also, several of the filters in Table 5.1 are not likely to be used in practice. Nevertheless, Table 5.1 is offered as an example of the performance of nonmatched filters. The usual "rule of thumb" when no other information is available, is to assume that a practical approximation to a matched filter has $B\tau \approx 1$ and a loss in SNR of about 0.5 dB.

Matched Filter for Nonwhite Noise   In the derivation of the matched-filter characteristic [Eq. (5.15)], it was assumed that the spectrum of the input noise accompanying the signal was white; that is, it was independent of frequency. When this assumption does not

Table 5.1   Efficiency of nonmatched filters compared with matched filters

Input signal	Filter	Optimum $B\tau$	Loss in SNR, dB
Rectangular pulse	Third-order Bessel filter	0.78	0.47
Rectangular pulse	Quadruply tuned (Butterworth)	1.06	0.48
Rectangular pulse	Double tuned (Butterworth)	0.81	0.46
Rectangular pulse	5 cascaded single-tuned stages	0.67	0.51
Rectangular pulse	2 cascaded single-tuned stages	0.61	0.56
Rectangular pulse	Single tuned	0.40	0.88
Rectangular pulse	Rectangular	1.37	0.85
Rectangular pulse	Gaussian	0.74	0.51
Gaussian pulse	Rectangular	0.74	0.51
Gaussian pulse	Gaussian	0.44	0 (matched)

hold and the noise is represented by a nonwhite power spectrum $[N_i(f)]^2$, the frequency-response function that maximizes the peak-signal-to-mean-noise power has been shown to be[6,7]

$$H(f) = \frac{G_a S^*(f) \exp(-j2\pi f t_m)}{[N_i(f)]^2} \qquad \text{[5.19]}$$

This is the frequency response function of the *nonwhite-noise matched (NWN) filter.* When the noise is white, $[N_i(f)]^2 = $ constant, and Eq. (5.19) reduces to that of Eq. (5.15) derived assuming white noise.

Equation (5.19) for nonwhite noise can be rewritten as

$$H(f) = \frac{1}{N_i(f)} \times G_a \left(\frac{S(f)}{N_i(f)} \right)^* \exp(-2\pi f t_m) \qquad \text{[5.20]}$$

From this the nonwhite-noise matched filter can be interpreted as the cascade of two filters. The first filter, with frequency-response function $1/N_i(f)$, makes the noise spectrum uniform, or white. It is sometimes called the *whitening filter.* The second is the matched filter given by Eq. (5.15) when the input noise is white and the signal spectrum is $S(f)/N_i(f)$.

It is seldom that noise is nonuniform over the bandwidth of the radar receiver. The nonwhite-noise matched filter is interesting, but it has had almost no application in radar.

Summary of the Matched Filter   The characteristics of the matched filter for an input signal $s(t)$ are summarized below in short notation, omitting realizability factors and constants. The symbols have been defined previously in this section.

1. Frequency response function: $S^*(f)$

2. Maximum output signal-to-noise ratio: $2E/N_0$

3. Magnitude of the frequency response: $|H(f)| = |S(f)|$

4. Phase of the frequency response: $\phi_m(f) = -\phi_s(f)$

5. Impulse response: $s(-t)$

6. Output signal waveform for large signal-to-noise ratio: autocorrelation function of $s(t)$

7. Relation between bandwidth and pulse width for a rectangular-like pulse and conventional filter: $B\tau \approx 1$

8. Frequency response function for nonwhite noise: $S^*(f)/[N_i(f)]^2$

The matched filter makes radar-signal detection quite different from detection in conventional communication systems. The detectability of signals with a matched-filter receiver is a function only of the received signal energy E and the input noise spectral density N_0. Detection capability and the range of a radar do not depend on the shape of the signal or the receiver bandwidth. The shape of the transmitted signal and its bandwidth therefore can be selected to optimize the extraction of information without, in theory, affecting detection. Also different from communications is that the signal out of the matched filter is not the same shape as the input signal. It should be no surprise that the output signal's shape is different from the input since the criterion for the matched filter states only that detectability is to be maximized, not that the shape of the signal is to be preserved.

5.3 DETECTION CRITERIA

Detection of signals is equivalent to deciding whether the receiver output is due to noise alone or to signal plus noise. This is the type of decision made (probably subconsciously) by a human operator from the information presented on a radar display. When the detection process is carried out automatically by electronic means without the aid of an operator, the detection criterion must be carefully specified and built into the decision-making device.

In Chap. 2, the radar detection process was described in terms of threshold detection. If the envelope of the receiver output exceeds a pre-established threshold, a signal is said to be present. The threshold level divides the output into a region of no detection and a region of detection. The radar engineer selects the threshold that divides these two regions so as to achieve a specified probability of false alarm, which in turn is related to the average time between false alarms. The engineer then determines the other parameters of the radar needed to obtain the signal-to-noise ratio for the desired probability of detection.

Neyman-Pearson Observer The usual procedure for establishing the decision threshold at the output of the radar receiver is based on the classical statistical theory of the *Neyman-Pearson observer*. This is described in terms of the two types of errors that might be made in the detection decision process.

One type of error is to mistake noise for signal when only noise is present. It occurs whenever the noise out of the receiver is large enough to exceed the decision-threshold level. In statistics this is called a Type I error. In radar it is a *false alarm*. A Type II error occurs when a signal is present, but is erroneously considered to be noise. The radar engineer would call such an error a *missed detection*. It might be desired to minimize both errors, but they both cannot be minimized independently. In the Neyman-Pearson observer, the probability of a Type I error is fixed, and the probability of a Type II error is minimized.

As discussed in Sec. 2.5, the threshold level is set by the radar engineer so that a specified false-alarm probability is not exceeded. This is equivalent to fixing the probability of a Type I error and minimizing the Type II error (or maximizing the probability of detection), which is the Neyman-Pearson test used in statistics for determining the validity of a specified statistical hypothesis.[8] In statistical terms it is claimed to be a uniformly most powerful test and an optimum one, no matter what the a priori probabilities of signal and noise. The Neyman-Pearson criterion is employed in most radars for making the detection decision, whether knowingly or not.

Likelihood-Ratio Receiver The *likelihood ratio* is a statistical concept that has been used in radar detection theory and information extraction theory to model optimum decision procedures. It is defined as the ratio of two probability density functions, with and without signal present, or

$$L_r(v) = \frac{p_{sn}}{p_n}$$ [5.21]

where p_{sn} is the probability-density function for signal plus noise and p_n is the probability-density function for noise alone. In Chap. 2 these two probability-density functions were given by Eqs. (2.27) and (2.21), respectively. The likelihood ratio is a measure of how likely it is that the envelope v of the receiver output is due to signal plus noise as compared with noise alone. If the likelihood ratio is sufficiently large, it would be reasonable to conclude that a signal is present.

The Neyman-Pearson observer is equivalent to examining the likelihood ratio and determining if $L_r(v) \geq K$, where K is a real, nonnegative number that depends on the probability of false alarm selected.

One does not find likelihood-ratio receivers in equipment catalogs. It is a statistical concept that models the basic nature of a receiver for maximizing the detectability of radar signals or a receiver that provides the most accurate measurement of radar parameter (such as range). The likelihood ratio is an analytical tool used to indicate optimum receiver and detector design. In most cases of practical interest a radar that employs a matched filter is equivalent to a likelihood-ratio receiver.[8]

Inverse Probability Receiver This is another statistical concept and is based on the relationship known as Bayes' rule for the probability of causes.[9,10] As with the likelihood ratio, inverse probability has been used as an analytical basis to model "optimum" receivers for detection and information extraction. *Inverse probability* is different from the more familiar *direct probability* that describes the chance of an event happening on a given hypothesis. If an event actually happens (such as a voltage appearing at the output of the radar receiver), the problem of forming the best estimate of its cause is a problem in inverse probability.

The operation of the inverse probability receiver will not be described here. (More detail can be found in earlier editions of this text and in the references thereto.) The inverse probability receiver, likelihood receiver, and matched-filter receiver, however, are all related to one another under certain conditions which are often met in most radar applications. The design information obtained from one can usually be obtained from the others as well. The inverse probability receiver differs from the likelihood-ratio receiver (and the matched filter) in that it requires knowledge of a priori probabilities. (An a priori probability is one that is known before the event occurs; e. g., the a priori probability that a flip of a coin results in a heads is 0.5.) It is not usually possible in radar to define quantitatively the a priori probability (for example, the probability of observing an aircraft echo signal at the output of the radar at a range of 110 nmi, azimuth of 75°, at 0630 tomorrow morning). Therefore, the inverse probability receiver has only been of academic interest. Sometimes it has been suggested that the problem of selecting the a priori probability can be satisfied by assuming it to be constant. If the a priori probability is constant, the inverse probability receiver reduces to the likelihood-ratio receiver. Thus one might as well start with the likelihood-ratio receiver in the first place.

The inverse probability receiver and the likelihood-ratio receiver are statistical models that have been used in the past to derive important relations in the theory of signal detection and information extraction. Although one does not build either type of receiver, they both have been useful since theoretical results derived from them indicate the best that can be achieved under the given assumptions.

Sequential Observer, Sequential Detection In a conventional radar based on the Neyman-Pearson observer, a fixed number of pulses, n, are obtained before a detection decision is made. When the signal-to-noise ratio is large, it might not be necessary to collect all n pulses before being able to make the decision that a target echo signal is present. Also, it might be possible to determine after only a few pulses that the receiver output is so low it is unlikely that, even with the remaining pulses, the integrated receiver output would cross the threshold. It should be possible, by taking advantage of the possibility of a quick decision, to make a detection decision with fewer pulses, on average, than would be needed for the Neyman-Pearson observer. This procedure is called the *sequential observer*.[11,12] It is an interesting detection method that, in some cases, can result in almost an order of magnitude decrease in power or revisit time when it can be applied. Unfortunately, its application in radar is limited.

After a single sample of the receiver output, the sequential observer makes one of three choices: (1) the sample is due to the presence of signal and noise; (2) it is due to noise alone; or (3) it cannot be determined whether it is due to noise alone or signal-plus-noise. If it can be decided that either No. 1 or No. 2 applies, the test is completed and the radar moves to the next resolution cell to repeat the operation. If the choice is No. 3, a decision cannot be made, and another observation is obtained and the choices examined again on the basis of the two observations. This procedure is repeated until a decision can be made as to whether noise alone or signal-plus-noise is present.

The sequential observer fixes the probability of errors beforehand and allows the number of observations (integration time) to vary. This procedure theoretically allows a significant reduction in the average number of pulses (samples) needed for making a decision. The sequential observer makes a relatively prompt decision when only noise is present. In one reported example,[13] the sequential observer can, on average, come to a decision with less than one-tenth the number of observations required for the Neyman-Pearson observer when only noise is present. When a threshold signal is present, the sequential observer requires, on average, about one-half the number of observations of the equivalent fixed-sample Neyman-Pearson observer.

A flexible phased array radar, or equivalent agile antenna, is required for the sequential observer in order to take advantage of the variable number of pulses to be integrated. Unfortunately, there is a severe limitation to its use. If there is only one range cell in each angular resolution cell, such as a "guard band," the sequential observer works as indicated above. At each angular position of a surveillance radar antenna, however, there can be a large number of range cells. The sequential observer must come to a decision in every one of these cells before moving on to the next angular position. Any savings in time to make a decision is lost when the number of cells is large, since the observation time at any angular position is determined by the time it takes for the slowest cell to make a decision.[12]

Although the sequential observer can, in principle, result in a saving in transmitter power or in revisit time, it is limited to applications such as a guard ring, the detection of border penetration,[14] or a radar with an omnidirectional transmit antenna and many contiguous fixed narrow-beam receiving antennas that look everywhere all the time.

The term *sequential detection* is sometimes used synonymously with the term sequential observer; but it is also used to describe a two-stage detection process that can be employed with a phased-array radar.[15,16] The radar transmits a pulse or a series of pulses in a particular direction, but with a lower threshold (and higher false-alarm probability) than normal. If no threshold crossings are obtained, the antenna beam moves to the next position. If a threshold crossing occurs, a second pulse or series of pulses is transmitted with higher energy, and with a higher threshold. A detection is declared if the threshold is crossed on both transmissions. It has been claimed that a second threshold is employed in about 4 percent of the beam positions and that there is a power saving of from 3 to 4 dB as compared with uniform scanning.

5.4 DETECTORS

The detector is that portion of the radar receiver that extracts the modulation from the carrier in order to decide whether or not a signal is present. It extends from the IF amplifier to the output of the video amplifier; thus, it is much more than a rectifying element. The conventional pulse radar as described in Chaps. 1 and 2 employs an *envelope detector* which extracts the amplitude modulation and rejects the carrier. By eliminating the carrier and passing only the envelope, the envelope detector destroys the phase information. There are other "detectors" in radar that are different from the above description. The MTI radar uses a *phase detector* to extract the phase of the radar echo relative to the phase of a coherent reference, as described in Chap. 3. In Chap. 4, the *phase-sensitive detector* employed in tracking radars for extracting angle information was mentioned.

Optimum Envelope Detector Law   The envelope detector consists of the IF amplifier with bandpass filter characteristic, a rectifying element (such as a diode), and a video amplifier with a low-pass filter characteristic. The detector is called a *linear detector* if the relation between the input and output signal is linear for positive voltage signals, and zero for negative voltage. (The detector, of course, is a nonlinear device even though it bears the name *linear.*) When the output is the square of the input for positive voltage, the detector is called *square law.* The detector law is usually considered the combined law of the rectifying element and the video integrator that follows it, if an integrator is used. For example, if the rectifying element has a linear characteristic and the video integrator has a square-law characteristic, the combination would be considered a square-law detector. There can be, of course, many other detector laws beside the linear and the square law.

The *optimum detector* law can be found based on the use of the likelihood-ratio receiver. It can be expressed as[17–19]

$$y = ln\, I_0(av) \qquad [5.22]$$

where y = output voltage of the detector

a = amplitude of the sinewave signal divided by the rms noise voltage

v = amplitude of the IF voltage envelope divided by the rms noise voltage

$I_0(x)$ = modified Bessel function of zero order

This equation specifies the form of the detector law that maximizes the likelihood ratio for a fixed probability of false alarm. A suitable approximation is[20]

$$y = ln\ I_0(av) \approx \sqrt{(av)^2 + 4} - 2 \qquad \textbf{[5.23]}$$

For large signal-to-noise ratios (a ≫ 1), this is approximately

$$y \approx av$$

which is a linear law. For small signal-to-noise ratios, the approximation of Eq. (5.23) becomes

$$y \approx (av)^2/4$$

which has the characteristic of a square-law detector. Hence, for large signal-to-noise ratio, the optimum $ln\ I_0$ detector may be approximated by a linear detector, and for small signal-to-noise ratios it is approximated by a square-law detector.

The linear detector usually is preferred in practice since it results in a higher dynamic range than the square law and is less likely to introduce distortion. On the other hand, the square-law detector usually is easier to analyze than the linear, so many analyses assume a square-law characteristic. Fortunately, the theoretical difference in detection performance between the square-law and linear detectors when performing noncoherent integration often is relatively insignificant.[21,22] Marcum[23] also showed that for a single pulse (no integration) the probability of detecting a given signal is independent of the detector law.

Logarithmic Detector   If the output of the receiver is proportional to the logarithm of the input envelope, it is called a *logarithmic detector,* or *logarithmic receiver.* It finds application where large variations of input signals are expected. Its purpose is to prevent receiver saturation and/or to reduce the effects of unwanted clutter echoes in certain types of non-MTI receivers (as in the discussion of the log-FTC receiver in Sec. 7.8). A logarithmic characteristic is not used with MTI receivers since a nonlinear characteristic can limit the MTI improvement factor that can be achieved.

There is a loss in detectability with a logarithmic receiver. For 10 pulses integrated the loss in signal-to-noise ratio is about 0.5 dB, and for 100 pulses integrated, the loss is about 1.0 dB.[24] As the number of pulses increase, the loss approaches a maximum value of 1.1 dB.[25]

I,Q Detector   The *I* and *Q,* or *in-phase* and *quadrature,* channels were mentioned in Sec. 3.5 in the discussion of the MTI radar. There it was noted that a single phase-detector fed by a coherent reference could produce a significant loss in signal depending on the relative timing (or "phase") of the pulse train and the doppler-shifted echo signal. In an MTI radar, the term *blind phase* (not a truly descriptive term) was used to describe this loss.

The loss due to blind phases was avoided if a second parallel detector channel, called the *quadrature*, or *Q* channel, were used with a reference signal 90° out of phase from the reference signal of the first channel, called the *in-phase*, or *I* channel. Most signal processing analyses now use *I* and *Q* channels as the receiver model especially when the doppler frequency is extracted.

The *I,Q* detector is more general than just for avoiding loss due to blind phases in an MTI radar. Figure 5.3 illustrates the *I,Q* detector. It is sometimes called a *synchronous detector.*[26] If the input is a narrowband signal having a carrier frequency f_0 (which could be the IF frequency) with a time-varying amplitude $a(t)$ and time-varying phase $\phi(t)$, then

$$\text{input signal: } s(t) = a(t) \sin [2\pi f_0 t + \phi(t)]$$

The output of the in-phase channel is $I(t) = a(t) \cos [\phi(t)]$ and the output of the quadrature channel is $Q(t) = a(t)\sin [\phi(t)]$. The input signal then can be represented as $s(t) = I(t) \sin 2\pi f_0 t + Q(t) \cos 2\pi f_0 t$. Thus the *I* and *Q* channels together provide the amplitude and phase modulations of the input signal.

If the outputs of the *I* and *Q* channels of Fig. 5.3 are squared and combined (summed), then the square root of the sum of the squares is the envelope $a(t)$ of the input signal. This describes an envelope detector. The phase $\phi(t)$ of the input signal is arctan (Q/I).

The *I,Q* representation is commonly used in digital signal processing.[27] The digitized signals are represented by complex numbers derived from the *I* and *Q* components. In each channel, the signal is digitized by an analog-to-digital (A/D) converter to produce a series of complex digital samples from the signal $I + jQ$. According to the sampling theorem, if the input signal has a bandwidth B there must be at least $2B$ samples per second (the Nyquist rate) to faithfully reproduce the signal. Because there are two channels in the *I,Q* detector, the A/D converter in each of the *I* and *Q* channels needs only to sample at the rate of B samples per second, thus reducing the complexity required of the A/D converters.

With a rate of B samples per second, there is a loss of about 0.6 dB compared to continuous sampling, since the sampling is not guaranteed to occur at the peak of the

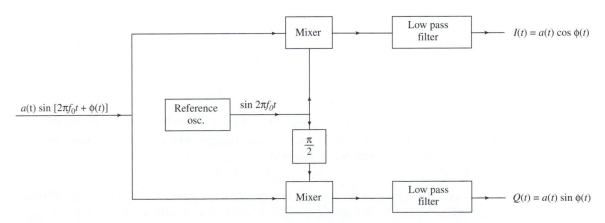

Figure 5.3   *I,Q* detector

output.[27] Much of this loss can be recovered by sampling at a rate of $2B$ samples per second. In some applications, further loss might occur due to the two channels not being precisely 90° out of phase, not being of equal gain, or if they are not perfectly linear.[28]

When I,Q channels are used for MTI processing, a doppler filter such as a delay-line canceler is included in each channel to separate moving targets from stationary clutter, as was discussed in Sec. 3.5.

Coherent Detector   The so-called "coherent detector" sometimes has been described in the past literature as a single-channel detector similar to the in-phase channel of the I,Q detector, but with the reference signal at the same exact frequency and same exact phase as that of the input signal. Compared to the normal envelope detector of Chap. 2, the signal-to-noise ratio from a coherent detector might be from 1 to 3 dB greater. Unfortunately, the phase of the received radar signal is seldom known, so the single-channel coherent detector as described generally is not applicable to radar. The I,Q detector of Fig. 5.3 can also be considered as a coherent detector, but without the limitation of the coherent detector described above.

5.5   AUTOMATIC DETECTION

An operator viewing a PPI display or an A-scope "integrates" in his/her eye-brain combination the echo pulses available from the target. Although an operator in many cases can be as effective as an automatic integrator, performance is limited by operator fatigue, boredom, overload, and the integrating characteristics of the phosphor of the CRT display. With automatic detection by electronic means, the operator is not depended on to make the detection decision. *Automatic detection* is the name applied to the part of the radar that performs the operations required for the detection decision without operator intervention. The detection decision made by an automatic detector might be presented to an operator for action or to a computer for further processing.

In many respects, automatic detection requires much better receiver design than when an operator makes the detection decision. Operators can recognize and ignore clutter echoes and interference that would limit the recognition abilities of some automatic devices. An operator might have better discrimination capabilities than automatic methods for sorting clutter and interference; but the automatic, computer-based decision devices can operate with far greater number of targets than an operator can handle.

Automatic detection of radar signals involves the following:

- *Quantization* of the radar coverage into range, and maybe angle, resolution cells.
- *Sampling* of the output of the range-resolution cells with at least one sample per cell, more than one sample when practical.
- *Analog-to-digital conversion* of the analog samples.
- *Signal processing* in the receiver to remove as much noise, clutter echoes, and interference as practicable before the detection decision is attempted.
- *Integration* of the available samples at each resolution cell.

- *Constant false-alarm rate* (CFAR) circuitry to maintain the false-alarm rate when the receiver cannot remove all the clutter and interference.
- *Clutter map* to provide the location of clutter so as to ignore known clutter echoes.
- *Threshold detection* to select target echoes for further processing by an automatic tracker or other data processor.
- *Measurement of range and angle* after the detection decision is made.

The automatic detection and tracking (ADT) system, which includes the above, was discussed in Sec. 4.9. We next consider the automatic integration of signals and the application of CFAR in the automatic detection process.

5.6 INTEGRATORS

A major part of an automatic detector that operates with more than one pulse is the *integrator* which integrates, or adds, the energy from the received pulses available from a target. The subject of predetection and postdetection integration was introduced in Sec. 2.6. In this section, several integration methods will be briefly reviewed. Integration of pulses in early radars often was performed by an operator viewing a cathode-ray tube display since automatic integration of pulses was seldom practical with the then existing analog technology. Modern radars almost always implement the integration of signals digitally. (Note that in the technical literature, some integration devices are called *detectors* even though they perform *integration.*)

Moving-Window Integrator[29]   The straightforward method for integrating the n pulses available from a target is to simply add them. It was not until advances in digital processing technology became available, however, that it became practical to do so. Continuous integration of the last n pulses at the output from a receiver from each range-resolution cell can be accomplished with a *moving-window integrator,* also called *moving-window detector.* The new output from the receiver is added to the previous sum, and the output received n pulses earlier is subtracted to achieve a running sum of n pulses. In a digital processor it is possible to apply weights to the outputs, based on the two-way gain of the antenna pattern, so as to provide increased signal-to-noise ratio. If uniform weighting is used instead (since it is easier to do), there is a loss in signal-to-noise ratio of about 0.5 dB compared to optimum weighting.[30]

The angular location of the target may be estimated by taking the midpoint between the first and last crossings of the detection threshold or by noting the location of the maximum value of the running sum. After correcting for the bias, the accuracy of the angular location measurement is only about 20 percent worse than theoretical.[30]

According to Trunk,[31] a disadvantage of the moving-window detector is that it is susceptible to large interference signals, a problem that can be minimized by using limiting. It also requires large storage since the last n pulses from each range cell must be put in memory. With increasing improvements in digital technology, this limitation has become less of a concern.

Binary Integration   This was the first automatic method developed to integrate pulses and make the detection decision without the aid of an operator.[32] It is still an important technique. Its chief advantage is that it can be implemented without the complexity of the moving-window integrator. It is, however, less efficient than ideal postdetection integration.

As a radar antenna scans by a target it will receive n echo pulses. If m of these expected n pulses exceed a predetermined value (threshold), a target is declared to be present. The use of a detection criterion that requires m out of n echo pulses to be present is a form of integration. It is called the *binary integrator,* but it is also well-known as the *binary detector, double-threshold detector,* and *m-out-of-n detector.*

A block diagram of the binary integrator is shown in Fig. 5.4. The radar video is passed through a threshold detector, whose level is lower than the normal threshold discussed previously in Chap. 2. It is the first of two thresholds in this system, hence the name *double-threshold detector.* The output of the first threshold is sampled by the quantizer at least once per range-resolution cell. A pulse with a standard amplitude is generated if the video waveform exceeds the first threshold, and nothing is generated if it does not exceed the threshold. These outputs are designated 1 and 0, respectively. Thus the output of the quantizer is a series of 1s and 0s The 1s and 0s from the last n *pulses* at each range cell are stored and counted in the binary counter. If there are at least m 1s within the last n sweeps, a target is said to be detected at that range. The number m is the second threshold to be passed in the double-threshold detector. The two thresholds must be selected jointly for best performance.

The optimum value of m/n for a nonfluctuating echo signal is shown in Fig. 5.5.[33] This curve is only approximate since there is a slight dependence on the false alarm probability, but it is said to be independent of the signal-to-noise ratio. A fluctuating Swerling Case 1 target has the same optimum value of m/n as a nonfluctuating target, but a fluctuating Swerling Case 2 has different optimum values.[34-36] The optimum value of m is not a sensitive selection. It can be quite different from the optimum without significant penalty.

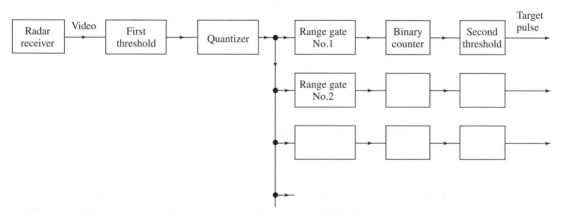

Figure 5.4   Block diagram of a binary integrator

Figure 5.5 Optimum number of pulses m_{opt} (out of a maximum of n) for a binary moving window detector, assuming a constant (nonfluctuating) target echo. | (After Swerling,[33] courtesy Rand Corp.)

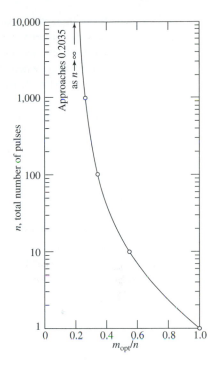

The loss in signal-to-noise ratio due to quantizing signals into two levels (1 or 0) in the binary integrator can vary from just under 1 dB to 2.5 dB.[37] For the particular case given in ref. 37 where the probability of detection is 0.9 and the probability of false alarm is 10^{-6}, the loss can reach 1.4 dB for a nonfluctuating target and 2.2 dB for a Swerling Case 1 fluctuating target, when n is about 8 hits per target. For larger values of n (n greater than about 100) the binary integrator asymptotically approaches a loss of 0.94 dB in all cases, as compared with optimum noncoherent integration. When the amplitude is quantized into more than two levels, the loss is less. V. Gregers-Hansen[37] states that quantization into four levels (two bits) reduces the loss to about one-third that of the two-level quantization.

The binary integrator is relatively simple as an automatic detector and is less sensitive to the effects of a single large interference pulse that might exist along with the target echo pulses. In the conventional integrator, the full energy of the interference pulse is included in the sum. This could result in a false-target indication even though only one interference pulse were present. In a binary integrator, however, a large interference pulse contributes no more to the sum than would any other pulse that crosses the first threshold. No matter what the energy in the pulse, the output of the first threshold is a "1". The same advantage occurs when the background is not receiver noise, but is nongaussian (or non-Rayleigh) clutter as in high-resolution sea clutter and many forms of land clutter. If the clutter statistics have high tails (which means a higher probability of having large values than the gaussian probability density function), these high values can result in false-target reports when a detection is based on gaussian statistics. The binary integrator treats these high values of clutter as any other first-threshold crossing, and it is not as likely to report a target when none is present as might a conventional detection criterion based on the total energy received in n pulses. The binary integrator is therefore *robust,* in that it can be used when the background noise or clutter is nongaussian.

An estimate of the target's angular position (beam splitting) also may be made by locating the center of the group of n pulses. For large n, the angular estimation error made with the binary integrator is about 25 percent greater than the theoretical lower bound.[31]

Batch Integrator[31]   A *batch integrator* is used when there is a large number of pulses available. If there are kn pulses received from the target, k pulses are summed (batched) and compared to a threshold to make a binary decision (0 or 1) as to whether the threshold has been crossed. The process is repeated for each of the remaining $n - 1$ sets of k pulses. The n 0s and 1s are summed and compared to a second threshold. The batch integrator, just as the binary integrator, is simpler to implement, is less affected by interference spikes, and is robust to the noise or clutter statistics. It is said to require less storage, have better detection performance, and provide a more accurate angle estimation than the binary integrator.

Feedback Integrators[31]   The advantage of the single delay-line feedback integrator is its simpler processing. As indicated in Fig. 5.6a, in this integrator the output of the delay line is recirculated so that the signals from each new sweep are added to the sum of all the previous sweeps. To prevent unwanted oscillations ("ringing") due to the positive feedback, the sum must be attenuated by an amount $k < 1$ after each pass through the delay line. The factor k is the gain of the loop formed by the delay line and the feedback path. It imparts an exponential weighting to the received pulses. The effective number of pulses integrated is equal to $(1 - k)^{-1}$.

The single delay-line feedback integrator has a loss of about 1.0 dB in signal-to-noise ratio compared to the ideal postdetection integrator that weights the received pulses in

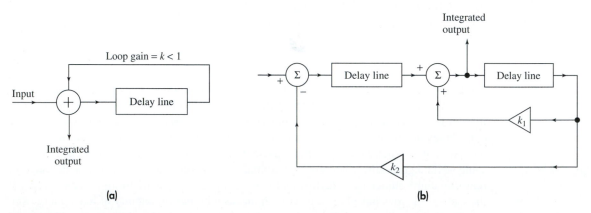

(a)   **(b)**

Figure 5.6   Recirculating delay-line integrator, or feedback integrator, k = loop gain < 1. (a) Single delay loop; (b) two-pole filter.

proportion to the two-way antenna gain. Estimating the angular location of the target based on threshold crossings produces a 20 percent error compared to the optimum. There is a bias, however, that must be estimated, and it can be large. The single delay-line integrator of Fig. 5.6a might have the advantage of simplicity, but its problems cause it to have only limited utility.

The two-pole filter of Fig. 5.6b requires more storage than the single delay-line feedback integrator, but its detection performance is only 0.15 dB less than the optimum. Its angular measurement accuracy has a standard deviation 15 percent greater than optimum, and the estimator based on the maximum value has a constant bias.[38] According to Trunk[31] the problems with this integrator are that it has rather high detector sidelobes (15 to 20 dB) and it is extremely sensitive to interference.

Other Types of Integrators/Detectors Some forms of integrators are also called detectors because the detection decision uses the *n* pulses received from a target. The *mean detector,* for example, is one that sets a threshold based on the mean, or average, of its *n* received pulses. It is therefore equivalent to the conventional method of setting the threshold on the basis of the addition (integration) of *n* pulses. The *median detector* sets a threshold based on the median value of its *n* expected pulses. It is more robust than a mean detector in that it is not as adversely affected by a large interference pulse that might be included among the *n* pulses. Also, it is not as degraded as the mean detector when the clutter or noise background is described by nongaussian statistics. There are also *censored mean detectors* in which one or more of the largest amplitude pulses of the *n* pulses received are eliminated from the detection decision on the assumption that they are likely to be from interference rather than from a target. The adaptive detector, nonparametric detector, and distribution-free detector are usually considered as forms of CFAR, which is discussed in the next section. Most of these detectors have been more of academic interest than candidates for application in operational radar systems.

5.7 CONSTANT-FALSE-ALARM RATE (CFAR) RECEIVER

As said before in this text, a target is detected when the output of the radar receiver crosses a predetermined fixed threshold level set to achieve a specified probability of false alarm. When noise at the receiver is due to internally generated noise of fixed level described by a gaussian probability density function, the procedure for setting the threshold is well established (Sec. 2.8). In many situations, however, clutter echoes and/or hostile noise jamming can be much larger than receiver internal noise. When this happens, the receiver threshold can be exceeded and many false alarms can occur, which cause havoc with radar detection and tracking.

A well-trained and alert operator viewing a PPI or other radar display is seldom misled into mistaking clutter or jamming for real targets, but an operator can lose effectiveness when there are many target echoes to be processed. An automatic detection and tracking (ADT) system can handle many targets, and will attempt to determine if clutter or jamming signals that cross the receiver threshold form realistic tracks. Eventually, a false alarm will not form a realistic track and will be discarded by the tracking computer. An automatic system, however, might be of limited capability and require too much time or computer capacity to recognize and discard false alarms. Although digital computers can have a high level of capability, the task of recognizing false echoes might cause them to be overloaded when there are a large number of real targets, a large number of clutter echoes, interference, and/or high external noise levels. Therefore, if ADT is to work properly, some method is necessary to keep clutter and external noise from reaching the automatic-tracking computer. One method has been CFAR, or *constant false alarm rate* receiver. CFAR automatically raises the threshold level to keep clutter echoes and external noise from overloading the automatic tracker with extraneous information. The need for CFAR was recognized when the early automatic detection and tracking systems were installed as add-ons to existing radars with no MTI or relatively poor MTI that did not have good clutter rejection. CFAR is achieved, of course, at the expense of a lower probability of detection of desired targets. In addition, CFAR can also produce false echoes when there is nonuniform clutter, suppress nearby targets, and degrade the range resolution.

Cell Averaging CFAR, or CA CFAR   The major form of CFAR has been the cell-averaging CFAR, due to Finn and Johnson,[39] and its variants. It is illustrated in Fig. 5.7. It uses an adaptive threshold whose level is determined by the clutter and/or noise in the vicinity of the radar echo. Two tapped delay-lines sample echo signals from the environment in a number of *reference cells* located on both sides of the test cell (the range cell of interest). The spacing between reference cells is equal to the radar range resolution (usually the pulse width). The output of the test cell is the radar video output, which is compared to the adaptive threshold derived from the sum of the outputs of the tapped delay lines defining the reference cells. This sum, therefore, represents the radar environment to either side of the test cell. It changes as the radar environment changes and as the pulse travels out in time. When multiplied by a predetermined constant k, the sum provides an adaptive threshold to maintain a constant false-alarm rate. Thus the threshold can adapt to the environment as the pulse travels in time.

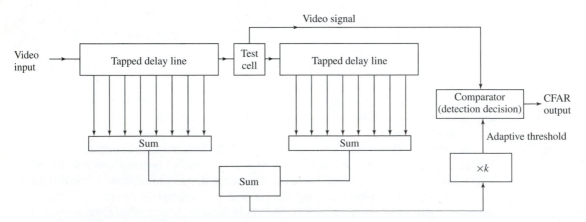

Figure 5.7   Cell averaging CFAR

If the radar output is noise or clutter described by the Rayleigh probability density function, the constant k which multiplies the sum of the tapped delay lines can be determined from classical detection theory, similar to that described in Chap. 2. When the statistics of the clutter are not known, which is often the case, the value of k can only be estimated or some form of nonparametric detector used.

CFAR Loss The greater the number of reference cells (delay-line taps) in the CA CFAR the better is the estimate of the background clutter or noise and the less will be the loss in detectability (signal-to-noise ratio). There is a limit, however, to the number of reference cells that can be used in practice since the clutter must be relatively homogeneous over the reference cells. A typical CFAR design for an aircraft-surveillance radar might have a total of 16 to 20 reference cells that sample the environment a half-mile to either side of the signal in the test cell. In a doppler processing radar, such as MTI or pulse doppler, reference cells can sometimes be taken from adjacent doppler filters as well as from adjacent range cells.

Since there are only a finite number of reference cells, the estimate of the noise or clutter is not precise and there will be a loss in detectability. Figure 5.8, derived from the publications of Mitchell and Walker[40] and from R. Nitzberg,[41] gives the theoretical CFAR loss as a function of the number of reference cells M, the probability of false alarm, and the number N of pulses integrated. (The CFAR loss is the signal-to-noise ratio required when CFAR is employed divided by the signal-to-noise ratio required for fixed-threshold

Figure 5.8 Theoretical CFAR loss.
(After R. L. Mitchell and J. F. Walker[40] and R. Nitzberg.[41])

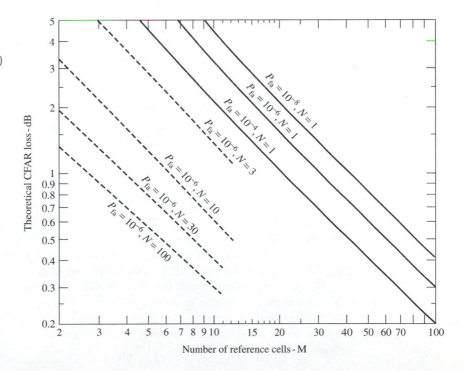

detection.) The solid curves apply for detection using only a single pulse. The dashed curves give the loss for a probability of false alarm of 10^{-6}, when the number of pulses N is greater than one. The curves of Fig. 5.8 apply to a nonfluctuating target as well as Swerling Case 1 and Case 2. When the number of reference cells is large, the CFAR loss is small. Nitzberg shows that the CFAR loss for single-pulse detection ($N = 1$) can be approximated by

$$\text{Loss (dB)} = -\frac{5}{M} \log P_{\text{fa}} \qquad \qquad \text{[5.25]}$$

where P_{fa} is the probability of false alarm.

Clutter Edges   The CA CFAR of Fig. 5.7 assumes that the statistics of the clutter or noise at each reference cell are independent, identical, and the same as the statistics at the test cell. This is not the case at the edges of clutter. As the reference cells pass over the leading and trailing edges of a patch of clutter, not all the cells contain clutter; so the threshold will be lower than when all reference cells contain clutter. False alarms (threshold crossings), therefore, can result at clutter edges. Threshold crossings from the clutter edges can be reduced by summing the leading and the lagging references cells separately, and using the greater of the two to determine the threshold.[42] The CFAR that uses the *greater of* the two sets of reference cells is known as GO-CFAR. It introduces an additional CFAR loss of from 0.1 to 0.3 dB.[43]

Effect of Multiple Targets   When there are one or more targets within the reference cells along with a primary target in the test cell, the threshold is raised (even in the absence of any clutter) and the detection of the primary target in the test cell of the CA CFAR might be suppressed. One method for reducing the effect of multiple targets is to remove (censor) the outputs of those reference cells that are much larger than the rest. A predetermined number J of reference cells (those with the largest outputs) are removed and the adaptive threshold determined by the outputs of the remaining $M - J$ cells.[44] This is known as a *censored mean-level detector* (CMLD). The number of censored cells should be equal to, or at least not smaller than, the number of interfering targets. The loss associated with the CLMD has been analyzed and has been said to be small,[45] but it can be 1 dB or greater.[46]

Another approach to handling multiple nearby targets is that of *ordered statistic,* or OS, CFAR, in which the CFAR threshold is determined from one single value selected from the so-called ordered statistic.[47,48] The outputs from M reference cells are put in order from smallest to largest, and the Kth ordered value when multiplied by a scalar is used to set the threshold. For example, if $M = 16$, K might be 10. In one particular analysis of the OS CFAR, the additional loss in signal-to-noise ratio is given as about 0.5 dB for one interfering target, and about 1 dB for two interfering targets.

The problem of dealing with the effect of additional targets within the reference cells has received much attention in the literature.[49–58] Still another method for dealing with loss in detectability when multiple targets are present within the reference cells is to employ log video in which a log detector is used ahead of a CA CFAR to suppress large echoes.[59]

Range Resolution   In general, two equal-amplitude targets can be resolved if they are separated in range by about 0.8 pulse width. The usual CFAR, however, considerably degrades the range resolution so that two equal-amplitude targets can be resolved only if they are spaced greater than about 2.5 pulse widths.[60] One reason for the poorer resolution is that the range cells adjacent to the test cell in Fig. 5.7 are not used as part of the reference cells since the target energy in the test cell spills over to nearby cells and affects the threshold. Another reason for degraded resolution is that in automatic detection systems, a large target can be detected in adjacent range cells, adjacent azimuth antenna beams, and adjacent elevation beams. Automatic detection systems have to merge the many detections of the same target into a single centroided detection. Trunk[61] has shown, using a generalized likelihood approach, that in theory it is possible to resolve two equal nonfluctuating targets with separations varying between 1/4 and 3/4 of a pulse width, depending on the relative phase between the two echo signals. This requires that the shape of the received echo pulse be known. (Trunk's result is a lower bound. It indicates what might be achieved, but does not necessarily apply to a specific CFAR configuration.)

Nonparametric Detectors   A common assumption in most CFARs is that the statistics of the clutter or noise in the reference cells are known (usually taken to be Rayleigh), except for a scale factor. In many cases, however, the form of the clutter probability density function is not known. A *nonparametric detector,* also known as a *distribution-free detector,* has been considered as a CFAR when the clutter statistics are not known.[62] Its operation is not described here since it is seldom used. The nonparametric detector has a relatively large CFAR loss and problems with correlated samples. In addition, it is fairly susceptible to target suppression from large targets in the reference cells, its implementation might not be simple, and there is loss of amplitude information.[63]

Clutter Map[64–66]   A clutter map divides the radar coverage area into range-azimuth cells on a polar or a rectangular grid. The clutter echo stored in each cell of the map then can be used to establish a threshold for that range and azimuth. It is, therefore, a form of CFAR.

The size of each clutter-map cell is equal to or greater than the radar resolution. At each of the range-azimuth cells of the clutter map, a number proportional to the amplitude of the clutter within the cell is stored in the map memory. Since clutter can change with time, the value of the clutter in each cell is updated periodically by averaging over a large number of scans (for example, 5 to 10 scans). The larger the number of scans the more accurate will be the estimate of the clutter, the lower the loss, and the less the effect of a target that moves through the cell. On the other hand, the averaging time (determined by the number of scans) should be shorter than the limited dwell time in which moving clutter (rain or chaff) is within the cell. A short averaging time also allows the threshold to recover to its proper state within a few scans after a target has passed through the cell.

A clutter map CFAR has an advantage over the CA CFAR in that it is not affected by nonhomogeneous clutter (edge effects). The response of the clutter map CFAR will be reduced when a target of slow speed remains within the cell long enough to affect the threshold. This effect can be reduced by making the map cell greater than the radar resolution cell.[65] Increasing the size of the clutter-map cell should not be carried too far, however, since it reduces the interclutter visibility.

The loss of signal-to-noise ratio in the clutter map will depend on the averaging time. The longer the time, the less the loss. In a particular example, Khoury and Hoyle[64] state that the loss is 0.8 dB when the averaging time is approximately 2 minutes.

Another attribute of the clutter map is the elimination of those resolution cells containing slowly moving objects such as birds.[67] Each threshold crossing is checked against a clutter map before initiating an acquisition. It has been said that even with bird densities as low as 0.1 to 0.2 birds/km^2 a radar tracker can be overloaded and waste much of its time on birds.

The clutter map used in the original MTD discussed in Sec 3.6 was not a true CFAR. It could be called a *blanking clutter map* since it passes targets whose amplitude exceeds that of the clutter.[64]

Other Forms of CFAR   Forms of CFAR that predate the CA CFAR include the following:

- *Siebert CFAR:* The output of a postdetection integrator (low-pass filter) provides an estimate of the average noise level which is then applied as a feedforward signal to control the threshold level to maintain the false-alarm rate constant.[68,69] This was one of the first attempts to provide a CFAR, and it was employed in the AN/FPS-23 bistatic CW radar installed by the U.S. Air Force on the DEW (Distant Early Warning) line in the middle 1960s for automatic detection of low-flying aircraft.

- *Hard limiter:* An example is the so-called *Dicke fix,* which consists of a broadband IF filter followed by a hard limiter (which is set low enough to limit receiver noise) and a narrowband matched filter.[69] The output is then unaffected by the amplitude of the noise. The Dicke fix is especially effective against impulse-like noise and broadband jamming. It would not normally be used with an MTI radar since, as mentioned in Sec 3.7, hard limiting restricts the improvement factor that can be achieved.

- *Log-FTC:* This is described in Sec. 7.8. It is a CFAR when the noise or clutter has a Rayleigh probability density function.

CFAR Use in Radar   CFAR is used in radars to maintain effectiveness when there are too many extraneous crossings of a fixed threshold caused by noise or clutter. Automatic tracking of targets can be seriously degraded if excessive false alarms occur.

CFAR is to a radar as crutches are to a person with a broken foot. The crutches allow the person to be mobile, but they are something the person would rather not need. CFAR may allow a radar to continue operation, but there are limitations in performance that accompany its use. CFAR automatically adjusts the threshold to prevent threshold crossings that tie up and overload the tracking computer. Increasing the threshold to maintain a constant probability of false alarm, however, lowers the probability of detection and results in missed detections of some targets. This loss of targets has to be tolerated when CFAR is used. In addition to missed target detections, CFAR can cause a loss in the signal-to-noise ratio when the statistics of the clutter or noise are not estimated accurately. The leading and trailing edges of some CFARs can produce undesired threshold crossings (false alarms). Target suppression can occur when one or more targets are within the reference cells. There is poor range resolution compared to a radar without CFAR. Furthermore, those CFAR designs that might be subject to spoofing by hostile electronic countermeasures have to be avoided in military radars. Thus CFAR is a "necessary evil," needed for maintaining operation of automatic detection and tracking systems that would cause excessive false alarms due to noise or clutter.

CFAR would not be required if the radar had good doppler processing to reject clutter, good ECCM (electronic counter-countermeasures) to reject hostile noise jamming, good EMC (electromagnetic compatibility) to reject interference, and a good tracking computer that recognizes (without overloading in the presence of a large number of threshold crossings) desired moving targets and rejects clutter echoes that break through the signal processing.

Doppler-Estimation False-Alarm Control   A quite different method of controlling false alarms is to estimate the target amplitude and the radial velocity of the target (from a measurement of the doppler frequency shift).[70] Noise or clutter are discriminated from targets by the variation in the radial velocity and amplitude over successive measurements. Consistency tests are applied to the measurements based on the assumption that clutter and noise will fluctuate in both amplitude and estimated doppler over successive measurements; but moving targets generally will not. Also, multiple measurements at different pulse repetition intervals and/or frequencies can be used to produce an unambiguous velocity estimate from which moving targets can be separated from stationary clutter to aid in the tracking process. No reference cells are required in this method, so that the problems of nonhomogeneous environments that degrade the CA CFAR (edge effects and multiple targets) do not appear.

5.8   THE RADAR OPERATOR

The discussion in this chapter assumed automatic detection without an operator making the detection decision. Modern radars usually make the detection decision automatically. An operator viewing a display can be a good detector of targets, as has been demonstrated in the past.[71] On the other hand, operators do not have the capacity to process large quantities of information as rapidly as do electronic circuits, and they can become fatigued.

It has been demonstrated experimentally that when an operator views a display in which the pulses received from successive sweeps are presented side-by-side without saturating the display and without fading, the integration improvement achieved by an operator is equivalent to what would be expected from classical detection theory.[72,73]

To obtain the benefits of both automatic detection and the capability of an operator to interpret unusual situations, some radar designers prefer to make the raw video information available to the operator along with the automatically processed information.

5.9 SIGNAL MANAGEMENT

This chapter has been concerned with the detection of desired radar signals. To close the chapter and introduce the next one on the extraction of radar information, we briefly list below the various parts of *signal management* that occur throughout the radar system. Signal management includes everything associated with the waveforms and their processing that is required for a radar to do its job of detecting and locating targets and determining something about their nature. Signal management starts with the design of a suitable waveform and its radiation into space, the collection by the receiver of echo signals reflected from targets and other objects, the use of signal processing to extract the desired signal and reject undesired echoes, the use of data processing to extract information about the detected signals, the coordinated control of these processes throughout the radar, and keeping within the resources and constraints that affect signals and their management. Some parts of signal management listed below apply to a conventional pulse radar with an envelope detector; some apply to a radar that extracts moving targets based on their doppler frequency shift; and some to both.

Figure 5.9 indicates the various factors that enter into radar signal management.

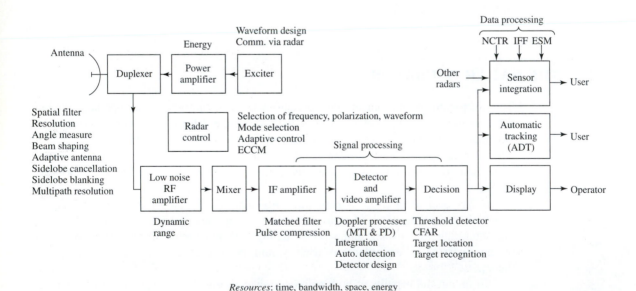

Figure 5.9 The various elements that enter into radar signal management.

Component Parts of Radar Signal Management

Signal Processing   This is processing for the purpose of detecting desired echo signals and rejecting noise, interference, and undesired echoes from clutter. It includes the following:

Matched filter: to maximize the signal-to-noise ratio at the output of the radar receiver, and thus maximize detectability of echo signals.

Detector/integrator: the means for processing in a convenient and efficient manner the number of pulses received from a target so as to take full advantage of the total signal energy received from a target.

Clutter reduction: to eliminate or reduce unwanted clutter by one or more methods, of which filtering of moving targets based on the doppler frequency shift is the most important.

CFAR: used to maintain a constant false-alarm rate at the output of the threshold detector when the radar cannot eliminate unwanted echoes.

Electromagnetic compatibility (EMC): the elimination of interference from other radars and other electromagnetic radiations that can enter the radar receiver.

Electronic counter-countermeasures (ECCM): in a military radar, those methods employed to reduce or eliminate the effectiveness of jamming, deception, and other hostile electronic active and passive measures whose purpose is to degrade radar performance. ECCM can be found throughout the radar, not just as part of the signal processing.

Threshold detection: the decision as to whether the output of the radar is a desired signal.

Data Processing   These are the processes that take place after the detection of the desired signals for the purpose of acquiring further information about the target.

Target location: in range, angle, and sometimes radial velocity (from the doppler shift). Location information is not generally thought of as either signal processing or data processing. It is usually obtained as part of the detection process (since detection without location is of no value).

Target trajectory: also called *target track,* which is the time history of the target's location. Usually a prediction of the target's future position is included.

Target recognition: the recognition of the type of target being viewed by the radar. It might include the recognition of aircraft from birds, one type of aircraft or ship from another, recognition of various types of weather, and information about the land and sea environment (remote sensing).

Weapon control: in military systems, the use of the radar output for the control and guidance of weapons.

Waveform Design   The selection of the waveform depends on what is required of the radar for detection in noise, clutter, interference, and electronic counter-countermeasures, as well as for the extraction of information from the radar signal. Waveform design will affect the signal and data processing. Multiple waveforms for different purposes can be an important aspect of high-performance radar. The radar signal can also be adapted to communicate with other radars.

Antenna　This is not just for radiating and collecting radar signals, but is the means by which angle information is obtained and by which the radar coverage is achieved. The antenna can act as a spatial filter that can affect the spectral properties of wideband signals. It can also provide, in some cases, the angle rate and extract a *spatial* doppler shift similar to the temporal doppler shift.[74] The target's tangential velocity obtained from the spatial doppler shift, along with the radial velocity obtained from the more common temporal doppler shift, can provide the vector velocity of the target.

Automatic Radar Control　A radar often employs multiple waveforms and various signal processing options to maximize performance under a variety of environmental conditions. Radar control involves the automatic selection of the proper waveform and signal processing according to the environment and interference (both natural and intentional) encountered by the radar.

Sensor Integration　The outputs from other radars covering the same region may be combined to form tracks. Information from the civil aviation Air Traffic Control Radar Beacon Systems (ATCRBS) or the military identification friend or foe (IFF), or other civil or military command and control information can be used to assist in identifying the target. Noncooperative target recognition (NCTR) based on special radar waveforms and processing, as well as signals and information obtained by electronic warfare (EW) methods, such as electronic support measures (ESM), may be used as part of an integrated military target-recognition system.

Resources for Signal Management　The radar engineer has available the following resources for pursuing the management of signals and extraction of information.

Energy　Sufficiently large transmitted energy is important for detection of weak signals in noise at long range and for obtaining accurate radar measurements.

Bandwidth　This is the classical measure of information and is especially important for accurate range measurement and the temporal resolution of targets.

Time　Time is necessary for accurate measurement of the doppler frequency. Time also is a means for obtaining increased energy when peak power is a limitation. It is important for achieving multiple functions from a single-beam radar within a required time, and for handling the processing of many echo signals.

Space　This applies to the physical aperture area required for an antenna. The larger the antenna aperture the greater the echo energy at the receiver and the more accurate the spatial measurements that can be obtained.

Constraints   It is not always possible or practical to obtain the desired energy, bandwidth, time, and spatial extent that might be required. Furthermore, the environment might cause clutter echoes that limit a radar's performance: at high microwave and millimeter wave frequencies atmospheric attenuation can be a nuisance; atmospheric refraction can produce both good and bad effects, as discussed in Chap. 8; and the curvature of the earth limits the range of a radar to targets within the line of sight. Military radars must be able to perform their mission in spite of hostile actions designed to degrade or eliminate their effectiveness. In most applications there are constraints imposed by size, space, weight, and perhaps primary power. Spectrum availability is always a consideration and can seriously limit what the engineer might do. There is also the ever-present constraint imposed by cost.

Engineers always have constraints on what they can do and almost never have everything they need to accomplish the desired task. The essence of successful engineering, however, involves compromise so as to provide in a timely manner a new and useful capability at an acceptable cost.

REFERENCES

1. North, D. O. "An Analysis of the Factors Which Determine Signal/Noise Discrimination in Pulsed-Carrier Systems." *Proc. IEEE* 51 (July 1963), pp. 1016–1027. Originally appeared as RCA Tech. Rept. PTR-6C, June 25, 1943 (ATI 14009).

2. Introduction to Matched Filters, *Special Issue on Matched Filters of the IRE Trans. on Information Theory* IT-6, no. 3 (June 1960).

3. Van Vleck, J. H., and D. Middleton. "A Theoretical Comparison of Visual, Aural, and Meter Reception of Pulsed Signals in the Presence of Noise." *J. Appl. Phys.* 17 (November 1946), pp. 940–971.

4. D'Aloisi, D., A. DiVito, and G. Galati. "Sampling Losses in Radar Signal Detection." *J. IERE* 56, no. 6/7 (June/July 1986), pp. 237–242.

5. Taylor, J. W., Jr. "Receivers." In *Radar Handbook,* 2nd ed., M. Skolnik, Ed. New York: McGraw-Hill, 1990, Chap. 3, Sec. 3.7.

6. Dwork, B. M. "Detection of a Pulse Superimposed on Fluctuation Noise." *Proc. IRE* 38 (July 1959), pp. 771–774.

7. Urkowitz, H. "Filters for the Detection of Small Radar Signals in Clutter." *J. Appl. Phys.* 24 (October 1952), pp. 1024–1031.

8. Peterson, W. W., T. G. Birdsall, and W. C. Fox. "The Theory of Signal Detectability." *IRE Trans.* PGIT-4 (September 1954), pp. 171–212.

9. Woodward, P. M. *Probability and Information Theory with Applications to Radar.* New York: McGraw-Hill, 1953.

10. Minkoff, J. *Signals, Noise, & Active Sensors.* New York: Wiley, 1992, Chap. 5.

11. Bussgang, J. J., and D. Middleton. "Optimum Sequential Detection of Signals in Noise." *IRE Trans.* IT-1 (December 1955), pp. 5–18.

12. Bussgang, J. J. "Sequential Methods in Radar Detection." *Proc. IEEE* 58 (May 1970), pp. 731–743.

13. Preston, G. W. "The Search Efficiency of the Probability Ratio Sequential Search Radar." *IRE Intern. Conv. Record* 8, pt. 4 (1960), pp. 116–124.

14. Kazovsky, L. G. "Sequential Radar Detection of Border Penetration." *IEE Proc.* 128, Pt. F, no. 5 (October 1981), pp. 305–310.

15. Brennan, L. E., and F. S. Hill, Jr. "A Two-Step Sequential Procedure for Improving the Cumulative Probability of Detection in Radars." *IEEE Trans.* MIL-9 (July–October 1965), pp. 278–287.

16. Nathanson, F. E. *Radar Design Principles,* 2nd ed. New York: McGraw-Hill, 1991, Sec. 4.7.

17. Marcum, J. "A Statistical Theory of Target Detection by Pulsed Radar, Mathematical Appendix." *IRE Trans.* IT-6 (April 1960), pp. 209–211.

18. Woodward, P. M., See Ref. 9, Sec. 5.5.

19. Skolnik, M. *Introduction to Radar Systems,* 2nd ed. New York: McGraw-Hill, 1980, Sec. 10.5.

20. This expression was suggested by Warren D. White, who reviewed the manuscript for the second edition of this text.

21. Marcum, J. Ref. 17, p. 189 and Fig. 42.

22. Bird, J. S. "Calculating the Performance of Linear and Square-Law Detectors." *IEEE Trans.* AES-31 (January 1995), pp. 39–51.

23. Marcum, J. Ref. 17, pp. 158–159.

24. Green, B. A., Jr. "Radar Detection Probability with Logarithmic Detectors." *IRE Trans.* IT-4 (March 1958), pp. 50–52.

25. Hansen, V. G. "Radar Detection Probability with Logarithmic Detectors." *IEEE Trans.* AES-8 (May 1972), pp. 386–388. See correction, AES-10 (January 1974), p. 168.

26. Eaves, J. L. and E. K. Reedy. *Principles of Modern Radar.* New York: Van Nostrand Reinhold, 1987, pp. 254, 270–272.

27. Nathanson, F. E. Ref. 16, Sec. 8.8.

28. Taylor, J. W., Jr. "Receivers." In *Radar Handbook.* 2nd ed., M. Skolnik, Ed. New York: McGraw-Hill, Shap. 3, Sec. 3.12.

29. Hansen, V. G. "Performance of the Analog Moving Window Detector." *IEEE Trans.* AES-6 (March 1970), pp. 173–179.

30. Trunk, G. V. "Radar Signal Processing," *Advances in Electronics and Electron Physics* 45, L. Marton, Ed. New York: Academic, 1978, pp. 203–252.

31. Trunk, G. V. "Automatic Detection, Tracking, and Sensor Intgration." *Radar Handbook,* 2nd ed. M. Skolnik, Ed. New York: McGraw-Hill, 1990, Chap. 8.

32. Harrington, J. V. "An Analysis of the Detection of Repeated Signals in Noise by Binary Integration." *IRE Trans.* IT-1 (March 1955), pp. 1–9.

33. Swerling, P. "The 'Double Threshold' Method of Detection." Rand Corp. Rept. RM-1081, Dec. 17, 1952, Santa Monica, CA.

34. Weiner, M. A. "Binary Integration of Fluctuating Targets." *IEEE Trans.* AES-27 (January 1991), pp. 11–17.

35. Walker, J. F. "Performance Data for a Double-Threshold Detection Radar." *IEEE Trans.* AES-7 (January 1971), pp. 142–146. See also comment by V. G. Hansen, p. 561, May, 1971.

36. Worley, R. "Optimum Thresholds for Binary Integration." *IEEE Trans.* IT-4 (March 1968), pp. 349–353.

37. Hansen, V. G. "Optimization and Performance of Multilevel Quantization in Automatic Detectors." *IEEE Trans.* AES-10 (March 1974), pp. 274–280.

38. Cantrell, B. H., and G. V. Trunk. "Angular Accuracy of a Scanning Radar Employing a Two-Pole Filter." *IEEE Trans.* AES-9 (September 1973), pp. 649–653.

39. Finn, H. M., and R. S. Johnson. "Adaptive Detection Mode with Threshold Control as a Function of Spatially Sampled Clutter-Level Estimates." *RCA Rev.* 29 (September 1968), pp. 414–464.

40. Mitchell, R. L., and J. F. Walker. "Recursive Methods for Computing Detection Probabilities." *IEEE Trans.* AES-7 (July 1971), pp. 671–676.

41. Nitzberg, R. "Analysis of the Arithmetic Mean CFAR Normalizer for Fluctuating Targets." *IEEE Trans.* AES-14 (January 1978), pp. 44–47.

42. Hansen, V. G. "Constant False Alarm Processing in Search Radars." *International Conference on Radar—Present and Future,* Oct. 23–25, 1973, pp. 325–332, IEE Publication No. 105.

43. Gregers-Hansen, V., and J. H. Sawyers. "Detectability Loss Due to 'Greatest Of' Selection in a Cell-Averaging CFAR." *IEEE Trans.* AES-16 (January 1980), pp. 115–116.

44. Rickard, J. T., and G. M. Dillard. "Adaptive Detection Algorithms for Multiple-Target Situations." *IEEE Trans.* AES-13 (July 1977), pp. 338–343.

45. Al-Hussaini, E. K. "Performance of the Greater-Of and Censored Greater-Of Detectors in Multiple Target Environments." *IEE Proc.* 135, Pt. F (June 1988), pp. 193–198.

46. Ritcey, J. A. "Performance Analysis of the Censored Mean-Level Detector." *IEEE Trans.* AES-22 (July 1986), pp. 443–454.

47. Rohling, H. "Radar CFAR Thresholding in Clutter and Multiple Target Situations." *IEEE Trans.* AES-19 (July 1983), pp. 608–621.

48. Levanon, N. *Radar Principles.* New York: Wiley, 1988, p. 263.

49. Weiss, M. "Analysis of Some Modified Cell-Averaging CFAR Processors in Multiple-Target Situations." *IEEE Trans.* AES-18 (January 1982), pp. 102–114.

50. Barboy, B., A. Lomes, and E. Perkalski. "Cell-Averaging CFAR for Multiple-Target Situations." *IEE Proc.* 133, Pt. F (April 1986), pp. 176–186.

51. Gandhi, P. P., and S. A. Kassam. "Analysis of CFAR Processors in Nonhomogeneous Background." *IEEE Trans.* AES-24 (July 1988), pp. 427–445.

52. Levanon, N. "Detection Loss Due to Interfering Targets in Ordered Statistics CFAR." *IEEE Trans.* AES-24 (November 1988), pp. 678–681.

53. Blake, S. "OS-CFAR Theory for Multiple Targets and Nonuniform Clutter." *IEEE Trans.* AES-24 (November 1988), pp. 785–790.

54. Barket, M., S. D. Himonas, and P. K. Varshney. "CFAR Detection for Multiple Target Situations." *IEE Proc.* 136 (October 1989), pp. 193–209.

55. Ritcey, J. A., and J. L. Hines. "Performance of MAX Family of Order-Statistic CFAR Detectors." *IEEE Trans.* AES-27 (January 1991), pp. 48–57.

56. Shor, M., and N. Levanon. "Performance of Order Statistics CFAR." *IEEE Trans.* AES-27 (March 1991), pp. 214–224.

57. Goldman, H., and I. Bar-David. "Analysis and Application of the Excision CFAR Detector." *IEE Proc.* 135, Pt. F (December 1988), pp. 563–575.

58. Minkler, G., and J. Minkler. *CFAR.* Baltimore, MD: Magellan, 1990.

59. Trunk, G. V. Ref. 31, pp. 8.17–8.18.

60. Trunk, G. V. "Range Resolution of Targets Using Automatic Detectors." *IEEE Trans.* AES-14 (September 1978), pp. 750–755.

61. Trunk, G. V. "Range Resolution of Targets." *IEEE Trans.* AES-20 (November 1984), pp. 789–797.

62. Trunk, G. V., B. H. Cantrell, and F. D. Queen. "Modified Generalized Sign Test Processor for 2-D Radar." *IEEE Trans.* AES-10 (September 1974), pp. 574–582.

63. Trunk, G. V., Ref. 31, pp. 8.19–8.20.

64. Khoury, E. N., and J. S. Hoyle. "Clutter Maps: Design and Performance." *Proc. of the 1984 IEEE National Radar Conference,* pp. 1–7, 84CH1963-8.

65. Farina, A., and F. A. Studer. "A Review of CFAR Detection Techniques in Radar Systems." *Microwave J.* 29, no. 9 (September 1986), pp. 115–128.

66. Nitzberg, R. "Clutter Map CFAR Analysis." *IEEE Trans.* AES-22 (July 1986), pp. 419–421.

67. Franzen, N. I. "The Use of a Clutter Map in the Artillery Locating Radar ARTHUR." *IEEE International Radar Conference,* Arlington, VA, May 7–10, 1990, pp. 207–210, IEEE Catalog No. 90CH-2882-9.

68. Siebert, W. M. "Some Applications of Detection Theory to Radar." *IRE Natl. Conv. Record* 6, pt. 4, pp. 5–14, 1958.

69. Hansen, V. G., and A. J. Zottl. "The Detection Performance of the Siebert and Dicke-Fix CFAR Detectors." *IEEE Trans.* AES-7 (July 1971), pp. 706–709.

70. Trunk, G. V., W. B. Gordon, and B. H. Cantrell. "False Alarm Control Using Doppler Estimation." *IEEE Trans.* AES-26 (January 1990), pp. 146–153.

71. Baker, C. H. *Man and Radar Displays.* New York: Macmillan, 1962.

72. Tucker, D. G. "Detection of Pulse Signals in Noise. Trace-to-Trace Correlation in Visual Displays." *J. Brit. IRE* 17 (June 1957), pp. 319–329.

73. Skolnik, M. I., and D. G. Tucker. "Discussion on 'Detection of Pulse Signals in Noise. Trace-to-Trace Correlation in Visual Displays.'" *J. Brit. IRE* 17 (December 1957), pp. 705–706.

74. Skolnik, M. "Radar Information from the Partial Derivatives of the Echo Signal Phase from a Point Scatterer." Naval Research Laboratory, Washington. D.C., Memorandum Rep. 6148, February 17, 1988.

PROBLEMS

5.1 (a) Find the matched-filter frequency response function $H(f)$ for a perfectly rectangular (video) pulse of duration τ, and amplitude A. (Assume the pulse extends in time from $-\tau/2$ to $+\tau/2$). (b) Sketch (roughly) its magnitude $|H(f)|$ for positive frequencies. (c) Sketch (roughly) the output of the video matched filter. (This can probably be done much easier by "inspection" than calculation.) All right to take $t_m = 0$.

5.2 (a) Find the matched-filter frequency response function $H(f)$ for a perfectly rectangular pulse of sinewave of duration τ, amplitude A, and frequency f_0. (Assume the pulse extends in time from $-\tau/2$ to $+\tau/2$). (b) Sketch (roughly) its magnitude $|H(f)|$ for positive frequencies. (c) Sketch (roughly) the output of the matched filter. (A rough sketch means it does not need to be precise or "artistic".) (d) Optional—In parts (a) and (b), your expression for $H(f)$ probably contained negative frequencies. What is the meaning of negative frequencies in $H(f)$ and what does one do about them? [Note that the answer to part (d) is not obvious or readily found in textbooks, and might require a little basic thinking about a Fourier transform and what it really is.]

5.3 From problem 5.1a you found that the output of a filter matched for a single *video* rectangular pulse of width τ and amplitude A is a triangular pulse with peak amplitude $A^2\tau$ and whose base has a width 2τ. (a) Sketch and label the output of a filter matched to a train of three equal video rectangular pulses with spacing T between pulses. (This can be done by inspection rather than by calculation.) (b) The more usual way to process a number of pulses, as discussed in Chap. 2, is to pass each pulse in sequence through a filter matched to a *single* pulse and integrate (add) the total number of pulses either coherently or noncoherently. Sketch and label the integrated output of a train of three equal video pulses when processed in this manner.

5.4 Find the ratio of the peak-signal-to-mean-noise power out of a matched filter designed for an RF signal

$$s(t) = Ae^{-at} \sin 2\pi f_0 t$$

where $0 < t < \tau$, and A and a are constants. The input noise is white and of spectral density N_0. You may assume there are many cycles of f_0 within the pulse duration τ, and that $e^{-a\tau}$ is small. (You should use integral tables.)

5.5 The input signal to its matched filter is $s(t) = (A/T)(T - t)$, where $0 \leq t \leq T$. Sketch the following: (a) the input signal, (b) the impulse response of the matched filter, and (c) the output signal from the matched filter. (d) Why is this particular waveform unrealizable?

5.6 What are the units of the constant G_a in the expression for the matched filter frequency response function given by Eq. (5.1)?

5.7 This problem involves finding the matched filter for fixed clutter modeled as nonwhite noise (NWN). The clutter power is assumed to be much larger than receiver noise so that the clutter echo rather than receiver noise determines signal detectability. It is assumed that the clutter is uniformly distributed and stationary so that the power spectral density $|N_i(f)|^2$ of the clutter echo signal can be considered to be the same as the power spectrum of the transmitted radar signal which is reflected from it. (a) Starting with Eq. (5.19) find the frequency response function $H(f)$ of the NWN matched filter for detecting a stationary point target in clutter as given by the above assumptions. (b) If the radar signal $s(t)$ were a perfectly rectangular pulse of width τ, sketch $|H(f)|$ for the NWN matched filter. (c) Why is this clutter matched filter not practical? (d) Optional—If you never heard of a matched filter, what type of radar waveform might you have selected to attempt to detect a stationary point target in uniform distributed clutter much larger than receiver noise? (e) If you answered part (d), how might your solution compare (better or worse) to the NWN matched clutter filter of part (b)?

5.8 This concerns the effectiveness of a nonmatched filter. (a) Find the peak-signal-to-mean-noise ratio (SNR) out of a one-stage low-pass RC network when the input is a rectangular pulse of width τ, amplitude $A = 1$, and the noise is white with a noise power per unit bandwidth of N_0. The normalized frequency response function is

$$\text{frequency response function of low-pass RC network} = H(f) = \frac{1}{1 + jf/B_v}$$

where B_v = bandwidth of the low-pass filter. Note that the maximum SNR occurs at a time equal to the pulse width τ. (b) Find the peak-signal-to-mean-noise power out of a filter that is perfectly matched to the rectangular pulse. (c) What is the loss in SNR (in dB) introduced by the nonmatched filter of (a) compared to the matched filter of (b)? (d) If the efficiency of the nonmatched filter relative to that of a matched filter is defined as

$$\rho_f = \frac{|s_0(t)|^2_{max}/N_{out}}{2E/N_0}$$

what is the value of $B_v\tau$ that maximizes the efficiency? [Note that a low-pass RC video network produces results for the above that are equivalent to what would be obtained with a single tuned RLC resonant network as might be used in the IF, assuming $B_v = B_{IF}/2$. Thus your answer to part (d) also applies to a single tuned RLC resonant network that could be in the IF portion of the receiver. A single-tuned circuit, however, is seldom found in radar receivers, so the answers you obtain in this problem might not be typical for radar.]

5.9 (a) Draw the block diagram of a correlation receiver. (b) Explain why the correlation receiver can be considered equivalent to the matched filter receiver in detection performance.

(c) Under what conditions, if any, might one choose to implement a correlation receiver rather than a matched filter receiver?

5.10   Sketch the matched-filter frequency response function when the waveform is just one RF cycle of sinewave in duration. You may start with the answer you found for problem 5.2(a). (A single cycle sinewave is an *ultrawideband* waveform.)

5.11   The matched filter of Eq. (5.1) assumed that the shape of the radar echo was the same as the shape of the transmitted radar signal. When a target is observed by a high-resolution radar (one with a range-resolution cell size much smaller than the target's radial extent), the target echo is not the same as that which was transmitted. It will consist of the superposition of echoes from the individual scattering centers of the target. (An example is a large ship 500 feet in length being observed head-on by a civil marine radar using a pulse width of 80 ns.) Discuss what has to be considered about the "matched filter" when attempting to detect a target that is much longer in radial size than the range-resolution cell so that the target echo is resolved into multiple scatterers. (Note that this question does not have a simple, unique answer.)

5.12   Show that the impulse response of a matched filter $[h(t) = G_a s(t_m - t)]$ is the inverse Fourier transform of its frequency response function $H(f) = G_a S^*(f) \exp(-j2\pi f t_m)$.

5.13   Why is a CFAR needed in some radars? What are the disadvantages of using CFAR?

5.14   What does one have to do in a radar system to avoid the use of a conventional CFAR?

5.15   In the VHF frequency region (30 to 300 MHz), the external noise at the receive-antenna terminals is generally higher than receiver internal noise. If one were to design an ultrawideband radar at VHF, qualitatively describe how the matched filter of Eq. (5.1), based on white noise, would have to be modified to allow for the large external noise levels that vary with frequency? (You might want to review Sec. 8.8 or other related sources on external noise.)

5.17   Show that the optimum detector law based on the criterion of the likelihood-ratio receiver is $y = \ln I_0(av)$, where y is the receiver output, a is the amplitude of the received sinewave signal normalized (divided) by the rms noise voltage, v is the amplitude of the IF voltage envelope normalized by the rms noise voltage, and $I_0(x)$ is the modified Bessel function of zero order. [The following outlines how you might work through the derivation. Start with the likelihood ratio of Eq. (5.21). Assume there are N independent pulses with normalized envelope-amplitudes $v_1, v_2, \ldots, v_N$ available from the radar receiver. The probability density function for the ith noise pulse $p_n(v_i)$ is found from Eq. (2.21), where v_i is the ratio $R/\psi_0^{1/2}$, R is the envelope amplitude of the ith output and $\psi_0^{1/2}$ is the rms noise level. The probability density function for the ith signal-plus-noise pulse $p_s(v_i)$ is found from Eq. (2.27), with a = ratio of the sinewave signal amplitude to rms noise. The likelihood ratio of the N pulses is

$$L_r(v) = \frac{\displaystyle\prod_{i=1}^{N} p_s(v_i)}{\displaystyle\prod_{i=1}^{N} p_n(v_i)} \geq K$$

where K is the receiver threshold level. After making the substitutions one should take the log of both sides so that the product becomes a more convenient sum. At this point, examination of the likelihood ratio will show how the signal should be processed and indicate the nature of the detector law.] How does this "optimum detector" relate to more conventional detectors?

5.18 What are the advantages and limitations of a binary integrator?

5.19 How does the performance of a radar operator making detection decisions by viewing the raw (unprocessed) video output of a radar display compare to the performance of an automatic (electronic) detector?

chapter

6

Information from Radar Signals

6.1 INTRODUCTION

This chapter includes the *basic measurements* that can be made by a radar; the *theoretical accuracy* of radar measurements; the *ambiguity diagram* that graphically illustrates the characteristics of radar waveforms in the time (range) and frequency (radial velocity) domains; *pulse compression,* which is used to achieve high range-resolution without the need for high peak power; and target *recognition* methods whereby a radar distinguishes one type of echo signal from another.

A radar obtains information about a target by comparing the received echo signal with the signal that was transmitted. It was said in Chap. 5 that the presence of a target is announced when the echo signal is strong enough to cross the receiver detection threshold. Knowing that a target is present, however, is almost never sufficient in itself. More must be known to be useful; thus a radar must provide information about the target, as discussed next.

6.2 BASIC RADAR MEASUREMENTS

A radar can obtain a target's location in range and azimuth, and sometimes elevation. After several observations of a moving target over a period of time, the target trajectory, or track, can be obtained. Radar can also do more than simply characterize the target as a "blob." In this section, the information available from a target first will be discussed as if the target were a point scatterer and then as a distributed scatterer. For purposes of this chapter, a *point scatterer*, or *point target*, is one with dimensions small compared to the size of the radar resolution cell in range, cross-range (angle), or both. The target's individual scattering features, therefore, are not resolved. A *distributed scatterer*, or *target* is one with dimensions large compared to the resolution-cell size, allowing the individual scatterers to be discerned. The resolution capabilities of a radar usually (but not always) determine whether a target is considered as a point scatterer (unresolved) or a distributed target (resolved). A *complex scatterer* is one that contains multiple scatterers. A complex scatterer can be either a point scatterer or a distributed scatter.

Measurements of a Point Target   The basic radar measurements that can be made for a point target when only a single observation is made are range, radial velocity, direction (angle), and, in some special cases, tangential velocity.

Range   It was said in Chap. 1 that the measurement of distance, or range, was obtained from the round-trip time T_R required for a radar signal to travel to the target and back. The range R is given by $cT_R/2$, where c = velocity of propagation. In many radar applications the target's range is the most significant measurement that is made. No other sensor has been able to compete with radar for determining range to a distant target, especially in accuracy, ability to make a measurement over very long or very short distances, and under adverse weather conditions. A long-range air-surveillance radar might measure range to an accuracy of many tens of meters, but accuracies of a few centimeters are possible with precision systems. In the most precise systems, the accuracy of a range measurement is limited only by the accuracy with which the velocity of propagation is known. The spectral bandwidth occupied by the radar signal is the fundamental resource required for accurate range measurement. The greater the bandwidth, the more accurate can be the range measurement.

Angle Measurement   Almost all radars utilize directive antennas with relatively narrow beamwidths. A directive antenna not only provides the large transmitting gain and large receiving aperture needed for detecting weak echo signals, but its narrow beamwidth allows the target's direction to be determined accurately. It can do this by noting the direction the antenna points when its received echo signal is a maximum. A typical microwave radar might have a beamwidth of one or a few degrees. The narrowest beamwidths of operational radars have been about 0.3°. This is not an absolute limit; but the narrower the beamwidth, the greater the mechanical and electrical tolerances that are required of the antenna.

   Angular accuracy can be much better than the antenna beamwidth, as described later in this chapter. Angle accuracy depends on the electrical size of the antenna (size as measured in wavelengths). With signal-to-noise ratios typical of those required for reliable detection, the angular location of a target can be determined to about 1/10th of a beamwidth. The best precision monopulse tracking radars used for range instrumentation can determine angle to about 0.1 mrad rms (0.006°) if the signal-to-noise ratio is large enough and if the proper efforts are taken to minimize errors.

Radial Velocity   Measurement of the radial component of velocity in many radars is obtained from the rate of change of range. This is known as the *range rate*. The classical method for finding the radial velocity is based on $v_r = (R_2 - R_1)/(T_2 - T_1)$. It is found from the range R_1 measured at time T_1 and the range R_2 at time T_2. However, this method of finding range rate (or any other derivative of a location measurement) is not considered here as a *basic* radar measurement even though it may be widely used. Instead, the doppler frequency shift is the basic method for obtaining radial velocity. It can be made on the basis of a single observation. Using the classical expression [Eq. (3.3)] for the doppler frequency shift, f_d, the radial velocity v_r is given as

$$v_r = \lambda f_d/2 \qquad\qquad\qquad [6.1]$$

where λ = wavelength. It can be shown from the theoretical accuracy expressions presented later in this chapter that the radial velocity accuracy derived from the doppler frequency shift can be much better than that found from the range rate, assuming the time between the two range measurements in the range-rate method is the same as the time duration of the doppler frequency measurement. (See Problem 6.5.)

It will be seen later that the accuracy of the doppler-frequency measurement depends on the time duration over which it is made. The longer the time, the better the frequency accuracy. Because of the relationship between radial velocity and wavelength in Eq. (6.1), the shorter the wavelength, the shorter can be the observation time to achieve a required velocity accuracy. (The shorter the wavelength, the higher the frequency.) Or, the shorter the wavelength, the better will be the velocity accuracy for a given observation time.

In spite of its good accuracy, the doppler frequency shift is not used as often as is the range-rate for obtaining the radial velocity since it can result in ambiguities in range and/or doppler when employed with a short or a medium pulse-duration radar.

Tangential (Cross-Range) Velocity   Just as the temporal doppler frequency shift can provide the radial velocity, there exists in the spatial (angle) domain an analogous *spatial doppler-frequency shift* from which the tangential velocity can be determined.[1] (If the radial velocity is $v_r = v \cos\theta$, the tangential velocity is $v_t = v \sin\theta$, where v is the target's speed and θ is the angle between the target's velocity vector and the radar line of sight.) The angle-rate times the range is equal to the tangential velocity. Together, the tangential velocity and the radial velocity can give the magnitude of the target's speed v and its direction θ. The measurement of tangential velocity has not been of practical interest in radar since it requires a long-baseline antenna system.

Measurements of a Distributed Target   With sufficient resolution in the appropriate dimension, the size and shape of a distributed target can be ascertained. It should be recalled that resolution and accuracy are not the same. Range *resolution* requires that the entire bandwidth be occupied continuously without gaps in the signal frequency spectrum. Range *accuracy,* however, only requires, as a minimum, that there be adequate spectral energy at the two ends of the spectral bandwidth. The spectral bandwidth need not be fully occupied. This assumes that there is only one scattering object present. Resolution requires a filled spectrum; accuracy can be achieved with a thinned, or sparse, spectrum. A similar description applies in the temporal (time) domain for frequency measurement, and in the spatial (antenna) domain for angle measurements. Generally, good resolution will provide good accuracy; but the reverse is not necessarily true since accurate measurements can be made with waveforms that do not provide good resolution.

Radial Profile   The target's profile (and size) in the range dimension can be obtained when the radar's range resolution cell is smaller in size than the target's dimensions (i.e., when the scattering centers of the target can be resolved). To obtain a radial profile of a target it is required that $c\tau/2 << D$, where D = the target's radial component and τ = pulse width. Good resolution in the range dimension requires large spectral bandwidth. The radial profile of a target sometimes can be employed to obtain limited "recognition" of one type of target from another.

Tangential (Cross-Range) Profile   With sufficient resolution in the angle dimension, the tangential (cross-range) profile of a distributed target can be determined. This can provide the angular size of the target and the location in angle of the scattering centers. If the range is known, the location of scatterers in the tangential dimension can be determined since the cross-range (tangential) dimension is equal to the product of range and angle (the latter in radians). Resolution in cross-range based on conventional angle measurements is generally not as good as the resolution that can be obtained in the range dimension. Synthetic aperture radar (SAR) and inverse synthetic aperture radar (ISAR), however, can provide excellent cross-range resolution without the need for large antennas. (In SAR and ISAR, the equivalent of resolution in angle may be thought of as being obtained because of resolution in the doppler frequency domain.)

Size and Shape   When the tangential profile is obtained at each range resolution cell, the target image (size and shape) is formed. Imaging radars, such as SAR, ISAR, and SLAR (side-looking airborne radar), have sufficient resolution in both range and cross-range to resolve the major scatterers of a distributed target. (SLAR achieves its tangential resolution by use of a narrow beam antenna directed perpendicular to the flight direction of the aircraft carrying the radar.)

Symmetry   The response of a target to changes in the polarization of the radar signal can provide a measure of the symmetry of the target. (The polarization of a radar signal is determined by the orientation of the electric field.) If a sphere (a perfectly symmetrical target) were directly viewed by a radar with a rotating linearly polarized signal, there would be no change of the echo signal when the polarization is changed. On the other hand, if the same rotating polarization radar were to view a long narrow rod, the echo would be maximum when the electric field (polarization) is parallel to the rod and minimum when it is perpendicular to the rod. By observing the variation of the amplitude of the echo signal as a function of polarization, the orientation and shape of the rod can be determined. Measurement of target symmetry using polarization is not widely used in radar; however, it is the basis for detection of aircraft (an asymmetrical target) in the presence of rain (symmetrical target) when using circular polarization (defined as the electric field rotating at the RF frequency, Sec. 7.8).

Change of Radial and Tangential Profiles   Here it is assumed that the pulse is long enough so that the individual scatterers of a complex target are not resolved. If the individual scatterers of this complex target change their relative locations in range (radial profile), the echo signal will experience a change in amplitude due to constructive and destructive interference among the echo signals from the individual scatterers. Changes in the target echo amplitude, therefore, indicate there are changes in the relative locations of the individual scatterers of the complex target.

Examples of target effects that might be recognized by the amplitude modulation of the echo signal include changes in target aspect, propeller modulation, jet engine modulation, and the time-varying separation of two closely spaced, unresolved targets (such as two aircraft or an aircraft and a missile).

Surface Characteristics   The dielectric constant of a target's surface material and the roughness of its surface can, in principle, be found from radar measurements. Surface roughness may be determined by varying the radar frequency and noting where the scattering changes from specular (a smooth surface) to diffuse (rough surface). This boundary depends on the size of the surface roughness relative to the radar wavelength. Surface roughness, such as the height of ocean waves (the sea state), can be found from a direct measurement with a high range-resolution radar, as has been done from space with a precision high-resolution altimeter.

The dielectric constant of the scattering surface can be found if the reflection coefficient can be measured and if the shape and roughness of the surface are known. This is practical under laboratory conditions, but difficult to apply with radar. Radar cross section measurements over a wide range of frequencies, however, were used to estimate the dielectric properties of the moon's surface (before astronauts landed on the moon and brought back rocks for laboratory analysis).[2,3]

The surface roughness and the dielectric constant are of interest for remote sensing with radar, especially from space. The former might indicate the sea state over the oceans of the world; and the latter, if it were practical, might be used to determine soil moisture, which is of interest for agriculture and hydrology. Although the radar determination of surface characteristics might be desirable, it has proven to be difficult to achieve except under limited circumstances.

6.3   THEORETICAL ACCURACY OF RADAR MEASUREMENTS

Noise is the fundamental limitation to accurate radar measurements. The theoretical aspects of the extraction of information from radar signals have benefited greatly from the theory of *statistical parameter estimation* just as the theory of detection has benefited from the statistical theory *of hypothesis testing.*[4] In this section, expressions for the theoretical accuracies of radar measurements will be presented. It is assumed that the signal-to-noise ratio is large. This is usually the case since it was found in Sec. 2.5 that large signal-to-noise ratios are required for detection of a signal. Detection must occur before meaningful information can be extracted about a target echo. It is further assumed that the measurement error associated with a particular parameter is independent of the errors in any of the other parameters, that accuracy is limited only by receiver noise, and that all bias errors are accounted for separately. The measure of error is the root mean square (rms) of the difference between the measured (estimated) value and the true value.

The expressions given in this chapter for the theoretical rms error δM of a radar measurement M have the following form:

$$\delta M = \frac{kM}{\sqrt{2E/N_0}} \qquad [6.2]$$

where k is a constant whose value is in the vicinity of unity, E is the received signal energy, and N_0 is the noise power per unit bandwidth. The following will be shown later in this section:

- For a time-delay (range) measurement, k depends of the shape of the frequency spectrum $S(f)$, and M is the rise time of the pulse (inversely proportional to bandwidth).

- For a measurement of radial velocity based on the doppler frequency, k depends on the shape of the time waveform $s(t)$, and M is the spectral resolution (inversely related to the time duration of the signal).

- For an angle measurement, k depends on the shape of the aperture illumination $A(x)$, and M is the beamwidth.

Theoretical radar accuracies may be derived by a variety of methods such as (1) simple geometrical relationships among the signal, noise, and the parameter to be measured;[5] (2) the likelihood ratio;[6] (3) the method of inverse probability;[7] (4) a suitably selected gating function preceded by a matched filter;[8] and (5) minimization of the mean square error.[9] The simple method (no. 1) for finding the rms error in the measurement of time delay when the waveform is a rectangular pulse will be illustrated next. This derivation takes some liberties, but it has the advantage of being easy to understand. Fortunately, the simple method and the more involved methods give similar answers for the rectangular pulse.

Time-Delay (Range) Accuracy—Simplified Method   The measurement of range R is the measurement of the round-trip time delay T_R for the radar signal (waveform) to travel out to the target and back. The rms error in range is $\delta R = (c/2)\, \delta T_R$, where c is the velocity of propagation, and δT_R is the rms error in time delay. The range, or time-delay, measurement to be described here is based on locating the leading edge of the video pulse, Fig. 6.1. The video pulse uncorrupted by noise is shown by the solid curve. Its shape is not perfectly rectangular, but has finite rise and fall times. (Zero rise or fall times require infinite bandwidth.) The effect of noise added to the pulse is to shift the time of threshold crossing as shown by the dashed curve. Since large signal-to-noise ratio is assumed, the slope of the leading edge of the noise-free pulse (solid curve) can be equated to the slope of the leading edge of the pulse with noise added (dashed curve). The slope of the leading edge of a pulse of amplitude A at the output of a video filter is A/t_r, where t_r is the rise time. From Fig. 6.1, the slope of the signal plus noise (dashed curve) can be

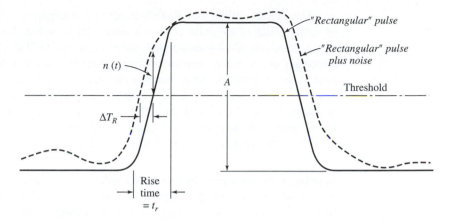

Figure 6.1   Measurement of time delay using the leading edge of the video pulse. Solid curve is the echo pulse uncorrupted by noise. Dashed curve represents signal plus noise.

written as $n(t)/\Delta T_R$, where $n(t)$ is the noise voltage at the threshold crossing of the pulse, and ΔT_R is the error in the time-delay measurement. Equating the slope of the leading edge of the pulse without noise to the slope of the pulse with noise gives

$$A/t_r = n(t)/\Delta T_R \qquad [6.3]$$

which leads to

$$[\overline{(\Delta T_R)^2}]^{1/2} = \delta T_R = \frac{t_r}{(A^2/\overline{n^2})^{1/2}} = \frac{t_r}{(2S/N)^{1/2}} \qquad [6.4]$$

where $A^2/\overline{n^2}$ is the video signal-to-noise power ratio. The last part of Eq. (6.4) follows from the fact that the signal-to-noise power ratio for a rectangular video pulse is equal to $2S/N$, where S/N is the signal-to-noise power ratio of a sinewave pulse in the IF portion of the receiver. This assumes a linear detector and a large signal-to-noise ratio. Equation (6.4) indicates that accurate measurements of time delay require video pulses with short rise times and large amplitudes. The width of the pulse does not enter explicitly in this expression.

If the rise time of the video pulse is limited by the spectral bandwidth B of the rectangular-shaped IF filter, then $t_r \approx 1/B$. Letting $S = E/\tau$ and $N = N_0 B$, the error in the time delay can be written

$$\delta T_R = \left(\frac{\tau}{2BE/N_0}\right)^{1/2} \qquad [6.5]$$

where τ = pulse width, B = spectral bandwidth of the rectangular filter, E = signal energy, and N_0 = noise power per unit bandwidth. If a similar measurement of time delay is made at the trailing edge of the video pulse, and if the noise at the trailing edge is independent of the noise at the leading edge, then the two measurements can be averaged to obtain an improvement in the time-delay accuracy of $\sqrt{2}$, which gives

$$\delta T_R = \left(\frac{\tau}{4BE/N_0}\right)^{1/2} \qquad [6.6]$$

The estimate of the time delay obtained by determining when either the leading edge or trailing edge of the pulse crosses a threshold, as in Fig. 6.1, will depend on the value of the threshold relative to the peak value of the pulse. The choice of threshold level is important, therefore, for consistency in accurate measurement when only the leading edge is used. It has been suggested that the bias error can be avoided by one of several methods.[10-12] One method is to use an adaptive threshold in which the level of the threshold is always a fixed fraction of the pulse amplitude. If the average of the two time delays found from both the leading and trailing edges of the pulse is used, there is no theoretical bias with change of amplitude if the pulse shape is symmetrical.

Time-Delay Accuracy and Effective Bandwidth   There are several methods[6-9] based on the likelihood ratio, inverse probability, and other statistical analyses, which all lead to the following expression for the rms error in the measurement of the time delay

$$\delta T_R = \frac{1}{\beta \, (2E/N_0)^{1/2}} \tag{6.7}$$

where E is the signal energy, N_0 is the noise power per unit bandwidth, and β is called the *effective bandwidth* and is defined as

$$\beta^2 = \frac{\displaystyle\int_{-\infty}^{\infty} (2\pi f)^2 |S(f)|^2 \, df}{\displaystyle\int_{-\infty}^{\infty} |S(f)|^2 \, df} = \frac{1}{E} \int_{-\infty}^{\infty} (2\pi f)^2 \, |S(f)|^2 \, df \tag{6.8}$$

It has also been called the *rms bandwidth*. The effective bandwidth β is such that $(\beta/2\pi)^2$ is the normalized second moment of $|S(f)|^2$ about its mean. Equation (6.8) assumes that the mean value of $S(f)$ is at $f = 0$, where $S(f)$ is the video spectrum with negative as well as positive frequencies. The effective bandwidth β is different from other bandwidths encountered in electronic engineering. It is not related to either the half-power bandwidth or the noise bandwidth. The more the spectral energy is concentrated at the two ends of the band, the larger will be β and the more accurate will be the measurement of time delay.

The first edition (1962) of this text discussed three different methods for deriving Eq. (6.8) based on statistical concepts.[6-8] The second edition (1980) described one method to derive this equation.[8] These methods are not included in the current edition since there now appears to be less interest in the mathematical aspects of the subject. The application of Eq. (6.8), however, is important and will be discussed next.

Rectangular Pulse   When the spectrum $S(f)$ of a perfectly rectangular pulse—one with zero rise time and zero fall time—is inserted in Eq. (6.8) for the effective bandwidth, the result is obtained that $\beta = \infty$. This means that $\delta T_R = 0$; hence, the measurement of the time delay can be made with zero error. It may seem strange, but it is correct for the perfectly rectangular pulse that was assumed. An infinite bandwidth implies zero rise time (infinite slope) so noise does not displace the threshold crossing in time (as it does in Fig. 6.1 for a finite rise time) and there will be no error in the delay. A perfectly rectangular pulse, however, requires infinite bandwidth, which is not possible. Thus the bandwidth of a practical "rectangular" pulse must be finite, there will be finite rise and fall times, and the rms time-delay error will not be zero.

To obtain the effective bandwidth β for a finite-bandwidth pulse, it will be assumed that the spectrum of a perfectly rectangular IF pulse of width τ_r is limited to a finite spectral bandwidth B_s. For the time-delay measurement that uses the envelope of the IF pulse, this is equivalent to a video spectrum $S(f) = (\sin \pi f \tau_r)/\pi f \tau_r$ that is limited to a spectral width $\pm B_s/2$, as in Fig. 6.2. (The video spectrum is shown here with both negative and positive frequencies, as is required in Fourier analysis.) Although the analysis considered here is based on the video pulse of width τ_r and a low-pass filter of video bandwidth $B_s/2$, the result is the same as taking the envelope of an IF pulse of width τ_r and an IF bandpass filter of bandwidth B_s. The value of β^2 for this case is found by setting the limits of the integration in Eq. (6.8) from $-B_s/2$ to $+B_s/2$ instead of from $-\infty$ to $+\infty$, which is then

$$\beta^2 = \frac{(2\pi)^2 \int_{-B_s/2}^{B_s/2} f^2 (\sin^2 \pi f \tau_r)/\pi^2 f^2 \, df}{\int_{-B_s/2}^{B_s/2} (\sin^2 \pi f \tau_r)/\pi^2 f^2 \, df} = \frac{1}{\tau_r^2} \frac{\pi B_s \tau_r - \sin \pi B_s \tau_r}{\mathrm{Si}(\pi B_s \tau_r) + (\cos \pi B_s \tau_r - 1)/\pi B_s \tau_r} \qquad \textbf{[6.9]}$$

where $\mathrm{Si}(x)$ is the *sine integral* function defined by $\int_0^x (\sin u)/u \, du$. For large $B_s \tau_r$ in Eq. (6.9), the product $\beta^2 \tau_r^2 \to 2 B_s \tau_r$, or

$$\beta^2 \approx \frac{2 B_s}{\tau_r} \approx \frac{2}{\tau_r t_r} \qquad \text{for large } B_s \tau_r \qquad \textbf{[6.10]}$$

It was assumed in Eq. (6.10) that the rise time t_r of the pulse is approximately the inverse of the spectral bandwidth B_s (the total width of the spectrum, not the half-power bandwidth).

Figure 6.2   Spectrum $[(\sin \pi f \tau_r)/\pi f \tau_r]$ of a rectangular pulse of width τ_r shown limited to a spectral extent of $\pm B_s/2$.

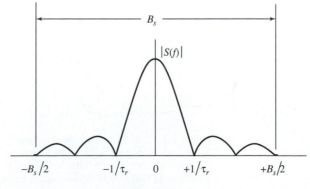

Frequency

Substituting this expression for β^2 into Eq. (6.7) gives the rms error in time delay as

$$\delta T_R = \left(\frac{\tau_r}{4B_s E/N_0} \right)^{1/2} = \left(\frac{t_r \tau_r}{4E/N_0} \right)^{1/2} \qquad \text{rectangular pulse, } B_s \tau_r \gg 1 \qquad \textbf{[6.11]}$$

This applies for large $B_s \tau_r$, or when the rise time is small compared to the pulse width. Note that Eq. (6.11), derived in a totally different manner, is the same as Eq. (6.6) except for denoting the spectral bandwidth in Eq. (6.11) as B_s instead of B and the pulse width as τ_r instead of τ.

The value of $\beta = (2B_s/\tau_r)^{1/2}$ for a long rectangular pulse of fixed bandwidth B_s (or fixed rise time) decreases with increasing pulse width τ_r. Thus if the total energy remains the same, the time-delay accuracy decreases (becomes worse) with increasing pulse width even though the rise time remains the same.

It is not often that $B_s \tau_r \gg 1$ in radar applications. Next, the more usual case is examined, where the product of half-power bandwidth and pulse width is approximately unity.

Quasi-Rectangular Pulse As before, we start with a perfectly rectangular pulse of width τ_r. The spectral extent B_s is assumed in Fig. 6.2 to be limited to the main portion of the rectangular video pulse spectrum that lies between the first nulls at $-1/\tau_r$ and $+1/\tau_r$ on either side of the spectrum peak at $f = 0$. (As mentioned, integration over negative as well as positive frequencies has to be considered in Fourier analysis). Thus the IF spectral bandwidth extent is $B_s = 2/\tau_r$. The half-power bandwidth is $B \approx B_s/2$; or $B \approx 1/\tau_r$. (The product of the half-power bandwidth B and the width τ_r of a rectangular pulse is actually equal to 0.886; but, for convenience, it is usually "rounded off" to unity. This is more like the usual case in radar where $B\tau \approx 1$.) The solid curve of Fig. 6.3 shows the pulse shape that

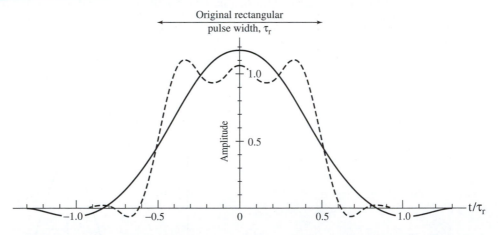

Figure 6.3 Solid curve is the theoretical waveform out of a low-pass rectangular filter of bandwidth $B_v = B_s/2 = 1/\tau_r$ when the input is a rectangular pulse of width τ_r. It is the same as the envelope out of an IF filter of bandwidth $B_s = 2/\tau_r$ when the input is a rectangular pulse of sinewave of width τ_r. In the text this is called a *quasi-rectangular pulse*. The dashed curve applies for $B_s = 6/\tau_r$.

emerges from the video low-pass filter of bandwidth $B_s/2$, which corresponds to an IF filter with bandwidth B_s. Although Fig. 6.3 doesn't resemble a perfectly rectangular pulse, it will be called the *quasi-rectangular pulse.* It is not unusual for a pulse radar to radiate a similar-looking waveform when it is thought that a "rectangular" pulse is being generated. (A rounded pulse is often used in radar since it produces less out-of-band interference to other users of the electromagnetic spectrum.) Substituting $B_s\tau_r = 2$ into Eq. (6.9), we get that $\beta\tau_r$ is very close to 2.1; which, when inserted into Eq. (6.7), gives the rms time-delay error as

$$\delta T_R = \frac{\tau_r}{2.1(2E/N_0)^{1/2}} \qquad \text{quasi-rectangular pulse, } B_s\tau_r \approx 1 \qquad \text{[6.12]}$$

This expression for the rms time-delay error is considered here to be representative of a conventional pulse typically used in radar. Since the rectangular IF filter is of bandwidth $B_s = 2/\tau_r$, then $\beta = 1.05B_s$. In terms of the half-power bandwidth B of the IF spectrum, $\beta = 2.4B$ when $B\tau_r = 0.886$, or $\beta \approx 2.1B$ when $B\tau \approx 1$. The dashed curve in Fig. 6.3 for $B_s\tau_r = 6$ is shown to indicate the shorter rise time of the pulse when the bandwidth B_s is much wider than the quasi-rectangular pulse.

The symbol τ_r was used for the pulse width in the above analysis instead of τ to indicate it was the width of the perfectly rectangular pulse before it is passed through a rectangular IF filter of bandwidth B_s (or video bandwidth $B_s/2$). If τ denotes the half-power pulse width that emerges from a filter of bandwidth $B_s = 2/\tau_r$, then $\tau = 0.625\tau_r$. The effective bandwidth in this case is $\beta = 1.3/\tau$.

Trapezoidal Pulse A rectangular pulse is sometimes approximated by a trapezoidal pulse with finite rise and fall times. If the width across the top of the trapezoid is τ_p and if the rise and fall times are t_r, then the rms error in time-delay measurement is

$$\delta T_R = \left(\frac{3\tau_p t_r + 2t_r^2}{12E/N_0}\right)^{1/2} \qquad \text{trapezoidal pulse} \qquad \text{[6.13]}$$

When the rise time t_r is small compared to the width τ_p of the top of the trapezoid, Eq. (6.13) becomes

$$\delta T_R = \left(\frac{\tau_p t_r}{4E/N_0}\right)^{1/2} \qquad t_r \ll \tau_p \qquad \text{[6.14]}$$

which is the same as Eq. (6.11) for the rectangular pulse with short rise time. Note that the larger the width τ_p, the poorer is the accuracy.

Triangular Pulse When $\tau_p = 0$ in Eq. (6.13), the pulse is triangular and its base = $2t_r = \tau_B$. The rms error becomes

$$\delta T_R = \frac{\tau_B}{12^{1/2}(2E/N_0)^{1/2}} \qquad \text{triangular pulse} \qquad \text{[6.15]}$$

In terms of the half-power bandwidth B of the $(\sin^2 x)/x^2$ spectrum of a triangular pulse (where $x = \pi f\tau_B/2$), the effective bandwidth $\beta = 2.72\,B$.

With a triangular or rounded pulse (such as the quasi-rectangular, gaussian, or $(\sin x)/x$ time waveforms) the time delay might be found by differentiating the pulse waveform and selecting the time at which $s'(t)$ goes through zero (which is the point at which the amplitude of $s(t)$ is a max).

Gaussian Pulse   Consider a pulse described by the gaussian function

$$s(t) = \exp\left(-\frac{1.38t^2}{\tau^2}\right) \tag{6.16}$$

where τ in this case is the half-power pulse width. Its rms time-delay error is

$$\delta T_R = \frac{\tau}{1.18(2E/N_0)^{1/2}} = \frac{1.18}{\pi B(2E/N_0)^{1/2}} \qquad \text{gaussian pulse} \tag{6.17}$$

where B is the half-power bandwidth of the gaussian-pulse spectrum.

Cosine Pulse   The positive half of one cycle of a cosine also can represent practical radar pulses. Its time waveform is $\cos(\pi t/\tau_B)$, where τ_B is the width of the base. Its effective bandwidth is $\beta = \pi/\tau_B = 2.64B$, where $B =$ half-power bandwidth. The parabolic pulse, given by $s(t) = 1 - 4t^2/\tau_B^2$, is similar in shape and has $\beta = 3.16/\tau_B$.

Pulse with Uniform Spectrum   The effective bandwidth of a waveform with a uniform spectrum of width B_u is $\beta = \pi B_u/\sqrt{3}$. The time waveform is of the form $(\sin \pi B_u t)/\pi B_u t$. The rms time-delay error is

$$\delta T_R = \frac{\sqrt{3}}{\pi B_u(2E/N_0)^{1/2}} \qquad \frac{\sin \pi B_u t}{\pi B_u t} \text{ waveform} \tag{6.18}$$

The above applies to a linear-FM pulse-compression waveform with large time-bandwidth product.

Minimum Error   Examination of the integral in the numerator of the expression for β^2 in Eq. (6.8) indicates that a large value of β requires that the spectral energy be concentrated at the extremes of the spectral bandwidth [note that the origin is at the mean value of $S(f)$]. Consider, therefore, a spectrum with its entire energy concentrated in delta functions at both ends of the video spectrum, such as $S(f) = \delta(f + B_s/2) + \delta(f - B_s/2)$. Substitution of this spectrum into Eq. (6.8) yields $\beta = \pi B_s$. This is the largest value of β that can be obtained for a spectrum occupying a bandwidth B_s. The time waveform corresponds to two sinewaves, each of infinite duration, separated in frequency by B_s. (The time duration of the waveform, of course, cannot be infinite. The result is approximately the same, however, if the time duration τ is such that $1/\tau << B_s$.) This waveform, however, results in range ambiguities that have to be resolved. It does not work when multiple unresolved targets are present.

Two CW waveforms separated in frequency by B have been considered in the past for measurement of range by determining the phase difference between them.[13,14] The phase difference $\Delta\phi$ between the two frequencies separated by B gives the time delay as

$$T_R = \frac{\Delta\phi}{2\pi B} \qquad\qquad [6.19]$$

The error in time delay, δT_R, based on the theoretical error in phase $\delta\phi$ [which is $\delta\phi = (2S/N)^{-1/2}$] from each of the two sine waves (with each having one half the total energy) results in the same value of $\beta = \pi B$ as found above by substituting the two delta-function spectrum into Eq. (6.8).

The two-frequency CW waveform might have a higher effective bandwidth than the others (and better time-delay accuracy), but it is inconvenient to use because of ambiguities. In practical embodiments of this method of range measurement, four or five frequencies have been used to resolve the ambiguities and achieve good accuracy.[15]

Other Considerations   In the expression for time-delay error as given by Eqs. (6.7) and (6.8), the energy E appears as a normalizing factor in the effective bandwidth β. It also appears in the signal-to-noise energy ratio ($2E/N_0$). The energy cancels, and the rms time-delay error can be written as

$$\delta T_R = \left(\frac{N_0}{2\int_{-\infty}^{\infty} (2\pi f)^2 |S(f)|^2 \, df} \right)^{1/2} \qquad\qquad [6.20]$$

The integral in the denominator also can be expressed as

$$\beta^2 E = \int_{-\infty}^{\infty} [s'(t)]^2 \, dt = -\int_{-\infty}^{\infty} s''(t)s(t) \, dt \qquad\qquad [6.21]$$

The energy E does not appear in the right-hand side of this equation for $\beta^2 E$ since $\beta^2 \sim 1/E$. When calculating β, the expressions of Eq. (6.21) sometimes are easier to compute than the integral of Eq. (6.8). Equation (6.21) implies that accurate time delay is obtained with waveforms having large first derivatives over the time duration of the signal.

Table 6.1 lists the effective bandwidths β for the various waveforms considered here. There is not much difference in β among the various waveforms in this table. One might, therefore, not be too concerned about which waveform ought to be chosen based on time-delay accuracy alone. The triangular waveform has a large theoretical accuracy, but the discontinuity of the slope at the middle of the pulse presents practical problems in its generation. The values of β for the "rounded" pulses (gaussian, cosine, and quasi-rectangular) are not much lower than the triangular, and they are better representations of the shape of practical radar pulses. Radar pulses are almost always band-limited and are, therefore, rounded rather than appear as the perfectly rectangular pulses as sometimes seen in textbooks. The cosine seems a good compromise choice as typical of these rounded waveforms. The uniform spectrum, such as that of the linear-FM pulse-compression waveform, is about $2/3d$ less accurate than the rounded pulse if its spectral extent B equals the half-power bandwidth of the rounded pulse, but this is not much of a difference.

Accuracy of Frequency and Radial Velocity   The measurement of frequency in radar is that of the doppler frequency shift. As mentioned previously, the radial component v_r of velocity can be found from the doppler frequency shift $f_d = 2v_r/\lambda$, where λ is the radar

Table 6.1   Effective Bandwidth β for Various Waveforms

Description	Time Waveform $s(t)$	Effective Bandwidth β
Gaussian pulse	$\exp(-1.38t^2/\tau^2)$	2.66 B, or 1.18/τ
Cosine pulse	$\cos(\pi t/\tau_B)$	2.64 B, or 3.14/τ_B
Triangular pulse	$(2t/\tau_B) + 1 \quad -\tau_B/2 < t < 0$ $-(2t/\tau_B) + 1 \quad 0 < t < \tau_B/2$	2.72 B, or 3.46/τ_B
Quasi-rectangular* $B_s\tau_r = 2$; $B\tau_r = 0.886$	$\text{Si}[\pi(B_st + 1)] - \text{Si}[\pi(B_st - 1)]$	2.38 B, or 2.1/τ_r, or 1.3/τ
Uniform spectrum of width B_u	$(\sin \pi B_u t)/\pi B_u t$	1.8 B_u, or 2.04/τ
Band-limited, rectangular pulse; $B_s\tau_r \gg 1$	$\text{Si}[\pi B_s(t + \tau_r/2)] - \text{Si}[\pi B_s(t - \tau_r/2)]$	$1.4\sqrt{B_s/\tau_r}$
2 CW sinewaves separated by B_s	video spectrum: $\delta(f + B_s/2) + \delta(f - B_s/2)$	3.14 B_s

B_s = spectral extent, B = half-power bandwidth, τ = half-power pulsewidth, τ_r = width of original rectangular pulse, τ_B = extent of pulse (at its base), Si [X] = sine integral function of X.

*Considered to be typical of many pulse waveforms.

wavelength. The rms error in radial velocity is $\delta v_r = (\lambda/2)\,\delta f_d$, where δf_d is the rms error in the doppler frequency.

Using the method of inverse probability, Roger Manasse[16] showed that the rms error in the measurement of frequency is

$$\delta f = \frac{1}{\alpha\,(2E/N_0)^{1/2}} \qquad \text{[6.22]}$$

where

$$\alpha^2 = \frac{\displaystyle\int_{-\infty}^{\infty} (2\pi t)^2 s^2(t)dt}{\displaystyle\int_{-\infty}^{\infty} s^2(t)dt} \qquad \text{[6.23]}$$

and $s(t)$ is the input signal as a function of time. Note the similarity of this expression for δf and that of δT_R in Eq. (6.7), as well as the similarity in the expressions for α and β. The parameter α is the *effective time duration* of the signal, and $(\alpha/2\pi)^2$ is the normalized second moment of $s^2(t)$ about the mean epoch, taken to be $t = 0$. [If the mean is not zero, but is some other value, t_0, the integrand in the numerator of Eq. (6.23) would be $(2\pi)^2(t - t_0)^2 s^2(t)$.]

Rectangular Pulse   The value of α^2 for a perfectly rectangular pulse of width τ is found to be $\pi^2\tau^2/3$; thus the rms frequency error is

$$\delta f = \frac{\sqrt{3}}{\pi\tau(2E/N_0)^{1/2}} \qquad \text{rectangular pulse} \qquad [6.24]$$

The longer the pulse, the more accurate is the frequency measurement. This expression can also be applied to a frequency measurement made by a CW radar since the observation time over which the CW measurement is made is equivalent to the pulse duration τ.

Quasi-Rectangular Pulse   For the bandwidth-limited rectangular pulse of Fig. (6.2), the value of α^2 is

$$\alpha^2 = \frac{\pi^2\tau_r^2}{3} \frac{\text{Si}(\pi B_s\tau_r) + \dfrac{\cos(\pi B_s\tau_r) - 3}{\pi B_s\tau_r} - \dfrac{2\sin\pi B_s\tau_r}{(\pi B_s\tau_r)^2} - \dfrac{8[\cos(\pi B_s\tau_r) - 1]}{(\pi B_s\tau_r)^3}}{\text{Si}(\pi B_s\tau_r) + [\cos(\pi B_s\tau_r) - 1]/\pi B_s\tau_r} \qquad [6.25]$$

where τ_r is the width of a rectangular pulse that is passed through a rectangular filter of bandwidth B_s, and Si (x) is the sine integral function of x. In the limit as $B_s\tau_r \to \infty$, α^2 approaches $\pi^2\tau_r^2/3$, which is the same as that obtained for the perfectly rectangular pulse.

In the discussion of the rms error in time delay given previously, a *quasi-rectangular pulse* with $B_s\tau_r = 2$ was considered. For this case, Eq. (6.25) gives $\alpha = 1.6\tau_r$, where τ_r is the width of the rectangular pulse before passing through a rectangular filter of bandwidth B_s. The half-power width τ after the pulse passes through the band-limited filter is $0.625\tau_r$, so that $\alpha = 2.6\tau$.

The value of α for a perfectly rectangular pulse is finite even though the value of β for a perfectly rectangular pulse was found to be infinite. The effective time duration α will be infinite, however, for a waveform that has a perfectly rectangular frequency spectrum of width B. Such a spectrum corresponds to a $(\sin x)/x$ time waveform of infinite duration, where $x = \pi Bt$. Any practical waveform must be limited in time, and α will therefore be finite. The frequency error for a time-limited waveform with a rectangular-like spectrum may be found in a manner similar to that which was employed for finding the time-delay error of a band-limited rectangular pulse. The $(\sin x)/x$ time waveform is limited to a duration T_s just as the $(\sin x)/x$ spectral bandwidth was limited to a bandwidth B_s. The frequency error can be found from Eq. (6.11) except the roles of bandwidth and frequency are reversed. In Eq. (6.11), replace the time-delay error δT_R with δf, replace the pulse width τ_r with the bandwidth B, and replace the bandwidth B_s with the signal duration T_s.

Trapezoidal Pulse   The theoretical rms error for the measurement of frequency with a trapezoidal pulse is

$$\delta f = \frac{(3\tau_p + 2t_r)^{1/2}}{2\pi\left(\dfrac{\tau_p^3}{4} + \dfrac{\tau_p^2 t_r}{2} + \dfrac{\tau_p t_r^2}{2} + \dfrac{t_r^3}{5}\right)^{1/2}\left(\dfrac{2E}{N_0}\right)^{1/2}} \qquad \text{trapezoidal pulse} \qquad [6.26]$$

where the rise and fall time is t_r and the width across the top of the trapezoid is τ_p. Assuming, for example, that $t_r = \tau_p/2$, the value of α would be $0.81\pi\tau_p$. When t_r is small compared to τ_p, the value of α approaches $\pi\tau_p/\sqrt{3}$, which is the value found in the above for the perfectly rectangular pulse.

Triangular Pulse This is obtained from the expression for the trapezoidal pulse by setting $\tau_p = 0$ and letting $2t_r = \tau_B$, where τ_B = width of the base of the triangular waveform. This results in

$$\delta f = \frac{(10)^{1/2}}{\pi\tau_B(2E/N_0)^{1/2}} \qquad \text{triangular pulse} \qquad \text{[6.27]}$$

Gaussian Pulse The rms error in frequency for a gaussian pulse is

$$\delta f = \frac{1.18}{\pi\tau(2E/N_0)^{1/2}} = \frac{B}{1.18(2E/N_0)^{1/2}} \qquad \text{gaussian pulse} \qquad \text{[6.28]}$$

where τ = half-power pulse width and B = half-power bandwidth.

Multiple Observations The error expressions for time delay and frequency presented here apply for a single observation. When more than one independent measurement is made, the resultant error is reduced and may be found by combining errors in the usual manner for gaussian statistics: the variance (the square of δT_R or δf) of the N independent observations is equal to $1/N$ of the variance of a single observation. Alternatively, the expression for a single pulse applies to multiple pulses if the energy E is the total energy of N pulses. The above assumes that the effective bandwidth β or the effective time-duration α remains the same for each of the N measurements.

Certainty of the Uncertainty Principle The product of the effective bandwidth and the effective time-duration α must be equal to or greater than π; that is,

$$\beta\alpha \geq \pi \qquad \text{[6.29]}$$

This relation may be derived from the definitions of β and α given by Eqs. (6.8) and (6.23) and by applying the Schwartz inequality Eq. (5.11). It is a consequence of the Fourier-transform relationship between a time waveform and its spectrum. The longer the time duration of a waveform, the narrower will be its spectrum. The wider the spectrum, the narrower will be the time waveform. Both the time waveform and its frequency spectrum cannot be made arbitrarily small simultaneously.

Equation (6.29) has sometimes been referred to as the *radar uncertainty principle* because of its supposed analogy to the important concept in quantum physics known as the Heisenberg uncertainty principle. The physics uncertainty principle states that both the position and velocity of an object (such as a subatomic particle) cannot be measured exactly at the same time.[17] Equation (6.29) actually has the opposite interpretation for radar signals, and it is inappropriate to refer to it as a radar uncertainty principle. It follows from this equation that there is no theoretical restriction on accuracy with which a radar can simultaneously locate the position of a target and determine its velocity. The product of the rms time-delay error, Eq. (6.7), and the rms frequency error, Eq. (6.22), is

$$\delta T_R \delta f = \frac{1}{\beta\alpha(2E/N_0)} \tag{6.30}$$

Substituting the inequality of Eq. (6.29) into the above gives

$$\delta T_R \delta f \leq \frac{1}{\pi(2E/N_0)} \tag{6.31}$$

This states that the time delay and the frequency may be simultaneously measured to as small a theoretical error as one desires by designing the radar to yield a sufficiently large ratio of signal energy (E) to noise power per unit bandwidth (N_0), or for fixed E/N_0, to select a waveform with a large $\beta\alpha$ product. Large $\beta\alpha$ requires waveforms with long time duration and wide spectral width. In terms of range accuracy δR and radial-velocity accuracy δv_r, the expression of Eq. (6.31) can be written

$$\delta R \delta v_r \leq \frac{c\lambda}{4\pi(2E/N_0)} \tag{6.32}$$

where λ = radar wavelength and c = velocity of propagation. This states that the shorter the wavelength, the better will be the accuracy that can be achieved in the simultaneous measurement of range and radial velocity.

There is nothing "uncertain" about the simultaneous radar measurement of range and radial velocity. The radar "uncertainly relation" and the physics uncertainty principle that describes quantum mechanical effects should not be confused. In quantum mechanics, the observer does not have control over the waveform with which the quantum particle is being observed. The radar engineer, on the other hand, can choose the value of the $\beta\alpha$ product, the signal energy E, and to some extent the noise density N_0. Any limits to classical radar measurement accuracies are practical ones.

The use of the Schwartz inequality in deriving Eq. (6.29) shows that the poorest waveform for obtaining accurate time-delay and frequency measurements simultaneously is the one for which $\beta\alpha = \pi$ (the smallest theoretical value allowed for $\beta\alpha$). This corresponds to the gaussian-shaped pulse. The quasi-rectangular pulse has $\beta\alpha = 1.22\pi$, and the trapezoidal pulse with rise time $t_r = \tau_p/2$ has $\beta\alpha = 1.4\pi$. Thus there is not much difference in the $\beta\alpha$ products of simple-shape waveforms. Large values of $\beta\alpha$ require internal modulation of the pulse to make the bandwidth much greater than the reciprocal of the pulse width. This is what is done in pulse compression waveforms, to be discussed later in this chapter (Sec. 6.5).

Angular Accuracy   The expression for the theoretical accuracy with which an antenna can measure the angle of arrival follows from the above discussion of accuracy for the time-delay measurement. This is possible because of the similarity of the mathematics in the spatial (angle) and spectral (frequency) domains. The one-dimensional electric-field strength pattern of an antenna in one plane is given in Chap. 9 as

$$g(\theta) = \int_{-D/2}^{+D/2} A(z) \exp\left(j2\pi \frac{z}{\lambda} \sin \theta\right) dz \qquad [6.33]$$

where the antenna is of length D and lies along the z-axis, $A(z)$ is the distribution of the current across the aperture (called the *aperture illumination*), λ = radar wavelength, and θ = angle measured from broadside ($\theta = 0$ is the perpendicular to the antenna). This is an inverse Fourier transform and resembles the inverse Fourier transform between the time waveform $s(t)$ and the spectrum $S(f)$.

$$s(t) = \int_{-\infty}^{\infty} S(f) \exp\left(j2\pi ft\right) df \qquad [6.34]$$

The antenna pattern $g(\theta)$ can be related to the time waveform $s(t)$, the aperture illumination $A(x)$ to $S(f)$, $\sin \theta$ to the time t, and the aperture coordinate z/λ can be related to the frequency f. What is learned about signals in the frequency domain often can be applied to signals in the spatial domain—and vice versa. Using the above analogies, the rms error for an angle measurement can be obtained from Eqs. (6.7) and (6.8) as

$$\delta\theta = \frac{1}{\gamma(2E/N_0)^{1/2}} \qquad [6.35]$$

where the effective aperture width γ is defined as

$$\gamma^2 = \frac{\displaystyle\int_{-\infty}^{\infty} (2\pi z/\lambda)^2 |A(z)|^2 \, dz}{\displaystyle\int_{-\infty}^{\infty} |A(z)|^2 \, dz} \qquad [6.36]$$

The effective aperture width is 2π times the square root of the normalized second moment of $|A(z)|^2$ about the mean value of z. The mean is at $z = 0$.

The theoretical angle-measurement error for an antenna with a uniform (rectangular) amplitude illumination across the aperture is

$$\delta\theta = \frac{\sqrt{3}\lambda}{\pi D(2E/N_0)^{1/2}} = \frac{0.628\theta_B}{(2E/N_0)^{1/2}} \qquad [6.37]$$

where the far-right expression uses the relation that the half-power beamwidth for this aperture illumination is $\theta_B = 0.88\lambda/D$. The units of $\delta\theta$ and θ_B are radians. With a cosine illumination $A(z) = \cos(\pi z/D)$ across the aperture of dimension D (where $|z| \le D/2$), the effective aperture width γ is $1.13D/\lambda$, or $1.37/\theta_B$. A triangular illumination has $\gamma = 0.99D/\lambda$; and for a parabolic illumination, $\gamma = 0.93D/\lambda$.

Commonality of Measurements   The measurements of range, angle, and radial velocity are all carried out differently in radar, but they share the concept of locating the maximum of a time waveform similar to that in Fig. 6.4. This figure might represent the radar echo time-waveform and the measurement of time delay (range); or it might represent a

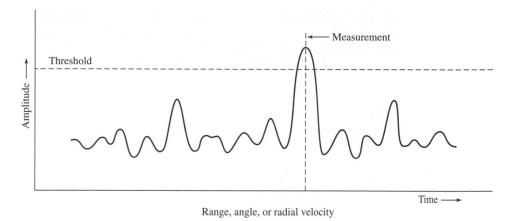

Figure 6.4 Representation of the radar measurement of range, angular location, or radial velocity by observing the waveform as a function of time, angle, or frequency.

scanning antenna pattern and the measurement of angle; or it might be the doppler frequency as observed at the output of a tunable filter that provides the radial velocity. The range, angle, or radial velocity is determined when the "signal" is a maximum.

6.4 AMBIGUITY DIAGRAM

It was mentioned in Sec. 5.2 that the output of the matched filter is the cross correlation between (1) the received signal plus noise and (2) the transmitted signal. When the signal-to-noise ratio is large (as it must be for detection), the output of the matched filter can usually be approximated by the *autocorrelation function* of the transmitted signal; that is, the noise is ignored. This assumes there is no doppler shift so that the received echo signal has the same frequency as the transmitted signal. In many radar applications, however, the target is moving so that its echo signal has a doppler frequency shift. The output of the matched filter, therefore, will not be the autocorrelation function of the transmitted signal. Instead, it must be considered as the cross correlation between the doppler-shifted received signal and the transmitted signal, with noise being ignored since the signal-to-noise ratio is assumed to be large.

The nature of the matched-filter output as a function of both time and doppler frequency is important for understanding the properties of a radar waveform, in particular its effect on measurement accuracy, target resolution, ambiguities in range and radial velocity, and the response to clutter. These aspects of the matched-filter output are examined next.

When the received echo signal is large compared to noise, the output of the matched filter [Eq. (5.17)] may be written as the following cross-correlation function:

$$\text{output of the matched filter} = \int_{-\infty}^{\infty} s_r(t)s^*(t - T_R') \, dt \qquad \text{[6.38]}$$

where $s_r(t)$ is the received echo signal, $s(t)$ is the transmitted signal, $s^*(t)$ is its complex conjugate, and T'_R is the estimate of the time delay (considered a variable). Complex notation is assumed, so that the transmitted signal can be written as $u(t) \exp [j2\pi f_0 t]$, where $u(t)$ is the complex modulation function whose magnitude $|u(t)|$ is the envelope of the real signal, and f_0 is the carrier frequency. The received echo signal $s_r(t)$ is assumed to be the same as that transmitted, except for a doppler frequency shift f_d and a delay equal to the true time delay T_0. Therefore,

$$s_r(t) = u(t - T_0) \exp [j2\pi(f_0 + f_d)(t - T_0)] \qquad \text{[6.39]}$$

(The change of amplitude due to the factors in the radar equation is ignored here.) With the above definitions, the output of the matched filter is

$$\text{output} = \int_{-\infty}^{\infty} u(t - T_0)u^*(t - T'_R)e^{j2\pi(f_0 + f_d)(t - T_0)}e^{-j2\pi f_0(t - T'_R)} \, dt \qquad \text{[6.40]}$$

For simplicity in understanding this equation, we take the origin to be the true time delay and the transmitted frequency; hence, $T_0 = 0$ and $f_0 = 0$. Then $T_0 - T'_R = -T'_R = T_R$. The output of the matched filter is then

$$\chi(T_R, f_d) = \int_{-\infty}^{\infty} u(t)u^*(t + T_R)e^{j2\pi f_d t} \, dt \qquad \text{[6.41]}$$

A positive time delay T_R indicates a target beyond the true target time delay T_0, and a positive doppler frequency f_0 indicates an approaching target.[18,19] The squared magnitude of Eq. (6.41), $|\chi(T_R, f_d)|^2$, is called the *ambiguity function*. Its three-dimensional plot as a function of time delay T_R and doppler frequency f_d is the *ambiguity diagram*.[20]

Properties of the Ambiguity Diagram   The ambiguity function $|\chi(T_R, f_d)|^2$ has the following properties:

$$\text{maximum value: } |\chi(T_R, f_d)|^2_{\max} = |\chi(0,0)|^2 = (2E)^2 \qquad \text{[6.42]}$$

$$\text{symmetry relation: } |\chi(-T_R, -f_d)|^2 = |\chi(T_R, f_d)|^2 \qquad \text{[6.43]}$$

$$\text{behavior on } T_R \text{ axis: } |\chi(T_R, 0)|^2 = \left| \int u(t)u^*(t + T_R) dt \right|^2 \qquad \text{[6.44]}$$

$$\text{behavior on } f_d \text{ axis: } |\chi(0, f_d)|^2 = \left| \int u^2(t)e^{j2\pi f_d t} \, dt \right|^2 \qquad \text{[6.45]}$$

$$\text{volume under surface: } \int\int |\chi(T_R, f_d)|^2 \, dT_R \, df_d = (2E)^2 \qquad \text{[6.46]}$$

Equation (6.42) states that the maximum value of the ambiguity function occurs at the origin, which is the true location of the target when the doppler shift $f_d = 0$. Its maximum value is $(2E)^2$, where E is the energy contained in the echo signal. Equation (6.43) is a symmetry relation. Equation (6.44) is the form of the ambiguity function on the time-delay axis. It is the square of the autocorrelation function of $u(t)$. Equation (6.45) describes the behavior on the frequency axis and is the square of the inverse fourier transform of $[u(t)]^2$. The total volume under the ambiguity diagram is given by Eq. (6.46) and is a constant, also equal to $(2E)^2$. (All limits in the above equations go from $-\infty$ to $+\infty$.)

"Ideal" Ambiguity Diagram   If there were no theoretical restrictions, the "ideal" ambiguity diagram would consist of a single peak of infinitesimal thickness at the origin and be zero everywhere else, as shown in Fig. 6.5. It would be an impulse function and have no ambiguities in range or doppler frequency (radial velocity). The infinitesimal (or very small) thickness at the origin would permit the time delay and/or frequency to be determined simultaneously to as high a degree of accuracy as desired. It would also permit the resolution of two very closely spaced targets and reject all clutter other than clutter at the origin. There would be no ambiguous responses. Such a highly desirable ambiguity diagram, however, is not theoretically allowed. It cannot be obtained because Eq. (6.42) requires that the maximum of the ambiguity diagram be equal to $(2E)^2$ and the volume under its surface as given by Eq. (6.46) also must be equal to $(2E)^2$.

The restrictions on the ambiguity diagram may be considered by imagining a box of sand.[21] The total amount of sand in the box is fixed, just as the volume under the ambiguity diagram is fixed at $(2E)^2$ by the signal energy E. The sand may be piled up in the center of the box (the origin of the ambiguity diagram), but its height can be no greater than $(2E)^2$. If one tries to pile sand at the center of the box in a very narrow pile (to obtain good accuracy and resolution), the sand that remains must be redistributed elsewhere in the box. Sand might then pile up in other parts of the box, which means ambiguities in range and/or doppler can result. Thus the nature of the ambiguity diagram indicates there have to be trade-offs made among the resolution, accuracy, and ambiguity.

An approximation to what might seem to be a good ambiguity diagram is shown in Fig. 6.6. The waveform does not result in ambiguities since there is but a single peak. In general, when a single peak is obtained, such as shown here, it might be so wide along the time-delay axis and the doppler-frequency axis that it might have poor accuracy and resolution. It appears that, in practice, waveforms generally have significant response somewhere in the ambiguity diagram outside the narrow region in the near vicinity of the origin. Practical waveforms do not approximate the ideal ambiguity diagram or even its more realistic version of Fig. 6.6.

Figure 6.5   Ideal, but unattainable, ambiguity diagram.

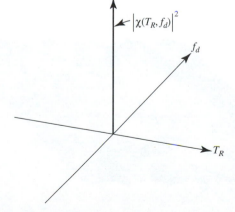

Figure 6.6 An approximation to the ideal ambiguity function with the restriction that the value at the origin is constant at $(2E)^2$ and the volume under the surface of $|\chi(T_R,f_d)|^2$ also is given by $(2E)^2$.

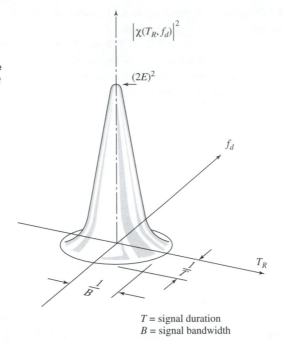

$$|\chi(T_R,f_d)|^2$$

$(2E)^2$

f_d

T_R

$\frac{1}{T}$

$\frac{1}{B}$

T = signal duration
B = signal bandwidth

Single Pulse of Sinewave and the Ridge Ambiguity Diagram A computer-generated 3-D plot of $\chi|T_R, f_d|$ for a single rectangular pulse of width τ is shown in Fig. 6.7, due to A. W. Rihaczek and R. L. Mitchell. (Note that it is the magnitude of the square root of ambiguity function that is plotted here.) The triangular shape of the output of the matched filter on the time axis ($f_d = 0$) can be seen as well as the $(\sin x)/x$ shape on the frequency axis.

The essence of the information found from an ambiguity diagram can usually be obtained from simpler two-dimensional plots, as in Fig. 6.8 for (a) a long pulse of width τ and (b) for a short pulse. Shading is used to indicate the regions where $|\chi(T_R, f_d)|^2$ is

Figure 6.7 Plot of $|\chi(T_R, f_d)|$ for a simple rectangular pulse of width τ.
(Due to A. W. Rihaczek and R. L. Mitchell[22])

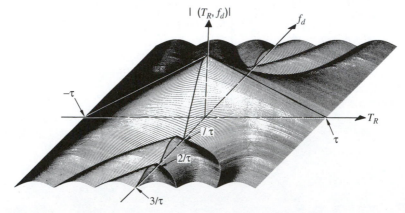

$|(T_R,f_d)|$

f_d

$-\tau$

$1/\tau$

T_R

τ

$2/\tau$

$3/\tau$

Figure 6.8 Two-dimensional representation of the ambiguity diagram for a single pulse of sinewave of width τ. (a) Long pulse; (b) short pulse.

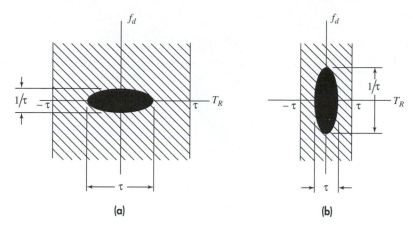

large (completely darkened area), where it is small (lightly shaded area), and regions of zero response (no shading). The plot for a single pulse shows most of the response as a completely shaded elliptically shaped region in which $|\chi|^2$ is large. No response occurs outside the time $\pm\tau$. The time-delay measurement error is proportional to τ and the frequency measurement error is proportional to $1/\tau$. The plot of Fig. 6.8a shows the good frequency measurement accuracy and the poor time-delay accuracy of a long pulse. Figure 6.8b shows the opposite occurs for a short pulse. The ambiguity diagram indicates that as the range accuracy of a simple pulse waveform is improved, the frequency accuracy worsens and vice versa. (This follows from the box of sand analogy mentioned previously.) The short pulse is *doppler tolerant* in that a single matched filter will produce a good output if there is a significant doppler shift. That is, the output from a filter matched to a zero doppler shift will not change much when there is a doppler shift. On the other hand, a long pulse will produce greatly reduced output for a doppler-frequency shift; so it is not tolerant to changes in doppler. This is the reason why in an MTI radar a bank of doppler filters sometimes is used to cover the expected range of doppler frequencies. The ambiguity diagram of the single pulse is known as a *ridge* or *knife edge*.

Single Linear Frequency-Modulated (FM) Pulse Another example of a ridge or knife edge ambiguity diagram is that produced by linearly frequency modulating a rectangular pulse over a bandwidth B, as shown by the 2-D plot of Fig. 6.9. The pulse width T is large compared to $1/B$. The frequency modulation increases the spectral bandwidth of the pulse so that $BT \gg 1$. The ridge is at an angle determined by the slope B/T. The time-delay measurement accuracy is proportional to $1/B$ and the frequency (if it can be measured) has an accuracy proportional to $1/T$. Since the pulse width T and the bandwidth B can be chosen independent of one another, the time-delay and frequency accuracies are independent of the other. This is unlike the time-delay and frequency accuracies of the simple, unmodulated rectangular pulse.

Periodic Pulse Train Consider, as shown in Fig. 6.10a, a series of five pulses, each of width τ, pulse repetition period T_p, and total duration T_d. The 2-D representation of its

Figure 6.9 Two-dimensional representation of the ambiguity diagram for a single linear frequency-modulated pulse of width T and bandwidth B.

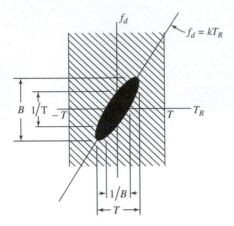

Figure 6.10 (a) Video pulse train of five pulses; (b) two-dimensional representation of the ambiguity diagram for (a).

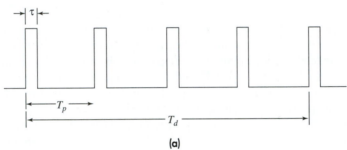

(a)

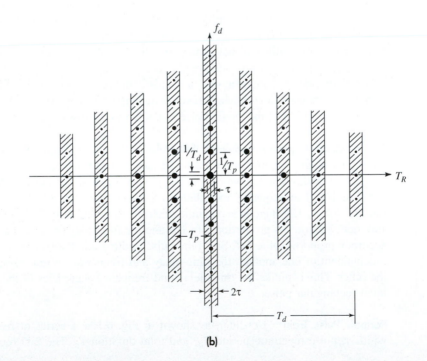

(b)

ambiguity diagram is shown in Fig. 6.10b. This type of ambiguity function is called a *bed of spikes*. Throughout this ambiguity diagram there are many possible ambiguities in range and doppler (blind speeds) such as were encountered before in this text. Ambiguities are a consequence of discontinuous waveforms, which is why they do not appear with a single pulse. It might seem at first glance that the many ambiguities produced by a pulse train result in it being a poor radar waveform. As seen in Chap. 3, a pulse train is widely used in pulse doppler and MTI radars, and it is often possible to adequately unravel, avoid, or (sometimes) ignore the ambiguities.

Examine the center (the origin) of the ambiguity diagram shown in Fig. 6.10b. The spike at the origin is of dimension τ on the time axis and $1/T_d$ on the frequency axis. The time-delay measurement accuracy (determined by the pulse width) and the frequency accuracy (determined by the duration of the signal) can be selected independently. If the pulse repetition period T_p is such that no radar echoes are expected with a time delay greater than T_p and no doppler-frequency shifts are expected greater than $1/T_p$, then the effective ambiguity diagram reduces to just a single spike at the origin whose dimensions are determined by τ and T_d. The pulse train, therefore, can be a good radar waveform. Ambiguities that occur can be resolved with different pulse repetition frequencies (Secs. 2.10 and 3.9). The fact that many radars employ this type of waveform attests to its usefulness far better than any analysis based on its ambiguity diagram.

Although this discussion applies to a pulse train, the signal processing employed with such waveforms is almost always different from that assumed for Fig. 6.10. The ambiguity function of a pulse train is the output of a matched filter designed for the *entire* pulse train. This is why the ambiguity diagram of Fig. 6.10 for five pulses has an output (in time) consisting of nine pulses of different amplitudes when the input is five equal amplitude pulses (the autocorrelation function of five pulses). Pulse radars, however, are not usually designed with a matched filter for N pulses. Instead, a pulse radar usually uses a filter matched to a single pulse and then integrates the N pulses.

Noiselike Waveforms and the Thumbtack Ambiguity Diagram   A noise waveform has an ambiguity function, called a thumbtack, similar to that of Fig. 6.11. Pure noise waveforms are seldom employed in radar; but constant-amplitude noiselike waveforms have been used to produce a thumbtack ambiguity diagram. Examples include nonmonotonic frequency modulation and pseudorandom variations of phase or frequency. These are discussed later in this chapter under the topic of pulse compression.

The thumbtack has the advantage that the time-delay and frequency measurement accuracies are independently determined, respectively, by the bandwidth of the modulation and the duration of the pulse. There are no apparent ambiguities and it resembles the approximation of the ideal ambiguity function of Fig. 6.6, except for the plateau on which the main response rests. This plateau extends over a dimension $2T$ along the time axis and $2B$ along the frequency axis. When the product BT is large, the volume under the peak response is small compared to the total volume, and almost all of the volume is in the plateau. Consequently, the average height of the plateau is approximately E^2/BT for large BT (based on Eq. 6.46).

The thumbtack ambiguity diagram, as illustrated in the sketch of Fig. 6.11, appears attractive; but it is an overly simplified sketch that can be misleading when practical

Figure 6.11 Ideal representation of the thumbtack ambiguity diagram as might be produced by a noiselike waveform or a pseudorandom coded pulse waveform (details of the sidelobe structure are not shown.)

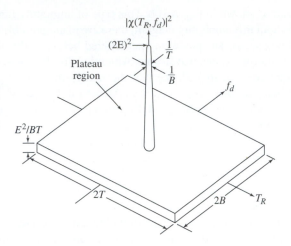

waveforms are examined. The plateau is not uniform as shown. Most of the volume under the ambiguity function is confined to the intervals $|T_R| < T/2$ and $|f_d| < B/2$.[23] Furthermore, when there is a doppler-frequency shift, the sidelobes can be relatively high and can cause false responses. There are some waveforms, as will be mentioned later, that can produce low sidelobes when $f_d = 0$; but such waveforms should be looked at with caution when good doppler filtering is required without false responses, as for the detection of moving targets.

Waveform Design and the Ambiguity Diagram   The waveform transmitted by a radar can affect (1) target detection, (2) measurement accuracy, (3) resolution, (4) ambiguities, and (5) clutter rejection. The ambiguity diagram may be used to assess qualitatively how well a particular waveform achieves these capabilities. Each of the above five capabilities will be briefly discussed.

Detection   It was said in Sec. 5.2 that if the receiver is designed as a matched filter, the output peak-signal-to-mean-noise ratio (which is related to the ability to detect a target) depends only on the ratio of the received signal energy E and the receiver noise power per unit bandwidth N_0. The requirements for detection do not place any demands on the *shape* of the transmitted waveform except that (1) the waveform be practical to generate and radiate, and (2) the matched filter required for the waveform be practical to achieve. The ambiguity diagram, therefore, is seldom used to assess the detection capability of a particular waveform, except to note if the signal contains sufficient energy.

Accuracy   The accuracy with which the range and radial velocity can be measured is indicated by the main response at the origin. The width along the time axis determines the range (time delay) accuracy, and the width along the frequency axis determines the radial-velocity accuracy.

Resolution   The width of the central response also determines the resolution ability of a radar waveform in range and radial velocity. In order to resolve closely spaced targets, the central response must be isolated. There cannot be extraneous peaks near the main response that would mask the echo from a nearby target.

Ambiguity   Ambiguities occur in radar measurements when the waveform is not continuous. For example, a pulse train is not continuous; hence, such a waveform can produce ambiguities in range (time delay) and in velocity (frequency). An ambiguous measurement is one in which there is more than one choice available, but only one is correct. Ambiguities appear in the ambiguity diagram as additional high responses similar in magnitude to the peak response at the origin. The correct response from the target cannot be readily recognized from these additional responses without taking some action to resolve the correct from the incorrect values.

The name *ambiguity function* for $|\chi(T_R, f_d)|^2$ can be misleading since this function describes more about the character of a waveform than just the ambiguities it produces. Woodward[24] coined the name for an entirely different reason than the ambiguities associated with discontinuous waveforms. The reader is advised not to be distracted by trying to understand the ambiguous use of the term "ambiguity" as the name for the function $|\chi(T_R, f_d)|^2$.

Clutter Attenuation   Resolution in range and velocity can enhance the target signal echo relative to nearby distributed clutter echoes. The ambiguity diagram can indicate the ability of a waveform to reject clutter by superimposing the locations of the clutter echoes (as stored in a *clutter map*) on the T_R, f_d plane of the ambiguity diagram. If the radar is to have good clutter rejection, the ambiguity diagram should have little or no response in regions of high clutter echoes. The short-pulse waveform, for example, will reduce stationary clutter that is extensive in range since the ambiguity function for this waveform has little response on the time (range) axis.

Waveform Synthesis   It is reasonable to ask if radar waveforms can be obtained by first defining a desired ambiguity diagram and then synthesizing the signal that yields this ambiguity diagram. Synthesis has not proven successful in the past. Instead, it is more usual to compute the ambiguity diagrams of various waveforms and determine which of them have suitable properties for the intended application.

6.5   PULSE COMPRESSION

High range-resolution, as might be obtained with a short pulse, is important for many radar applications, as indicated by the list of capabilities in Table 6.2. There can be limitations, however, to the use of a short pulse. Since the spectral bandwidth of a pulse is inversely proportional to its width, the bandwidth of a short pulse is large. Large bandwidth can increase system complexity, make greater demands on the signal processing, and increase the likelihood of interference to and from other users of the electromagnetic

Table 6.2 Capabilities of short-pulse, high range-resolution radar

---

Range resolution. Usually easier to separate (resolve) multiple targets in range than in angle.

Range accuracy. A radar capable of good range resolution is also capable of good range accuracy.

Clutter reduction. Increased target-to-clutter ratio is obtained by reducing the amount of distributed clutter with which the target echo signal must compete.

Interclutter visibility. With some types of "patchy" land and sea clutter, a high-resolution radar can detect moving targets in the clear areas between the clutter patches.

Glint reduction. Angle and range tracking errors introduced by a complex target with multiple scatterers are reduced when high range-resolution is employed to isolate (resolve) the individual scatterers that make up the target.

Multipath resolution. Range resolution permits the separation of the desired target echo from the echoes that arrive at the radar via scattering from longer propagation paths, or multipath.

Multipath height-finding. When multipath due to scattering of radar energy from the earth's surface can be separated from the direct-path signal by high range-resolution, target height can be determined without a direct measurement of elevation angle.

Target classification. The range, or radial, profile of a target in some cases can provide a measure of target size in the radial dimension. From the range profile one might be able to sort one type of target from another based on size or distinctive profile, especially if the cross-range profile is also available.

Doppler tolerance. With a short-pulse waveform, the doppler-frequency shift from a moving target will be small compared to the receiver bandwidth; hence, only a single matched filter is needed for detection, rather than a bank of matched filters each tuned for a different doppler shift.

ECCM. A short-pulse radar can negate the effects of certain electronic countermeasures such as range-gate stealers, repeater jammers, and decoys. The wide bandwidth of the short-pulse radar can, in principle, provide some reduction in the effects of broadband noise jamming and reduce the effectiveness of some electronic warfare receivers and their associated signal processing.

Minimum range. A short pulse allows the radar to operate with a short minimum range. It also allows reduction of blind zones (eclipsing) in high-prf radars.

---

spectrum. Another limitation is that in some high-resolution radars the limited number of resolution cells available with conventional displays might result in overlap of nearby echoes when displayed, which results in a collapsing loss (Sec. 2.12) if the detection decision is made by an operator. Wide bandwidth can also mean less dynamic range in the receiver because receiver noise power is proportional to bandwidth. Also, a short-pulse waveform provides less accurate radial velocity measurement than if obtained from the doppler-frequency shift. It spite of such limitations, the short-pulse waveform is used because of the important capabilities it provides.

A serious limitation to achieving long ranges with short-duration pulses is that a high peak power is required for a large pulse energy. The transmission line of a high peak power radar can be subject to voltage breakdown (arc discharge), especially at the higher frequencies where waveguide dimensions are small. If the peak power is limited by breakdown, the pulse might not have sufficient energy. Consider, for example, a more or less conventional radar with a pulse width of one microsecond and one megawatt peak power, as might be found in a medium-range air-surveillance radar. In this example, the energy contained in a single pulse is one joule. (The energy per pulse and the number of pulses integrated determine the detectability of a target.) A one microsecond pulse has a range resolution of 150 m. If it were desired to have a resolution of 15 cm (one-half foot), the pulse width would have to be reduced to one nanosecond and the peak power increased to *one gigawatt* (10^9 W) in order to maintain the same pulse energy of one joule. This is an unusually large peak power that cannot be propagated without breakdown in the usual types of transmission lines employed at microwave radar frequencies.

A short pulse has a wide spectral bandwidth. A long pulse can have the same spectral bandwidth as a short pulse *if* the long pulse is modulated in frequency or phase. (Amplitude modulation can also increase the bandwidth of a long pulse, but is seldom used in radar because it can result in lower transmitter efficiency.) The modulated long pulse with its increased bandwidth B is compressed by the matched filter of the receiver to a width equal to $1/B$. This process is called *pulse compression*. It can be described as the use of a long pulse of width T to obtain the resolution of a short pulse by modulating the long pulse to achieve a bandwidth $B \gg 1/T$, and processing the modulated long pulse in a matched filter to obtain a pulse width $\tau \approx 1/B$. Pulse compression allows a radar to simultaneously achieve the energy of a long pulse and the resolution of a short pulse without the high peak power required of a high-energy short-duration pulse. It is used in high-power radar applications that are limited by voltage breakdown if a short pulse were to be used. Airborne radars might experience breakdown with lower voltages than ground-based radars, and might be candidates for pulse compression. It is almost always used in high-power radars with solid-state transmitters since solid-state devices, unlike vacuum tubes, have to operate with high duty cycles, low peak power, and pulse widths much longer than normal. Pulse compression is also found in SAR and ISAR imaging systems to obtain range resolution comparable to the cross-range resolution.

The pulse compression ratio is defined as the ratio of the long pulse width T to the compressed pulse width τ, or T/τ. The bandwidth B and the compressed pulse width τ are related as $B \approx 1/\tau$. This would make the pulse compression ratio approximately BT. If amplitude weighting of the received waveform (but not the transmitted waveform) is used to reduce the time sidelobes of the linear-FM waveform (as will be discussed later in this section), the pulse compression ratio defined as BT usually is a little larger than T/τ. It is better, therefore, to define the pulse compression ratio by T/τ (the ratio of the before and after pulse widths) rather than the bandwidth-time product BT when weighting on receive is used. (In spite of this caution, the pulse compression ratio is often given as BT in this text as well as in other radar literature. The reader should be aware of which value is used, especially with linear-FM waveforms.) The pulse compression ratio in practical radar systems might be as small as 10 (although 13 is a more typical lower value) or greater than 10^5. Values from 100 to 300 might be considered typical.

There are two ways to describe the operation of a pulse compression radar. One is based on the ambiguity function of Sec. 6.4. A long pulse is modulated to increase its bandwidth. On reception the modulated long-pulse echo signal is passed through the matched filter. Its resolution in range can be found from examination of the ambiguity diagram. The constant-amplitude linear-FM pulse, whose ambiguity diagram was shown in Fig. 6.9, is an example of a widely used pulse compression waveform. Its ambiguity diagram shows that the long pulse of width T provides a compressed pulse width equal to $1/B$.

The other method for describing pulse compression is based on how linear-FM pulse compression was presented in the original patent of R. H. Dicke,[25] before the concept of the ambiguity function was known. The modulation applied to a long pulse can be considered as providing distinctive "marks" in either the frequency or the phase along the various portions of the pulse. For instance, the changing frequency of a linearly frequency-modulated pulse is distributed along the pulse so that each small segment of the pulse corresponds to a different frequency. By passing the modulated pulse through a dispersive delay line whose delay time is a function of frequency, each part of the pulse experiences a different delay, so that it is possible for the trailing edge of the pulse to be speeded up in a dispersive* delay line and the leading edge slowed down so that they "come together" to effect a compression of the long pulse.

There have been two general classes of pulse-compression waveforms used in the past. The most popular has been the *linear FM* (also known as chirp) to be discussed next. *Binary phase-coded* pulses is the other type. Also to be mentioned here are *polyphase codes* whose phase quantizations are less than π radians; *Costas codes* in which the frequencies of the subpulses are changed in a prescribed fashion; *nonlinear FM, nonlinear binary phase-coding*; *complimentary codes* that in principle produce zero time-sidelobes; and codes with very low or no sidelobes that require amplitude modulation of the subpulses on transmit. No one type of pulse-compression waveform can do everything that might be required; but linear FM probably has been the most widely used.

Linear Frequency Modulation (LFM) Pulse Compression   The basic concept of the linear frequency-modulated pulsed compression radar was described by R. H. Dicke in a patent filed in 1945 and issued in 1953.[25] Figure 6.12, which is derived from Dicke's patent, is a block diagram of such a radar. It is similar to the block diagram of a conventional radar except that the transmitter is shown here as being frequency modulated and the receiver

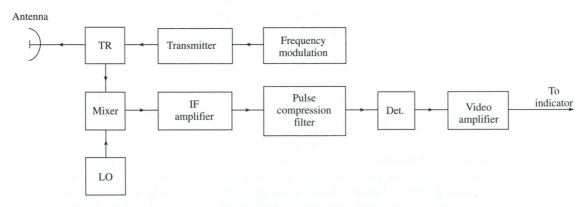

Figure 6.12   Block diagram of an FM pulse compression radar.

| *"Dispersive" in this case means that the velocity of propagation is a function of frequency.

contains a pulse-compression filter (which is identical to a matched filter). More usually, however, the FM waveform is generated at low power and amplified by a power amplifier rather than by frequency modulating a power oscillator as indicated in this figure. The transmitted waveform (Fig. 6.13a) consists of a rectangular pulse of constant amplitude A and duration T. The frequency increases linearly from f_1 to f_2 over the duration of the pulse (Fig. 6.13b). This is sometimes known as an *up-chirp*. Alternatively, the frequency could just as well decrease with time, and it is then called *down-chirp*. The time waveform is represented in Fig. 6.13c. On reception, the frequency-modulated signal is passed through the pulse-compression filter, which is a delay line whose velocity of propagation is proportional to frequency. It speeds up the higher frequencies at the trailing edge of the pulse relative to the lower frequencies at the leading edge so as to compress the signal to a width $1/B$, where $B = f_2 - f_1$ (Fig. 6.13d). The pulse compression filter is a matched filter; hence, its output envelope (neglecting noise) is the autocorrelation function of the input. In this case, the output is proportional to (sin πBt)/πBt.[26,27] The peak power of the pulse is increased by the pulse compression ratio $BT \approx T/\tau$ after passage through the filter.

The ambiguity diagram for a linear-FM pulse-compression waveform (Fig. 6.9) shows that a large doppler shift in the echo signal can result in the indicated range not being the true range. This is known as *range-doppler coupling*. In many cases, the range error due to the doppler shift is small and can be tolerated. If the range error is large, the effect of the doppler can be removed by averaging the two range indications obtained with both a rising FM (up-chirp) and a falling FM (down-chirp) similar to the manner in which the range error due to doppler is eliminated in an FM-CW radar with triangular modulation.[28]

Reduction of Time (Range) Sidelobes   The (sin πBt)/πBt envelope out of a matched filter when the input is a linear-FM sinewave has relatively high peak time-sidelobes of -13.2 dB adjacent to the main response. This is usually not acceptable since high sidelobes can be mistaken for targets or can mask nearby weaker targets. The time sidelobes can be reduced by transmitting a pulse with nonuniform amplitude; that is, by amplitude weighting the long pulse over its duration T. (This is similar to *tapering* the aperture illumination of an antenna, Sec. 9.3, or *windowing* to reduce the spectral sidelobes of a digital filter.) Unfortunately, it is often not practical in high-power radar to have a transmitted waveform whose amplitude varies over the pulse duration. High-power transmitters such as klystrons, traveling wave tubes, and crossed-field devices should be operated saturated to obtain maximum efficiency. They don't like to be operated with amplitude modulation and they should be either full-on or full-off. Solid-state transmitters can have a linear input-output relation and be amplitude modulated if operated Class-A; but they are almost always operated Class-C because of the much higher efficiency of Class-C.[29] It is seldom that high-power microwave radar transmitters are operated with a deliberate change in amplitude over the pulse duration.

As a compromise, the time sidelobes with a linear-FM pulse-compression signal usually are reduced by applying amplitude weighting on receive (to the dispersive delay-line pulse compression filter) and not on transmit. Since the pulse compression filter is the matched filter, applying the weighting only on receive results in a mismatched filter and a loss in signal-to-noise ratio.[30] This is the price paid to reduce the time sidelobes. Table 6.3 gives examples of weightings, the peak sidelobe that results, and other properties of

Figure 6.13 Linear FM pulse compression. (a) Transmitted waveform; (b) frequency of the transmitted waveform as a function of time; (c) representation of the linear FM waveform; (d) theoretical output from the pulse-compression filter.

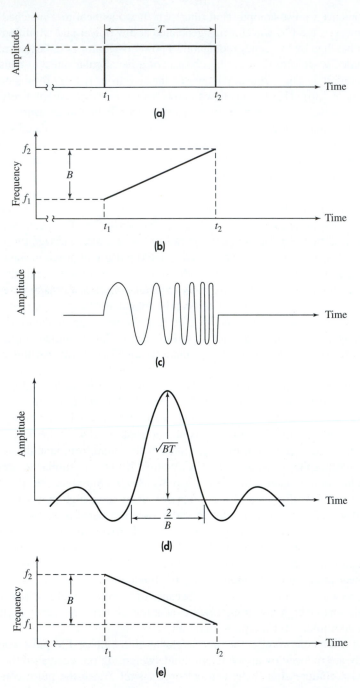

Table 6.3   Properties of weighting functions to reduce time sidelobes

Weighting Function	Peak sidelobe dB	Loss dB	Mainbeam width (relative)
Uniform	−13.2	0	1.0
$0.33 + 0.66 \cos^2 (\pi f/B)$	−25.7	0.55	1.23
$\cos^2 (\pi f/B)$	−31.7	1.76	1.65
$0.16 + 0.84 \cos^2 (\pi f/B)$	−34.0	1.0	1.4
Taylor $(\bar{n} = 6)$	−40.0	1.2	1.4
$0.08 + 0.92 \cos^2 (\pi f/B)$ (Hamming)	−42.8	1.34	1.5

the output waveform. The mismatched-filter loss can generally be kept to about 1 dB when the peak sidelobe level is reduced to 30 dB below the peak response. It is a loss that is tolerated in order to achieve lower time-sidelobe levels.

It is possible to have no theoretical loss in signal-to-noise ratio and still achieve low time-sidelobes with a uniform-amplitude transmitted waveform if a nonlinear frequency-modulated waveform is used, as discussed later in this section.

Stretch[31,32]   This is a variant of linear-FM pulse compression in which narrow bandwidth processing can be employed if the extent of the range interval where high-range resolution occurs can be reduced. Stretch starts with a linear-FM waveform that is of much smaller bandwidth B_1 than required for the desired resolution. Before being transmitted, it is heterodyned (mixed) with a wideband linear FM of bandwidth B_2 to produce a signal with the desired bandwidth $B_1 + B_2$, which is radiated. On receive, the signal of bandwidth $B_1 + B_2$ is heterodyned again with the linear-FM sweep of wide bandwidth B_2. The difference signal has the narrow bandwidth B_1 and is processed as a normal (narrowband) pulse-compression signal. The heterodyne operation on receive results in a time expansion by a factor $\alpha = (B_1 + B_2)/B_1$ (time expansion means smaller bandwidth). The range resolution of a signal, however, corresponds to a signal with bandwidth $B_1 + B_2 = B$. The processing of the linear-FM signal and its generation are both done with narrow bandwidth circuitry of bandwidth $B_1 = B/\alpha$. The reduced bandwidth in waveform generation and processing is the advantage of Stretch.

Because of the "stretching" of the waveform during processing, the range interval over which high resolution is obtained is $1/\alpha$ that which would normally be obtained by a conventional linear-FM radar of bandwidth B. One trades processing bandwidth for the size of the high-resolution range interval. This capability might not be appropriate for a surveillance radar that has to look everywhere in range, but it is well suited to a tracking radar or a high range-resolution radar used for target classification.

The Stretch waveform was used in the Cobra Dane (AN/FPS-108), a large *L*-band phased array radar located in the Aleutian Islands of Alaska.[33] Its purpose was to gather information about Soviet/Russian intercontinental ballistic missiles (ICBM). It also has been used for space surveillance and ICBM warning. It performs high range-resolution observation of ICBM reentry vehicles at ranges of about 1000 nmi. For this purpose it uses a 1000 μs duration linear-FM wideband waveform with 200 MHz bandwidth extending from 1175 to 1375 MHz. Its time-bandwidth product is 200,000, a relatively large value. Stretch processing examines a preselected range interval 250 ft in extent. The inherent range resolution is 2.5 ft, but the actual resolution is degraded to 3.75 ft since Taylor amplitude-weighting is employed on receive to reduce its time sidelobes to -30 dB.

Linear FM Pulse-Compression Filters   In the first edition of the *Radar Handbook,* which appeared in 1970, the chapter on pulse compression[34] listed nine different devices that might be used as the pulse compression filter for linear FM. Since that time, the practice has narrowed to mainly two choices: the surface acoustic wave (SAW) dispersive delay line and digital processing. The 1970 edition briefly discussed the SAW device (then called a *surface-wave dispersive delay line*) but it didn't mention digital processing for pulse compression. Digital processing is usually preferred when the A/D converter can provide the very wide bandwidths required of high-resolution pulse-compression radar. The SAW device has been the method of choice, however, when large-bandwidth signal processing cannot be obtained digitally. Digital methods are therefore used when they are applicable and the analog SAW delay line is used for very high resolution when digital methods cannot compete.

There are two general technical areas of application for pulse compression. One is when the compressed pulse is very small, perhaps of the order of a nanosecond, but the original uncompressed pulse is of conventional width, perhaps a microsecond. High resolution might be of interest for synthetic-aperture imaging radars (SAR), radars designed to detect a person swimming in the water, or for radars required to recognize one class of ship from another based on inverse synthetic aperture radar. SAW devices have been appropriate as the pulse compression filter for such applications. The other area is when one starts with a long pulse, say a few hundred microseconds, and compresses it to a conventional width, for example, of the order of a microsecond. Digital processing can be used in such situations. An important example is when a solid-state transmitter is used that requires long pulses for efficient operation.

SAW Devices[35–37]   The concept of a surface acoustic wave delay line is shown schematically in Fig. 6.14. It consists of a very smooth piezoelectric substrate, such as a thin slice of quartz, lithium niobate, or lithium tantalate. The function of the piezoelectric substrate is to support propagation of acoustical waves along the surface. The low velocity of acoustic waves (approximately 3500 m/s) compared to electromagnetic waves means that significant delay times can be achieved with a relatively small device. The input/output devices arranged on the surface are known as *interdigital transducers* (IDT). They are metallic thin films, usually aluminum, that are deposited on the substrate using photolithographic methods. The transducers are the means by which electrical signals are converted to acoustical signals, and vice versa. They determine the impulse response of the SAW delay line and the length of the IDT determines the duration of the signal. Since the IDT can launch waves in both the forward and back directions, the acoustic energy propagating in the opposite direction has to be attenuated so that it is not reflected and cause interference. One method for attenuating the unwanted signal is to use acoustic absorbers located at each end of the device, as indicated in Fig. 6.14.

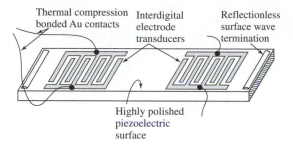

Thermal compression bonded Au contacts

Interdigital electrode transducers

Reflectionless surface wave termination

Highly polished piezoelectric surface

Figure 6.14   Schematic of a simple surface acoustic wave (SAW) delay line.
(From T. W. Bristol: Surface Acoustic Wave Devices—Technology and Applications, IEEE 1976 WESCON, paper 245/1. Courtesy IEEE.)

Efficient electric-to-acoustic coupling occurs when the comb fingers, or electrodes, of the IDT are spaced one-half wavelength of the acoustic signal that propagates along the substrate material. The frequency response of the delay line depends, therefore, on the spacings between the electrodes. A dispersive delay line whose delay is a function of frequency can be obtained with varying electrode spacing as illustrated in Fig. 6.15a or b. The SAW delay-line configuration in 6.15a is called an *in-line single dispersive chirp filter.* The received linear-FM signal sketched below the delay line is applied to the broadband IDT on the left. The output from the dispersive IDT structure on the right is the compressed pulse. (Either IDT can be used as the input or output since the SAW delay line is reciprocal.)

The *in-line double dispersive chirp filter* of Fig. 6.15b is capable of larger *BT* products (50 to 1000) than the single configuration of 6.15a.[35] The *slanted-array compressor* (SAC) sketched in Fig. 6.15c also is capable of large pulse compression ratios. An advantage the inclined IDTs have over the in-line configurations is that they avoid the distortion that occurs when the low-frequency components have to propagate under the high-frequency electrodes, and vice versa. Another advantage of the SAC configuration is that corrections in the phase characteristic of the device can be made by inserting a "phase plate" between the input and output IDTs. An example of a SAC linear-FM filter described by Cambell[37] (but attributed to S. Jen and C. F. Shaffer) operated at a center frequency of 1.4 GHz with a bandwidth of 1.1 GHz. This is a relative bandwidth of almost 80 percent, and can be called ultrawideband. The uncompressed pulse width was 0.44 μs, the compressed pulse was about 0.9 ns, and the bandwidth-time product was 484. Amplitude weighting was used to reduce the time sidelobes to 26.8 dB below the peak.

The *reflective-array compressor* (RAC) shown schematically in Fig. 6.15d is another form of SAW device. It is capable of larger pulse compression ratios (500 to 10,000) than other configurations. Shallow grooves are etched in the delay path that result in reflections to form a delay that depends on frequency. The structure is less sensitive to fabrication tolerances than conventional transducers. Compensation for phase errors can also be inserted between the oblique-angle grooves, as can attenuation for weighting the amplitude.

Figure 6.15
Configurations of interdigital transducers for linear-FM pulse-compression dispersive delay lines. (a) In-line single dispersive SAW delay line; (b) in-line double dispersive SAW delay line with dispersion in both transducers; (c) slanted-array compressor (SAC); and (d) reflective-array compressor (RAC).
(Figs. (a), (b), and (d) are from Maines and Paige,[36] courtesy of Proc. IEEE.)

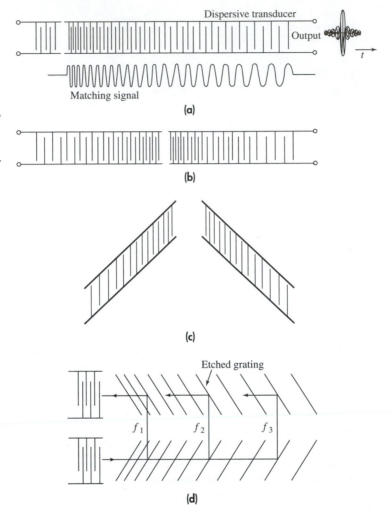

In summary, linear-FM SAW devices might have pulse compression ratios (BT) greater than 10,000, uncompressed pulse durations up to 150 μs, bandwidths greater than 1 GHz, insertion loss from 20 to 60 dB, and operate at center frequencies from 100 MHz to 1.5 GHz. They are usually found in the IF portion of the radar receiver. SAW devices provide wide bandwidth pulse-compression filters in small size packages. They are readily reproducible, easy to manufacture in large quantities, and of relatively low unit cost.

A device related to SAW devices is the IMCON, which is a reflection-mode delay line configured similar to the RAC SAW device, but fabricated on steel acoustic media. IMCONs are more suited for narrow bandwidths and low center frequencies (below 30 MHz) and when uncompressed pulse durations greater than 50 μs are desired. Bandwidths are from 0.5 to 12 MHz. Pulse durations of 600 μs are possible and units can be cascaded to obtain longer pulse durations.[38] For example, a pulse duration of 10 ms was obtained by cascading 18 IMCONs with amplifiers. It operated at a center frequency of 7.5 MHz and bandwidth 2.5 MHz. The pulse compression ratio was 25,000 and its dimensions were 15 by 15 by 12 inches.[39]

Amplitude Weighting   Amplitude shaping of the frequency response of a SAW filter can be obtained by the amount of overlap of the electrodes of the IDT, as in Fig. 6.16. This is sometimes called *apodization.*

Digital Processing for Frequency-Modulated Pulse Compression   The linear-FM pulse-compression waveform, as well as most other pulse compression waveforms, can be processed and generated at low power levels by digital methods when A/D converters are available with the required bandwidth and number of bits.[40,41] Digital methods are very stable and can handle long-duration waveforms. The same basic digital system implementation can be used when one wishes to employ multiple bandwidths and pulse durations, different types of pulse compression modulations, good phase repeatability, low time-sidelobes, or when flexibility is desired in waveform selection.

Generation of the Pulse-Compression Waveform   The analog and digital methods used to obtain the linear-FM pulse-compression filter can also be applied to generate the transmitted waveform. The waveforms may be generated passively or actively. An example of the former is the SAW device; an example of the latter is a voltage-controlled oscillator. Other examples of each can be found in Ref. 40.

Single Transmit-Receive Filter   It will be recalled (Sec. 5.2) that the impulse response of a matched filter is the time inverse of the signal for which it is matched. If an up-chrip linear-FM is transmitted, the impulse response of its matched filter will be a down-chirp. To employ the same filter for both transmit and receive, the local oscillator frequency of the radar superheterodyne receiver should be greater than the frequency of the received signal. When the difference signal is taken from the mixing operation, the linear-FM waveform will be inverted; that is, an up-chirp becomes a down-chirp, which is the signal required for the matched filter when a transmitted up-chirp is generated using the same matched filter.

Binary Phase-Coded Pulse Compression   Changes in phase can also be used to increase the signal bandwidth of a long pulse for purposes of pulse compression. A long pulse of

Figure 6.16   Interdigital transducer (nondispersive) showing overlap of the comb fingers to provide an amplitude weighting along the pulse.

duration T is divided into N subpulses each of width τ. An increase in bandwidth is achieved by changing the phase of each subpulse (since the rate of change of phase with time is a frequency). A common form of phase change is *binary phase coding,* in which the phase of each subpulse is selected to be either 0 or π radians according to some specified criterion. If the selections of the $0, \pi$ phases are made at random, the waveform approximates a noise-modulated signal and has a thumbtack-like ambiguity function as in Fig. 6.11. The output of the matched filter will be a compressed pulse of width τ and will have a peak N times greater than that of the long pulse. The pulse compression ratio equals the number of subpulses $N = T/\tau \approx BT$, where the bandwidth $B \approx 1/\tau$. The matched filter output extends for a time T on either side of the peak response. The unwanted, but unavoidable, portions of the output waveform other than the compressed pulse are known as *time sidelobes.* When the selection of the phases is made at random, the expected maximum (power) sidelobe is about $2/N$ below the peak of the compressed pulse.

Barker Codes   Some random selections of the $0, \pi$ phases are better than others (where *better* means a lower maximum sidelobe level). Completely random selection of the phases, therefore, is not a good idea if compressed waveforms with low time-sidelobes are desired—and they usually are. One criterion for selecting the subpulse phases is that all the

Figure 6.17 (a) Barker code of length 13; a long pulse with 13 equal subdivisions whose individual phases are either 0° (+) or 180° (−). (b) Autocorrelation function of (a), which represents the output of the matched filter. (c) Tapped delay line for generating the Barker code of length 13.

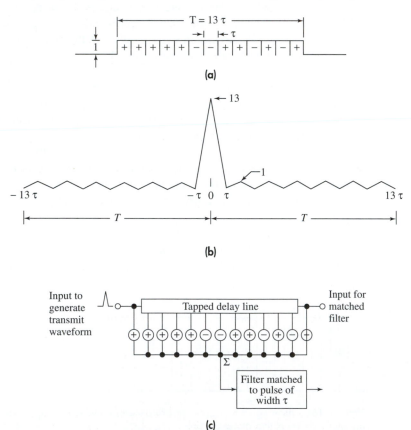

Table 6.4   Barker codes

Code length	Code elements	Sidelobe level, dB
2	+−, ++	−6.0
3	++−	−9.5
4	++−+, +++−	−12.0
5	+++−+	−14.0
7	+++−−+−	−16.9
11	+++−−−+−−+−	−20.8
13	+++++−−++−+−+	−22.3

time-sidelobes of the compressed pulse should be equal. (The reasoning here is that one can allow the small sidelobes to increase if it results in a lowering of the high sidelobes, and having all sidelobes equal is an indication that this has happened.) The $0, \pi$ binary phase-codes that result in equal time-sidelobes are called *Barker codes*. The Barker code of length $N = 13$ is shown in Fig. 6.17a. The (+) indicates 0 phase and (−) indicates π radian phase. Its autocorrelation function, which is the output of the matched filter, is shown in (b). There are six equal time-sidelobes to either side of the peak, each at a level 22.3 dB below the peak. (The sidelobe level of the Barker codes is $1/N^2$ that of the peak signal.) In (c) is shown schematically a tapped delay line that generates the Barker code of length 13 when the input is from the left. The same tapped delay-line filter can be used as the receiver matched filter if the received signal is applied from the right. The Barker codes are listed in Table 6.4. There are none greater than length 13; hence, the greatest pulse-compression ratio for a Barker code is 13. This is a relatively low value for pulse-compression applications.

Linear Recursive Sequences, or Shift-Register Codes   When a pulse compression ratio larger than 13 is required, some other criterion for selecting the $0, \pi$ phases is needed. One method for obtaining a set of random-like phase codes is to employ a shift register with feedback and modulo 2 addition that generates a pseudorandom sequence of zeros and ones of length $2^n - 1$, where n is the number of stages in the shift register.[42−44] An n-stage shift register consists of n consecutive two-state memory units controlled by a single clock. The two states considered here are 0 and 1. At each clock pulse, the state of each stage is shifted to the next stage in line. Figure 6.18 shows a seven-stage shift register used to generate a pseudorandom sequence of zeros and ones of length 127. In this particular case, feedback is obtained by combining the output of the 6th and 7th stages

Figure 6.18   Seven-bit shift register for generating a pseudorandom linear recursive sequence of length 127.

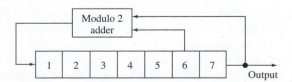

in a modulo-2 adder. (In a modulo-two adder the output is *zero* when both inputs are the same [(0,0) or (1,1)] and the output is *one* when they are not the same. It is equivalent to base-two addition with only the least-significant bit carried forward.) An n-stage binary device has a total of 2^n different possible states. The shift register cannot, however, employ the state in which all stages are zero since it would produce all zeros thereafter. Thus an n-stage shift register can generate a binary sequence of length no greater than $2^n - 1$ before repeating. The actual sequence obtained depends on both the feedback connection and the initial loading of the shift register. When the output sequence of an n-stage shift register is of period $2^n - 1$, it is called a *maximal length sequence*, or *m-sequence*.

This type of waveform is also known as a *linear recursive sequence* (LRS), *pseudo-random sequence*, *pseudonoise* (PN) sequence, or *binary shift-register sequence*. They are linear since they obey the superposition theorem. When applied to phase-coded pulse compression, the *zeros* correspond to zero phase of the subpulse and the *ones* correspond to π radians phase.

There can be more than one maximal length sequence, depending on the feedback connection. For example, 18 different maximal length sequences, each of length 127, can be obtained with a seven-stage shift register by using different feedback connections. With the proper code, the highest (power) sidelobe can be about $1/2N$ that of the maximum compressed-pulse power. A 24-dB sidelobe can be available with a sequence of length 127. Not all maximal length codes, however, have this low a value of peak sidelobe. For example,[45] with $N = 127$, the highest sidelobe of the various maximal length sequences can vary from 18 to 24 dB below the peak. It is generally said that the more usual maximum sidelobe of a "typical" maximal-length shift register sequence is approximately $1/N$ that of the peak response. In the above example with $N = 127$, this is 21 dB. As mentioned above, a completely random selection of the phases usually results in a sidelobe approximately $2/N$ below the peak; the typical maximal-length shift-register sequence might have a sidelobe of $1/N$, and the best of the maximal-length sequences might approach 1/2N. By comparison, the Barker codes have a peak sidelobe $1/N^2$ below the peak.

Sometimes the term *code* is used and at other times the term *sequence* is used to describe the phases of the individual subpulses of a phase-coded waveform. Both terms are found in the literature and are often interchangeable when discussing pulse compression, as is the practice in this section.

The shift-register codes fit several of the tests for randomness. They are called *pseudo random* since they may appear to be random, but they are actually deterministic once the shift-register length and feedback connections are known. The fact that a pulse compression sequence is random or pseudorandom does not mean it will produce the lowest time-sidelobes at the output of the matched filter. For instance, the Barker code of length 13 in Table 6.4 is a good sequence (for its length) in that it produces a -22.3-dB peak sidelobe, but it is not what is usually thought of as "random." It does not satisfy the *balance* property of a random sequence (the number of ones differs from the number of zeros by at most one); nor does it satisfy the *run* property (among the number of runs of ones and zeros in each sequence, one half are of length two, one quarter of length three, and so forth); nor does it satisfy the *correlation* property (when the sequence is compared term by term with any cyclic shift of itself, the number of agreements differs from the number of disagreements by at most one). Thus the Barker codes are not random in the above sense, but they produce the lowest sidelobes for their length. It should be no surprise, therefore, to find that there are better binary sequences than the shift-register sequences.

Other Binary Sequences   Computer search has shown that the longest code with side-lobe level of 2 is of length 28;[46,47] the longest code with sidelobe level 3 is 51;[48] and the longest codes for levels 4 and 5 are 69 and 88 respectively.[48] It should be noted that the above sidelobe levels are almost 25 dB for code lengths varying from 51 to 88. These sidelobe levels are better than the $1/2N$ values of the best maximal-length sequences.

Doppler Effects   The binary phase-coded waveform produces a thumbtack ambiguity diagram so that a bank of matched filters (with each filter tuned to a different doppler frequency) will be needed when the doppler shift of the target echo signal is large. There is, however, a more serious problem with these waveforms when there are large doppler shifts. The sidelobes in the plateau region of the thumbtack can be relatively high and a single target can result in responses from more than one filter to cause ambiguities and/or false target reports at incorrect doppler shifts. Thus the binary phase-coded waveforms might not be suitable when there is a significant doppler shift.

Quadriphase Code[49]   Binary phase (biphase) coded signals using rectangular subpulses can result in poor fall-off of the radiated spectrum, mismatch loss in the receiver pulse-compression filter, and loss due to range sampling when the pulse compression is digital. These effects can be reduced by modifying the biphase codes to produce what is called a *quadriphase code.*

   To generate the quadriphase code, one first starts with a good biphase code. The binary-code phases, designated as W_k for each subpulse, are transformed to the quadriphase codes, designated V_k, by means of the following transformation:

$$V_k = j^{s(k-1)} W_k \qquad \text{[6.47]}$$

where s is fixed and is either $+1$ or -1. Since the phases of the subpulses of a binary code are either 0 or π radians, the above equation shows that the phases of the quadriphase subpulses will be either 0, $\pi/2$, π, or $3\pi/2$ radians. Between subpulses the phase change will be either $+\pi/2$ or $-\pi/2$ radians. Each subpulse of the quadriphase code has a half-cosine shape rather than a rectangular shape. The spectrum of a cosine subpulse decreases more rapidly than that of a rectangular shape, and is therefore less likely to cause interference. The subpulse width τ is measured from the half-power points of the half-cosine. (The base-width of the half-cosine is therefore 2τ.) The overlap of the cosine-shaped subpulses results in an uncompressed pulse that has constant amplitude except for the leading and trailing edges. As mentioned previously in this section, constant amplitude output is desired of high-power transmitters since variations in the amplitude of the pulse can cause loss of transmitter efficiency. The half-cosine shape of the subpulses and the crossover at the half-power points result in constant amplitude and also eliminate phase transients that can cause spectral splatter.

   The maximum value of the compressed biphase waveform (the autocorrelation function) is preserved when converted to the quadriphase compressed pulse. The peak sidelobe level of the quadriphase code, however, can be larger than that of the biphase signal from which it was derived. The increase approaches 1.5 dB when the biphase side-lobe is very large. If, however, the amplitude of the peak sidelobe of the compressed pulse of the biphase code is unity, as with Barker codes, then according to Taylor and Blinchikoff[49] there is no increase in the sidelobe amplitude when transforming to the quad-riphase code. When the peak sidelobe of the compressed pulse from a biphase code is two, the peak sidelobe of the compressed quadriphase code is increased by 0.41 dB.

If digital processing is employed, a loss in signal-to-noise ratio occurs when the sampling straddles the peak of the compressed pulse rather than being exactly at the peak. When losses due to random sampling are averaged over all possible locations within the pulse, the range-straddling loss is about 2.3 dB for binary phase codes, but is only 0.8 dB for quadriphase codes.[49] (In both the binary phase and the quadriphase codes, it is assumed in the above that the spacing between samples is the same as the subpulse spacing τ, but can occur anywhere within the subpulse. If the binary phase codes are double sampled, however, the 2.3-dB loss is reduced to 0.8 dB.) The doppler behavior of quad-riphase codes is said by Taylor and Blinchikoff to be the same as the doppler behavior of biphase codes. Levanon and Freedman,[50] however, give examples where the ambiguity diagram with a nonzero doppler shift can be significantly different for a quadriphase code than for the biphase code from which it was derived. For example, the ambiguity diagram of a quadriphase code derived from a Barker code of length 13 has a diagonal ridge more like that of a linear-FM ambiguity function rather than the thumbtack ambiguity function of the Barker code. The diagonal ridge, however, is not a general feature of quadriphase codes.

Polyphase Codes   The phases of the subpulses in phase-coded pulse compression need not be restricted to the two levels of 0 and π. When other than the binary phases of 0 or π, the coded pulses are called *polyphase codes*. They produce lower sidelobe levels than the binary phase codes and are tolerant to doppler frequency shifts if the doppler frequencies are not too large. An example is the Frank polyphase code[51,52] defined by an M by M matrix as shown on the left side of Table 6.5. The numbers in the matrix are each multiplied by a phase equal to $2\pi/M$ radians (or $360/M$ deg). The polyphase code starts at the upper left-hand corner of the matrix, and a sequence of length M^2 is obtained. The pulse compression ratio is $M^2 = N$, the total number of subpulses. An example of the phases for each of the 25 pulses of a Frank code of $M = 5$ is shown on the right side of Table 6.5. The basic phase increment in this example is $360/5 = 72°$. The phases are shown modulo 360°.

Frank conjectured that for large N, the highest sidelobe of a polyphase code relative to the peak of the compressed pulse is $\pi^2 N \approx 10 \times$ (pulse compression ratio). In the above example with $N = 25$, the peak sidelobe is 23.9 dB. (By comparison, the closest maximal-length shift register sequence, of length 31, has a peak sidelobe of 17.8 dB.)

Since the rate of change of phase is a frequency, examination of the matrix of Table 6.5 indicates that the frequencies of the Frank code change linearly with time in a discrete fashion. The Frank codes can be thought of as approximating a stepped linear-FM waveform. The ambiguity diagram for a polyphase code is similar to that of a linear-FM waveform, but there can be a 3- to 4-dB loss in signal at doppler frequencies that are

Table 6.5   Frank Polyphase Code

$M \times M$ Matrix Defining Frank Polyphase Code	Example of Frank Matrix with $M = 5$ and pulse compression ratio $N = M \times M = 25$
0 0 0 0... 0	0, 0, 0, 0, 0,
0 1 2 3... $N - 1$	0, 72, 144, 216, 288,
0 2 4 6.... $2(N - 1)$	0, 144, 288, 72, 216,
0 3 6 9.... $3(N - 1)$	0, 216, 72, 288, 144,
.	0, 288, 216, 144, 72.
.	
0............ $(N - 1)^2$	

The phases of each of the M^2 subpulses are found by starting at the upper left of the matrix and reading each row in succession from left to right.

odd-integer multiples of π radians per pulse.[53] The doppler response of polyphase codes should be satisfactory with aircraft targets but might be a problem when detecting high-speed targets such as satellites and ballistic missiles.

B. Lewis and F. Kretschmer have described variants of the Frank polyphase codes, which they called P-codes.[54] They devised four P-codes that they claim to be more tolerant than the Frank code to receiver bandlimiting prior to pulse compression.

Lewis[55] has also shown that the range, or time, sidelobes of the polyphase codes can be reduced significantly after reception by following the polyphase pulse-compression network with a two-sample sliding-window subtractor for the Frank and P1 codes, and by a two-sample sliding-window adder for the P3 and P4 polyphase codes. The sliding-window subtractor is a one-sample delay whose input and output drive the subtractor. With this additional processing, the sidelobes of a polyphase code of length N are uniform and are $(2/N)^2$ relative to the peak. The sidelobe level is of unit magnitude, just as with the Barker codes. The width of the compressed pulse, however, is doubled so that the pulse compression ratio is reduced to $N/2$. There is a loss because of the doubling that Lewis estimates to be about 1 dB. For example, a polyphase waveform with an original pulse-compression ratio of 400 will have an effective pulse-compression ratio of 200 after the two-sample sliding-window subtractor or adder, and its sidelobe level will be 46 dB below the peak. This is the sidelobe level that would be achieved if there existed a Barker code with a pulse compression ratio of 200. Lewis also states that "tests of the sidelobe suppression with doppler as would be encountered in radar where the codes were useful revealed that the doppler did not significantly reduce the effect of the sidelobe suppression." There exists, however, a range-doppler coupling that is characteristic of linear FM and FM-derived polyphase codes.

Costas Codes   A *frequency hopping,* or *time-frequency coded,* waveform is generated by dividing a long pulse of width T into a series of M contiguous subpulses (Fig. 6.19a). The frequency of each subpulse is selected from M contiguous frequencies within a band B (Fig. 6.19b).[56] The frequencies are separated by the reciprocal of the subpulse width (or

Figure 6.19 Discrete frequency-coded pulse-compression waveform. (a) Long pulse divided into M subpulses; (b) M contiguous uniformly increasing frequencies covering the band B; (c) a frequency-hopping code.

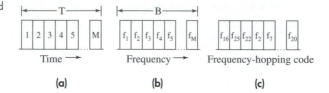

$\Delta B = M/T$); there are B/M different frequencies to choose for the subpulses; and the width of each subpulse is T/M. If the frequencies of the subpulses were to be selected so as to be monotonically increasing (or decreasing) in frequency from subpulse to subpulse, it would be a stepped-frequency waveform that approximates the linear-FM waveform, especially if the frequency and time steps are small. Its ambiguity diagram will be a ridge, like that of the linear FM. When the frequencies are selected at random, as in Fig. 6.19c, the result is a thumbtack ambiguity diagram. The pulse compression ratio is $BT = (M \Delta B)T = M(M/T)T = M^2$. Only $M = \sqrt{BT}$ subpulses are needed instead of the BT subpulses required for binary phase-coded pulses.

Some random selections, however, are better than others for producing an ambiguity diagram with low sidelobes; so it is not wise to choose the frequencies haphazardly. With M choices of frequencies for M subpulses, there are $M!$ different sequences. An exhaustive blind search for the best sequences is not practical except for very small values of M.

J. P. Costas has suggested a procedure for selecting the order of frequencies so as to provide well-controlled range and doppler sidelobes.[57] Costas codes attempt to have a sidelobe no greater than one unit high, so that the maximum (voltage) sidelobe level of the thumbtack ambiguity diagram is $1/M$th the central peak (voltage), where M is the number of subpulses. The sidelobes of the ambiguity diagram (in terms of relative power) are about $(1/M)^2$ relative to the central peak in those regions of the ambiguity diagram away from the central peak, and can be close to $(2/M)^2$ near the central peak.

The waveforms based on the criterion of Costas are described by an array of M rows representing frequency and M columns representing time intervals (subpulses). There is exactly one mark in each row and each column. Such an array is called an $M \times M$ Costas array if the *coincidence function* satisfies $C(r,s) \le 1$ for all integer pairs $(r,s) \ne (0,0)$ with $|r| \le M - 1$, $|s| \le M - 1$. The coincidence function gives the number of coincidences of marks between the original array and its translation along the time and/or frequency axis. It may be regarded as a discrete version of the (unnormalized) ambiguity function. The parameters r and s define the amount of the translation: r is the number of integer shifts to the right or left (translated by columns), and s is the number of integer shifts up or down (translated by rows).

Consider, for example, the 6×6 Costas array of Fig. 6.20.[58] The dashed lines in this figure indicate the translation of the array two time intervals to the right and three frequency intervals up. The Xs locate the frequencies transmitted in the original array, and the Os the same frequencies in the shifted array. From this discrete representation, one point on the ambiguity diagram can be obtained. In the example of Fig. 6.20 there is only one cell in the original array and the shifted array where the X and O marks (signals) coincide. Hence the value for this point (2,3) on the ambiguity diagram is 1. The complete

Figure 6.20 A 6 × 6 Costas array indicated by Xs along with its translation two to the right and three up (2,3) shown dashed and indicated by Os. A single coincidence occurs in cell (3,6).
(After Chang and Scarbrough.[58] copyright 1989 IEEE.)

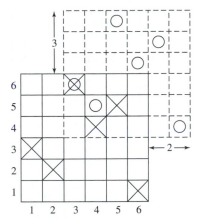

ambiguity diagram based on the discrete array is obtained by shifting the two arrays through all values of r and s.

A Costas code is therefore a frequency-hopped signal where there is no more than one coincidence between the original and the translated array. With no translation, there will be M coincidences, which is the peak at the center of the ambiguity diagram. Thus the maximum sidelobe (voltage) ratio is $1/M$. This is only approximate since discrete values are used. A more exact calculation of the ambiguity function, therefore, might have sidelobe levels that can be greater than $1/M$ in the vicinity of the central peak.

Golomb and Taylor[59] have analyzed the Costas codes and conjecture that $M \times M$ Costas arrays exist for every positive integer M. They show that Costas arrays exist when $M = p - 1$, $M = q - 2$, $M = q - 3$, and sometimes when $M = q - 4$ and $M = q - 5$, where p is a prime number and q is any power of a prime number. If lines were to be drawn connecting pairs of marks in the distinct cells of a Costas array, no two of these lines would be equal in both length and slope. Golomb and Taylor list the known Costas arrays for 271 values of M up to 360. They also indicate that there can be a large number of different Costas arrays for a given value of M. If $C(M)$ represents the total number of $M \times M$ Costas arrays, they give $C(M)$ for values from $M = 1$ to 13. For example, when $M = 7$ (pulse-compression ratio of 49) there are 200 different 7×7 arrays that meet the Costas criterion of only one coincidence in the discrete ambiguity diagram. For $M = 13$, there are 12,828 Costas arrays. The probability that a randomly chosen $M \times M$ permutation matrix will be a Costas array is $C(M)/M!$. Golomb and Taylor state that this probability is less than 10^{-21} when $M = 32$. Thus trial and error search is not practical for large M. Figure 6.21 is an example of a Costas array for $M = 24$ and pulse-compression ratio of 576. For this example, the maximum sidelobe in regions of the ambiguity diagram is predicted to be about 27.6 dB below the central peak when not too close to the central peak, and a bit higher in the vicinity of the central peak.

Nonlinear FM Pulse Compression[60,61] Nonlinear FM offers the advantage over linear FM of producing low time-sidelobes using a constant-amplitude waveform and a theoretically lossless matched filter. It does not experience the loss in signal-to-noise ratio associated

Figure 6.21   Example of frequency-time sequences for a Costas code with $M = 24$, which gives a pulse compression ratio of 576.
| (After Golomb and Taylor.[59])

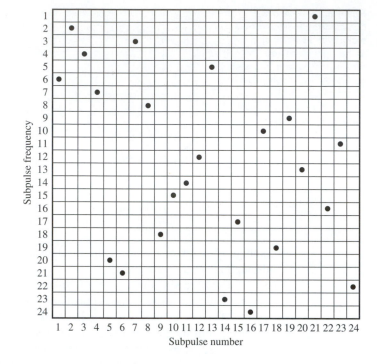

with the mismatched filter used to reduce sidelobes in a linear-FM pulse-compression system. A constant-amplitude envelope allows efficient generation of high power. The nonlinear rate of change of frequency performs the same role as amplitude weighting of the spectrum. If less time is spent over some part of the spectrum, it is equivalent to reduced amplitude of the spectrum. In addition, there is no significant widening of the compressed pulse. When a symmetrical nonlinear FM is used, the ambiguity diagram is that of a thumbtack; that is, it has a single peak rather than a ridge. (A symmetrical waveform is one where the frequency increases during the first half of the pulse and decreases in a similar manner during the second half of the pulse, or vice versa.) Hence, symmetrical nonlinear FM is more sensitive to large doppler shifts and is not doppler tolerant. A nonsymmetrical waveform utilizes only one half of the symmetrical waveform and has some of the range-doppler coupling characteristic of the linear FM.

Nonlinear FM waveforms result in more system complexity than linear FM. The surface acoustic wave (SAW) dispersive delay line and digital methods can be used to generate and process nonlinear FM. Nonlinear FM waveforms are designed to produce the equivalent of the classical amplitude-weighting functions mentioned previously. Examples of nonlinear FM based on 40-dB Taylor, Hamming, truncated gaussian, and cosine-squared-on-a-pedestal weightings that give low sidelobes (35 to 40 dB, or better) can be found in the literature.[61,62]

Doppler-Tolerant Pulse-Compression Waveforms   A *doppler-tolerant waveform* is one whose signal-to-noise ratio out of its matched filter is relatively independent of the doppler frequency shift (over a wide range of doppler shifts). Thus a single matched filter can be used rather than multiple filters (as in a filter bank). Such waveforms have also been called *doppler invariant*. Several of these have already been encountered in this section.

Short Pulse and Linear-FM   A short pulse is doppler tolerant, as can be seen from its ridge ambiguity diagram of Fig. 6.8b. The long pulse of Fig. 6.8a, however, is not doppler tolerant. A bank of matched filters, each tuned to a different doppler shift, is required in order to detect the received signal when its doppler is unknown. The linear-FM is also doppler tolerant, as long as the absolute value of $2(v_r/c)BT$ is not greater than unity, where v_r = radial velocity of the target, c = velocity of propagation, B = bandwidth, and T = pulse duration.[62] When this product exceeds unity the peak-signal out of the matched filter will be significantly reduced and the compressed pulse widened. Except for very high-speed targets and very large values of BT, this restriction is not a concern in most radar applications; but it can be important in sonar since the velocity of acoustic propagation is much lower than that of electromagnetic waves.[63]

Linear Period Modulation   The linear-FM waveform is only an approximation to a "true" doppler-tolerant waveform that is not limited by the restriction given in the above paragraph. The doppler-tolerant waveform is[64–66]

$$s(t) = A(t) \cos \left[\frac{2\pi f_0^2 T}{B} \, ln \left(1 - \frac{Bt}{f_0 T} \right) \right]$$ [6.48]

The amplitude $A(t)$ represents modulation by a rectangular pulse of width T [sometimes written as rect (t/T)]. The bandwidth is B and the carrier frequency is f_0. This expression for the doppler-tolerant waveform is difficult to interpret as it stands, but if the natural log is expanded using the series $ln(1 - x) = -(x + x^2/2 + x^3/3 + \cdots)$, for $x^2 < 1$, Eq. (6.48) becomes

$$s(t) = A(t) \cos \left(2\pi f_0 t + \frac{\pi B t^2}{T} + \frac{2\pi B^2 t^3}{3 f_0 T^2} + \cdots \right)$$ [6.49]

When terms greater than the first two can be neglected, Eq. (6.49) is the same as the classical linear-FM (LFM) waveform. Thus the linear FM, or chirp, pulse-compression waveform is a practical approximation to the theoretical doppler-tolerant waveform given by Eq. (6.48).

Differentiating the argument of Eq. (6.48) with respect to time, gives the frequency of the doppler-tolerant waveform as $f_0^2 T/(f_0 T - Bt)$. Inverting to obtain the period, it can be seen that the doppler-tolerant waveform is one whose period is linearly modulated with time; that is, period $= [(1/f_0) - mt]$, where $m = B/f_0^2 T$. This doppler-tolerant waveform has been called *linear-period modulation* (LPM). Because of the logarithm in the argument of Eq. (6.48), it has also been called *logarithmic phase modulation*. It is also known as *hyperbolic frequency modulation* (HFM), since the relationship between frequency and time is hyperbolic. Both the LFM and the LPM waveforms are doppler tolerant in the sense that the signal-to-noise ratio and the compressed pulse width are not significantly changed by a doppler frequency shift. However, with both waveforms there is coupling between the range and the doppler; and the peak response will be shifted in time so that a range error can result, as evident in the ambiguity diagram of Fig. 6.9 for the LFM waveform.

The time sidelobes of the LPM waveform, after passing through the matched filter, are higher than those of the LFM waveform.[66] It has been said that weighting can reduce the LPM sidelobes to the range of −20 to −30 dB.[67] The spectrum of the LPM is not uniform but falls off exponentially as the frequency increases. The amount of exponential decay depends on the ratio of the highest-to-the-lowest frequency.

LPM pulse-compression is more likely to be important for acoustic (sonar) echo-location applications than for radar. Since the velocity of sound is much smaller than the velocity of light, the acoustic LFM waveform might result in an unacceptable widening of the compressed pulse and a lowering of the output signal-to-noise ratio. Various species of bats have been found to employ acoustic echolocation with LPM waveforms. The LPM is needed rather than LFM since the bandwidth-pulsewidth products are generally high enough to cause the absolute value of $2(v_r/c)BT$ to exceed unity.[68] The bat, which depends on acoustic echolocation to fly through dark caves with impunity and to find and catch insects in flight, has evolved a remarkable pulse compression system during its many years on earth, long before radar or sonar existed.

An accelerating target causes a frequency shift of an LPM signal.[69] The LPM waveform can be made to accommodate accelerating targets if a bank of filters is employed that is matched to the frequency-shifted versions of the received LPM waveform. It has been said, however, that wideband LFM signals "may be quite capable of achieving high acceleration tolerance without the need for separate acceleration-processing channels," even though this waveform is less doppler tolerant than the wideband LPM.[70]

Other Pulse Compression Waveforms There are several other pulse-compression waveforms that have been considered for radar. They each have some interesting characteristics; but as is true of almost everything, they also have some limitations.

Nonlinear Binary Phase-Coded Sequences[71,72] Linear recursive sequences, or shift-register codes, were discussed earlier in this section. They are formed with an n-stage shift register with a feedback logic consisting of modulo-2 additions. The number of different maximal-length sequences that can be obtained with an n-stage shift register with linear feedback logic is approximately $2^n/n$. If nonlinear feedback logic is used instead, the number of maximal length sequences increases to $2^{2^{n-1}}/2^n$. This large number is of interest when many different codes are desired for such purposes as minimizing mutual interference, providing more security to the code, or making deception jamming more difficult. With a five-stage shift register, for example, only six 31-symbol maximal-length sequences can be obtained when linear feedback logic is used. With nonlinear feedback, 2048 different 32-symbol pseudorandom sequences are available. The length of a nonlinear sequence from an n-stage shift register can be 2^n rather than the $2^n - 1$ for linear sequences.

Complementary Codes It is possible to find pairs of equal-length phase-coded pulses in which the sidelobes of the autocorrelation function of one are the negative of the other. If the autocorrelation functions from the outputs of the two matched filters are added, the algebraic sum of the sidelobes will be zero and the main response will be 2*N*, where *N* is the number of elements in each of the two codes. These are called *complementary codes,* or *Golay codes* after the person who first reported their existence and described how to construct them.[73] Theoretically, there are no sidelobes on the time axis when complementary codes are employed. Complementary codes can be obtained with either binary or polyphase sequences.[74]

There are two problems, however, that limit the use of complementary codes.[75] The first is that the two codes have to be transmitted on two separate pulses, detected separately, and then subtracted. Any movement of the target or instability in the system that occurs during the time between the two pulses can result in incomplete cancellation of the sidelobes. Transmitting the two codes simultaneously at two different frequencies does not solve the problem since the target response can vary with frequency. The second problem is that the sidelobes are not zero after cancellation when there is a doppler frequency shift so that the ambiguity diagram will contain other regions with high sidelobes. Thus this method of obtaining zero sidelobes has serious practical difficulties and is not as attractive as it might seem at first glance.

Welti Codes[76]   These are related to Golay complementary codes in that they are used in pairs that are subtracted from one another to obtain autocorrelation functions with zero time sidelobes. They use four phases (0,90,180,270°) rather than the two phases (0, 180) of the Golay codes. They are a class of polyphase codes but have ambiguity diagrams more like those of biphase codes than the Frank polyphase codes.[77] Welti codes can present the same problems as other complementary codes.

Huffman Codes[78]   So far, every pulse-compression waveform discussed in this section is of constant amplitude across the uncompressed pulse. The signal bandwidth is increased by phase or frequency modulation rather than by amplitude modulation. The Huffman codes, on the other hand, consist of elements that vary in amplitude as well as in phase. When the doppler shift is zero, they produce autocorrelation functions with no sidelobes on the time axis except for a single unavoidable sidelobe at both ends of the compressed waveform. The level of these two end-sidelobes is a design tradeoff. In one example,[79] a Huffman code of length 64 with no doppler shift has a sidelobe at each end that is 56 dB below the peak. As with other methods for obtaining zero or low sidelobes, the volume under the ambiguity diagram must remain constant [Eq. (6.46)], which means that higher sidelobes will appear elsewhere in the doppler domain. The sidelobes also degrade if the tolerances in amplitude and phase are not maintained sufficiently high or if there are too few bits in the A/D converter.

Variants of the Barker Code   The pulse compression ratio of a conventional Barker code can be increased (beyond the maximum of 13) by making each element of the Barker code itself a Barker code. For example, a Barker code of length 13 in which each element is also a length 13 Barker gives a combined code length of 169 with a pulse compression ratio of 169 and a maximum sidelobe 22.3 dB below the peak.[80] Thus the pulse compression ratio is increased by increasing the length of the uncompressed pulse, but there is no decrease in absolute sidelobe level. This is called a *combined Barker code,* but it has also been known as a *compound Barker* or a *concatenated Barker.*

The sidelobe level of a conventional Barker code can be decreased by extending the time-sidelobe region beyond $2T$, where $T =$ the uncompressed pulse width. This method, suggested many years ago by Key, Fowle, and Haggarty,[81,82] makes use of the fact that the equal-amplitude sidelobes and the main lobe of the autocorrelation function of the Barker codes are of similar shape. Because of this geometric similarity, it is possible to suppress the sidelobes by adding properly weighted and time-shifted replicas of the matched-filter output to the original output of the matched filter. That is, the autocorrelation function of the Barker-13 code, as was shown in Fig. 6.17b, is passed through a transversal filter with delays and weighting designed to eliminate the original sidelobes. If, for example, the six sidelobes that are on each side of the main lobe of a 13-bit Barker are to be removed, there can be six weighted (reduced amplitude) replicas of the Barker code autocorrelation function applied at six different times ahead of the main lobe and six different times behind the main lobe. To accomplish this the Barker code matched filter would be followed by a 13-tap transversal filter with the proper weightings at each tap to completely eliminate (in theory) the original sidelobes. As a result, new sidelobes will appear farther out in time on either side of the original Barker-13 sidelobes. These new sidelobes will be lower than the original. The new maximum sidelobe generated because of the transversal filter will be 32.4 dB below the peak instead of the 22.3 dB of the original. The loss in signal-to-noise ratio is about 0.25 dB, which is small.

Sidelobe Reduction for Phase-Coded Pulse-Compression Signals Suggestions have been made to reduce the sidelobes of phase-coded signals by following the matched filter with a transversal filter with weights that reduce the time sidelobes, similar to that described by Key et al. in the above. The weights are selected according to one of two criteria: (1) minimize the total energy in the sidelobes or (2) minimize the peak sidelobe. The number of bits (taps) in the delay line of the transversal filter can be greater than the length (number of bits) of the phase-coded signal. When the number of bits in the transversal filter is greater than that of the matched filter, the phase-coded signal is zero-padded to the same length as the filter.

One of the first papers to discuss this method was by Ackroyd and Ghani,[83] who determined the weights of a transversal filter so as to give a compressed output signal that approximates the desired output signal in the least-squares sense. The output to be minimized is a sum that represents the "energy" of the difference between the actual sidelobe levels and the desired levels. This is sometimes known as minimizing the integrated sidelobes, or ISL.[84] Starting with a Barker-13 code, Ackroyd and Ghani show that with least-squares filters of lengths 13, 41, 53, and 66, the maximum sidelobe levels below the main peak are 24 dB, 40 dB, 50 dB, and 60 dB, respectively. It is said that when the Barker-13 matched filter is replaced with the least-squares filter, the loss in signal-to-noise ratio is 0.2 dB.

This technique has also been applied to reducing the sidelobe levels of the combined Barker code.[85] A 52-element combined Barker code, for example, can be generated by combining a 4-element Barker code with a 13-element Barker code. When passed through a 200-bit minimum square-error transversal filter, a maximum sidelobe level 40.7 dB below the peak is produced. This mismatched filter results in a computed loss of 1.86 dB.

Similarly, the weights of a transversal filter can be chosen to minimize the peak side-lobe.[86] Baden and Cohen[87] show that the peak sidelobe obtained using this criterion is from one to seven dB less than the peak sidelobe obtained using the minimization of the integrated sidelobes, depending on the particular code and filter length. Loss in signal-to-noise ratio is said to be less than one dB. Compared to the sidelobes of the original output of the matched filter, reductions of 15 dB or more "are readily achieved for reasonable filter lengths." This sidelobe-weighting technique can also be used for generating mismatched filters that produce sidelobe-free regions in the vicinity of the main lobe.

Although there are several methods described for reducing the sidelobes of the output of a matched filter for binary phase-coded signals, apparently they have not had as much application as the nonmatched weighting filter used to reduce the time sidelobes of the linear-FM pulse compression waveform.

Other Aspects of Pulse Compression

Generic Compressed Signal   By way of review, the general nature of the compressed pulse produced by a matched filter is sketched in Fig. 6.22. The peak of the compressed pulse is equal to $2E$, where E is the energy of the input signal. (Note, however, that the units of the peak output is in volts, not joules.) The peak value depends only on the input-signal energy and not on the signal shape. The width of the compressed pulse is approximately $1/B$, where $B =$ signal bandwidth, regardless of how the bandwidth is obtained. When the total duration of the uncompressed signal is T, the sidelobes of the compressed pulse extend over a time $2T$. The nature of the sidelobes is determined by the shape of the uncompressed signal, $s(t)$. One of the most important aspects of pulse compression design is the selection of $s(t)$ to minimize the sidelobe level.

The major difference between pulse compression and a short pulse of the same resolution and energy is that the short pulse has no time sidelobes, whereas the pulse compression signal has sidelobes that can be mistaken for real targets or mask the presence of small targets located within the sidelobe region that extends over a time $2T$ in extent.

Figure 6.22   Sketch of a generic compressed pulse whose uncompressed waveform is $s(t)$, with bandwidth B, time duration T, and energy E.

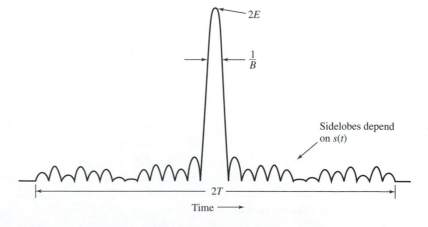

Limiting in Pulse Compression   Since the amplitude of uncompressed radar waveforms is constant, limiting is sometimes used in the receiver before the matched filter to insure that the input signal to the matched filter is of uniform amplitude or to provide a constant false-alarm rate using CFAR. It was said in Chap. 3 that hard limiting should not be used with MTI radar since it can significantly decrease the improvement factor. If pulse compression is used with MTI, and if hard limiting is desired for pulse compression, the limiter and the pulse compression matched filter can follow the doppler processing rather than precede it. On the other hand, when pulse compression follows the MTI, the MTI processor does not benefit from the reduction in clutter obtained with pulse compression. If there are significant instabilities in the MTI transmitter and oscillators, the use of limiters with pulse compression is different from that assumed here, as will be described later under the heading "Compatibility with Other Processing."

If MTI precedes the pulse-compression matched filter or if MTI is not used at all in the radar, then limiting can be employed with pulse compression. Even so, there are several concerns with the use of a limiter in pulse compression. These include (1) the loss incurred in the signal-to-noise ratio, (2) the suppression of nearby small target signals, and (3) the possibility of spurious (false) targets.

For phase-coded waveforms, the loss in signal-to-noise ratio decreases with the length of the code and the number of pulses integrated. Table 6.6 summarizes the loss due to limiting as given by Castella and Rudie.[88] A hard limiter is one in which the limit level is set low enough (well into the noise) so that there is no variation in the amplitude of the output signal. Phase and frequency modulations, however, are preserved. The hard limiter listed in the table corresponds to analog limiting in the IF. A digital soft limiter is defined by Castella and Rudie as one where the I and Q signal components are quantized to three bits at baseband, with two of these bits allotted to ± 2 sigma of the noise. The table indicates that one should try to have long code lengths and a large number of pulses integrated if hard limiting is used.

Table 6.6   Loss with phase-coded waveforms due to limiting[88] (probability of detection = 0.5; probability of false alarm = 10^{-6})

Type of Limiter*	Code Length	Number of Pulses Integrated	Loss in Signal-to-Noise Ratio
Hard limiter	16	1	6.4 dB
	16	8	1.8 dB
	128	1	1.4 dB
	128	8	1.1 dB
Digital soft limiter	16 to 128	1 to 8	0.4 to 0.5 dB
Digital hard limiter	16	1	8.6 dB
	16	8	3.0 dB
	128	1	2.4 dB
	128	8	2.0 dB

| *See text for descriptions of limiters

The second concern mentioned above, small-signal suppression, involves two uncompressed pulses that overlap. The effect of the limiter is to cause the smaller of the two signals to be suppressed and be less detectable than if limiting were not employed. This can occur with either a linear FM or a phase-coded signal. In the overlap region the weaker signal can be suppressed by as much as 6 dB if the stronger signal has a large signal-to-noise ratio.[89] The noise is also reduced. The net result can be a reduction of the detection probability of the weaker signal, which deteriorates rapidly depending on the amount of overlap.[90]

The third concern is the possibility of generating false targets in addition to the real targets when there is overlap of the uncompressed pulses. Woerrlein[91] states that with two overlapped linear-FM signals there can be an array of false targets that appear on both sides of the two target returns. If the two signals are only partially overlapped, there will be a consequent reduction in the number of false targets. There is, on the other hand, no evidence of any false targets when two binary-phase signals overlap; but when three binary-phase signals overlap, there is a fourth signal that is false. His analysis applies only to large signal-to-noise ratios.

In the above, limiting was assumed to be intentionally included in the radar processing; however, there can be unintentional limiting that can occur when the receiver is driven to saturation by large clutter echoes or interference signals.

Cross-Correlation Properties   It is sometimes desirable for more than one radar to operate simultaneously at the same frequency within the same region of coverage. Each radar can employ different pulse-compression waveforms to minimize crosstalk (when one radar accepts signals from another radar as true targets when they are really false alarms). The potential that various waveforms will cause mutual interference or false target-reports can be determined by examining the cross-correlation function (the cross-ambiguity function) between two different signals. In general, the maximum output produced in one radar by a signal from another radar with a different waveform covering the same bandwidth will usually be greater than the sidelobes of the radar's own autocorrelation function (the output of its matched filter).

If there are only two radars each using linear-FM pulse compression, one could have an up-chirp and the other a down-chirp to reduce crosstalk. Nonlinear FM waveforms can also be considered for avoiding crosstalk among various radars.

It is possible to find two Costas (frequency hopped) codes so that the sidelobe power level of the response of one in the other is no greater than $(2/M)^2$ that of the main lobe peak, where M is the number of frequencies and M^2 is the pulse compression ratio.[92] This compares with a maximum sidelobe level of $(1/M)^2$ for the ambiguity function of a Costas code, as mentioned earlier in this section. When more than two signals with Costas codes are present, the maximum cross-ambiguity sidelobes can be much greater than $(2/M)^2$.

For binary phase-coded pulses, H. Deng[93] estimated that the maximum sidelobe (power) level of the cross-correlation function for a set of k binary sequences ($k \geq 2$) is $kM/3(k-1)$ below the peak of the compressed pulse. Each of the k binary sequences is of the same length M. For $k = 2$, this expression predicts a maximum sidelobe $(M/1.5)^2$ below the peak. Thus the cross-correlation sidelobes are not as low as those of the autocorrelation function [which can be $(M)^2$ below the peak]. Using the optimization technique known as *simulated annealing,* Deng finds sequences of codes with cross-correlation lobes close to that predicted when $k = 2$ or 3. (In his optimization he allowed the autocorrelation maximum sidelobe level to be weighted equally with the cross-correlation maximum sidelobe level.) With increasing k, the expression for estimating the cross-correlation sidelobes predicts better sidelobes than actually obtained by the use of simulated annealing, which is attributed to it being difficult to satisfy the approximations used in the estimate when k is large.

Compatibility with Other Processing   Pulse compression may be used in conjunction with MTI radar, as was mentioned previously in this section. Care needs to be taken when they are used together so that if limiting is used for pulse compression it does not degrade the performance of the MTI. The increased range resolution of pulse compression can reduce the amount of clutter the radar sees so that less MTI improvement factor is required. This may be true for an ideal system; but as pointed out by Shrader and Gregers-Hansen,[94] the performance of the MTI may be no better, or even worse, than a system transmitting the same-length uncoded pulse if there are significant system instabilities that cause increased sidelobes.

The sidelobes produced in a pulse compression waveform because of system design or components that are nonlinear with frequency do not vary with time and will therefore cancel in the MTI processor just as would clutter. Instabilities due to noise from the local oscillators, transmitter power supplies, time jitter, and other transmitter noise, however, result in noiselike time-varying sidelobes that are proportional to the clutter amplitude. These noiselike time-sidelobes are not canceled by the MTI and can produce residual clutter that can cross the detection threshold and appear as targets. Shrader and Gregers-Hansen suggest a method for dealing with these noiselike sidelobes. The pulse-compression filter precedes the MTI processor, but a limiter is placed ahead of it. The limiter's dynamic range at its output is set equal to the difference between the peak transmitter power and the transmitter noise within the system bandwidth. A second limiter is placed after the pulse compression filter. The dynamic range at its output is set equal to the expected MTI improvement factor. The MTI processor follows. The two limiters cause the residue due to transmitter noise and other instabilities to be equal to the front-end thermal noise at the input to the MTI processor. The two limiters are adjustable so that when the radar is placed in the field, they can be compensated to allow for differences in clutter and the condition of the radar equipment.

Special consideration has to be given to pulse compression when used in multichannel radars, such as the three receiving channels of a monopulse tracker or the many thousands of channels of an active-aperture phased array radar. Each channel has to be highly matched (almost identical) to all the rest. The reproducibility and precision of SAW devices makes them attractive for applications where precision multiple pulse-compression units are required.[95]

Spread Spectrum   Spread spectrum communication systems[96] employ coded waveforms similar to those used in pulse compression radar. The purpose of coded waveforms in communications is different, however, from their purpose in radar pulse compression. Spread spectrum communication waveforms allow multiple simultaneous use of the same frequency band by coding each transmitted signal differently from the others. In military communications, spread spectrum waveforms also have the capability of rejecting interference as well as reducing the probability of being detected by a hostile electronic-warfare receiver. Sometimes pulse compression radars that use waveforms similar to those of spread spectrum communications have been called *spread spectrum radars*. This terminology, however, is misleading since the purpose of coded waveforms in pulse compression is entirely different from their purpose in spread spectrum communications. It is suggested that the use of the term *spread spectrum* for describing a pulse compression radar be avoided.

Comparison of Pulse Compression Waveforms   As seen in this section, there are a number of different pulse compression waveforms with different advantages and limitations. Table 6.7 compares the theoretical sidelobe levels that might be achieved with various

Table 6.7   Maximum sidelobe level for various pulse compression waveforms. (Maximum sidelobe is in dB down from the peak of the compressed signal.)

Pulse Compression Ratio	Pseudorandom Sequences	Computer-Search Binary Phase	Polyphase[1]	Costas[2]
13		22.3		
15	14.0			
16			21.2	12.0
25			23.9	14.0
28		22.9		
31	17.8			
49			26.8	16.9
63	20.4			
64			28.0	18.0
73		25.2		
88		24.7		
100			29.9	20.0
112		25.4		
121			30.8	20.8
127	24			
129		25.3		
144			31.5	21.6
255	25.9			
256			34.0	24.1

[1]For polyphase codes, the maximum sidelobe is taken here to be $\pi^2 N$ down from the central peak, where $N = M \times M$ = pulse compression ratio, and M is the dimension of the matrix.

[2]For Costas codes the maximum sidelobe away from the central peak is taken to be N down from the peak, where $N = M \times M$ = pulse compression ratio and M = number of subpulses (equal to the number of frequencies). Near the central compressed peak, the sidelobes can be larger than indicated in the table.

pulse compression waveforms. The polyphase codes have the lowest predicted sidelobes in this table; but the greater the pulse compression ratio, the smaller the phase increment and the greater must be the precision. A pulse compression ratio of 900, for instance, requires a phase increment of $360/\sqrt{900} = 12°$, and a phase tolerance that is a small fraction of this increment.

Linear FM has been the waveform used most in the past in radar pulse compression. It is less complex than some others, especially if the application permits the use of Stretch. It usually requires weighting on receive to reduce the -13.2 dB sidelobes to the order of -30 dB, with a loss of about one dB. The range-doppler coupling that causes an error in the range measurement when there is a doppler frequency shift is sometimes of little consequence. If not, the true range can be obtained by use of both an up-chirp and a down-chirp. The ridge ambiguity diagram of the linear-FM waveform means that it is doppler tolerant and that a single filter can be used when there is a large doppler shift.

The linear-period waveform is related to the linear FM and is in theory a true doppler-tolerant waveform. In most radar applications, it does not seem necessary to use linear period instead of linear FM.

The nonlinear FM waveform might be more complex than linear FM, but it can give low sidelobes without the loss caused by a mismatched weighting filter. Its thumbtack ambiguity diagram means that a bank of matched filters is needed if there are large doppler shifts, further complicating the processing. In long-range radars where it is important to minimize loss, the nonlinear FM might be considered when low sidelobes are needed.

Binary phase-coded pseudorandom waveforms were sometimes considered in the past for military pulse compression radar when it was originally believed they could provide some degree of security from deception jamming or spoofing. Shift-register codes might appear random, but by examining only a portion of the code, the rest of the code can be readily predicted. Geffe[97] pointed out that the connections of an n-stage shift register that generates binary coded waveforms can be determined by elementary methods from a knowledge of $2n - 1$ successive digits of the shift-register sequence. Thus they have no inherent security. Nonlinear shift-register codes have many more options than linear shift-register codes and for this reason might be a little better for security purposes. Truly random codes might not have the limitation of pseudorandom codes; but even if these codes were fully crypto secure, they would not possess as low sidelobes as other pulse compression waveforms. Processing of binary phase-coded signals is more complex than for linear FM, and they require a filter bank to be employed when the doppler shifts are large.

A brief comparison of the linear FM and the binary phase-coded pulse compression waveforms is given in Table 6.8.

Polyphase codes have lower sidelobes than binary phase codes. They are not very doppler tolerant for large doppler-frequency shifts, but appear to be suitable for detection of targets with aircraft velocities. They could be of interest for pulse compression applications, but have not been widely used. The sliding-window modification suggested by B. Lewis[55] appears to provide significantly lower sidelobes than any other pulse compression method but with a small loss in signal-to-noise ratio.

Costas (frequency-hopping) codes achieve a particular pulse compression ratio with fewer subpulses than phase-coded waveforms. Their sidelobes appear to be almost the same as ordinary binary phase-coded waveforms. Many more different Costas codes of a

Table 6.8   Comparison of Linear Frequency Modulation and Binary Phase-Coded Pulse Compression Waveforms

Property	Linear FM	Binary Phase-Coded Pulse
Time sidelobes	Good (~30 dB) when weighting on receive, and when a loss of about 1 dB can be tolerated	Can be equal to $1/2N$, and are not easy to improve; poor doppler sidelobes
Doppler	Doppler tolerant	Requires filter bank
Ambiguity diagram	Ridge	Thumbtack (but with high sidelobes in plateau)
Pulse compression filter	Single filter can be used for transmit and receive; usually analog for high resolution	Single filter can be used for transmit and receive, but with input at opposite end; usually digital
Complexity	Less complex, especially if Stretch can be used	More complex, (requires filter bank)
Application	High resolution (wide bandwidth)	Long pulses
Other	Range-doppler coupling; has been more widely used than other pulse compression	Bandwidth limited by availability of A/D converter; erroneously thought to be less susceptible to ECM spoofing

given length are available than can be obtained with binary phase codes. This property might be of interest in military radars concerned with operating against some forms of electronic countermeasures.

Complementary codes and Hoffman codes that are supposed to produce zero sidelobes along the zero-doppler time axis have interesting theoretical properties, but have serious practical limitations that make their use in radar less likely.

In engineering design when there is more than one possible method for accomplishing some desired objective, there is seldom one solution that is best for all applications. This applies as well to pulse compression. The radar designer should keep an open mind and examine the options carefully to determine the type of pulse compression waveform to be used for any particular radar application.

6.6   TARGET RECOGNITION

Early radars were "blob" detectors in that they detected the presence of a target and gave its location in range and angle, but could not provide much else about the type of target being detected. As radar resolution in range and cross-range improved over the years, it became possible to resolve the individual scattering centers of a target and infer something about its nature. Radar began to be more than a blob detector and could provide recognition of one type of target from another.

Even without high resolution, radar has been able to recognize the general nature of a target, or scattering object, based on such information as its behavior in space and time. The frequency dependence of the cross section can also be a useful discriminant in some cases.

The various degrees of target recognition are listed below in increasing order of information required for a decision:

- *General nature of target:* Recognition that the echo on a radar display is that of an aircraft, ship, motor vehicle, bird, person, rain, chaff, clear-air turbulence, land clutter, sea clutter, bare mountains, forested areas, meteors, aurora, ionized media, or other natural phenomena. A trained and experienced radar operator with the right type of radar should be able to sort these broad classes of target echoes from one other.

- *Target type:* This includes recognizing a fighter aircraft from a multi-engine bomber aircraft, a cargo ship from a tanker, a tracked military vehicle from a truck, chaff rather than a ship, a buried rock instead of a mine; or a surface-to-air missile site from a dump site. Sometimes this coarse form of recognition has been called *perceptual classification.*

- *Target class:* This involves determining the particular class to which a target belongs among the many possible classes. For example, if the radar believes it is detecting an aircraft, is it an F-18, F-22, MIG-31, B-2, A-6, Rafale-2000, or something else? If the target is a ship, does it belong to the Aegis destroyer Class DDG-51, Aegis cruiser CG-47, Kara, Sovermeney, or so forth, or is the echo that of a chaff decoy? If it is a bird, is it a starling, mallard, or what else? The process of determining the class of the target is known as *target classification.*

The above definitions are not standardized, so one needs to be careful when hearing or reading such terms to make sure their meanings are clear. Unfortunately, this is not always the case in the literature of target recognition.

The ultimate form of target recognition is *target identification,* which involves determining the actual name of the target, its serial number, or its side number. Identification of a target usually requires a cooperative system; that is, the target must cooperate in some manner with the identification sensor. The target has to have some form of communication system, data link, or transponder system that allows it to identify itself on a regular basis or when asked for its identification by an interrogator. *Noncooperative target recognition* (NCTR) systems, of which radar is an example, obtain target recognition information without any cooperation from the target itself. In the case of radar NCTR, recognition is based on examining the characteristics of the radar echo signal received from the target.

Military cooperative identification systems are called IFF, or *Identification Friend or Foe.* They have also been known as CAI, or *Cooperative Aircraft Identification.* In civil air-traffic control, cooperative identification methods are called ATCRBS, or *Air-Traffic Control Radar Beacon System.* IFF and ATCRBS are examples of *question and answer systems,* in that an interrogator asks the question, "who are you?" and a transponder on the target automatically answers "I am a friend and my name is . . . ."

The ability to perform noncooperative target recognition with radar depends on the type of target. Ships, for example, are easier to sort by class than aircraft since ships contain many more individual scatterers with which to recognize one class from another. Target classification generally requires greater signal-to-noise ratio than does target detection since detection depends on the total signal energy, but target recognition depends on discerning the details of the target echo signature. The many small scatterers on a target can sometimes be more important for recognition than the few large scatterers that are more important for detection. The need for large signal-to-nose ratio to detect the small scatterers means that reliable target recognition usually occurs at a shorter range than does detection.

Noncooperative target recognition methods are mainly of interest to the military. By contrast, civilian needs for aircraft target recognition are usually satisfied by cooperative methods, such as the ATCRBS. NCTR sensors, such as radar designed for that purpose, have the task of recognizing one class of target from many others that might be present. Both NCTR and cooperative methods are jointly used for the important military function of *Combat Identification.* When the target is a spacecraft or satellite, target recognition is sometimes called *Space Object Identification,* or SOI. *Automatic Target Recognition,* or ATR, is a name that could apply equally well to NCTR or combat identification, but it appears to be used mainly to describe automatic methods used for the recognition of military land targets. In the civilian sector, the use of radar and other sensors to determine the nature of the natural environment is known as *remote sensing.* Remote sensing radars observe precipitation, atmospheric effects, wind shear, birds and insects; determine the earth's surface topography; explore planets and their moons (such as probing the surface of Venus beneath its ever-present cloud cover); and monitor ice conditions, the mean sea level, and the winds that drive the ocean surface.[98]

There are two aspects to target recognition. The first is to separate the target echo from its surroundings (such as clutter) and extract from the radar echo-signal information about the unique features of a target that can help distinguish one target from another. The second aspect is the method by which one makes the actual decision as to which class or type of target a particular radar signature belongs. The decision is usually, but not always, made automatically and is often based on classical pattern recognition methods or similar decision-making methods. For example, known target signatures can be stored in a computer memory (a library) and when an unknown signature is measured, it is compared with the library of signatures to see which it matches best. (This is an overly simplified statement. The actual algorithms for target recognition can be quite sophisticated.) Only the first aspect, that of information extraction, is considered here.

Target recognition, whether by radar, optics, or the human eye, is not 100 percent accurate. Even cooperative systems do not have accuracies approaching 100 percent.* A target recognition method must be able to recognize one class of target out of many tens of classes with an accuracy of perhaps 85 percent or better before it can be taken seriously. Good target recognition methods might have accuracies of 95 percent or better, which is

---

*The accuracy of target recognition or identification can be no better than the "availability" of the equipment (where availability is the fraction of the time the equipment can be operable). Thus if a particular IFF or an NCTR has an availability of, for example, 95 percent, the accuracy of target recognition or identification obtained from such an equipment can be no better than 95 percent.

one error out of 20 decisions. There have been many radar target recognition methods proposed and explored; but only a few have been able to achieve the accuracy and reliability required, especially when the total number of targets to be recognized at any one time is large. For this reason, more than one type of recognition method is generally employed in combination to increase the overall probability of a correct decision and approach 100 percent accuracy of a correct recognition decision.

When a target is moving in the presence of clutter, the clutter echoes can be filtered out by doppler processing, as in MTI (moving target indication) radar. If the target is stationary in the midst of clutter, MTI is not applicable and some other technique must be used to separate the target from its surrounding clutter. Some of the target recognition methods mentioned here can be applied for such a purpose. Detection of nonmoving targets is sometimes called *stationary target indication* (STI).

In the remainder of this section, several target recognition methods based on radar will be briefly reviewed. All targets are assumed to be in the clear; or, if clutter is present, the targets are separated by some means from the clutter.

One-Dimensional Imaging with High Range-Resolution Radar A radar with sufficiently high range-resolution can resolve the individual scattering centers of a target and provide the radial profile (the one-dimensional image) of the target. The radial profile might provide a measure of the length of a target in the range dimension, but the true physical target-length usually cannot be determined accurately in this manner. The ends of the target might not always provide large enough echoes to be detected, one of the ends might be masked by other parts of the target, or the aspect angle of the target with respect to the radar might not be known accurately. Even if the length could be measured with accuracy, it is not usually a good means for recognizing the particular class of most targets of interest.

The radial profile of a 757 aircraft obtained with an *L*-band air-surveillance radar having 1-m range resolution is shown in Fig. 6.23.[99] In the upper portion of the figure are superimposed seven time-aligned pulse-to-pulse radial profiles. The radial velocity of the target can be obtained by measuring the target movement from the first pulse to the last. The average of the seven time-aligned radial profiles is displayed in the lower part of the figure, from which a target dimension can be obtained. With a knowledge of aspect angle, a wingspan or aircraft length can then be estimated. Figure 6.23 is typical for a jet aircraft in that the individual resolved scatterers are relatively constant. A propeller aircraft, on the other hand, can have constant returns from the nose and the tail, but there can be large pulse-to-pulse fluctuations from the propellers, which make it possible to distinguish a propeller aircraft from a jet.

Figure 6.24 is a radial profile of a large naval ship using an *X*-band radar with a resolution of about 0.3 m. As can be seen, the radial profile of this target remained relatively the same when measurements were obtained a year later.

The radial profile has often been considered a potential method of aircraft target classification. A serious difficulty exists, however, in that the details of the radial profile can change with only a small change in aspect. Masking of one part of the target by another can occur. If there is more than one scattering center within a radar's resolution cell, the relative phases of each scatterer can change with a change in aspect. This causes

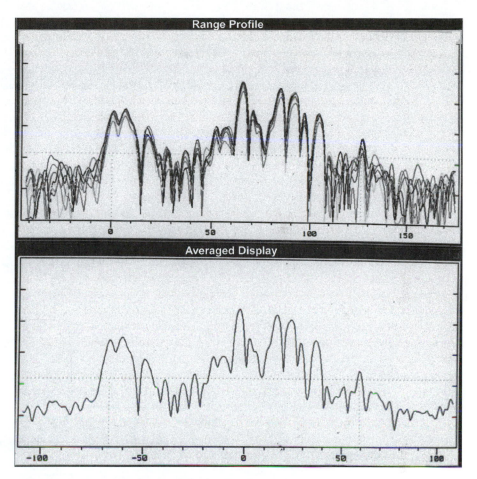

Figure 6.23 Radial profile of a commercial *757* aircraft obtained with an L-band radar with a range resolution of about 1 meter. Upper: the superposition of seven time-aligned pulse-to-pulse radial profiles. Lower: average of the seven profiles. Abscissa is in feet.
I (Provided by G. Linde of the Naval Research Laboratory.)

constructive and destructive interference and a change in the resultant cross section of the scatterers within the resolution cell. When creating a library of radial profiles to be used to match to an unknown profile, each target in the library has to be characterized by many reference profiles corresponding to different aspect angles. To make use of this library of reference profiles, the aspect angle of the unknown target must be estimated. (Observing the target's trajectory is one method for estimating the aspect.) There might need to be a large number of reference profiles stored in computer memory for each target. Furthermore, there can be many tens or even hundreds of classes of targets that need to be considered. The use of high range-resolution for target recognition starts out simple but quickly becomes complex. The result is that the use of high-resolution radial profiles for target classification is not easy to achieve in practical situations.[100,101]

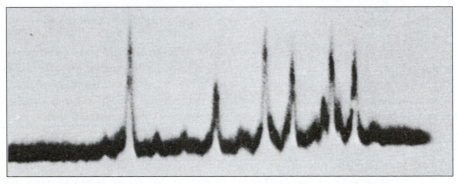

Figure 6.24 Radial profile of the former U.S. Navy gun cruiser USS Baltimore obtained with an X-band radar of 1-ft range resolution. (a) Outline of ship; (b) stern aspect; (c) bow aspect taken one year earlier than in (b).
| (Provided by I. W. Fuller, formerly of the Naval Research Laboratory)

Perceptual Classification   Although it is difficult to use one-dimensional radial profiles of a target to recognize one class of target from another, it is possible to separate targets into a simpler set of general classes. This has sometimes been called *perceptual classification.* In the case of aircraft, G. Linde and C. Platis[99] were able to show from the radial profile obtained with an L-band radar, having 1-m range resolution, that targets could be separated into the following general categories: small or large jet-engine aircraft, small or large propeller aircraft, helicopters, and missiles. The perceptual classification of aircraft does not always need a dedicated special purpose radar system. It can be achieved, for example, as demonstrated by Linde, by employing a wideband air-surveillance radar using Stretch pulse compression to obtain high range-resolution as the antenna scans by the target.

When employing high-resolution range-only information for the one-dimensional imaging of ships, it is possible to separate the ships into such categories as large ships, small ships, military ships, commercial ships, tankers, and aircraft carriers. Although perceptual classification is limited in its utility, the requirements on the radar are not as large as when determining the class of target. For reliable target classification other information usually is needed in addition to the radial profile. This added information can be the high-resolution cross-range profile, as discussed next.

Two-Dimensional Radar Imaging of Targets   The two-dimensional image of a target (in range and cross-range) can be obtained by use of an imaging radar such as *synthetic aperture radar* (SAR), *inverse synthetic aperture radar* (ISAR), or more conventional high-resolution radars such as the *side-looking airborne radar* (SLAR).

Synthetic Aperture Radar   SAR produces a high-resolution image of a scene of the earth's surface in both range and cross-range.[102] It can produce images of scenes at long range and in adverse weather that are not possible with infrared or optical sensors. The theoretical cross-range resolution of SAR is $\Delta x = D/2$, where D is the horizontal dimension of the SAR's real antenna aperture. SAR does not, however, accurately image moving targets. Moving targets can be seriously distorted and displaced from their true location. For instance, a railroad train, if it can be imaged at all, will be displaced from the track it is riding on. The motion of a ship target can be complicated because of its roll, pitch, and yaw. Also, all parts of the ship might not be moving rigidly together. Thus SAR is restricted to the recognition of stationary objects. One application of SAR is its military use for airborne surveillance of the battlefield and for imaging of fixed targets, as in the X-band Joint STARS. An X-band SAR image of a B-52 aircraft sitting on a runway, taken by Metratek, Inc., is shown in Fig. 6.25. Its resolution is about 1 ft. This might be compared with the ISAR image in Fig. 2.18 of a similar aircraft at VHF which has less resolution. (The resolution of published images, such as those in Figs. 6.25 and 2.18, is often degraded by the printing process and is not as good as that available from the original image that emerges from the processor.)

Inverse Synthetic Aperture Radar   ISAR can be considered as a radar in which the cross-range resolution is obtained by means of high resolution in the doppler-frequency domain. Each part of a moving target can have a different relative velocity with respect to the radar, especially if there is a large rotational component of the target's motion. Resolution in doppler frequency will allow the various parts of a moving target to be resolved in the cross-range dimension. Resolution in range is obtained with either a short pulse or pulse compression so that a two-dimensional image is obtained. The cross-range resolution can be shown to be $\Delta x = \lambda/(2\,\Delta\theta)$, where $\Delta\theta$ is the change in aspect angle during the ISAR observation time and λ is the radar wavelength. Thus cross-range resolution depends on the amount of angular rotation of the target during the radar observation time. Unlike SAR, ISAR takes advantage of the target's motion to provide a two-dimensional image.

Figure 6.25 X-band SAR image of a B-52 aircraft sitting on a runway. Resolution is about 1 ft. The outline of the aircraft is shown for comparison.
| (Courtesy of Ray Harris of Metratek, Inc.)

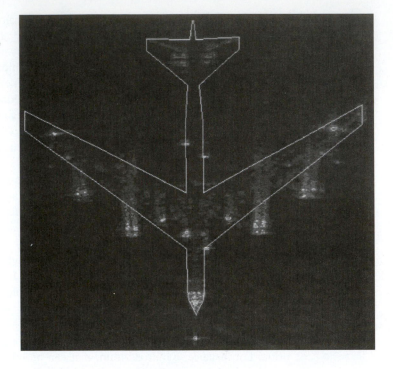

SAR and ISAR are related in that they both require a change in the aspect of a target.* In SAR, the target is assumed stationary and the radar is in motion. In ISAR, the target motion provides the changes in relative velocity that cause different doppler shifts to occur across the target. The doppler shifts from the individual scatterers are resolved by filtering. ISAR generally requires that a single large scatterer from the target be tracked to act as the reference.

Figure 6.26 is an example of an ISAR image of a commercial ship obtained with an X-band radar with about 2 m resolution in range and cross range.[103] The *pitch* motion of a ship causes the top of the masts to have a higher velocity than the bottom of the masts or the superstructure. These differences in velocity cause different doppler shifts. Resolution in doppler then allows the mast to be imaged. Along with conventional range resolution, the pitching motion of the ship gives a vertical profile of the target along its length dimension. *Roll* motion also provides height information, but in the plane that includes the width of the ship. Roll is not that significant for recognition since the width of a ship is small compared to its length. *Yaw* motion of the ship gives a plan view of the target. As the ship pitches, rolls, and yaws, ISAR might produce a vertical profile along the major axis of the ship, or a vertical profile along the minor axis of the ship, or a horizontal plan

---

*SAR and ISAR are variations of the same phenomenon. This is seen in ISAR by requiring a change in the target aspect of $\Delta\theta = \lambda/2\Delta x$. In SAR the change in fixed-target aspect is determined by the beamwidth of the real radar aperture $\Delta\theta \approx \lambda/D$. Using this expression for beamwidth in the expression for the SAR cross-range resolution $\Delta x = D/2$, it is found that $\Delta x = \lambda/2\Delta\theta$, which is the same as the ISAR cross-range resolution.

Figure 6.26   ISAR image, before image processing, of a commercial ship (17,000 ton) obtained with an *X*-band radar having 2 m resolution. The vertical scale in this image is slightly exaggerated. Note that "radar eyes" are not like "optical eyes," yet useful information can be obtained from a series of such images.
(Provided by Ronald Lipps of the Naval Research Labortary)

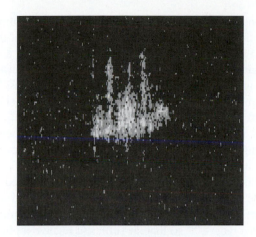

view, or some combination of the three that results in the image appearing as a perspective view. Since the angular rates of pitch, roll, and yaw are not generally known, an ISAR image does not represent true distances in cross-range, as it does in range. This doesn't seriously limit, however, the ability to recognize one class of ship target from the other.

Since different ISAR images are obtained as the ship pitches, rolls, and yaws it might require many tens of seconds of observation to acquire images suitable for classification. For ship targets, relatively long observation times are not necessarily a serious problem since time is not as urgent as it would be for aircraft recognition. Sometimes an experienced operator can recognize the ship by simply viewing its ISAR image. In most cases, however, the operator has to employ a more structured technique, especially when there are a large number of possible ship classes to which a target might belong. An operator can use a combination of three different techniques: (1) *measurement of the relative locations* of the major scatterers along the bow-stern axis, such as masts, superstructure breaks, guns, and missile launchers; (2) *feature descriptions* which characterize scattering features by their degree of match to descriptive templates such as the shape of the stern (straight, curved, or rounded) or the type of masts (pole, lattice, or solid); and (3) *shape correlation* by visually comparing on the same display a single ISAR image with a superimposed wire-frame model of the candidate target which has been transformed to match the orientation of the target image produced by the ISAR.

Ships are generally large radar targets so that with a good radar the range at which ship recognition can be performed is basically limited by the horizon. It can be tens of miles with a ship radar, 200 miles with an aircraft radar, or much greater ranges with a spaceborne radar.

The classification of aircraft with ISAR is much more difficult than the ISAR classification of ships. First, an aircraft has many fewer scattering centers than does a ship. With fewer scattering centers or features presented by the unknown target, the recognition decision will be less accurate. (ISAR images of an aircraft have seldom been shown in the past without an outline of the aircraft superimposed to allow the viewer to recognize what is being seen.) Second, an aircraft does not usually experience the relatively large pitch, roll, and yaw motions of a ship, so that the cross-range resolution might not be good enough to isolate the scattering features important for target recognition. Since resolution depends on having a sufficient change of aspect angle, the aircraft must be observed for a relatively long time if it is moving on a straight-line path. Usually a deliberate maneuver by the target aircraft is needed in order to obtain the required change in viewing aspect ($\Delta\theta$) needed for acceptable high-resolution ISAR imaging. Third, aircraft have much smaller radar cross sections than do ships. Therefore, the ranges at which aircraft recognition can be performed, even if the other two limitations are not present, will be much less than those obtained with ships. These difficulties combine to make ISAR imaging of aircraft less attractive than the ISAR imaging of ships.

Although microwave ISAR might not provide aircraft target recognition as well as it does ship recognition, it has been successfully used as a diagnostic tool for understanding the nature of radar scattering from aircraft and the design of low radar cross-section aircraft. Radar images of aircraft in flight can be obtained from both ground-based and airborne radars,[104] so long as sufficient relative motion is achieved between the radar and the target. An ISAR image of a KC 135 jet aircraft (a military version of the Boeing 707) was shown in Fig. 2.18.[105] The aircraft to be imaged flies behind the aircraft carrying the imaging radar in its tail. The radar aircraft maneuvers from one side to the other so that its radar can obtain the relative velocity (and doppler shift) needed to produce an image. Whether one calls this radar an ISAR or a SAR is immaterial. It relies on the relative motion between the imaging radar and the various scatterers that make up the aircraft to be imaged. The advantage of this high-resolution imaging method is that the individual scatterers that contribute to the backscatter echo can be readily recognized and their contributions to the total target cross section can be determined.

An innovative, experimental aircraft-recognition radar is that reported by Steinberg[106,107] to observe commercial aircraft flying into Philadelphia International Airport. It employed a ground-based ISAR to image aircraft passing at relatively close range (typically 3 km). Target aircraft were viewed by the radar in the vicinity of broadside where the rate of change of angle was large so that good cross-range resolution was obtained. A single transmitter was used with two receivers. One receiver was colocated with the transmitter, the other was separated by 25 m. Both were operated as monostatic (single-site) radars. Their physical separation allowed two different target images to be obtained. A third image was obtained by operating the two receiving antennas as an interferometer. The resulting highly processed image of an L-1011 commercial aircraft, with 47-m wingspan and 54-m length, is shown in Fig. 6.27a, along with a plan view drawing of the L-1011 (Fig. 6.27b) for comparison. Figure 6.27a is not that of a single image, as is Fig. 6.27c, but is the superposition of the three images mentioned above (one from each receiver and the two as an interferometer) as well as the inverted images of the three so as to take advantage of the known symmetry of the aircraft about the longitudinal axis. This is an exceptionally good "ISAR" image because of the short range, large signal-to-noise ratio, large change in aspect angle, diversity overlay of multiple images, lack of clutter, and inclusion of the inverted images so as to depict scatterers that might have been masked by the fuselage or the tail. For comparison, several range profiles of the same L-1011 aircraft are shown in Fig. 6.28 at different aspect angles to illustrate the dramatic change in profile with aspect.

Generally, aircraft in normal (nonmaneuvering) flight do not change their aspect angle sufficiently to make good ISAR images with X-band radar. Also there are few

  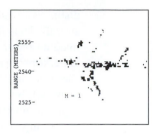

Figure 6.27 (a) ISAR radar image of an L-1011 aircraft made up by superimposing three independent images along with their individual inverted images (see text). (b) Outline drawing of the same aircraft shown for comparison. (c) One of the single images that was used as a part of (a). (Courtesy of Prof. Bernard Steinberg of the Univ. of Pennsylvania.)

Figure 6.28 One-dimensional profiles of the Lockheed L-1011 as a function of aspect angle, obtained from Fig. 6.27a. Head-on aspect occurs at $\alpha = 0$ deg; positive angles view the port side of the aircraft.
(Courtesy of Prof. Bernard Steinberg of the Univ. of Pennsylvania.)

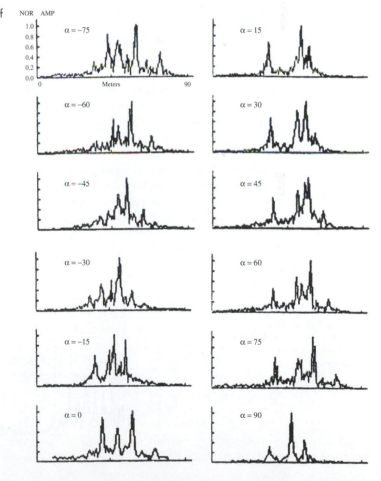

distinctive features in an aircraft image that allow it to be recognized from other similar aircraft. Thus ISAR imaging of aircraft in normal flight has been disappointing. The ISAR imaging of aircraft at *W* band (94 GHz), however, requires only one tenth the change of target aspect necessary at *X* band for the same resolution. Because scattering occurs from variations in the target surface that are comparable to the radar wavelength, it is also likely there will be more scatterers imaged at millimeter waves than at lower frequencies if the signal-to-noise ratio is high enough. In one experimental investigation,[108] it was found that, compared to ISAR images at *X* band, ISAR images of a small Piper Navajo aircraft made at 49 GHz "showed scattering from small details that tended to fill in the target shape and produce an outline view of the target." Furthermore, the target cross section at 49 GHz, when averaged over 360°, was 8 dB greater than the 9-GHz cross section. The largest increase was 19 dB, and was at nose-on incidence. Millimeter-wave ISAR, therefore, should produce better results than does ISAR at the lower frequencies.

Earth-based radars were used in the past to make ISAR images of the moon, Venus, and Mercury; but this technique has been replaced by planetary SAR radars that can orbit close to a planet and obtain superior resolution than an earth-based radar. ISAR for planetary exploration was called *delay-doppler mapping* by the radar astronomer.[109]

Radar Cross-Section Modulations   Radar echoes from moving parts of a target can help recognize one type of target from another or determine the class to which a target belongs. Modulations of the radar backscatter due to propellers, helicopter rotors, jet engines, tank treads, rotating antennas, rotating machinery, the wing beat of birds and insects, and even the heartbeat of a human are examples that might be employed for target recognition. Narrowband (long pulse) as well as wideband (short pulse) waveforms can be used for extracting these modulations.

Propeller Modulation   Aircraft propellers cause a distinctive modulation of the echo signal. The modulation depends on the rotation rate of the engine, the aspect angle, the number of blades on the propeller, and the shape of the propeller. An example of the doppler spectrum obtained with a coherent S-band radar is shown in Fig. 6.29 for a DC-7 aircraft,

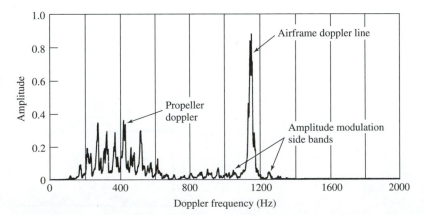

Figure 6.29   Spectrum of the propeller modulation from a DC-7 four-engine commercial aircraft taken with an *S*-band radar and one-second dwell time.
(From R. Hynes and R. E. Gardner of the Naval Research Laboratory.[110,116])

a four-engine propeller-driven commercial aircraft.[110,111] The "airframe doppler-line" is the doppler frequency shift from the aircraft itself. The low-level amplitude modulation of the airframe doppler signal is caused by the propeller blade "chopping" a portion of the radar energy reflected from the airframe. At a frequency lower than the airframe line, there are spectral lines due to the radar echo from the rotating propeller blades themselves.

Propeller modulation has not been used to recognize one type of propeller aircraft from another, but its absence can be used to readily distinguish a jet-engine aircraft from a propeller-driven aircraft. In a high-resolution imaging radar, propeller modulation might locate the position of the engines, which could be helpful for target recognition.

Helicopter Blade Modulation   Helicopters can be distinguished from fixed-wing aircraft by the characteristic modulation of the radar echo that comes from the rotating blades and the rotating hub structure supporting the blades.

A "blade flash" occurs every time one of the rotating blades is perpendicular to the radar beam direction.[112,113] In this position the radar cross section is a maximum. For a two-blade rotor, there are two flashes per revolution. Rotors with an even number of blades produce a number of flashes per rotation equal to the number of blades. An odd number of blades produce twice the number of flashes per revolution as the number of blades. (This can be verified by drawing a sketch and noting that there will be a separate flash from the front and rear when there are an odd number of blades, but a simultaneous flash from a front and a rear with an even number of blades.) The linear speed at the tip of the blade of helicopter is not highly dependent on the type of helicopter. It varies from about 210 to 230 m/s. If L = the length of the blade in meters, N = number of blades, and the speed of the tip is taken to be 210 m/s, the blade period in seconds for an even number of blades is

$$T_B = \frac{2\pi L}{210N} \qquad [6.50]$$

The period for an odd number of blades is one half that for an even number. Collot[112] points out that the blade period can be used to separate one type of helicopter from another since the rotation speed is constant whatever the type of flight.

Collot also states that the radar cross section, in square meters, for the blade flash is approximately aL^2/λ, where L = length of the blade, λ = wavelength, and the constant a = 0.5 for the front edge of the blade and 0.1 for the trailing edge. (The front edge has a higher cross section than the trailing edge since it is blunt and the trailing edge is sharp.[114]) Because of the difference between front and rear cross sections, a rotor with an odd number of blades will have alternating values for the blade flash (since the front edge and rear edge are not seen simultaneously). The echoes from the blade flash with an even number of blades, however, will be the same strength each time. The duration of the blade flash depends on the rotor speed, the length of the blade, and the radar wavelength. Assuming that the reflection from the blades can be approximated by a $(\sin x)/x$ relation and that the angular extent of the reflected echo is λ/L radians, the duration of the blade flash is approximately $t_f = \lambda/420$ for an even number of blades and twice this for an odd number (again assuming that the velocity of the blade tip is 210 m/s). At X band, with λ = 3.2 cm, the duration of the flash with an even number of blades is about 75 μs.

The intermittent nature of the short-duration helicopter blade flash might go unnoticed if the time on target is less than the time between flashes [which is the blade period of Eq. (6.50)]. In order that the helicopter blade flash be intercepted by the radar, a high prf and a long time on target are required.

The frequency spectrum of the radar echo from a helicopter has energy about the airframe doppler line due to reflections from the rotating hub that holds the blades. The helicopter spectrum will be asymmetrical since the echo from the approaching blade will be at a higher frequency than the airframe line and be of a larger amplitude than the echo from the receding blade at a lower frequency. Collot states that recognizing that an echo is from a helicopter and not a fixed wing aircraft is "immediate," and the recognition "of the type of helicopter is given in a very short time with a probability better than 95 percent." (He does not state, however, how many different types of helicopters were involved in obtaining this probability.)

In addition to the characteristic features mentioned above, the classification of helicopters might also include the echo characteristics of the tail rotor, whether there are single or twin rotors, and the configuration of the rotating hub.[115] Bullard and Dowdy[113] point out that to recognize a helicopter based on the spectral characteristics of its echo signal, the radar signal sampling must be at the Nyquist rate or higher to prevent aliasing. For X-band radar they state that a minimum prf of 30 kHz is required.

Jet-Engine Modulation (JEM) The radar echo from a jet-engine aircraft will be modulated by the engine's rotating compressor when looking in the vicinity of the nose. There might also be radar echo modulations from the turbine when looking from the rear of the aircraft; but these usually have a much smaller echo than the echo from the compressor.[110] The characteristic modulations of the radar echoes from aircraft can be used for recognizing one type of aircraft from another, or more correctly, one type of aircraft engine from another. Even though the jet engines are set back at a distance from their air intake and are entirely enclosed except for intake and exhaust ducts, there usually is sufficient propagation down the duct at microwave frequencies to obtain an echo signal from the compressors. Radar echoes from the compressors are sometimes obtained out to angles 60 deg from the head-on.[116]

An example of the *jet-engine modulation* (JEM) produced by a multiple-engine jet viewed head-on by an X-band radar is shown in Fig. 6.30.[110] Shown in this spectrum are the airframe line and the lower frequency sidebands of the compressor modulation. The component labeled c is displaced from the airframe line by a frequency $\Delta f = nb_c$, where n is the number of revolutions per second of the compressor and b_c is the number of blades on the first-stage compressor. There are other engine spectral components that are displaced from the airframe doppler line by integer multiples of Δf. The upper sidebands (not shown here) are symmetrical in frequency with the lower sidebands (with reference to the airframe line), but are lower in amplitude. Because of the high rotation speed of the engine components, aircraft jet-engine modulations are likely to be as high as ten to twenty kilohertz at X band. This requires high sampling frequencies (prfs) to avoid aliasing.

Mathematical analysis of jet-engine modulations indicates that the overall JEM signal can be decomposed into an amplitude modulation component and an angle (phase or frequency) modulation component that can be considered separately. The details of the line spectrum of

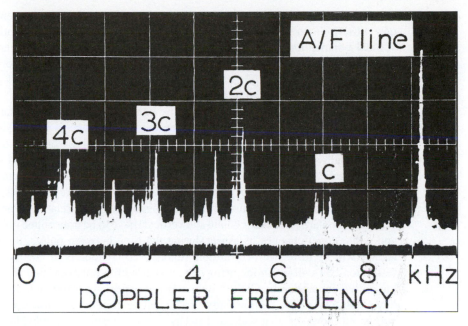

Figure 6.30 Spectrum of a multi-engine jet aircraft taken at 0 deg aspect angle showing the airframe line due to the aircraft's relative velocity and modulations from the compressor. (From R. Hynes and R. E. Gardner of the Naval Research Laboratory.[110,116])

the echo signal depend on many things including the number of blades on the first and second stages of the compressor. A minimum time of observation is needed to obtain a meaningful spectrum suitable for target recognition, which is said to be at least 25 ms.[117]

Polarization Response and Target Recognition   The radar echo from a target can be affected by the polarization of the radar signal. Many research engineers have hoped that the differences in the echo from targets viewed with different polarizations might be used as a means for distinguishing one from another.[118] This is possible in simple cases, as, for example, distinguishing between a long thin rod and a sphere by using a rotating linear polarization (rotating electric field vector).

In the more general situation with realistic targets, such as aircraft, there have been many attempts to use the polarization matrix of the target echo and other polarization descriptors[119–121] for target recognition. Target recognition by means of polarization generally requires measurement of what is called the *polarization matrix*. This is obtained from the radar echo signals received on both horizontal and vertical polarization when horizontal is transmitted, and on the received horizontal and vertical polarization signal when vertical is transmitted. The polarization matrix is a 2×2 complex scattering operator (phase as well as amplitude) that characterizes a target's scattering properties. It may be expressed as

$$S = \begin{bmatrix} S_{HH} & S_{VH} \\ S_{HV} & S_{VV} \end{bmatrix}$$

[6.51]

where the first subscript represents the polarization of the receiving antenna and the second subscript is the transmitted polarization, H is for horizontal polarization and V is for vertical polarization. For example, S_{HV} is the complex target backscatter signal (amplitude and phase) received with a vertically polarized antenna when a horizontally polarized signal is transmitted. The HH and VV are both *co-polar* components and the HV and VH are both *cross-polar* components. In general $HV = VH$.

Although theory might indicate that the polarization response of a target depends on the nature of the target, there has not been the desired success in applying polarimetric methods for practical target recognition. There are several reasons why this has been so: (1) Multipath signals from the surface of the ground or from clutter and structures along the propagation path can modify the polarization of the signal. (2) Practical antennas are limited in the purity of their polarization response, and there is always some finite response to the cross polarization which sometimes might be relatively large. (3) The two orthogonal polarizations (horizontal and vertical) have to be transmitted on different pulses; and if the target changes its aspect during the time between pulses, the polarization response will be distorted. (4) Multiple (unresolved) scatterers or targets within the radar resolution cell will result in a composite signal with a polarization different from any of the individual scatterers. Giuli[122] in his classic paper on polarization diversity states that low-resolution radars "intrinsically provide quite limited classification properties." He points out, however, that wideband radars that can resolve the individual scatterers of a target might make exploitation of the polarization information more profitable by avoiding the changes in echo polarization caused by unresolved scatterers.

Even though the use of polarimetry for target recognition has excited much interest, one can conclude that as a means for recognizing complex targets its effectiveness in the past has been disappointing.

Other Radar Target Recognition Methods   Several other radar methods for the recognition of targets are briefly mentioned below. Most can be thought of in terms of aircraft recognition, although they are not limited to aircraft.

High Range-Resolution with Monopulse[123]   The use of monopulse radar for target tracking in angle was discussed in Sec. 4.2. If the monopulse radar has sufficiently high range-resolution so as to isolate the major scattering centers of a target, then a measurement of the angular location of each scatter in azimuth and elevation can be made. Thus high range-resolution with monopulse can provide a 3D-like "image" of a target that might be used for target recognition. In addition to requiring high range-resolution, a limitation is that the range must be short enough so that angle measurements can be made with enough accuracy to obtain adequate cross-range measurements of the major scatterers.

Fluctuations of Radar Cross Section with Aspect Angle   Section 2.7 indicated that small changes in the aspect angle of complex (multiple scatterer) targets can cause major changes in the radar cross section. This has been considered in the past as a means for target recognition, but it has not had much success. In some respects this method is related to the imaging of moving targets by means of ISAR, discussed previously, except that the highly successful ISAR technique employs both phase and amplitude. Only amplitude information is available when cross-section variations are observed.

Resonance Region Response   There have been several methods proposed for target recognition based on using a number of distinct frequencies in the resonance region of target scattering. (In the resonance region the dimensions of the target are comparable to the radar wavelength.) One method is derived from the ramp response of a target and develops a so-called *feature space* based on the amplitude, phase, and/or polarization of each distinct frequency that observes the target.[124,125] A related method employs the complex *natural resonances,* or *poles,* of the target to characterize the target.[126,127] The *Singularity Expansion Method* has also been applied to the target recognition method that employs frequencies in the resonance region.[128]

These methods have a good theoretical basis; they generally do not depend on knowing the target aspect angle; and they have been tested with computer simulations and experimental model measurements at scaled frequencies. They are not practical, however, for most applications of target recognition since they require frequencies in the resonance region, which for aircraft targets would be in the HF portion of the spectrum. For ships, even lower frequencies would be required. These recognition methods depend on the so-called *late-time scattering* due to creeping waves that travel around the target. Even if long radar wavelengths were no problem, it has been pointed out[129] that the effect of noise can mask the creeping waves that occur beyond the first creeping wave so that there might not be enough information to determine the complex resonances.

Another related method is to transmit a radar waveform that is based on knowing ahead of time the natural resonances of the target so that the desired target echo signal can be readily recognized from the echo signals from targets with a different set of natural resonances. One such method is called *K-pulse;* another is called *E-pulse.*[130] They require that the radar transmit a special signal for each target to be recognized. Rather than transmit special waveforms it might be better to transmit a more normal wideband radar pulse and perform target recognition by convolution processing in the receiver to recognize a particular target signal.[131]

Since they have been difficult to apply in practice, the above resonance region methods have been more of academic interest than for practical target recognition applications.

Nonlinear Scattering Effects   When a metal comes into contact with another metal, oxides can form and the junction can act as a nonlinear diode. A radar echo signal reflected from such a metal-to-metal junction will contain higher harmonic components, with the third harmonic usually being the largest.[132–134] Thus if a frequency f_1 is transmitted, detection of the third harmonic, $3f_1$, indicates that the scatterer has metal-to-metal contacts. This nonlinear response has been of interest as a possible method for recognizing certain types of metallic targets. If the third harmonic of the echo signal is used, the transmitted signal must not have significant third-harmonic radiation of its own. A method for avoiding the third harmonic problem is to simultaneously transmit two frequencies f_1 and f_2, and tune the receiver to a strong cross-product such as $2f_1 \pm f_2$. The use of nonlinear target properties for radar has sometimes been called METRRA, which stands for *Metal Reradiating Radar.*

The amount of signal returned from a nonlinear contact at a harmonic frequency is a nonlinear function of the incident field strength. Thus the nonlinear target cross section depends on the peak power. The normal radar equation does not apply.[135,136] The range dependence in the radar equation might be as the sixth power or higher, instead of the usual fourth power. Because of the nonlinearity, in this specialized case high peak power is more important for detection than is high average power.

Although radar detection based on harmonics produced by nonlinear metal-to-metal contacts has some interesting attributes for applications (such as detection of stationary targets in clutter), it is difficult to apply since the echo signals at frequencies other than those transmitted can be very weak.

A related effect is based on the modulation of the scattered signal when metal-to-metal contacts are opened and closed so that the current distribution on a metal target is modified.[137] An example is the intermittent contacts due to the moving wheels of a train. This effect is sometimes called RADAM, which means *Radar Detection of Agitated Metals*.[138] The echo signals from RADAM are usually too weak to be of interest for satisfying the needs of target recognition.

Target Track History   In military operations it is important to determine whether a target held in track by a radar is a friend or is a hostile threat. There are several military recognition or identification methods that might be used for separating friend from foe; but the radar itself might be able to assist in the recognition task based on the history of a target's track. If one knew where an aircraft or ship originated, where it has been, and what it is currently doing, it ought to be possible in many cases to determine whether the target is potentially hostile or not. A radar with a good automatic tracking system and with adequate memory is required, as well as logic to recognize the trajectories of potentially hostile targets from ones that are not.

In some cases, military aircraft recognition is accomplished by having friendly aircraft fly in pre-designated corridors in a known manner that the radar is able to observe.

Signal Intercept and Direction Finding Combined with Radar   So far, this section has been concerned with radar methods for recognizing one target from another. Sometimes, however, the use of radar with other sensors can have a synergistic effect on the quality of target information. An example is the use of an electronic warfare intercept receiver (a part of ESM, or Electronic Support Measures*) and direction finder (DF) in combination with a radar. The intercept receiver might be able to recognize the type of target by the characteristic signals it radiates. The direction to the target can be obtained by the DF, but electronic warfare equipment usually cannot obtain the target's location in range. Radar, on the other hand, provides target location and track. The two together can locate, track, and obtain target recognition if the radar target track can be properly associated with the DF angle track. The angle accuracy of ESM/DF is often much poorer than the angle accuracy of radar. Nevertheless, algorithms can be developed that allow the ESM/DF information to be associated with a target under track by the radar so that target recognition provided by the ESM can be applied to the radar target under track.[139–141]

| *ESM is also known as Electronic Warfare Support, or ES.

Target Recognition Applications   There are many areas of application where the radar recognition of a target echo signal is important. Some have already been mentioned previously in this subsection, others are briefly described below. All of these require the intelligent extraction of information from a radar signal, a radar capability which this chapter has tried to introduce.

Military Combat Identification   Studies of fratricide, which is the unintentional infliction of casualties on one's own forces, show that fratricide occured in past military hostilities at rates that typically vary from 10 to 20 percent, or greater.[142] Casualties from "friendly fire" have been considered by the military as an unwelcome consequence of war that needs to be kept to a minimum. To minimize the fratricide rate, military forces employ a number of recognition and identification methods as well as strict rules of engagement. Reliable, accurate, and secure combat identification is necessary to reduce fratricide. Both cooperative and noncooperative methods are employed in combination to increase the probability of a correct decision in a timely manner. The particular methods differ depending on whether the target is an aircraft, ship, ground vehicle, or the soldier on the ground. Radar target-recognition methods include engine modulations, track history, SAR, and ISAR.

Ballistic Missile Target Discrimination   Defending against a single ballistic missile is a demanding task, but it can be done. The difficulty in ballistic missile defense escalates tremendously when the reentry vehicle carrying the warhead is accompanied by *penetration aids* that might consist of the spent booster tank or the many fragments from the tank after it has been deliberately exploded, the very many pieces of chaff that can be launched in space to accompany the reentry vehicle, solid fuel fragments, separation debris, and the deliberate use of inflated decoys that resemble the reentry vehicle. One of the fundamental challenges in ballistic missile defense, therefore, is to determine which element in the multi-element threat complex is the lethal object to be destroyed or made ineffective.[143] If the reentry vehicle is engaged by the defense after it reenters the earth's atmosphere, the drag introduced by the atmosphere causes the lightweight penetration aids to slow down much faster than the heavier high-speed reentry body. Thus the atmosphere separates the warhead from the penetration aids. A serious limitation of this tactic is that it results in a relatively small defended area. If, on the other hand, the ballistic missile threat is engaged outside of the atmosphere so as to achieve a large defended area, then there needs to be some way in which the threatening warhead can be distinguished from the "junk" that accompanies it. The use of radar in such a role is called *ballistic missile target discrimination.*

Meteorological Observation   The original weather radars used for many years by the National Weather Service determined the presence of rainfall and estimated the rainfall rate. These were replaced in the early 1990s with the *S*-band Nexrad (WSR-88D) doppler weather radar,[144] which was designed to obtain much more information about the weather, other than it is raining somewhere. In addition to determining rainfall as a function of azimuth and height, Nexrad employs the doppler frequency shift to estimate wind speed as a function of direction and height in order to detect and measure damaging winds, severe turbulence, dangerous wind shear, and recognize the onset of tornadoes. Nexrad differentiates hail from heavy rainfall, determines the tops of thunderstorms (an indicator of the intensity of such storms), and detects and tracks severe storms and mesocyclones. There are about 40 different weather products that can be generated by the WSR-88D radar.[145] These are processed automatically and displayed to an operator for further action. The extensive processing is such that the trained operator need not be a professional meteorologist to readily employ the information presented by the radar. Nexrad is a good example that radar need no longer be a "blob" detector!

Battlefield Surveillance   The *X*-band Joint STARS airborne radar system employs a SAR mode for generating a maplike image of the battlefield and a GMTI mode for detecting moving targets. These two modes allow the recognition of terrain features, roads, fixed structures, military forces, artillery, and stationary as well as moving vehicles. Detection of moving vehicles is readily accomplished with modern GMTI methods. The radar detection of stationary ground targets of military interest is a classical radar application, but recognizing what type of target is being seen by the radar can be difficult. To be useful for target recognition, a radar should be able to distinguish tanks from trucks and from artillery. It would be even better if the type of tank or type of artillery could be determined.

Vehicles are difficult to recognize with radar since they have only a relatively few distinctive scattering centers. This is in contrast to a ship target that has many scattering centers, which makes recognizing one class of ship from many other ship classes achievable based on its ISAR image. Although the writer is not aware of a documented source defining the minimum number of distinctive scatterers needed for classifying targets, it appears that perhaps a dozen (more or less) individual scatterers might be necessary in a target image for reliable classification. The minimum number will depend, of course, on the type of target and the number of different target classes that have to be sorted. The number of scatterers defining a target might have to be greater if there are a large number of different classes of targets that have to be distinguished from one another. With both military aircraft and ships, the number of different classes that might be encountered range from many tens of target classes to more than a hundred if worldwide operation is expected.

As mentioned previously, determining the type of military land target seen by a SAR is called *automatic target recognition,* or ATR.[146] Radar ATR of land targets of military interest is complicated by the military tactic of hiding targets among the trees and other vegetation that can contaminate the target's echo signature as well as interfere with detection when the clutter echo is large. Thus ATR involves some method of STI (stationary target indication) to detect targets in the midst of clutter as well as foliage penetration. Foliage penetration requires operation at the lower radar frequencies, such as VHF, where the attenuation in propagating through trees and underbrush is lower than at microwave frequencies. Recognition of targets also requires very wide bandwidths to isolate the target's scatterers. Wide bandwidth combined with VHF results in ultrawideband (UWB) SAR.[147]

The interpretation of SAR images of terrain and populated areas has also been used in the civilian world for adding to the base of Geographic Information Systems (GIS), which use such information for land usage and planning, urban development, natural resource management, and topography.

Other Applications of Radar Information Extraction   Radar has been used for recognizing one species of birds from another by means of the modulation on the bird's radar echo signals thought to be due to the characteristic beating of its wings.[148] Doppler radar can be used to detect the heartbeat and breathing of humans, even when located behind a cinder-block wall.[149,150] Radars can readily detect buried land mines, but the challenge has been to recognize the echo of a mine from the echoes from the many other underground objects that can be present, such as rocks, tree roots, debris, and changes in the underground surface characteristics.

REFERENCES

1. Skolnik, M. I. "Radar Information from the Partial Derivatives of the Echo Signal Phase from a Point Scatterer." Naval Research Laboratory, Washington, D.C., Memorandum Rep. 6148, February 17, 1988.

2. Evans, J. V., and T. Hagfors. *Radar Astronomy,* New York: McGraw-Hill, 1968, Sec. 5.9.

3. Jurgens, R. F. "Earth-Based Radar Studies of Planetary Surfaces and Atmospheres." *IEEE Trans.* GE-20 (July 1982), pp. 293–305.

4. McDonough, R. N., and A. D. Whalen. *Detection of Signals in Noise.* San Diego, CA: Academic, 1995.

5. Goldman, S. *Frequency Analysis, Modulation, and Noise.* New York: McGraw-Hill, 1948, p. 281.

6. Slepian, D. "Estimation of Signal Parameters in the Presence of Noise." *IRE Trans.* no. PGIT-3 (March 1954), pp. 68–89.

7. Manasse, R. "Range and Velocity Accuracy from Radar Measurements," unpublished internal report dated February 1955, MIT Lincoln Laboratory, Lexington, MA. (Not generally available.)

8. Mallinckrodt, A. J., and T. E. Sollenberger. "Optimum Pulse-Time Determination." *IRE Trans.* no. PGIT-3 (March 1954), pp. 151–159.

9. Weiss, A. J. "Composite Bound on Arrival Time Estimation Errors." *IEEE Trans.* AES-22 (November 1986), pp. 751–756.

10. Torrieri, D. J. "Arrival Time Estimation by Adaptive Thresholding." *IEEE Trans.* AES-10 (March 1974), pp. 178–184.

11. Torrieri, D. J. "Adaptive Thresholding Systems." *IEEE Trans.* AES-13 (May 1977), pp. 273–280.

12. Ho, K. C., Y. T. Chan, and R. J. Inkol. "Pulse Arrival Time Estimation Based on Pulse Sample Ratios." *IEE Proc.* 142, Pt. F (August 1995), pp. 153–157.

13. Varian, R. H., W. W. Hansen, and J. R. Woodyard. "Object Detecting and Locating System." U.S. Patent 2,435,615, February 10, 1948.

14. Skolnik, M. *Introduction to Radar Systems,* 2nd ed. New York: McGraw-Hill, 1980, Sec. 3.5.

15. Wadley, T. L. "Electronic Principles of the Tellurometer." *Trans. South African Inst. Elec. Engrs.* 49 (May 1958), pp. 143–161; discussion pp. 161–172.

16. Manasse, R. "Range and Velocity Accuracy from Radar Measurements." Internal memo dated February 1955, MIT Lincoln Laboratory, Lexington, MA. (Not generally available.)

17. *Encyclopaedia Britannica* vol. 12, 15th ed., Chicago, IL, 1991, p. 125.

18. Sinsky, A. I., and C. P. Wang. "Standardization of the Definition of the Radar Ambiguity Function." *IEEE Trans.* AES-10 (July 1974), pp. 532–533.

19. IEEE Standard 686-1990, Radar Definitions, April 20, 1990.

20. Woodward, P. M. *Probability and Information Theory, with Applications to Radar.* New York: McGraw-Hill, 1953, chap. 7.

21. Siebert, W. McC. "A Radar Design Philosophy." *IRE Trans.* IT-2 (September 1956), pp. 204–221.

22. Deley, G. W. "Waveform Design." In *Radar Handbook,* 1st ed., M. Skolnik (Ed.). New York: McGraw-Hill, 1970, p. 3-17.

23. Rihcazek, A. W. *Principles of High-Resolution Radar.* New York: McGraw-Hill, 1969, Sec. 5.3.

24. Woodward, P. M. Ref. 20, Sec. 7.2.

25. Dicke, R. H. "Object Detection System." U.S. patent no. 2,624,876, issued Jan. 6, 1953.

26. Klauder, J. R., et al. "The Theory and Design of Chirp Radars." *Bell System Technical J.* 39 (July 1960), pp. 745–808.

27. Cook, C. E., and M. Bernfeld. *Radar Signals.* Norwood, MA: Artech House, 1993.

28. Skolnik, M. *Introduction to Radar Systems.* 2nd ed. New York: McGraw-Hill, 1980, Sec. 3.3.

29. Borkowski, T. T. "Solid-State Transmitters." In *Radar Handbook,* 2nd ed. M. Skolnik (Ed.) New York: McGraw-Hill, 1990, Chap. 5.

30. Cook, C. E., and M. Bernfeld. Ref. 27, Sec. 7.3.

31. Caputi, W. J., Jr. "Stretch: A Time-Transformation Technique." *IEEE Trans.* AES-7 (March 1971), pp. 269–278.

32. Holt, D. J., and M. B. Fishwick. "Analog Waveform Generation and Processing." *Electronic Progress* 17, no. 1 (Spring 1975), pp. 2–16. (Published by Raytheon, Lexington, MA.)

33. Brookner, E. *Aspects of Modern Radar.* Norwood, MA: Artech House, 1988, pp. 25–28.

34. Farnett, E. C., T. B. Howard, and G. H. Stevens. "Pulse-Compression Radar." In *Radar Handbook,* 1st ed., M. Skolnik (Ed.). New York: McGraw-Hill, 1970, chap. 20.

35. Hartmann, C. S. "SAW Device Technology: Recent Advances and Future Trends." *Microwave J.* 1990 State of the Art Reference, pp. 73–89.

36. Maines, J. D., and E. G. S. Paige. "Surface-Acoustic-Wave Devices for Signal Processing Applications." *Proc. IEEE* 64 (May 1976), pp. 639–652.

37. Cambell, Colin. *Surface Acoustic Wave Devices and Their Signal Processing Applications.* New York: Academic, 1989.

38. Information obtained from Anderson Laboratories, Bloomfield, CT, brochure titled "Dispersive Delay Lines," 1986.

39. Martin, T. A. "Low Sidelobe IMCON Pulse Compression." *Proc. 1976 IEEE Ultrasonics Symposium,* pp. 411–414, IEEE Cat. #76 CH1120-5SU.

40. Farnett, E. C., and G. H. Stevens. "Pulse Compression Radar." In *Radar Handbook,* 2nd ed. M. Skolnik (Ed.). New York: McGraw-Hill, 1990, chap. 10.

41. Wehner, D. R. *High-Resolution Radar,* 2nd ed. Boston, MA: Artech House, 1995, Sec. 4.7.

42. MacWilliams, F. J., and N. J. A. Sloan. "Pseudo-Random Sequences and Arrays." *Proc. IEEE* 64 (December 1976), pp. 1715–1729.

43. Farnett, E. C., and G. H. Stevens. Ref. 40, Sec. 10.6.

44. Golomb, S. W. *Shift Register Sequences.* rev. ed. Laguna Hills, CA: Aegean Park, 1982.

45. Taylor, S. A., and J. L. MacArthur. "Digital Pulse Compression Radar Receiver." *APL Technical Digest* 6 (March/April 1967), pp. 2–10.

46. Turin, R. "Sequences with Small Correlation." In *Error Correcting Codes.* H. B. Mann (Ed.). New York: John Wiley, 1968, pp. 195–228.

47. Linder, J. "Binary Sequences Up to Length 40 With Best Possible Autocorrelation Function." *Electron. Lett.* 11 (October 10, 1975), p. 507.

48. Kerdock, A. M., R. Mayer, and D. Bass. "Longest Binary Pulse Compression Codes with Given Peak Sidelobe Levels." *Proc. IEEE* 74 (February 1986) p. 366.

49. Taylor, J. W., Jr., and H. J. Blinchikoff. "Quadriphase Code—A Radar Pulse Compression Signal with Unique Characteristics." *IEEE Trans.* AES-24 (March 1988), pp. 156–170.

50. Levanon, N., and A. Freedman. "Ambiguity Function of Quadriphase Coded Radar Pulse." *IEEE Trans.* AES-25 (November 1989), pp. 848–853.

51. Frank, F. L. "Polyphase Codes with Good Nonperiodic Correlation Properties." *IEEE Trans.* IT-9 (January 1963), pp. 43–45.

52. Cook and Bernfeld, Ref. 27, Sec. 8.4.

53. Nathanson, F. E. *Radar Design Principles.* 2nd ed. New York: McGraw-Hill, 1991, Sec. 12.5.

54. Lewis, B. L., F. F. Kretschmer, Jr., and W. W. Shelton. *Aspects of Radar Signal Processing.* Norwood, MA: Artech House, 1986, Chap. 2.

55. Lewis, B. L. "Range-Time Sidelobe Reduction Technique for FM-Derived Polyphase PC Codes." *IEEE Trans.* AES-29 (July 1993), pp. 834–840.

56. Nathanson, F. E. Ref. 53, Sec. 13.5.

57. Costas, J. P. "A Study of a Class of Detection Waveforms Having Nearly Ideal Range-Doppler Ambiguity Properties." *Proc. IEEE* 72 (August 1984), pp. 996–1009.

58. Chang, W., and K. Scarbrough. "Costas Arrays with Small Number of Cross-Coincidences." *IEEE Trans.* AES-25 (January 1989), pp. 109–112.

59. Golomb, S. W., and H. Taylor. "Constructions and Properties of Costas Arrays." *Proc. IEEE* 72 (September 1984), pp. 1143–1163.

60. Nathanson, F. E. Ref. 53, Sec. 13.11.

61. Farnett, E. C., and G. H. Stevens. Ref. 40, Sec. 10.4.

62. Cook, C. E., and M. Bernfeld. Ref. 27, Sec. 6.8.

63. Minkoff, J. *Signals, Noise, and Active Sensors.* New York: John Wiley, 1992, Sec. 9.2.

64. Thor, R. C. "A Large Time-Bandwidth Pulse-Compression Technique." *IEEE Trans.* MIL-6 (April 1962), pp. 169–173. Reprinted in *Radars* vol. 5, Pulse Compression, D. K. Barton (Ed.). Boston, MA: Artech House, 1875.

65. Kroszcynski, J. J. "Pulse-Compression by Means of Linear-Period Modulation." *Proc. IEEE* 57 (July 1969), pp. 1260–1266.

66. Minkoff, J. Ref. 63, Sec. 9.3.

67. Rowlands, R. O. "Detection of a Doppler-Invariant FM Signal by Means of a Tapped Delay Line." *J. Acoust. Soc. Am.* 37 (April 1965), pp. 608–615.

68. Altes, R. A., and E. L. Titlebaum. "Bat Signals as Optimally Doppler Tolerant Waveforms." *J. Acoust. Soc. Am.* 48 (1970), pp. 1014–1020.

69. Altes, R. A. "Radar/Sonar Acceleration Estimation with Linear-Period Modulated Waveforms." *IEEE Trans.* AES-26 (November 1990), pp. 914–923.

70. Kramer, S. A. "Doppler and Acceleration Tolerances of High-Gain, Wideband Linear FM Correlation Sonars." *Proc. IEEE* 55 (May 1967), pp. 627–636.

71. Golomb, S. W. Ref. 44, Chap. VI.

72. Belyayev, V. S. "A New Pseuodorandom, Phase-Controlled Signal Based Upon a Nonlinear Sequence and the Possibilities for Generating It." *Radiophysics and Quantum Electronics* 34 (March 1991), pp. 285–287.

73. Golay, M. J. E. "Complementary Series." *IRE Trans.* IT-7 (June 1960), pp. 82–87.

74. Levanon, N. *Radar Principles.* New York: John Wiley, 1988, pp. 159–162.

75. Cloke, J. A. "Ambiguity Function of Complementary Series." *IEE International Conf., Radar-82,* October 18–20, 1982, pp. 477–481.

76. Welti, G. R. "Quaternary Codes for Pulse Radar." *IRE Trans.* IT-7 (June 1960), pp. 400–408.

77. Nathanson, F. E. Ref. 53, p. 564.

78. Cook, C. E., and M. Bernfeld. Ref. 27, pp. 264–269.

79. Kretschmer, F. F., and F. C. Lin. "Huffman-Coded Pulse Compression Waveforms." Naval Research Laboratory, Washington, D.C., Report 8894, May 23, 1985.

80. Eves, J. L., and E. K. Reedy. *Principles of Modern Radar.* Van Nostrand Reinhold, New York, 1987, Sec. 15.3.2.

81. Key, E. L., E. N. Fowle, and R. D. Haggarty. "A Method of Sidelobe Suppression in Phase-Coded Pulse Compression Systems." MIT Lincoln Laboratory, Lexington, MA, TR-209, August 28, 1959.

82. Nathanson, F. E. Ref. 53, Sec. 12.4.

83. Ackroyd, M. H., and F. Ghani. "Optimum Mismatched Filters for Sidelobe Suppression." *IEEE Trans.* AES-9 (March 1973), pp. 214–218.

84. Eves and Reedy, Ref. 80, Sec. 15.5.2.

85. Morgan, G. B., P. Dassanayake, and O. A. Liberg. "The Design and Performance of Transversal Filters for Sidelobe Reduction of Pulses Compressed from Combined Barker Phase Codes." *The Radio and Electronic Engineer* 51 (June 1981), pp. 272–280.

86. Nathanson, F. E. Ref. 53, pp. 556–557.

87. Baden, J. M., and M. N. Cohen. "Optimal Peak Sidelobe Filters for Biphase Pulse Compression." *Record of the 1995 IEEE International Radar Conf.,* IEEE Catalog No. 90CH2882-9, pp. 249–252.

88. Castella, F. R., and S. A. Rudie. "Detection Performance of Phase-Coded Radar Waveforms with Various Types of Limiting." *IEE Proc.* 136, Pt. F (June 1989), pp. 118–121.

89. Cahn, C. R. "A Note on Signal-to-Noise Ratio in Band-Pass Limiters." *IEEE Trans.* IT-7 (January 1961), pp. 39–43.

90. Bogotch, S. E., and C. E. Cook. "The Effect of Limiting on the Detectability of Partially Time-Coincident Pulse Compression Signals." *IRE Trans.* MIL-9 (January 1965), pp. 17–24.

91. Woerrlein, H. H. "Capture and Spurious Target Generation Due to Hard Limiting in Large Time-Bandwidth Product Radars." Naval Research Laboratory, Washington, D.C., Report 7001, December 22, 1969.

92. Maric, S. V., I. Seskar, and E. L. Titlebaum. "On Cross-Ambiguity Properties of Welsh-Costas Arrays." *IEEE Trans.* AES-30 (October 1994), pp. 1063–1071.

93. Deng, H. "Synthesis of Binary Sequences with Good Autocorrelation and Cross-correlation Properties by Simulated Annealing." *IEEE Trans.* AES-32 (January 1996), pp. 98–107.

94. Shrader, W. W., and V. Gregers-Hansen. *Radar Handbook,* M. Skolnik (Ed.). New York: McGraw-Hill, 1990, "MTI Radar," Chap. 15, pp. 15.55–15.57.

95. Arthur, J. W. "SAW Pulse Compression in Modern Multi-Channel Radar Applications." *Microwave J.* 29 (January 1986), pp. 159–169.

96. Dixon, R. C. *Spread Spectrum Systems.* New York: Wiley Interscience, 1976.

97. Geffe, P. R. "Open Letter to Communications Engineers." *Proc. IEEE* 55 (December 1967), p. 2173.

98. Skolnik, M. I. "Radar's Environmental Role." *IEEE Potentials* 10 (April 1991), pp. 13–16.

99. Linde, G. J. "Use of Wideband Waveforms for Target Recognition with Surveillance Radars." *Record of the IEEE 2000 International Radar Conf.* May 7–12, 2000, Washington, D.C., pp. 128–133. See also Linde, G. J., and C. V. Platis. "Target Recognition with Surveillance Radar." *NRL Review.* Naval Research Laboratory, Washington, D.C., pp. 118–120, 1995.

100. Hudson, S., and D. Psaltis. "Correlation Filters for Aircraft Identification from Radar Range Profiles." *IEEE Trans.* AES-29 (July 1993), pp. 741–748.

101. Zyweck, A., and R. E. Bogner. "Radar Target Classification of Commercial Aircraft." *IEEE Trans.* AES-32 (April 1996), pp. 598–606.

102. Cutrona, L. J. "Synthetic Aperture Radar." *Radar Handbook,* M. Skolnik (Ed.). New York: McGraw-Hill, 1990, Chap. 21.

103. Kerr, D., S. Musman, and C. Bachmann. "Automatic Recognition of ISAR Ship Images." *IEEE Trans.* AES-32 (October 1996), pp. 1392–1404.

104. Jain, A., and I. Patel. "Dynamic Imaging and RCS Measurements of Aircraft." *IEEE Trans.* AES-31 (January 1995), pp. 211–226.

105. Harris, et al. "Dynamic Air-to-Air Imaging Measurement System." *Conf. Proceedings of the 14th Annual Meeting of the Antenna Measurements Techniques Association,* October 19–23, 1992, pp. 6-11 to 6-16.

106. Steinberg, B. D. "Microwave Imaging of Aircraft." *Proc. IEEE* 76 (December 1988), pp. 1578–1592.

107. Steinberg, B. D., D. L. Carlson, and W. Lee. "Experimental Localized Radar Cross Sections of Aircraft." *Proc. IEEE* 77 (May 1989), pp. 663–669.

108. Dinger, R., et al. "Measurements of the Radar Cross Section and Inverse Synthetic Aperture Radar (ISAR) Images of a Piper Navajo at 9.5 GHz and 49 GHz." Naval Command Control and Ocean Surveillance Center (NRaD), Tech. Rep. 1569, January 1993.

109. Pettengill, G. H. "Radar Astronomy." In *Radar Handbook,* 1st ed. M. Skolnik (Ed.). New York: McGraw-Hill, 1970, chap. 33.

110. Hynes, R., and R. E. Gardner. "Doppler Spectra of S-Band and X-Band Signals." Supplement to *IEEE Trans.* AES-3 (November 1967), pp. 356–365. Also, *Report of NRL Progress* (January 1968), pp. 1–10.

111. Dunn, J. H., and D. D. Howard. "Target Noise." *Radar Handbook,* 1st ed. M. Skolnik (Ed.). New York: McGraw-Hill, 1970, Chap. 28, Sec. 28.5.

112. Collot, G. "Fixed/Rotary Wings Classification/Recognition." *Proc. of the CIE International Conf. on Radar,* Beijing, China, October 22–24, 1991, pp. 610–612.

113. Bullard, B. D., and P. C. Dowdy. "Pulse Doppler Signature of a Rotary-Wing Aircraft." *Proc. 1991 IEEE National Radar Conf.,* Los Angeles, CA, March 12–13, 1991, pp. 160–163.

114. Fliss, G. G., and D. L. Mensa. "Instrumentation for RCS Measurements of Modulation Spectra of Aircraft Blades." *Proc. IEEE 1986 National Radar Conf.,* pp. 95–99.

115. Kulpa, K., Z. Czekala, J. Misiurewicz, and J. Falkiewicz. "Parametric Detection of the Helicopter Hub Echo." *Proc. 1999 IEEE Radar Conf.,* Waltham, MA, pp. 262–266, IEEE Catalog No. 99CH36249.

116. Gardner, R. E. "Doppler Spectral Characteristics of Aircraft Radar Targets at S-Band." Naval Reserach Laboratory, Washington, D.C., Report 5656, August 3, 1961.

117. Bell, M. R., and R. A. Grubbs. "JEM Modeling and Measurement for Radar Target Identification." *IEEE Trans.* AES-29 (January 1993), pp. 73–87.

118. Copeland, J. R. "Radar Target Classification by Polarization Properties." *Proc. IRE* 48 (July 1960), pp. 1290–1296.

119. Holm, W. A. "Polarimetric Fundamentals and Techniques." *Principles of Modern Radar,* J. L. Eaves and E. K. Reedy (Eds.). New York: Van Nostrand Reinhold, 1987, chap. 20.

120. Evans, D. L., T. G. Farr, J. J. Van Zyl, and H. A. Zebker. "Radar Polarimetry: Analysis Tools and Applications." *IEEE Trans.* GRS-26 (November 1988), pp. 774–789.

121. Boerner, W. M., W-L Yan, A-Q Xi, and Y. Yamaguchi. "On the Basic Principles of Radar Polarimetry: the Target Characteristic Polarization State Theory of Kennaugh, Huynen's Polarization Fork Concept, and Its Extension to the Partially Polarized Case." *Proc. IEEE* 79 (October 1991), pp. 1538–1550.

122. Giuli, D. "Polarization Diversity in Radars." *Proc. IEEE* 74 (February 1986), pp. 245–269.

123. Howard, D. D. "High Range-Resolution Monopulse Radar." *IEEE Trans.* AES-11 (September 1975), pp. 749–755.

124. Ksienski, A. A., Y. T. Lin, and L. J. White. "Low-Frequency Approach to Target Identification." *Proc. IEEE* 63 (December 1975), pp. 1651–1660.

125. Lin, H., and A. A. Ksienski. "Optimum Frequencies for Aircraft Classification." *IEEE Trans.* AES-17 (September 1981), pp. 656–665.

126. Chuang, C. W., and D. L. Moffatt. "Natural Resonances of Radar Targets via Prony's Method and Target Discrimination." *IEEE Trans.* AES-12 (November 1976), pp. 583–589.

127. Moffatt, D. L., and R. K. Mains. "Detection and Discrimination of Radar Targets." *IEEE Trans.* AP-23 (May 1975), pp. 358–367.

128. Baum, C. E., E. J. Rothwell, K-M Chen, and D. P. Nyquist. "The Singularity Expansion Method and Its Application to Target Identification." *Proc. IEEE* 79 (October 1991), pp. 1481–1492.

129. Dudley, D. G. "Progress in Identification of Electromagnetic Systems." *IEEE Ant. and Prop. Society Newsletter* (August 1988), pp. 5–11.

130. Fok, F. Y. S., and D. L. Moffatt. "The K-Pulse and E-Pulse." *IEEE Trans.* AP-35 (November 1987), pp. 1325–1326.

131. Chen, K-M, D. P. Nyquist, E. J. Rothwell, L. L. Webb, and B. Drachman. "Radar Target Discrimination by Convolution of Radar Return with Extinction-Pulse and Single-Mode Extraction Signals." *IEEE Trans.* AP-34 (July 1986), pp. 896–904.

132. Optiz, C. L. "Metal-Detecting Radar Rejects Clutter Naturally." *Microwaves* 15 (August 1976), pp. 12–14.

133. Harger, R. G. "Harmonic Radar Systems for Near-Ground In-Foliage Nonlinear Scatterers." *IEEE Trans.* AES-12 (March 1976), pp. 230–245.

134. Flemming, M. A., F. H. Mullins, and A. W. D. Watson. "Harmonic Radar Detection Systems." *IEE RADAR-77 International Conf.,* October 25–28, pp. 552–554, 1977.

135. Powers, E. J., J. Y. Hong, and Y. C. Kim. "Cross Sections and Radar Equation for Nonlinear Scatterers." *IEEE Trans.* AES-17 (July 1981), pp. 602–605.

136. Hong, J. Y., and E. J. Powers. "Detection of Weak Third Harmonic Backscatter from Nonlinear Metal Targets." *IEEE Eascon-83,* September 19–21, 1983, pp. 169–175, 83CH1967-9, ISSN:0531-6863.

137. Bahr, A. J., and J. P. Petro. "On the RF Frequency Dependence of the Scattered Spectral Energy Produced by Intermittent Contacts Among the Elements of a Target." *IEEE Trans.* AP-25 (July 1978), pp. 618–621.

138. Newburgh, R. G. "Basic Investigation of the RADAM Effect." Rome Air Development Center, Rome, N.Y., Report RADC-TR-151, June 1978.

139. Trunk, G. V., and J. D. Wilson. "Association of DF Bearing Measurements With Radar Tracks." *IEEE Trans.* AES-23 (July 1987), pp. 438–447.

140. Saha, R. K. "Analytical Evaluation of an ESM/Radar Track Association Algorithm." *SPIE* 1698, "Signal and Data Processing of Small Targets." (1992), pp. 338–347.

141. Farina, A., and B. La Scala. "Methods for the Association of Active and Passive Tracks for Airborne Sensors." *International Radar Symposium,* Munich, Germany, September, 1998.

142. Hawkins, C. F. "Friendly Fire: Facts, Myths and Misperceptions." *Proc. U.S. Naval Institute* 120 (June 1994), pp. 54–59.

143. Silberman, G. L. "Parametric Classification Techniques for Theater Ballistic Missile Defense." *Johns Hopkins APL Technical Digest* 19, no. 3 (1998), pp. 322–339.

144. Heiss, W. H., D. L. McGrew, and D. Sirmans. "Nexrad: Next Generation Weather Radar (WSR-88D)." *Microwave J.* 33 (January 1990), pp. 79–98.

145. Crum, T. D., and R. L. Alberty. "The WSR-88D and the WSR-88D Operational Support Facility." *Bulletin of the American Meteorological Society* 74 (September 1993), pp. 1669–1687.

146. Dudgen, D. E., and R. T. Lacoss. "An Overview of Automatic Target Recognition." MIT Lincoln Laboratory Journal, "Special Issue on Automatic Target Recognition," vol. 6, pp. 3–10, Spring 1993.

147. Sheen, D. R. et al. "The P-3 Ultra-Wideband SAR: Description and Examples." *Proc. 1996 IEEE National Radar Conf.,* pp. 50–53, IEEE Catalog no. 96CH35891.

148. Vaughn, C. R. "Birds and Insects as Radar Targets: A Review." *Proc. IEEE* 73 (February 1985), pp. 205–227.

149. Chen, K-M, et al. "An X-Band Microwave Life-Detection System." *IEEE Trans.* BME-33 (July 1986), pp. 697–701.

150. Geisheimer, J. "A Radar System for Monitoring Human Vital Signs." *IEEE Potentials* 17 (January 1999), pp. 21–24.

PROBLEMS

6.1   (a) Sketch the rms range error (in meters) for a quasi-rectangular pulse with a half-power pulse width of 2 μs, as a function of the peak-signal-to-mean-noise ratio ($2E/N_0$) over the range of values from 10 to 60 dB. (b) Why might it not be appropriate to consider signal-to-noise ratios below 10 dB and above 60 dB?

6.2   Derive the rms error in measuring the time delay for a gaussian pulse of half-power width τ [Eq. (6.17)].

6.3   Based on the measurement of doppler frequency shift, sketch the rms error of the radial velocity (in meters) as a function of the width τ of a rectangular pulse when the pulse width varies from 1 μs to 10 ms for (a) constant pulse energy and (b) constant peak power. The frequency is 5400 MHz. Assume in both cases that $2E/N_0 = 36$ when the pulse width is 1 μs.

6.4   (a) What is the minimum width τ of a rectangular pulse that can be used with an X-band radar (9375 MHz) if it is desired to achieve a 10 kt radial velocity accuracy (based on the doppler frequency measured by a single pulse), when $2E/N_0 = 23$ dB? (b) What is the minimum range (in nautical miles) that corresponds to this pulse width? (c) In part (a) of this question, what should be the value of $2E/N_0$ (in dB) to achieve a 10 kt radial velocity accuracy if the pulse width can be no longer than 10 μs? (d) What would be the minimum pulse width in (a) if the radar operated at W band (94 GHz)? (e) Comment on the utility of accurately measuring the velocity with a single short pulse.

6.5   There are two methods for finding the radial velocity of a target. One is based on the doppler shift $f_d = 2v_r/\lambda$; the other is based on the rate of change of range with time $\Delta R/\Delta t$. They give different measurement accuracies. (a) What is the expression for the radial velocity error, δv_d, found by measuring the doppler frequency shift of a long quasi-rectangular pulse of width τ and RF frequency f_0? (b) What is the expression for the radial velocity error, δv_r, found from the rate of change of range based on two range measurements R_1 and R_2 separated by a time τ, so that the velocity is $v_r = (R_2 - R_1)/\tau$, and τ is the same as the pulse width of the doppler measurements? The pulses in this range-rate measurement are of gaussian shape with a half-power bandwidth B [use right-hand side of Eq. (6.17)]. Assume the total value of $2E/N_0$ in each of these two methods for radial velocity measurement [(a) and (b)] is the same. (c) What is the value of $\delta v_d/\delta v_r$? (d) Based on your answer in (c), which is the more accurate method of velocity measurement, the doppler method or the range-rate method? (e) Under what conditions will these two methods give comparable accuracies (assuming the same total $2E/N_0$)? (f) Why do you think the doppler method has not been used very often for a velocity measurement?

6.6 (a) What value of $2E/N_0$ (in dB) is required to achieve an angular accuracy of 0.3 mrad when the antenna beamwidth is one deg, assuming the antenna has a cosine aperture illumination? (b) If the signal received by this radar antenna has a value of $2E/N_0 = 23$ dB for a particular target at a range of 150 nmi, at what range will the radar first be capable of obtaining an angular accuracy of 0.3 mrad?

6.7 Determine $\beta\alpha$ (the product of the effective pulse width β and the effective time duration α) for the following: (a) gaussian pulse, (b) triangular pulse, and (c) quasi-rectangular pulse. (Use the expressions for β and α already given in the text.) (d) Based on the above answers, how much flexibility is there in selecting one of these simple waveforms to achieve a large value of $\beta\alpha$ for the purpose of making an accurate measurement of both time delay and frequency? (e) What option, other than a large $\beta\alpha$, is available for obtaining an accurate measurement of both time delay and frequency simultaneously?

6.8 Derive the rms error in measuring frequency for a rectangular pulse of width τ (Eq. 6.24).

6.9 How is the rms error in measuring frequency affected when the width of a rectangular pulse is increased by a factor of four, with the peak power remaining constant?

6.10 (a) A K-band (24.15 GHz) radar speed-gun, used for measuring the speed of an auto or a baseball, has a claimed accuracy of 0.1 statute mile/hour. If the signal-to-noise ratio ($2E/N_0$) is 17 dB, what observation time is needed to achieve this accuracy? (b) How far does a car with a speed of 60 mph travel during this time? (c) Assuming the same obervation time and value of $2E/N_0$ as in part (a), what would be the accuracy of an X-band (10.525 GHz) radar speed-gun?

6.11 Using the first part of Eq. 6.21, $\beta^2 E = \int_{-\infty}^{\infty} [s'(t)]^2\, dt$, show that the rms error in measuring the time delay of a trapezoidal pulse is the same as that given by Eq. (6.14).

6.12 Note that the sine integral function $Si(4\pi) = 1.492$. (The energetic reader might try to find the time waveform $s(t)$ out of a rectangular filter when $B_s\tau_r = 4$, plot $s(t)$ to find its half-power width τ, and express β and α in terms of the half-power width τ.) Find the values of the effective bandwidth β and the effective time duration α for a different quasi-rectangular pulse formed by taking the bandwidth B_s [in Fig. 6.2, Eq. (6.9), and Eq (6.25)] extending from $-2/\tau_r$ to $+2/\tau_r$, where τ_r is the width of the original rectangular pulse. Thus $B_s\tau_r = 4$.

6.13 Using the definition of β [Eq. (6.8)], definition of α [Eq. (6.23)], Eq. (6.21), Parseval's relation [Eq. (5.13)], Schwartz inequality, [Eq. (5.11)], and performing integration by parts, show that $\beta\alpha \geq \pi$ and that the equality holds for the gaussian function. (Only for those who enjoy a mathematical exercise.)

6.14   Show that $\int (2\pi f)^2 |S(f)|^2 \, df = -\int s''(t)s(t)dt = \int [s'(t)]^2 \, dt$, which is Eq. (6.21). The limits of all integrals are from $-\infty$ to $+\infty$. (Start with $s(t)$ expressed as an inverse Fourier transform of $S(f)$, and differentiate $s(t)$ twice.)

6.15   What is the message of the radar "uncertainty principle"?

6.16   Show that the rms error for the measurement of the phase of a sinewave is $\delta\phi = (2S/N)^{-1/2}$, where S/N is the signal-to-noise ratio. [This can be obtained from the error in making a measurement of the time when a sinewave crosses the time axis. Note that phase ϕ is equal to $2\pi ft$, so that $\delta\phi = 2\pi f \, \delta t$. The derivation is similar to that given for the rectangular pulse that led to Eq. (6.4), except that the time is measured when the sinewave crosses the time axis once.]

6.17   What information about the radar waveform can be obtained from the ambiguity diagram?

6.18   The amplitude of the peak of the signal out of a matched filter is $2E$, where E is the signal energy. This can be seen from Eq. (6.42). A peak output of $2E$ implies that the output has units of energy [(volts)^2 × time], but this cannot be so since the output of a filter should have volts as the unit. Show that the unit for the output of the matched filter is actually volts and not energy. (This is related to problem 5.6.)

6.19   Explain qualitatively (no derivation needed) why the ambiguity diagram shown in Fig. 6.9 is that of a down-chirp; i. e., frequency of the linear FM waveform decreases with time.

6.20   (a) Why might one want to use both an up-chirp and a down-chirp waveform on two successive pulses in a linear-FM pulse-compression radar system? (b) When might one use a waveform with the following three contiguous parts: (1) an unmodulated CW waveform, (2) a down-chirp and (3) an up-chirp? (c) Why might one want a radar to operate with an up-chirp when another similar nearby radar operates with a down-chirp?

6.21   A C-band (5.5 GHz) ballistic-missile detection radar employs a linear-FM pulse-compression radar with a 1.0-ms uncompressed pulse-width having a down-chirp covering a bandwidth of 200 MHz. If the target has a radial velocity of 2 km/s, what is the error in range (km) due to the doppler shift of the target? What is the range error in terms of the resolution of the compressed pulse?

6.22   Determine the shift in the time delay ΔT_r with a linear-FM pulse-compression waveform whose bandwidth is B and time duration is T when the target experiences a doppler frequency shift f_d for the following two cases: (a) Ballistic-missile-detection radar with $B = 1$ MHz and $T = 1$ ms, when $f_d = 100$ kHz. (b) Aircraft-detection radar with $B = 100$ MHz and $T = 10$ μs, when $f_d = 1$ kHz. In each of the two cases: (c) What is the doppler-caused time-shift error, measured in meters of range? (d) What is the ratio of the doppler-caused time shift and the resolution in time delay of the waveform (which you may assume to be $1/B$)?

6.23   How can one find the true range and true doppler frequency shift of a target when using linear-FM pulse-compression waveforms?

6.24   In this problem two nearby aircraft are both located within the antenna beam, but one is trailing slightly behind the other. The trailing aircraft has a slower speed than the leading aircraft. They are being detected by a linear-FM pulse-compression radar. Describe why, in this case, it is better to use an up-chirp rather than a down-chirp when it is desired to provide a wide separation between the two output echo signals.

6.25 (a) Show that the tapped delay line that generates a Barker code of length 5 can be used as a matched filter if the receive signal is inserted from the opposite end. (This can be shown by sketching the output of the matched filter.) (b) How can the dispersive filter used to generate a linear-FM (chirp) signal be used as the matched filter to receive the echo signal?

6.26 Show (simple diagrams are okay) that the two different Barker codes of length 4 in Table 6.4 are also complementary codes.

6.27 *Background:* Generally, one can assume that the doppler shift is constant across the spectrum of the echo signal. There are cases, however, where this is not so, as in the current problem. *Problem:* A ballistic missile detection radar is attempting to detect a target moving with a radial velocity of 2 nmi/s. The radar employs a linear-FM pulse-compression waveform with a bandwidth of 100 MHz. What is the longest pulse width that can be used before significant degradation of the compressed pulse width occurs? (See the discussion in the subsection *Doppler-Tolerant Pulse-Compression Waveforms.*)

6.28 In the previous problem, one might have used the criterion that $2BTv_r/c$ must not exceed unity. This criterion is based on the doppler at the leading edge of a long FM pulse of length T and bandwidth B being significantly different from the doppler at the trailing edge of the pulse when the target radial velocity is v_r. Derive the expression $2BTv_r/c < 1$. [One criterion that might be applied is to require the spread in the time-delay measurements caused by the difference in the doppler shifts at the leading and trailing edges of the pulse of duration T to be no greater than the spread $(1/B)$ in the time-delay measurement when the target is stationary (no doppler shift).]

6.29 A linear-FM pulse-compression radar has a bandwidth B and an uncompressed pulse duration T. The matched filter is followed by a sidelobe-reduction filter whose weighting function is

$$W(f) = \cos(\pi f/B) \, \text{rect}(f/B)$$

where rect $(x) = 1$ for $|x| < 1/2$ and equals 0 for $|x| > 1/2$. The signal spectrum at the output of the matched filter is rectangular and is given as

$$S_m(f) = \sqrt{T/B} \, \text{rect}(f/B)$$

and the noise spectrum at the matched filter output is

$$N(f) = (N_0/2) \, \text{rect}(f/B)$$

Find:

a. The signal waveform $s_w(t)$ at the output of the sidelobe-reduction filter.
b. The loss in signal-to-noise ratio due to the sidelobe-reduction filter.
c. The level of the first sidelobe of the output waveform $s_w(t)$.

(Acknowledgment: This problem was generously given to the author by an instructor of a radar course in Huntsville, Alabama many years ago. He used it as part of a take-home exam. Unfortunately, over the years I lost his name, but wish to acknowledge, with gratitude, my use of his problem here and throughout the years in my own radar course.)

6.30   An X-band (9.5 GHz) ground-based ISAR (inverse synthetic aperture radar) is imaging an aircraft at a range of 20 nmi. The aircraft is traveling at a speed of 250 kt and is on a tangential trajectory (i.e., it is perpendicular to the radar line of sight). The ISAR image is obtained as the aircraft is viewed broadside to the radar line of sight, so that the viewing aspect is 90° or in close vicinity to it. What must the total radar observation time be in order to obtain an image with a cross-range resolution of 1 m?

6.31   What effect might the echo from a rapidly rotating antenna located within a range-resolution cell of an ISAR have on the ISAR image of the target?

6.32   A ground-based S-band (3.2 GHz) radar is observing a helicopter. One may assume that the velocity of the tip of the blade of a helicopter is 210 m/s and that the length of a blade is 6 m.

    a. What is the time between blade flashes for a two-blade and a three-blade helicopter?

    b. What is the time duration of the blade flash for a two-blade and a three-blade helicopter?

    c. If the radar antenna has an azimuth beamwidth of 3°, at what rpm must the antenna be rotated in order to insure that the blade flash will be detected on each scan of the antenna?

    d. What must the prf of the radar be to insure that there are at least five pulses received from the blade flash?

    e. What is the radar cross section of the blade flash for an odd number of blades?

    f. If the helicopter is hovering, what is the maximum doppler shift that might be obtained from the helicopter blades?

    g. What is the maximum forward speed (kt) that a helicopter can have if the actual velocity of the tip is to be less than 0.8 of Mach 1? (Mach 1 can be assumed to be 343 m/s.) What is the maximum doppler shift of the radar echo from the helicopter blade?

6.33   (a) What options are available for reliably recognizing one aircraft target from another by noncooperative methods? (b) Describe how one might reliably recognize a ship target by noncooperative methods. (c) Describe how one might reliably recognize a helicopter target by noncooperative methods.

6.34   It has sometimes been said that two pulse signals, one at frequency f_1 and the other at frequency f_2, each of width τ, can be resolved (separated by filters) by a radar if the number of cycles within the pulse width τ at frequency f_1 differs by at least one from the number of cycles within the pulse width τ at frequency f_2. (a) Show that this criterion for resolution is equivalent to the more usual criterion that $|f_1 - f_2| \geq 1/\tau$. (b) A long-range UHF radar (440 MHz) for the detection of satellites is assumed to have a pulse width of

2 ms. What is the minimum difference in radial velocity of two targets that will allow them to be separated by doppler filtering?

6.35 A linear-FM waveform and a $(\sin \pi ft)/\pi ft$ waveform both have a uniform power spectrum. Does that mean they both can use the same pulse compression filter? (More than a simple yes or no answer would be nice.)

6.36 It was said in Sec. 6.5 in the discussion of linear-FM pulse-compression that the true range of a target can be found by using both an up-chirp and a down-chirp waveform, and taking the average of the two time delays. It is also possible to obtain the true doppler frequency shift using the same two measurements of time delay from the up-chirp and down-chirp waveforms. (This was Problem 6.23.) Derive an expression for the rms accuracy of the frequency measurement found from the two time delay measurements when the frequency extent of the linear FM is B and the time duration is T.

chapter

7

Radar Clutter

7.1 INTRODUCTION TO RADAR CLUTTER

Clutter is the term used by radar engineers to denote *unwanted* echoes from the natural environment. It implies that these unwanted echoes "clutter" the radar and make difficult the detection of wanted targets. Clutter includes echoes from land, sea, weather (particularly rain), birds, and insects. At the lower radar frequencies, echoes from ionized meteor trails and aurora also can produce clutter. The electronic warfare technique known as *chaff,** although not an example of the natural environment, is usually considered as clutter since it is unwanted and resembles clutter from rain. Clutter is generally distributed in spatial extent in that it is much larger in physical size than the radar resolution cell. There are also "point," or discrete, clutter echoes, such as TV and water towers, buildings, and other similar structures that produce large backscatter. Large clutter echoes can mask echoes from desired targets and limit radar capability. When clutter is much larger than receiver noise, the optimum radar waveform and signal processing can be quite different from that employed when only receiver noise is the dominant limitation on sensitivity.

Radar echoes from the environment are not always undesired. Reflections from storm clouds, for example, can be a nuisance to a radar that must detect aircraft; but storm clouds

---

*Chaff is an electronic countermeasure that consists of a large number of thin passive reflectors, often metallic foil strips. When released from an aircraft they are quickly dispersed by the wind to form a highly reflecting cloud. A relatively small bundle of chaff can form a cloud with a radar cross section comparable to that of a large aircraft.

containing rain are what the radar meteorologist wants to detect in order to measure rain-fall rate over a large area. The backscatter echoes from land can interfere with many applications of radar, but they are the target of interest for ground-mapping radar, synthetic aperture radars, and radars that observe earth resources. Thus the same environmental echo might be the desired signal in one application and the undesired clutter echo in another. The observation of land, sea, weather and other natural phenomena by radar and other sensors for the purpose of determining something about the environment is known as *remote sensing of the environment,* or simply *remote sensing.* All radars, strictly speaking, are remote sensors; but the term is usually applied only to those radars whose major function is to observe the natural environment for the purpose of extracting information about the environment. A prime example of a radar used for remote sensing is the doppler weather radar.

Echoes from land or sea are examples of *surface clutter.* Echoes from rain and chaff are examples of *volume clutter.* The magnitude of the echo from distributed surface clutter is proportional to the area illuminated. In order to have a measure of the clutter echo that is independent of the illuminated area, the *clutter cross section per unit area,* denoted by the symbol σ^0, is commonly used to describe surface clutter. It is given as

$$\sigma^0 = \frac{\sigma_c}{A_c} \qquad \qquad [7.1]$$

where σ_c is the radar cross section of the clutter occupying an area A_c. The symbol σ^0 is spoken, and sometimes written, as *sigma zero.* It has also been called the *scattering coefficient, differential scattering cross section, normalized radar reflectivity, backscattering coefficient,* and *normalized radar cross section (NRCS).* The zero is a superscript since the subscript is reserved for the polarization employed. Sigma zero is a dimensionless quantity, and is often expressed in decibels with a reference value of one m^2/m^2.

Similarly, a cross section per unit volume is used to characterize volume clutter. It is defined as

$$\eta = \frac{\sigma_c}{V_c} \qquad \qquad [7.2]$$

where σ_c in this case is the radar cross section of the clutter that occupies a volume V_c. Clutter cross section per unit volume, η, is sometimes called the *reflectivity.*

In the next section, the radar range equation for targets in surface clutter is derived along with a brief review of the general character of scattering from surface clutter. This is followed by descriptions of radar echoes from land, sea, weather, and the atmosphere. The chapter concludes by describing methods that might be used to enhance the detection of targets in clutter.

7.2 SURFACE-CLUTTER RADAR EQUATION

The radar equation describing the detection of a target in surface clutter is different from the radar equation discussed in Chap. 2, where it was assumed that detection sensitivity was limited by receiver noise. The radar equation for detection of a target in clutter leads to different design guidelines than the radar equation for detection of a target when limited by receiver noise.

Low Grazing Angle   Consider the geometry of Fig. 7.1 which depicts a radar illuminating the surface at a grazing angle ψ. Assume the grazing angle is small. A small grazing angle usually implies that the extent of the resolution cell in the range dimension is determined by the radar pulse width τ rather than the elevation beamwidth. The width of the cell in the cross-range dimension is determined by the azimuth beamwidth θ_B and the range R. From the simple radar equation [such as Eq. (1.7) of Sec. 1.2] the received echo power P_r is

$$P_r = \frac{P_t G A_e \sigma}{(4\pi)^2 R^4}$$

[7.3]

where

P_t = transmitter power, W

G = antenna gain

A_e = antenna effective aperture, m^2

R = range, m

σ = radar cross section of the scatterer, m^2

When the echo is from a target (rather than from clutter), we let $P_r = S$ (received target signal power) and $\sigma = \sigma_t$ (target cross section). The signal power returned from a target is then

$$S = \frac{P_t G A_e \sigma_t}{(4\pi)^2 R^4}$$

[7.4]

Figure 7.1
Geometry of radar surface clutter. (a) Elevation view showing the extent of the surface illuminated by the radar pulse, (b) plan view showing the illuminated clutter patch (or resolution cell) consisting of individual, independent scatterers.

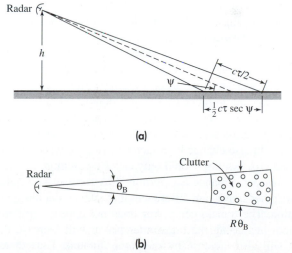

When the echo is from clutter, the cross section σ becomes $\sigma_c = \sigma^0 A_c$, where the area A_c of the radar resolution cell is given from Fig. 7.1 as

$$A_c = R\theta_B (c\tau/2) \sec \psi \qquad \text{[7.5]}$$

θ_B = two-way azimuth beamwidth, c = velocity of propagation, τ = pulsewidth, and ψ = grazing angle (defined with respect to the tangent to the surface). The extent of the area A_c in the range coordinate (the range resolution) is $c\tau/2$, where the factor of 2 in the denominator accounts for the two-way propagation of radar. With these definitions, the radar equation for the surface-clutter echo-signal power C is

$$C = \frac{P_t G A_e \sigma^0 \theta_B (c\tau/2) \sec \psi}{(4\pi)^2 R^3} \qquad \text{[7.6]}$$

The received echo power C from surface clutter is seen to vary inversely as the cube of the range rather than the fourth power, as was the case for point targets in free space.

When the echo from surface clutter is large compared to receiver noise, the signal-to-clutter ratio is Eq. (7.4) divided by Eq. (7.6), or

$$\frac{S}{C} = \frac{\sigma_t}{\sigma^0 R\theta_B(c\tau/2) \sec \psi} \qquad \text{[7.7]}$$

If the maximum range $R_{\max}$ corresponds to the minimum discernible signal-to-clutter ratio $(S/C)_{\min}$, then the radar equation for the detection of a target in surface clutter at low grazing angle is

$$R_{\max} = \frac{\sigma_t}{(S/C)_{\min}\sigma^0 \theta_B(c\tau/2) \sec \psi} \qquad \text{[7.8]}$$

If pulse compression is used, the pulse width τ is that of the compressed pulse.

The azimuth beamwidth θ_B in the above equations is the two-way beamwidth. When the gaussian function can be used to approximate the beam shape (which is usually a good assumption), the two-way beamwidth is smaller than the one-way beamwidth by $\sqrt{2}$. (If the one-way beamwidth is used to calculate σ^0, the value of σ^0 will be lower by 1.5 dB than if the two-way beamwidth were used.)

The radar equation for surface clutter, Eq. (7.8), is quite different from the noise-dominated radar equation derived in Sec 1.2. The range appears as the first power rather than the fourth power as in the usual radar equation [Eq. (7.3)]. This results in greater variation of the maximum range of a clutter-dominated radar than a noise-dominated one when there is uncertainty or variability in the parameters of the radar equation. For example, if the target cross section in Eq. (7.8) were to change by a factor of two, the maximum range would also change by a factor of two. However, the same change of a factor of two in target cross section would only cause a variation in range of 1.2 (the fourth root of 2) when radar performance is determined by receiver noise alone.

There are other significant differences that affect radar design when clutter is the dominant limitation. The transmitter power does not appear explicitly in the surface-clutter radar equation. Increasing the transmitter power will increase the target signal, but the clutter echo will also increase by the same amount. Thus there is no net gain in the detectability of desired targets. The only requirement on the transmitter power when Eq. (7.8) is used is that it be great enough to cause the clutter power at the radar receiver to be large compared to receiver noise.

Neither the antenna gain nor effective aperture enters explicitly, except as affected by the azimuth beamwidth θ_B. Equation (7.8) indicates that the narrower the azimuth beamwidth, the greater the range. Also, the narrower the pulse width, the greater the range, which is just the opposite of conventional radar detection of target echo signals in noise. A long pulse is desired when the radar is dominated by noise so as to increase the signal-to-noise energy ratio. A long pulse, on the other hand, decreases the signal-to-clutter ratio.

If the statistics of the clutter echoes are similar to the statistics of receiver noise (gaussian probability density function), the minimum signal-to-clutter ratio in Eq. (7.8) can be selected similarly to that for the signal-to-noise ratio as described in Chap. 2. Gaussian statistics, however, are rarely applicable to sea clutter and seldom for land clutter, unless the resolution cell is large. Therefore, the selection of the minimum signal-to-clutter ratio required for detection of a target in sea clutter can be difficult. When no other information is available concerning the statistics of clutter, many engineers will cautiously use the gaussian model that is used for receiver noise—and hope for the best. Even if the clutter can be described by gaussian statistics, clutter echoes do not vary with time in the same manner as receiver noise. The temporal correlation of successive clutter echoes has to be taken into account.

Radars integrate (add together) a number of echoes from a target to enhance detection, as was discussed in Sec. 2.6. Integration of pulses is generally much less effective when detection is limited by clutter than when limited by receiver noise. The statistics of receiver noise are independent (uncorrelated) in a time equal to $1/B$, where B = receiver bandwidth (usually measured in the IF portion of the radar receiver). When pulses are integrated, the sum of the noise-power fluctuations do not increase as rapidly as does the sum of the target echoes so that increased signal-to-noise ratio is obtained as the number of pulses integrated is increased. On the other hand, with perfectly stationary clutter (rocks or fence posts, for example), the clutter echo does not fluctuate from pulse to pulse. It builds up at the same rate as the target signal. There is no increase in signal-to-clutter ratio, so that integrating pulses provides no benefit. Sea clutter, however, changes with time, but slowly. At X band, for example, the decorrelation time of sea clutter is about several milliseconds. One might talk about an effective number, n_{eff}, of pulses integrated for incorporating in the numerator of Eq. (7.8), but it is likely to be much smaller than the n_{eff} for a noise-limited radar with the same number of received pulses. It is much more difficult to determine n_{eff} for clutter than for noise. Because of the uncertainties in determining n_{eff} for clutter, the conservative engineer might omit any gain due to pulse integration when detection is dominated by stationary or slowly changing clutter rather than receiver noise.

System losses also are not included explicitly in the radar equation of Eq. (7.8). Many of the system losses mentioned in Sec. 2.12 affect the target and clutter echo signals the same. Losses, therefore, have less effect on detection when signals are limited by clutter than when signals are limited by receiver noise. As long as losses do not make invalid the assumption that the clutter echo is large compared to receiver noise, losses have a lesser effect than in a noise-dominated radar.

High Grazing Angle   Next consider the case where the radar observes surface clutter near perpendicular incidence. The clutter area viewed by the radar is determined by the antenna beamwidths θ_B and ϕ_B in the two principal planes. The clutter illuminated area A_c in Eq. (7.1) is $(\pi/4)R\theta_B R\phi_B/2 \sin \psi$, where ψ = grazing angle and R = range. The factor $\pi/4$ accounts for the elliptical shape of the illuminated area, and the factor of 2 in the denominator is necessary since in this case θ_B and ϕ_B are the one-way beamwidths. Substituting $\sigma = \sigma^0 A_c$ in Eq. (7.3), letting $P_r = C$ (the clutter echo power), and taking $G = \pi^2/\theta_B\phi_B,$[1] the clutter radar equation in this case is

$$C = \frac{\pi P_t A_e \sigma^0}{128 R^2 \sin \psi} \tag{7.9}$$

The clutter power is seen to vary inversely as the square of the range. This equation applies to the echo power received from the ground by a radar altimeter or the remote sensing radar known as a scatterometer. An equation for detecting a target at high grazing angles could be derived, but this represents a situation not often found in practice.

Grazing, Incidence, and Depression Angles   In most of this chapter, the grazing angle is used to describe the aspect at which clutter is viewed. There are two other angles that are sometimes used instead of the grazing angle (Fig. 7.2). The *incidence angle* is defined with respect to the normal to the surface; the *grazing angle* is defined with respect to the tangent to the surface; and the *depression angle* is defined with respect to the local horizontal at the radar. The incidence angle is the complement of the grazing angle. When the earth's surface can be considered smooth and flat, the depression angle and the grazing angle are the same. When the earth's curvature must be taken into account, as in space-borne radars, the depression angle can be quite different from the grazing angle. The incidence angle is usually used when considering earth backscatter at near perpendicular incidence, as in the altimeter and the scatterometer. The grazing angle is preferred in most other applications. Some engineers (as in references 3 and 4) prefer to use the depression angle when a rough or varying earth's surface is viewed at low grazing angles since it might be easier to determine than the grazing angle when the earth is not a flat surface.

Variation of Surface Clutter with Grazing Angle   The general form of surface clutter as a function of grazing angle is shown in Fig. 7.3. There are three different scattering regions. At high grazing angles, the radar echo is due mainly to reflections from clutter that can be represented as a number of individual *planar facets* oriented so that the incident

Figure 7.2   Angles used in describing geometry of the radar and surface clutter.

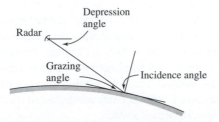

Figure 7.3 General nature of the variation of surface clutter as a function of grazing angle, showing the three major scattering regions.

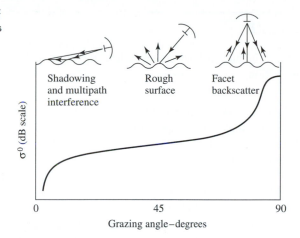

energy is directed back to the radar. The backscatter can be quite large at high grazing angles. At the intermediate grazing angles, scattering is somewhat similar to that from a *rough surface*. At low grazing angles, back scattering is influenced by *shadowing* (masking) and by *multipath propagation*. Shadowing of the trough regions by the crests of waves prevents low-lying scatterers from being illuminated. Multipath reduces the energy propagating at low angles because of cancellation of the direct energy by the out-of-phase surface-reflected energy (similar to the multipath phenomenon described in Sec. 8.2). The curve drawn in Fig. 7.3 is descriptive of the general character of both land and sea scattering; but there are significant differences in the details depending on the particular type of clutter. The difference between the maximum clutter at perpendicular incidence and the minimum clutter at grazing incidence can be many tens of dB.

Mean and Median Values The characteristics of clutter echoes are usually given in statistical terms. It is often described either by the mean (average) value of σ^0 or the median (the value exceeded 50 percent of the time). For a Rayleigh probability density function, the difference between the mean and the median is small (a few percent); but for non-Rayleigh distributions, the mean and the median can be quite different. In some cases, the accuracy of clutter measurements might not be sufficient to make a distinction between the mean and the median. Nevertheless, when using clutter data, it is best to know whether it is the mean or the median that is being used. According to Nathanson,[2] the mean value is much better behaved than the median. The median can vary depending on the siting, pulse duration, masking, and other factors.

For proper radar design, the probability density function of the clutter echoes should be known so that the receiver detector can be designed appropriately. Unfortunately, when the statistics of the clutter cannot be described by the classical Rayleigh probability density function, it is often difficult to define a specific quantitative statistical description. For this reason, many radar designs have been based on the mean value of the clutter σ^0 rather than some statistical model. Further discussion of surface-clutter probability density functions is given later in Sec. 7.5.

7.3   LAND CLUTTER

The general nature of land clutter at low, medium, and high grazing angles is described in this section.

Land Clutter at Low Grazing Angles   An extensive multiple-frequency database of land clutter at low angles was acquired by the MIT Lincoln Laboratory and reported by J. B. Billingsley and J. F. Larrabee.[3] This is one of the few collections of land clutter data that have been obtained over a wide variety of terrain, with good "ground truth," good calibration, and observations over a long period of time. The Lincoln Laboratory data provides much better understanding than previously, and it has caused some earlier notions about the nature of land clutter to be modified.

Forty-two different sites widely dispersed geographically across North America were measured, with most of the sites in western Canada. Measurements were made over a period of almost 3 years with an average time of 17 days at a site. Some sites were visited more than once to determine seasonal variations. Measurements were made at five frequencies: VHF (167 MHz), UHF (435 MHz), L (1.23 GHz), S (3.24 GHz), and X (9.2 GHz) bands. The radars were mobile with antennas mounted on a tower that could be extended to heights of 30, 60, or 100 ft. Range resolution was either 150 m or 36 m at VHF and UHF. At the other three frequencies, the resolution was either 150 m or 15 m. Both vertical and horizontal polarizations were employed. The rms accuracy of the clutter echo measurements over all sites was said to be 2 dB, a very good value for field operations.

The radar measured $\sigma^0 F^4$ (called *clutter strength* in the Lincoln Laboratory Report), where σ^0 is the clutter cross-section per unit area and F is the *propagation factor* that sometimes appears in the radar equation to account for propagation effects such as multipath reflections, diffraction, and attenuation. The authors define the propagation factor F as the "ratio of the incident field that actually exists at the clutter cell being measured to the incident field that would exist there if the clutter cell existed by itself in free space." Even though one might wish to separate clutter backscatter (σ^0) from propagation effects (F), it is difficult to do so and it is not generally done. (Many of the measurements of σ^0 reported in the literature are measurements of $\sigma^0 F^4$ even though they might be said to be of σ^0.)

Clutter observations were made at low depression angles, at ranges from 1 to 25 or 50 km or more. The depression angle was used rather than the grazing angle since it was difficult to define the grazing angle over a non-flat surface such as natural terrain. The depression angle in the Lincoln Laboratory report is "the complement of the incidence angle at the backscattering terrain point under consideration." This definition of depression angle includes the effect of earth curvature on the angle of illumination but not the effect of the local terrain slope.

The results of the Lincoln Laboratory measurements are summarized in Fig. 7.4 and Table 7.1. Clutter strength is given as the median value of the measured means by terrain type and frequency. The values in the figure and table were averaged over both vertical and horizontal polarizations, and with both 150 m and 15 or 36 m range resolution. This

Figure 7.4   Mean clutter strength as a function of frequency for various terrain types.

(From J. B. Billingsley and J. F. Larrabee.[3] Reprinted with permission of MIT Lincoln Laboratory, Lexington, Massachusetts.)

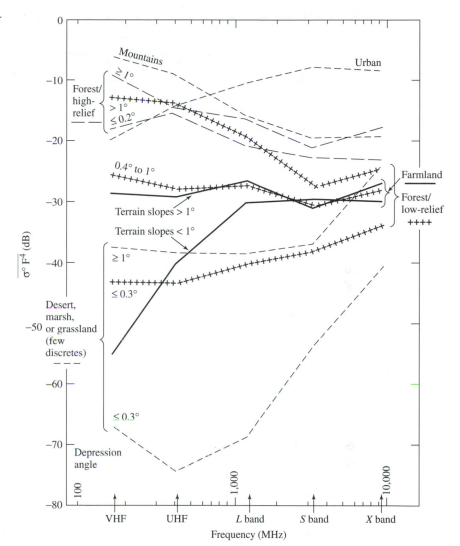

averaging was done since the variations of the mean clutter echo with both polarization and resolution were small, generally about 1 or 2 dB.

Figure 7.5 shows the mean value of rural clutter strength (no urban areas are included) as a function of frequency, measured over 36 sites. The mean values are almost independent of frequency. The average of the means is 29.2 dB, and all five values lie within 1.7 dB of this value. The standard deviation is indicated by the vertical bars, and the extreme horizontal bars in the figure represent the extremes of the measurements at each frequency. The values in Fig. 7.5 are *before* terrain classification. The corresponding one-sigma patch-to-patch variation in mean clutter strength by frequency band *after* terrain classification is 3.9, 3.8, 2.9, 2.7, and 2.3 dB at VHF, UHF, *L*, *S*, and *X* bands,

Table 7.1   Median value of mean land clutter strength over many measurements* by terrain type and frequency

Terrain Type	Median Value of $\sigma^0 F^4$ (dB)				
	Frequency Band				
	VHF	**UHF**	**_L_ Band**	**_S_ Band**	**_X_ Band**
URBAN	−20.9	−16.0	−12.6	−10.1	−10.8
MOUNTAINS	−7.6	−10.6	−17.5	−21.4	−21.6
FOREST/HIGH-RELIEF (Terrain Slopes > 2°)					
High depression angle (>1°)	−10.5	−16.1	−18.2	−23.6	−19.9
Low depression angle (≤0.2°)	−19.5	−16.8	−22.6	−24.6	−25.0
FOREST/LOW-RELIEF (Terrain slopes < 2°)					
High depression angle (>1°)	−14.2	−15.7	−20.8	−29.3	−26.5
Intermediate depression angle (0.4° to 1°)	−26.2	−29.2	−28.6	−32.1	−29.7
Low depression angle (≤0.3°)	−43.6	−44.1	−41.4	−38.9	−35.4
AGRICULTURAL/HIGH-RELIEF (Terrain Slopes > 2°)	−32.4	−27.3	−26.9	−34.8	−28.8
AGRICULTURAL/LOW-RELIEF					
Moderately low-relief (1° < terrain slopes < 2°)	−27.5	−30.9	−28.1	−32.5	−28.4
Very low-relief (terrain slopes < 1°)	−56.0	−41.1	−31.6	−30.9	−31.5
DESERT, MARSH, OR GRASSLAND (Few discretes)					
High depression angle (≥1°)	−38.2	−39.4	−39.6	−37.9	−25.6
Low depression angle (≤0.3°)	−66.8	−74.0	−68.6	−54.4	−42.0

*Medianized (central) values within groups of like-classified measurements in a given frequency band, including both V- and H-polarizations and high (15 or 36 m) and low (150 m) range resolution.

respectively. If the terrain can be separated by class (as it is in Fig. 7.4 and Table 7.1), the variability in all bands is substantially reduced.

Some of the other findings of this comprehensive measurement program were:

• Most of the significant land clutter echoes at low angles come from spatially localized or discrete vertical features associated with the high regions of the visible landscape (such as trees, tree lines, buildings, fences, or high points of the terrain). Low regions of terrain are shadowed at low grazing angles so that the clutter is spatially patchy. Clutter occurs "within kilometer-sized macroregions of general geometric visibility, each containing hundreds or thousands of spatial resolution cells." Groups of cells that produce a strong return are often separated by cells with weak echoes or just receiver noise. (The patchiness of the clutter is what gives rise at low grazing angles to interclutter visibility, Sec. 7.8).

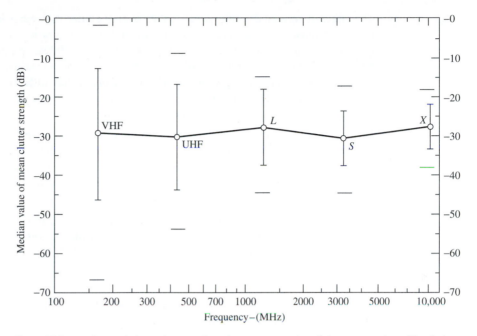

Figure 7.5   General dependence at low depression angles of the mean value of land clutter (circles), standard deviation (vertical lines), and the extreme measured values (horizontal bars) as a function of frequency at 36 rural sites.
| (After J. B. Billingsley and J. F. Larrabee.[3])

- Over the range of depression angles employed in these measurements, the mean clutter strengths increase and the cell-to-cell fluctuations decrease with increasing angle. This is attributed to the reduction in masking with increasing angle. At the lower angles where masking occurs, the statistics can be represented by the Weibull probability density function. At depression angles of 6 to 8°, clutter is no longer spiky and is represented by the Rayleigh probability density.

- The difference between vertical and horizontal polarization is small, typically about 1 or 2 dB. The median difference between polarizations over all frequencies is 1.5 dB and the standard deviation is 2.8 dB, with little apparent dependence on frequency. (An exception is at VHF in steep mountainous terrain where the measurements show that the mean clutter strength is 7 to 8 dB greater with vertical polarization than with horizontal.) There was also little apparent dependence of the polarization on the range resolutions that were used. Even though the nominal calibration accuracy is 2 dB over all sites, the authors of reference 3 "accept the conclusion that, on average, the mean ground clutter strength is often 2 dB or so stronger at vertical polarization than at horizontal." They suggest that this may be due to the vertical orientation of many discrete clutter scatterers.

- The variation of clutter echo due to weather or season was found to be small, usually less than 1.5 dB for weather and 3 dB for changes in season. This was believed to be related to low-angle clutter being dominated by discrete scatterers.

- The effect of vertical discrete objects on the overall clutter strength is large even when they are relatively sparse. They contribute to clutter in a major way and are relatively unchanged with season. For example, the telephone lines, trees, and the fence around a wheat field are much more significant scatterers than the wheat itself. Even though the wheat field changes appearance with the seasons, it has little effect on the overall clutter strength. The echo from a tree line contributes strong clutter echoes no matter whether the trees are in leaf or bare, wet or dry.

- When a ground-based radar experiences strong clutter from visible terrain at ranges of 100 km or more, it is generally due to echoes from mountains that rise high enough to be within the line-of-sight of the radar. The echoes from mountains at long range are significantly greater when viewed with VHF than with microwave frequencies.

- Forests provide mean clutter echoes at VHF that are 10 to 15 dB stronger than at microwaves. This is attributed to the decreased loss at the lower frequencies when propagating radar energy through forests. Clutter from farmland, however, is 20 or 30 dB less at VHF than at microwaves since multipath (Sec. 8.2) is more of a factor with relatively flat, smooth surfaces. (Multipath, when it is present, reduces the energy at low angles.)

- Since forests tend to destroy the multipath, forest clutter echoes at high illumination angles are more likely to be the intrinsic value of σ^0 rather than $\sigma^0 F^4$.

- The median difference of the mean ground clutter strength between the low (150 m) and high resolutions (15 or 36 m) used in these measurements is less than 2 dB. Thus, on average, over many measurements, there is no significant difference expected between low and high range-resolution.

- Although the mean value of the clutter echo signal does not depend on resolution, the *variation* of the clutter echo amplitudes is a function of resolution. The smaller the resolution cell, the less the averaging within the cell and the greater is the variability from cell to cell. Reference 3 provides measurements of the ratio of the standard deviation to the mean, the skewness (third central moment), and kurtosis (fourth central moment) for the seven major terrain types at the five frequencies. Tables of the 50-, 70-, and 90-percentile levels are also given.

- Many past radar designs were based on only the mean value of the expected clutter. Different types of clutter, however, might have the same mean value, but quite different statistical variations. Figure. 7.6a is a histogram of the clutter for a particular region of forest at X band. Its mean is about the same as the histogram from farmlands shown in Fig. 7.6b. The distributions, however, are seen to be quite different. It should not be expected that the performance of a radar design based only on the mean value of clutter will be the same in these two regions. The mean-to-median ratio for forest in this case was 8 dB and for level farmland it was a very large value of 33 dB. The histograms for VHF over these same two sites are shown in Figs. 7.6c and d. The means are quite different in these two cases. The mean-to-median ratio was 4 dB for forest and 15 dB for farmland.

- The most likely day-to-day difference in the mean clutter strength was 0.2 dB, the average difference was about 1 dB, and the one-sigma range of variability beyond

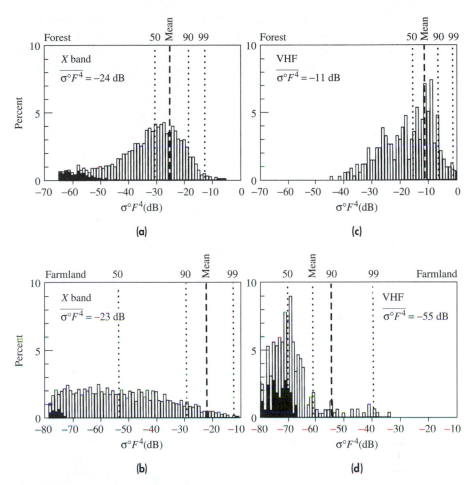

Figure 7.6 Examples of histograms of *X*-band land clutter at low depression angles in (a) a particular region of forest and (b) a particular level-farmland; both with the same approximate mean value of clutter strength, but with different statistical distributions. Histograms for the same terrain, but at VHF, are shown in (c) and (d). Black values are receiver noise.

(From J. B. Billingsley and J. F. Larrabee.[3] Reprinted with permission of MIT Lincoln Laboratory, Lexington, Massachusetts.)

the mean was 2.5 dB. Occasionally when the day-to-day difference was large (near 10 dB), it was attributed to measurement equipment malfunctions or singular events such as a train passing through the scene.

- It is said that land clutter's most salient attribute is its variability. This needs to be kept in mind when designing a radar to detect targets in clutter. The design should be conservative in order to provide reliable detection over a wide range of possible clutter values.

Figure 7.7 is an *X*-band PPI display of the clutter seen at Gull Lake West, in Manitoba, Canada, one of the sites measured by Lincoln Laboratory. The maximum range is 7 km, range resolution is 15 m, and the polarization is horizontal. Cells with $\sigma^0 F^4 \geq -40$ dB are shown in white. The area outlined in black lines in the northwest is a sector in which much of the data was recorded. This particular area is a forested wetland out to 3.5 km range, followed by a swampy open pond from 3.5 to 5 km, then to a higher sand dune along the shore of Lake Winnipeg between 5 and 6 km. Beyond that is the water of Lake Winnipeg. This PPI display illustrates that land clutter consists of many different kinds of textures and degrees of spatial correlation. It shows that clutter does not occur as "random salt and pepper."

The cumulative amplitude distributions of the *X*-band echo from rural low-relief and rural high-relief terrain are shown in Fig. 7.8 as a function of depression angle.[4] By low relief is meant slopes <2° and variations in height <100 ft; and by high relief, slopes >2° and variations in height >100 ft. At the higher depression angles, the slope of the distribution approaches that of the Rayleigh, which indicates that microshadowing of the clutter is small at higher angles.

Figure 7.9, derived from Billingsley,[4] shows the variation of the mean and median values for low- and high-relief rural land clutter. At the top of this figure are values of the Weibull skewness parameter, which is mentioned later in Sec. 7.5 and Eq. (7.18).

Further information about the Lincoln Laboratory ground-clutter measurements can be found in the detailed reports by Billingsley.[3-5]

Figure 7.7   PPI clutter map at Gull Lake West, Manitoba, Canada. Cells with $\sigma^0 F^4 \geq -40$ dB are shown in white.
(From J. B. Billingsley and J. F. Larrabee.[3] Reprinted with permission of MIT Lincoln Laboratory, Lexington, Massachusetts.)

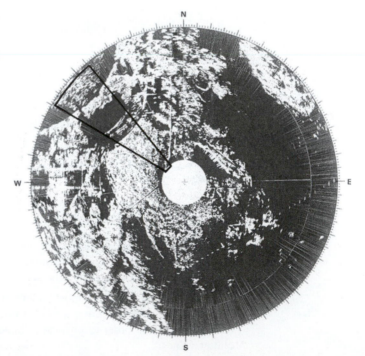

Figure 7.8
Cumulative amplitude probability distributions as a function of depression angle for X-band clutter from rural terrain of low and high relief. A straight line on this graph indicates Weibull clutter. The slope for a Rayleigh distribution (Weibull with $\alpha = 2$) is shown for comparison.
(From J. B. Billingsley.[4] Reprinted with permission of MIT Lincoln Laboratory, Lexington, Massachusetts.)

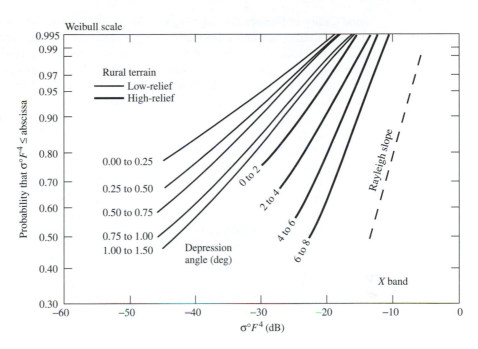

Figure 7.9 Mean and median values of low- and high-relief rural terrain as a function of depression angle. The upper curves give the Weibull skewness parameter.
(After J. Billingsley.[4])

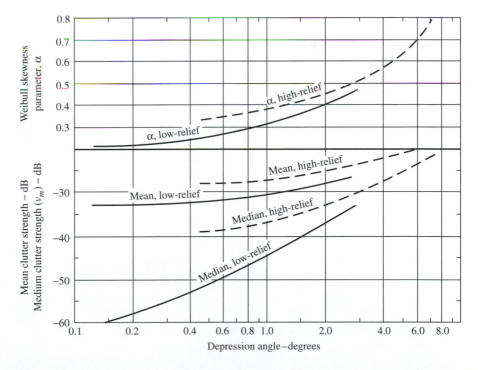

Land Clutter at Medium Grazing Angles There exist many measurements of land clutter at medium grazing angles, from a few degrees to about 70° (see, for example, references 2,6,9,10,12,15,17–19). Much of the information in this angle regime has been obtained from aircraft or spacecraft, where there is poor ground truth about the nature of the clutter being viewed by the radar. Radars mounted on portable "cherry pickers" also have been used to acquire data at medium grazing angles, especially by those interested in remote sensing. Ground truth is easier to obtain with this type of instrumentation since the radar is a short distance from the clutter and one can readily discern its nature. Radars mounted on cherry pickers, however, do not obtain the highly averaged values possible with airborne or spaceborne sensors.

An example of the average value of clutter that might be typical of North America in the summer is shown in Fig. 7.10. This figure was derived from data presented by Moore et al.[6] It represents a combination of measurements made by the Skylab S-193 scatterometer (at 13.9 GHz) and the University of Kansas "microwave-active-spectrometer system"* mounted on a cherry picker. The ground-based cherry-picker radars could operate from 1 to 18 GHz. The data from the cherry-picker radars were combined by Moore et al. by comparing the 13.8-GHz ground-based data with the 13.9-GHz Skylab measurements. The absolute level of the model was determined by the Skylab measurements, but the relationships among the other frequencies were set by the ground-based measurements. The curve in Fig. 7.10 for *L* band combines the horizontal and vertical polarizations, and represents an average over the two years for which measurements were reported. The *X*-band values at low grazing angles were taken from the means averaged over all terrain types, as presented by Billingsley.[4] The dashed portion of the curve is a bold interpolation.

Many experimental measurements of land clutter seem to indicate that the value of sigma zero at grazing angles from a few degrees to perhaps 70° is approximately

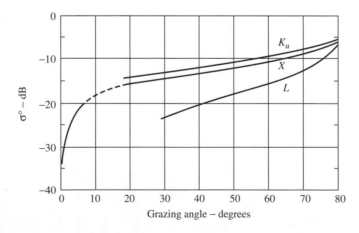

Figure 7.10 Clutter sigma zero that might be typical of North America in the summer.

(After Moore et al.[6] Courtesy of the IEEE. The *X*-band values at low angles are taken from Billingsley.[4])

| *A radar by any other name is still a radar.

proportional to the sine of the grazing angle ψ. For this reason, land clutter is sometimes described by the parameter γ, defined as

$$\gamma = \frac{\sigma^0}{\sin \psi} \tag{7.10}$$

The parameter γ is said to be almost independent of the grazing angle in this angular regime. (Note that a perfectly rough surface that reradiates in the direction of the source a power per unit solid angle independent of the direction of the incident energy will result in $\sigma^0 = 2 \sin \psi$.[7] Substituting in Eq. (7.10), γ in this case is 2, or 3 dB.) Nathanson[2] states that the maximum value of γ for grazing angles from 6 to 70°, all frequencies from 0.4 to 35 GHz, and for all polarizations is -3 dB. The median is -14 dB, and the minimum is -29 dB. In this oversimplification, these values are a maximum of various experiments rather than a peak value in time. Barton[8] adds that for rural terrain covered by crops, bushes, and trees, the value of γ lies between -10 and -15 dB, and that urban and mountain clutter might have a value of γ that approaches -5 dB. Billingsley's X-band data for all terrain types at depression angles from 1 to 8° can be approximated by $\gamma = -11$ dB.[4]

These values of γ are said to apply over a wide range of frequencies; but it is known that they are frequency dependent, as was indicated in Fig. 7.10. Nathanson suggests that since these values are based to large extent on X-band data, they might be considered as applicable for X band and that the frequency dependence can be represented by

$$\gamma = \gamma_{10} + 5 \log \left(\frac{f}{10} \right) \tag{7.11}$$

where f is the frequency in GHz and γ_{10} is the value of γ at 10 GHz. For example, if $\gamma_{10} = -14$ dB at X band, γ will be -18.5 dB at L band. Equation (7.11) is shown by the dashed curve in Fig. 7.11.

The solid curves of Fig. 7.11 are derived from Table 7.16 of Ref 2. They plot the maximum value of γ for various terrains. These are an indication of the decibel average

Figure 7.11 Dashed curve is the mean value of γ as given by Eq. (7.11) with $\gamma_{10} = -14$ dB at X band. The solid curves are the maximum values of γ as given by Table 2.16 of Ref. 2.

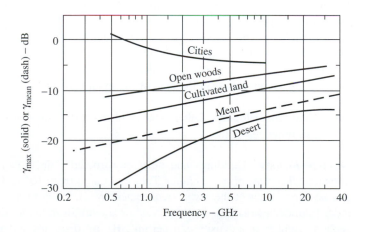

of the maximum values reported by various experimenters rather than an indication of the maximum expected return. (Because of the wide variation in the data, some liberties have been taken in smoothing the values from which Fig. 7.11 was derived.)

Land Clutter at High Grazing Angles (Near Vertical Incidence)   The values of σ^0 can be large near vertical incidence. As mentioned previously, scattering in this regime is due to those facets (surfaces that are "flat" relative to a wavelength) that reflect incident energy back to the radar. The values of σ^0 near vertical incidence can also be affected by the antenna pattern and the antenna gain.

When determining the values of σ^0 at perpendicular incidence, the effect of the antenna pattern shape must be taken into account. Because of the finite antenna beamwidth, measured values of σ^0 at vertical incidence are sometimes lower than actual. This results from the averaging done over a finite beamwidth where the clutter echo is changing rapidly with angle. With a finite beamwidth antenna, the large value at the peak (90° grazing angle) is averaged with lower values at angles off normal so that the value obtained in this manner is less than it really is at 90°.

The antenna gain can influence, in some cases, the value of σ^0 near vertical incidence. Schooley[7] has shown that the value of σ^0 is $4/(\theta_{2B})^2$ at normal incidence (grazing angle = 90°) for an infinite perfectly smooth reflecting surface when the antenna is a pencil beam of two-way beamwidth θ_{2B}. The one-way beamwidth $\theta_B = \sqrt{2}\ \theta_{2B}$. Since the maximum gain G of an antenna is approximately equal to $\pi^2/(\theta_B)^2$, the value of sigma zero in this case is

$$\sigma^0 = \frac{8}{\theta_B^2} = \frac{8G}{\pi^2} \approx G \qquad\qquad [7.12]$$

Thus the antenna gain G can have a significant effect on the value of σ^0 at normal incidence and in the vicinity of normal incidence for a perfectly smooth surface. This probably applies, however, more to a perfectly smooth sea than to land. Over a perfectly flat earth with a reflection coefficient less than unity, the value of σ^0 will be reduced by the (power) reflection coefficient. If the surface is slightly rough (roughness small compared to a wavelength), the value of σ^0 will also be reduced similar to that for an antenna with errors in the aperture distribution.

Examples of the values of σ^0 near vertical incidence may be found in Chapter 12, Ground Echo, in the *Radar Handbook*.[9]

The value of σ^0 at normal incidence is generally larger than unity. That is, the radar cross section is greater than the area of the radar resolution cell.

Other Land Clutter Topics   Other subjects of interest in the radar scattering from land are briefly summarized below.

Discrete Echoes   Buildings and other constructed objects can result in large echoes, known as *discretes* or *point clutter.* They can range from 10^4 to 10^5 m^2 at S band.[10] They can be even larger at the lower frequencies. W. H. Long et al.[11] state that at the higher radar frequencies (assumed to be S band or above) discrete clutter echoes of 10^4 m^2 cross section might have a density of 1 per mi^2; 10^5 m^2 discretes, a density of 0.1 per mi^2; and

10^6 m^2 discretes, only 0.01 per mi^2. It takes a good MTI radar to eliminate large discrete echoes. Other techniques (such as blanking based on a clutter map) might be used to reduce the effects of discrete echoes when the MTI cannot eliminate them.

Snow   The effect of snow covering the ground depends on its thickness, water content, and frequency. Dry snow (no liquid water) has a dielectric constant ranging from 1.4 to 2, so that significant transmission of energy can take place across the air-snow boundary over a wide range of angles.[12] Dry snow is of low loss so that radar energy is not highly attenuated when propagating through the snow. The ground beneath the snow, therefore, can have a major effect on σ^0, especially at low frequencies and shallow snow cover where the attenuation is quite low. It was found, for example, that 15 cm of dry powder-like snow had no effect on the measured backscatter over the frequency range from 1 to 8 GHz.[13] Any backscatter from snow under these conditions was completely dominated by the contribution from the underlying surface. With 12 cm of wet snow, however, σ^0 decreased approximately 5 to 10 dB at grazing angles between 30 and 80°.

The attenuation in propagating through snow depends on the amount of water in the snow. When the sun shines, it melts the snow and produces water, which increases attenuation. At night this water freezes. Thus there can be large variations in radar scattering, depending on whether it is night or day.[6] For example, it was reported that in the morning as the snow began to melt and produce water, the radar return decreased as much as 10 dB within half an hour (at the higher microwave frequencies, 40° grazing angle, and 48 cm snow depth). Figure 7.12 shows the effect of dry and wet snow as a function of frequency at a grazing angle of 40°.[14] At *L* band, the snow is essentially transparent and the echo is primarily from the underlying ground. At the upper end of the microwave frequency spectrum, at K_a band, the snow is likely to have a dominant effect on the backscatter.

Other examples of radar measurements of snow can be found in *Microwave Remote Sensing,* vol III.[15]

Snow and Altimeters   The effect of snow (or ice) over the ground is dramatically illustrated by the problem it can introduce in some radar altimeters. Serious, life-threatening

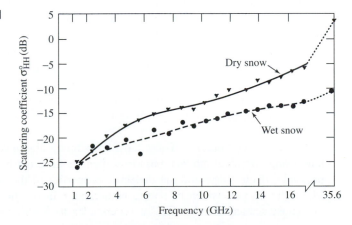

Figure 7.12   Effect of wet and dry snow as a function of frequency for a grazing angle of 40°.

(Reprinted from Ulaby and Stiles, Ref. 14, Copyright 1981, with permission from Elsevier Science.)

errors in the measurement of the aircraft altitude can result, especially at UHF or lower frequencies where the propagation loss in snow is low.[16] In some cases, the echo from the air-snow interface can be so small that it might be missed and the altitude measurement is falsely determined by the reflection from the ground underneath the snow. To make matters worse, the dielectric constant of snow compared to air results in a greater altitude measurement for a given time delay if the velocity of propagation in air rather than snow is assumed. (The velocity of propagation in snow is about 0.8 that in air; in ice it is 0.55 times that in air.) It can result in a pilot being misled as to the altitude of the aircraft above the snow and might crash into the snow, thinking that the plane's altitude was safely above the surface. This can be a serious problem with UHF altimeters when flying over a region such as the Greenland ice cap. Altimeters are now operated at the higher microwave frequencies (4.2–4.4 GHz) where the propagation loss is higher than that at UHF so that altimeter measurements are more likely to indicate the distance to the air-snow interface rather than to the ground beneath the snow.

Remote Sensing   Much information on land backscatter reported in the literature has been obtained for purposes of developing radar for the remote sensing of the environment. The goal of the remote sensing radar designer is different from that of other radar designers. In remote sensing the interest is in acquiring information about the environment (such as soil moisture or crop census) while other radar engineers are interested in detecting targets in unwanted clutter. The data of interest to the remote sensing designer, therefore, might not always be useful to the radar designer, and vice versa. Examples of the backscatter from land obtained mainly for purposes of remote sensing design are given by Ulaby and Dobson[17] as well as in a computer database at the University of Massachusetts.[18]

Variability of Land Clutter   As mentioned earlier in this chapter, the chief characteristic of clutter is its variability. There is seldom precise agreement among similar data taken by different investigators. Long[19] points out that "two flights over apparently the same type of terrain may at times differ by as much as 10 dB in σ^0." Another analysis of the problem of the variability of the value of σ^0 from eleven sets of measurements of agricultural crops under "similar" conditions and for approximately the same radar parameters (frequency, polarization, and grazing angle) found maximum differences of approximately 17 dB.[20] Some of these differences are due to the absence of good ground truth (that is, the measurements are not from similar terrain), accurate system calibration, or a lack of accurate data processing (including accounting for the effect of range variation of clutter as well as the effect of the antenna pattern shape). The variation and uncertainty in the value of land clutter is something the radar designer has learned to accept and compensate for by conservative system design.

Theory of Land Clutter   There have been many attempts to describe scattering from land using empirical models; but these attempts have not been completely successful. In principle, if the shape and the dielectric properties of the scattering surface are known, Maxwell's equations can be solved numerically. In practice, however, the land features that cause scattering are generally too complex to describe simply.

Radar and Land Clutter   Information about radar backscatter from land is required for several diverse applications, each with its own special needs that differ from the others. These applications include:

- *Detection of aircraft over land.* Clutter echoes might be 50 to 60 dB, or more, greater than aircraft echoes. MTI or pulse doppler radar (Chap. 3) is commonly used for this application to remove the large unwanted clutter.

- *Detection of moving surface-targets over land.* Vehicles or personnel can be separated from clutter by means of properly designed doppler processing radar.

- *Altimeters.* A large echo from the ground or the sea is desired for the measurement of the height of an aircraft or spacecraft since "clutter" is the target. The altimeter has also been used in "map matching" for missile guidance, as well as for remote sensing.

- *Detection and height measurement of terrain features.* This provides warning to an aircraft that it is approaching high ground so that the aircraft can fly around it (terrain avoidance) or follow the contour of the land (terrain following).

- *Mapping, or imaging, radars.* These utilize high resolution to recognize ground objects by their shape and by contrast with their surroundings. Such radars might be used by the military for navigation or target recognition. The synthetic aperture radar (SAR), sidelooking airborne radar (SLAR), and military air-to-surface radar are examples.

- *Remote sensing.* Imaging radars (such as SAR), altimeters, and scatterometers (a radar that measures sigma zero as a function of elevation angle) are used to obtain specific information about the characteristics of the earth's surface.

7.4   SEA CLUTTER

The radar echo from the sea when viewed at low grazing angles is generally smaller than the echo from land. It usually does not extend as far in range as land clutter and is more uniform over the oceans of the worlds than typical land clutter. It has been difficult, however, to establish reliable quantitative relationships between sea echo measurements and the environmental factors that determine the sea conditions. Another difficulty in dealing with sea echo is that the surface of the sea continually changes with time. Nevertheless, there does exist a large body of information regarding the radar echo from the sea that can be used for radar design and provide a general understanding of its effect on radar performance.

The nature of the radar echo (clutter) from the sea depends upon the shape of the sea surface. Echoes are obtained from those parts of the sea whose scale sizes (roughness) are comparable in dimension to the radar wavelength. The shape, or roughness, of the sea depends on the wind. Sea clutter also depends on the pointing direction of the radar antenna beam relative to the direction of the wind. Sea clutter can be affected by contaminants that change the water surface-tension. The temperature of the water relative to that of the air is also thought to have an effect on sea clutter.

The *sea* generally consists of waves that result from the action of the wind blowing on the water surface. Such waves, called *wind waves,* cause a random-appearing ocean-height profile. *Swell waves* occur when wind waves move out of the region where they were originally excited by the wind or when the wind ceases to blow. Swell waves are less random and sometimes appear to be somewhat sinusoidal. They can travel great distances (sometimes thousands of miles) from the place where they originated. The echoes from an *X*-band radar viewing swell at low grazing angles will be small if there is no wind blowing, even if the swell waves are large. If a wind occurs, the surface will roughen and radar echoes will appear.

Sea state is a term used by mariners as a measure of wave height, as shown in Table 7.2. The sea state description shown in this table is that of the World Meteorological Organization. Sea state conditions can also be described by the Douglas scale, the Hydrographic Office scale, and the Beaufort scale. The Beaufort is actually a wind-speed scale.[21] Each gives slightly different values, so when a sea state is mentioned one should check which scale is being used.

Although sea state is commonly used to describe the roughness of the sea, it is not a complete indicator of the strength of sea clutter. Wind speed is often considered a better measure of sea clutter, but it is also limited since the effect of the wind on the sea depends on how long a time it has been blowing (called the *duration*) and over how great a distance (called the *fetch*). Once the wind starts to blow, the sea takes a finite time to grow and reach equilibrium conditions. When equilibrium is reached it is known as a *fully developed sea.* For example, a wind speed of 10 kt with a duration of 2.4 h and a fetch of at least 10 nmi produces a fully developed sea with a significant wave height (average height of the one-third highest waves) of 1.4 ft.[22] It corresponds to sea state 2. A 20 kt

Table 7.2 World Meteorological Organization sea state

| Sea State | Wave Height | | Descriptive Term |
	Feet	Meters	
0	0	0	Calm, glassy
1	$0-\frac{1}{3}$	0–0.1	Calm, rippled
2	$\frac{1}{3}-1\frac{2}{3}$	0.1–0.5	Smooth, wavelets
3	2–4	0.6–1.2	Slight
4	4–8	1.2–2.4	Moderate
5	8–13	2.4–4.0	Rough
6	13–20	4.0–6.0	Very rough
7	20–30	6.0–9.0	High
8	30–45	9.0–14	Very high
9	over 45	over 14	Phenomenal

wind blowing for 10 h over a fetch of 75 nmi results in a significant wave height of about 8 ft and corresponds to sea state 4.

Average Value of σ^0 as a Function of Grazing Angle   A composite of sea clutter data from many sources is shown in Fig. 7.13. This figure was derived from data for winds varying from approximately 10 to 20 kt, and can be considered representative of sea state 3. (Sea state 3 is roughly the medium value of sea state; i.e., about half the time over the oceans of the world, the sea state is 3 or less.) It is believed that Fig. 7.13 is representative of the average behavior of sea clutter, but there is more uncertainty in the data than is indicated by the thin lines with which the curves were drawn.

The curves of Fig. 7.13 provide the following conclusions for sea clutter with winds from 10 to 20 kt:

- At high grazing angles, above about 45°, sea clutter is independent of polarization and frequency.

- Sea clutter with vertical polarization is larger than with horizontal polarization. (At higher wind speeds the differences between the two polarizations might be less.)

- Sea clutter with vertical polarization is approximately independent of frequency. (This seems to hold at low grazing angles even down to frequencies in the HF band.)

- At low grazing angles, sea clutter with horizontal polarization decreases with decreasing frequency. This is apparently due to the interference effect at low angles between the direct radar signal and the multipath signal reflected from the sea surface.

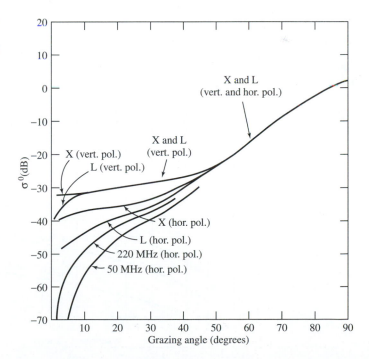

Figure 7.13   Composite of averaged sea clutter σ^0 data from various sources for wind speeds ranging from 10 to 20 kt.

There is no simple law that describes the frequency dependence of sea clutter with horizontal polarization.

Effect of Wind   The wind is the most important environmental factor that determines the magnitude of the sea clutter. At low grazing angles and at microwave frequencies, backscatter from the sea is quite low when the wind speed is less than about 5 kt. It increases rapidly with increasing wind from about 5 to 20 kt, and increases more slowly at higher wind speeds. At very high winds, the increase is small with increasing wind.

Figure 7.14 was derived from experimental data of John Daley et al. using the Naval Research Laboratory Four-Frequency Airborne Radar.[23-25] As mentioned, sea clutter at the higher microwave frequencies and low grazing angles increases with increasing wind speed, but begins to level off at winds of about 15 to 25 kt. When viewed at vertical incidence (grazing angle of 90°) and with zero or low wind speed, the sea surface is flat and a large echo is directed back to the radar. As the wind speed increases and the sea surface roughens, some of the incident radar energy is scattered in directions other than back to the radar, so that σ^0 will decrease. According to Daley et al., the value of σ^0 at vertical incidence in Fig. 7.14 can be represented by $25\,w^{-0.6}$, where w is the wind speed in knots.

At low grazing angles, less than about one degree, it is difficult to provide a quantitative measure of the effect of the wind on sea clutter. This is due to the many factors that influence σ^0 at low angles, such as shadowing of parts of the sea by waves, multipath interference, diffraction, surface traveling (electromagnetic) waves, and ducted propagation.

Figure 7.14   Effect of wind speed on sea clutter at several grazing angles. Radar looking upwind. Solid curves apply to X band, dashed curve applies to L band.
| (After J. C. Daley, et al.[23-25])

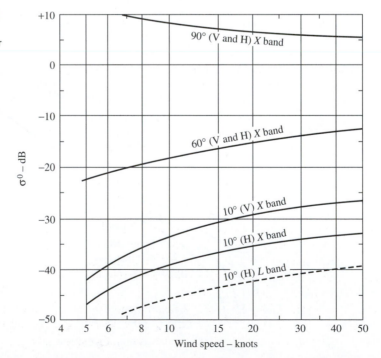

Another factor affecting the ability to obtain reproducible measurements of sea clutter is that a finite time and a finite fetch are required for the sea to become fully developed. Unfortunately, few measurements of sea clutter mention the fetch and duration of the wind.

Sea clutter is largest when the radar looks into the wind (upwind), smallest when looking with the wind (downwind), and intermediate when looking perpendicular to the wind (crosswind). There might be as much as 5 to 10 dB variation in σ^0 as the antenna rotates 360° in azimuth.[26] Backscatter is more sensitive to wind direction at the higher frequencies than at lower frequencies; horizontal polarization is more sensitive to wind direction than vertical polarization; the ratio of σ^0 measured upwind to that measured downwind decreases with increasing grazing angle and sea state; and at UHF the backscatter is practically insensitive to wind direction at grazing angles greater than 10°.

The orthogonal component of polarization from sea clutter (cross polarization response) at grazing angles from 5 to 60° appears to be about 5 to 15 dB less than the echo from the same polarization (co-pol) as transmitted.[25]

Sea Clutter with High-Resolution Radar (Sea Spikes)   The use of a clutter density, σ^0 (clutter cross section per unit area), to describe sea clutter implies that the clutter echo is independent of the illuminated area. When sea clutter is viewed by a high-resolution radar, especially at the higher microwave frequencies (such as X band), sea clutter is not uniform and cannot be characterized by σ^0 alone. High-resolution sea clutter is spiky. The individual echoes seen with high-resolution radar are called *sea spikes*. They are sporadic and have durations of the order of seconds. They are nonstationary in time, spatially nonhomogeneous, and have a probability density function that is non-Rayleigh. Sea spikes are important since they are the major cause of sea clutter at the higher microwave frequencies at low grazing angles, with any radar resolution.

Fig. 7.15 is an example of the time history of sea spikes in sea state 3 for pulse widths varying from 400 to 40 ns and with vertical polarization.[27,28] With 40-ns pulse width (6-m range resolution), Fig. 7.15 shows that the time between sea spikes can be several tens of seconds and the duration of each spike is of the order of one or a few seconds. As the pulse width is increased, more sea spikes appear within the larger resolution cell of the radar and the time between spikes decreases. The peak radar cross section of sea spikes in this example is almost 10 m^2. (At times they have been observed to be of even higher cross section.) The relatively large cross section and time duration of sea spikes can result in their being mistaken for small radar targets. This is a major problem with sea spikes; they can cause false alarms when a conventional detector based on gaussian receiver noise is used.

Under calm conditions, sea state 1 or less, the echo signals still have the same spiky appearance as in Fig. 7.15, but are as much as 40 dB lower in cross section. A time history similar to that of Fig. 7.15, but for horizontal polarization, would show that sea spikes with this polarization occur slightly less frequently and are sharper (of briefer duration) than with vertical polarization.

Based on airborne radar measurements made at L, S, and X bands with pulse widths ranging from 0.5 to 5 μs, clutter is more spiky when the radar looks upwind or downwind rather than crosswind and with low rather than high grazing angles.[29]

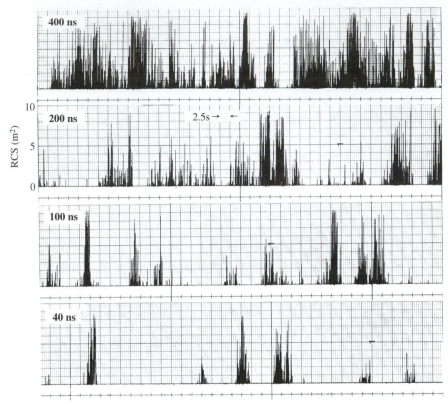

Figure 7.15 Amplitude as a function of time at a fixed range-resolution cell for low grazing angle X-band (9.2 GHz) sea clutter obtained off Boca Raton, Florida, with pulse widths ranging from 400 ns to 40 ns in a windblown sea with many white caps (sea state 3). Cross-range resolution is approximately 9 m, grazing angle of 1.4°, and vertical polarization.
| (From J. P. Hansen and V. F. Cavaleri.[27])

Sea spikes are evident when the radar resolution is less than the water wavelength. The physical size of sea spikes usually is less than the resolution of the radar which observes them, and it is said they appear to move at approximately the surface wave velocity.[30] Sea spikes obviously are present with low as well as high resolution. At low resolution, the sum of many individual sea spikes within the resolution cell produces an almost continuous noiselike echo. At the lower frequencies where the radar wavelength is large compared to the sea-surface features that give rise to sea spikes, it would be expected that the backscatter is no longer characterized by sea spikes.

Another characteristic of sea spikes is a relatively rapid and high-percentage pulse-to-pulse amplitude modulation. According to Hansen and Cavaleri,[27] measured modulation frequencies at X band vary from 20 to 500 Hz. The modulating frequency seems to be affected by the type of physical surface (breaking water, sharp wave crests, ripples, etc.), the relative wind speed and direction, and polarization of the radar. The characteristic amplitude modulations can be used for recognizing and rejecting sea spike echoes from true target echoes.[31]

In addition to changes in clutter characteristics with narrow pulse widths, as illustrated in Fig. 7.15, similar effects have been noted with narrow antenna beamwidths.[32] In some cases, target detection in clutter can actually be enhanced when a smaller antenna with a lower resolution is used rather than one with a high resolution that is bothered by sea spikes.[32] On the other hand, with very high resolution, targets can be seen in the clear regions between the sea spikes, so that subclutter visibility in the traditional sense might not be required.

Origin of Sea Spikes   Sea spikes are associated with breaking waves or waves about to break. Visible whitecaps are also associated with breaking waves, but whitecaps themselves do not appear to be the cause of sea spikes since whitecaps are mainly foam with entrapped air (which does not result in significant backscatter). It has been reported[33] that about 50 percent of the time the whitecap is seen visually either simultaneously or a fraction of a second later than the appearance of a sea spike on the radar display. About 35 to 40 percent of the time a spike is observed when the waves have a very peaked crest, but with no whitecap developed. A sea spike echo can appear without the presence of a whitecap, but no whitecap is seen in the absence of a radar observation of a spike. Thus it can be concluded that the whitecap is not the cause of the sea spike echo.

Wetzel[34] has offered an explanation for the origin of sea spikes based on the entraining plume model of a spilling breaker.[35] In this model "a turbulent plume emerges from the unstable wave crest and accelerates down the forward face of the breaking wave, entraining air as it goes." Wetzel further assumes "that the breaking 'event' involves a cascade of discrete plumes emitted along the wave crest at closely spaced times." Based on this model and some assumptions about the characteristics of the plume, Wetzel was able to account for the peak radar cross section, frequency dependence, the spikier nature of horizontal polarization, similarity of the appearance of sea spikes with both horizontal and vertical polarizations at very low grazing angles, appearance of the characteristic modulation, and other properties. He has pointed out that this model is based on a simplistic hypothesis that requires further elaboration. It does not account, however, for the scattering when the radar views the sea downwind and it does not adequately explain the internal amplitude modulations of the sea spikes.

Detection of Signals in High-Resolution Sea Clutter   The nature of sea spikes as seen by high-resolution radar results in a probability density function (pdf) that is not Rayleigh. Therefore, conventional methods for detection of signals in gaussian noise do not apply. (A Rayleigh pdf for clutter power is equivalent to a gaussian pdf for receiver noise voltage.) Non-Rayleigh pdfs have "high tails"; that is, there is a higher probability of obtaining a large value of clutter than when it is Rayleigh. A receiver detector designed as in Chap. 2 on the basis of gaussian noise, or Rayleigh clutter, will result in a high false-alarm rate when confronted with sea spikes. The statistics of non-Rayleigh sea clutter are not easily quantified and can vary with resolution and sea state. Thus conventional receiver detector-design based on the assumption of gaussian noise cannot be applied. To avoid excessive false alarms with non-Rayleigh sea clutter, the detection-decision threshold might have to be increased (perhaps 20 to 30 dB). The high thresholds necessary to avoid false alarms reduce the probability of detecting desired targets. When sea spikes are a concern, detection criteria other than those based on gaussian or Rayleigh statistics must be used if a severe penalty in detection capability is to be avoided.

One method for dealing with a sea spike is to recognize its characteristic amplitude modulations and delete the sea spike from the receiver.[31] Another method is to employ a receiver with a log-log output-input characteristic whose function is to provide greater suppression of the higher values of clutter than a conventional logarithmic receiver. In a log-log receiver, the logarithmic characteristic progressively declines faster than the usual logarithmic response by a factor of 2 to 1 over the range from noise level to +80 dB above noise.[36]

Conventional pulse-to-pulse integration does not improve the detection of targets in spiky clutter because of the long correlation time of sea spikes. However, a high antenna scan rate (several hundred rpm) allows independent observations of the clutter to be made, so that scan-to-scan integration can be performed.[37] Similarly, if the target is viewed over a long period of time before a detection decision is made, the target can be recognized by its being continuously present on the display while sea spikes come and go. One method for achieving this is *time compression,* as mentioned later in Sec. 7.8.

The effect of sea spikes is less important for radars that are to detect ships since ship cross sections are large compared to the cross sections of sea spikes. Sea spikes might interfere, however, with the detection of small targets such as buoys, swimmers, submarine periscopes, debris, and small boats. With ultrahigh resolution (ultrawideband radar) where the range resolution might be of the order of centimeters, sea spikes are relatively sparse in both time and space so that it should be possible to see physically small targets when they are located in between the spiky clutter.

Sea Clutter at Very Low Grazing Angles[38]   Clutter at very low grazing angles differs from that at higher angles because of shadowing, ducted propagation (Sec. 8.5), and the changing angle at which the radar ray strikes the fluctuating sea surface.

The surface of the sea is seldom perfectly flat. It usually has a time-varying angle with respect to the radar. A grazing angle might be defined with respect to the horizontal, but it is difficult to determine the angle made with respect to the dynamic sea surface. Refraction by the atmosphere can also change the angle the radar ray makes with the surface.

Shadowing of wave troughs by wave crests can occur. Scatterers as seen by the radar are thought to be mainly those from the crests, especially for horizontal polarization. Attempts to compute the effect of shadowing by simple geometrical considerations (to determine how much of the sea surface is masked) have not proven successful. One reason for failure of geometrical shadowing is that scattering features (such as sloshes, plumes, and other effects of breaking waves) are not uniformly distributed and are more likely to be near the crests of the waves. Diffraction effects can occur which complicate shadowing calculations. Thus the effect of shadowing is more complicated than just simple masking.[38,39]

Another factor to consider when searching for a theoretical understanding of sea spikes (and microwave sea clutter in general) is the effect of a surface traveling radar wave that is launched when the incident wave has a component of electric field in the plane of incidence.[40] The effect was mentioned in Sec. 2.7 and shown in Fig. 2.10 for scattering from a long thin road. The surface traveling wave and its reflection from a discontinuity can be a reason why microwave sea clutter seen with vertical polarization usually is larger than sea clutter seen with horizontal polarization.

At very low grazing angles (less than about one degree), sea clutter should decrease rapidly with decreasing grazing angle, especially for horizontal polarization. Figure 7.3, which is a generic plot of clutter echo strength as a function of grazing angle, attempts to show this behavior. The decrease of clutter at low angles results from the cancellation of the direct and surface scattered waves as illustrated in the ideal representation given in Sec. 8.2. Measurements[41] have demonstrated this rapid decrease in sea clutter below some *critical angle* (the angle at which clutter changes from a R^{-3} dependence at short range to an R^{-7} at longer range and low grazing angle, where R = range). Not all low-angle sea clutter measurements, however, show this effect. For example, the curve for X-band sea clutter in Fig. 7.13 does not show the presence of a critical angle below which the clutter decreases rapidly.

One reason for the lack of a critical angle in some cases is that at very low grazing angles ducted propagation can occur.[42] Erroneous measurements of σ^0 can be made unless propagation effects are separated from sea-surface scattering. This requires that ducting propagation be accounted for using the proper propagation models.[43] The commonly encountered evaporation duct, which occurs a large portion of the time over most of the oceans of the world, probably is the dominant factor that gives rise to larger values of sea clutter than expected with normal (nonconducting) refractive conditions.

It is difficult to obtain reliable empirical information on sea clutter at very low grazing angles that is correlated with environmental conditions. It is also difficult to obtain a theoretical understanding of the nature of the scatterers and the propagation medium at low angles. Fortunately, sea clutter is quite low at very low grazing angles and might not be a serious factor in most radar applications that have to detect targets in a sea clutter background. For example, at X band and sea state 3, σ^0 is less than -40 dB at $1.0°$, less than -45 dB at $0.3°$, and less than -50 dB at $0.1°$ grazing angle.[44] At lower frequencies, sea clutter is even less than at X band when the polarization is horizontal.

Sea Clutter at Vertical (Normal) Incidence   Although sea clutter usually is much lower than land clutter at low grazing angles, it is higher than land clutter at perpendicular incidence. In the discussion of clutter at vertical (normal) incidence over a flat, perfectly reflecting land surface, it was mentioned that the value of σ^0 is approximately equal to the antenna gain G. The same is true over the sea. Thus when examining measured values of sea (or land) clutter at or near normal incidence, it needs to be kept in mind that the antenna has a significant effect on the value of the backscatter clutter power per unit area when the surface is flat or even when it is slightly rough.

Theory of Sea Clutter   Many theoretical models have been proposed over the years to explain sea clutter. Most of the discussion here applies to moderate or low grazing angles. Scattering near vertical incidence generally requires a different theoretical model than at the lower grazing angles.

Past attempts to explain sea echo have been based on two different approaches. In one, the clutter is assumed to originate from scattering features at or near the sea surface. Examples include reflection from a corrugated surface,[45] backscatter from droplets of spray thrown into the air above the sea surface,[46] and backscatter from small facets, or patches, that lie on the sea surface.[41] Each of these can be applied to some limited aspect of sea echo, but none has provided an adequate explanation that accounts for the experimental evidence.

The other approach is to derive the scattering field as a boundary-value problem in which the sea surface is described by some kind of statistical process. One of the first attempts assumed that the surface disturbances could be described by a gaussian probability density function. It has been said[47] that gross observation of the sea characterized by large water wavelengths shows that the surface can be considered to be approximately gaussian, but observation of the sea's fine structure (which is what is of interest for radar backscatter) shows it is not so. Calculations of the scattering of the sea based on a gaussian surface[48] produce results that appear at first glance to be reasonable, but on close examination do not match experimental data. The conclusion that the sea cannot be represented as a gaussian statistical surface was also found by application of the theory of chaos in nonlinear dynamical systems to experimental sea clutter data.[49]

Bragg Scatter   A model that has been successfully used to describe sea echo at long radar wavelengths (HF and VHF) is Bragg resonance, or Bragg scatter. This type of backscatter is so named because of its similarity to X-ray diffraction in crystals as originally put forward by Sir Lawrence Bragg and his father, Sir William Bragg, for which they jointly received the 1915 Nobel Prize in Physics. Bragg scatter is based on the coherent reinforcement of scattering from a periodic scatterer. Although real seas driven by the wind do not appear periodic, the sea can be considered as made up of a large number of individual sinewaves of different wavelengths and directions, similar to the manner in which an electrical engineer describes a noiselike voltage waveform by its frequency spectrum (also a collection of individual sinewaves of different frequencies). The Fourier transform is the means by which one can convert the voltage waveform to the frequency spectrum, and vice versa. The sea spectrum is two-dimensional (wave height as a function of frequency and direction), which differs from the one-dimensional frequency spectrum usually considered by the electrical engineer.

A rough sea surface can be described by its vertical displacement from the mean. A Fourier transform of this surface displacement gives a spectrum. Scattering from such a surface can be characterized as scattering from that particular (sinewave) frequency component of the surface spectrum that is *resonant* with the radar frequency. By resonant is meant backscattering such that the echo energy from each cycle of the sinewave reinforces coherently, as depicted in Fig. 7.16. In this figure, the radar wavelength λ_r is twice that of the spectral component λ_w (water wavelength) so that coherent addition takes place. Echoes, from any other spectral component that is not at the resonant frequency, add noncoherently and result in much smaller amplitude than the echo that experiences coherent addition from each cycle. Hence, the major scattering effect is from the resonant component and not from the other components of the sea-surface spectrum.

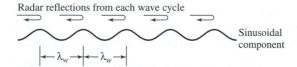

Radar reflections from each wave cycle

Sinusoidal component

λ_w λ_w

Figure 7.16   Representation of "resonant," or Bragg, scatter from a sinusoidal component of the sea-surface spectrum. Grazing angle = 0°. When the radar wavelength λ_r equals twice the wavelength λ_w of the resonant spectral component of the sea spectrum, then in-phase (coherent) addition of scattering takes place.

In general, the grazing angle is not zero, as was shown in Fig. 7.16. The Bragg resonant condition relating the radar wavelength λ_r and the water wavelength λ_w for a grazing angle ψ is given as

$$\lambda_r = 2\lambda_w \cos \psi \qquad [7.13]$$

The first to successfully apply the Bragg scattering model to sea clutter was Douglas Crombie,[50] who used it to explain the distinctive nature of the sea clutter doppler spectrum from HF radar. Bragg scatter forms the theoretical basis for remote sensing of sea and wind conditions over the sea with HF radar.[51,52] The Bragg model, however, has not been successful in explaining experimental observations of sea echo at the higher microwave frequencies.

Composite Surface Model   At the higher microwave frequencies, the resonant water waves of the classical Bragg model that might contribute to backscatter have wavelengths of the order of centimeters. These short water waves are said to ride on the higher-amplitude long waves. The short-wavelength water waves are primarily responsible for the backscattering, and the long-wavelength water waves act to tilt the shorter waves. The effect of tilt can be taken into account by averaging over the tilt angles. The sea surface is thus modeled as a composite surface with two scales of roughness.[53] It has been called the *two-roughness* model or the *composite surface* model.

Long-wavelength water waves are called *gravity waves* because their velocity of propagation is determined primarily by gravity. *Capillary waves,* also called *ripples,* are small waves whose velocity is determined mainly by the surface tension of the water. Waves less than about 1.73 cm are capillary waves; waves of longer wavelength are considered gravity waves. At *X* band, the composite surface model is said to consist of short capillary waves riding on longer gravity waves. (There are, however, other surface effects in addition to capillary waves that can give rise to scattering at the higher frequencies; and these surface effects can produce greater backscatter than capillaries.)

The combination of the Bragg theory of sea echo and the composite-surface model has had considerable acceptance by some theorists since it is one of the few theories that provides an analytical basis for computation. Bragg theory, however, has serious limitations at microwave frequencies, which make its validity suspect. These include:

- The composite-surface model for microwave sea echo does not explain many of the important experimental observations of sea clutter, especially at low grazing angles.

- The theoretical formulation of Bragg scattering as a global boundary-value problem is based on the assumption that the sea-surface displacements are small compared to the radar wavelength, an assumption usually not satisfied at microwave frequencies.[54]

- Examination of sea echo obtained with high-resolution radar indicates that the dominant scattering features (sea spikes) are of relatively large amplitude (order of a few square meters radar cross section) and have a time duration of seconds. When observed with sufficient resolution, they are seen by the radar intermittently. This is far different from the view that small waves ride on top of big waves as in the composite model. Most proponents of Bragg scatter tend to consider the capillaries or ripples as the small waves. In photographs of rough sea that show capillaries or ripples, however, there are almost always present other, more significant, sea-surface features. These are generally larger than capillaries and have sharp features that can give rise to larger echoes than the capillaries. Any theory of microwave sea clutter has to be based on the observed nature of the sea and what the radar sees, which at low grazing angle is predominantly (maybe even exclusively) sea spikes.

Scattering by Surface Features   One of the limitations to obtaining a satisfactory theory of sea echo is the lack of a quantitative description of the sea surface. If the exact nature of the sea surface is described, the calculation of electromagnetic scattering is achieved by solution of Maxwell's equations using more or less standard methods. Oceanographers and hydrodynamicists, however, have not been very interested in the small-scale features of the sea surface that give rise to radar scattering at microwave frequencies. The lack of a full understanding of these small-scale features has been a limitation in formulating a suitable model.

At the beginning of this discussion on sea echo theory, it was mentioned that early theorists tried to explain sea echo by postulating rather simple surface features (sinewaves, facets, and spray). Wetzel[55] has enumerated several other sea-surface features that could account for sea scatter. These are:

- *Plumes:* which slide down the front faces of breaking waves
- *Sloshes or hydraulic shocks:* structures associated with localized wind-puffs or the passage of a steep wave
- *Pools of surface roughness:* associated with breaking waves
- *Wedgelike structures*[56]
- *Pyramidal cusplike structures*

In addition, scattering from the tops of the disturbed crests might be included in the list. If the size, shape, distribution, and transient behavior of these surface features were known, one might begin to formulate a theory of sea echo.

Sea Spikes   Whatever its cause and nature, the scattering feature known as the sea spike has to be a major part of any theory of sea clutter at low grazing angles at the higher microwave frequencies. Experiments show that the sea spike is all that a high-resolution radar sees. There are no other major scatterers present in the echo. Thus the sea spike should be the basis for any radar sea scatter theory at microwave frequencies (*L* band and higher), and Bragg scatter is the basis for sea scatter theory at the lower radar frequencies (HF and VHF).

Sea Clutter Models and Radar Design   As indicated, it has been difficult to formulate a satisfactory theory of microwave sea clutter that starts from fundamental principles and fully describes the experimental observations of electromagnetic scatter from the sea. A satisfactory theory should allow the radar engineer to determine optimum waveforms and related signal processing to maximize radar performance when sea clutter limits detection. Lacking a suitable theory, radar engineers have been able to design radars that operate satisfactorily in the presence of sea clutter by using simplified models and conservative design. Empirical observations are still a widely used basis for specifying a sea clutter model for engineering design or procurement specification. This is an engineering expedient that will be used until a theory suitable for reliable design is formulated.

There exist several collections of averaged values of σ^0 used to describe sea clutter. Two of the most popular are the tables given in Nathanson's book[57] and the formulas of Georgia Tech.[58] Other models have been described by Morchin.[59]

The values of σ^0 in the literature are usually the mean or the median. There is less information on the statistical properties of sea clutter. Design of the radar signal processor and detector by the methods of classical detection theory requires knowledge of the probability density function of the clutter. The spatial as well as temporal variations of clutter are also necessary, especially with high radar-resolution. These descriptions have been difficult to obtain. It has been even more difficult to relate them to the environmental parameters that affect sea clutter. There has been only limited success in devising a theory of signal processing and detector design that can be used by the radar designer when the clutter with which the radar echo must compete is time varying (nonstationary) and spatially nonhomogeneous.

If there were a satisfactory model for sea clutter, it would also provide information about the physical mechanisms occurring at the sea surface. In remote sensing, the radar is used as an instrument to measure some characteristic of the sea, such as the sea state, surface currents, and the winds that drive the sea. One can obtain empirical relations between the radar echo and some environmental parameters of the sea to be measured; but a satisfactory theory of sea clutter could advance one's ability to perform remote sensing of the environment, as well as improved detection of aircraft, missiles, ships, and other targets.

Backscatter from Sea Ice[60–62]   At low grazing angles, as might occur with a shipboard radar, there is little backscattered energy from smooth, flat ice. On a PPI or similar display the areas of ice will be dark except perhaps at the edges. If areas of water are present along with the ice, the sea clutter seen on the display will be relatively bright in contrast with the ice echoes. Backscatter from rough ice, such as floes and pack ice, can produce an effect on the radar display similar to sea clutter. Clutter echoes from rough ice, however, can be distinguished from sea clutter since its pattern will remain stationary from scan to scan but the sea clutter echo pattern will change with time.[63] Measurements of ice fields near Thule, Greenland indicated that the radar backscatter (σ^0) varied linearly with frequency over a range of grazing angles from 1° to 10°, was proportional to the grazing angle from 2° to 10°, but was inversely proportional to grazing angle from 1° to 2°.[64] The scatterometer, a radar that measures σ^0 as a function of elevation angle in the region near vertical incidence, has been used as an ice sensor to differentiate between first-year ice and multiyear ice (ice that doesn't completely melt in the summer). Shipboard radars with multiple polarizations are also able to distinguish between first-year ice, multiyear ice, and icebergs.[65] Multiyear ice is harder, and more difficult for a ship to penetrate, than first-year ice. Imaging radars, such as SAR and SLAR, have been used to image ice fields to determine the nature of the ice and inform shipping of the best routes to travel through the ice.[66]

Radar can also detect icebergs, especially if they have faces that are nearly perpendicular to the radar direction of propagation.[63] Icebergs with sloping surfaces can have a small echo even though they may be large in size. Icebergs can be readily identified in radar images by the characteristic shadows (absence of echo) they produce when viewed at low grazing angles.[67] Growlers, which are small icebergs that are large enough to be of danger to ships, are poor radar targets because of their small size and shape.

Oil Slicks[68]   Oil on the surface of the sea has a smoothing effect on breaking waves. Oil slicks, therefore, are readily detectable since they appear dark on a radar PPI compared to the surrounding sea. Vertical polarization produces greater contrast in radar images of oil slicks than horizontal polarization.

7.5   STATISTICAL MODELS FOR SURFACE CLUTTER

Because of the highly variable nature of clutter echoes it is often described by a probability density function (pdf) or a probability distribution (Sec. 2.4). This section describes several statistical models that have been suggested for characterizing the fluctuations of the surface-clutter cross section per unit area, or σ^0. They can apply to both sea and land clutter. The term *distribution,* as in *Rayleigh distribution,* is used here to indicate the statistical nature of the phenomenon and applies to either the pdf or the probability distribution function. In this chapter, however, the pdf rather than the probability distribution is usually used to describe clutter statistics.

Rayleigh Distribution[69]   This popular model is based on the assumption that there are a large number of randomly located independent scatterers within the clutter surface area illuminated by the radar. (The assumption of a large number is usually satisfied with as few as ten scatterers.) It is further assumed that none of the individual scatterers is significantly larger than the others.

If the radar receiver uses a linear detector, the probability density function of the voltage envelope of the Rayleigh distributed clutter at the receiver output is[17]

$$p(v) = \frac{2v}{m_2} \exp\left(-\frac{v^2}{m_2}\right) \qquad v \geq 0 \qquad \text{[7.14a]}$$

where m_2 is the mean-square value (second moment) of the envelope v. (This equation differs slightly from the form given in Ref. 17.) The mean m_1 (first moment) is $(m_2\pi/4)^{1/2}$, and the median is $(m_2 \, ln \, 2)^{1/2}$ so that the mean-to-median ratio is $[\pi/(4 \, ln \, 2)]^{1/2} = 1.06$, or 0.27 dB. The mean of the Rayleigh distribution is proportional to the standard deviation, or

$$\text{standard deviation} = \sqrt{\frac{4}{\pi} - 1} \times \text{mean} = 0.523 \times \text{mean}$$

The Rayleigh pdf of v normalized to its median v_m, rather than the mean-square value, is

$$p(v_n) = 2(ln \, 2)v_n \exp \left[-(ln \, 2) \, v_n^2 \right] \qquad v_n \geq 0 \qquad \text{[7.14b]}$$

where $v_n = v/v_m$.

The Rayleigh pdf also describes the envelope of the output of the receiver when the input is gaussian noise, as was discussed in Chap. 2 for the detection of signals in noise. The theory given in Chap. 2 for signals in noise applies to the detection of signals in clutter when the clutter statistics are Rayleigh and the clutter echoes are independent pulse to pulse (as is receiver noise). However, this is not often true with clutter. The noise voltage from a receiver is independent from pulse to pulse since it is decorrelated in a time $1/B$, where $B =$ bandwidth. The decorrelation time of clutter can be much longer. This is important when attempting to integrate pulses for improving detection since clutter echoes are generally not decorrelated pulse to pulse. (Correlation of clutter echoes was discussed in relation to the surface-clutter radar equation in Sec. 7.2.) To apply the methods of Chap. 2 based on Rayleigh statistics, the clutter might be decorrelated by changing the frequency pulse to pulse or by waiting a sufficiently long time between pulses to allow the temporal decorrelation of the clutter to occur (if there is clutter motion). However, these methods for decorrelating clutter are not practical in all applications. For example, frequency agility cannot be used with doppler processing. Waiting for the clutter to decorrelate due to its own internal motion might take too long.

If the receiver uses a square-law detector, the output voltage is proportional to the input signal power P. The probability density function in this case is the exponential, or[17]

$$p(P) = \frac{1}{\overline{P}} \exp \left(-\frac{P}{\overline{P}} \right) \qquad P \geq 0 \qquad \text{[7.15]}$$

where $\overline{P}$ is the mean value of the power. The standard deviation of the exponential pdf is equal to the mean.

Although the output of a linear receiver is described by the Rayleigh pdf, as in Eq. (7.14), and the output of the square-law receiver is given by the exponential pdf of Eq. (7.15), they both are considered to belong to the Rayleigh model. The term *Rayleigh clutter* derives from the basic nature of the clutter model rather than whether a linear or a square-law receiver is used.

Log-Normal Distribution   As mentioned, the Rayleigh clutter model usually applies when the radar resolution cell is large so that it contains many scatterers, with no one scatterer dominant. It has been used to characterize relatively uniform clutter. However, it is not a good representation of clutter when the resolution cell size and the grazing angle are small. Under these conditions, there is a higher probability of getting large values of clutter (higher "tails") than is given by the Rayleigh model.

One of the first models suggested to describe non-Rayleigh clutter was the log-normal pdf since it has a long tail (when compared to the Rayleigh).[70] In the log-normal pdf the clutter echo power expressed in dB is gaussian. The log-normal pdf for the echo power when the receiver has a square-law detector is[17]

$$p(P) = \frac{1}{\sqrt{2\pi}\, sP} \exp\left[-\frac{1}{2s^2}\left(ln\frac{P}{P_m}\right)^2\right] \qquad P \geq 0 \qquad \text{[7.16]}$$

where s = standard deviation of $ln\, P$, and P_m = median value of P. The ratio of the mean to the median is $\exp(s^2/2)$. With a linear receiver, the pdf for the normalized output voltage amplitude $v_n = v/v_m$, with v_m = median value of v, is

$$p(v_n) = \frac{2}{\sqrt{2\pi}\, sv_n} \exp\left[-\frac{2}{s^2}(ln\, v_n)^2\right] \qquad v_n \geq 0 \qquad \text{[7.17]}$$

where s remains the standard deviation of $ln\, P$.

The log-normal distribution is specified by two parameters (the standard deviation and the median), whereas the Rayleigh is specified by only one parameter (the mean square value). Log-normal clutter is often characterized by the ratio of its mean to median.[71] Based on measurements by different experimenters, as reported in Ref. 71, the mean-to-median ratio for sea clutter varies from about 0.6 dB for a grazing angle of 4.7° and low sea state, to 5.75 dB for a high sea state and low grazing angle of 0.5°. For a particular ground clutter measurement at low grazing angles, it had a value of about 2.6 dB.

It should be expected that because of its two parameters the log-normal pdf can be made to fit experimental data better than can the one-parameter Rayleigh. However, it is sometimes said that the log-normal tends to predict higher tails for the pdf than normally experienced with most non-Rayleigh clutter, just as the Rayleigh model tends to predict lower values.

Weibull Distribution[17,72]   The Weibull distribution is a two-parameter family that can be made to fit clutter measurements that lie between the Rayleigh and the log-normal. The Rayleigh is actually a special case of the Weibull; and with appropriate selection of the distribution's parameters, the Weibull can be made to approach the log-normal.

If v is the amplitude of the voltage out of a linear detector, the Weibull pdf for the normalized amplitude $v_n = v/v_m$ is

$$p(v_n) = \alpha(ln\, 2)v_n^{\alpha-1} \exp\left[-(ln\, 2)v_n^{\alpha}\right] \qquad v_n \geq 0 \qquad \text{[7.18]}$$

where α is a parameter that relates to the skewness of the distribution (sometimes called the *Weibull skewness parameter*), and v_m is the median value of the distribution. When $\alpha = 2$, the Weibull takes the form of the Rayleigh; and when $\alpha = 1$, it is the exponential pdf. The mean-to-median ratio is $(ln\, 2)^{-1/\alpha}\Gamma(1 + 1/\alpha)$, where $\Gamma(\cdot)$ is the gamma function. With a square-law detector, the Weibull pdf for the power $P = v^2$ is

$$p(P_n) = \beta(ln\, 2)P_n^{\beta-1} \exp\left[-(ln\, 2)P_n^{\beta}\right] \qquad P_n \geq 0 \qquad \text{[7.19]}$$

where $\beta = \alpha/2$, $P_n = P/P_m$, and $P_m = v_m^2$ is the median value of P.

The Weibull distribution has been applied to land clutter,[73,74] sea clutter,[75,76] weather clutter,[77] and sea ice.[78] Table 7.3 lists examples of the Weibull skewness parameter α for several types of land and sea clutter.[79]

Table 7.3   Weibull clutter parameters.[79]

Terrain/Sea State	Frequency	Beamwidth (deg)	Pulse Width (μs)	Grazing Angle	Weibull Parameter
Rocky mountains	S	1.5	2	....	0.52
Wooded hills	L	1.7	3	~0.5	0.63
Forest	X	1.4	0.17	0.7	0.51–0.53
Cultivated land	X	1.4	0.17	0.7–5.0	0.61–2.0
Sea state 1	X	0.5	0.02	4.7	1.45
Sea state 3	K_u	5	0.1	1.0–30.0	1.16–1.78

With regard to land clutter at small depression angles, Billingsley[4] states: "Weibull formulations of clutter amplitude statistics represent better engineering approximations to clutter spatial amplitude distributions than do log normal formulations. Log normal formulations of clutter amplitude statistics tend to provide too much spread in the statistics."

K-Distribution   This two-parameter probability distribution for modeling the statistics of clutter, proposed by Jakeman and Pusey,[80] is described as a compound distribution that consists of two components. The K-distribution probability density function describing the voltage amplitude x is

$$p(x) = \frac{2b}{\Gamma(\nu)} \left(\frac{bx}{2} \right)^{\nu} K_{\nu-1}(bx) \qquad \text{[7.20]}$$

where b is a scale parameter[81] that relates only to the mean of the clutter or σ^0, $\Gamma(\cdot)$ is the gamma function, ν is the shape parameter which depends on the higher moments in relation to the mean, and K_ν is the modified Bessel function of the second kind. (The parameter x has been used in the above instead of the more usual v for voltage so as to not confuse it with the Greek letter ν which is the shape parameter.) The statistical moments of the K-distribution lie between those of the Rayleigh and the log-normal distributions, which is sometimes used as a justification for the K-distribution since experimental sea clutter measurements seem to have the same property. Although it can be applied to land,[82] there appears to have been more interest in its application to the sea.

For sea clutter, the K-distribution is said to be made up of two components that might be associated with experimental observations.[83] There is a fast varying component, with a correlation time on the order of 5 to 10 ms. (This applies for a particular set of data obtained at X band, with a 30 ns pulse duration and a two-way 3-dB beamwidth of 1.2°.) The fast component can be decorrelated pulse to pulse by means of frequency agility. It is sometimes called the *speckle component* and its statistics can be represented by a Rayleigh distribution. The other component has a longer decorrelation time, of the order of seconds, and is unaffected by frequency agility. The slowly varying component can be represented by a gamma distribution.[84] Thus the model assumes a Rayleigh-distributed rapidly fluctuating component modulated by a slowly fluctuating gamma-distributed component which results in the compound K-distribution of Eq. (7.20).

In the literature it has been said that the rapid component "occurs because of the multiple nature of the clutter in any illuminated patch."[85] The slowly varying component is "thought to be associated with the sea swell structure." There is not universal agreement about this description. (It is also possible to describe sea clutter as consisting of two components with different correlation times because of the observed behavior of sea spikes, as was discussed in Sec. 7.4.)

Based on measurements with an X-band radar having a 30-ns pulse width, and a 1.2° (two-way) beamwidth (resulting in a cross-range resolution of 100 to 800 m), the shape parameter ν in the K-distribution generally falls within the range 0.1 and ∞.[81] The value of ν increases with increasing cell size; and when $\nu = \infty$, the K-distribution is the same as the Rayleigh. An empirical estimate of the shape parameter ν, based on the above radar characteristics, is

$$\log \nu = \frac{2}{3} \log \psi + \frac{5}{8} \log \Delta + \zeta - k \qquad [7.21]$$

where ψ = grazing angle in degrees (from 0.1 to 10°); Δ = cross-range resolution in meters; $\zeta = -1/3$ for up or down swell directions, $+1/3$ for across swell directions, and 0 for intermediate directions or when no swell waves exist; and k gives the effect of polarization with $k = 1$ for vertical and $k = 1.7$ for horizontal polarization. No significant statistical trend was noted for variations in the sea state, wind speed, or aspect angle relative to the wind direction. It has been suggested[86] that changes in the range resolution might have a similar effect on ν as the cross-range resolution Δ, and that the parameter ζ can be expressed as $-1/3 \cos 2\theta$, where the angle θ is zero when the radar antenna points in the direction of the swell. [There is also a slightly different version of Eq. (7.21).[86]] As the range-resolution cell becomes smaller, the clutter distribution deviates from Rayleigh and ν becomes smaller. The smaller the ν, the higher the threshold for a given false-alarm probability. However, the higher the resolution, the smaller will be the clutter echo. In one analysis of range resolution,[87] it was concluded that better performance will generally be achieved with the higher resolution radar, provided the K-distribution model is maintained.

It should be cautioned that Eq. (7.21) was obtained with measurements from only a single radar and might not be universally applicable to other situations. (It would seem that ν should vary to some extent with sea state, since experiments show that the probability density function of sea clutter changes with the sea state.[88]) Although most of the above information is based on X-band data, similar behavior of the parameter ν was also observed at S band and K_u band.[89]

The above discussion has applied mainly to the noncoherent detection of targets in K-distributed clutter. The model has been extended to coherent detection,[90,91] the addition of thermal noise,[92] and the performance of CFAR.[93]

Other Statistical Distributions   Other distributions that have been proposed for describing the statistics of clutter include contaminated normal,[88] Ricean,[94] Rice squared,[95] gamma,[96] and log-Weibull.[97] Shlyakhin[98] provides an extensive listing of probability density functions for describing non-Rayleigh fluctuating radar signals. He lists twelve two-parameter, six three-parameter, and seven multiparameter distributions.

Application of the Statistical Descriptions of Clutter Probability distributions such as those given in this section have been of interest for providing a concise description of the amplitude statistics of clutter. As mentioned previously, except for the Rayleigh, there is no physical basis for the use of these distributions. They are curve-fitting models used to describe experimental data. Analyses in the literature of the detection of targets in non-Rayleigh clutter can be found that employ log-normal clutter,[99,100] Weibull clutter,[88,101] and K-distribution clutter.[102,91]

The probability distributions that contain two parameters, such as the Weibull and the K, can be made to fit experimental data better than can a single-parameter distribution. Generally, the Weibull and K-distributions have similar curve-fitting capability; but it has been said that the Weibull is easier to use. However, curve fitting of a particular distribution is not always of sufficient help to the radar engineer trying to determine how to design a signal processor to detect signals in non-Rayleigh clutter since a single statistical expression does not do well in fitting the wide variety of real-world clutter.

A comparison of the Rayleigh pdf with examples of log-normal and Weibull is shown in Fig. 7.17. (The equivalence of the Rayleigh and the Weibull with $\alpha = 2$ is noted in the figure.) The ability of a two-parameter distribution like the Weibull to fit experimental data is illustrated by the quite different shapes for the $\alpha = 0.5$ and $\alpha = 2$ curves.

Figure 7.17
Probability density functions of the envelope v of the output of a linear detector, normalized by the median value v_m, for Rayleigh [Eq. (7.14b)], log-normal [Eq. (7.17), with a mean-to-median ratio of 5 dB], and Weibull [Eq. (7.18) with a skewness parameter of 0.5].

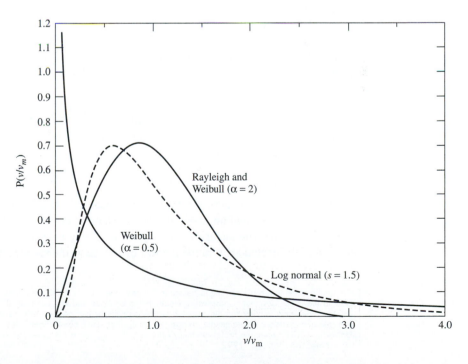

The theory for the detection of signals in clutter that is non-gaussian, partially correlated in time, nonhomogeneous, and nonstationary (which is the nature of real-world clutter) is a difficult problem. Classical detection-theory methods have generally been inadequate. They either contain restricting assumptions that might not be applicable or are difficult to apply to the varying nature of real-world clutter. The radar designer, when confronted with such a problem, has to design conservatively and from experience.*

7.6 WEATHER CLUTTER

Radar is much less affected by weather than are optical and infrared sensors; but weather can seriously degrade radar performance at the higher microwave and millimeter wave frequencies. Backscatter from precipitation that masks the desired target echo is the chief problem. Attenuation of the electromagnetic signal in propagating through precipitation also occurs; but at most microwave frequencies it is seldom strong enough to be of serious concern for the detection of targets.

Radar Equation for Detection of Rain[1,103–105]   The radar equation describing the echo from meteorological particles such as rain, hail, sleet, and snow is of interest to the meteorologist as well as the radar engineer. The engineer needs to understand the effect weather has on the detection of targets. The meteorologist, on the other hand, is interested in using radar to determine the rate of fall of precipitation and the direction and speed of the wind. The radar equation derived below is based on the detection of rain, but it can be extended to other forms of precipitation.

Within the radar resolution cell there are many individual rain drops, each with some radar cross section σ_i. The total radar cross section, σ_c, from volume-distributed rain is the summation of the echoes from all the individual scatterers, and can be represented as

$$\sigma_c = V_c\eta = V_c \sum_i \sigma_i \qquad [7.22]$$

where η = radar cross section per unit volume, and V_c = volume of the radar resolution cell. The summation is taken over the unit volume. The volume V_c is

$$V_c = \frac{\pi}{4}(R\theta_B)(R\phi_B)\left(\frac{c\tau}{2}\right)\frac{1}{2\,ln\,2} \qquad [7.23]$$

where R = range, θ_B = horizontal half-power beamwidth of the radar antenna, ϕ_B = vertical half-power beamwidth, τ = pulse duration, c = velocity of propagation, and ln is the natural logarithm. The factor $\pi/4$ is included to account for the elliptical shape of the antenna-beam projected area. In the interest of accuracy, the radar meteorologist reduces the

---

*A similar statement has been made before in this text, referring to those times when full information is not available to the radar designer. *Conservative* means being cautious and the inclusion of "safety factors" where appropriate. *Experience* implies that the designer learns from his or her own experience and the experience of others. The role of experience in engineering design, especially that which comes from failure, has been described by Duke University Civil Engineering Professor Henry Petroski in *To Engineer is Human—The Role of Failure in Successful Design*. New York: Vintage, 1992.

volume of Eq. (7.23) by a factor of 2 *ln* 2 to account for the fact that the effective volume of uniform rain illuminated by the *two-way* radar antenna pattern (described by a gaussian beamshape) is less than that indicated if the half-power (one-way) beamwidths are used in Eq. (7.23).

The echo power P_r, received from rain of radar cross section σ_c, is given by the simple radar equation of Sec. 1.2:

$$P_r = \frac{P_t G^2 \lambda^2 \sigma_c}{(4\pi)^3 R^4} \qquad [7.24]$$

where P_t = transmitted power, G = antenna gain, λ = wavelength, and R = range. Using Eqs. (7.22) and (7.23), the radar equation becomes

$$P_r = \frac{P_t G^2 \lambda^2 \theta_B \phi_B \, c\tau}{1024(ln\ 2)\pi^2 R^2} \sum_i \sigma_i \qquad [7.25]$$

The antenna gain is approximated by $G = \pi^2/\theta_B \phi_B$, if the antenna pattern can be described by a gaussian function.[1] With this substitution, the radar equation becomes

$$\overline{P}_r = \frac{P_t G \lambda^2 c\tau}{1024(ln\ 2)R^2} \sum_i \sigma_i \qquad [7.26]$$

The bar over P_r denotes that the received power is averaged over many independent samples to smooth the fluctuations of the rain echo. (The radar meteorologist might average 30 or 40 samples to make the standard deviation of the received signal power less than one dB.)

The above radar equation assumes that the volume of the radar resolution cell is completely filled with uniform rain. If not, a correction must be made by introducing a dimensionless beam-filling factor, which is the fraction of the cell filled by rain. It is difficult to accurately estimate this correction. The resolution cell is not likely to be completely filled at long range or when the beam is viewing the edge of a precipitation cell.

If a single drop of rain can be considered as a sphere of diameter D_i, with circumference small compared to the radar wavelength (Rayleigh scattering region), the radar cross section is given by

$$\sigma_i = \frac{\pi^5 D_i^6}{\lambda^4}|K|^2 \qquad [7.27]$$

where $|K|^2 = (\epsilon - 1)/(\epsilon + 2)$, and ϵ = dielectric constant of water. The value of $|K|^2$ for water varies with temperature and wavelength. At 10°C and 10 cm wavelength, it is approximately 0.93. Equation (7.27) also applies for ice. The value of $|K|^2$ for ice is about 0.197 and is independent of frequency in the centimeter-wavelength region. Substituting Eq. (7.27) into (7.26) yields

$$\overline{P}_r = \frac{\pi^5 P_t G c\tau}{1024(ln\ 2)R^2 \lambda^2}|K|^2 \sum_i D_i^6 \qquad [7.28]$$

Since the diameter D_i appears as the sixth power, in any distribution of drop sizes the small number of large drops will dominate the contribution to the radar echo.

Equation (7.28) does not include the attenuation of radar energy by rain. Attenuation usually is not of concern at microwave frequencies, except when accurate measurements of rainfall rate are required. (It is for this reason that the meteorological radar known as Nexrad is at S band, where the effects of attenuation are small.) The two-way attenuation of the radar signal in traversing the range R and back through uniform rain is $\exp(-2\alpha R)$, where α is the one-way attenuation coefficient. If rain is not uniform, the total attenuation must be expressed as the integrated value of $\exp(-2\alpha R)$ over the two-way path. When the radar equation has to include the effect of attenuation, it cannot be readily solved for the range R. (In practice, it can be solved by trial and error.)

The parameter $\sum D_i^6$ in Eq. (7.28) is called the *radar reflectivity factor* by the radar meteorologist and is denoted by Z. Experimentally, Z can be related to the rainfall rate r by

$$Z = \sum_i D_i^6 = ar^b \qquad [7.29]$$

where a and b are empirically determined constants. Equation (7.29) allows the echo signal power P_r to be related to the rainfall rate r. A number of experimenters have attempted to determine the constants in Eq. (7.29), but considerable variability exists among the reported results.[103] One form of Eq. (7.29) that has had wide acceptance is

$$Z = 200r^{1.6} \qquad [7.30]$$

where Z is in units of mm^6/m^3 and the rainfall rate r is in mm/h. (Sometimes the expression $Z = 300 R^{1.4}$ has been used for the Nexrad WSR-88D weather radar system for rainfall estimation.) The units of Z must be properly accounted for when substituting into the radar equation. Equation (7.30) applies to stratiform rain which, in temperate climates, is relatively uniform over distances of 100 km or more (except for embedded showers that might have dimensions of the order of 10 km).[106] Intensities rarely exceed 3 mm/h except in the imbedded showers that can have rates up to 20 mm/h or more. Their vertical extent is from 4 to 6 km, and durations are of several hours. For orographic rain $Z = 31r^{1.71}$ and for thunderstorm rain $Z = 48r^{1.37}$. (Orographic rain is the rain produced when a mountain deflects moisture-laden wind upward.)

When Eq. (7.30) is substituted into Eq. (7.28) the radar equation becomes

$$\overline{P}_r = \frac{2.4P_tG\tau r^{1.6}}{R^2\lambda^2} \times 10^{-8} \qquad [7.31]$$

where r is in mm/h, R and λ in m, τ in seconds, and P_t in watts. This is the form of the equation used to measure rainfall. The received power varies as $1/R^2$ rather than the $1/R^4$ found in the usual radar equation for conventional "point" targets.

The backscatter cross section of rain per unit volume, η, as a function of wavelength and rainfall rate is shown plotted in Fig. 7.18. The dashed lines apply to Rayleigh scattering when $D_i \ll \lambda$, and were obtained by summing the values of σ_i given by Eq. (7.27) over the unit volume and using Eqs. (7.29) and (7.30) to give

$$\eta = \sum_i \sigma_i = 7f^4r^{1.6} \times 10^{-12} \quad \text{m}^2/\text{m}^3 \qquad [7.32]$$

where f is the radar frequency in GHz and r the rainfall rate in mm/h. The solid curves are exact values computed by Haddock.[107] Rayleigh scattering is seen to be a good approximation over most of the frequency range of interest to radar.

Figure 7.18 Exact (solid curves) and approximate (dashed curves) backscattering cross section per unit volume of rain at a temperature of 18°C. (From Gunn and East,[104] Quart. J. Roy. Meteor. Soc.)

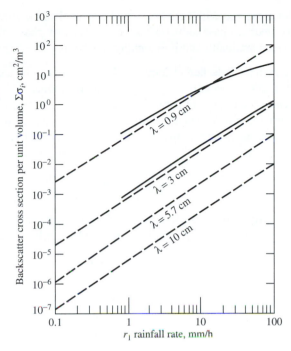

The radar meteorologist prefers to describe the output of a weather radar by the parameter Z [Eq. (7.29)] rather than rainfall rate r. It would not be unusual for a weather radar to observe a rain cloud filled with rain that had not yet reached the ground or which evaporated before hitting the ground. Therefore a radar measurement of r in mm/h from such a cloud would have no meaning to an observer on the ground with a conventional rain gauge. By using Z rather than r, the meteorologist avoids this problem. The parameter Z is usually given in dB and denoted as dBZ. Using the relationship of Eq. (7.30), a rainfall rate of 1 mm/h equals 23 dBZ, 4 mm/h equals 33 dBZ, and 16 mm/h equals 42 dBZ. (This may be an incorrect usage of the precise definition of decibels as a power ratio, but it is the jargon of the radar meteorologist.) With the parameter Z instead of rainfall rate, Eq. (7.31) becomes (ignoring attenuation)

$$\overline{P}_r = \frac{1.2 P_t G \tau Z}{R^2 \lambda^2} \times 10^{-10} \qquad [7.33]$$

Sometimes the radar meteorologist uses, instead of Z, an equivalent radar reflectively factor Z_e when the assumption of Rayleigh scattering does not apply. It is defined as[108]

$$Z_e = \lambda^4 \eta / (\pi^5 |K|^2) \qquad [7.34]$$

The measurement of rainfall based on the Z-r relationship discussed in this section is not always sufficiently accurate for the meteorologist. Other microwave methods for obtaining potentially more accurate measurements are based on attenuation measured at one or two frequencies, or at two polarizations; differential reflectivity (comparison of reflectivity measured with both horizontal and vertical polarizations); polarization differential propagation phase shift (difference in rates of phase change with increasing distance between vertically and horizontally polarized waves); and others.[109]

Radar Equation for Detection of Targets in Rain   Rain can be a serious limitation to the detection of targets, especially at *L*-band and higher frequencies. The radar equation derived here indicates the important parameters that affect detection of targets in rain.

We derive the radar equation for detection of a target in rain by taking the ratio of the echo from the target and the echo from rain. It is assumed that the rain echo is much larger than receiver noise. The received signal power *S* from the target is (Sec. 1.2)

$$S = P_r = \frac{P_t G^2 \lambda^2 \sigma_t}{(4\pi)^3 R^4} \qquad [7.35]$$

where σ_t = target radar cross section, and the other parameters are defined as in Eq. (7.24). The rain clutter *C* is similar to that of Eq. (7.33), and is written

$$C = \frac{K_1 P_t G \tau Z}{R^2 \lambda^2} \qquad [7.36]$$

where K_1 is a constant = 1.2×10^{-10}. (The constant K_1 includes the velocity of propagation which has units of m/s.) The ratio of Eqs. (7.35) and (7.36) gives the signal-to-clutter ratio for a *single* pulse. If the maximum range R_{max} corresponds to the minimum detectable signal-to-clutter ratio $(S/C)_{min}$, then

$$R_{max}^2 = \frac{K_2 G \lambda^4 \sigma_t}{\tau Z (S/C)_{min}} \qquad [7.37]$$

where $K_2 = 4.2 \times 10^6$. Attenuation has not been included. (For *Z*, one can substitute $200r^{1.6}$, or any other suitable ar^b relationship.) It should be noted that the statistics of rain echo can be different from those of receiver noise, and the value of $(S/C)_{min}$ might not be easy to determine. We see from Eq. (7.37) that for long-range detection of targets in rain, the radar wavelength should be large (low frequency), the pulse width small, and the beamwidths small (high antenna gain).

The radar equation derived above for detection of targets in rain applies for one pulse. When a number of pulses are available from a target, they may be added together (integrated) to get larger signal-to-clutter ratio if the echoes are not correlated. It was mentioned that for land and sea clutter the pulses might not be decorrelated pulse to pulse, and the use of an effective number of pulses n_{eff} must be done with caution. Rain clutter, however, is likely to be decorrelated quicker than other clutter echoes and have the statistics of the Rayleigh pdf if the radar resolution cell is not too small and the prf is not too high. Therefore, it might be appropriate to include an effective number of pulses in the numerator of Eq. (7.37) when the conditions for independent pulses apply. The decorrelation (or independence) time of rain in seconds has been said to be

$$T_I = 0.2\lambda/\sigma_v \qquad [7.38]$$

where λ = radar wavelength in meters and σ_v = standard deviation of the velocity spectrum of the rain echo in m/s.[110] For example, at *S* band (λ = 10 cm) and σ_v = 1 m/s, the decorrelation time is 0.02 s, which means there are only 50 independent samples of rain echo available per second. (The value of σ_v depends on wind shear, turbulence, and the terminal fall velocities of the precipitation.[111] It varies from 0.5 m/s for snow to 1 m/s for rain. In convective storms it might reach 5 m/s.)

Scattering from Snowfall Dry snow particles are essentially ice crystals, either single or aggregated. The relationship between Z and snowfall rate r is of the same form as that given by Eq. (7.29) for rain, but with different values of a,b. This allows the radar equation derived for rain, Eq. (7.28) or (7.31), to be applied to snow with the proper value of $|K|^2$ inserted (typically 0.197) and the snowfall rate r at the ground in millimeters of water measured when the snow is liquid (melted).

There have been fewer measurements of the Z-r relationship for snow than rain. The following two expressions have been suggested[112,113]

$$Z = 2000 \, r^2 \qquad\qquad\qquad \text{[7.39]}$$

$$Z = 1780 \, r^{2.21} \qquad\qquad\qquad \text{[7.40]}$$

Other values have been proposed, but there does not seem to be universal agreement.[114,115]

A radar is less affected by snow and ice than by rain since $|K|^2$ is smaller for ice than rain, and the snowfall rates (equivalent water content) are generally smaller than rainfall rates.

Scattering from Water-Coated Ice Spheres Moisture at altitudes where the temperature is below freezing takes the form of ice crystals, snow, or hail. As frozen particles fall to the warmer lower altitudes, they melt and form water-coated ice before becoming rain. Radar scattering by water-coated ice or snow is similar in magnitude to that of water drops of the same size and shape. Therefore, when an ice particle begins to change to liquid, radar backscatter increases since water-coated particles reflect more strongly than ice. Even for comparatively thin coatings of water, the composite particle scatters nearly as well as an all-water particle. One effect of this phenomenon is to give rise to what is called the *bright band*.

Radar observations of light precipitation show a horizontal bright band at an altitude at which the temperature is just above 0°C. It can be seen with a RHI display (range-height indicator). The center of the bright band might be 100 to 400 m below the 0°C isotherm. The echo in the center of the bright band is typically about 12 to 15 dB greater than the echo from the snow above it and about 6 to 10 dB greater than the rain beneath it.[103]

The bright band is due to changes in snow falling through the freezing level.[104] As snow just begins to melt, it changes from flat or needle-shaped particles which scatter feebly to similarly shaped particles which scatter relatively strongly due to their water coating. As melting progresses, the particles lose their extreme shapes and their velocity of fall increases. This results in a decrease in the number of particles per unit volume, and a reduction in the backscatter.

Scattering from Clouds and Fog Most water droplets in fair-weather (cumulus) clouds do not exceed 100 μm in diameter (1 μm = 10^{-6} m); consequently Rayleigh scattering applies. Since the diameter of cloud droplets is small compared to the radar wavelength, echoes from fair-weather clouds are not of concern.

It might be possible to obtain weak echoes from a thick, intense fog at millimeter wavelengths; but at most radar frequencies, echoes due to fog may generally be regarded as insignificant.

Attenuation in Precipitation When precipitation particles are small compared to the radar wavelength (Rayleigh region), the attenuation due to absorption is small. This is the case for frequencies below S band. Since rain attenuation is usually small and unimportant at the longer wavelengths, the relative simplicity of the Rayleigh scattering approximation is of limited use for predicting attenuation through rain. The computation of rain attenuation must therefore be based on a more exact formulation, the results of which are shown in Fig. 7.19 as a function of wavelength and rainfall rate.[104] The attenuation produced by ice particles in the atmosphere, whether occurring as hail, snow, or ice-crystal clouds, is less than that caused by rain of an equivalent rate of precipitation.[116]

Effect of Weather on Radar Because the echo from precipitation varies as f^4, where f = frequency, UHF radars (420–450 MHz) are seldom bothered by weather effects. At L band weather echoes can be a problem, and some method for seeing aircraft targets in weather is usually needed. A radar at S band will have its range considerably reduced in modest rainfall if nothing is done to reduce the effect of rain. Radars at higher frequencies are even further degraded by rain. (Airborne weather-avoidance radars at X band, for example, can be severely degraded by heavy rain and prevent the radar from seeing hazardous weather.)

A typical specification for an air-surveillance radar might be that it has to detect its target when rainfall in the vicinity of the target is at the rate of 4 mm/h. This is called a

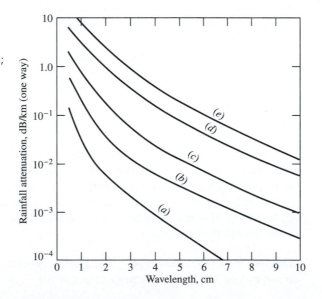

Figure 7.19 One-way attenuation in rain at a temperature of 18°C. (a) Drizzle—0.25 mm/h; (b) light rain—1 mm/h; (c) moderate rain—4 mm/h; (d) heavy rain—16 mm/h; (e) excessive rain—40 mm/h. In Washington, D.C., a rainfall rate of 0.25 mm/h is exceeded 450 h/yr, 1 mm/h is exceeded 200 h/yr, 4 mm/h is exceeded 60 h/yr, 16 mm/h is exceeded 8 h/yr, and 40 mm/h is exceeded 2.2 h/yr.

moderate rain. At Washington, D.C., this rainfall rate is equaled or exceeded approximately 60 h/yr.[117] As has been mentioned, attenuation is generally not a problem at frequencies below X band, unless the precipitation is very heavy.

7.7   OTHER SOURCES OF ATMOSPHERIC ECHOES

Radar echoes are sometimes obtained from regions of the atmosphere where no apparent reflecting sources seem to exist. Before the causes of these echoes were understood they were called *angels*. Today we know that discrete or "point" angel echoes are usually due to birds or insects. Even though the echo from a single bird or insect is small, it can be detected by radar if the range is short enough. For example, a radar that can detect a target with a radar cross section of 1 m^2 at a range of 200 nmi can detect a 10^{-2} m^2 target (a large bird) at a range greater than 60 nmi, and a 10^{-4} m^2 target (a large insect) at a range of 20 nmi. Angel echoes also can be caused by large flocks of birds, especially during migration season, or by large numbers of insects.

Distributed angel echoes can be due to inhomogeneities in the atmosphere such as clear-air convective cells and other turbulent atmospheric effects. Echoes from atmospheric turbulence are weak and can only be detected by powerful radars or at short range. Echoes from birds and insects, on the other hand, can be quite substantial in numbers and can be seen within a large area of the radar coverage. They can present a problem to an unprepared radar designed to detect aircraft. Operation at VHF or UHF can reduce the backscatter of insects and, to some extent, the echo from small birds when their cross sections are within the Rayleigh scattering region where the cross section is proportional to the fourth power of the frequency.

Knowledge of the scattering characteristics of birds and insects is important for understanding their effect on radar performance. In addition, radar has proven to be an important instrument for the study of the flight characteristics of insects (entomology) and birds (ornithology). It can obtain information not possible by other sensing means.

Birds   Table 7.4 gives examples of the radar cross sections of birds taken at three frequencies with vertical polarization.[118] The largest values occur at S band. It is difficult, however, to be precise about the radar cross section of birds since echoes can fluctuate over quite large values, with the maximum and minimum differing at times by more than two orders of magnitude.[118]

The effect of birds on radar depends on the total cross sections that can be expected from large flocks. Table 7.5 gives representative estimates of the area over which bird echoes might be expected, the number of birds within the area, their density, individual cross section, and the volume reflectivity η (or radar cross section per unit volume). The value of η can be used in Eq. (7.2) to find the clutter cross section σ_c for the radar equation. This table is taken from an excellent review paper[119] published in the *Proceedings of the IEEE* by Charles Vaughn of NASA, Wallops Island, VA. (References to the sources of this information can be found in Vaughn's paper.) According to Table 7.5, most birds fly at altitudes less than a few hundred meters; but at times they have been observed much

Table 7.4   Radar cross sections of birds.[118]

Bird	Frequency Band	Mean Radar Cross Section (cm²)	Median Radar Cross Section (cm²)
Grackle	X	16	6.9
	S	25	12
	UHF	0.57	0.45
Sparrow	X	1.6	0.8
	S	14	11
	UHF	0.02	0.02
Pigeon	X	15	6.4
	S	80	32
	UHF	11	8.0

higher.[120] Over the ocean, the altitudes of migrating land birds have been reported as high as 4000 m.[121] Birds generally fly at speeds of from 15 to 40 kt,[122] and some have been observed at greater than 50 kt. These speeds are too high to be rejected by conventional microwave MTI radars without special design consideration.

The radar echo from a bird exhibits a periodic modulation that is caused by the beating of the bird's wings. A wing-beat fluctuation in radar cross section of 10 dB is common. A single tracked bird has sometimes been observed to fluctuate as much as 30 to 40 dB at the wing-beat rate.[119] The wing-beat frequency might vary from a few Hz to 10 to 20 Hz. The nonsinusoidal nature of the wing-beat modulation results in a spectrum with significantly higher harmonics. One estimate of the wing-beat frequency is given by $fL^{0.827} = 572$, where f is the wing-beat frequency in hertz and L is the length of the bird's wing in millimeters.[123] The spectral components of the wing-beat modulation has been said to be remarkably stable and suited for recognition.[124]

Williams and Williams[121] state that the great majority of radar observations of birds over the ocean are straight tracks at altitudes of more than 100 m, and are found at the season of peak land-bird migration on nearby continents. Migrating land and sea birds move over the oceans in massive waves that can extend up to a thousand kilometers wide and a hundred kilometers deep. The maximum density of radar echoes over the ocean is said to be significantly less than that detected with similar radars at continental sites. Williams and Williams also report that radar will often detect no bird echoes over the ocean for several days and then suddenly the display will be filled with small echoes, all moving at about the same direction and speed. Sea birds are usually seen close to the surface of the sea and move in curving paths rather than in the large number of parallel tracks typical of migrating land birds. Land birds cannot feed or rest over the sea and must make nonstop flights. Sea birds, on the other hand, feed from the sea and make use of dynamic soaring.

A statistical prediction of radar bird clutter was reported for the Distant Early Warning (DEW) line of aircraft surveillance radars that were located in northern Canada and

Table 7.5 Representative order of magnitude values for some high average areal number densities of birds (integrated over altitude), most of which are averaged over wide geographical areas. The cross section values apply for frequencies from S to X bands.

Type of Concentration	Area Affected (km^2)	No. of Birds	n (Birds/m^3)	σ (cm^2)	η (cm^{-1})
Blackbird roosts lower Miss. Valley (winter)	10^5	10^8	$10^{-6(1)}$	5–50	10^{-11}–10^{-10}
Single blackbird roost (Winter feeding area)	10^3–10^4	10^7	10^{-6}–$10^{-5(1)}$	5–50	10^{-11}–10^{-9}
Crows, gulls, geese, and ducks costal areas, wildlife refuges (winter)	10^3	10^4–10^6	10^{-9}–$10^{-6(2)}$	10–500	10^{-13}–10^{-9}
Shearwater migration, California coast	10^3	$> 10^6$	$10^{-5(3)}$	50–500	10^{-9}–10^{-8}
Greater Shearwater (*Puffinus gravis*) Georges Banks, non-breeding	10	10^5	$10^{-4(3)}$	50–500	10^{-9}–10^{-8}
Wading and seabird (breeding) U.S. Gulf Coast	10^5	10^6	$10^{-9(2)}$	50–500	10^{-13}–10^{-12}
Nocturnal fall migration U.S. east of Rocky Mts.	10^7	10^9	$10^{-7(4)}$	1–50	10^{-13}–10^{-12}
Nocturnal fall migration, one location, eastern U.S.	$> 10^3$	?	$10^{-6(4)}$	1–50	10^{-12}–10^{-10}
Quelea (*Quelea quelae*) breeding colony, semi-arid African savannahs	2	$> 10^7$	$10^{-2(1)}$	5–50	10^{-7}–10^{-6}
Quelea, semi-arid African Savannahs	10^9	10^9–10^{11}	10^{-9}–$10^{-7(1)}$	5–50	10^{-14}–10^{-11}

| $^{(1)}$10 percent within 100 m of the ground.
| $^{(2)}$50 percent within 200 m of the ground.
| $^{(3)}$50 percent within 20 m of the surface.
| $^{(4)}$50 percent within a 500 m interval.
| Sources of this information can be found in Vaughn,[119] from which this table was taken. (*Copyright 1985 IEEE*)

Alaska.[125] Birds in this region can be relatively large, and can have mean cross sections of about 0.02 m^2 at L band and 0.05 m^2 at UHF. A migrating flock might contain between 25 and 100 birds, but at times can be as large as 1000. The probability that a flock of Elders, a common northern bird, will have a radar cross section greater than 0.1 m^2 was said to be 90 percent, and the probability that the cross section will be greater than 1.0 m^2 is 50 percent. In northern Alaska, for example, from July to September, approximately 1.5 million large size birds such as ducks, swans, and geese can be expected to migrate across the length of the DEW line. It was estimated that the maximum number of birds that might be seen on a particular radar screen at various locations along the DEW line during the peak of the migratory season could vary from 8,000 to 30,000.

There are times when birds or insects can cause serious degradation of radar performance if the radar operator and/or designer are not aware of the effects that birds can have as moving clutter echoes. Radars that have to operate reliably under all conditions need to account for clutter from birds and insects. Most civil air-traffic control air-surveillance radars employ sensitivity time control (to be described later in this chapter) to turn down the receiver gain at short ranges so as to reduce the clutter from nearby birds and insects. Military air-surveillance radars cannot reduce their gain by means of sensitivity time control in order to avoid low cross section birds and insects, since a military radar's target of interest can also have low cross section. During migratory seasons, bird echoes can extend over a wide area and can cause great concern to those who are not aware of what is happening. In addition to times of migration, Vaughn[119] states that mass bird concentrations can occur in the vicinity of breeding colonies, nocturnal roosts, premigratory staging areas, and concentrated feeding areas.

Vaughn summarizes the effect of birds on radar as follows:

> "From spring through fall, birds and/or insects are generally common to abundant in the atmosphere to an altitude of 1 or 2 km over most land areas of the world—especially at night. Some migrant bird species regularly fly above 4-km altitude and more than 1000 km from the nearest land. From the surface to 2-km altitude a density of 10^{-7} to 10^{-6} m^{-3} birds is not uncommon; in areas with social gatherings, 10^{-5} m^{-3} birds can be encountered during the day. A radar resolution cell of 10^{6} to 10^{7} m^3 (which will occur within 20 km of many radars) will result in a significant fraction of these cells occupied by at least one bird. Single birds typically have mean σ [radar cross section] from 10^{0} to 10^{2} cm^2, while migrant flocks, which often occur during the day, can have a mean σ greater than 10^{4} cm^2."

Insects   Despite their small size, insects can be detected by radar, especially at short range. In sufficient numbers, insects can *clutter* the display and prevent the detection of desired targets. It has been said that the density of airborne insects can be many orders of magnitude greater than the density of birds.[119]

Figure 7.20 is an approximate plot of the X-band radar cross section of insects as a function of their mass. This particular curve is taken from E. F. Knott,[126] who based it on a much more complete presentation of insect cross sections given by J. R. Riley.[127] The reader is referred to Riley's paper for the complete list of insects that make up this plot. Riley states that the data indicate that "the radar cross section of an insect may be very approximately represented by that of a spherical water droplet of the same mass, and that this representation holds true over a mass range of 10,000:1." This is only approximate since the cross section of insects is aspect dependent. Insects have been observed to have echoes at broadside aspect of from 10 to 1000 times greater than when viewed end-on.[128] Insects generally travel with the wind. Because of the aspect dependence, when a radar looks into or away from the wind, the radar echoes from a swarm of insects will be smaller than the echo when looking broadside to the wind direction. Thus the configuration of a large area of insects on a PPI display will not be uniform with angle, but more like a figure-eight. (The orientation of the figure-eight can indicate the approximate direction of

Figure 7.20   Sample of measured radar cross section of insects as a function of insect mass, as drawn by E. Knott[126] from data presented in a paper by J. Riley.[127] The dashed curve is the calculated cross section of spherical water drops, and is shown for comparison.
| (Copyright, 1985 IEEE.)

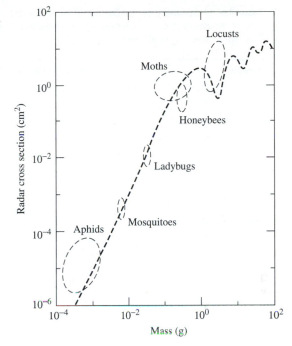

the wind, with a 180° ambiguity.) The wing beat of insects is also observed to produce a distinctive modulation on the radar echo. It has been said,[128] "Male and female grasshoppers and locusts usually differ significantly in size and hence in wing-beat frequency, making it possible to distinguish sexes under favorable circumstances."

Radar echoes can be produced by insect concentrations that would scarcely be noticed visually. Insect echoes are more likely to be found at the lower altitudes, near dawn and twilight. The majority of insects do not fly at temperatures below 40°F (4.5°C) or above 90°F (32°C), so that large concentrations of insect echoes would not be expected outside this temperature range. As with clutter due to birds, sensitivity time control can reduce the adverse effects of clutter caused by insects.

Vaughn[119] summarizes his discussion of insects as follows:

"Insect densities can easily exceed 10^{-5} to 10^{-4} m^{-3}, while 10^{-3} to 10^{-2} m^{-3} insects seem to be regularly encountered when a wind field convergence brings them together. The highest densities are very local, less than 100 km^2 probably being a typical area of concentration. Insects also trace many atmospheric boundary layer features, thus complicating data interpretation by radar meteorologists . . . . At a wavelength of less than 3–5 cm, a volume reflectivity due to insects of 10^{-8} cm^{-1} might reasonably be expected in a region of wind convergence; 10^{-11} to 10^{-10} cm^{-1} should be regularly expected elsewhere."

Table 7.6 is a summary of some relevant number densities compiled by Vaughn from a number of sources. His paper gives the references for the values shown in the table.

Table 7.6 Summary of number densities of insects; as presented by Vaughn[119] based on data from R. Rainey and C. G. Johnson. The altitude in the table is that at which the sample was taken. The cross section is at X band.

Sample	Altitude (m)	Insects (m^{-3})	σ (cm^2)	η (cm^{-1})
Medium sized butterflies (perids), heavy migration	surface	10^{-2}	10^{-2}–10^0	10^{-10}–10^{-8}
Major aphid migration	surface	10^0	10^{-3}–10^{-2}	10^{-9}–10^{-8}
Noctuid moths	surface	10^{-2}	10^{-1}–10^0	10^{-9}–10^{-8}
All insects in 1 h	surface	10^1	10^{-3}–10^{-1}	10^{-8}–10^{-6}
Stratified desert locust swarm	near surface	10^0	10^{-1}–10^0	10^{-7}–10^{-5}
Suction trap during passage of intertropical front, Africa	15	10^{-2}	10^{-3}–10^0	10^{-11}–10^{-8}
Aircraft flying across intertropical front	300	10^{-3}	10^{-3}–10^0	10^{-11}–10^{-8}
Mean over 50-km traverse. Mostly *Aiolopus simulatrix*	450	10^{-3}	10^{-2}–10^0	10^{-11}–10^{-9}
Spodoptera exempta, average over 12-h period	2	10^{-2}	10^{-2}–10^0	10^{-10}–10^{-8}

I *(Copyright 1985 IEEE.)*

Clear-Air Turbulence Radar echoes from clear air (no birds, insects, clouds, dust, or precipitation) are caused by inhomogeneities in the refractive index (index of refraction) of the atmosphere that result from turbulence.[129,130] (The index of refraction is the square root of the dielectric constant.) The refractive index of the atmosphere is a function of the water vapor (humidity), temperature, and pressure. At microwave frequencies, inhomogeneities in the water vapor are usually more important than inhomogeneities in the temperature or pressure. Reflections from clear-air turbulence are small in terms of radar cross section and are detectable only by large radars or at close range.

The atmosphere can be considered turbulent everywhere, but its intensity varies widely both in space and time. It is only when turbulence is concentrated into regions of greater or lesser intensity than its surroundings (so that there is a contrast that can be detected) is it of interest as an electromagnetic scatterer. There are at least two types of turbulent atmospheric phenomena that result in clear-air echoes. One is the *convective cell,* or *plume,* that occurs in the lower part of the atmosphere. Atmospheric convective cells, also called *thermals,* are the means by which birds and gliders soar. The other turbulent phenomenon is the *atmospheric horizontal layer,* which can occur at any altitude and is the turbulent effect that aircraft encounter at the higher altitudes. To radar, the atmospheric layer is generally weaker than the convective cell.

The basic theory for scattering from homogeneous, isotropic turbulence was first given by Tatarskii.[131] The radar reflectivity (cross section per unit volume) of a homogeneous turbulent medium is given as

$$\eta = 0.38C_n^2\lambda^{-1/3} \qquad [7.41]$$

where C_n^2 is the *structure constant* (a measure of the intensity of refractive-index fluctuations), and λ is the radar wavelength. At altitudes of several hundred meters, values of C_n^2 can be between 10^{-9} and 10^{-11} m$^{-2/3}$, which correspond to a volume reflectivity of about 0.8×10^{-9} to 0.8×10^{-11} m^{-1} at S band ($\lambda = 10$ cm). This is quite low, as can be seen by comparison with the volume reflectivity of rain in Fig. 7.18. The value of C_n^2 is variable and decreases with increasing height. At an altitude of 10 km, it can have values in the vicinity of 10^{-14} m$^{-2/3}$. An S-band radar with a 1° beamwidth and a 1-μs pulse width viewing a turbulent medium with $\eta = 10^{-9}$ (a very high value) sees a radar cross section at 10 km of about 3×10^{-3} m^2. Thus even with large values of C_n^2, echoes from clear-air turbulence are not likely to bother most radars.

Doppler weather radars, such as Nexrad[133] and the Terminal Doppler Weather Radar,[134] are used to detect wind shear, including that caused by the dangerous microburst, or downburst.* Wind shear caused by a downburst can be a hazard to aircraft during landing or take-off. It has sometimes been said that these radars detect clear-air wind shear, but the atmospheric wind-shear pattern is usually inferred from the radar echo of wind-driven rain rather than clear-air turbulence. When a downburst occurs without the presence of rain, wind-blown dust and insects provide the echoes that allow wind shear to be detected.

Other "Angel" Echoes   Radar has been reported to detect the passage of an invisible sea breeze as it moves toward the shore.[103] Echoes have also been reported from the vicinity of forest fires, dump fires, and oil fires.[135,136] The reflectivity of smoke from such fires is too small to account for these echoes. Such echoes might be caused by numerous large particles and debris lifted into the air above the fires and/or by the generation of atmospheric turbulence due to the high heat from the fire.

Another example of clear-air echoes that are detectable by radar is gas seepage from underground oil and gas deposits.[137,138] Although these echoes are weak and the radar used for this purpose is small (a modified civil marine radar), the observations are performed at short range (500 to 1000 ft). This radar technique has been used commercially on many occasions for oil and gas exploration. It is speculated that the radar is detecting turbulent water vapor included with the gas seepage.

Ring echoes, or ring angels, are sometimes observed on PPI displays.[139] They start as a point and form a rapidly expanding ring. After one ring grows to a diameter of several miles, a second ring forms. Other rings can form in succession. They expand at velocities typically ranging from 20 to 30 kt and can attain diameters of 30 km or more. Over time, their appearance is similar to the outward moving ripples that occur when a pebble is dropped in a pond of water. These ring echoes have been associated with birds, such as starlings, flying away from their roosting areas at sunrise.

7.8 DETECTION OF TARGETS IN CLUTTER

Chapters 2 and 5 discussed the radar detection of targets when radar sensitivity is limited by receiver noise. When clutter is larger than receiver noise, different radar design methods

---

*Wind shear in radar meteorology is the local variation of wind velocity in range, altitude, or azimuth (or x, y, and z in a rectangular coordinate system). According to Theodore Fujita of the University of Chicago, a downburst is a strong downdraft that induces an outburst of damaging winds on or near the ground.

need to be employed for target detection. This section discusses the various techniques that might allow the detection of desired targets in the presence of undesired clutter echoes.

There is no one method that works equally well for the detection of all targets in various types of clutter. In brief, the major techniques available to the radar designer may be summarized as follows:

- The *doppler frequency shift* in MTI and pulse doppler radar is a powerful method for separating moving targets from stationary clutter.

- *High resolution in range and/or angle* reduces the amount of clutter with which the target must compete, thus increasing the signal-to-clutter ratio.

- Generally, for many types of clutter the *lower radar frequencies* produce smaller echo power. (An exception is an ionized media such as meteor trails and aurora).

- The use of *polarization discrimination* can increase the target-to-clutter ratio from some types of clutter, such as rain.

- *Clutter-echo decorrelation* by observing the target and clutter at different times or with different frequencies has some limited potential for allowing improved detectability.

- There are also techniques that help to *avoid saturation* of the receiver due to large clutter echoes. These can be important to good radar performance even though they do not increase the target-to-clutter ratio.

Doppler Frequency Shift (Target Radial Velocity)   The doppler frequency shift of the echo signal is widely used for separating a moving target's echo signal from large, unwanted stationary clutter echoes. Doppler filtering allows echoes of moving targets to be separated from those of stationary clutter even though the clutter might be greater by many orders of magnitude. This is the basis for moving target indication radar (MTI), pulse doppler radar, and CW radar (Chap. 3). In spite of limitations, doppler filtering works well and can provide far greater suppression of the clutter echoes than any other technique.

Since radar echoes from land usually are much larger than those from the sea, the use of doppler filtering to detect aircraft is more demanding over land than over the sea. In the absence of ducted propagation or mountainous terrain, clutter usually does not extend more than a few tens of miles when the radar is at or near the surface. Airborne radars, however, can look down on extensive land or sea clutter and can experience surface clutter at relatively long ranges compared to what is observed from land or ship-based radars. When doppler filtering is employed to detect moving targets in rain, the rain itself is moving and the doppler processor must be designed differently from a doppler processor designed for stationary clutter. The MTD radar described in Sec. 3.6 is an example of how aircraft can be detected in the presence of moving rain clutter.

High Resolution   The smaller the radar resolution cell, the less will be the competing clutter echo. The benefit of small resolution cells is indicated in the radar equations for detection of targets in surface clutter [Eq. (7.8)] and in rain [Eq. (7.37)]. Short pulses (or the equivalent pulse compression) and narrow-beamwidth antennas, therefore, can reduce the amount of clutter with which the target echo signal must compete. High resolution usually is the preferred method for detection of ships.

With large resolution cells, the clutter statistics can be described by the Rayleigh probability density function (pdf). With small resolution cells, however, sea clutter is no longer Rayleigh and is likely to produce higher values (such as from sea spikes) than predicted by Rayleigh clutter. The result is that serious false alarms can occur when small targets are to be detected if the clutter is non-Rayleigh but the receiver is designed on the basis of Rayleigh statistics. Increasing the detection threshold can reduce the false alarms from non-Rayleigh clutter, but it results in the undesired lowering of the probability of detection. Other solutions are required.

A knowledge of clutter statistics is necessary for receiver design based on classical detection theory. It is also needed to design CFAR (Sec. 5.7) when an optimum detector is not practical. Unfortunately, the statistics of the sea clutter are variable, and it is not always possible to specify a single statistical description of the clutter when the statistics are not Rayleigh.

High resolution is usually a good method for enhancing the signal-to-clutter ratio, even when the statistics of the clutter are not Rayleigh. It is one of the few methods available for increasing the target-to-clutter ratio when the target is stationary (no doppler shift).

The clutter echo must be large compared to receiver noise in order for the clutter radar equations [Eqs. (7.8 and 7.37)] to apply. If this condition is not met, it is not necessarily bad since it means that radar detection will be noise limited rather than clutter limited—which is to be preferred. It is cautioned that the resolution cell size cannot be decreased without limit when considering detection of signals in clutter. With a small enough resolution cell size, the radar becomes noise limited and there is no need for a further decrease.

High-Resolution Clutter and Detector Design   With sea clutter, non-Rayleigh clutter statistics might be expected to appear when the beamwidth is less than about one degree and the pulse width is less than about one μs (as indicated in the previous discussion of sea spikes in this chapter). Non-Rayleigh sea clutter is especially obvious when pulse widths are on the order of nanoseconds. Attempts have been made to model the statistics of high-resolution sea clutter, but without total success. The Rayleigh pdf generally underestimates the range of values obtained from real sea clutter. On the other hand, the log-normal pdf tends to overestimate the range of values. Several other analytical pdfs that lie between these two extremes have been suggested for modeling the statistics of sea clutter, some of which were mentioned in Sec. 7.5.

Any particular probability density function might be able to fit some particular set of data; but there does not seem to exist a single analytical form of probability density function that fits all the observed data and can be correlated with environmental parameters. For this reason it has not been practical to design a detector in a classical manner when sea clutter is dominant. (*Classical manner* means that the optimum detector is found based on the statistics and correlation properties of the background clutter and target signals.)

Billingsley[140] has said that on the basis of many measurements of various types of land clutter at low depression angles, "measured amplitude distributions almost never pass rigorous statistical hypothesis tests for belonging to Weibull, log normal, or other theoretical distributions that have been tried." Thus the fitting of experimental data to arbitrary statistical models has not been successful as a method for describing clutter to aid in radar design.

When the clutter cannot be described by a specific pdf, *robust detection* and *distribution-free detection* are often considered. A robust detector is one that works satisfactorily when the clutter (or noise) statistics can be described as one of a family of pdfs or a number of different classes of pdf. It is not necessarily optimum for any of them. A distribution-free detector makes as few assumptions as possible regarding the statistics of the clutter or noise.[141] Its probability of false alarm is supposed to be constant under weak assumptions on the statistical character of the clutter. Distribution-free detectors have not been used much in practice because they are not easy to implement and the loss in detectability is larger than might be desired. The term *nonparametric detector* is sometimes used interchangeably with the term *distribution-free detector.*

The *median detector* and the *m-out-of-n detector* (Sec. 5.6) are examples of robust detectors that have been considered for detecting targets in high-resolution sea clutter. A median detector is one that bases its detection decision on the median value* of n received pulses. Trunk and George[142] showed that the *median detector* gave better performance in non-Rayleigh clutter than the conventional receiver (which can be called a *mean detector*). A median detector is robust in that the threshold values and detection probabilities do not depend on the detailed shape of the clutter pdf, but only on its median value. For instance, if one of the n received pulses happens to be very large, it can have a significant effect on the mean value; but it would have no greater effect on the median value than a pulse that is only slightly larger than the median. Trunk[143] has shown that the *trimmed-mean detector* is even better than the median detector for non-Rayleigh clutter. If there are n pulses available from which to make a detection decision, the trimmed-mean detector discards the smallest and the largest of the n pulses before finding the mean and making a detection decision. Although its performance may be attractive, the trimmed mean is more difficult to implement in practice since the n pulses must be rank-ordered to find the smallest and the largest.

According to Schleher,[144,145] the median detector is inferior to both the *m-out-of-n* detector and the logarithmic detector when clutter can be described by the Weibull pdf. The optimum value of m for an *m-out-of-n* detector will depend on the nature of the pdf. When the optimum m is used with non-Rayleigh clutter, Schleher states that the *m-out-of-n* detector is better than other detection criteria.

The logarithmic detector has an output voltage whose amplitude is proportional to the logarithm of the input envelope. A logarithmic characteristic provides a constant false-alarm rate (CFAR) when the clutter is described by the Rayleigh pdf (see the log-FTC discussion presented later in this section). The performance of the logarithmic detector in non-Rayleigh clutter is almost as good as the *m-out-of-n* detector and is probably easier to implement.[145] A variant sometimes useful for reducing the false-alarm rate in non-Rayleigh clutter is the *log-log detector.* In one implementation, this is a logarithmic receiver in which the slope of the logarithmic characteristic progressively declines by a

---

| *The middle value of a statistical distribution, above and below which lie an equal number of values.

factor of 2 to 1 over the range from noise level to $+80$ dB.[146,147] In the log-log detector, the higher clutter values (tails of the distribution) are subjected to greater suppression.

Land Interclutter Visibility   The patchiness of land clutter, when viewed with high-resolution radar at low grazing angles, usually is beneficial for target detection. Patches of large clutter are separated by areas with little or no clutter. Typical sizes of clutter patches can be several kilometers on a side. Billingsley[140] states that the macropatches of clutter might typically occupy about one half of the resolution cells of discernible clutter. However, for level terrain in which only isolated discrete clutter echoes appear, the percentage of cells with clutter might be as low as 25 percent. For high radar sites and/or steep terrain "in which relatively full illumination exists, it can approach 100%." Interclutter visibility is a result of masking caused by hills or higher levels of terrain which create shadow areas where clutter echoes are absent. Aircraft targets can be detected in regions with low or no clutter even though they might be undetectable when within patches of high clutter.[148] It has been said that at low grazing angles over land, interclutter visibility can be expected with beamwidths less than a degree and pulse widths less than one μs.[149]

Weather   Reducing radar beamwidth and/or pulse width is generally considered good for decreasing the weather clutter with which the target must compete. There is, however, little information on the characteristics of rain with high-resolution radar.[150] Most forms of rain are not uniform when viewed with high resolution; but for purposes of radar design, it is usually assumed so. Reducing the radar resolution cell to improve signal-to-clutter ratio is an acceptable technique, but the designer should be aware that a small resolution cell size might cause the statistics of the clutter to become non-Rayleigh and increase the probability of false alarm.

Target Break-up with High-Resolution Waveforms   In some cases, the range resolution cell might be smaller than the dimensions of the target, resulting in "break-up" of the target echo. This means that the cross section of each resolved area of the target will be less than the total cross section as seen by a lower resolution radar. A serious reduction of detectability need not occur because of target break-up. Reduction of clutter by high range-resolution is still a good tactic.

If echoes from the various parts of the target viewed with high-resolution radar are properly displayed to an operator, and if the operator knows the general shape of the target or even that it is a distributed target, there appears to be little, if any, loss in detectability. The operator seems to be able to provide a form of noncoherent integration. Even if the recognition ability of an operator is not utilized, detectability will likely be reduced only slightly with high resolution since many targets tend to consist of one or a few large scatterers and a large number of small scatterers.[151] The echo in the largest resolution cell and in surrounding cells might still be of significant value to produce adequate detection.

Hughes[152] has investigated detection strategies when the radar resolution is less than the target dimensions. He assumed, for sake of discussion, that a range cell was 10 ns (5 ft) and that a 100-ns window was examined. The distributed target was completely within the 100-ns window. Clutter was homogeneous and described by a gaussian distribution. In one strategy, a threshold decision was made every resolution cell (10 ns), and any threshold crossing within the 100 ns window was declared a detection. This is similar to an *m*-out-of-*n* detector with *m* = 1. The other strategy, called the integrated detector, was to integrate 10 cells and make a detection decision every 100 ns. As reference case with which to judge these two procedures, Hughes took a 100-ns range cell with a detection decision made every 100 ns. The reference case was always significantly worse than either of the two strategies. Hughes concluded that the strategy that is better depends on the a priori knowledge of the target's nature that is available to the designer. If about two-thirds or more of the target echo energy comes from a single "flare" point, or individual scatterer, the *m*-out-of-*n* detector will be better. He stated that the correct strategy to use can be chosen if detailed knowledge of the target is known. Without such knowledge or when a variety of targets must be detected, the second strategy—integration before detection—is better.

Frequency   Radar echoes from rain, sea (with horizontal polarization), and (to some extent) land decrease with decreasing frequency. Although this favors the lower frequencies, it is more difficult at the lower frequencies to achieve narrow beamwidths and high range-resolution. Thus some of the benefits of reduced clutter at the lower frequencies might be partially offset by the poorer resolution that results.

Rain   The radar equation for detection of targets in rain, Eq. (7.37), shows that for a constant-gain antenna the range varies as the square of the wavelength. (For an antenna with a constant effective aperture, however, the range is directly proportional to wavelength.) There can be great benefit in using lower frequencies to reduce rain clutter. It is unlikely, for instance, that a radar at UHF or lower frequencies will be degraded by rain clutter. Unfortunately, as indicated above, in some applications it is not always possible to operate radars at frequencies low enough to make rain clutter negligible because of other undesirable constraints such as broad beamwidth, narrow bandwidth, a crowded frequency spectrum, external noise, and poor low altitude coverage.

Radars at L band or higher frequencies are likely to be affected by rain clutter. Some higher frequency radars might even be rendered useless in rain if no corrective measures are taken.

Sea   The plot of σ^0 versus grazing angle that was shown in Fig. 7.13 for sea clutter indicates that at low grazing angles and horizontal polarization, the lower the frequency the lower will be the echo from the sea when everything else remains the same. There are some reservations, however, about how low in frequency one might wish to go. The lower the frequency, the more difficult it is to direct the radar energy at low angles. This is illustrated in Sec. 8.2, which shows that the elevation angle of the lowest interference lobe is $\lambda/4h_a$, where λ = wavelength and h_a = antenna height above the water. If the target of interest is a small boat, buoy, submarine periscope, swimmer, or low altitude (sea skimming) missile, the higher microwave frequencies might be preferred even though the sea clutter might be greater.

Another reason higher frequencies might be preferred in some overwater applications in spite of higher sea clutter is the better range and angle resolution that can be obtained. In addition to providing less clutter, the higher resolutions obtainable at the higher frequencies might be required for particular applications. An example is the civil marine radar, which is commercially available at both X and S bands. When radars at both frequencies are operated on a ship, it is the higher frequency radar (X band) that is generally depended upon, except in bad weather when rain clutter degrades the higher frequency much more than S band. The fact that sea clutter is less at the lower frequencies is but one of several factors to be considered in the selection of the best frequency for a particular radar application.

Land   Although it was shown in Fig. 7.10 that land clutter at high grazing angles decreases with decreasing frequency, the frequency dependence is not usually a major consideration in radar design. At low grazing angles, Fig. 7.5 and the discussion of Sec. 7.3 show that for any particular type of terrain there might be significant variation of clutter with frequency, but the mean value of clutter taken over all types of terrain (excluding urban area) is essentially independent of frequency.

Polarization

Rain   Circular polarization is often employed to reduce weather clutter in microwave air-surveillance radars. Discrimination of a target located in the midst of rain results because the echo from an aircraft (an asymmetrical scatterer) differs from that of a raindrop (a circularly symmetric scatterer) when using circular polarization. This difference in scattering can be used to enhance the echo from the target and suppress echoes from rain.

A circularly polarized wave incident on a spherical scatterer will be reflected as a circularly polarized wave of opposite sense of rotation (when viewed in the direction of propagation) and will be rejected by the same antenna that originally radiated it. On the other hand, when a circularly polarized wave is incident on an asymmetrical target such as an aircraft, it has been found experimentally that the reflected energy is more or less equally divided between the two senses of polarization rotation.[153] The two senses of circular polarization are (1) right-hand circular, when the electric field rotates clockwise when viewed in the direction of propagation and (2) left-hand circular, when the electric field rotates counterclockwise. Since about half of the reflected energy is of the same sense of circular polarization as that transmitted, it will be accepted by the antenna that originally radiated it. The circularly polarized echo from spherical raindrops is of the opposite sense of polarization (when viewed in the direction of its propagation) and will be rejected by the radar antenna; but an aircraft will have a significant amount of its reflected energy with a polarization that is accepted by the antenna. This then provides target-to-clutter enhancement.

Raindrops are seldom perfect spheres, especially when they are large. They deviate from a sphere when their radius is greater than about 1 mm, and have a shape more like that of an oblate spheroid with a flattened base (a hamburger bun).[154,155] This particular raindrop shape, which is quite different from that usually depicted by an illustrator or cartoonist, is a result of aerodynamic forces as the drop falls.[156] The greater its deviation from a spherical shape, the less the echo signal will be rejected. Thus circular polarization is limited in the amount of rain cancellation it can provide and becomes less as the rain becomes heavier.

In addition to the nonspherical shape of raindrops, the ability to cancel rain echo using circular polarization is limited by practical considerations. It is difficult to obtain an antenna with pure polarization that does not accept energy from the orthogonal polarization. The cancellation of an orthogonally polarized signal by an exceptionally well designed, well maintained antenna might be limited to about 40 dB, a very good value.[157] Most antennas are not this good, and they can be even poorer because of the depolarizing effects of the nearby environment. Depolarization is also introduced when propagating through rain located between the radar and the target. The different reflection coefficients of horizontal and vertical polarization on reflection from the earth's surface also will cause a change in the signal's polarization when multipath reflection occurs from the surface. This factor has been said to limit the cancellation of rain clutter in certain cases to 13 dB over water[158] and 34 dB over desert.[159]

The various forms of depolarization that cause circular polarization to become elliptical suggest that better rain cancellation might be achieved if the optimum form of elliptical polarization were used rather than circular. With the optimum elliptical polarization, cancellation in some regions of heavy rain might be increased by as much as 12 dB.[160] However, the polarization that is optimum for one particular region of rain might not be optimum for some other region of rain, and might prove to be worse than the cancellation obtained with circular polarization. The optimum polarization will depend on the distance traveled and the nature of the rain. Thus the polarization on receive should be adaptive and made variable with range (time). Nathanson suggested that an improvement in cancellation of 6 to 9 dB can be obtained with adaptive polarization.[161] Adaptive polarization, however, has not been easy to implement successfully.

The radar cross section of aircraft is generally less with circular polarization than with linear polarization. As mentioned above, experimental measurements indicate that when an aircraft is illuminated with one sense of circular polarization, the echo power on a statistical basis is divided more or less equally between right-hand and left-hand circular polarization. Other experiments, however, have shown the circular polarized echo to be smaller than the linearly polarized echo by about 5 dB for aircraft and even greater reductions for missiles and satellites.[162]

An alternative to circular polarization is to transmit with a linear polarization (horizontal or vertical) and receive with the orthogonal linear polarization. Spherical scatterers will not produce an echo at the orthogonal linear polarization; but asymmetrical scatterers will have energy in the orthogonal component. It has been suggested that crossed linear polarization might give better rain rejection than conventional circular polarization.[163] Some measurements show, however, that the backscatter from aircraft with the orthogonal linear polarization can be as much as 7 dB below that reflected at the transmitted linear polarization.[164]

Although the use of circular polarization is often employed in surveillance radars during rainfall, at the higher microwave frequencies where the echo from rain can be a problem, cancellations of heavy rain might be 15 dB or less. In heavy rain that accompanies thunderstorms cancellations of only 5 dB have been observed.[161] If detection of targets in rain is important, then circular polarization might not provide all that is needed. MTI designed for moving clutter will obtain much greater clutter rejection. Lowering the frequency by a factor of 2 or using a pulse-compression waveform with a 13-bit Barker code could likely provide in heavy rain as much cancellation as circular polarization.

Sea   Over the sea, horizontal polarization at low and moderate sea states results in less sea clutter than vertical, as was indicated in Fig. 7.13. The majority of the radars that operate over the sea employ horizontal polarization. If a target, such as an oil slick or a swimmer's head, lies on the water, vertical polarization is preferred even though sea clutter may be larger than with horizontal polarization. Sea clutter is large with vertical polarization at low angles for the same reason the echo energy from targets on the water surface is large: these targets and the sea surface are illuminated with more energy than if horizontal polarization were used because of the behavior of the vertical polarization reflection coefficient (Sec. 8.2).

Land   It was mentioned in Sec. 7.3 that there is no significant difference between horizontal and vertical polarizations from land clutter when viewed at low grazing angles. Although there are indications that horizontal polarization produces a lower clutter echo than vertical, the difference is not significant enough to affect the choice of polarization in most radar applications over land.

Time Decorrelation   Unlike receiver noise, clutter echoes are generally correlated from pulse to pulse, and sometimes even from scan to scan. The techniques of *rapid antenna scan* for detection of small targets in the sea, and *time compression* for detection of moving targets in patchy land or sea clutter are examples of detection techniques that try to take advantage of the nature of correlated clutter echoes.

Antenna Scan Rate (Sea Clutter)   As mentioned previously in this chapter, when clutter is stationary (not moving) there is no gain to be had by integrating pulses. The sea is always changing, but over a short time interval it might not change sufficiently for the radar echo to be decorrelated pulse to pulse. For a medium-resolution X-band radar observing the sea at low grazing angles, the time required for the sea to decorrelate is about 10 ms.[165] Any pulses received during this time will be correlated and no improvement in signal-to-clutter ratio will result by integrating pulses. The sea surface, however, will often change by the time of the next antenna scan so that sea clutter is generally decorrelated from scan to scan. To take advantage of the decorrelation of sea clutter from scan to scan, the antenna rotation rate can be increased, as was first demonstrated by J. Croney for civil marine radar.[166]

Consider, for example, a civil marine X-band radar with a 1.2° beamwidth, pulse repetition frequency of 6000 Hz, and a 20-rpm antenna rotation rate. There will be 60 pulses in the 10 ms it takes for the beam to scan past a target. No integration improvement is obtained from adding these 60 pulses together since they occur within a time that is less than the decorrelation time of X-band sea clutter. If the antenna rotation rate is speeded up to 600 rpm, two pulses will be received per scan. These will be correlated, but the two echoes received on the next scan, one tenth of a second later, will be decorrelated from the two echoes from the previous scan. In 30 scans at 600 rpm, there will be 30 sets of two pulses each, and each set will be decorrelated from the others. Thus there can be an integration improvement corresponding to that obtained with 30 independent samples. Experimental measurements with a civil marine radar having a 1° beamwidth and 5000 Hz pulse repetition frequency show that increasing the antenna rotation rate from 20 to 420 rpm improves the target-to-clutter ratio from about 4 to 8 dB, depending on the sea state.[167] The smaller improvement corresponds to the higher sea state.

Radars that use a high antenna-rotation-rate to improve detection of small targets in sea clutter include the AN/APS-116,[168] and its successor the AN/APS-137. The antenna scan rate in these radars is 300 rpm (five scans per second). Scan-to-scan integration is performed with an m-out-of-n detector. With spiky clutter, analysis shows that the best ratio of m/n was 0.6, which is larger than the optimum value used in an m-out-of-n detector when only receiver noise is present.

The decorrelation time of sea clutter is approximated by the reciprocal of the width of the clutter doppler spectrum. Normally, the decorrelation time might be expected to be inversely proportional to frequency, if the decorrelation is due solely to doppler effects. However, experimental measurements show that when the decorrelation time is 10 ms at X band, it will be about 60 ms at L band instead of the 70 to 75 ms predicted on the basis of an inverse relationship of decorrelation time and frequency. Therefore, it is found that when clutter spectra are given in units of velocity rather than frequency, the velocity spectra are broader at the lower frequencies.[169]

Time Compression   Even if a target seen with a high-resolution non-MTI radar can be resolved in the midst of patchy land clutter or spiky sea clutter, it might be difficult to recognize the target from the clutter on the basis of a single observation. If the target is in motion, however, an operator can eventually distinguish it from the stationary clutter. One method that has been proposed to accomplish this is time compression, in which the last N scans of a radar are stored and then repeatedly displayed speeded up to accentuate the target motion. (The number of scans N might be from 5 to 10.) An operator's attention is called to the moving targets by the flickering of the repeated speeded-up scans. Even with a fixed target, time compression can separate the short duration sea spikes from the more persistent target echoes. As a target at long range comes within the coverage of the radar, there will be a delay of N scans before time compression can be fully effective. In the time needed to accumulate the N scans, however, a trained and alert operator might be expected to detect the moving target even without time compression.

Frequency Agility   If the RF frequency of a pulse of duration τ is changed by more than $1/\tau$, the echo from uniformly distributed Rayleigh type clutter will be decorrelated. This can be shown by calculating the change in frequency Δf that causes the difference in the phase between the clutter echo from the leading edge of the range-resolution cell and the clutter echo from the trailing edge of the range-resolution cell to be greater than 2π radians (as in Problem 7.15). Pulse-to-pulse frequency changes (called *frequency agility*) greater than $1/\tau$ Hz will therefore decorrelate the clutter and permit an increase in target-to-clutter ratio when the decorrelated pulses are integrated. This assumes the target's physical size is small compared to the size, $c\tau/2$, of the radar resolution cell. When the clutter is non-Rayleigh, as when there are one or a few dominate scatterers, there might be little benefit in the use of frequency agility. Spiky sea clutter, for example, is highly non-Rayleigh and frequency agility is not expected to offer much improvement. If land clutter is dominated by only one or a few large clutter scatterers in a resolution cell, the benefit of frequency agility also is lessened. Weather clutter is probably more uniform than either sea or land clutter, and frequency agility might be more effective than with other forms of clutter.[170]

The effect on the target and the clutter by changing frequency from pulse to pulse is illustrated in the A-scope presentations of Fig. 7.21.[171] The left-hand trace is that of a single frequency at X band. The clutter is shown on the left side of each figure. Its amplitude decreases with increasing range. A single target is shown on the right side. Its appearance indicates that it is composed of multiple scatterers. The right-hand trace is with pulse-to-pulse frequency agility. Two changes are noted: The clutter is smoothed in appearance and the target echo is larger. The increase in target echo is due to some of the individual frequencies returning large echoes from the target, at least as compared to the particular single frequency used in the upper trace. If a different frequency were chosen in the upper trace, it might have been larger or it might have been smaller. But by using multiple frequencies, the composite of the echoes is not as likely to produce a small echo as might a single frequency. If the target on the right side of the A-scope presentation were located in the midst of the clutter in the left-hand trace, it might not be detected. However, the target signal in the right-hand trace would likely be detected. It seems from this example that the real benefit of frequency agility is in the greater target cross section rather than the suppression of clutter.

It should be cautioned that conventional doppler filtering, as in MTI or pulse doppler radar, is not possible with pulse-to-pulse frequency agility.

Although frequency agility is sometimes claimed to be an attractive method for obtaining increased target-to-clutter ratio, it has not had the full success that its proponents

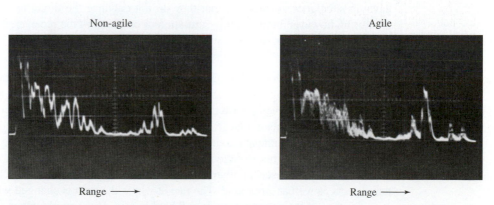

Non-agile   Agile

Range ⟶   Range ⟶

Figure 7.21   Effect of frequency agility on clutter and target. This A-scope presentation of clutter (left-hand side of each illustration) and a target (right-hand side of each illustration) was made with an X-band radar and a 0.2-μs pulse width. Left-hand trace is for a single frequency. Right-hand trace is for pulse-to-pulse frequency agility over a 100-MHz bandwidth.
I (Courtesy of CPI Beverly Microwave Division, Beverly, MA.)

have desired since most clutter is non-Rayleigh. When a large frequency band is available, the bandwidth is usually better used to provide high range-resolution to decrease the clutter rather than to attempt to decorrelate the clutter by pulse-to-pulse frequency change.

Clutter Fence   Reflections from nearby mountains and other large clutter might not be fully suppressed by conventional clutter reduction techniques. These echoes can be so large that they can enter the radar via the antenna sidelobes. At a fixed radar site it is possible to reduce the effects of such clutter by surrounding the radar with an electromagnetically opaque fence to prevent direct viewing of the clutter. The two-way attenuation provided by a typical clutter fence with a straight edge might be about 40 dB. Greater attenuation can be achieved by incorporating two continuous slots close to, and parallel with, the upper edge of the fence to cancel a portion of the energy diffracted by the fence. The increase in two-way attenuation by this method is 20 dB or more.[172]

A clutter fence can produce effects that are not always desirable. It can limit the accuracy of the elevation-angle measurement because of blockage and because of radar energy diffracted by the fence. Energy diffracted by the fence will interact with the direct energy and cause lobing of the radiation pattern in the angular region just above the fence. Radar energy backscattered by the fence towards the radar can sometimes damage the receiver front-end. In one design, the fence was tilted 15° away from the vertical to prevent damage to the receiver from occurring.[173]

Clutter fences are useful only when the targets are at higher elevation angles than the clutter. They have not been an attractive method to enhance target-to-clutter ratios for most applications.

Methods to Avoid Receiver Saturation by Clutter Echoes   The finite dynamic range of radar receivers means that they can be saturated by large clutter signals, with the result that target echoes can be prevented from being detected even though they might be larger than clutter. Several methods have been used to minimize the effect of large clutter but they have no subclutter visibility in that they do not enhance the target-to-clutter ratio. They are useful, however, in preventing saturation or overload of the receiver, automatic processor, automatic tracker, or display.

Sensitivity Time Control (STC)   The fourth-power relationship between the received echo-signal power and range means that clutter echoes at close ranges will be large and can saturate a radar receiver. A solution is to reduce the receiver gain at short ranges where the clutter echo signals are large, continually increase the gain as the pulse travels out in range, and finally operate at maximum gain (maximum sensitivity) at ranges beyond which clutter is expected. This is called *sensitivity time control* (STC). It has also been called *swept gain.* STC is used in air-traffic control radars to reduce nearby clutter from birds and insects as well as from land. The rate at which the gain is changed with time depends on the nature of the environment. It will be different, for example, when the radar is looking over water or desert than when the radar is looking over rugged or mountainous terrain. An air-surveillance radar might have more than one STC gain-versus-time characteristic that can be selected depending on the type of clutter the radar encounters. The variation of gain with range might be as R^2 for rain, R^3 or R^4 for surface clutter, to R^7 over water at long range (as in a civil marine radar for targets below the lowest interference lobe[174]).

Even without the presence of clutter, STC can be used to compensate for the large change in the magnitude of the received target echo signal as a function of range. The change of target cross section with range has been said to overshadow other causes of echo variation.[175]

The use of STC causes the sensitivity of a 2D air-surveillance radar with a cosecant-squared antenna elevation radiation pattern (Secs. 2.11 and 9.11) to be reduced for targets at high angles and short range since the gain is lowered in these directions just as it is at low angles and long range. This can cause echoes from aircraft at short range and high altitude to be too weak to be detected. By modifying the cosecant-squared antenna pattern to direct more energy at the higher angles, it is possible to see these close-in aircraft even with STC applied. Multiple-elevation-beam (stacked beam) 3D radars can have a separate STC variation with range to match the clutter conditions found as a function of elevation angle.

STC usually cannot be used with pulse doppler and other radars that have prfs high enough to result in ambiguous range echoes. It can also cause degradation to pulse compression radars that employ very long uncompressed pulses, such as required with solid-state transmitters.

Raising the Antenna Beam   Surface clutter can be decreased by raising the antenna beam in elevation and not illuminating the clutter. This technique can be applied only if the targets of interest (e.g., aircraft) are at a higher altitude than the clutter. Tilting an antenna beam upward reduces coverage at long range and low altitudes. A better technique is to use two beams in elevation, with one pointing higher than the other. Most civil air-traffic control radars use this method to reduce unwanted echoes from cars and trucks. In the ARSR-3 long range (nominally 200 nmi) air-traffic control radar, the gain of the upper beam is 16 dB less along the horizon than the gain of the lower beam along the horizon. The radar signal is transmitted on the lower beam and received only on the higher beam at short range so that surface clutter at short range is illuminated with less energy. After the pulse travels beyond the range of expected surface targets (typically 50 nmi in the ARSR-3), the receiver is shifted to the low beam to allow detection of aircraft at long range.

Log-FTC   This is a receiver with a logarithmic input-output characteristic followed by a high-pass filter (sometimes called a *fast time-constant,* or FTC). When the input clutter or noise is described by the Rayleigh probability density function, the output clutter or noise is constant, independent of the input amplitude. Thus it acts to provide a constant false alarm rate, or CFAR.

The Rayleigh pdf can be written

$$p(v) = \frac{2v}{m_2} \exp\left(-\frac{v^2}{m_2}\right) \qquad v > 0 \qquad\qquad \text{[7.41b]}$$

where m_2 is the mean square value of v. The rms amplitude of the fluctuations about the mean (denoted here by δv_{in}) is proportional to the mean $\overline{v_{in}}$ or $\delta v_{in} = k\overline{v_{in}}$. (This is Problem 2.4.) A logarithmic receiver has the characteristic

$$v_{out} = a \log (bv_{in}) \qquad [7.42]$$

The slope of the logarithmic receiver characteristic at $\overline{v_{in}}$ is

$$\frac{\Delta v_{out}}{\Delta v_{in}} = \frac{a}{v_{in}} \qquad [7.43]$$

If the input clutter fluctuations δv_{in} are small compared to the total range of the logarithmic characteristic, the output fluctuations δv_{out} are approximately

$$\delta v_{out} = \text{slope} \times \delta v_{in} = \frac{a}{v_{in}} \delta v_{in} = ak \qquad [7.44]$$

Thus the output fluctuations are constant, independent of the input mean.

Although the output fluctuations about the mean are constant, the output mean is not. A high-pass filter removes the mean value of the output, leaving the fluctuation at a constant value on the display. The high-pass filter is equivalent to a differentiation.

The noise or clutter fluctuations that appear at the output of a logarithmic receiver are not symmetrical in amplitude since large amplitudes are suppressed more than normal. To make the output more like that of a linear receiver, the log-FTC may be followed by an amplifier with the inverse of the logarithmic characteristic (antilog).

A true logarithmic characteristic cannot be maintained down to zero input voltage since $v_{out} \rightarrow -\infty$ as $v_{in} \rightarrow 0$. At some point the receiver characteristic must deviate from logarithmic and go through the origin. Therefore, the practical receiver will have a linear characteristic at low signal levels and logarithmic at higher levels. This is called a *lin-log receiver*. The logarithmic receiver must be maintained at about 20 dB below the receiver rms noise level.[176]

Log-FTC prevents clutter from saturating the limited dynamic-range radar display and obscuring the presence of desired targets even though the targets may be of greater amplitude than the clutter. It is an example of a CFAR, but its effectiveness depends on the clutter having a Rayleigh pdf, which is not always the case.

The log-FTC was originally considered for the detection of targets in sea clutter;[176] however, it is probably more useful for operation in precipitation clutter. Precipitation is more likely to be described by a Rayleigh pdf than is sea or land clutter. When log-FTC when used for reducing the effect of rain, it has sometimes been called *weather fix*.

CFAR   Circuitry for maintaining the false alarm rate constant (CFAR) has been described in Sec. 5.7 and by Taylor[177] in the *Radar Handbook*. As mentioned, CFAR does not provide an improvement in signal-to-clutter ratio, and it maintains the false-alarm constant at the expense of probability of detection. It also has the disadvantage of poorer range resolution and results in a loss in signal-to-noise ratio. It should be used only when necessary.

REFERENCES

1. Probert-Jones, J. R. "The Radar Equation in Meteorology." *Quart. J. Roy. Meteor. Soc.* 88 (1962), pp. 485–495.

2. Nathanson, F. E. *Radar Design Principles.* 2nd ed. New York: McGraw-Hill, 1991, p. 316.

3. Billingsley, J. B., and J. F. Larrabee: "Multifrequency Measurements of Radar Ground Clutter at 42 Sites," MIT Lincoln Laboratory, Lexington, MA, Technical Report 916, November 15, 1991, Vol. 1, Principal Results (ESD-TR-91-061); Vol. 2 Appendices A through D (ESD-TR-91-175); Vol. 3 Appendix E (ESD-TR-91-176). A condensation appears in Billingsley, J. B. "Radar Ground Clutter Measurements and Models, Part 1, Spatial Amplitude Statistics," paper No. 1 in *Target and Clutter Scattering and Their Effects on Military Radar Performance,* NATO Advisory Group for Aerospace Research and Development, AGARD Conf. Proceedings 501, September 1991. AD-A244 893.

4. Billingsley, J. B. "Ground Clutter Measurements for Surface-Sited Radar." MIT Lincoln Laboratory, Lexington, MA, Tech. Rep. 786, Revision 1, February 1, 1993.

5. Billingsley, J. B. "A Handbook of Multifrequency Land Clutter Coefficients for Surface Radar." MIT Lincoln Laboratory, Lexington, MA, Tech. Rep. 958. This will appear in the book *Low-Angle Radar Land Clutter* by J. B. Billingsley.

6. Moore, R. K., K. A. Soofi, and S. M. Purduski. "A Radar Clutter Model: Average Scattering Coefficients of Land, Snow and Ice." *IEEE Trans.* AES-16 (November 1980), pp. 783–799.

7. Schooley, A. H. "Some Limiting Cases of Radar Sea Clutter Noise." *Proc. IRE* 44 (August 1956), pp. 1043–1047.

8. Barton, D. K. *Modern Radar System Analysis.* Norwood, MA: Artech House, 1988.

9. Moore, R. K. "Ground Echo." In *Radar Handbook,* 2nd ed., M. Skolnik (Ed.). New York: McGraw-Hill, 1990, chap. 12.

10. Ref. 2, Table 7.17.

11. Long, W. H., D. H. Mooney, and W. A. Skillman. "Pulse Doppler Radar." In *Radar Handbook,* 2nd ed., M. Skolnik (Ed.). New York: McGraw-Hill, 1990, chap. 17, pp. 17.11–17.16.

12. Ulaby, F. T., R. K. Moore, and A. K. Fung. *Microwave Remote Sensing,* vol. II, Reading, MA: Addison-Wesley, 1982, Sec. 11-4.2.

13. Ulaby, F. T., W. H. Stiles, L. F. Dellwig, and B. C. Hansen. "Experiments on the Radar Backscatter of Snow." *IEEE Trans.* GE-15 (October 1977), pp. 185–189.

14. Ulaby, F. T., and W. H. Stiles. "Microwave Response of Snow." *Adv. Space Res.* (1981), pp. 131–149. (Also found in Refs. 15 and 17.)

15. Ulaby, F. T., R. K. Moore, and A. K. Fung. *Microwave Remote Sensing,* vol. III. Dedham, MA: Artech House, 1986, Sec. 21-7.

16. Waite, A. H., and S. J. Schmidt. "Gross Errors in Height Indication from Pulsed Radar Altimeters Operating over Thick Ice or Snow." *Proc. IRE* 50 (June 1962), pp. 1515–1520.

17. Ulaby, F. T., and M. C. Dobson. *Handbook of Radar Scattering Statistics for Terrain.* Norwood, MA: Artech House, 1989.

18. Borel, C. C., R. E. McIntosh, R. M. Narayanan, and C. T. Swift. "File of Normalized Radar Cross Sections (FINRACS)—A Computer Program for Research of the Scattering of Radar Signals by Natural Surfaces." *IEEE Trans.* GE-24 (November 1986), pp. 1020–1022.

19. Long, M. W., *Radar Reflectivity of Land and Sea,* 2nd ed. Norwood, MA: Artech House, 1983.

20. Bush, T. F., F. T. Ulaby, and W. H. Peake. "Variability in the Measurement of Radar Backscatter." *IEEE Trans.* AP-24 (November 1976), pp. 896–899.

21. Bowditch, *American Practical Navigator,* U. S. Hydrographic Office, H. O. Publication No. 9, 1966, Appendix R.

22. Pierson, W. J., G. Neumann, and R. W. James. *Observing and Forecasting Ocean Waves,* U. S. Naval Oceanographic Office, H. O. Publication No. 603, 1955.

23. Daley, J. "An Empirical Sea Clutter Model." Naval Research Laboratory, Washington, D.C., Memorandum Report 2668, October 1973.

24. Daley, J. C., J. T. Ransone, Jr., and W. T. Davis. "Radar Sea Return—JOSS II." Naval Research Laboratory, Washington, D.C., Rep. 7534, February 21, 1973.

25. Daley, J. C., J. T. Ransone, Jr., and J. A. Burkett. "Radar Sea Return—JOSS I." Naval Research Laboratory, Washington, D.C., Report 7268, May 11, 1971.

26. Daley, J. C., et al. "Upwind-Downwind-Crosswind Sea Clutter Measurements." Naval Research Laboratory, Washington, D.C., Report 6881, April 14, 1969.

27. Hansen, J. P., and V. F. Cavaleri. "High-Resolution Radar Sea Scatter, Experimental Observations and Discriminants." Naval Research Laboratory, Washington, D.C., Rep. 8557, March 5, 1982.

28. Lewis, B. L., and I. D. Olin. "Experimental Study and Theoretical Model of High-Resolution Radar Backscatter from the Sea." *Radio Science.* 15 (July–August 1980), pp. 815–828.

29. Macdonald, F. C. "Characteristics of Radar Sea Clutter, Pt. 1—Persistent Target-Like Echoes in Sea Clutter." Naval Research Laboratory, Washington, D.C., Report 4902, March 19, 1957.

30. Ewell, G. W., M. T. Tuley, and W. F. Horne. "Temporal and Spatial Behavior of High Resolution Sea Clutter 'Spikes.'" *Proc 1984 IEEE National Radar Conf.* pp. 100–104, IEEE Catalog no. 84CH1963-8.

31. U. S. Patent 3,971,997: "Sea Spike Suppression Technique," issued to B. L. Lewis and I. D. Olin, July 27, 1976.

32. Williams, P. D. L. "Limitations of Radar Techniques for the Detection of Small Surface Targets in Clutter." *The Radio and Electronic Engineer* 45 (August 1975), pp. 379–389.

33. Long, M. W. Ref. 19, Sec. 5.6.

34. Wetzel, L. B. "On Microwave Scattering by Breaking Waves." In *Wave Dynamics and Radio Probing of the Ocean Surface.* O. M. Phillips and K. Hasselmann (Ed.). New York: Plenum, 1986, pp. 273–284.

35. Longuet-Higgins, M. S., and J. S. Turner. "An 'Entraining Plume' Model of a Spilling Breaker." *J. Fluid Mech.* 63 (1974), pp. 1–20.

36. Croney, J., A. Woroncow, and B. R. Gladman. "Further Observations on the Detection of Small Targets in Clutter." *The Radio and Electronic Engineer* 45 (March 1975), pp. 105–115.

37. Croney, J. "Improved Radar Visibility of Small Targets in Sea Clutter." *The Radio and Electronic Engineer* 32 (September 1966), p. 135–148.

38. Wetzel, L. B. "Electromagnetic Scattering from the Sea at Low Grazing Angles." In *Surface Waves and Fluxes.* Vol. II Netherlands: Kluwer Academic, 1990, chap. 12, pp. 109–171.

39. Wetzel, L. B. "A Model for Sea Backscatter Intermittency at Extreme Grazing Angles," *Radio Science* 12 (September–October 1977), pp. 747–756.

40. Knott, E. F., J. F. Shaffer, and M. T. Tuley. *Radar Cross Section,* 2nd ed. Norwood, MA: Artech House, 1993, Secs. 5.9 and 6.3.

41. Katzin, M. "On the Mechanisms of Radar Sea Clutter." *Proc. IRE* 45 (January 1957), pp. 44–54.

42. Helmken, H. H., and M. J. Vanderhill. "Very Low Grazing Angle Radar Backscatter from the Ocean Surface." *Record of the IEEE 1990 International Radar Conf.* pp. 181–188, IEEE Catalog No. 90CH-2882-9

43. Dockery, G. D. "Method for Modelling Sea Surface Clutter in Complicated Propagation Environments." *IEE Proc.* 137, Pt. F (April 1990), pp. 73–79.

44. Nathanson, F. E. Ref. 2, Sec. 7.2

45. Kerr, D. E. (Ed.): *Propagation of Short Radio Waves.* MIT Radiation Laboratory Series. New York: McGraw-Hill, 1951, vol. 13.

46. Goldstein, H. "Frequency Dependence of the Properties of Sea Echo." *Phys. Rev.* 70 (Dec. 1 and 15, 1946), pp. 938–946.

47. Kinsman, B. *Wind Waves.* Upper Saddle River, NJ: Prentice Hall, 1965, Sec. 7.4.

48. Beckmann, P., and A. Spizzichino. *The Scattering of Electromagnetic Waves from Rough Surfaces.* New York: Macmillan, 1963, Secs. 5.3 and 18.2.

49. Leung, H., and S. Haykin. "Is There a Radar Clutter Attractor?" *Appl. Phys. Lett.* 56 (February 1990), pp. 593–595.

50. Crombie, D. D. "Doppler Spectrum of Sea Echo at 13.56 Mc/s." *Nature* 175 (1955), pp. 681–682.

51. Maresca, J. W., and T. M. Georges. "Measuring RMS Wave Height and the Scalar Ocean Wave Spectrum with HF Skywave Radar." *J. Geophys. Res.* 85 (1980), pp. 2759–2771.

52. "Special Issue on High-Frequency Radar for Ocean and Ice Mapping and Ship Location." *IEEE J. Oceanic Engr.* OE-11, No. 2 (April 1986).

53. Wright, J. W. "A New Model for Sea Clutter." *IEEE Trans.* AP-16 (March 1968), 217–223.

54. Wetzel, L. B. "Sea Clutter." In *Radar Handbook,* 2nd ed., M. Skolnik (Ed.). New York: McGrall-Hill, 1990, Chap. 13, Sec. 13.4.

55. Wetzel, L. B. Ref. 38, Chap. 12.

56. Lyzenga, D. R., A. L. Moffett, and R. A. Shuchman. "The Contribution of Wedge Scattering to the Radar Cross Section of the Ocean Surface." *IEEE Trans.* GE-31 (October 1983), pp. 502–505.

57. Nathanson, F. E: Ref. 2, Chap. 7.

58. Horst, M. M., F. B. Dyer, and M. T. Tuley. "Radar Sea Clutter Model." *Inter. Conf. on Ant. and Prop.* IEE Pub. No. 169, Pts. 1 and 2, London, 1978. (Also available in Ref. 2, pp. 307–308.)

59. Morchin, W. *Radar Engineer's Sourcebook.* Norwood, MA: Artech House, 1993, Sec. 3.3.2.

60. Ulaby, F. T., R. K. Moore, and A. K. Fung. Ref. 15, Sec. 20–4.

61. Lewis, E. O., B. W. Currie, and S. Haykin. *Detection and Classifcation of Ice.* New York: John Wiley, 1987.

62. Haykin, et al. *Remote Sensing of Sea Ice and Icebergs.* New York: John Wiley, 1994.

63. Wylie, F. J. *The Use of Radar at Sea,* 5th ed. Annapolis, MD: Naval Institute Press, 1978.

64. Ringwalt, D. L., and F. C. Macdonald. "Terrain Clutter Measurements in the Far North." *Report of NRL Progress,* Naval Research Laboratory, Washington, D.C., pp. 9–14, December 1956.

65. Orlando, J. R., R. Mann, and S. Haykin. "Classification of Sea-Ice Images Using a

71. Schleher, D. C. "Radar Detection in Log-Normal Clutter." *Record of the IEEE 1975 International Radar Conf.,* pp. 262–267. Reprinted in *Automatic Detection and Radar Data Processing,* D. C. Schleher (Ed.). Dedham, MA: Artech House, 1980.

72. Sekine, M., and Y. Mao. *Weibull Radar Clutter.* London: Peter Peregrinus, 1990.

73. Boothe, R. R. "The Weibull Distribution Applied to the Ground Clutter Backscatter Coefficient." U.S. Army Missile Command Report No. RE-TR-69-15, June, 1969; reprinted in *Automatic Detection and Radar Data Processing,* D. C. Schleher (Ed.). Dedham, MA: Artech House, 1980, pp. 435–450.

74. Sekine, et al. "Weibull Distributed Ground Clutter." *IEEE Trans.* AES-17 (July 1981), pp. 596–598.

75. Fay, F. A., J. Clarke, and R. S. Peters. "Weibull Distribution Applied to Sea Clutter," *Radar 77, IEE Conf. Publ.* 155, 1977, pp. 101–104; reprinted in *Advances in Radar Techniques.* London: Peter Peregrinus, 1985, pp. 236–239.

76. Sekine, M., et al. "Weibull Distributed Sea Clutter." *IEE Proc.* 130, Pt. F (1983), p. 476.

77. Sekine, M., et al. "On Weibull Distributed Weather Clutter." *IEEE Trans.* AES-15 (November 1979), pp. 824–830.

78. Ogawa, H., et al. "Weibull-Distributed Radar Clutter Reflected from Sea Ice." *Trans. IEICE* E70 (1987), pp. 116–120.

79. Schleher, D.C. "Radar Detection in Weibull Clutter." *IEEE Trans.* AES-12 (November 1976), pp. 736–743.

80. Jakeman, E., and P. N. Pusey. "A Model for Non-Rayleigh Sea Echo." *IEEE Trans.* AES-24 (November 1976), pp. 806–814.

81. Ward, K. D., and S. Watts. "Radar Sea Clutter." *Microwave J.* 28 (June 1985), pp. 109–121.

82. Jao, J. K. "Amplitude Distribution of Composite Terrain Radar Clutter and the K-Distribution." *IEEE Trans.* AES-32 (October 1984), pp. 1049–1062.

83. Ward, K. D. "A Radar Sea Clutter Model and Its Application to Performance Assessment." *International Conf. Radar-82* October 18–20, 1982. IEE Conference Publication No. 216, pp. 204–207.

84. Ward, K. D., C. J. Baker, and S. Watts. "Maritime Surveillance Radar Part 1: Radar

89. Hair, T., T. Lee, and C. J. Baker. "Statistical Properties of Multifrequency High-Range-Resolution Sea Reflections." *IEE Proc.* 138, Pt. F (April 1991), pp. 75–79.

90. Baker, C. J. "K-Distributed Coherent Sea Clutter." *IEE Proc.* 138, Pt. F (April 1991), pp. 89–92.

91. Pentini, F. A., A. Farina, and F. Zirilli. "Radar Detection of Targets Located in a Coherent K Distributed Clutter Background." *IEE Proc.* 139, Pt. F (June 1992), pp. 239–245.

92. Watts, S. "Radar Detection Prediction in K-Distributed Sea Clutter and Thermal Noise." *IEEE Trans.* AES-23 (January 1987), pp. 40–45.

93. Armstrong, B. C., and H. D. Grifiths. "CFAR Detection of Fluctuating Targets in Spatially Correlated K-Distributed Clutter." *IEE Proc.* 138, Pt. F (April 1991), pp. 139–152.

94. Trunk, G. V. "Radar Properties of Non-Rayleigh Sea Clutter." *IEEE Trans.* AES-8 (March 1972), pp. 196–204.

95. Tough, R. J. A., C. J. Baker, and J. M. Pink. "Radar Performance in a Maritime Environment: Single Hit Detection in the Presence of Multipath Fading and Non-Rayleigh Sea Clutter." *IEE Proc.* 137, Pt. F (February 1990), pp. 33–40.

96. Oliver, C. J. "Representation of Radar Sea Clutter." *IEE Proc.* 135, Pt. F (December 1988), pp. 497–500.

97. Sekine, et al. "Log-Weibull Distributed Sea Clutter." *IEE Proc.* 127, Pt. F (June 1980), pp. 225–228.

98. Shlyakhin, V. M. "Probability Models of Non-Rayleigh Fluctuations of Radar Signals (A Review)." *Soviet J. Communications Technology and Electronics.* 33 (January 1988), pp. 1–16.

99. Schleher, D. C. "Radar Detection in Log Normal Clutter." *Record of the IEEE 1975 International Radar Conf.* pp. 262–267, IEEE Publication 75 CHO 938-1 AES.

100. Farina, A., A. Russo, and F. A. Studer. "Coherent Radar Detection in Log-Normal Clutter." *IEE Proc.* 133, Pt. F (February 1986), pp. 39–54.

101. Farina, A., et al. "Theory of Radar Detection in Coherent Weibull Clutter." *IEE Proc.* 134, Pt. F (April 1987), pp. 174–190.

102. Sangston, K. J. "Coherent Detection of Radar Targets in K-Distributed, Correlated Clutter." Naval Research Laboratory, Washington, D.C., Report 9130, August, 1988.

103. Battan, L. J. *Radar Observation of the Atmosphere.* Chicago, IL: Univ. of Chicago, 1973.

104. Gunn, K. L. S., and T. W. R. East. "The Microwave Properties of Precipitation Particles." *Quart. J. Roy. Meteor. Soc.* 80 (October 1954), pp. 522–545.

105. Sauvageot, H. *Radar Meteorology.* Norwood, MA: Artech House, 1992.

106. Personal communication from Raymond Wexler.

107. Haddock, F. T. "Scattering and Attenuation of Microwave Radiation Through Rain." Naval Research Laboratory, Washington, D.C., (unpublished manuscript), 1948. (Mentioned in Gunn and East, Ref. 104)

108. Smith, P. L. Jr., K. R. Hardy, and K. M. Glover. "Applications of Radar to Meteorological Operations and Research." *Proc. IEEE.* 62 (June 1974), pp. 724–745.

109. Jameson, A. R. "A Comparison of Microwave Techniques for Measuring Rainfall." *J. Applied Meteorology* 30 (January 1991), pp. 32–54.

110. Nathanson, F. E. Ref. 2, Sec. 3.6.

111. Sauvageot, H. Ref. 105, Sec. 1.5.3.

112. Gunn, K. L. S., and J. S. Marshall. "The Distribution with Size of Aggregate Snowflakes." *J. Meteor.* 15 (1958), pp. 452–466.

113. Sekhon, R. S., and R. C. Srivastava. "Snow Size Spectra and Radar Reflectivity." *J. Atmos. Sci.* 27 (1970), pp. 299–307.

114. Puhakka, T. "On the Dependence of the Z-R Relationship on the Temperature in Snowfall." *Preprints 16th Radar Meteorology Conf.* Am. Meteor. Soc., April 22–24, 1975, Houston, TX, pp. 504–507.

115. Austin, P. M. "Measurements of the Distribution of Precipitation in New England Storms." *Proc. 10th Weather Radar Conf.,* Am. Meteor. Soc., pp. 247–254, 1963.

116. Saxton, J. A. "The Influence of Atmospheric Conditions on Radar Performance." *J. Inst. Navigation* (London) 11, pp. 290–303, 1958.

117. Nathanson, F. E., Ref. 2, Fig. 6.4.

118. Konrad, T. G., J. J. Hicks, and E. B. Dobson. "Radar Characteristics of Brids in Flight." *Science* 159 (Jan. 19, 1968), pp. 274–280.

119. Vaughn, C. R. "Birds and Insects as Radar Targets: A Review." *Proc. IEEE* 73 (February 1985), pp. 205–227.

120. Eastwood, E. *Radar Ornithology.* London: Methuen, 1967.

121. Williams, T. C. and J. M. Williams. "Open Ocean Bird Migration." *IEE Proc.* 137, Pt. F (April 1990), pp. 133–137.

122. Houghton, E. W., F. Blackwell, and T. A. Wilmot. "Bird Strike and the Radar Properties of Birds." *Int. Conf. on Radar—Present and Future,* Oct. 23–25, 1973, pp. 257–262, IEE Conference Publication no. 105.

123. Flock, W. L., and J. L. Green. "The Detection and Identification of Birds in Flight, Using Coherent and Noncoherent Radars." *Proc. IEEE* 62 (June 1974), pp. 745–753.

124. Blackwell, F., and E. W. Houghton. "Radar Tracking and Identification of Wild Duck During the Autumn Migration." *Proc. World Conf. on Bird Hazards to Aircraft, Canada* (1969), pp. 361–376.

125. Antonucci, J. "A Statitical Model of Radar Bird Clutter at the DEW Line." Rome Laboratory (EEAS), Air Force Systems Command, Hanscomb AFB, MA, Report no. RL-TR-91-85, May, 1991.

126. Knott, E. F. *"Radar Cross Section."* In *Radar Handbook*, 2nd ed. M. Skolnik (Ed.) New York: McGraw-Hill, 1990, Chap. 11.

127. Riley, J. R. "Radar Cross Section of Insects." *Proc. IEEE* 73 (February 1985), pp. 228–232.

128. Schaefer, G. W. "Radar Observations of Insect Flight," *Insect Flight,* R. C. Rainey (Ed.) London: Blackwell Scientific Publications, 1976, Chap. 8.

129. Gossard, E. E., and R. G. Strauch. *Radar Observations of Clear Air and Clouds.* New York: Elsevier, 1983.

130. James, P. K. "A Review of Radar Observations of the Troposphere in Clear Air Conditions." *Radio Sci.* 15 (March–April 1980), pp. 151–175. (The entire March-April 1980 issue of *Radio Sci.* is devoted to "Radar Investigations of the Clear Air.")

131. Tatarskii, V. T. *Wave Propagation in a Turbulent Medium.* New York: McGraw-Hill, 1961.

132. Skolnik, M. "Atmospheric Turbulence and the Extension of the Radar Horizon." Naval Research Laboratory, Washington, D.C., Memorandum Rep. 2903, October 1974.

133. Heiss, W. H., D. L. McGrew, and D. Sirmans. "Nexrad: Next Generation Weather Radar (WSR-88D)." *Microwave J.* 33 (January 1990), pp. 79–98.

134. Michelson, M., Shrader, W. W., and J. G. Wieler. "Terminal Doppler Weather Radar." *Microwave J.* 33, (February 1990), pp. 139–148.

135. Plank, V. G. "A Meteorological Study of Radar Angels." U.S.A.F. Cambridge Research Center Geophys. Resarch Papers, no. 52, July, 1956, AFCRC-TR-56-211, AD 98752.

136. Rogers, R. R., and W. O. J. Brown. "Radar Observations of a Major Industrial Fire." *Bull. Am. Meteorological Soc.* 78 (May 1997), pp. 803–814.

137. Skolnik, M., D. Hemenway, and J. P. Hansen. "Radar Detection of Gas Seepage Associated with Oil and Gas Deposits." *IEEE Trans.* GRS-30 (May 1992), pp. 630–633.

138. Skolnik, M., and T. C. Bailey. "A Review of Radar for the Detection of Gas Seepage Associated with Underground Oil and Gas Deposits." *Proc. of the Conf. on Applications of Emerging Technologies: Unconventional Methods in Exploration for Petroleum and Natural Gas V,* Institute for the Study of Earth and Man, Southern Methodist Univ., Dallas, Texas, 1997, pp. 207–228.

139. Eastwood, E. Ref. 120, Chap. 9; also V. G. Plank, Ref. 135.

140. Billingsley, J. B. "Ground Clutter Measurements for Surface-Sited Radar," MIT Lincoln Laboratory, Lexington, MA, Tech. Rep. 786, Revision 1, February 1, 1993.

141. Caspers, J. W. "Automatic Detection Theory." In *Radar Handbook,* 1st ed. M. Skolnik (Ed.). New York: McGraw-Hill, 1970, chap. 15, Sec.15.6.

142. Trunk, G. V., and S. F. George. "Detection of Targets in Non-Gaussian Sea Clutter." *IEEE Trans.* AES-6 (September 1970), pp. 620–628.

143. Trunk, G. V. "Further Results on the Detection of Targets in Non-Gaussian Sea Clutter." *IEEE Trans.* AES-7 (May 1971), pp. 553–556.

144. Schleher, D. C. "Radar Detection in Weibull Clutter." *IEEE Trans.* AES-12 (November 1976), pp. 736–743. Correction in AES-13 (July 1977), p. 435.

145. Schleher, D. C. "Radar Detection in Log-Normal Clutter." *IEEE International Radar Conf.* Arlington, VA, April 21–23, 1975, pp. 262–267.

146. Croney, J., A. Woroncow, and B. R. Gladman. "Further Observations on the Detection of Small Targets in Sea Clutter." *The Radio and Electronic Engineer* 45 (March 1975), pp. 105–115.

147. Williams, P. D. L. "Observations on the Further Optimization of Radar Signal Processing for the Display and Detection of Targets in Sea Clutter." *Int. J. Remote Sensing* 5 (1984), pp. 489–496.

148. Tonkin, S. P., and R. A. McCulloch. "Gross Spatial Structure of Land Clutter." *IEE Proc.* 138, Pt. F. (April 1991), pp. 99–108.

149. Shrader, W. W., and V. Gregers-Hansen. "MTI Radar." *Radar Handbook,* M. Skolnik (Ed.). New York: McGraw-Hill, 1990, Chap. 15, p. 15.13.

150. Gordon, W. B. "Analysis of Rain Clutter Data from a Frequency Agile Radar." *Radio Science* 17 (July-August 1982), pp. 801–816.

151. Queen, F. D., and J. J. Alter. "Results of a Feasibility Study for Determining the Yaw Angle of a Landing Aircraft." Naval Research Laboratory, Washington, D.C., Rep. 8480, May 27, 1981.

152. Hughes, P. K., II. "A High-Resolution Radar Detection Strategy." *IEEE Trans.* AES-19 (September 1983), pp. 663–667.

153. Gent, H., I. M. Hunter, and N. P. Robinson. "Polarization of Radar Echoes, Including Aircraft, Precipitation, and Terrain." *IEE Proc.* 110 (December 1963), pp. 2139–2148.

154. Sauvageot, H. *Radar Meteorology.* Norwood, MA: Artech House, 1992, Sec. 2.2.7.

155. Oguchi, T. "Electromagnetic Wave Propagation and Scattering in Rain and Other Hydrometeors." *Proc. IEEE* 71, (September 1983), pp. 1029–1078.

156. McDonald, J. E. "The Shape of Raindrops." *Scientific American* (February 1954).

157. Schneider, A. B., and P. D. L. Williams. "Circular Polarization in Radars." *The Radio and Electronic Engineer* 47, no. 1/2 (January/February 1976), pp. 11–29.

158. Kalafus, R. M. "Rain Cancellation Deterioration Due to Surface Reflections in Ground-Mapping Radars Using Circular Polarization." *IEEE Trans.* AP-23 (March 1975), pp. 269–271.

159. Beasley, E. W. "Effect of Surface Reflections on Rain Cancellation in Radars Using Circular Polarization." *Proc. IEEE* 54 (December 1966), pp. 2000–2001.

160. Hendry, A., and G. C. McCormick. "Deterioration of Circular-Polarization Clutter Cancellation in Anisotropic Precipitation Media." *Electronics Letters* 10 (May 16, 1974), pp. 165–166.

161. Nathanson, F. E. "Adaptive Circular Polarization." *IEEE 1975 International Radar Conf.,* April 21–23, 1975, pp. 221–225.

162. Nathanson, F. E. Ref. 2, Sec. 5.2.

163. Schneider, A. B., and P. D. L. Williams. Ref. 157.

164. Olin, I. D., and F. D. Queen. "Dynamic Measurement of Radar Cross Sections." *Proc. IEEE* 53 (August 1965), pp. 954–961.

165. Kerr, D. E. (Ed.). *Propagation of Short Radio Waves.* MIT Radiation Laboratory Series. New York: McGraw-Hill, 1951, vol. 13.

166. Croney, J. "Improved Radar Visibility of Small Targets in Sea Clutter. *The Radio and Electronic Engineer* 32 (September 1966), pp. 135–148.

167. Croney, J., and A. Woroncow. "Dependence of Sea Clutter Decorrelation Improvments Upon Wave Height." *IEE Int. Conf. On Advances in Marine Navigational Aids,* July 25–27, 1972, IEE (London) Conf. Publication no. 87, pp. 53–59.

168. Smith, J. M., and R. H. Logan. "AN/APS-116 Periscope Detecting Radar." *IEEE Trans.* AES-16 (January 1980), pp. 66–73.

169. Valenzuela, G. R., and M. B. Laing. "Study of Doppler Spectra of Radar Sea Echo." *J. Geophys. Res.* 75 (January 20, 1970), pp. 551–563.

170. Nathanson, F. E. Ref. 2, Sec. 6.6.

171. Fuller, J. B., and J. R. Martin. "Radar Subsystems," Varian Co. brochure, Beverly Division, Beverly MA, no date, Fig. 2. (Now known as CPI Beverly Microwave Division.)

172. Becker, J. E., and R. E. Millet. "A Double-Slot Radar Fence for Increased Clutter Suppression." *IEEE Trans.* AP-16 (January 1968), pp. 103–108.

173. Ruze, J., F. I. Sheftman, and D. A. Cahlander. "Radar Ground-Clutter Shields." *Proc. IEEE* 54 (September 1966), pp. 1171–1183.

174. Harrison, A. "Marine Radar Today—A Review." *The Radio and Electronic Engineer* 47 (April 1977), pp. 177–183.

175. Taylor, J. W., Jr. "Receivers." In *Radar Handbook,* M. Skolnik (Ed.). New York: McGraw-Hill, 1990, Chap. 3, Sec. 3.6.

176. Croney, J. "Clutter on Radar Displays." *Wireless Engr.* 33 (April 1956), pp. 83–96.

177. Taylor, J. W., Jr. Ref. 175, Sec. 3.13.

PROBLEMS

7.1 Find the range at which the radar echo from low grazing angle surface clutter equals receiver noise for the following radar: peak power = 100 kW, azimuth beamwidth = 1°, elevation beamwidth = 20° [see Eq. (9.5c)], wavelength = 3 cm, pulse width = 0.1 μs, and receiver noise figure = 4 dB. The radar is located at a height of 1 km over a flat sea surface. You may assume the following variation of σ^0 with grazing angle Ψ:

$$\text{sigma zero} = -46 \text{ dB at } 1.0° \text{ grazing angle}$$

$$= -42 \text{ dB at } 3.0°$$

$$= -37 \text{ dB at } 10.0°$$

7.2 The ARSR-3 is a long-range, ground-based, fan-beam, L-band radar that was used by the FAA for enroute air-traffic control. It had an azimuth beamwidth of $1.25°$ and a pulse width of 2 μs. (a) If the clutter cross section per unit area, σ^0, for surface land clutter is -20 dB, what is the MTI clutter attenuation required to detect a 2 m^2 target at a range of 30 nmi with an output signal-to-clutter ratio of 15 dB? (Clutter is considered to be much greater than receiver noise.) (b) What will be the radar cross section of rain clutter (in m^2) seen by the ARSR-3 radar at a range of 30 nmi when the rainfall rate is 4 mm/h? You may assume a flat earth and that the rain uniformly fills the radar resolution cell from the ground up to a height h of 3 km. Take the radar frequency to be 1.3 GHz. (The elevation angle coverage should not be needed here.)

7.3 Derive a radar equation for the detection of a target in surface clutter when the grazing angle is $90°$ (normal incidence). Assume the antenna employs a pencil beam. (This is not a usual radar detection situation because the clutter echo is large when the antenna beam is perpendicular to the surface.)

7.4 An S-band pencil-beam radar (3.2 GHz) with a $1.5°$ beamwidth and 3-μs pulse width can detect a 2 m^2 target at a range of 200 nmi in a clear atmosphere (no rain). If a single-pulse signal-to-clutter ratio of 10 dB is required, what would be the range of the radar when observing a 2-m^2 target in rain of 4 mm/h? (Assume that clutter is uniformly distributed and is much greater than receiver noise, attenuation in rain at this frequency can be neglected, the antenna beam is pointing at a low elevation angle but doesn't strike the ground, and rain completely fills the radar resolution cell.)

7.5 Sketch as a function of rainfall rate (from 1 to 40 mm/h) the variation of the pulse echo power from rain for a radar with the following characteristics (which are those of Nexrad):

peak power = 1 MW

frequency = 3.0 GHz

pulse width = 1.57 μs

antenna gain = 45.5 dB (pencil beam)

effective number of hits integrated = 80

range = 150 km

(Assume that the beam is completely filled with rain, the beam is low to the ground but does not intercept it, and the attenuation due to rain can be neglected.)

7.6 What will be the attenuation due to rain uniformly distributed throughout the radar coverage and what will be the radar cross section of rain at a range of 20 km for a radar pointing at a low elevation angle with a $2°$ by $4°$ beamwidth and a 2-μs pulse width at frequencies of 3, 10, and 35 GHz? The rain falls at a rate of 4 mm/h.

7.7 Briefly comment on how the radar parameters listed below affect radar performance when detection is limited by (1) surface clutter and by (2) receiver noise. The radar parameters are: (a) pulse width, (b) antenna gain, (c) transmitter power, (d) number of pulses returned from the target, (e) system losses, and (f) the sensitivity of the maximum detection range to changes in the radar cross section. (It may help to make a table.)

7.8   Show that the radar cross section per unit area, σ^0, for a radar with a narrow pencil-beam antenna viewing a large perfectly reflecting flat surface at perpendicular incidence is approximately equal to G, where G = antenna gain. The flat surface is much larger than the extent of the footprint of the radar beam hitting the surface. In other words, derive Eq. (7.12). [The key to this problem is solving for the received power in two different ways, and then equating them to find σ^0. Start with the simple form of the radar equation for the received echo signal power P_r from a radar directed perpendicular to a flat surface. The cross section of the target is $\sigma^0 A_c$, where A_c is the area of that portion of the flat surface which is illuminated by the radar beam. Find an expression for A_c. The half-power beamwidth is θ_B, but you have to account for the change in illuminated area due to the two-way beamwidth. The second way to find the received power is recognize that the flat plate creates an image of the radar antenna a distance R behind the plate. The receive signal power is found by considering the geometry as a one-way communication link between the radar antenna and its image located a distance R behind the surface. Derive a one-way communication equation for the signal received at the image antenna when the radar antenna transmits to its image a distance $2R$ away. The two values of P_r, found from the two different models of propagation path, are the same and may be equated. You might need to use Eq. (9.5b) for the antenna gain.]

7.9   This problem concerns finding the radar cross section (in square meters) of various forms of distributed volumetric clutter at low altitudes as might be seen at a range of 10 km by an X-band (10 GHz) radar with a 1° beamwidth and a 1-μs pulse width. The cross section per unit volume of the volumetric clutter is denoted by η. The clutter is assumed to occupy the entire radar resolution cell. (Be careful of units.) (a) Rain falling at 4 mm/h. (b) Migrating Shearwater birds with $\eta = 3 \times 10^{-9}$ cm^{-1}. Compare your answer with the radar cross section of a single Shearwater bird as given in Table 7.5. (c) Insects with $\eta = 10^{-10}$ cm^{-1}. (d) Clear-air turbulence with $C_n^2 = 10^{-10}$ m$^{-2/3}$.

7.10 List five options available to the radar designer for enhancing the detection of aircraft in rain. Briefly describe the operation of each and their chief limitation.

7.11 If an S-band radar is capable of detecting a 1 m^2 target at a range of 200 nmi, at what range will it be able to detect a single sparrow? (See Table 7.4.)

7.12 Assume an X-band radar ($\lambda = 3.2$ cm) has a range of 45 km in the absence of rain. If it is further assumed that the attenuation in rain is the only factor affecting range, what will be the radar range when rain is falling throughout the region at a rate of 4 mm/h? (You will probably have to solve for the range by trail and error. The ultimate in precision is not vital in the answer to this question.)

7.13 What radar parameters are under the control of the radar designer for increasing the range at which a target may be detected in rain? Which parameter do you think is most important for increasing the detectability of a target? Explain why.

7.14 The value of the cross section per unit volume of rain, η, varies as f^4, where f = radar frequency. Assume the rain completely fills the radar resolution cell. (a) What is the variation of the echo from rain when the antenna gain G is independent of frequency? (b) What is the variation of the echo from rain when the antenna aperture area A_e is independent of frequency?

7.15   Show that the clutter echo seen by a pulse of width τ is decorrelated if the radar frequency is changed by an amount greater than $1/\tau$. [A decorrelated echo due to a change in frequency Δf means that the phase difference between the radar echo from the leading edge of the resolution cell and the trailing edge of the resolution cell changes by more than 2π radians when the frequency is changed an amount Δf. This occurs since phase is modulo 2π. Start by obtaining an expression for the two-way phase difference ϕ in terms of the radar resolution cell ΔR and wavelength λ. Note that the resolution cell is $\Delta R = c\tau/2$. The change in phase $\Delta\phi$ has to be greater than 2π radians when the frequency is changed by Δf in order to decorrelate the echo. The answer is now staring at you.]

7.16   Explain why frequency agility is not compatible with MTI (doppler) radar processing for detection of moving targets in clutter. (You may take as a model the single delay-line canceler.)

7.17   Why is the Bragg-scatter model not suitable for describing sea clutter at the higher microwave frequencies?

7.18   Briefly indicate what you think might be the preferred radar method or methods for detecting the targets listed below when in the presence of clutter. Assume the radar is located on the surface of the earth.

  a. Aircraft over normal land clutter.
  b. Aircraft over normal sea clutter.
  c. Aircraft in the presence of large numbers of birds and insects.
  d. Aircraft in the presence of moving ground targets.
  e. Aircraft in rain clutter but beyond the range of surface clutter.
  f. Ships and large buoys with corner reflectors.
  g. Small boats, periscopes, buoys, and swimmers.
  h. Stationary targets in distributed clutter.

8

Propagation of Radar Waves

8.1 INTRODUCTION

Most of the discussion of the radar equation in Chap. 2 considered the radar energy to propagate in free space. In the real world, however, the earth's surface and atmosphere can have major effects on radar performance. In Sec. 2.12 we briefly mentioned a few of the propagation factors that can influence the range and coverage of a radar. Since propagation effects might extend the radar range significantly or reduce it drastically, it is important to account for the earth's environment when attempting to predict radar performance.

Free-space radar performance is modified by the following propagation effects:

- *Forward scattering* (reflection) of the radar energy from the surface of the earth, which enhances the radiated energy at some elevation angles and decreases it at others.

- *Refraction* (bending) of the radar energy by the earth's atmosphere, which can cause the radar energy to deviate from straight-line propagation.

- *Ducting* (trapping) of the radar energy, a form of severe refraction, which causes extended radar ranges (and, surprisingly, might not always be a good thing).

- *Diffraction* of radar waves by the earth's surface that causes energy to propagate beyond the normal radar horizon. It applies mainly at the lower frequencies that are seldom used for radar applications.

- *Attenuation* of radar waves by the clear atmosphere, which generally has little or no effect on microwave propagation.
- *External noise* that enters the radar receiver and increases the receiver noise level.
- *Backscatter from land, sea, and weather clutter, and attenuation in rain and other hydrometeors.* These are not discussed here since they were included in Chap. 7.

Most of the significant propagation effects at microwave radar frequencies occur within the line of sight of the radar. Diffraction effects when the radar is at a sufficiently low frequency, and ducting at almost any microwave frequency, can cause radar waves to bend around the surface of the earth and extend the radar range beyond the normal line-of-sight horizon.

Although the basic theory of radar propagation may be well understood, accurate quantitative predictions for a particular place and for a specific time in the future are not always easy to obtain because of the difficulty in acquiring the necessary information about the environment in which a radar operates. In some respects, predicting the effects of propagation is a little like forecasting the weather. The radar system designer is usually interested in a long-term statistical description of propagation effects so that the radar can be designed to fulfill its mission satisfactorily. Sometimes, however, the radar designer has to be content with only a qualitative knowledge of "average" propagation effects. The military tactical commander or the air-traffic controller is not as interested in the long-term statistical effects of propagation or average conditions, but is more interested in the current or short-term forecast of radar propagation conditions that might be encountered. For example, based on current measurements of the environmental factors that affect radar propagation, a military tactical planner might be able to determine a flight profile of an attacking aircraft that would minimize the range where it is first detected by a defender's radar.

8.2   FORWARD SCATTERING FROM A FLAT EARTH

To determine the type of effects the earth's surface has on radar propagation, we initially assume a plane, smooth, perfectly reflecting flat earth. The results obtained with this simplification are indicative of what is obtained with more realistic models. The geometry is shown in Fig. 8.1a. The radar antenna is located at a height h_a above the planar surface. Its antenna radiation pattern is assumed to be uniform in elevation angle. The target is at a height h_t and at a range R from the radar. The ground distance between the radar and the target is D (not shown in the figure). Energy radiated by the radar antenna arrives at the target via two separate paths. One is the direct path (*AB*) from radar to target; the other is the path (*AMB*) from the radar to the target that includes a forward-scatter reflection from the surface. The signal reflected by the target also arrives back at the radar by these same two paths. The magnitude of the resultant echo signal back at the radar antenna depends on the amplitudes and relative phases of the signals that propagate via the direct and surface-scattered paths. The modification of the field strength η (measured in volts/meter) caused by the presence of the earth's surface may be expressed by the ratio

$$\eta = \frac{\text{field strength at target in presence of earth's surface}}{\text{field strength at target if in free space}} \qquad [8.1]$$

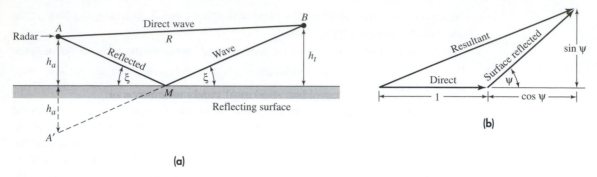

Figure 8.1   (a) Geometry illustrating radar propagation over a plane reflecting surface. (b) Vector addition of direct and surface-reflected signals, each of unity amplitude, with a phase difference of ψ.

It is assumed in this analysis that the path lengths of the direct and surface-reflected signals are almost (but not quite) equal so that the amplitudes of the two signals are essentially the same, except for any loss of signal suffered on reflection from the earth's surface. That is, if the two signals differ in amplitude from one another it is due to the surface reflection-coefficient being less than unity, and is not due to a significant difference in the $1/R^2$ factor. The slight range difference between the direct and surface-reflected paths, however, results in a difference in phase between the two which, when they combine at the target or at the radar, affects their sum. There is also a change in phase of the signal when it is reflected from the surface. This is represented by a reflection coefficient, which is a complex quantity $\Gamma = \rho e^{j\psi_r}$. The magnitude ρ describes the change in amplitude on reflection, the argument ψ_r describes the phase shift.

In this particular analysis we take the reflection coefficient to be $\Gamma = -1$. Thus the surface-reflected wave does not change its amplitude on reflection, but its phase is shifted by an amount $\psi_r = \pi$ radians. A reflection coefficient of $\Gamma = -1$ applies to a perfectly smooth, perfectly conducting surface if the radiation is horizontally polarized and the grazing angle is small.

The problem is easier to analyze if we replace the surface-reflected signal with the signal radiated from the image of the radar antenna that is below the surface, at A' in Fig. 8.1a. Instead of the surface-reflected path AMB, we consider the equivalent straight-line path $A'MB$. The path length AB is

$$AB = [D^2 + (h_t - h_a)^2]^{1/2} = D\left[1 + \frac{(h_t - h_a)^2}{D^2}\right]^{1/2}$$

$$\approx D\left[1 + \frac{1}{2} \cdot \frac{(h_t - h_a)^2}{D^2}\right] = D + \frac{(h_t - h_a)^2}{2D} \tag{8.2}$$

In the above we assumed $|h_t - h_a| \ll D$. The surface-reflected path length AMB, or its equivalent, $A'MB$ is similarly

$$AMB = A'MB = [D^2 + (h_t + h_a)^2]^{1/2} \approx D + \frac{(h_t + h_a)^2}{2D} \tag{8.3}$$

where we have assumed that $h_t + h_a << D$. Subtracting Eq. (8.2) from Eq. (8.3) results in the difference between the two paths $AMB - AB$, which is $\Delta = 2h_t h_a/D$. We make the assumption that the horizontal distance D can be replaced by the range R, so that the difference between the two paths is

$$\Delta \approx \frac{2h_a h_t}{R} \qquad\qquad \textbf{[8.4]}$$

To recapitulate, the geometrical assumptions we have made are that $(h_t \pm h_a) << D \approx R$. The phase lag ψ_Δ associated with this difference is found by multiplying Eq. (8.4) by $2\pi/\lambda$. The total phase difference is then

$$\psi = \psi_\Delta + \psi_r = \frac{4\pi h_a h_t}{\lambda R} + \pi \qquad\qquad \textbf{[8.5]}$$

At the target there are two signals, the direct and surface-reflected, which are of the same approximate amplitude with a phase difference between them of ψ, as given in Eq. (8.5). To obtain η, the vector addition of these two signals is divided by the signal amplitude that would have appeared if in free space. The value of η is found by applying the Pythagorean theorem to the sum of the two signal vectors, Fig. 8.1b, both of the same amplitude (normalized to unity) but with relative phase ψ. This gives

$$\eta = [(1 + \cos \psi)^2 + (\sin \psi)^2]^{1/2} = [2(1 + \cos \psi)]^{1/2} \qquad\qquad \textbf{[8.6a]}$$

The value η^2 is the ratio of the signal power density (W/m^2) at the target to the power density that would have been at the target if it were in free space, which becomes

$$\eta^2 = 2\left(1 - \cos \frac{4\pi h_a h_t}{\lambda R}\right) = 4 \sin^2 \left(\frac{2\pi h_a h_t}{\lambda R}\right) \qquad\qquad \textbf{[8.6b]}$$

Because of reciprocity in propagation, the path from target to radar is the same as that from radar to target. The echo signal power density received at the radar, relative to what would have been received in free space, is the fourth power of η, or

$$\eta^4 = 16 \sin^4 \left(\frac{2\pi h_a h_t}{\lambda R}\right) \qquad\qquad \textbf{[8.7]}$$

Lobing   The radar equation describing the received echo power is multiplied by the factor η^4 as given by Eq. (8.7). Since the sine function varies from 0 to 1, the factor η^4 varies from 0 to 16. The effect of the earth's surface in this simplified example is to increase the received signal power at some elevation angles by as much as 16. At other elevation angles it can be zero. Because of the fourth-power relation between range and received echo-signal power, the radar range will vary from 0 to 2 times the range the radar would have if it were in free space. The result is that the radiation in elevation is broken up into lobes which increase the range at some elevation angles and decrease it at others, as shown in Fig. 8.2. This effect is sometimes called *lobing*.

Figure 8.2   Vertical (elevation) lobe structure of the radar radiation caused by the presence of a planar reflecting surface.

$\lambda/4h_a$

The field strength in the presence of the earth's surface is a maximum when the argument of the sine term in Eq. (8.7) is equal to $\pi/2$, $3\pi/2$, . . . , $(2n + 1)\pi/2$, where $n = 0, 1, 2, ...$ The peaks of the lobes occur when

$$\frac{4h_a h_t}{\lambda R} = 2n + 1 \qquad \text{maxima} \qquad [8.8]$$

and the nulls, or minima, occur when the sine term is zero, or when

$$\frac{2h_a h_t}{\lambda R} = n \qquad \text{minima} \qquad [8.9]$$

From Eq. (8.8), the angle of the peak of the first (lowest) lobe ($n = 0$), is at

$$\theta_1 \approx h_t/R = \lambda/4h_a \qquad [8.10]$$

Thus if it is desired to see targets at low angles, the wavelength must be small (high frequency) and/or the antenna height must be large.

To illustrate the effects of a flat, smooth, perfectly reflecting earth's surface on radar performance we include η^4 in the simple form of the radar equation [Eq. (1.6)], which then becomes

$$P_r = \frac{P_t G^2 \lambda^2 \sigma}{(4\pi)^3 R^4} \cdot 16 \sin^4\left(\frac{2\pi h_a h_t}{\lambda R}\right) \qquad [8.11a]$$

When the argument of the sine is small,

$$P_r \approx \frac{4\pi P_t G^2 \sigma (h_a h_t)^4}{\lambda^2 R^8} \qquad [8.11b]$$

This represents the region below the peak of the first lobe. For targets at small angles (lower than the first lobe), the signal power is seen to vary as the inverse eighth power of the range, rather than as the inverse fourth power as occurs in free space.* The gain and wavelength appear in Eq. (8.11b) as the factor G/λ instead of $G\lambda$. The above applies for an antenna with a constant gain as a function of frequency. A different result is obtained if the effective aperture of the antenna is maintained constant with change in frequency.

*As shall be seen later, the variation of signal strength with range at low angles can be much more complicated than that given by Eq. (8.11b), especially under conditions of anomalous propagation.

Surface Reflection Coefficient   In the above several simplifying assumptions were made. One was that the antenna elevation pattern was uniform. With an actual antenna the antenna gain as a function of elevation angle has to be taken into account since the gain in the direction of the target can be different from the gain in the direction of the surface-reflected ray. Another assumption was that the surface was smooth and perfectly reflecting. This is not always realistic since the reflection coefficient depends on the surface roughness, the dielectric properties of the surface, polarization of the radar energy, and the frequency. Figure 8.3a gives the magnitude of the reflection coefficient as a function of the grazing angle and frequency for vertical polarization over smooth sea water. Figure 8.3b is the phase of the reflection coefficient for vertical polarization and a smooth sea. The magnitude of the reflection coefficient for horizontal polarization is given in Fig. 8.3c. The phase of the reflection coefficient for horizontal polarization is approximately

Figure 8.3
Reflection coefficient for a smooth sea as a function of grazing angle, frequency, and polarization. (a) Magnitude of the reflection coefficient for vertical polarization; (b) phase of the reflection coefficient for vertical polarization (phase of the reflected wave lags the phase of the incident wave); (c) magnitude of the reflection coefficient for horizontal polarization. The phase of the reflection coefficient for horizontal polarization is 180°, and is approximately independent of grazing angle and frequency.
(From Lamont Blake, *Radar Handbook*.[20])

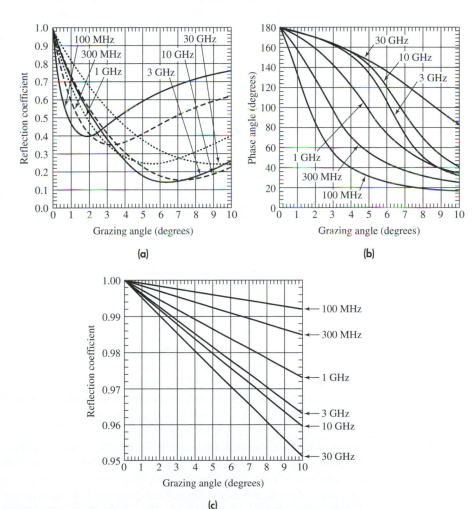

π radians, and doesn't vary much with frequency or grazing angle. The magnitude of the reflection coefficient is generally less for vertical polarization than for horizontal.

The minimum reflection coefficient for vertical polarization occurs at a grazing angle known as the *Brewster's angle.* When the reflection coefficient is less than unity and/or the phase of the surface reflection is not 180°, the nulls in the radiation pattern due to multipath will not be as deep and the peak value of the various lobes will decrease.

Surface roughness depends on the physical roughness *relative* to the radar wavelength. The lower the radar frequency (longer the wavelength), the smoother a surface will appear to the radar and the more likely the lobing pattern due to multipath will be highly pronounced. For example, the range in the direction of the lowest lobe of a VHF radar that is suitably sited over a sea surface might be increased almost by the theoretical factor of two indicated by the simple flat-earth model. At the higher microwave and millimeter wave frequencies, the less pronounced will be the effects of lobing.

The different values of reflection coefficients shown in Fig. 8.3 for vertical and horizontal polarization can result is different coverage patterns. The nulls with vertical polarization are not as deep and the maxima are not as great as with horizontal polarization. Vertical polarization might be specified when more uniform vertical coverage is desired and horizontal polarization might be preferred when greater range in the direction of the lobes is more important than more uniform coverage. Almost all air-surveillance radars, however, seem to employ horizontal polarization. The greater range at some elevation angles is a benefit that many radar manufacturers take advantage of when advertising the capabilities of their radars. The fact that there also are holes in the "long range" coverage with horizontal polarization is seldom deliberately mentioned.

Rough Surface Reflection Coefficient   The theoretical curves of Fig. 8.3 assume a smooth reflecting surface. A smooth surface is sometimes defined by the *Rayleigh roughness criterion* which considers a surface to be smooth if $h \sin \psi < \lambda/8$, where ψ is the grazing angle and h is the difference between the extremes of the surface height. (Some take h to be approximately $4\sigma_h$, and some take it to be $3\sigma_h$, where σ_h is the standard deviation of the gaussian distribution of the surface heights.) A surface is smooth or rough to a radar signal depending on the grazing angle and the physical roughness relative to the radar wavelength.

The surface roughness can affect the reflection coefficient more than the electrical properties of the surface which enter into the reflection coefficients given by Fig. 8.3. Measurements by many workers have shown that the reflection coefficient for normal (nonsmooth) ground terrain is in the range 0.2 to 0.4 and is seldom greater than 0.5 at frequencies above 1500 MHz at low grazing angles.[1] An expression for the reflection coefficient ρ_r of a rough, perfectly conducting surface, such as the sea, was originally given by Ament[2] as

$$\rho_r = \rho_0 \exp \left[-2k^2 \overline{h^2} \sin^2 \psi \right] \qquad \textbf{[8.12]}$$

where ρ_0 is the complex reflection coefficient for a smooth surface (Fig. 8.3), $k = 2\pi/\lambda$, λ = wavelength, $\overline{h^2}$ is the mean square surface height, ψ is the grazing angle, and h has a mean value of zero. Experimental data taken over the sea fit the expression of Eq. (8.12) for small values of the *surface roughness parameter* defined as $(\sigma_h \sin \psi)/\lambda$, where σ_h is the rms value of the surface height h. The Ament theory underestimates the experimental data of Beard[3] when the roughness parameter is greater than 0.1. Miller, Brown, and Vegh[4,5] extended the theoretical analysis of Ament and obtained the following expression for the rough-surface reflection coefficient

$$\rho_r = \rho_0 \exp\left[-2k^2\overline{h^2}\sin^2\psi\right] \cdot I_0(k^2\overline{h^2}\sin^2\psi) \qquad \text{[8.13]}$$

where $I_0(z)$ is the modified Bessel function of zero order. Eq. (8.13) is Ament's equation multiplied by the I_0 factor. It fits experimental data for roughness parameters less than 0.3. Figure 8.4 plots Ament's expression along with its modification, Eq. (8.13), and a set of experimental data.

In the above the *specular,* or *coherent,* component of the scattered signal has been discussed. There is also a *diffuse,* or *incoherent,* component of reflection.[6] Its phase and amplitude are random, and scattering occurs over a wider range of angles than does the specular component. It has been reported to increase linearly with increasing surface roughness parameter, level off to a maximum, and then decrease as the inverse square root of the roughness parameter.

The surface-scattered energy that causes lobing of the antenna elevation pattern not only affects the coverage of a radar but it also can introduce serious errors in height-finding methods as well as degrade low-angle tracking (as was discussed in Sec. 4.5).

The effect of the surface-scattered wave on the coverage of the radar is indicated in the radar equation by the propagation factor F^4, where F is defined as

$$F = \frac{E_s}{E_0} \qquad \text{[8.14]}$$

where E_s is the field strength of the signal at the target (it includes the effects of the antenna pattern normalized to unity gain), and E_0 is the electric field strength that would occur in free space with loss-free isotropic antennas. It is similar to the parameter η defined by Eq. (8.1) except that F includes the effects of the antenna pattern on the elevation coverage. The propagation factor F was included in the numerator of the radar equation given in Chap. 2 as Eq. (2.61)

Figure 8.4   Surface reflection coefficient ρ_r as a function of the surface roughness parameter defined as $(\sigma_h \sin\psi)/\lambda$, where σ_h = rms value of the surface height h, ψ = grazing angle, and λ = radar wavelength. Top curve is the theoretical expression given by Eq. (8.13), bottom curve is the original expression given by Ament as in Eq. (8.12), and the middle curve is the experimental data of Beard. (After A. R. Miller and E. Vegh, Naval Research Laboratory Report 8898, July 31, 1985.)

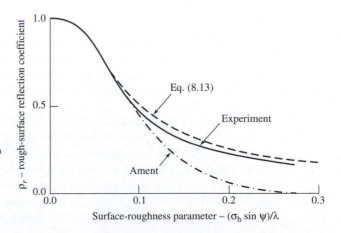

8.3 SCATTERING FROM THE ROUND EARTH'S SURFACE

The use of a flat-earth model is quite suitable for understanding the general nature of the modifications that occur in the antenna coverage. The earth, of course, is not flat, and precise predictions of the effect of the surface on the antenna coverage must consider the round earth. This is especially true for coverage at low elevation angles near the horizon.

The reflection coefficient from a round earth is less than from a flat-earth surface because of the divergence, or widening, of the beam when scattered from a round surface. The so-called *divergence factor* describes the decrease in the scattered signal. The divergence of the beam, however, means that the reflected energy will be spread over a wider angular region than specular scattering from a flat surface. The grazing angle of specular reflection is easy to determine for a perfectly flat surface, but this same angle from a spherical surface is more difficult to compute. In the past it has been found using either approximations or numerical computations; but Miller and Vegh have provided a deterministic method for obtaining the grazing angle from a spherical surface.[7]

There exists in the literature the necessary information, graphs, and nomographs to compute the coverage of a radar when lobing occurs due to the presence of the earth's surface.[8,9] This can be tedious when done by hand, especially when there are many lobes generated in the coverage. Computer programs are available that considerably ease the burden of calculating and plotting the coverages, one of the first of which was by Lamont Blake.[10]

An example of a calculated coverage diagram for an *L*-band radar over sea water is shown in Fig. 8.5 for both horizontal and vertical polarization. At low angles the maximum range with vertical polarization in this example is decreased only slightly relative to the range with horizontal polarization. The effect of the Brewster angle on vertical polarization, however, can be seen in the reduced ranges at higher elevation angles. One of the major consequences of the lobed elevation pattern due to multipath is that tracking of an aircraft flying at a constant altitude will not be continuous. Echoes will be received when the target is in one of the lobes, but the target might not be detected when it is in a null between the lobes. For instance, a target flying at a constant height of 30,000 ft will first be seen by a radar, whose coverage pattern is given by Fig. 8.5a, at a range of 170 nmi. The target will be lost at 160 nmi, reappear at 146 nmi, be lost again at 136 nmi, and so forth. The coming and going of the radar echo can create problems with automatic detection and tracking systems. In such systems, allowance has to be made for the track to coast whenever the target is momentarily lost, rather than immediately drop the track as soon as it leaves the coverage of a lobe and have to initiate a new track when the target echo reappears.

As has been mentioned, the effect of the interference between the direct and surface-scattered waves is to cause the peak of the lowest lobe of the elevation coverage to be at an angle higher than zero degrees, as was seen by Eq. (8.10) for the example of a flat earth. The lowest lobe with a round earth likewise will be at some angle above the horizontal. The result will be a lack of coverage of low-altitude targets. Figure 8.6a is a sketch of the elevation coverage of a long-range enroute air-traffic control radar. The details of the lobing are not shown. (There are two beams, an upper and a lower. We need only be

Figure 8.5   Example of a calculated vertical-plane coverage diagram for (a) horizontal polarization and (b) vertical polarization. Frequency = 1300 MHz, antenna height = 50 ft, antenna vertical beamwidth = 12° with the beam maximum pointing on the horizon, a sea surface with 4 ft wave height and free-space radar range of 100 nmi.

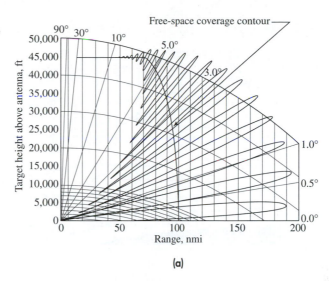

(a)

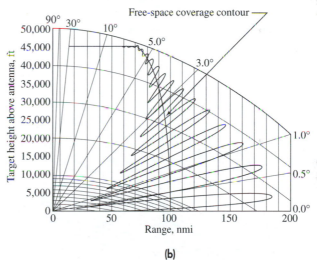

(b)

concerned with the lower.) The lower beam is tilted so that its half-power point, rather than its maximum, is on the horizon. This makes the lobes (which are not shown) less pronounced, but it also decreases the coverage at low altitudes. The radar is seen to have a maximum range of about 235 nmi. This is a good range for an air-surveillance radar, until it is noted that this range occurs at an altitude in the vicinity of 85,000 ft, which is much higher than commercial aircraft fly. The range for an aircraft flying at 30,000 ft, according to this coverage diagram, is about 185 nmi, and is 165 nmi when the altitude

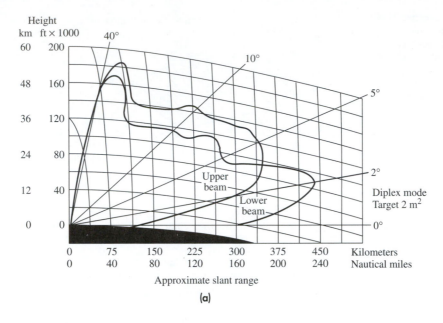

(a)

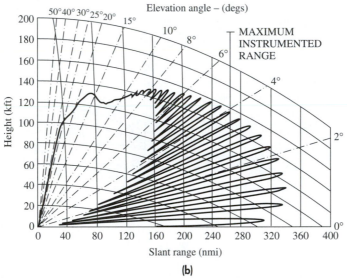

(b)

Figure 8.6 (a) Vertical coverage diagram of an *L*-band long-range en-route air-traffic control radar, illustrating the reduced range of the lower beam at very low angles due to the multipath from the earth's surface. (From the brochure *The ARSR-3 Story*, Westinghouse Defense and Electronic System Center, Baltimore, MD. Courtesy of Northrup Grumman Corporation.) (b) Calculated vertical coverage diagram of the Raytheon AN/SPS-49 radar. Long-range mode with antenna rotation rate of 6 rpm, 0.50 probability of detection, 10^{-6} probability of false alarm, Swerling Case 1 target with 1 m^2 radar cross section, sea state 3, and antenna height of 75 ft. This radar operates in the band from 850 to 942 MHz, with an antenna gain of 29 dB and an average transmitter power of 13 kW. (Courtesy of Vilhem Gregers-Hansen and the Raytheon Co.)

is 20,000 ft. Thus the quoted range of a radar might be much greater than the actual range at which an aircraft is first detected when flying at its more usual altitudes. If low-altitude performance at long range is important, the radar should be located on high ground so that its first lobe will be lowered in angle.

Another example of radar coverage, shown in Fig. 8.6b, is that of a long-range ship-board air-surveillance radar known as the AN/SPS-49.

Methods for Minimizing Lobing Effects   In many instances, the added range produced by the multipath lobing is a desirable attribute in spite of the nulls which can cause loss of targets. In other cases, lobing might not be desired. This can be true for automatic tracking radars or when the target undertakes maneuvers that require the radar to obtain information at a high data rate. One method to reduce the effects of lobing is to tilt the antenna beam upward so that the radiated energy illuminating the surface is reduced. Thus it is customary in many air-surveillance radars to tilt the beam so that its lower half-power point lies on the horizon.

We have seen that the location of multipath nulls depends on the frequency and the antenna height. By changing the height of the antenna (switching the radar between antennas at different heights) or by changing the radar frequency, the nulls can be filled-in when the radar data from the two or more antennas at different heights are combined. The utility of *height diversity* is limited by the need for antennas at different heights and by the difficulty in obtaining more than two different heights. *Frequency diversity* has been demonstrated to effectively fill in the nulls and allow continuous tracking of the target.[11] This capability, however, requires a wide frequency band. A wide frequency range is highly desired in many applications since it has other advantages than just filling in the nulls.

A radar with *polarization diversity* utilizes two orthogonal polarizations, such as horizontal and vertical. It could, in principle, provide some filling in of the nulls, but it has seldom been used. The nature of the Brewster's angle seems to limit its utility.

A *fence* surrounding the radar antenna can prevent radiated energy from illuminating the ground and can thus reduce lobing. This has seldom been employed since a fence can be large and expensive, diffraction from the edge of the fence limits the amount of attenuation of ground clutter that can be achieved, and there can be a loss of detection of desired low altitude targets.

The land-based military air-surveillance radars used in World War II did not employ doppler processing for separating the unwanted clutter from the doppler-shifted echoes from moving targets. For this reason, when conditions allowed, the radars were sited where there might be natural shielding to attenuate the energy radiated in the direction of the surface. Modern military radars, however, have to see targets at low altitude and would not benefit from this technique.

Not much can be done to enhance the signals from targets that lie below the first lobe. This is why surface-based radars that might have large free-space ranges have significantly reduced detection ranges against targets low on the horizon.

The lobing effect was put to good use in World War II for obtaining the height of an aircraft by using VHF radars that were capable of measuring only azimuth angle and not elevation angle. The range of first detection seen by the lowest lobe was used as an indication of target height. This required good calibration of the radar's sensitivity, and calibration flights with known aircraft to verify the relation between aircraft height and range of first detection. It also required the enemy to cooperate by not introducing new aircraft whose echo signals might be of quite different strength than those on which the radar was calibrated. In spite of its limitations, this method of height finding did what was required at the time.

8.4 ATMOSPHERIC REFRACTION—STANDARD PROPAGATION

Radar waves travel in straight lines in free space. Propagation in the earth's atmosphere, however, is not in free space. The atmosphere is not uniform; hence, it causes electro-magnetic waves to be bent, or refracted. Normally the effect of bending caused by at-mospheric refraction is favorable, in that it causes the radar horizon to be extended and increases the coverage of a radar beyond the geometrical horizon, Fig. 8.7a. On the other hand, the bending of the rays by the atmosphere can introduce an error in the measure-ment of the elevation angle, Fig. 8.7b.

Refractivity   Refraction of radar waves in the atmosphere is due to the variation of the velocity of propagation with altitude. The *index of refraction* is a measure of the velocity of propagation and is defined as the velocity in free space divided by the velocity in the medium in question, the atmosphere in this case. (The index of refraction is the square root of the dielectric constant, a parameter that might be more familiar to the electrical engineer.) The difference in velocity of propagation in the atmosphere compared to that

Figure 8.7 (a) Extension of the radar horizon due to refraction of radar rays by the atmosphere; (b) angular error caused by atmospheric refraction.

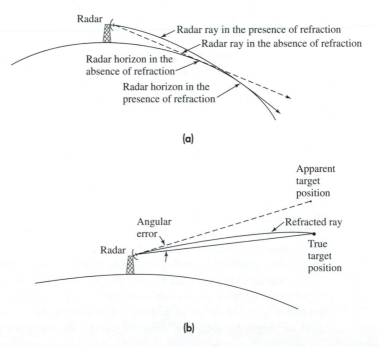

in free space is very small. According to the International Telecommunications Union, the average value of the surface index of refraction at mid-latitudes is 1.000315.[12] (Hitney,[13] on the other hand, gives 1.000350 as a typical value for the index of refraction at the earth's surface.) Rather than use the index of refraction, *n*, it is more convenient to use a modified parameter called the *refractivity, N,* which is defined as $N = (n - 1)10^6$. Thus an index of refraction $n = 1.000315$ corresponds to a refractivity $N = 315$.

At microwave frequencies, the refractivity *N* for air is given by the empirical relation[14,15]

$$N = (n - 1) \cdot 10^6 = \frac{77.6}{T}\left[p + \frac{4810e}{T}\right] \qquad \text{[8.15]}$$

where

  p = barometric pressure, mbar (1 mm Hg = 1.3332 mbar)

  e = partial pressure of water vapor, mbar

  T = absolute temperature, K

Atmospheric refractivity depends on the pressure, temperature, and water vapor. Of these, water vapor is the most important at microwave frequencies. It strongly affects the speed of microwave propagation. Temperature variations are more significant than pressure variations. (It might be mentioned that although refractivity is generally not a function of frequency within the microwave region, atmospheric refraction at optical frequencies differs from that at microwave frequencies since it is more dependent on temperature than on water vapor.)

Since the barometric pressure *p* and the water-vapor content *e* decrease rapidly with height above the earth's surface, while the temperature decreases slowly, the refractivity normally decreases with increasing altitude. The decrease of *N* means that the velocity of propagation increases with altitude, causing the radar rays to bend downward. (Refraction of radar waves in the atmosphere is analogous to bending of light rays by an optical prism.) The result is an increase in the radar coverage, as was illustrated in Fig. 8.7a. The magnitude of the atmospheric refractivity at some particular altitude is not as important in determining the effect of refraction on propagation as is the small change of refractivity with height; that is, it is the *gradient* of refractivity that causes the rays to bend.

The major changes in atmospheric refractivity occur in the vertical dimension. There may be changes in the horizontal dimension as well; but these are generally small (especially over water) so that radar propagation can be considered independent of the azimuth direction, unless the radar ranges are very large. The path of radar waves in the atmosphere may be plotted using ray-tracing[16] techniques, provided the variation of refractivity with altitude is known.

Effective Earth's Radius   A simple method to account for the effects of atmospheric refraction is to assume that the gradient of the index of refraction is constant with height, at least over the lower part of the atmosphere. This assumption allows the actual earth of radius *a* (*a* = 3440 nmi) and its nonuniform atmosphere to be replaced with an earth having a different radius (*ka*) and a uniform atmosphere in which radar waves propagate in straight

Figure 8.8   (a) Bending of the antenna beam due to refraction by the earth's atmosphere; (b) shape of the beam in the equivalent-earth representation with radius *ka*.

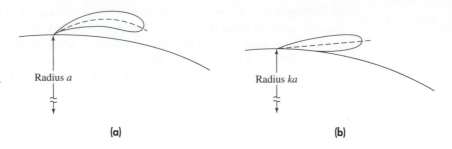

Radius *a*   Radius *ka*

(a)   (b)

lines rather than along curved paths, Fig. 8.8. The factor *k* depends on the refractivity gradient at the surface. From Snell's law in spherical geometry, the value of *k* by which the earth's radius must be multiplied in order to plot the propagation paths as straight lines is

$$k = \frac{1}{1 + a(dn/dh)}$$ [8.16]

where *dn/dh* is the rate of change of the refractive index with height. The vertical gradient of the refractive index is normally negative. The gradient of refractivity usually can vary from −79 to 0 *N* units per km of height.[12,13] The long-term average value of the gradient of *N* over the United States is approximately −39 *N*/km. When *N* is converted to *n* and substituted into the above equation, we get *k* = 4/3. The use of the *k* = 4/3 effective earth's radius to account for normal atmospheric refraction is convenient and widely used. It is only an approximation, however, and might not yield correct results when precise predictions are required. The term *standard refraction* is sometimes used to signify a value of *k* = 4/3 with the index of refraction decreasing uniformly with altitude with a gradient of $dn/dh = -39 \times 10^{-9}$/m.

The 4/3rd earth radius represents an average and should not be used where precision is important. The correct value of *k* depends on meteorological conditions and can be found by measurement. Bean[17,18] states that the average value of *k* measured at an altitude of 1 km varies from 1.25 to 1.45 over the continental United States during the month of February and from 1.25 to 1.90 during August. In general, higher values of *k* occur in the southern part of the country.

Equation (8.17) is often used as a measure of the line-of-sight coverage of a radar. This can lead to optimistic results since the propagation loss at the range d given by Eq. (8.17) can be quite high, as mentioned later in the discussion of diffraction in Sec. 8.6. The actual coverage of a radar can be less than given by the above simple geometric relation because of the large diffraction losses at the horizon. In spite of this, Eq. (8.17) has been widely used. It should be replaced, however, by diffraction calculations when it is important to know the maximum range at which a radar can detect low altitude targets.

Exponential Model of Refractivity   A limitation in the use of the effective earth's radius model is that the gradient of refractive index dn/dh is not linear with altitude, but is better approximated by an exponential model, especially in the troposphere above an altitude of 1 km. The exponential decrease of refractivity with height has been given as

$$N = N_s \exp\left(-h/H_s\right) \tag{8.18}$$

where

$$N_s = \text{refractivity at the surface of the earth}$$
$$h = \text{height above sea level in km}$$
$$H_s = \text{scale height in km}$$

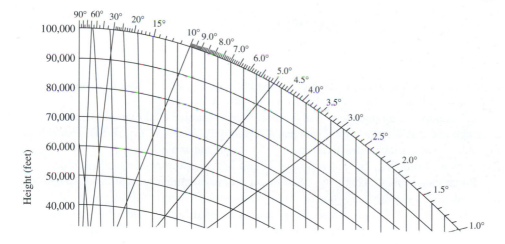

Figure 8.10 Radar range-height-angle diagram when the antenna is at an altitude of 30,000 ft and a U.S. Standard Atmosphere at 45° N latitutde in spring or fall. Due to W. G. Tank.[21]

(Reprinted with permission of Artech House, Inc., Norwood, MA. www.artechhouse.com.)

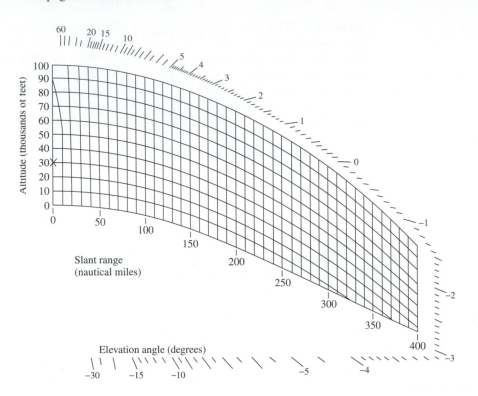

At mid-latitudes, the average value of N_s has been said to be 315 and the average value of H_s as 7.35 km.[19] Other values have been proposed, however.

An example of a range-height-angle chart (due to Lamont Blake) used to plot antenna coverage patterns based on the exponential refractivity model of Eq. (8.18) is shown in Fig. 8.9.[20] The surface refractivity is $N_s = 313$, and the scale height $H_s = 6.95$ km. In charts such as this, it is customary to plot height in feet and range in nautical miles.

Range-height-angle charts are often based on an antenna at ground level. When the radar antenna is elevated, as when on a tall mountain or when performing as an Airborne Early Warning (AEW) radar, these charts have to be modified. Figure 8.10 is an example due to W. G. Tank[21] for a radar at an altitude of 30,000 ft and a U.S. Standard Atmosphere. This can be thought of as a nomogram for determining the angle of arrival by first selecting the radar altitude (30 kft in this example), and then laying a straight edge from the radar location to the target point. The angle of arrival is read on the angle scales around the edges of the figure. A slightly different approach for plotting the range-height-angle charts when the antenna is not at ground level was suggested by Bauer,[22] based on the CRPL exponential atmosphere.

Standard Atmosphere A *U.S. Standard Atmosphere* is a hypothetical vertical distribution of atmospheric temperature, pressure, and density which by international agreement and for historical reasons, is roughly representative of year-round mid-latitude (45°N) conditions.[23,24] At microwave frequencies, a model of the moisture as a function of altitude must be added, which results in a refractivity $N = 316 \exp(-Z/26.5)$, for $Z \leq 25$, where Z = altitude in thousands of feet.

Radar Measurement Errors Due to Refraction   As was illustrated in Fig. 8.7b, refraction causes the radar rays to bend and results in an apparent target height that is different from the true height. An example of the angular error as a function of height and elevation angle as calculated by Shannon[25] is shown in Fig. 8.11. He assumed $N_s = 313$ and $H_s = 7$ km. At an altitude of 40 kft and an elevation angle of 3° the angle error is 2.67 mrad (17.45 mrad equals 1°).

When precision measurements are required, corrections should be made to the radar data to obtain accurate elevation angle, target height and range.[26] Surface observations often are sufficient for ascertaining the effects of refraction, but there can be cases when the variation of refractivity with height is not simple (as in the case of ducting discussed in the next section). In these cases, the variation of refractivity with height has to be measured and ray-tracing methods used to determine the measurement errors.

Even without ducting or other nonlinear refractivity profiles, errors can be significant. A comparison of heights obtained with the 4/3rd earth-radius model and the exponential model can be found in Brown[27] (and repeated in Murrow[28]). For example, at a range of 100 nmi with $N_s = 315$, the exponential model predicts a height 200 ft greater than the 4/3rd earth-radius model, when the elevation angle is 0.5°. It results in a 500 ft greater height compared to the 4/3rd earth model when the elevation angle is 2.0°. At the same range of 100 nmi and an elevation angle of 2.0°, the exponential model gives a

Figure 8.11   Calculated angle error (abscissa) due to atmospheric refraction for a standard atmosphere as a function of elevation angle and target height (ordinate.)
| (After Shannon.[25])

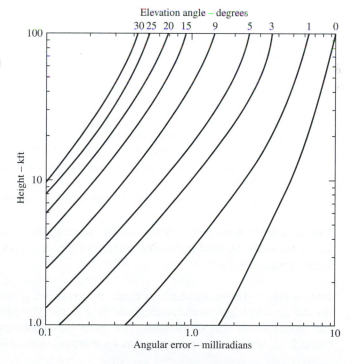

height error of 2.32 kft when $N_s = 370$ and an error of 1.38 kft when $N_s = 280$. The need for highly accurate knowledge of atmospheric refractivity limits the accuracy with which target height can be determined. This is the reason radar has not been used in the past when precision height measurements are needed.

In measurements to determine how well one can predict the effect of refraction, the accuracy of an AN/FPS-16 tracking radar was checked against photo theodolites at the U. S. Army Electronic Proving Ground, Fort Huachuca, Arizona.[29] The AN/FPS-16 is a highly precise tracker with rms mechanical and electronic measurement errors of ± 0.1 mrad for azimuth and elevation angles and ± 15 ft for range. At 40 nmi, the three-dimensional rms precision of this radar is ± 33 ft. Location measurements were made on aircraft at a range of 40 nmi and elevation angles from 0.7 to 2.8°. It was found that at an elevation angle of 2.5°, the rms elevation error was equal to the inherent ± 0.1 mrad precision of the radar. Above 2.5°, the predicted elevation errors were smaller than the inherent precision. Below 2.5°, the angle errors due to refraction were larger than the inherent radar precision.

There is also an additional time delay in propagating within the atmosphere that leads to an error in range. Shannon calculated this for the same conditions as for the angle error in Fig. 8.11. At 40 kft and an elevation angle of 3°, the range error is 97.5 ft. Shannon's plot can be found reproduced in Nathanson.[30]

A correction for the range error due to refraction, suggested by G. Robertshaw,[31] is given by the expression

$$R_c \cong 0.42 + 0.0577\ R_t \left(\frac{N_s}{h} \right)^{1/2} \qquad [8.19]$$

where

R_c = range correction, m
R_t = radar range, km
N_s = surface refractivity in N-units
h = radar altitude in kft

This expression was derived from ray-trace computations performed for radar altitudes of 15 kft, 35 kft, and 65 kft, based on a total of three months of measured atmospheric data over a period of three years at two sites in Germany, one in Saudi Arabia, and one in South Korea. It was said to have "the advantage of great simplicity with a tolerable loss of accuracy" and that "it will provide adequate range correction for airborne radar."

Measurement of Refractivity   There are two basic methods for determining the refractivity profile of the atmosphere: one is the radiosonde, or equivalent, and the other is the refractometer.

Radiosonde   The atmospheric refractivity profile may be obtained indirectly by the use of the empirical relationship given by Eq. (8.15) which equates refractivity with the properties of the atmosphere. The three measurements of water vapor pressure (humidity), atmospheric pressure, and temperature can be obtained with conventional weather-observation instruments launched into the atmosphere on a balloon. Such an instrument is called a *radiosonde*. Radiosonde measurements are routinely obtained daily throughout the world. The accuracy of radiosonde weather measurements, however, is generally not as good as might be desired for radar propagation predictions. Furthermore, the temperature, pressure, and humidity are sampled sequentially rather than simultaneously, and the sampling interval for each measurement is approximately 100 m in altitude. This may be satisfactory for weather observations, but it can sometimes cause the radiosonde to miss the sharp changes in refractivity that are characteristic of the strong, but shallow atmospheric layers that give rise to nonstandard propagation conditions.

Helicopter Probes[32]   In addition to having less than desired resolution in altitude, the conventional balloon-borne radiosonde makes one vertical pass and obtains only one vertical profile. The wind can cause it to move over a considerable horizontal distance during its flight and its measurements might not be representative of a true vertical profile. Furthermore, the measurements produced by conventional radiosondes launched from ships at sea might be contaminated by the microclimate created by the ship. A better method to probe the lower atmosphere where nonstandard propagation is prevalent is to mount instruments on a helicopter which can simultaneously provide the water vapor, pressure, and temperature as a function of range as well as altitude. A helicopter equipped by the Johns Hopkins University Applied Physics Laboratory (APL) employed more precise meteorological sensors than are normally found with weather radiosondes to obtain in almost real time the atmospheric refractivity. All three meteorological measurements were made at a data rate of 0.5 s. The altitude could be determined with a resolution of 0.3 m from 0 to 1000 m. These measurements were fed to a computer that calculated and displayed a continuously updated plot of modified refractivity versus altitude. The helicopter traveled with an airspeed greater than 30 m/s. Soundings were made with the helicopter ascending or descending at a rate of 1.5 to 3 m/s.

Small Rocket Probes[32]   The meteorological measurements needed to determine refractivity profiles can be obtained with a simple, low-cost expendable rocket. It can be used when it is not practical or convenient to use a helicopter. Such a rocketsonde developed and employed by APL carried an instrument package weighing 453 g to an altitude of 150 to 800 m, where it deployed the instruments using a 1 m diameter parachute. Measurements can be made with a vertical resolution of 2 m, which are then telemetered back.

Refractometer   An alternative to using meteorological measurements to indirectly derive the atmospheric refractivity is to use the more accurate and more responsive *microwave refractometer.* This instrument directly measures the index of refraction (or its square, the dielectric constant) by comparing the resonant frequencies of two identical microwave cavities. The resonant frequency of a cavity depends on its dimensions and its contents. One cavity is vented so as to sample the atmosphere; the other is hermetically sealed and acts as the reference. The cavities are fed by the same microwave source that is swept in frequency. The difference between the resonant frequencies of the sampled and reference cavities is a measure of the index of refraction. The microwave refractometer has much greater accuracy than the indirect measurement based on Eq. (8.15). It can measure changes in refractivity of less than 0.1 N unit; and when used to determine variations about an undetermined mean, it can have a time constant that allows detection rates of up to 100 Hz.[33] Although the refractometer may provide excellent refractivity measurements, it might not be suitable for all applications since it requires an aircraft or helicopter. It is usually too costly to be expendable, as is a radiosonde or a rocketsonde.

8.5 NONSTANDARD PROPAGATION

The previous section described the effect on radar propagation of standard, or normal, refractive conditions. Refractive effects, however, can be much more complex than described by the standard exponential model, and can cause significant changes in radar propagation. Such conditions are known as *anomalous,* or *nonstandard, propagation.* As a rough generalization, when nonstandard propagation conditions occur, the maximum ranges of a surface radar for detecting low-altitude or surface targets might be extended from two to five times what would be expected with a uniform atmosphere.[34]

Normal refraction occurs when the refractive gradient with height, dN/dh, is between 0 and -79 N units per km of height. (Note that we have said previously that the long-term mean gradient over the continental United States is about -39 N/km.) When the gradient equals -157 N/km, the effective earth's radius as given by Eq. (8.16) becomes infinite. Rays that are initially horizontal will then follow the curvature of the earth. Under such conditions, the radar range is significantly increased and detection beyond the radar horizon can result. Refractive gradients between -79 and -157 N/km result in what is called *superrefraction.* When the gradients exceed -157 N/km, the curvature of the propagating ray exceeds the curvature of the earth and ducts can form that trap the radar energy. The trapped energy within the duct can propagate to ranges well beyond the normal horizon.

If the refractive gradient were to increase with height, instead of the more usual decrease, the propagating rays would curve upward and the radar range would decrease as compared to normal conditions. This is called *subrefraction.* Its occurrence is rare; but when it does occur, its effect can be serious. It has been suspected of causing ship accidents when using marine radar. The term nonstandard propagation, or anomalous propagation, applies to any of the above propagation conditions other than normal. Table 8.1 summarizes these refractive conditions.

There are three general classes of ducts that will be briefly described. These are evaporation ducts, which occur at the surface of the sea; surface-based ducts; and elevated ducts. The later two can occur over land as well as water. To propagate energy within the duct, the angle the ray makes with the duct should be small, usually less than about one-half degree. Therefore, only those rays launched nearly parallel to the duct are trapped.

Evaporation Ducts   In a maritime environment, standard refractive conditions seldom appear, so that nonstandard conditions of some form are often present.[35] The most common type of anomalous propagation over the ocean and other large bodies of water is the evaporation duct. It is found at the surface, and is a relatively common occurrence. The air in

Table 8.1   Summary of Refractive Propagation Conditions

Refractive Condition	Gradient: N units per km
Subrefraction	Positive gradient
No refraction (uniform atmosphere)	0
Standard refraction (4/3rd earth radius)	-39
Normal refraction	0 to -79
Superrefraction	-79 to -157
Trapping, or ducting	-157 to $-\infty$

contact with the sea surface is usually saturated with water vapor so that its relative humidity is almost 100 percent. The air several meters above the sea surface is not usually saturated so there will be a decrease in humidity from the surface value to the ambient value determined by the general meteorological conditions well above the surface. The rapid decrease of water vapor causes a rapid decrease of refractivity that results in the formation of a low-lying duct that traps the radar energy so that it propagates close to the sea surface. Ducting can cause the radar ranges for targets at or near the sea surface to be considerably greater than the free-space range.

Duct Height   The "height" used to characterize an evaporation duct is not the height below which an antenna must be located in order to obtain extended propagation. It is more a measure of the strength of the duct. Evaporation duct heights, which typically might have values from 6 to 30 m, vary with the geographic location, season, time of day, and wind speed. Of the factors that can affect the strength of a duct, the wind speed seems to be of special importance. In one set of experimental observations conducted in the Atlantic trade wind region off the east coast of Antigua with X- and S-band radars, it was said that the wind was the only meteorological factor that was correlated with the rate of attenuation in the duct.[36] Wind speeds from 8 to 15 kt produced a moderately strong duct of low height. Winds from 20 to 30 kt produced a greater duct height, but in this case the duct was weaker according to the meteorological predictions. The attenuation in the duct, however, decreased with increasing wind. Passing squalls and rain showers did not affect the duct or decrease the propagated signal strength. The duct heights varied from 20 to 50 ft and were found to exist all the time during these experiments.

Evaporation duct heights cannot be readily determined by standard radiosondes or refractometers. They can be inferred, however, from theoretical models based on meteorological measurements. One such model due to H. Jeske[37,38] and modified by Paulus[39] utilizes the sea-surface temperature, air temperature, relative humidity, and wind speed—the last three being obtained at some reference altitude (often taken to be 6 m).[38,40] Table 8.2 gives the calculated average duct height, for various areas of the world, based on this formulation using meteorological measurements from a 15-year subset of data from the National Climatic Data Center.[41] The histograms for evaporation-duct heights for three of the areas in Table 8.2, plus the worldwide average are shown in Fig. 8.12. (These were originally given in Ref. 41 in 2-m increments of height, but here a smooth curve was drawn.)

The predicted evaporation-duct height using the Jeske-Paulus model does not always agree with actual observations of radar propagation. This might be due, in part, to the difficulty in making the relevant meteorological observations at sea with sufficient accuracy. Also, the theory and the accompanying simplifying assumptions on which the predictions are based might not be applicable under all conditions, especially when large duct heights are predicted. The nature of an evaporation duct can be more complex and varying than are accounted for in a simple model.

An improved method of determining the evaporation duct height has been proposed by Babin, Young, and Carton, which they call *Model A*.[42] They take advantage of the availability of high-speed desktop computers, not available when the Jeske-Paulus model was first formulated, to eliminate some of the assumptions that were originally made, incorporate more atmospheric boundary-layer physics, and decrease the use of empirical relationships. This model not only provides a more accurate duct height, but it also gives the standard deviation of duct heights based on the accuracy of the sensors used to obtain the atmospheric data.

Frequency Dependence   The thicker the duct, the lower the frequency that can be propagated; but there is a limit. The lower frequency limit for propagation within an evaporation duct is said to be about 3 GHz.[13] An approximate model for propagation in an evaporation duct is that of a waveguide with the sea as the bottom wall and a leaky top wall. Thus there is a low-frequency cutoff for propagation in this type of leaky waveguide.

According to Hitney,[13] a rough guide to the lowest frequency that can be trapped by a duct of a given height is as follows: 3 GHz requires a duct height of at least 25 m; 7 GHz, 14 m; 10 GHz, 10 m; and 18 GHz, 6 m. As the frequency increases, however, the propagation loss increases due to attenuation caused by the high concentration of water

Table 8.2   Calculated average height of the evaporation duct[41]

Area	Average Duct Height, m
North Atlantic	5.3
Canadian Atlantic	5.8
East Atlantic	7.4
North Pacific	7.8
Mediterranean	11.8
West Atlantic	14.1
Persian Gulf	14.7
Indian Ocean	15.9
Tropics	15.9
World Average	13.1

Figure 8.12
Examples of the statistics of evaporation-duct height for three areas of the world and the worldwide average, as given by Hitney et al.[41] These were calculated based on historical average meteorological measurements.

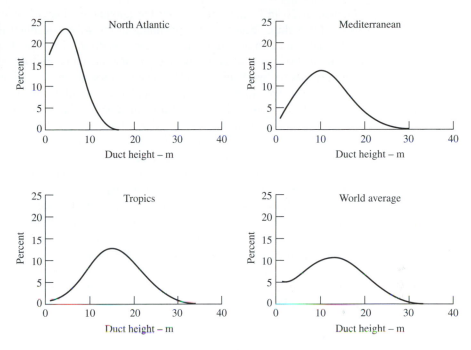

vapor within the duct as well as the attenuation due to a rough sea surface. For this reason Hitney suggests that the optimum frequency for duct propagation is around 18 GHz. Significant ducting has been observed, however, at frequencies as high as 94 GHz. Experiments at 94 GHz conducted over a 40.6 km over-the-horizon path along the Southern California coast found that the median loss with ducted propagation was 60 dB less than the loss that would have been obtained if there were no ducting (as with a standard atmosphere).[43]

Multiple-Mode Propagation   If the duct height is large enough, more than one mode can be propagated. (A *mode* is a configuration of electric and magnetic field distribution, similar to the modes of propagation in a conventional waveguide.) Multiple propagation modes have two consequences: (1) the signal strength will not vary uniformly throughout the duct and (2) there can be more than one antenna height suitable for low-loss propagation. In normal propagation within a uniform or a standard atmosphere, the propagation loss decreases with increasing antenna height. The higher the antenna the better. On the other hand, with a single mode of propagation in an evaporation duct, the loss will increase with height within the duct. Propagation is better with a low-sited antenna. With a duct height that supports multiple modes of propagation, there can be more than one choice of antenna height that provides low-loss propagation. In one particular *X*-band experiment, the minimum attenuation occurred with an antenna height of 2 m.[36] Increasing the antenna height increased the loss until a maximum loss was found at about 10 m. Further increases of height decreased the attenuation again until a secondary minimum was obtained at 20 m height, after which the attenuation increased (at least to a height of 30 m). On a ship, because of the pounding of the waves, it might not be practical to site a radar antenna 2 m over the sea if it were desired to obtain the benefits of ducted propagation. Instead, in this particular case, the antenna could be located at 20 m height, with a slightly greater loss being the price to be paid for the more convenient antenna location. These values apply to a particular location at a particular time. On most ships one probably would not want to have a variable antenna height to maximize propagation in the duct. What this example illustrates is that a fixed high-sited antenna might be a suitable compromise, rather than try to place an antenna in the duct at a low height where it can be subjected to destruction by the force of the waves.

Theoretical models of ducted propagation confirm the general nature of the above experimental observations.[44] Each mode of propagation of a guided wave has a particular configuration of electric field strength. When multiple modes are present, the fields of each mode can interfere constructively or destructively with the others. Consequently, the field strength along the duct or across the duct might not be uniform, which can lead to variations in the radar echo signal strength. When multiple modes of propagation are present in the duct, theory predicts considerable variation of the signal strength, called *fades,* because of the interference among the several modes. Fades can be of the order of 20 dB. In one example, it was shown that fading was strongly dependent on the height of the radar and the target, and for centimeter wavelengths the fading occurred at intervals of from two to three miles. Because of the strong dependence with height, the negative effect of holes in the coverage might be reduced by having more than one antenna, each at a different height, to provide height diversity. Frequency diversity might also reduce the effects of the holes in the coverage, if the frequencies are widely separated.

Ducting Within or Near the Horizon   Most of the above has been concerned with propagation in a duct beyond the normal horizon. Within the horizon, the refractive effects of the duct can lead to a modification of the normal lobing pattern caused by the interference of the direct and surface-reflected rays discussed in Sec. 8.2.[45] The relative phase between the direct and surface-reflected rays can be different in the presence of the duct, and focusing[46] can change the relative amplitudes of the two components. Focusing by the atmosphere might even cause the amplitude of the surface-reflected ray to sometimes exceed that of the direct ray. The effect of the duct on line-of-sight propagation is to reduce the angle of the lowest lobe, bringing it closer to the horizon.

Although the evaporation duct provides a significant increase in signal strength for ranges well beyond the horizon, compared to what would be received if there were no ducting, it results in reduced signal strength at or near the horizon. This is illustrated in Fig. 8.13 which shows the propagation loss as a function of range for an *X*-band radar (where propagation loss is defined here as the ratio of transmitted to received power assuming that the antenna pattern is normalized to unity gain).[47] The presence of a duct increases the signal at the longer range compared to no duct, but it decreases the signal at the first lobe (about 10 dB for the 16 m duct) and it shifts the first null to longer range than when no duct is present. This confirms what was said in the previous paragraph about the evaporation duct lowering the lowest lobe.

Figure 8.13 Propagation loss between radar and target as a function of range for an X-band radar located 23.5 m above the sea with a point target 4.9 m above the sea. The radar has a free-space range of 7.5 km. The dashed horizontal line represents the detection threshold for this radar. The "0 m duct" corresponds to a standard atmosphere. (From Anderson,[47] Copyright 1995 IEEE.)

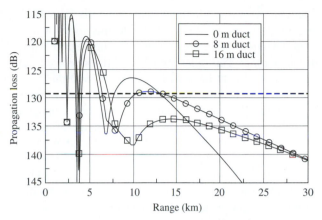

Surface-Based Ducts[48,49]   A surface-based duct is one whose base is at the earth's surface. There are three types of such ducts depending on the relationship of the trapping layer to the earth's surface. One example is the evaporation duct discussed in the above. The evaporation duct, however, is usually considered separately (as was done here) because of its unique characteristics and because it is a nearly permanent worldwide feature.[38] The second is a surface duct created from an elevated trapping layer, and the third is a surface duct created from a surface-based trapping layer. In this subsection, only the latter two will be considered, since the first has been discussed.

Surface-based ducts are formed when the upper air is exceptionally warm and dry compared with the air at the surface. There are several meteorological conditions that may lead to their formation. Over land, a surface-based duct can be caused by the radiation of heat from the earth on clear nights, especially in the summer when the ground is moist. The earth loses heat and its surface temperature falls, but there is little or no change in the temperature of the upper atmosphere.[50] This leads to a temperature inversion at the ground and a sharp decrease in moisture with height. Thus over land, ducting is most noticeable at night and usually disappears during the warmest part of the day.

Another cause of surface-based ducts is the movement (advection) of warm dry air, from land, over cooler bodies of water. Examples of such advections are the Santa Ana wind of Southern California, the sirocco of southern Mediterranean, and the shamal of the Persian Gulf.[38] Warm dry air that is blown out over the cooler sea is cooled at its lowest layers to produce a temperature inversion. At the same time, moisture from the sea is added to produce a moisture gradient, and a surface-based duct can be formed. This type of ducting tends to be on the leeward side of land masses, it occurs either during the day or night (but more likely to occur in the late afternoon or evening when the warm afternoon air drifts out over the sea), it might extend out over the ocean for several hundreds of kilometers, and it can last for long periods of time (several days).

The height of surface-based ducts generally does not exceed a few hundred meters. Propagation within a surface-based duct is relatively insensitive to frequency, and long-range propagation can occur at frequencies exceeding 100 MHz. Ducted propagation has even been reported for frequencies down to 20 MHz.[51] Most of the reports of unusual long-range radar detection are due to this form of duct. A classic example is the often cited detection in the Indian Ocean during World War II by a 200 MHz radar located at Bombay, India. This radar, located 225 ft above sea level, frequently detected echoes from points in Arabia at ranges from 1000 to 1500 miles during the hot season. It was quite common to plot ships out to ranges of 200 miles, and cases were reported to 700 miles. By contrast, during the monsoon season when propagation conditions were more normal, the radar was able to plot ships out to a range of only about 20 miles.

An example of the effects of surface-based ducts is shown in Fig. 8.14. This is a plot of the two-way propagation factor (ordinate) seen in the vertical plane at a range of 50 nmi. The abscissa is the height above sea level. The radar frequency is 900 MHz and antenna height is 85 ft. The dashed curve represents the calculated propagation factor for a standard atmosphere. (According to Eq. (8.10) the peak of the lowest lobe due to multipath is predicted to be at 0.18°, and the peak of the second lobe is at 0.55°.) The solid curve was calculated for a nonstandard atmosphere whose refractivity profile was measured in the Persian Gulf at local noon on a day in August. The duct height was of the order of 700 to 800 ft. The signal strength below the duct height is considerably increased compared to the no-ducting standard-atmosphere situation. The increase in energy at the lower altitudes is accompanied by a decrease in energy in the lowest lobe below about 0.5°. The maximum of the two-way radar signal in the lowest lobe is seen to decrease by about 10 dB with ducting present.[52]

Surface-based ducts may also be formed by the diverging downdraft of air under a thunderstorm.[50] The relatively cool air which spreads out from the base of a thunderstorm results in a temperature inversion in the lowest few thousand feet. The moisture gradient is also appropriate for the formation of a duct. Although duct formation by a thunderstorm might not be as frequent as other ducting mechanisms, it may be used as a means of detecting the presence of a storm. An operator carefully observing a radar display can detect the storm by the sudden increase in the number and range of ground echoes. The conditions appropriate for the formation of a thunderstorm duct might have a duration of from one-half hour to a few hours.

Figure 8.14   The solid curve is the two-way propagation factor as a function of height above sea level (ASL) for a surface-based duct taken in the Middle East at a range of 50 nmi measured at local noon on a day in August. The dashed curve represents the propagation factor calculated for a standard atmosphere.
(Courtesy of John Walters and Vilhelm Gregers-Hansen, NRL Radar Division.)

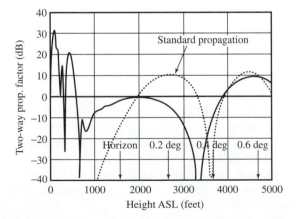

Elevated Ducts[38]   The base of an elevated duct lies above the surface of the earth. Elevated ducts can occur in the tradewind regions between the mid-ocean high-pressure cells and the equator. Semipermanent high-pressure areas are centered at approximately 30° north and 30° south latitudes over the ocean. The region between these high-pressure areas and the equator are called the tradewinds. Two such tradewind areas that have been studied lie between Brazil and the Ascension Islands[53] and between Southern California and Hawaii.[54] Within the high-pressure regions, there is a slow-sinking of high-altitude air (called large-scale *subsidence*) which meets low-level maritime air (marine boundary layer) flowing toward the equator. The general sinking of air from high altitudes causes compression which results in adiabatic heating and a decrease in moisture content. This leads to warmer, drier air lying above cooler, moist air to produce a temperature inversion (an increase in temperature that causes a decrease in refractivity with height). The result is a strong duct along the interface of the temperature inversion along the top of the marine boundary layer.

In some cases, a stratus cloud layer will form at the base of the temperature inversion, and the duct altitude can be identified by the height of the cloud tops that are suppressed by the temperature inversion.[55] When the temperature inversion occurs below the altitude at which clouds are formed, a haze layer in the air below the temperature inversion can be observed.

The altitude of the tradewind duct varies from hundreds of meters at the eastern part of the tropical oceans to thousands of meters at the western end. Thus, the height rises gradually in going from east to west. Over the coast of Southern California, elevated ducts were found to occur an average of 40 percent of the time and have an average maximum elevation of about 600 m. On the western side of the Pacific Ocean along the coast of Japan, elevated ducts are said to occur 10 percent of the time and have an average maximum height of 1500 m.[38] The thickness of elevated ducts can range from near zero to several hundred meters. Frequencies as low as 100 MHz can propagate in the thicker elevated ducts. Since the elevated duct is due to meteorological effects, there can be seasonal as well as diurnal variations. It has been said, however, that elevated ducts give rise to strong persistent anomalous propagation throughout most of the year over at least one third of the oceans.

Ray-optics theory indicates that both the radar antenna and the target must be within the duct to obtain the benefits of the low propagation loss provided by the duct. In practice, it has been observed that this is not always necessary since enhanced propagation can occur with the radar and/or the target outside the duct (as defined by its classical duct thickness). This is likely due to the oversimplification of the duct model when defined by a smooth surface. Both the upper and lower boundaries of a duct can be irregular, allowing energy to "leak" or scatter into or out of the duct. It has been reported[56] that the presence of a very strong evaporation duct (height less than 100 ft) along with an elevated duct (at 2000 ft) can result in a significant increase in signal strength of a 3 GHz signal beyond the horizon at heights (3000 ft) which are an order of magnitude or more greater than the evaporation duct height.

The capability of elevated ducts to propagate to long range in the tradewind region is illustrated by a particular flight of an instrumented Naval Research Laboratory aircraft from San Diego, California, to Oahu, Hawaii. These were communications, rather than radar, experiments. A 220-MHz signal was transmitted from an antenna located at an elevation of about 800 ft near San Diego. The transmitting antenna was within the duct. The signal was received by an aircraft throughout the entire path and was even detected after the aircraft was on the ground in Hawaii.[54] This long range, however, occurred only once during the fourteen runs conducted during these experiments. In all these runs, significant increases in range were achieved compared to what would be expected from tropospheric scatter propagation.

Meteorological conditions necessary for an elevated duct are similar to those for a surface-based duct. Under the proper conditions, one can turn into the other.[38]

Although enhanced propagation can occur when the target and the radar are properly located with respect to the duct, it is also possible to obtain reduced or no coverage above or below the duct, compared to that expected with a standard atmosphere. This lack of coverage due to the duct is called a *radar hole.*

Subrefraction   The gradient of refractivity may, at times, be such as to bend electromagnetic rays upward rather than downward, causing a decrease in range when compared with standard refractive conditions.[57] This is called *subrefraction,* or *substandard propagation.* It occurs when the index of refraction increases with altitude, instead of decreasing as is the more usual situation. Subrefraction can occur when warm, moist air flows over a cool ocean surface or a over a cooler air mass just above the ocean surface.

An interesting example of subrefraction on radar performance has been reported by Brookner et al.[58] for an S-band marine radar located near the entrance to the Delaware Bay between the States of Delaware and New Jersey. The radar was operated by the Pilots Association of the Bay & River Delaware. Subrefraction was said to occur throughout the year in the Delaware Bay region. The radar typically had a range of 20 nmi for ships of interest, but when subrefraction occurred the range was reduced in half to 10 nmi. A one-way loss was observed that was greater than 20 dB relative to free-space propagation. Reduced conditions could last for several hours before returning to normal. Subrefraction results in an effective earth's radius less than one, and can be one half of the actual earth's radius. Some subrefraction conditions can produce a skip zone. For example, an approaching ship might be first detected at 20 nmi range, be undetected starting at about 12 nmi, and not seen again until it is within 6 to 8 nmi of the radar. It has been said that operating the radar antenna at a higher height above the surface can reduce the effects of subrefraction, but frequency diversity has little effect.

In some cases fog can lead to subrefraction. When fog forms, part of the water in the air changes from the gaseous to the liquid state, but the total amount of water remains unchanged. Water in liquid form contributes far less to the refractive index than water in the gaseous (vapor) form. The formation of fog near the surface results in a reduction of water vapor and a corresponding lowering of refractivity in the region of fog. The result is an upward bending of the radar rays, and a shortening of the radar range. Because other factors can enter, the presence of fog is neither a necessary nor a sufficient condition for the occurrence of substandard propagation.

It was mentioned in both references 57 and 58 that under some subrefraction conditions (and in the absence of fog), a ship that is not seen by radar might be seen visually.

This results because water vapor makes a significant contribution to the atmospheric index of refraction at microwave frequencies, but has little effect at optical frequencies.

Nonstandard Propagation over Land   Much of the discussion about nonstandard propagation in this section has been based on overwater paths. Similar effects can occur over land; but there has been less written about overland effects than the effects that occur over water.

Over land, nonstandard propagation can be caused by radiation of heat from the earth on clear nights, especially in summer when the ground is moist. When the earth loses heat, its surface temperature falls, but there is little or no change in the temperature of the upper atmosphere. This leads to conditions favorable to superrefraction; that is, a temperature inversion at the ground and a sharp decrease of moisture with height. Thus over land masses, the phenomena of superrefraction and ducting are most noticeable at night and disappear during the warmest part of the day. Ducting observed over the Arizona desert during the winter when the atmosphere was clear and dry (low moisture content) has been attributed to this type of temperature inversion caused by nocturnal cooling of the ground.[59] In these experiments, the duct increased in height and intensity as the night progressed. It was also stated that "The cyclic variation in the meteorological conditions occurs with the same general character night after night and, in fact, year after year" and that "The field-strength variations reflect this consistency."

Terrain generally consists of high and low regions. Diffraction (forward scattering) from the high regions can have a significant effect on radar propagation and sometimes can be the dominant propagation mechanism.[13] In other cases, ducting may be dominant, but the profile of the terrain along the propagation path might reduce the strength of the duct.

The computation of radar propagation over irregular land surfaces is much more difficult than computations of propagation over the sea. The variation of refractivity with range must be taken into account when radar waves propagate over irregular terrain. The irregular features, including trees and other structures, are not easy to model realistically. The mathematical techniques must include diffraction as well as refraction effects. One approach has been to map the irregular terrain boundary into a rectangular domain where a numerical solution can be generated by applying well-established numerical methods.[60] The parabolic equation method, mentioned later in this section, is one of the computation techniques that have been applied to predict propagation over land since it can account for the variation of refractivity with range.

Effect of Ducting on Surface Clutter Measurements   Most clutter measurements do not take account of the effects of nonstandard propagation. Ducting effects, when they exist, are usually inherent in the clutter data since it is difficult to separate them. Ducting can be, therefore, a major source of inaccuracy when trying to make quantitative measurements of the radar cross section per unit area (σ^0) of surface clutter.

Another reason why measurements of clutter are suspect when ducted propagation exists is that the precise grazing angle of the radar ray is not known. Even if the grazing angle were known, it is not easy to accurately determine the attenuation caused by ducted propagation (which is needed to determine the radar cross section of clutter).

Unfortunately, most of the available clutter data does not account for nonstandard propagation conditions. If the refractivity conditions of the atmosphere were accurately known (as with extensive measurements), it is possible in principle to extract the value of grazing angle* and the clutter cross section during ducted propagation.[61,62] This, however, is not easy to do in practice.

In the curves of sea clutter data shown in Fig. 7.13, the value of sigma zero (clutter cross section per unit area) for surface-based X-band radars did not drop off sharply at low gazing angles, as might usually be expected due to multipath from the surface. On the other hand, it was found that the X-band clutter echo did drop off with decreasing grazing angle, as expected when the radar was in an aircraft. It is suggested that this behavior might be due to the surface-based radars experiencing ducted propagation while the airborne radars flying above the evaporation duct did not.

It has also been observed in some cases with large radars looking at low angles over the ocean under ducted conditions, that echoes at long range from atmospheric clear-air turbulence can be much larger than the echoes from sea clutter.[63]

Occurrence of Ducting   Ducting is essentially a fine-weather phenomenon (with the exception of thunderstorm ducts). Since tropical climates, other than at the equator, are noted for their fine weather, it is not surprising to find the most intense ducting occurring in such regions.[64] In temperate climates, ducting is more common in summer than in winter. It does not occur when the atmosphere is well mixed, a condition generally accompanying poor weather. When it is cold, stormy, rainy, or cloudy, the lower atmosphere is well stirred up and propagation is likely to be normal. Both rough terrain and high winds tend to increase atmospheric mixing, reducing the occurrence of ducting. Although windy weather which causes the atmosphere to be well mixed can inhibit the formation of ducts, experiments in the Atlantic trade wind region indicated that the wind was the most important meteorological factor required for the appearance of an evaporation duct.[36] Thus, evaporation ducts might be weak or even not exist if there were no wind at all.

Consequences of Ducted Propagation on Radar Performance   Ducts can provide extended ranges against surface targets or low-flying aircraft considerably beyond the ranges that would be expected from a radar within a standard atmosphere. As previously mentioned, the radar antenna and the target must be in, or near to, the duct to experience extended range. Although ducts can significantly increase the range of a radar, the consequences of ducted propagation are not necessarily good. In fact, in most cases the negatives tend to outweigh the positives.

The ability to see at long ranges because of ducted propagation cannot be readily predicted in advance, and the increased ranges cannot be relied upon since they are not always present. One would certainly not want to depend on ducted propagation to extend the range of a radar when the proper ducting conditions for long range might not be available when needed. Furthermore, increased ranges in some directions are balanced by decreased ranges in other directions. There can be significant "radar holes" which prevent

---

*The grazing angle for sea clutter that can be found in this manner is the angle to the undisturbed sea surface. In practice, the angle the radar ray makes with the sea will depend on the slope of the large water waves on which ride the shorter waves as well as other surface disturbances.

a radar from seeing targets in some direction that would normally be detected if surface-based or elevated ducts were not present. The loss of detectability caused by radar holes can affect airborne radars as well as ground-based and shipborne radars. An aircraft or missile flying just above a duct might not be detected until it is at short range. A radar in an aircraft, such as for airborne air-surveillance, might not be able to detect targets that are below the duct even if they were well within the range of the radar. This can be avoided by proper control of the altitude of the aircraft carrying the radar, but it requires knowledge in real time of the local refractive conditions that can affect radar propagation.

Ducted propagation can make possible the detection of unwanted clutter echoes at long ranges that might otherwise not be detected in a normal atmosphere. This can place a severe burden on MTI radars designed on the assumption that clutter will not appear beyond a certain range. Also, multiple-time-around clutter echoes that arrive from beyond the maximum unambiguous range might not be eliminated if the doppler processing employs pulse-to-pulse staggered repetition periods.

In areas of the world where surfaced-based ducts can significantly extend the radar horizon for a large portion of the time, the greatly increased clutter echoes that are obtained can seriously degrade the performance of a radar not designed to cope with it. Such radars that experience the detrimental effects of ducting should be designed with a large dynamic range so as to avoid receiver saturation by large clutter echoes, the radar should have additional MTI or pulse doppler improvement factor to eliminate the larger than normal clutter, and the radar waveforms and processing should be designed to cancel multiple-time-around clutter echoes that originate from long ranges. The last mentioned is accomplished by using a constant prf (instead of pulse-to-pulse prfs) with processing that uses the required number of "fill pulses."[65] Fill pulses are those given zero weight in the digital MTI processor (that is, they are discarded) so as to eliminate pulse repetition intervals that do not have multiple-time-around clutter.

Again it should be mentioned that ducted propagation theoretically requires that the radar and target be within or close to the duct. It is seldom that a conventional surface-based radar, for example, will experience serious ducting effects if its beam is pointed to an elevation angle greater than about 0.5°.

Modified Refractivity   In this section the refractivity, denoted by N, has been used to describe the property of the atmosphere to bend radar waves. It is sometimes convenient, however, to employ a modified refractivity, defined as

$$M = N + (h/a) \times 10^6 \qquad \text{[8.20]}$$

where h = height above the earth's surface, and a = earth's radius (in the same units as h). The modified refractivity M takes account of the curvature of the earth. It is useful in identifying ducting, since the trapping of radar waves occurs for all negative gradients of M. The modified refractivity, rather than N, is commonly employed by propagation engineers for determining the effects of refraction.

Prediction of Refractive Effects   The theory of ducted propagation is not as complete as one might like. Available theoretical models sometimes fail to adequately describe what is taking place in nature. Nevertheless, there have been developed usable prediction methods based on local meteorological measurements that provide estimates of how ducted propagation might affect the radar coverage and how the radar platform might be positioned to minimize the effects of nonstandard propagation. In the following are briefly described the several methods mentioned by Hitney[66] for determining the effects of refraction by a nonstandard atmosphere.

Ray Tracing   This uses geometrical optics to determine the paths taken by radar waves as they propagate through the atmosphere.[67] In ray tracing the modified refractivity profile is assumed to vary only with height. The height is divided into small increments Δh, and Snell's law is applied with the small-angle approximation to determine the bending of a ray as it leaves a region of refractivity N at a height h and passes into a region of refractivity $N + \Delta N$ at height $h + \Delta h$. The refractivity at each increment of Δh is assumed to be linear. Since refractivity does not depend on frequency (within the normal range of radar frequencies), a single ray-tracing diagram can also be considered to be independent of frequency. Ray tracing is relatively simple compared to other models; but it does not provide the magnitude of the field strength. It also requires that the refractive index not change significantly in a distance comparable to a wavelength and the spacing between neighboring rays must be small in order to produce correct results when the rays diverge, converge, or cross. Also, diffraction effects are not taken into account.

Waveguide Model   The trapping layer can be considered as a waveguide with multiple modes of propagation.[68] Those readers familiar with microwave propagation in metallic waveguides know that the dimensions of the waveguide usually are chosen so that only one dominant "mode" can propagate. In a rectangular guide this occurs when the broad dimension is slightly greater than a half-wavelength. If the dimensions are much larger than this, more than one mode of propagation can take place. Each mode has a different field configuration along the guide than the others. The various modes interfere or reinforce with one another as the energy travels along the guide, which is why a single mode of propagation usually is preferred. The theory and practice of propagation in waveguides was developed primarily in the 1930s and 1940s. The theory of guided wave propagation in other than metallic waveguides, however, is much older. It was first devised to explain how long-wavelength radio waves can propagate around the earth along the structure formed by the earth's surface and the ionosphere. Such a situation results in the energy propagating in more than one mode. A similar type of propagation occurs when sound waves travel in water when one surface of the guide is the bottom of the sea and the other is the sea surface. In a metallic waveguide, the walls present sharp boundaries that reflect the propagating waves incident upon them. Waves can also be reflected from a stratified medium in which the refractivity varies continuously and which has no sharp boundary. This is what happens in ducted propagation.

The theory of guided propagation in a layered medium has been applied to radar propagation in atmospheric ducts. This is a physical optics approach that takes into account propagation loss and diffraction. The theory predicts a cutoff frequency, below which no propagation can take place in the guide medium. The waveguide model can be employed when the vertical refractivity profile is independent of range (a vertically stratified, horizontally homogeneous atmosphere). Thus it is not suited when the atmospheric refraction varies with range. It can be used for ducted propagation beyond the horizon. The mode theory applies best when only a few modes are present, and when both the radar and the target are well within the duct. It tends to predict greater loss than indicated by actual measurements when the radar, target, or both, are not within the duct. One reason for this is that the duct is usually *leaky*. The upper boundary of the atmospheric duct is not necessarily smooth, but can be ragged so that the simple model of a plane surface is not realistic. Another limitation is that mode theory when applied over land does not usually include the effects of scattering from the nonsmooth terrain which is characteristic of most of the earth's surface.

The effects of multiple modes on the propagation within a duct has been mentioned earlier in this section, in the subsection "Multiple-Mode Propagation." It was said there that fades due to multiple modes can be of the order of 20 dB and can occur at intervals of 2 to 3 mi in range.

Parabolic Equation Model   Unlike ray tracing or the waveguide model, the parabolic equation method can handle refractive index changes that are inhomogeneous in both the horizontal and vertical directions. Thus it is useful whenever the refractivity profile varies with range; such as at land/ocean interfaces and in propagation over irregular terrain. This approach solves the Helmholtz wave equation with a parabolic equation approximation, such as a numerical method called the Fourier split-step algorithm.[69] The *split-step parabolic equation* provides an efficient method for modeling atmospheres where the vertical refractivity profile changes along the propagation path. It works well within the horizon and over the horizon as well as near the horizon, so that a single model can be used to make computations in all regions of interest. It has considerable advantage over prior methods, but it might require large computer resources of memory and run time. It has been noted[66] that rough surface effects are difficult to handle rigorously with this model. In addition to being applied over the ocean, the parabolic equation method has been applied to propagation over terrain.[70–72]

Hybrid Method   The purpose of a hybrid method is to provide the benefits of the split-step parabolic equation (PE) method without the extensive computations. One example is the Radio Physical Optics (RPO) model that uses a combination of ray optics and split-step PE methods.[73] These two are complementary in that split-step PE works well for small angles and the ray optics works well at the higher angles not covered by the PE method. Hitney states that this hybrid method can be 100 times faster than the pure split-step PE method for stressful cases.

Computer-Based Propagation Assessment Methods   There exist several computer software programs for determining the effects of propagation based on knowledge of the environmental factors that influence atmospheric refractivity. These have been used mainly by the military to determine the actual coverage of their radars under nonstandard propagation conditions. Much of the pioneering work in this area was performed by the SPAWAR Systems Center, San Diego (formerly called NRaD).

Figure 8.15 illustrates in a simple manner one example why the military tactical planner would be interested in knowing the radar propagation conditions to be expected.[74] The

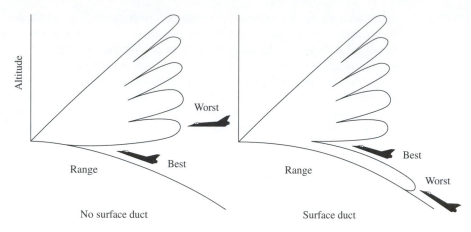

Figure 8.15   Sketch illustrating the effect of atmospheric ducting on the flight altitude of an attacking aircraft.
| (From J. H. Richter.[74])

left-hand side of the figure shows the radar coverage that might be obtained under standard propagation conditions with no surface duct present. An attack aircraft could avoid detection at long range by flying at low altitude. If, on the other hand, a surface duct were present, as depicted on the right-hand side of the figure, a low-flying attacker would be detected at much longer range because of the ducted propagation. A better tactic under such circumstances would be to fly at an altitude just above the duct.

IREPS   One of the first computer-based propagation prediction programs for operational use was *IREPS,* which stands for *Integrated Refraction Effects Prediction System.* It was developed by the SPAWAR Systems Center as an operational assessment tool for the U.S. Navy and was installed on major ships of the Fleet.[75,76] It was used with a PC, and had an interactive display. IREPS provided the following products:

1. *Propagation conditions summary.* The plot of refractivity as a function of height was shown graphically for the location, date, and time of day, along with a plain-language narrative assessment of what effects might be expected over surface-to-surface, surface-to-air, and air-to-air paths. The presence and vertical extent of any ducts were indicated, and wind speed and evaporation duct height were listed numerically. A correction for the measurement error of the target elevation angle due to bending by refraction was also given.

2. *Display of radar coverage.* These were vertical coverage diagrams (range versus height) for specific radars and other electromagnetic systems.

3. *Display of one-way path loss with range.*

4. *AEW aircraft stationing aid.* This showed the distortion of normal propagation caused by refractive effects at a given range for various combinations of radar and target altitudes. It provided the optimum location for the radar, so as not to lose targets

because of radar holes and allowed the aircraft to minimize detection by a hostile intercept receiver.

5. *Surface-search radar-range table.* This provided predictions of the detection range for an operator-selected surface-search radar against an operator-selected table of surface targets.

6. *Electronic support measures (ESM) intercept-range table.* A display was given of the maximum intercept range for an operator-selected ESM system (intercept receiver) against various operator-selected radars or other emitters.

In addition, IREPS could display the 50 percent probability of detection range of various size targets for an operator-selected forward-looking infrared (FLIR) system operating at various altitudes.

There were several methods by which IREPS obtained information about refractive conditions, depending on what was available. Refractivity could be obtained from radiosonde measurements (balloon borne instruments to determine upper atmosphere pressure, temperature, and water vapor pressure); aircraft-borne microwave refractometers; or inputs from surface meteorological measurements made on board the ship on which IREPS was located. When none of these were available, IREPS utilized a stored library of historic refractivity and climatology statistics as a function of the latitude, longitude, season, and time of day. When historic data was used, the output was a prediction of propagation performance in probabilistic terms. Also stored were the necessary system parameters for the various radars, communications, electronic warfare, and other electromagnetic systems whose predicted propagation performance was required. IREPS could make propagation predictions for frequencies from 100 MHz to 20 GHz.

Hitney[66] indicated that one of the most important uses of IREPS was for selecting the best flight profile for an attack aircraft attempting to penetrate the coverage of a hostile radar. Under normal (nonducting) propagation conditions, an attack aircraft flies most of the distance to its target at low altitude in order to be below the coverage of the hostile radar. If, on the other hand, a surface-based duct was predicted by IREPS, the defending radar detection range against low-altitude targets might be greater than when ducting was absent. Under such circumstances, it is usually better for the attacker to fly at an altitude slightly above the duct. Hitney has said that "the use of IREPS coverage diagrams in strike [attack] warfare flight profile selection has been verified operationally to be effective 85% of the time."

TESS, or Tactical Electronic Support System[66]   This is designed for naval tactical decision making and is similar to IREPS in that it uses the same basic propagation assessment models and displays, but with better environmental information and some improved propagation models. For example, it employs the hybrid RPO model mentioned previously so as to take account of range-dependent refractive effects. It uses real-time satellite data, and has the ability to overlay this data with other meteorology analyses and forecasts. TESS is designed for surface ships that have officers trained in the environmental sciences, including aircraft and helicopter carriers and amphibious assault ships.

EREPS, or Engineer's Refractive Effects Prediction System[38]   This is also derived from IREPS, but is designed for the use of engineers instead of naval tactical decision makers.

For example, the engineer is more interested in the long-term performance of a radar, usually in statistical terms, rather than in single-event performance prediction that IREPS was designed to provide as a tactical decision aid. EREPS is more flexible than IREPS in that it increases the user's ability to edit the various parameters and determine how changes in radar parameters affect performance. It also allows the use of a high-fidelity range-dependent propagation model such as the RPO program, including the use of the binary files of propagation loss versus range as generated by the RPO program.

AREPS or Advanced Refractive Effects Prediction Program (AREPS)   This software program is an advanced version of IREPS that computes and displays radar probability of detection, electronic support measures (ESM) vulnerability, UHF/VHF communications capability, and simultaneous radar detection and ESM (intercept) vulnerability.[77] It is Windows based and is available from the SPAWAR Systems Center as a CD or from the Internet.

Other versions of computer-based propagation prediction methods have been reported by Ferranti Computer Systems, Ltd. of the United Kingdom[78] and by the Ukraine.[79] The original Ferranti system was called IMP, or Identification of Microwave Propagation. As described, it appears to have less capability than IREPS. The Ukrainian system is said to incorporate scattering from atmospheric turbulence.

8.6   DIFFRACTION

In the previous section we discussed how radar waves can propagate beyond the geometrical horizon of the earth by means of atmospheric refraction. Another mechanism that permits electromagnetic waves to extend beyond the geometrical horizon is *diffraction*. Radio waves are diffracted around the curved earth in a manner similar to the way light is diffracted by a straight edge (a topic usually covered in college physics courses). The ability of electromagnetic waves to propagate beyond the horizon by diffraction depends upon the frequency (the lower the better). At microwave radar frequencies there is very little energy diffracted by the earth's surface so that microwave radar coverage cannot be significantly extended beyond the line of sight by this propagation mechanism. Diffraction is important, however, for understanding HF surface-wave radar and for predicting the signal strength at or near the radar horizon at any radar frequency.

Frequency Dependence of Diffraction   Figure 8.16 is a theoretical plot of the electric field strength (relative to free space) incident on a target as a function of the distance from the radar transmitting antenna. Both the transmitting antenna and the target are at a height of 100 m, and there is no refraction to extend the horizon. The geometrical horizon at 71.4 km is also the horizon distance for optical frequencies ($\lambda \rightarrow 0$). It represents the approximate boundary between propagation and no propagation for optics and infrared. As the frequency decreases (wavelength increases), energy propagates farther into the region beyond the geometrical horizon. As the wavelength increases, however, it is noted that there is a reduction of energy at the geometrical horizon as well as just within the

Figure 8.16 Theoretical field strength (relative to free-space field strength) as a function of the distance from the transmitting antenna. Vertical polarization, h_a = h_t = 100 m, k = 1, ground conductivity = 10^{-2} mho/m, dielectric constant = 4. (After Burrows and Attwood,[8] courtesy Academic Press, Inc.)

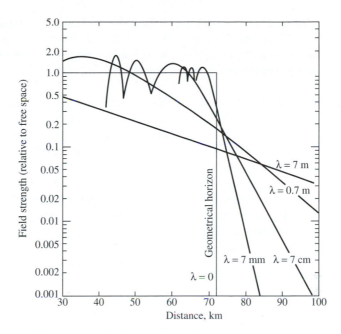

horizon. In the absence of refraction effects, microwave radar seldom has the ability to detect low-altitude targets beyond the geometrical horizon, or even at the horizon, unless it has sufficient power to overcome the loss caused by diffraction.

If low-altitude radar coverage is desired beyond the geometrical horizon in the diffraction region, the frequency should be as low as possible and there should be excess power to compensate for the diffraction loss. As an example, the loss in signal strength in the diffraction region at a frequency of 500 MHz is roughly 1 dB/mi at low altitudes. (It is even greater at higher frequencies.) Therefore, to penetrate 10 mi beyond the horizon within the diffraction region, the radar power at 500 MHz must be increased by 20 dB over that required for free-space propagation. Even if lower frequencies were available for radar to take advantage of the lower diffraction loss, the range resolution is poorer (because of narrower bandwidths), beamwidths are wider, the spectrum is crowded, and external noise increases with decreasing frequency.

If, on the other hand, the low-altitude coverage is to be optimized within the horizon in the interference region, Fig. 8.16 shows that the radar frequency should be as high as practical (consistent with other constraints on the choice of radar frequency).

Coverage at the Horizon The maximum "line-of-sight" distance d between a radar at a height h_a and a target at height h_t is given by the expression

$$d = \sqrt{2kah_a} + \sqrt{2kah_t} \qquad\qquad [8.21]$$

where k is the effective earth radius, and a is the earth actual radius. In this equation, the line of sight between the radar and the target is assumed to be tangent to the earth's surface. This equation is often used to describe the maximum range at which a target at an altitude h_t can be seen above the radar horizon. The diffraction loss at the horizon, however, might be from 10 to 30 dB below that if in free space.[80] Thus caution should be used when employing the above equation to describe the low-altitude coverage of a radar when propagation is near the horizon.

Surface-Wave HF Radar   The diffraction loss at HF frequencies (3 to 30 MHz) is much lower than that at microwave frequencies. For this reason, the potential of extended coverage at low altitudes beyond the horizon has been examined many times in the past using HF radar and the surface wave, or ground wave, mode of propagation where the diffraction loss is low. It has been said[81] that for every nautical mile increase in radar range beyond 75 nmi over the sea in the diffraction region requires an increase of radar energy of approximately 0.3, 0.5, and 0.6 dB per nautical mile at frequencies of 5, 10, and 15 MHz, respectively. The loss depends on the surface conductivity. Over land the loss is much higher than over the sea; which is why HF surface-wave radars are seldom considered for operation over land.

The lower the frequency the lower will be the propagation loss. The lower the frequency, however, the greater will be the external noise. (External noise can be many orders of magnitude greater than receiver noise). External noise is due to either atmospheric, cosmic, or anthropogenic noise. The radar cross section of most aircraft at HF is usually much larger than at microwave frequencies; but at a sufficiently low frequency (depending on the size of the target), the cross section falls in the Rayleigh region where it varies as the fourth power of the radar frequency. It then decreases rapidly with decreasing frequency. For example, the decrease in cross section of a fighter aircraft signifying the Rayleigh region might begin somewhere around 15 to 20 MHz. For a large bomber aircraft, the Rayleigh region might begin at a frequency of 3 to 6 MHz.

There will be an optimum frequency for a surface-wave radar. Below the optimum frequency there is an increase in radar power because of increased external noise and lower radar cross section. Above this frequency, increased power is also needed because of the increase in diffraction loss with increasing frequency. The optimum frequency will depend on the type of target and the variability of external noise with time of day and season. Because of the variability of external noise and the need to detect small as well as large targets, the choice of HF radar frequency is often a compromise. In one analysis of the optimum frequency for a particular HF radar, the region from 5 to 10 MHz seemed to be the best place to operate.[81]

Vertical polarization is used in HF surface-wave radar since its energy extends down to the surface, which is what is desired to detect targets at or near the surface. The energy radiated by a horizontally polarized antenna, on the other hand, decreases as the surface is approached. When surface-wave radars are considered for shore-based operation looking out over the sea, the antennas can be of large extent in the horizontal dimension. They might be from 300 to 1000 m (more or less). A 500-m antenna at 10 MHz has a beamwidth of about 4°. The antenna is an array which is either electronically steered in azimuth angle or which forms a number of multiple fixed receiving beams covering the area of interest. With multiple receiving antenna beams, the transmitting antenna is much smaller in size than the receiving array antenna since it utilizes one broad beam to cover the region viewed by the multiple, narrow receiving beams. The advantage of this antenna arrangement is that it is easier to make a large receiving antenna than a large transmitting antenna. The multiple beams of the receiving antenna allow the simultaneous processing of the radar echo signals from each beam, which allows a faster data rate. Because of expense, HF radar antennas seldom have significant directivity in the elevation plane other than that obtained from a monopole radiator in front of a backscreen. These radars need doppler processing in order to detect desired moving targets in the midst of the large echo from the land or sea.

Because of the exponential diffraction loss, land-based HF surface-wave radars that look over the sea might have average powers of from several tens of kW to well over 100 kW. The pulse repetition frequencies are usually low so that ground echoes received via skywave propagation from long range do not interfere with target echoes from the near-range. The ranges of such radars might be from 50 to 150 nmi depending on the type of target. (Most of the numbers presented here are not meant to be interpreted as rigid bounds.)

HF surface-wave radars for shipboard operation have to be much smaller in size than land-based systems and be satisfied with lower power transmitters.[82] Their ranges would be correspondingly reduced. The main naval application for such radars might be to detect low-altitude antiship missiles at greater ranges than can be detected by microwave radar.

In spite of some attractive attributes of HF surface-wave radar there are many difficulties in their application (which mainly have been for military purposes). They have poor resolution in range and angle, no indication of target elevation angle (or height), they require large antennas, and on board ship, they can cause mutual interference with HF communications. If the chief interest is in detecting targets at over-the-horizon ranges, the radar cannot immediately differentiate between a target that is at the same range as the over-the-horizon target but at a higher altitude and within line of sight. Military radars for air defense require some form of target identification, usually a cooperative IFF (identification friend or foe) system that operates at microwave frequencies. Thus there is no convenient means for target recognition when using HF radar. Also, the powerful HF signals radiated by a military radar can be detected by a hostile intercept receiver at very long distances. Thus there are reasons why the HF surface-wave radar has not seen operational application even though it has the capability of seeing over the horizon.

In general, the long-range HF over-the-horizon radar that employs skywave propagation has greater range and greater coverage than a surface-wave radar, and is not that much larger than the largest surface-wave systems.[83] Much smaller HF surface-wave radars have also been operated at within-the-horizon ranges for remote sensing of sea state,[84] surface currents,[85,86] and icebergs.[87,88]

8.7   ATTENUATION BY ATMOSPHERIC GASES

Water vapor and oxygen in the clear atmosphere can attenuate radar energy when the radar frequency is at or in the vicinity of one of the resonant frequencies of these molecules. Figure 8.17 shows the attenuation due to both water vapor and oxygen as a function of frequency. There is a resonance peak of water vapor at a frequency of 22.2 GHz, and another in the millimeter wave region at 184 GHz.[89,90] The attenuation due to water

Figure 8.17 Attenuation of electromagnetic energy by atmospheric gases in an atmosphere at 76 cm pressure. Dashed curve is absorption due to water vapor in an atmosphere containing 1 percent water vapor molecules (7.5 g water/m^3). The solid curve is the absorption due to oxygen. (From Burrows and Attwood[8] and Straiton and Tolbert.[89])

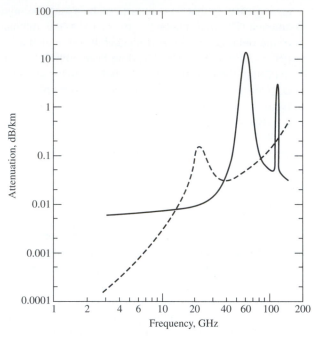

vapor will depend on the amount of moisture in the atmosphere, which can vary with time and place. Although the attenuation is only about 0.2 dB/km at a frequency of 22 GHz, absorption can be sufficient to deteriorate the effectiveness of radars that operate at the original K-band frequency of 24 GHz. When radars were first developed at K band during World War II, it was not realized that there was a nearby absorption band. To avoid this problem, the original K band was split into a lower band, K_u, and an upper band, K_a, Table 1.1. Radars are almost never found any more at the original K band. The oxygen molecule has resonances at 60 GHz and 118 GHz. The 16 dB/km attenuation at 60 GHz makes this region unusable except for very short range radars and radars that operate in space outside the atmosphere.

Atmospheric attenuation generally has negligible effect on radar performance at the normal microwave frequencies. It begins to be increasingly important at frequencies above 10 GHz. The large attenuations experienced at millimeter wavelengths is one of the chief reasons why long-range radars are seldom found above 40 GHz.

The effect of attenuation, when it is large enough to be a problem, is accounted for in the radar range equation by inserting into the numerator the multiplicative factor exp $[-2\alpha R]$, where α is the one-way attenuation coefficient measured in units of distance^{-1}, and R is the range to the target. Instead of α it is more usual to express the one-way attenuation, especially when plotted in graphs, as decibels per unit distance—which is what is done here. This is equivalent to the quantity 4.34α, where the constant 4.34 accounts for the conversion from the natural logarithm to the base 10 logarithm.

If the attenuation per unit distance α is not constant, exp $[-2\alpha R]$ should be replaced by exp $[-2\int \alpha(R)dR]$, with the integration taken from 0 to the target range R.

Atmospheric attenuation decreases with increasing altitude (there are fewer molecules to absorb the radar energy). When the antenna beam is pointed at some elevation angle, the variability of attenuation with altitude must be taken into account when determining the total attenuation along the propagation path. With a ground-based radar the attenuation is greatest when the antenna points along the horizon, and is least when it points to the zenith. For example, at the water vapor absorption line located at 22.2 GHz, when the energy is directed at the horizon (elevation angle of 0°), the total attenuation in propagating completely through the earth's troposphere and back again is 80 dB, a formidable number. When the energy propagates at the zenith (elevation angle of 90°), the total two-way attenuation through the entire troposphere is only 1.3°. If the elevation angle were greater than 10°, the total attenuation is less than 7 dB, so that when the radar is looking at the higher elevation angles, attenuation might not be important. Figure 8.18 gives

Figure 8.18 Two-way atmospheric attenuation as a function of range and frequency for (a) 0° elevation angle and (b) 5° elevation angle.
| (From L. V. Blake.[91])

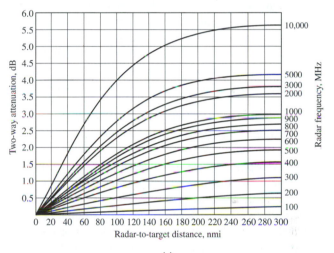

(a)

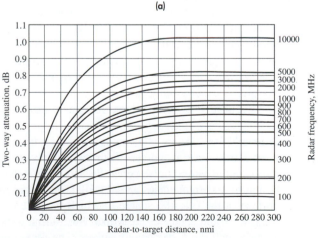

(b)

examples of the two-way attenuation in the atmosphere as a function of range and frequency for elevation angles of 0 and 5°.[91] It might be noted that even if there were no loss at 0° elevation, the null at the horizon due to multipath prevents significant propagation of radar energy at or near this angle (except under ducting conditions or at low frequencies with vertical polarization.)

8.8 ENVIRONMENTAL, OR EXTERNAL, NOISE

The inherent internal noise of the radar receiver is what usually limits the detectability of targets by microwave radars (in the absence of clutter echoes). At frequencies at either end of the microwave spectrum, however, the limitation on sensitivity is usually the external noise that appears at the antenna terminals from some outside source. The reradiation noise due to atmospheric absorption usually determines the receiver sensitivity at the upper end of the microwave spectrum and at millimeter waves. At VHF and lower frequencies the receiver sensitivity is usually set by cosmic noise, the noise due to the combined effects of lightning strokes throughout the world, and anthropogenic noise. The minimum noise occurs at the middle of the microwave region, in the vicinity of S band. Generally, external noise is not a factor in radar performance unless the radar frequencies are outside the range of the usual microwave frequencies. Harmful external noise at the antenna terminals due to deliberate hostile jamming, however, can cause serious degradation to an unprepared military radar system, but this is not the subject of this chapter.

Atmospheric Absorption Noise   It is known from the theory of blackbody radiation that any body that absorbs energy reradiates the same amount of energy that it absorbs, else it would increase in temperature. As mentioned in Sec. 8.7, water vapor and oxygen absorb (attenuate) radar energy. This absorbed energy must then reradiate as thermal noise. If L is the loss of radar energy when propagating through the atmosphere, and T_a is the ambient temperature of the absorbing atmosphere, the effective noise temperature of the reradiated energy is $T_e = T_a(L - 1)$. (The effective temperature is defined in Sec. 11.2.) Atmospheric absorption noise, just like atmospheric attenuation, is of potential concern only at the higher radar frequencies. Figure 8.19, shown later, is a composite plot of the several sources of electromagnetic noise as a function of frequency. Atmospheric absorption noise is the dominant effect at the right side of the figure. The maximum value of atmospheric absorption noise occurs for an elevation angle of 0°, the minimum values for an elevation angle of 90° (looking straight up).

Cosmic Noise   There is a continuous background of noiselike electromagnetic radiation from extraterrestrial sources in our own galaxy (the Milky Way), extragalatic sources, and radio stars. Cosmic noise generally decreases with increasing frequency and can usually be ignored at frequencies above UHF. The magnitude of cosmic noise depends upon the portion of the celestial sphere to which the antenna points. It is a maximum when looking toward the center of the Milky Way, and a minimum when observing along the pole about which the Milky Way revolves. The maximum and minimum *brightness temperature* due to cosmic noise is shown in the left-hand portion of Fig. 8.19. The brightness temperature can affect the system noise temperature and the sensitivity of the radar receiver (especially at the lower frequencies). In the absence of any radio stars, the background cosmic noise left over from the "big bang" at the start of the universe is the minimum noise level that might be expected. Its value is 2.7 K, which is too small to bother any radar receiver.

The sun is a relatively strong emitter of noise if the radar antenna beam directly views the solar disk. It might also be detectable with radars having poor sidelobes and very sensitive receivers. The minimum level of solar noise is due to blackbody radiation at a temperature of 6000 K. Solar storms (sunspots and flares), however, can increase the solar-noise level several orders of magnitude over that of the quiet or undisturbed sun. Radar stars are too weak to be a serious source of interference. Both radio stars (in conjunction with sensitive receivers) and the sun have been used as sources to calibrate the beam-pointing (boresight) of large antennas.[92–94]

Atmospheric Noise (Lightning)   A single lightning stroke radiates considerable RF noise power, especially at the lower frequencies. At any one moment there are an average of 1800 thunderstorms in progress in different parts of the world. From these storms about 100 lightning strokes take place every second somewhere in the world.[95] The combined effect of all the lightning strokes is to give rise to a noise spectrum that is especially large at broadcast and shortwave frequencies. Noise radiated by lightning strokes throughout the world is called *atmospheric noise* (not to be confused with the noise produced by atmospheric absorption, as mentioned previously). The spectrum of atmospheric noise falls off rapidly with increasing frequency and is usually of little consequence above 50 MHz. It is seldom, therefore, an important consideration in radar design; except possibly for radars in the lower portion of the VHF region.

Anthropogenic Noise   In preparing the previous (second) edition of this text, the publisher reminded me on several occasions to avoid the use of sexist terminology. I believe I was successful in doing so, except for one term: that of *man-made noise*. My reason for wanting to continue to use it was not that men usually make such noise (which they do), but I could not find a suitable substitute. Terms such as "human-made noise," "people-made noise," or "population-made noise" might be nonsexist, but they just didn't sound right. The publisher took pity on my inability to find a suitable substitute and relented; so on p. 463 of the 2d edition, the term "man-made noise" was allowed to appear. After the book was published, I came upon an excellent substitute, which is "anthropogenic noise." *Anthropogenic* is an adjective that means *relating to, or resulting from the influence of humans on nature*. It is a proper replacement for the no longer acceptable term *man-made noise*.

Electromagnetic emissions that appear as noise or interference to other electromagnetic services, originate from many possible sources: higher harmonics and other incidental radiation from transmitters, automobile ignition, electric razors, power tools, automatic garage door openers, fluorescent lights, industrial processing equipment, and power transmission lines.[96,97] Anthropogenic noise is more prevalent in urban and industrial areas than in rural areas. It decreases with increasing frequency and is seldom a factor in the design of microwave radars. It can be of concern, however, for VHF and lower-UHF systems. Because of its variability in time and space, it is difficult to be precise about the quantitative nature of this form of noise. In the UHF radar frequency band, one source indicates[98] that the noise temperature of anthropogenic noise from a business region (industrial park, large shopping center, busy street, or highway) might be about 500 K, and for a residential region (at least two dwelling units per acre and no nearby highways) it might be about 200 K. At VHF and UHF, anthropogenic noise varies almost inversely as the cube of the frequency.

Interference and noise can also be experienced from other users of the electromagnetic spectrum, such as other radars and communications of all varieties. These sources of interference are generally different from anthropogenic noise, and is considered a problem in electromagnetic compatability (EMC).

Composite Graph   Figure 8.19 is a composite graph of the several forms of environmental noise that might affect radar.[99] Only the minimum and maximum resultants from the various component factors are shown. Atmospheric noise and cosmic noise dominate at the lower frequencies, and atmospheric absorption noise dominates at the higher frequencies. The minimum noise levels occur from about 1 to 5 GHz (*L* to *C* band). Anthropogenic noise is not included on this graph since it is so variable and generally is of little consequence to radars that operate within the usual microwave radar bands. It can be quite important, however, for radars that operate at VHF or lower frequencies.

Earth Thermal-Noise   The temperature of the earth is nominally 290 K; hence, it will radiate thermal noise. If an antenna beam illuminates the ground, it will receive a portion

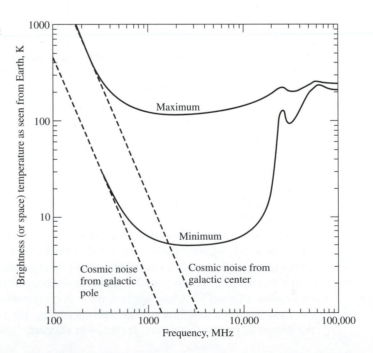

Figure 8.19   Maximum and minimum brightness temperatures of the sky as seen by an ideal single-polarization antenna on earth.
| (After Green and Lebenbaum.[99])

of the thermal noise radiated by the earth. The actual noise temperature seen by the receiver depends on whether the entire main beam of the antenna views the ground or if only a portion of the beam does so. The radiated noise depends on the emissivity of the ground as well as its temperature. Thus the brightness temperature seen by the antenna will be less than the actual temperature when the emissivity of the ground is less than unity, as will happen if the antenna views a water surface. The effect of the thermal noise radiation from the ground will affect only those systems with very sensitive receivers, generally those with noise figures of a fraction of a decibel. Radar receivers are seldom that sensitive; hence, the thermal noise radiated by the ground usually does not bother conventional radars.

8.9   OTHER PROPAGATION EFFECTS

Radar Siting   Minimizing of environmental noise effects sometimes can be an important consideration in determining where a radar should be located. The siting of a radar also depends on the masking (obscuring or screening of targets) due to terrain and the backscatter from terrain features. It is especially important to know how the terrain affects the performance of ground-based radars when there is either severe masking, severe clutter echoes, or when low altitude targets have to be detected with high reliability.

The air-surveillance radars employed during World War II had no doppler processing, so that an aircraft target located in the same radar resolution cell as a clutter echo was likely to be undetected because of the large clutter echoes. As mentioned previously, when siting military air-surveillance radars at that time, it was desired that a site be chosen such that the surrounding terrain would shield clutter echoes from the radar. In effect, the terrain was used to mask the low-lying clutter echo and prevent it from entering the radar receiver so that aircraft at high altitudes could be seen without the presence of clutter. That was satisfactory in World War II, since heavy bomber aircraft generally flew at high altitudes. Modern aircraft and missiles, however, fly at low altitudes to deliberately avoid a radar's coverage. The military radar designer can respond by siting the antenna as high as practical and employing doppler processing to separate moving targets from fixed clutter echoes. Thus the siting of ground-based military air-defense systems has to take account of the local terrain so as to minimize the regions where the detection ranges of low-altitude targets are significantly reduced because of terrain masking.

The effect of terrain is an important part of the siting of civil air-traffic control radars. Computer software has been developed for providing detailed information about the terrain and its effect on civil air-traffic control radars so that the performance of a potential radar site can be determined before the radar is installed.[100] This program, called the Radar Support System (RSS), provides information that allows the radar designer to optimize the performance of a site-specific sensor by selecting the optimum radar height, optimum beam-tilt-angle, sensitivity time control (STC) characteristics, and to determine the likelihood of false alarms due to echoes from highway traffic or large clutter echoes. The input to RSS includes a digital model of the terrain obtained from databases available from the United States Geological Survey (USGS) or the Defense Mapping Agency. The height resolution is one meter and the steps in latitude and longitude are three seconds. There is also a USGS database of land use and land cover with the same latitude/longitude quantization. It identifies the land as one of ten possibilities; such as urban, agricultural, forest, wet land, barren land, and so forth. Also included is a cultural database that provides three-dimensional models of the significant buildings near the radar site (such as an airport), and includes an estimate of the construction material of each structure. The RSS determines plots of the line-of-sight visibility from the radar, areas which are screened from the radar, the radar cross section of all visible terrain cells, and the probability of detection for specified targets.

Similar computer models and terrain databases have been applied to a military overland situation when the radar is located off shore.[101] Based on the Defense Mapping Agency's Digital Terrain Elevation Data and propagation programs like those mentioned in the previous section for standard and nonstandard atmospheric refraction conditions, it determines terrain and target visibility, masking, clutter echo strength, and plots the results on a maplike presentation to show how radar coverage is affected by the environment.

Atmospheric Lens-Effect Loss   Weil[102,103] has shown there is another effect of atmospheric refraction on radar propagation, in addition to what was discussed in Secs. 8.4 and 8.5. The variation of standard refractivity with altitude causes the atmosphere to act as a negative lens that decreases the radiated energy density incident on a target. This loss is independent of radar frequency. Unlike atmospheric attenuation, the lens-effect loss continues to increase at ranges beyond the sensible atmosphere, but it approaches a limiting value asymptotically. For a CRPL exponential atmosphere with surface refractivity equal to 313, the loss is less than 1 dB at a range of 200 nmi and 0° elevation angle, and less than 0.18 dB at an elevation angle of 5°. The limiting values at very long range (10,000 nmi) are 2.9 dB at 0° elevation and 0.27 dB at 5° elevation. The lens-effect loss is an additional loss that is additive to the atmospheric attenuation. It is usually small enough to be neglected, except at low elevation angles and long range.

Ionospheric Propagation at Microwave Frequencies   The ionosphere is a partially ionized region of the upper atmosphere that extends from about 50 km to 2000 km in altitude.[104] It is generated by high-energy particles that travel from the sun to ionize the atoms of the thin upper atmosphere. The refraction, or bending, of electromagnetic radiation by the ionosphere allows long-range propagation by shortwaves (the HF region) that is well know to radio amateurs. It is also the basis for HF over-the-horizon radar which allows the detection of aircraft, ships, and ballistic missiles at ranges extending out to 2000 nmi and beyond. The ionosphere is usually considered transparent to microwave radiation, but this is not fully correct. It can adversely affect in several ways the propagation of microwave radiation that travels through it.

Faraday Rotation of Polarization   An electromagnetic wave experiences a rotation of its plane of polarization when traveling in an ionized medium (the ionosphere) and a magnetic field (that of the earth). This is known as *Faraday rotation.* If a ground-based radar is to detect satellites and other space objects, and if the radar employs linear polarization, the polarization of the echo signal will be different from that transmitted. A loss in signal can result. If, for example, the polarization were rotated by 90° because of

Faraday rotation, the received signal would be zero since the polarization is orthogonal to that of the transmitting antenna (which is assumed to be the same antenna as that used for receiving). The amount of polarization rotation varies inversely as f^2, where f = radar frequency. The rotation is determined by the total electron content of the ionosphere seen over the radar propagation path, which depends on the radar location, time of day, time of year, and the sunspot cycle. The effect is greater when the radar beam is pointing north or south and minimum when pointing east or west (directions are relative to magnetic north).

Radars for the detection of extraterrestrial targets that operate at UHF or lower frequencies can encounter loss due to the large polarization rotations of their echo signals. One solution for radars that are subject to Faraday rotation is to transmit on a single linear polarization (for example, vertical) and receive on two orthogonal linear polarizations (horizontal and vertical) in order to avoid loss in signal because of Faraday rotation. The echo signals in each polarization receive-channel are processed separately and then combined. This technique has been used in UHF radars for the detection of space objects such as BMEWS (ballistic missile early warning system), the AN/FPS-85 space surveillance radar, and the Pave Paws missile warning radar. At one time it was thought that radars at L band did not experience sufficient polarization rotation to require receiving with dual polarizations. Faraday rotation, however, can sometimes be significant enough at L band to require compensation.[105] The maximum one-way polarization rotation at a frequency of 1 GHz, for example, has been said to be 108° for a radar in the United States that views a target at an elevation angle of 30°.[106] Faraday rotation can also affect spaceborne radars viewing ground targets.

Other Ionospheric Effects   The ionosphere will introduce a time delay that is inversely proportional to f^2. At 1 GHz, the maximum delay is said to be 0.25 μs.[105] There can also be refraction of the beam, loss by absorption, and frequency dispersion that introduces distortion into wideband signals. These effects are generally small at microwave frequencies for most radar applications that propagate through the ionosphere. There is one important example, however, where compensation had to be made in a microwave radar to avoid degradation of performance because of dispersion when propagating through the ionosphere. This occurred for the Cobra Dane high resolution L-band radar (1175 to 1375 MHz).[107]

The Cobra Dane radar, located on the island of Shemya at the southern tip of the Aleutian island chain in Alaska, was designed to gather intelligence information on Soviet ballistic missile systems. Its high range-resolution waveform employed a 1000-μs linear FM pulse with a 200-MHz bandwidth. Stretch pulse compression was incorporated to achieve a range resolution of about one meter. The time delay in propagating through the ionosphere was different at the low-frequency end of the 200-MHz bandwidth compared with the time delay at the high end of the bandwidth. This difference in propagation time was sufficient to introduce phase distortion and broaden the compressed pulse width unless compensated. The necessary corrections were obtained by predistorting the transmitted pulse by the inverse of the distortions introduced by the ionosphere. In addition to the ionospheric corrections, there had to be a correction applied to account for the fact that the doppler-frequency shift also was not constant over the 200-MHz bandwidth of the high-resolution waveform.

REFERENCES

1. Bachynski, M. P. "Microwave Propagation Over Rough Surfaces." *RCA Review,* 20, no. 2 (June 1959), pp. 308–335.

2. Ament, W. S. "Toward a Theory of Reflection by a Rough Surface." *Proc. IRE,* vol. 41 (January 1953), pp. 142–146.

3. Beard, C. I. "Coherent and Incoherent Scattering of Microwaves from the Ocean." *IRE Trans.* AP-9 (September 1961), pp. 470–483.

4. Miller, A. R., R. M. Brown, and E. Vegh. "New Derivation for the Rough-Surface Reflection Coefficient and the Distribution of Sea-Wave Elevations." *IEE Proc.,* 131, Pt. H (1984), pp. 114–116.

5. Miller, A. R., and E. Vegh. "Family of Curves for the Rough Surface Reflection Co-efficient." *IEE Proc.* 133, Pt. H (December 1986), pp. 483–489.

6. Barton, D. K. "Low-Angle Radar Tracking." *Proc. IEEE* 62 (June 1974), pp. 687–704.

7. Miller, A. R., and E. Vegh. "Exact Result for the Grazing Angle of Specular Reflection from a Sphere." Naval Research Laboratory, Washington, D.C., Memorandum Rep. 6867, August 9, 1991.

8. Burrows, C. R., and S. S. Attwood. *Radio Wave Propagation.* New York: Academic, 1949.

9. Shibuya, S. *A Basic Atlas of Radio-Wave Propagation.* New York: John Wiley, 1987.

10. Blake, L. "Machine Plotting of Radio/Radar Vertical-Plane Coverage Diagrams." Naval Research Laboratory, Washington, D.C. Rep. 7098, June 25, 1970 (AD 709897).

11. Skolnik, M. "Improvements for Air-Surveillance Radar." *Proc. 1999 IEEE Radar Conf.* pp. 18–21, IEEE Catalog no. 99CH36249.

12. ITU-R Recommendation P.453–5, "The Radio Refractive Index: Its Formula and Refractivity Data," 1995.

13. Hitney, H. V. "Refractive Effects from VHF to EHF, Part A: Propagation Mechanisms." Paper no. 4A in *Propagation Modeling and Decision Aids for Communications, Radar, and Navigation Systems.* AGARD-LS-196, NATO, September 1994.

14. Bean, B. R., and E. J. Dutton. "Radio Meteorology," National Bureau of Standards Monograph 92, March 1, 1966.

15. Shibuya, S. Ref. 9, p. 28.

16. Bean and Dutton. Ref. 14, Secs. 3.1 and 3.2.

17. Bean, B. R. "The Geographical and Height Distribution of the Gradient of Refractive Index." *Proc. IRE,* 41 (April 1953), pp. 549–550.

18. Bean, B. R. et al. "A World Atlas of Atmospheric Radio Refractivity," U. S. Dept. of Commerce, ESSA Monograph 1, 1966.

19. Hall, M. P. M., L. W. Barclay, and M. T. Hewitt. *Propagation of Radiowaves.* London: Institution of Electrical Engineers, 1996, Sec. 6.2.

20. Blake, L. V. "Prediction of Radar Range." *Radar Handbook,* 2nd ed. M. Skolnik (Ed.). New York: McGraw-Hill, 1990, Chap. 2, Fig. 2.18.

21. Tank, W. G. "Atmospheric Effects." *Airborne Early Warning.* W. C. Morchin (Ed.). Boston, MA: Artech House, 1990, Chap. 3. See also Morchin, W. *Radar Engineer's Sourcebook.* Boston, MA: Artech House, 1993, Sec. 15.4.3.

22. Bauer, K. W. "Range-Height-Angle Charts with Lookdown Capability." *Microwave J.* 24 (October 1981), pp. 89–92.

23. Jursa, A. S. (Ed.). *Handbook of Geophysics and the Space Environment.* U.S. Air Force Geophysics Laboratory, 1985. (Available from National Technical Information Service, Springfield, VA, 22161.)

24. Valley, S. L., (Ed). *Handbook of Geophysics and Space Environments.* New York: McGraw-Hill, 1965.

25. Shannon, H. H. "Recent Refraction Data Corrects Radar Errors." *Electronics* 35, no. 49 (Dec. 7, 1962), pp. 52–56.

26. Blake, L. V. "Ray Height Computation for a Continuous Nonlinear Atmospheric Refractive-Index Profile." *Radio Science* 3 (January 1968), pp. 85–92.

27. Brown, B. P. "Radar Height Finding." *Radar Handbook,* 1st ed., M. Skolnik (Ed.). New York: McGraw-Hill, 1970, Chap. 22, Sec. 22.3.

28. Murrow, D. J. "Height Finding and 3D Radar." In *Radar Handbook,* 2nd ed. M. Skolnik (Ed.). New York: McGraw-Hill, 1990, Chap. 20, Sec. 20.2.

29. Barnett, K. M., and S. H. Brown. "Accuracy of Calculated Radar Refraction Errors." *IEEE Trans.* AP-13 (November 1965), p. 986.

30. Nathanson, F. E. *Radar Design Principles,* 2nd ed. New York: McGraw-Hill, 1991, Fig. 6.26.

31. Robertshaw, G. "How Accurate is Range Correction?" *Microwaves & RF* 25 (March 1986), pp. 129–132.

32. Rowland, J. R., and S. M. Babin. "Fine-Scale Measurements of Microwave Refractivity Profiles with Helicopter and Low-Cost Rocket Probes." *Johns Hopkins APL Tech. Dig* 8, no. 4 (1987), pp. 413–417.

33. Bean, B. R., and E. J. Dutton. Ref. 13, Sec. 2.3.

34. Anderson, L. J., and E. E. Gossard. "Prediction of Oceanic Duct Propagation from Climatological Data." *IRE Trans.* AP-3 (October 1955), pp. 163–167.

35. Anderson, K. D. "Radar Measurements at 16.5 GHz in the Oceanic Evaporation Duct." *IEEE Trans..* AP-37 (January 1989), pp. 100–106.

36. Katzin, M., R. W. Bauchman, and W. Binnian. "3- and 9-Centimeter Propagation in Low Ocean Ducts." *Proc. IRE* 35 (September 1947), pp. 891–905.

37. Jeske, H. "State and Limits of Prediction Methods of Radar Wave Propagation Conditions Over the Sea." In *Modern Topics in Microwave Propagation and Air-Sea Interaction,* A. Zancla (Ed.). D. Reidel Publishing, 1973.

38. Patterson, W. L., et al. "Engineer's Refractive Effects Prediction System (EREPS)." Version 3.0, Naval Command, Control, and Ocean Surveillance Center, RDT&E Division, San Diego, CA, Tech. Document 2648, May, 1994.

39. Paulus, R. A. "Practical Applications of an Evaporation Duct Model." *Radio Science* 20 (July–August 1985), pp. 887–896.

40. Jeske, H. "The State of Radar-Range Prediction Over Sea." *AGARD Conf. Proceedings* 70(2) (1971), pp. 50.1–50.10.

41. Hitney, H. V., A. E. Barrios, and G. E. Lindem. "Engineer's Refractive Effects Prediction System (EREPS) User's Manual." Naval Ocean Systems Center, San Diego, CA, July 1988, (AD 203443)

42. Babin, S. M., G. S. Young, and J. A. Carton. "A New Model for the Oceanic Evaporation Duct." *J. Applied Meteorology* 36 (March 1997), pp. 193–204.

43. Anderson, K. D. "94-GHz Propagation in the Evaporation Duct." *IEEE Trans.* AP-38 (May 1990), pp. 746–753.

44. Joseph, R. I., and G. D. Smith. "Propagation in an Evaporation Duct: Results in Some Simple Analytic Models." *Radio Science* 7 (April 1972), pp. 433–441.

45. Früchtenicht, H. W. "Notes on Duct Influences on Line-of-Sight Propagation." *IEEE Trans.* AP-22 (March 1974), pp. 295–302.

46. Giger, A. J. *Low-Angle Microwave Propagation.* Norwood, MA: Artech House, 1991, Sec. 2.5.

47. Anderson, K. D. "Radar Detection of Low-Altitude Targets in a Maritime Environment." *IEEE Trans.* AP-43 (June 1995), pp. 609–613.

48. Hitney, H. V. Ref. 13, Sec. 6.2.

49. Patterson, W. L., et al. Ref. 38, pp. 12–14.

50. Battan, L. J. *Radar Observation of the Atmosphere.* Chicago: University of Chicago Press, 1973.

51. Pappert, R. A., and C. L. Goodhart. "A Numerical Study of Tropospheric Ducting at HF." *Radio Science* 14 (September–October 1979), pp. 803–813.

52. Appreciation is expressed to John Walters of the U. S. Naval Research Laboratory for supplying this information, and to Vilhelm Gregers-Hansen who originally created the figure.

53. Ringwalt, D. L., and F. C. McDonald. "Elevated Duct Propagation in the Tradewinds." *IRE Trans.* AP-9 (July 1961), pp. 377–383.

54. Guinard, N. W., J. Ransone, D. Randall, C. Purves, and P. Watkins. "Propagation Through an Elevated Duct: Tradewinds III." *IEEE Trans.* AP-12 (July 1964), pp. 479–490.

55. Purves, C. G. "Geophysical Aspects of Atmospheric Refraction." Naval Research Laboratory, Washington, D.C. Rep. 7725, June 7, 1974.

56. Hitney, H. V., R. A. Pappert, C. P. Hattan, and C. L. Goodhart. "Evaporation Duct Influences on Beyond-the-Horizon High Altitude Signals." *Radio Science* 13 (July–August 1978), pp. 669–675.

57. Kerr, D. E. (Ed.). *Propagation of Short Radio Waves.* MIT Radiation Laboratory Series, vol. 13 New York: McGraw-Hill, 1951.

58. Brookner, E., E. Ferraro, and G. D. Ouderkirk. "Radar Performance During Propagation Fades in the Mid-Atlantic Region." *IEEE Trans.* AP-46 (July 1998), pp. 1056–1064.

59. Day, J. P., and L. G. Trolese. "Propagation of Short Radio Waves Over Desert Terrain." *Proc. IRE* 38 (February 1950), pp. 165–175.

60. Donohue, D. J., and J. R. Kuttler. "Modeling Radar Propagation Over Terrain." *Johns Hopkins APL Tech. Dig.* 18, no. 2 (1997), pp. 279–287.

61. Dockery, G. D. "Method of Modeling Sea Surface Clutter in Complicated Propagation Environments." *IEE Proc.* 137, no. 2, Pt. F (April 1990), pp. 73–79.

62. Reilly, J. P., and G. D. Dockery. "Influence of Evaporation Ducts on Radar Sea Return." *IEE Proc.* 137, no. 2, Pt. F (April 1990), pp. 80–88.

63. Heimken, H. F. "Low-Grazing-Angle Radar Backscatter from the Ocean Surface." *IEE Proc.* 137, no. 2, Pt. 7 (April 1990), pp. 113–117.

64. Booker, H. G. "Elements of Radar Meteorology: How Weather and Climate Cause Unorthodox Radar Vision Beyond the Geometrical Horizon." *J. IEE* 93, Pt. IIIA (1946), pp. 69–78.

65. Nathanson, F. E. Ref. 30, Sec. 9.5.

66. Hitney, H. V. "Refractive Effects from VHF to EHF, Part B: Propagation Models." Paper no. 4B in *Propagation Modeling and Decision Aids for Communications, Radar, and Navigation Systems.* AGARD-LS-196, NATO, September 1994.

67. Kerr, D. E. Ref. 57.

68. Budden, K. G. *The Wave-Guide Mode Theory of Wave Propagation.* Englewood Cliffs, NJ: Prentice-Hall, 1961.

69. Dockery, G. D. "Modeling Electromagnetic Wave Propagation in the Troposphere Using the Parabolic Equation." *IEEE Trans.* AP-36 (October 1988), pp. 1464–1470.

70. McArthur, R. J. "Propagation Modeling Over Irregular Terrain Using the Split-Step Parabolic Equation Method." *International Conf. Radar 92, IEE Conf. Publication 365,* London, 1992, pp. 54–57.

71. Barrios, A. E. "Parabolic Equation Modeling in Horizontally Inhomogeneous Environments." *IEEE Trans.* AP-40 (July 1992), pp. 791–797.

72. Barrios, A. E. "A Terrain Parabolic Equation Model for Propagation in the Troposphere." *IEEE Trans.* AP-42 (January 1994), pp. 90–98.

73. Hitney. H. V. "Hybrid Ray Optics and Parabolic Equations Methods for Radar Propagation Modeling." *Int. Conf. Radar 92, IEE Conf. Publication 365,* London (1992), pp. 58–61.

74. Richter, J. H. "Electromagnetic Wave Propagation Assessment." *AGARD Highlights* 92/1 (March 1992), pp. 6–14.

75. Hitney, H. V., and J. H. Richter. "Integrated Refractive Effects Prediction System (IREPS)." *Naval Engineers J.* 2 (April 1976), pp. 257–262.

76. Patterson, W.L., et al. *IREPS 3.0 User's Manual.* Naval Ocean Systems Center (NOSC), San Diego, CA, Tech. Document 1151, September 1987. See also Revision PC-2.0, Tech. Document 1874, August 1990.

77. Patterson, W. L. *Advanced Refractive Effects Prediction Sysem (AREPS), Version 1.0 User's Manual.* SPAWAR Systems Center, San Diego, Technical Document 3028, April 1998.

78. Gelsenheyner, S. "Computerized Data Help Predict Anomalous Propagation." *Microwave System News* 11 (April 1982), pp. 45–46.

79. Belobrova, M. V., et al. "Software Suite for Diagnosing USW Propagation Over the Sea." *Radiophysics and Quantum Electronics* 33 (June 1991), pp. 961–965.

80. Skolnik, M. I. "Radar Horizon and Propagation Loss." *Proc. IRE* 45 (May 1957), pp. 697–698.

81. Millman, G. H., and G. R. Nelson. "Surface Wave HF Radar for Over-the-Horizon Detection." *Record of the IEEE 1980 International Radar Conf.* pp. 106–112, IEEE Publication 80CH1493–6 AES, New York.

82. Powers, R. L., L. M. Lewandoski, and R. J. Dinger. "High Frequency Surface Wave Radar—HFSWR." *Sea Technology* 37 (November 1996), pp. 25–32.

83. Headrick, J.M., and M. I. Skolnik. "Over-the-Horizon Radar in the HF Band." *Proc. IEEE* 62 (June 1974), pp. 664–673.

84. Lipa, B. J., and D. E. Barrick. "Extraction of Sea State from HF Radar Sea Echo: Mathematical Theory and Modeling." *Radio Science* 21 (January–February 1986), pp. 81–100.

85. Lipa, B. J., D. E. Barrick, J. Isaacson, and P. M. Lilleboe. "CODAR Wave Measurements From a North Sea Semisubmersible." *IEEE J. Oceanic Engineering* 15 (April 1990), pp. 119–125.

86. Graber, H. C., B. K. Haus, R. D. Chapman, and L. N. Shay. "HF Radar Comparisons with Moored Estimates of Current Speed and Direction: Expected Differences and Implications." *J. Geophys. Res.* 102, no. C8 (August 15, 1997), pp. 18,749–18,766.

87. Walsh, J., B. J. Dawe, and S. K. Srivastava. "Remote Sensing of Iceburgs by Ground-Wave Doppler Radar." *IEEE J. Oceanic Engineering* OE-11 (April 1986), 276–284.

88. Srivastava, S. K., and J. Walsh. "Over-the-Horizon Radar." Chapter 7 in *Remote Sensing of Sea Ice and Ice Bergs,* S. Haykin, E. O. Lewis, R. K. Raney, and J. R. Rossiter (Eds.). New York: John Wiley, 1994.

89. Straiton, A. W., and W. Tolbert. "Anomalies in the Absorption of Radio Waves by Atmospheric Gases." *Proc. IRE* 48 (May 1960), pp. 898–903.

90. Straiton, A. W. "The Absorption and Reradiation of Radio Waves by Oxygen and Water Vapor in the Atmosphere." *IEEE Trans.* AP-23 (July 1975), pp. 595–597.

91. Blake, L. V. "Prediction of Radar Range." In *Radar Handbook,* 1st ed. M. Skolnik (Ed.). New York: McGraw-Hill, 1970, Chap. 2, Sec. 2.7.

92. Baars, J. W. M. "The Measurement of Large Antennas with Cosmic Radio Sources." *IEEE Trans.* AP-21 (July 1973), pp. 461–474.

93. Graf, W., R. N. Bracewell, J. H. Deueter, and J. S. Rutherford. "The Sun as a Test Source for Boresight Calibration of Microwave Antennas." *IEEE Trans.* AP-19 (September 1971), pp. 606–612.

94. Evans, G. E. *Antenna Measurement Techniques.* Boston, MA: Artech House, 1990, Secs. 2.6 and 3.3.3.

95. Schonland, B. F. J. *The Flight of Thunderbolts.* 2nd ed. Oxford, London: Clarendon Press, 1964.

96. Skomal, E. N. "Man-Made Noise in the M/W Frequency Range." *Microwave J.* 18 (January 1975), pp. 44–47.

97. Skomal, E. N. *Man-Made Radio Noise.* New York: Van Nostrand, 1978.

98. Ralston, J., J. Heagy, and R. Sullivan. "Environmental/Noise Effects on VHF/UHF UWB SAR." Institute for Defense Analyses, Alexandria, VA, IDA Paper, P-3385, September 1998.

99. Greene, J. C., and M. T. Lebenbaum. Letter in *Microwave J.* 2 (October 1959), pp. 13–14.

100. Pieramico, A. F., D. A. Rugger, and L. R. Moyer. "The Radar Support System (RSS): A Tool for Siting Radars and Predicting their Performance." *ATC Systems* 2, no. 1 (January/February 1996), pp. 32–40.

101. Lin, C. C., and J. P. Reilly. "A Site-Specific Model of Radar Terrain Backscatter and Shadowing." *Johns Hopkins APL Tech. Digest* 18, no. 3 (July–September 1997), pp. 432–447.

102. Weil, T. A. "Atmospheric Lens Effect; Another Loss for the Radar Range Equation." *IEEE Trans.* AES-9 (January 1973), pp. 51–54.

103. Shrader, W. W., and T. A. Weil. "Lens-Effect Loss for Distributed Targets." *IEEE Trans.* AES-23 (July 1987), pp. 594–595.

104. Goodman, J. M., and J. Aarons. "Ionospheric Effects on Modern Electronic Systems." *Proc. IEEE* 78 (March 1990), pp. 512–528.

105. Brookner, E., Hall, W. M., and R. H. Westlake. "Faraday Loss for L-band Radar and Communcations Systems." *IEEE Trans.* AES-21 (July 1985), pp. 459–469.

106. Flock, W. L. "Propagation Effects on Satellite Systems at Frequencies Below 10 GHz." *NASA Reference Publication 1108* (December 1983), Sec. 2.4.

107. Filer, E., and J. Hartt. "Cobra Dane Wideband Pulse Compression System." *IEEE EASCON '76,* pp. 61-A to 61-M.

PROBLEMS

8.1   In this problem, you may assume a flat earth. (a) What are the elevation angles (in degrees) of the two lowest elevation-pattern multipath interference lobes for an *L*-band (1300 MHz) radar antenna located at a height 50 ft above a perfectly conducting flat surface? (b) What is the height (in meters) of the peak of the first (lowest) lobe above the earth's surface at a range of 3 nmi? (c) Repeat (a) and (b) for an *X*-band (9375 MHz) radar antenna. (d) What can you conclude from the above about the detection of low-altitude targets and radar frequency? (e) When might the *X*-band ship navigation-radar of part (c) have trouble detecting navigation buoys because of multipath lobing, especially if the ship is sailing in calm waters? (You may assume that the echo from the buoy is due to a corner reflector mounted at the top of the buoy 6 m above mean sea level.)

8.2   Under what conditions might the received echo-signal power from a point target located over a flat conducting surface (such as a smooth sea) vary inversely as the eighth power of the range?

8.3   Figure 8.1 illustrates the two paths between the radar and a point target (the direct path and the surface-reflected path). Assume a radar transmits a single short-pulse with pulse duration much less than the time difference between the signals transiting these two paths; i. e., the pulse width is small compared to $2\Delta/c$, where Δ was given by Eq. (8.4). (You may think of the pulse as a delta function that propagates in space.) (a) Sketch the nature of the echo signal received back at the radar after reflection by the point target. (There will be more than one echo returned to the radar.) (b) Derive an expression for the time separation between the pulses? (c) How can this type of short-pulse transmission be used to measure the height of a target?

8.4   The lobes in the elevation pattern due to multipath (such as in Fig. 8.6b) cause loss of target signal when the target is within the null regions of the pattern. What might the radar system designer do to avoid the loss of signal due to the multipath nulls?

8.5   Many air-surveillance radars operate with the antenna beam pointed slightly upward so that the lower half-power point of the elevation pattern is directed along the horizon rather than have the maximum antenna gain along the horizon. Discuss the pros and cons of having the antenna half-power point at the horizon instead of the maximum antenna gain at the horizon.

8.6   What radar characteristics are important for detecting targets at low altitudes?

8.7   The caption of Fig. 8.6b states that this is a plot of the elevation pattern of a 900-MHz radar at an antenna height of 75 ft. Using Eq. (8.8) verify that this is correct based on the antenna pattern shown in Fig. 8.6b. Assume a flat earth and an effective earth's radius of 4/3. (Use of the lowest lobe, however, will probably not give as correct an answer as will the next higher lobe.)

8.8   (a) Show that the distance *d* to the horizon from a radar at a height *h* above a spherical earth of effective radius *ka* is $d = \sqrt{2kah}$. (b) If $k = 4/3$ and the actual radius of the earth $a = 3440$ nmi, what is the distance in nautical miles to the horizon for a radar at a height of 10,000 ft? (c) How much would the distance to the horizon in (b) be increased if the atmospheric refraction were such that $k = 1.8$ instead of 1.33? (d) If a radar had a free-space range exactly equal to the distance to the horizon, d, why might it not be able to see a target located at the horizon?

8.9 The factor k that describes the modification to the earth's radius to account for atmospheric refraction was given by Eq. (8.16) as

$$k = \frac{1}{1 + a(dn/dh)}$$

where a = radius of the earth and dn/dh is the rate of change of the index of refraction with height. (a) What value of dn/dh results in $k = \infty$? (b) What does it mean when $k = \infty$?

8.10 In Sec. 8.4 of the text it is reported[25] that the range error due to atmospheric refraction is 97.5 ft for a target at 40,000 ft and an elevation angle of 3°. Compare this to the answer you get for the range error when using Eq. (8.19), based on different experimental data. (You might need to use Fig. 8.9. Take the surface refractivity to be 313.)

8.11 (a) Why are radars seldom operated at or near a frequency of 22 GHz or near 60 GHz? (b) What is the two-way attenuation of a radar signal (in dB) in the clear atmosphere at a frequency of 5 GHz when propagating 200 nmi (and back) at an elevation of 0°? (c) What is the two-way attenuation when the elevation angle is increased to 5°? (d) Why are aircraft targets not likely to be detected at long range at zero degrees elevation angle?

8.12 A shipboard military radar for detecting low-altitude missile targets over water might be based on either (1) microwave ducted propagation or (2) HF surface-wave propagation to extend the detection range beyond the horizon. Compare these two radar methods with respect to their effectiveness in performing this task. Include in your comparison the effect of their relative size, reliability for detection under all conditions, accuracy, and anything else you think is appropriate.

8.13 Equation 8.11b assumes that the antenna gain remains constant with frequency (which means the beamwidths remain constant), so the received echo signal power P_r varies as λ^{-2}, where λ = wavelength. How would the echo signal power P_r vary with wavelength if the antenna aperture A_e remained constant with frequency?

9

The Radar Antenna

9.1 FUNCTIONS OF THE RADAR ANTENNA

The radar antenna is a distinctive and important part of any radar. It serves the following functions:

- Acts as the transducer between propagation in space and guided-wave propagation in the transmission lines.

- Concentrates the radiated energy in the direction of the target (as measured by the antenna gain).

- Collects the echo energy scattered back to the radar from a target (as measured by the antenna effective aperture).

- Measures the angle of arrival of the received echo signal so as to provide the location of a target in azimuth, elevation, or both.

- Acts as a spatial filter to separate (resolve) targets in the angle (spatial) domain, and rejects undesired signals from directions other than the main beam.

- Provides the desired volumetric coverage of the radar.

- Usually establishes the time between radar observations of a target (revisit time).

In addition, the antenna is that part of a radar system that is most often portrayed when a picture of a radar is shown. (More can be learned about the nature of a radar from a picture of its antenna than from pictures of its equipment racks.)

With radar antennas, big is beautiful (within the limits of mechanical and electrical tolerances and the constraints imposed by the physical space available on the vehicle that carries the antenna). The larger the antenna, the better the radar performance, the smaller can be the transmitter, and the less can be the total amount of prime power needed for the radar system. The transmitting antenna gain and the receiving effective aperture are proportional to one another [as given by Eq. (1.8) or Eq. (9.9)] so that a large transmitting gain implies a large effective aperture, and vice versa. As was mentioned in Chap. 1, in radar a common antenna generally has been used for both transmission and reception.

Almost all radar antennas are directive and have some means for steering the beam in angle. Directive antennas mean narrow beams, which result in accurate angular measurements and allow closely spaced targets to be resolved. An important advantage of microwave frequencies for radar is that directive antennas with narrow beamwidths can be achieved with apertures of relatively small physical size.

In this chapter, the radar antenna will be considered as either a transmitting or a receiving antenna, depending on which is more convenient for explaining a particular antenna property. Results obtained for one may be readily applied to the other because of the reciprocity theorem of antenna theory.[1]

Antenna designers have a variety of directive antenna types from which to choose including the reflector antenna in its various forms, phased arrays, endfire antennas, and lenses. They all have seen application in radar at one time or other. These antennas differ in how the radiated beam is formed and the method by which the beam is steered in angle. Steering the antenna beam can be done mechanically (by physically positioning the antenna) or electronically (by using phase shifters with a fixed phased array). The relatively simple *parabolic reflector,* similar to the automobile headlight or the searchlight, in one form or other has been a popular microwave antenna for conventional radars. As will be discussed later in this chapter, a parabolic reflector can be a paraboloid of revolution, a section of a paraboloid, a parabolic cylinder, Cassegrain configuration, parabolic torus, or a mirror scan (also called polarization-twist Cassegrain). There have also been applications of spherical reflectors, but only for special limited purposes.

The mechanically rotating array antenna was the basis for most of the lower frequency air-surveillance radars that saw service early in World War II. They were eventually replaced by parabolic reflector antennas when radar frequencies increased to the microwave region during and just after World War II. In the 1970s the mechanically scanned planar array antenna reappeared, but at microwave frequencies with slotted waveguide radiators or printed-circuit antennas rather than dipoles. The mechanically scanned planar array is found in almost all 3D radar antennas, low sidelobe antennas, and in airborne radars where the antenna is fitted behind a radome in the nose of the aircraft. (A planar aperture allows a larger antenna to be used inside the radome than is practical with a parabolic reflector.) An example of a very low sidelobe rotating planar array used in the AWACS radar is shown later in Fig. 9.49.

Starting in the mid-1960s the electronically steered phased array antenna began to be employed for some of the more demanding military radar applications. It is the most interesting and the most versatile of the various antennas, but it is also more costly and more complex.

9.2 ANTENNA PARAMETERS

Several of the important antenna parameters were introduced in the discussion of the radar equation in Chapters 1 and 2. Here we review and expand on those we have introduced previously and add some other parameters not yet discussed.

Directive Gain   *Gain* is a measure of the ability of an antenna to concentrate the transmitted energy in a particular direction. There are two different, but related, definitions of antenna gain. One is the *directive gain,* sometimes called *directivity.* The other is the *power gain,* and is often simply called *gain.* Both gain definitions need to be understood by the radar systems engineer since both are used. The directive gain is descriptive of the nature of the antenna radiation pattern, and is usually the definition of gain that interests the antenna engineer. The power gain is related to the directive gain, but it takes account of loss in the antenna itself. The power gain is more appropriate for use in the radar range equation and is therefore of more interest for the radar engineer.

We will denote the directive gain by G_D. (In other literature it is sometimes denoted by D.) The directive gain of a transmitting antenna may be defined as

$$G_D = \frac{\text{maximum radiation intensity}}{\text{average radiation intensity}} \qquad [9.1]$$

where the radiation intensity is the *power per unit solid angle* radiated in the direction (θ, ϕ), and is denoted $P(\theta, \phi)$. Its units are watts per steradian. A plot of the radiation intensity as a function of the angular coordinates is called a *radiation-intensity pattern.* The power density, or *power per unit area,* when plotted as a function of angle is called the *power pattern.* The power pattern and the radiation-intensity pattern are identical when each is plotted on a relative basis; that is, when the maximum is normalized to a value of unity. When plotted on a relative basis, they are called the *antenna radiation pattern,* or simply *radiation pattern.*

Since the average radiation intensity over the entire solid angle of 4π steradians is equal to the total power radiated by the antenna divided by 4π, the directive gain of Eq. (9.1) can be written as

$$G_D = \frac{4\pi(\text{maximum power radiated per unit solid angle})}{\text{total power radiated by the antenna}} \qquad [9.2]$$

This equation indicates the procedure by which the directive gain may be found from the antenna radiation pattern. The maximum power radiated per unit solid angle is obtained by inspection, and the total power radiated is found by integrating the volume under the radiation pattern. From Eq. (9.2) the directive gain can be expressed as

$$G_D = \frac{4\pi\, P(\theta, \phi)_{\text{max}}}{\iint P(\theta, \phi)\, d\theta d\phi} = \frac{4\pi}{B} \qquad [9.3]$$

where B is called the *beam area* and is defined by

$$B = \frac{\iint P(\theta, \phi)\, d\theta d\theta}{P(\theta, \phi)_{\text{max}}} \qquad [9.4]$$

The beam area B is the solid angle through which all the radiated power would pass if the power per unit solid angle over the entire beam area were equal to $P(\theta,\phi)_{max}$. It defines, in effect, an equivalent antenna pattern. If θ_B and ϕ_B are the half-power (radian) beamwidths in the two orthogonal planes, the beam area B is approximately equal to $\theta_B\phi_B$. Substituting into Eq. (9.3) gives

$$G_D \approx \frac{4\pi}{\theta_B\phi_B} \qquad\qquad \text{[9.5a]}$$

This is only an approximation and should be used with caution. Another approximation, an improvement on the above, is

$$G_D \approx \frac{\pi^2}{\theta_B\phi_B} \qquad\qquad \text{[9.5b]}$$

This expression assumes a gaussian beamshape with θ_B and ϕ_B defined as the half-power beamwidths.[2] (This form of the directive gain has been popular with radar meteorologists.) Beamwidths in Eqs. (9.5a and b) are in radians. If the beamwidths are in degrees in Eq. (9.5a), the 4π in the numerator is replaced by 41,253. Equation (9.5a), however, is overly optimistic in that it provides too high a value of directive gain. It applies for a rectangular beam with no sidelobes.

Warren Stutzman,[3] however, recommended that for practical antennas the following "is an excellent approximation for general use:"

$$G \approx \frac{26,000}{\theta_B\phi_B} \qquad\qquad \text{[9.5c]}$$

where the half-power beamwidths are in degrees. He states that "gain [G] is used here instead of directivity, not because of losses . . . [but] to indicate that the formula is appropriate to real antenna hardware, where gain is the parameter used in performance descriptions." This equation states that a one-degree pencil-beam antenna has a gain of 44 dB. It should not be a substitute for more exact analysis, calculation, or measurement; but it is far better than Eqs. (9.5a and b) when nothing further is known about the antenna other than its beamwidths in the principal planes.

Power Gain The power gain, which we denote by G, is similar to the directive gain except that it takes account of dissipative losses in the antenna. (It does not include loss arising from mismatch of impedances or loss due to polarization mismatch.) It can be defined similarly to the definition of directive gain, Eq. (9.2), if the denominator is the net power accepted by the antenna from the connected transmitter, or

$$G = \frac{4\pi(\text{maximum power radiated per unit solid angle})}{\text{net power accepted by the antenna}} \qquad\qquad \text{[9.6a]}$$

An equivalent definition is

$$G = \frac{\text{maximum radiation intensity from subject antenna}}{\text{radiation intensity from a lossless isotropic radiator with the same power input}} \qquad\qquad \text{[9.6b]}$$

Whenever there is a choice, the power gain should be used in the radar equation since it includes the dissipative losses introduced by the antenna. The directive gain, which is always greater than the power gain, is more closely related to the antenna beamwidth. The difference between the two antenna gains is usually small for reflector antennas. The power gain and the directive gain are related by the radiation efficiency ρ_r as follows

$$G = \rho_r G_D \qquad \text{[9.7]}$$

The radiation efficiency is also the ratio of the total power radiated by the antenna to the net power accepted by the antenna at its terminals. The distinction between the two definitions of gain often can be ignored in practice, especially when the dissipative loss in the antenna is small.*

The definitions of power gain and directive gain described in the above were given in terms of a transmitting antenna. Because of reciprocity the pattern of a receiving antenna is the same as the pattern of a transmitting antenna, so the receiving antenna can be described by a gain just as can the transmitting antenna. This is why one can talk of a receiving gain even though gain was defined in terms of a transmitting antenna. The effective aperture of a receiving antenna, on the other hand, has no similar attribute in a transmitting antenna.

It should be kept in mind that the accuracy with which the gain of a radar antenna can be measured is usually about ± 0.5 dB.[4] Thus one should not specify or quote antenna gains to an accuracy much better than this unless there is a reason to be more accurate.

Antenna Radiation Pattern   In the above, antenna gain meant the maximum value. It is also common to speak of gain as a function of angle. Quite often the ordinate of a radiation pattern is given as the gain as a function of angle, normalized to unity. It is then known as *relative gain*. Unfortunately the term *gain* is used to denote both the maximum value and the gain as a function of angle. Uncertainty as to which usage is meant can usually be resolved from the context.

An example of an antenna radiation pattern for a paraboloidal reflector antenna is shown in Fig. 9.1.[5] This particular pattern might not be representative of a well-designed modern high-gain antenna, but it does illustrate the various features that a simple reflector-antenna radiation pattern might have. The *main beam* is shown at zero degrees. The remainder of the pattern outside the main beam is the *sidelobe* region. As the angle increases from the direction of maximum gain, there is an irregularity in this particular radiation pattern at about 22 dB below the peak. This is called a *vestigial lobe* or "shoulder" on the side of the main beam. It does not appear in all radiation patterns and is not desired since it is indicative of phase errors in the aperture illumination. Normally, when errors in the aperture illumination are small, the first sidelobe appears near where this vestigial lobe is indicated rather than where the first sidelobe is indicated in the figure.

The near-in sidelobes generally decrease in magnitude as the angle increases. The decrease is determined by the shape of the aperture illumination (as described in the next

---

*In some types of phased array antennas, the losses in the phase shifters and the power dividing networks can be quite high so that the difference between the power gain and the directive gain can be significant. In such cases, the directive gain is what is usually quoted and the losses are accounted for elsewhere.

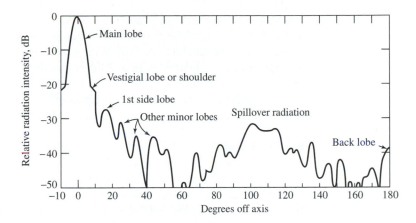

Figure 9.1 Radiation pattern for a particular paraboloid reflector antenna illustrating the main beam and sidelobe radiation.
| (After Cutler et al.,[5] Proc. IRE.)

section). Eventually, the sidelobes due to the aperture illumination are masked by sidelobes due to the random errors in the aperture [Sec. (9.12)]. With a conventional reflector antenna, there usually will be spillover radiation from that part of the feed radiation pattern that is not intercepted by the reflector (in the example of Fig. 9.1, this appears from about 100 to 115°). This radiation pattern also has a pronounced lobe in the backward direction (180°) due to diffraction around the edges of the reflector as well as direct leakage through the mesh reflector surface (if the surface is not solid).

The radiation pattern shown in Fig. 9.1 is plotted as a function of one angular coordinate, but the actual pattern is a plot of the radiation intensity $P(\theta,\phi)$ as a function of two angles. The two angle coordinates commonly employed with a ground-based radar antenna are azimuth and elevation, but other appropriate angle coordinates also can be used.

A complete three-dimensional plot of the radiation pattern can be complicated to display and interpret, and is not always necessary. For example, an antenna with a symmetrical pencil-beam pattern can be represented by a single plot in one angle coordinate because of its circular symmetry. The radiation intensity pattern for rectangular or rectangular-like apertures can often be written as the product of the radiation-intensity patterns in the two coordinate planes; for instance,

$$P(\theta,\phi) = P(\theta,0)\, P(0,\phi) \tag{9.8}$$

Thus when the pattern can be expressed in this manner, the complete radiation pattern in two coordinates can be determined from the two single-coordinate patterns in the θ and in the ϕ planes.

Effective Aperture The effective aperture of a receiving antenna is a measure of the effective area presented to the incident wave by the antenna. As was given previously as Eq. (1.8), the transmitting gain G and receiving effective area A_e of a lossless antenna are related by

$$G = \frac{4\pi A_e}{\lambda^2} = \frac{4\pi \rho_a A}{\lambda^2} \tag{9.9}$$

where λ = wavelength, ρ_a = antenna aperture efficiency, A = physical area of the antenna, and $A_e = \rho_a A$. The aperture efficiency depends on the nature of the current illumination across the antenna aperture. With a uniform illumination, $\rho_a = 1$. The advantage of high efficiency obtained with a uniform illumination is tempered by the radiation pattern having a relatively high peak-sidelobe level. An aperture illumination that is maximum at the center of the aperture and tapers off in amplitude towards the edges has lower sidelobes but less efficiency than the uniform illumination.

Sidelobe Radiation   Sidelobe radiation is radiation from an antenna that is not radiated by the main beam. It is possible in theory to have an antenna radiation pattern with only a single main beam and no sidelobes, but not only is this impractical to achieve, it is not desirable since the main beam would be unusually wide (as mentioned later in Sec. 9.13). In general, the lower the sidelobes, the lower will be the antenna gain and the aperture efficiency, and the greater will be the width of the main beam.

   The sidelobe level of an antenna may be described by the value of the peak sidelobe, the rms value of all the sidelobe radiation (usually of the dB values), or some other suitable measure. The peak sidelobe is a good measure of sidelobe behavior for purposes of the radar system engineer. For a line-source aperture with a uniform illumination, the peak sidelobe (which is the first sidelobe) is 13.2 dB down from the maximum value of the main beam, or a value of -13.2 dB. This is usually too high for most radar applications even though its aperture efficiency is unity, that is, 100 percent.

   Low antenna sidelobes are desired in a radar so as to avoid detecting large targets when they are illuminated by the antenna sidelobes. Any echoes received from the sidelobes will not be indicated by their true angle. (The angle assigned to an echo from a sidelobe will be the angle at which the main beam points at the time of detection rather than the angle of the sidelobe which illuminates the target.) Low sidelobes are also useful for minimizing the effect of strong jamming and interference and to reduce the large clutter echoes that can enter the antenna sidelobes of a high-prf pulse doppler radar (as was discussed in Sec. 3.9).

   The highest sidelobe of an antenna is usually the first sidelobe adjacent to the main beam. A typical parabolic reflector antenna fed from a waveguide horn might have a peak sidelobe of -23 to -28 dB. Peak sidelobes of -40 to -50 dB are possible with specially designed array antennas. Peak sidelobes less than -50 dB are sometimes called *ultralow sidelobes*.

Aperture Efficiency   The aperture efficiency is based on the maximum radiation intensity, which usually occurs at the center of the main beam. It is not like the radiation efficiency that is a measure of the energy dissipated as the signal travels through the antenna. A radiation efficiency less than unity means that energy is lost. On the other hand, an aperture efficiency less than unity means that the radiated energy is redistributed in angle rather than be dissipated. For example, consider a line-source antenna aperture-illumination proportional to $\cos^2(\pi z/2)$, where z is the distance from the center of the aperture, $-D/2 \leq z \leq +D/2$, and D is the aperture dimension. In this example the amplitude of the aperture illumination is one-half cycle of the square of the cosine over the aperture. Its radiation pattern has a first sidelobe of -32 dB compared to the -13.2 dB of a uniform illumination. The gain is reduced by 0.67 and its beamwidth is increased by 1.63 compared to the radiation pattern from a uniformly illuminated antenna. With a scanning antenna, as might be used in a surveillance radar, the reduction in gain is compensated in part by the increased number of pulses received because of the wider beamwidth.

There are other antenna properties that might be more important to the radar systems engineer than the aperture efficiency. Thus the aperture efficiency might be reduced in order to obtain other benefits. These might include low-sidelobe levels, a radar antenna beam which maximizes the radiated energy within a specified angular region,[6,7] shaped beams such as the cosecant-squared pattern, and monopulse antenna patterns optimized for good angle-tracking accuracy. The aperture efficiency might be important to the antenna designer, but to the radar systems engineer it is often something to be traded in order to achieve some more important antenna characteristic.

Polarization   The polarization of an electromagnetic wave is defined by the orientation of the electric field. Most radar antennas are *linearly polarized,* with the orientation of the electric field being either horizontal or vertical. Air-surveillance radars generally employ horizontal polarization. Most tracking radars are vertically polarized. *Circular polarization* occurs when the electric field rotates at a rate equal to the RF frequency. It is sometimes used to enhance the detectability of aircraft targets in the midst of rain (Sec. 7.8). There is also *elliptical polarization,* where the electric field also rotates at the RF frequency; but unlike circular polarization, the amplitude of the elliptically polarized electric field varies during the rotation period. Circular and linear polarizations are special cases of elliptical polarization. Although some radar applications seem to prefer a particular polarization (based sometimes on tradition), in many applications there is often not a strong requirement for one polarization over the other. Even the use of circular polarization for rain is not absolutely necessary since orthogonal linear polarizations can be used instead.

9.3   ANTENNA RADIATION PATTERN AND APERTURE ILLUMINATION

The electric-field intensity $E(\phi)$ (units of volts per meter) produced by the electromagnetic radiation emitted from a line-source antenna is a function of the amplitude and phase of the distribution of current across the aperture. The angle ϕ, shown in Fig. 9.2, is with respect to the normal to the center of the antenna aperture. $E(\phi)$ may be found by adding vectorially the individual contributions radiated from the various current elements that constitute the line-source antenna aperture. The mathematical summation at a point in space of all the contributions radiated by the current elements contained within the aperture gives the field intensity in terms of an integral that is difficult to evaluate in the general case.[8] It reduces, however, to a conventional inverse Fourier transform when the distances from the antenna are large enough for the radiation to be considered a plane wave. This occurs in the so-called *far field* of the antenna, when the range $R > D^2/\lambda$, where D is the size of the aperture and λ is the radar wavelength, with D and λ being in the same units. (Although antenna engineers call this region the far field, optical physicists call it the *Fraunhofer region.*) In radar, the target is almost always in the far field.

Figure 9.2 Coordinate system for a line source lying along the z axis. Field-intensity pattern E(φ) lies in the yz plane.

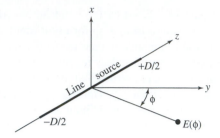

Electric-Field Intensity and the Fourier Transform   In Fig. 9.2, the width of the aperture in the z coordinate is D and the angle in the y,z plane as measured from the y axis is ϕ. The aperture is a line source, or linear antenna, in that its dimension in the z-coordinate is much larger than its dimension in the x-coordinate, and the latter (x-coordinate dimension) is small compared to a quarter wavelength. We are interested in the electric-field intensity in the y,z plane. Assuming $D >> \lambda$ and $R > D^2/\lambda$, the variation of the electric field intensity with angle ϕ in the far field is proportional to

$$E(\phi) = \int_{-D/2}^{+D/2} A(z) \exp\left(j2\pi\frac{z}{\lambda}\ \sin\ \phi\right) dz \qquad [9.10]$$

where the *aperture illumination* $A(z)$ is the current at a distance z from the center of the radiating line-source antenna. The aperture illumination, also called the *current distribution,* can be a complex quantity including both an amplitude $|A(z)|$ and a phase $\psi(z)$, so that $A(z) = |A(z)| \exp [j\psi(z)]$. The phase of the aperture illumination becomes important if the beam is to be steered in a direction other than broadside or if the antenna is focused (which is rare in radar applications). Here we assume $\psi(z) = 0$.

The electric-field intensity given by Eq. (9.10) applies to a one-dimensional radiating antenna lying along the z axis. If the aperture were two-dimensional and situated in the x,z plane, the aperture illumination, $A(z)$, would be the integral of $A(x,z)$ over the variable x. The electric field in the far field is a function only of the angle ϕ and does not depend on the range R except for the usual $1/R$ factor. The expression of Eq. (9.10) can be extended to two dimensions by considering the aperture illumination to be a function of x as well as z. The plot of the magnitude of the electric-field intensity $|E(\theta,\phi)|$ is called the *electric-field intensity pattern* of the antenna. The plot of the square of the field intensity $|E(\theta,\phi)|^2$ normalized to unity is the *power pattern* or the *radiation pattern.*

The integral describing the electric-field intensity of a radiating source in the far field is an inverse Fourier transform, which most electrical engineers are familiar with since the Fourier transform relates the frequency spectrum and waveform of a temporal signal. The Fourier transform of a time waveform $s(t)$ is the frequency spectrum

$$S(f) = \int_{-\infty}^{+\infty} s\ (t) \exp\ (-j2\pi ft)\ dt \qquad [9.11]$$

and the inverse Fourier transform of the spectrum $S(f)$ is the time waveform

$$s(t) = \int_{-\infty}^{+\infty} S(f) \exp{(j2\pi ft)}\, df \qquad [9.12]$$

This happens to be of the same form as the field-intensity expression of Eq. (9.10). Since the aperture illumination is zero beyond $z = \pm D/2$, the limits of Eq. (9.10) can extend from $-\infty$ to $+\infty$ to make it consistent with the inverse Fourier transform of Eq. (9.12). Thus the mathematical model that relates the time waveform and its spectrum is analogous to the mathematical model that relates the radiated field-intensity and the aperture illumination. The (spatial) antenna pattern $E(\phi)$ is related mathematically to the (temporal) waveform $s(t)$, and the aperture illumination $A(z)$ is related to the spectrum $S(f)$. What is known from signal theory about the role of the spectrum $S(f)$ in determining the nature of the signal $s(t)$ is applicable to how the aperture illumination $A(z)$ affects the radiation in space $E(\phi)$. The converse is also true.

In the above, the antenna was viewed as transmitting. As was stated earlier, the property of antenna reciprocity means that the variation of the radiated field on transmit as a function of angle will be similar to the variation of the received signal as a function of angle when the same antenna is used for both transmit and receive.

In the remainder of this section, the antenna field intensity will be examined for various analytical aperture illuminations. The phase distribution is assumed zero or constant so that only the effects of the amplitude variation across the aperture need be considered. The aperture over which the integral of Eq. (9.10) is taken is defined as the projection of the antenna surface on a plane perpendicular to broadside. In this formulation of antenna radiation based on the inverse Fourier transform of Eq. (9.10) it does not matter whether the illumination is produced by a reflector antenna, a lens, or an array so long as the illumination is that in the plane of the aperture.

One-Dimensional Aperture Illumination   Consider in Fig. 9.2 a uniform (constant) aperture illumination extending from $-D/2$ to $+D/2$, and zero outside. This represents the illumination across a line source or the projected illumination in one of the principal planes of a uniformly illuminated rectangular aperture. If the constant value of the amplitude of the aperture illumination is A_0, the variation of the electric-field intensity as a function of angle ϕ is computed from Eq. (9.10) as

$$E(\phi) = A_0 \int_{-D/2}^{+D/2} \exp{\left(j2\pi\frac{z}{\lambda}\sin\phi\right)}\, dz = \frac{A_0 D \sin{[\pi(D/\lambda)\sin\phi]}}{\pi(D/\lambda)\sin\phi} \qquad [9.13]$$

Normalizing to make $E(0) = 1$ results in

$$E(\phi) = \frac{\sin[\pi(D/\lambda)\sin\phi]}{\pi(D/\lambda)\sin\phi} \qquad [9.14]$$

This is of the form $(\sin x)/x$. The square of the above is the antenna radiation pattern or power pattern. It is shown by the solid curve in Fig. 9.3. The first sidelobe adjacent to the main beam is 13.2 dB below the peak value of the main beam. The angular distance between the two nulls defining the main beam is $2\lambda/D$ radians, and the beamwidth as

Figure 9.3 Solid curve is the radiation pattern produced by a uniform aperture illumination of a line source of dimension D. The dashed curve is the radiation pattern of an aperture illumination proportional to the cosine function. Both curves are normalized to unity maximum gain.

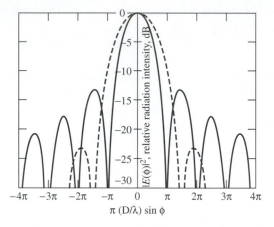

measured between the half-power points is $0.88\lambda/D$ radians, or $51\lambda/D$ degrees. (Quite often the half-power beamwidth for this uniform illumination is approximated by λ/D radians.) The field-intensity pattern of Eq. (9.14) is positive over the entire main lobe, but changes sign in passing through the first zero, returning to a positive value in passing through the second zero, and so on. The odd-numbered sidelobes are therefore out of phase with the main beam, and the even-numbered ones are in phase.

The normalized field-intensity pattern for an aperture illumination $A(z)$ proportional to one-half cycle of the cosine function, given by $\cos(\pi z/D)$, with $|z| \leq D/2$, is from Eq. (9.10)

$$E(\phi) = \frac{\pi}{4}\left[\frac{\sin(\psi + \pi/2)}{\psi + \pi/2} + \frac{\sin(\psi - \pi/2)}{\psi - \pi/2}\right] = \frac{\pi^2 \cos\psi}{\pi^2 - 4\psi^2} \tag{9.15}$$

where $\psi = \pi(D/\lambda)\sin\phi$. The square of the above is shown as the dashed curve in Fig. 9.3. In this figure, the peak gains of both patterns (for the uniform and the cosine illuminations) are normalized to unity. In reality, however, the maximum gain of the pattern from the cosine function is 0.9 dB less than that of the maximum gain of the uniform illumination. Notice that the peak sidelobe of the pattern from the cosine function is much lower than the peak sidelobe from the uniform illumination. Its beamwidth, however, is increased and its maximum gain decreased. The greater the taper of the aperture illumination as it approaches the edges of the antenna aperture, the lower will be the sidelobe level, but at the cost of a wider beamwidth and a lower maximum gain.

Table 9.1 lists some on the characteristics of the radiation patterns produced by various one-dimensional (line-source) antenna aperture-illuminations.[8] These aperture illuminations are expressed in analytic form so that the solution of the inverse Fourier transform of Eq. (9.10) can be conveniently determined. They are not necessarily what might be employed by the antenna designer, but they do illustrate how variations in the form of the aperture illumination affect the antenna pattern. More complicated distributions, which cannot be found from available tables of Fourier transforms, may be determined by computer computation. Using the Schwartz inequality, Silver[8] showed that the uniform

Table 9.1   Radiation-pattern characteristics produced by various aperture distributions

$\lambda = wavelength;\ D = aperture\ wideth$

Type of distribution, $\|z\| < 1$	Relative gain	Half-power beamwidth, deg	Intensity of first sidelobe, dB below maximum intensity
Uniform; $A(z) = 1$	1	$51\lambda/D$	13.2
Cosine; $A(z) = \cos^n(\pi z/2)$:			
$n = 0$	1	$51\lambda/D$	13.2
$n = 1$	0.810	$69\lambda/D$	23
$n = 2$	0.667	$83\lambda/D$	32
$n = 3$	0.575	$95\lambda/D$	40
$n = 4$	0.515	$111\lambda/D$	48
Parabolic; $A(z) = 1 - (1 - \Delta)z^2$:			
$\Delta = 1.0$	1	$51\lambda/D$	13.2
$\Delta = 0.8$	0.994	$53\lambda/D$	15.8
$\Delta = 0.5$	0.970	$56\lambda/D$	17.1
$\Delta = 0$	0.833	$66\lambda/D$	20.6
Triangular; $A(z) = 1 - \|z\|$	0.75	$73\lambda/D$	26.4
Circular; $A(z) = \sqrt{1 - z^2}$	0.865	$58.5\lambda/D$	17.6
Cosine-squared plus pedestal;			
$0.33 + 0.66 \cos^2(\pi z/2)$	0.88	$63\lambda/D$	25.7
$0.08 + 0.92 \cos^2(\pi z/2)$, Hamming	0.74	$76.5\lambda/D$	42.8

aperture illumination produces the maximum gain. When either the $\cos^n$ (cosine raised to the nth power) or the parabolic distributions shown in this table are examined, it is seen that, as has mentioned before, the more tapered the illumination the lower is the peak sidelobe, the wider the beamwidth, and the lower the maximum gain. Note that relative gain in Table 9.1 is the same as the aperture efficiency ρ_a defined previously by Eq. (9.9).

The cosine-squared on a pedestal listed one line from the bottom of the table is a representative illumination for conventional antennas. This is close to the so-called 25-dB Taylor illumination discussed later in Sec. 9.11. The Hamming illumination produces the lowest peak sidelobe for a cosine-squared on a pedestal illumination (in this case the peak lobe is not the one closest to the main beam). The reduction of the spatial sidelobes of an antenna by a tapered aperture illumination is similar to the windowing employed in digital processing to reduce filter sidelobes and to the filter weighting in pulse compression to reduce the time sidelobes of the compressed pulse (Sec. 6.5).

Having the proper aperture illumination is an important requirement for achieving suitable antenna radiation patterns. There is more to consider, as will be discussed in Sec. 9.13, which is on the subject of very low sidelobes.

Two-Dimensional Aperture Illumination   To extend the above discussion of radiation from a line source to an aperture with two dimensions, the angles θ, ϕ are defined by the coordinate system shown in Fig. 9.4. The antenna is in the x,y plane. This is the coordinate

Figure 9.4 Coordinate system for a planar antenna lying in the *xy* plane.

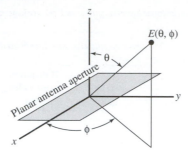

system usually used by antenna theorists, and differs from the azimuth-elevation coordinates preferred by radar engineers. With the coordinate system of Fig. 9.4, the two-dimensional field-intensity pattern from an aperture is given by[8]

$$E(\theta,\phi) = \iint A(x,y) \exp[j(2\pi/\lambda) \sin\theta (x\cos\phi + y\sin\phi)] \, dxdy \qquad [9.16]$$

This integral is not easy to solve analytically, so that numerical techniques are sometimes used. Equation (9.16) is easier to use when the aperture illumination is separable; that is, when $A(x,y) = A_1(x) A_2(y)$, where $A_1(x)$ is the projection of the aperture illumination along the x axis and $A_2(y)$ is the projection along the y axis. Silver[9] showed that when aperture illuminations are separable, the field-intensity patterns are also separable. Thus the two-dimensional pattern of Eq. (9.16) can be written as the product of the one-dimensional patterns in the two principal planes, as was indicated earlier by Eq. (9.8). One principal plane occurs when $\phi = 0$, and the other when $\phi = 90°$ in the coordinate system of Fig. 9.4. When the patterns are separable, the pattern in the xz plane is the same as would be produced by a linear antenna with aperture illumination $A_1(x)$, and the pattern in the yz plane is the same as that produced by the linear aperture illumination $A_2(y)$.

Circular Aperture[10–12] Instead of the rectangular coordinates used in Eq. (9.16), polar coordinates are used to describe the aperture illumination $A(r,\theta)$ of a circular aperture. The radial distance from the center of the circular aperture is r, and θ is the angle measured in the plane of the aperture with respect to a reference. When the aperture illumination depends only on the radial distance and is independent of the angle θ, the field intensity is proportional to

$$E(\theta) = 2\pi \int_0^{r_0} A(r) J_0[2\pi(r/\lambda) \sin\phi] \, rdr \qquad [9.17]$$

where $A(r)$ is the aperture illumination as a function of the radial distance, r_0 is the radius of the circular aperture, and ϕ is the angle with respect to the normal to the circular aperture. If the aperture illumination is uniform [$A(r) = 1$], this reduces to

$$E(\theta) = 2\pi r_0^2 J_1(\xi)/\xi \qquad [9.18]$$

where $\xi = 2\pi(r_0/\lambda) \sin\phi$ and $J_1(\xi)$ is the first-order Bessel function. A normalized plot of the square of this equation is shown in Fig. 9.5. The first sidelobe is 17.5 dB below

Figure 9.5 Radiation pattern for a uniformly illuminated circular aperture of radius r_0.

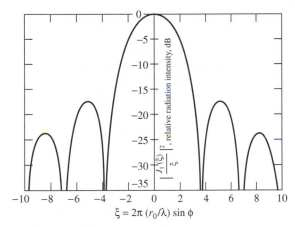

$$\xi = 2\pi \,(r_0/\lambda)\,\sin\phi$$

the main-beam maximum, and the beamwidth in degrees is $58.5\lambda/D$. Note that this is the same as from a one-dimensional (line source) antenna with a circular aperture-illumination (as was listed in Table 9.1) since the projection of the uniform circular-aperture-illumination on its diameter is a circular one-dimensional illumination.

Tapering of the amplitude illumination in the radial dimension of a circular aperture reduces the peak sidelobe, but at the expense of broader beamwidth and less antenna gain. Consider the family of circularly symmetrical aperture illuminations[8] given by

$$A(r) = [1 - (r/r_0)^2]^P \qquad\qquad \text{[9.19a]}$$

where $p = 0, 1, 2,\ldots$. This aperture illumination depends only on r and not on θ. The field-intensity pattern is[12]

$$E(\theta) = \pi r_0^2 \, 2^p p! \; \frac{J_{p+1}(\xi)}{\xi^{p+1}} \qquad\qquad \text{[9.19b]}$$

where ξ is defined as it was for Eq. (9.18). When $p = 0$, the illumination is uniform and the radiation pattern reduces to that given by Eq. (9.18). For $p = 1$, the gain is 0.75 of the gain of a uniformly illuminated aperture, the half-power beamwidth is broadened to $72.6\lambda/D$, and the first sidelobe is 24.6 dB below the maximum. The sidelobe level is 30.6 dB down for $p = 2$, and the gain is 0.56 relative to that of the uniform illumination.

Aperture Blocking[13]   An obstacle in front of an antenna can alter the effective aperture illumination and distort the radiation pattern. This is called *aperture blocking* or *shadowing*. Examples are the feed and its supports in a reflector antenna; masts on board a ship; and nearby buildings, trees, and other obstructions to a land-based radar. The subreflector, as well as the feed, of a Cassegrain antenna (Sec. 9.4) also blocks the aperture illumination. Aperture blocking lowers the antenna gain, raises the sidelobes, and fills in the nulls. It would not be unusual for a low-sidelobe antenna with -30 to -40 dB peak sidelobe level to be increased to a sidelobe level of from -15 to -20 dB when its beam is obstructed.

The effect of aperture blocking can be approximated by subtracting from the antenna pattern of the undisturbed aperture the antenna pattern produced by the shadow of an obstacle. This procedure is possible because of the linearity of the Fourier transform that relates the aperture illumination and the radiation pattern. An example of the effect of aperture blocking caused by the feed in a paraboloid-reflector antenna is illustrated in Fig. 9.6.[14] This relatively simple method for determining the effect of blocking a portion of the radiated energy is only an approximation.

The reduction of the antenna gain η due to the blockage of a circular obstacle of radius r_b placed in front of a circular aperture of radius r_0 whose aperture illumination is given by Eq. (9.19a) is

$$\eta = \left[1 - \delta^2 \{ [(1 - \delta^2)^p] p + 1 \} \right]^2 \qquad [9.20]$$

where $\delta = r_b/r_0$ and p is defined by Eq. (9.19a). This equation, due to Sciambi,[15] would apply to a circular feed at the focus of a parabolic reflector or to the subaperture of a Cassegrain antenna. (Sciambi included in his paper the aperture illumination of Eq. (9.19a) on a pedestal, but it was omitted here for simplicity.) When $\delta = r_b/r_0$ is small, then $\eta \approx [1 - (p + 1)\delta^2]^2$. Based on this approximation, the new sidelobe level due to aperture blocking can be written as

$$sl_b = \left(\frac{\sqrt{sl} + (p + 1)\delta^2}{1 - (p + 1)\delta^2} \right)^2 \qquad [9.21]$$

where sl is the original sidelobe level relative to the main-beam peak when there is no aperture blockage. The sidelobes sl_b and sl are power ratios less than one. In obtaining the above, the maximum value of the obstacle pattern based on Eq. (9.19b) was taken as $\pi r_b^2/2$ (with $p = 0$ and replacing r_0 with r_b).

As an example, consider a parabolic aperture illumination as in Eq. (9.19a) with $p = 1$ and $\delta = r_b/r_0 = 0.1$ (one-percent of the antenna area is blocked). The reduction in gain due to blockage from Eq. (9.20) is 0.96 (about 0.2 dB), and the peak sidelobe of the antenna from Eq. (9.21) is increased from -24.6 dB to -21.9 dB. When $\delta = 0.2$ (4 percent blockage), the peak sidelobe is increased to -16.4 dB. Thus with antennas having

Figure 9.6   Effect of aperture blocking caused by an obstacle (such as a feed) in a parabolic-reflector antenna.
I (From C. Cutler[5] Proc. IRE.)

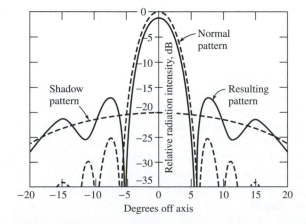

conventional sidelobes of −23 to −28 dB, the aperture blockage should not be more than about 1 percent in order to maintain decent sidelobes. With low peak sidelobes of −40 dB, the aperture blockage should be less than 0.05 percent, and with ultralow sidelobe levels (Sec. 9.13), no aperture blockage at all can be tolerated.

9.4   REFLECTOR ANTENNAS

The parabola, sketched in Fig. 9.7, works well as a reflector of electromagnetic energy and has been the basis for many radar antennas. The parabolic surface is illuminated by a source of radiated energy called the *feed,* which is placed at the focus of the parabola. The parabola converts the spherical wave radiated from the feed to a plane wave because (1) any ray radiated from the focus that intersects the parabolic surface is reflected in a direction parallel to the axis of the parabola, and (2) the distance traveled by any ray from the focus to the parabola and by reflection to a plane perpendicular to the parabola's axis is the same for all rays no matter what angle they emanate from the focus. (This description is only an approximation based on geometric optics. In practice a plane wave does not emerge after reflection until the wave travels a sufficient distance to be in the far field of the antenna, but this need not be of concern at present.)

Paraboloid   There are several ways in which the parabola is used for antennas. Rotating the parabolic curve shown in Fig. 9.7 about its axis produces a surface which is a parabola of revolution called a *circular parabola;* or, more usually, a *paraboloid.* When properly illuminated by a source at the focus, the paraboloid generates a nearly symmetrical pencil-beam antenna pattern. This has been a popular antenna for tracking radars. (The paraboloid reflector is sometimes called a *dish.*)

Section of a Paraboloid   Instead of a circular shape, consider the reflector antenna to have an elliptical shape (as though an elliptical section were cut from the symmetrical paraboloid). This produces an asymmetrical beam shape known as a *fan beam.* It is often used

Figure 9.7   Contour of a parabolic-reflector antenna.

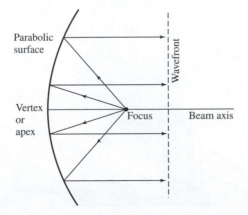

for two-dimensional (range and azimuth angle) air-surveillance radars, as shown in the example of Fig. 9.8. Sometimes this type of asymmetrical antenna has a different curvature in the horizontal and vertical planes so as to shape the beams differently in the two planes. This might be the case for an antenna used with an air-surveillance radar where the azimuth beamwidth is required to be narrow, and the vertical beam is shaped to provide a broader coverage, such as the cosecant-squared pattern discussed in Sec. 9.11.

Feeds for Paraboloids   The "ideal" feed for a paraboloid reflector would be a source at the focus with a radiation pattern that (1) had no phase variation with angle, (2) produced on the reflector surface the desired aperture amplitude illumination, and (3) had a directivity that allowed all of the feed radiation to be intercepted by the aperture without spillover. The ideal may be approximated but never fully accomplished. The radiation pattern produced by the feed is called the *primary pattern* and that radiated by the aperture is called the *secondary pattern*.

A simple half-wave dipole with a parasitic reflector to direct most of its energy towards the antenna aperture can be used as the feed for a paraboloid, Fig. 9.9a. A dipole, however, is of limited utility as a reflector feed since it is difficult to shape the primary pattern and it is limited in power handling, especially at the higher microwave frequencies. An open-ended waveguide is usually preferred over a dipole for microwave-radar reflector feeds. A circular paraboloid, for example, might be fed by a circular, open-ended waveguide operating in the TE_{11} mode. A rectangular guide operating in the dominant TE_{10} mode, however, does not result in a perfectly symmetrical secondary pattern since its dimensions in the E and H planes are different.

When more directivity is required from the feed than is available from an open-ended waveguide, some form of waveguide horn can be used. A horn can be made to provide the asymmetrical feed illuminations (the primary pattern) for a fan beam generated by a

Figure 9.8   Signaal LW08 2D (Jupiter) *L*-band fan-beam air and surface surveillance radar with an elliptical shape fan-beam antenna fed by a horn (on the left). The reflector surface is a mesh, the horizontal antenna mounted on the top is for IFF (identification friend or foe). At the middle right can be seen a fin which is added to the back of the antenna to counterbalance the wind forces on the reflector in the position of the worst yawing moment.
(Courtesy Hollandse Signaalapparten B. V., The Netherlands.)

Figure 9.9   Examples of the placement of the feeds in parabolic reflectors. (a) Rear feed using half-wave dipole with parasitic reflector; (b) rear feed using horn; (c) front feed using horn.

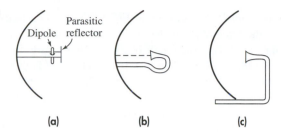

(a)   (b)   (c)

section of a paraboloid. Feeds for reflector antennas can come in many varieties in addition to the simple horn and open-ended waveguide.[16]

As an approximate rule of thumb, the intensity of the radiation from the feed toward the edge of the reflector should be about 10 dB down from the maximum radiation. The aperture illumination at the edges of the reflector surface will be even less than this because of the longer path length from the feed to the edge compared to the path length from the feed to the center of the reflector. When the primary feed pattern is 10 dB down toward the edges, the first sidelobe in the secondary pattern usually is in the vicinity of 22 to 25 dB.

The *f/D ratio* of a reflector antenna is the focal length *f* divided by the aperture diameter *D*. Most practical reflector antennas have *f/D* ratios ranging from 0.3 to 0.5. A small ratio means a deep reflector that is difficult to illuminate properly. A large *f/D* ratio results in a shallow reflector. The shallow reflector is easier to support and to position mechanically, but the feed must be supported farther from the reflector. The farther the feed is from the reflector, the narrower must be the primarily pattern (to avoid spillover loss) and the larger must be the feed. A large *f/D* is preferred for tracking radars and when the beam must be offset in angle by displacing the feed from the focus.

Calculations of the antenna efficiency based only on the aperture illumination established by the primarily pattern from a feed as well as the spillover indicate theoretical efficiencies of about 80 percent compared to an ideal uniformly illuminated aperture without spillover. In practice, however, phase variations across the aperture, poor polarization characteristics, and antenna mismatch reduce the overall antenna efficiency to the order of 55 to 65 percent for ordinary paraboloidal-reflector antennas.

Feed Support   The dipole and the waveguide horn (or open-ended waveguide) can be arranged to feed the paraboloid from the rear as shown in Fig. 9.9a and b. Other types of rear-feed systems have also been used. Figure 9.9c illustrates what is called a front feed using a horn radiator at the focus. It is well suited for supporting horn feeds, but the supports obstruct the aperture.[17] These obstructions due to the feed and its supports reduce the antenna gain, increase the sidelobes, and cause some of the radiated energy to be cross polarized. Analytical expressions and design curves for determining the adverse effects of aperture blockage have been proposed.[18] There is also an impedance mismatch at the feed due to some of the energy reflected by the antenna surface re-entering the feed and its transmission line. Both aperture blockage and mismatch due to reflection can be eliminated by the offset feed.

Offset-Fed Reflector[19,20]   As seen in Fig. 9.10, the feed in this arrangement is placed at the focus of the parabola, but the horn is tipped (upwards in the figure) with respect to the parabola's axis. The lower half of the parabolic surface is removed, leaving that portion shown by the solid curve in the figure. The feed is therefore outside the path of the energy reflected from the antenna surface. There is no pattern deterioration due to aperture blocking nor is there any significant amount of radiation intercepted by the feed to produce an impedance mismatch (high VSWR).

Although the offset feed eliminates aperture blockage and mismatch of rear and front feeds, it introduces problems of its own. Its *f/D* ratio (focal length divided by diameter) is greater than that of conventional paraboloids so that the feeds are larger. Furthermore, this type of antenna is generally more difficult to support mechanically. Because of the increased asymmetry of this geometry, when illuminated by a conventional linearly polarized feed, cross-polarized radiation lobes are produced which can reduce radar system performance by indicating false targets. It has been said[19] that when circular polarization is used, the offset-fed reflector does not depolarize the radiated field, but the beam will be squinted relative to the electrical boresight of the antenna. With the increased importance of operating satellite communications with dual orthogonal polarizations, there have been improvements made in the cross-polarization properties of offset-fed reflector antennas.[21,22] Cross-polarized sidelobes of a single-reflector offset-fed antenna can be made comparable to the co-polarized sidelobes, and can be much lower if a dual-reflector antenna is used.

Cassegrain Antenna   This is a dual-reflector antenna, Fig. 9.11, with the feed at or near the vertex of the parabola rather than at its focus. The larger (primary) reflector has a parabolic contour and the (secondary) subreflector has a hyperbolic contour. One of the two foci of the hyperbola is the real focal point of the system. The feed is located at this point, which can be at the vertex of the parabola or, more usually, in front of it. The other focus is a virtual focal point and is located at the focus of the primary parabolic surface. Parallel rays coming from a target are reflected by the parabola as a convergent beam and are re-reflected by the hyperbolic subreflector so as to converge at the position of the feed. There exists a family of hyperbolic surfaces that can serve as the subreflector. The larger the subreflector, the nearer it will be to the primary reflector and the shorter will be the axial dimension of the antenna assembly. A large subreflector, however, results in large aperture blocking, which may not be desirable. A small subreflector reduces aperture blocking, but it has to be supported at a greater distance from the primary reflector.

The chief advantage of the Cassegrain configuration is that the feed at or near the apex of the parabola does away with the need for long transmission lines out to a feed at

Figure 9.10   Parabolic reflector with offset feed.

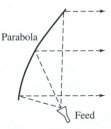

Parabola

Feed

Figure 9.11   (a) Cassegrain antenna showing the hyperbolic subreflector, the feed at the vertex of the main parabolic reflector, and the paths of the rays from the feed; (b) geometry of the Cassegrain antenna.

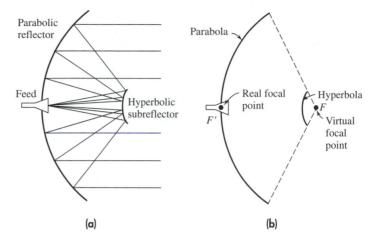

the normal focus of the parabola. Furthermore, it allows greater flexibility in the size of the feed system. It is popular for monopulse tracking radars since the microwave hardware for generating the sum and difference patterns can be located behind the reflector without increasing aperture blocking. It has also been a good structure for experimental systems that use radar and other electromagnetic systems for different purposes, such as in the MIT Lincoln Laboratory Haystack Hill microwave research system. In that system there were separate RF systems for radar, radiometer, and space communications which operated at various frequencies. Each was constructed in replaceable modules, 8 by 8 by 12 ft in size, which were mounted directly behind the primary reflector.[23]

The antenna noise temperature (Sec. 11.2) of a Cassegrain configuration is usually smaller than that of a conventional front-focus antenna since there are no lossy transmission lines between the receiver and the feed. Also, the sidelobes caused by the spillover of the feed radiation from the subreflector illuminate the cold sky rather than the warm earth. Low antenna noise temperature is important for antennas used for radio astronomy or space communications, but it is generally not an issue in radar since extremely low-noise receivers are not always desirable, especially for military applications.

Aperture Blocking in the Cassegrain Antenna   The hyperbolic subreflector of the Cassegrain antenna causes aperture blocking. Aperture blocking can be reduced by decreasing the size of the subreflector. This requires that the feed be made more directive or moved closer to the subreflector in order to minimize the spillover from the subreflector. A more directive feed means a larger feed that partially shadows the primary reflector and contributes to blockage. Thus blockage includes the obstacle presented by the feed as well as the subreflector. Minimum total aperture blocking occurs when the area of the subreflector and the projected area of the feed are equal.[24]

Polarization-Twist Reflector[24]   The technique diagrammed in Fig. 9.12 can reduce aperture blocking if the application permits the antenna to operate with only a single polarization. The subreflector consists of a horizontal grating of wires, called a *transreflector*.

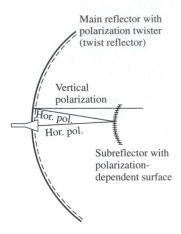

Figure 9.12 Polarization-twist Cassegrain antenna. Aperture blocking by the subreflector is reduced with this design.

It will pass vertically polarized radiation with negligible attenuation but will reflect horizontal polarization radiated by the feed. At the primary reflector the horizontally polarized radiation reflected by the subreflector is rotated 90° by the *twist reflector*. The twist reflector consists of wires oriented 45° to the incident polarization and placed one-quarter wavelength from the reflector's surface. Half the energy incident on the wires oriented at 45° passes through the grating; the other half is reflected. The one-quarter wavelength spacing of the wire grid from the reflector surface results in half-wavelength total travel of the component reflected from the surface. When combined with the component reflected from the wire grid, the resultant polarization is rotated 90°, and is therefore vertically polarized. This vertically polarized component is perpendicular to the horizontal-wire grid of the subreflector and passes through with negligible attenuation. The twist reflector as described above is narrowband, but it can be made to have very wide bandwidths.[25,26]

Gregorian Antenna The Gregorian antenna uses a dual-reflector similar to the Cassegrain except that the subreflector is an ellipsoid with one of its foci at the focus of the primary paraboloidal reflector. The ellipsoid lies beyond the focus of the paraboloid, instead of closer to it as does the subreflector of the Cassegrain. Also in the Gregorian configuration, the concave side of the secondary reflector ellipsoid faces the primary reflector, which differs from the Cassegrain in which the secondary reflector has its convex side facing the feed. The Gregorian has not seen as much application to radar as has the Cassegrain. There are other multireflector antennas; but they also have not had significant radar application.

Parabolic Cylinder Another method for obtaining an asymmetrical antenna pattern is to use a *parabolic cylinder,* shown in Fig. 9.13. This antenna surface is generated by moving the parabolic contour parallel to itself. A line source, such as a linear array, located at the focus of the cylinder is used to illuminate the parabolic-cylinder reflector (the focus is a line rather than a point). The beamshape and beamwidth in the plane containing

Figure 9.13   Example of a vertically oriented parabolic cylinder reflector antenna with a linear-array feed. This is the antenna used for the U.S. Marine Corps AN/TPS-63 air-surveillance radar.

| (Courtesy Northrop Grumman Corp.)

the linear feed are determined by the illumination of the line-source feed, while the beamwidth in the perpendicular plane is determined by the illumination across the parabolic profile. The reflector is usually made slightly longer than the linear feed to avoid spillover and diffraction effects.

An advantage of the parabolic cylinder is that the large number of individual radiators on its linear-array (line-source) feed provides more control of the aperture illumination than does a single point-source feeding a paraboloid. The aperture illuminations required for low-sidelobe radiation patterns are more readily achieved with a parabolic cylinder than a paraboloid or a section of a paraboloid because of the control that can be applied at each of the radiating elements of the linear-array feed. The line feed, however, shapes the radiated beam in one plane only. Shaping of the beam in the orthogonal plane is determined by the reflector. Precise elevation-beam shaping is the purpose of the parabolic cylinder shown in Fig. 9.13, where the cylindrical antenna is oriented in the vertical so that the elevation radiation pattern can be shaped to minimize the radiation that strikes the ground.

The parabolic cylinder can generate an asymmetrical fan beam with a much larger ratio of the two orthogonal beamwidths than can a section of a paraboloid. Aspect ratios greater than 8:1 are practical with a parabolic cylinder but are difficult to achieve with a section of a paraboloid. Also, there is usually less depolarization on reflection from a parabolic cylinder than from a paraboloid.

9.5   ELECTRONICALLY STEERED PHASED ARRAY ANTENNAS

Background   A phased array is a directive antenna made up of a number of individual antennas, or radiating elements. Its radiation pattern is determined by the amplitude and phase of the current at each of its elements. The phased array antenna has the advantage of being able to have its beam electronically steered in angle by changing the phase of the current at each element. The beam of a large fixed phased-array antenna therefore can be rapidly steered from one direction to another without the need for mechanically positioning a large and heavy antenna. A typical phased array radar for microwave radar might have several thousand individual radiating elements using, for example, ferrite or diode phase shifters that allow the beam to be switched from one direction to another in several microseconds, or less.

Electronically steerable phased arrays are of interest because they can provide:

*   Agile, rapid beam-steering.

*   Potential for large peak and large average power. Each element can have its own transmitter. The power-aperture product can be large, especially at the lower frequencies.

*   Multiple-target tracking. This can be accomplished either by generating multiple, simultaneous, independent beams or by rapidly switching a single beam to view more than one target in sequence.

*   A convenient means to employ solid-state transmitters.

*   Convenient shape for flush mounting or for blast hardening.

*   Control of the aperture illumination because of the many antenna elements available.

*   A lower radar cross section, if properly designed.

*   Operation with more than one function (a multifunction radar), especially if all functions are best performed at the same frequency.

The chief disadvantages of a phased array radar are that it is complex and can be of high cost. Although an advantage of a phased array is that it can perform multiple functions in a sequential (time-shared) manner, its ability to employ multiple functions requires serious compromises for some applications.

A *linear array* consists of antenna elements arranged in a straight line in one dimension. It was mentioned in the last section that a linear array can be used as the feed for a parabolic cylinder antenna. A *planar array* is a two-dimensional configuration of antenna elements arranged to lie in a plane. In both the linear and planar arrays, the element spacings usually are uniform (equal spacing). The planar array may be thought of as a linear array of linear arrays. Most phased arrays of interest for radar are planar, but in this section we will start with the linear array as the model since it is simpler to analyze. A *broadside array* is one in which the direction of maximum radiation is perpendicular to, or almost perpendicular to, the plane (or line) of the antenna. An *endfire array* has its maximum radiation parallel to the array or at a small angle to the plane of the array.

Radiation Patterns of Phased Arrays   Consider, as in Fig. 9.14, a receiving linear array made up of N elements equally spaced a distance d apart. The elements are assumed to be isotropic radiators in that they have uniform response for signals from all directions. Although isotropic radiators are not realizable in practice, they are a convenient concept in array theory. The outputs received from all N elements are summed via lines of equal length to produce a sum output voltage E_a. Element 1 will be taken as the reference with zero phase. From simple geometry, the difference in path length between adjacent elements for signals arriving at an angle θ with respect to the normal to the antenna, is

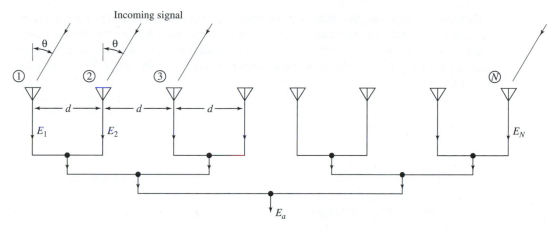

Figure 9.14 *N*-element receiving, parallel-feed, linear array, with equal lengths of transmission lines between each antenna element and the antenna output (at the bottom on the figure).

$d \sin \theta$. This gives a phase difference between adjacent elements of $\phi = 2\pi(d/\lambda) \sin \theta$, where $\lambda =$ wavelength of the received signal. It is assumed that there is no further amplitude or phase weighting of the received signals. For convenience, we take the amplitude of the received signal at each element to be unity. The sum of all the voltages from the individual elements, when the phase difference between adjacent elements is ϕ, can be written

$$E_a = \sin \omega t + \sin(\omega t + \phi) + \sin(\omega t + 2\phi) + \cdots + \sin[\omega t + (N - 1)\phi] \qquad [9.22]$$

where ω is the angular frequency of the signal. The sum can be written[27]

$$E_a = \sin\left[\omega t + (N - 1)\frac{\phi}{2}\right] \frac{\sin (N\phi/2)}{\sin (\phi/2)} \qquad [9.23]$$

The first factor is a sinewave of frequency ω with a phase shift $(N - 1)\phi/2$. (If the phase reference were taken at the center of the array instead at the left-hand side, this phase shift would be zero. In any event this factor is not as important as the second factor.) The second factor is an amplitude of the form $(\sin NX)/(\sin X)$. The magnitude of Eq. (9.23) represents the *field-intensity pattern,* or

$$|E_a(\theta)| = \left| \frac{\sin [N\pi(d/\lambda) \sin \theta]}{\sin[\pi(d/\lambda) \sin \theta]} \right| \qquad [9.24]$$

The field-intensity pattern has zeros when the numerator is zero. This occurs when $N\pi(d/\lambda) \sin \theta = 0, \pm\pi, \pm2\pi, \ldots, \pm n\pi$, where $n =$ integer. The denominator, on the other hand, is zero whenever $\pi(d/\lambda) \sin \theta = 0, \pm\pi, \pm2\pi, \ldots, \pm n\pi$. When the denominator is zero, it is seen that the numerator is also zero, and the value of $|E_a(\theta)| = 0/0$ is indeterminate. By applying L'Hopital's rule (differentiating numerator and denominator separately) it is found that $|E_a(\theta)|$ is a maximum and is equal to N when $\sin \theta = \pm n\lambda/d$. The maximum at $\theta = 0$ defines the main beam of the field-intensity pattern. The other maxima are called *grating lobes* and are of the same magnitude as the main beam. They are generally undesirable in that they can cause ambiguities by being mistaken for the response of a target in the main beam. Grating lobes can be avoided if the spacing *d* between elements is equal to or less than λ. (There is still a grating lobe at $\theta = \pm 90°$ when $d = \lambda$, but practical radiating elements are not isotropic and can have negligible radiation $\pm 90°$.)

Equation (9.24) indicates that $E_a(\theta) = E_a(\pi - \theta)$; which means that an array of isotropic elements has a similar pattern in the rear of the antenna as in the front. The same is true for an array of dipole antennas. To avoid ambiguities between echoes from the front and the rear, the backward radiation can be eliminated by placing a reflecting screen behind the array so that only radiation over the forward half of the array antenna ($-90° \le \theta \le +90°$) need by considered. The field-intensity pattern with a back screen will be different from that of Eq. (9.24).

The normalized radiation pattern of an array of isotropic elements, which is sometimes called the *array factor,* is

$$G_a(\theta) = \frac{|E_a|^2}{N^2} = \frac{\sin^2 [N\pi(d/\lambda) \sin \theta]}{N^2 \sin^2 [\pi(d/\lambda) \sin \theta]} \qquad [9.25]$$

If Nd = D, the antenna dimension, and if the sine in the denominator can be replaced by its argument (implying that the angle θ is small), the pattern of the uniformly illuminated array is similar to the pattern of a uniformly illuminated line-source antenna, as was given by Eq. (9.14). The half-power beamwidth of this uniformly illuminated array of *N* elements when d = λ/2 is approximately

$$\theta_B = \frac{102}{N} \qquad [9.26]$$

When *N* is sufficiently large, the first (and largest) sidelobe is 13.2 dB below the main-beam maximum value.

When the radiating elements are not isotropic, the antenna radiation pattern of Eq. (9.24) has to be modified by the radiation pattern $G_e(\theta)$ of an individual directive element, so that

$$G(\theta) = G_e(\theta) \frac{\sin^2 [N\pi(d/\lambda) \sin \theta]}{N^2 \sin^2 [\pi(d/\lambda) \sin \theta]} = G_e(\theta) G_a(\theta) \qquad [9.27]$$

This is the product of the *element factor* $G_e(\theta)$ times the *array factor* $G_a(\theta)$, the latter being the pattern of an array composed of isotropic elements. Grating lobes caused by element spacings greater than half-wavelength may be eliminated by using directive elements whose pattern is zero or small in directions of undesired grating lobes. For example, if the element spacing $d = 2\lambda$, grating lobes occur at $\pm 30°$ and $\pm 90°$, in addition to the main beam at $\theta = 0°$. If the individual radiating elements, for example, have a radiation pattern whose null width (defining its main beam) is less than 60°, the grating lobes produced by the array factor will be suppressed. When this occurs, the antenna beam cannot be steered beyond the coverage of the individual elements that make up the array.

Equation (9.27) assumes that the radiation pattern of each element is the same. This is not true in practice, however. The radiation from an element in an array will be affected by the mutual coupling among elements and the coupling due to the outward-traveling wave. An element in the center of the array sees a different electromagnetic environment from an element at the edge of the array. The radiation patterns of the elements will not be the same and will depend on the mutual coupling. Thus the pattern of an individual element depends on where it is located within an array. In order to obtain a more accurate representation of the radiation pattern of an array antenna, the pattern of each element within the array should be measured (or otherwise determined) in the presence of all others. Because the element pattern is not the same for each element, the radiation pattern of Eq. (9.27) is only an approximation, but one which has been widely employed.

Two-Dimensional Radiation Pattern   In a two-dimensional, rectangular planar array whose aperture illumination can be separated into two orthogonal planes such as the horizontal and the vertical planes, the radiation pattern may then be written as the product of the radiation patterns in these two planes (sometimes called *principal planes* of the antenna). If the radiation patterns in the two principal planes are $G_1(\theta_a)$ and $G_2(\theta_e)$, the two-dimensional antenna pattern in this case is

$$G(\theta_a, \theta_e) = G_1(\theta_a) \, G_2(\theta_e) \qquad \text{[9.28]}$$

The angles θ_a and θ_e are not necessarily the elevation and azimuth angles normally associated with radar antennas. The normalized radiation pattern of a uniformly illuminated rectangular array of isotropic elements with spacing d is

$$G(\theta_a, \theta_e) = \frac{\sin^2 [N\pi(d/\lambda) \sin \theta_d]}{N^2 \sin^2 [\pi(d/\lambda) \sin \theta_a]} \ \frac{\sin^2 [M\pi(d/\lambda) \sin \theta_e]}{M^2 \sin^2 [\pi(d/\lambda) \sin \theta_e]} \qquad \text{[9.29]}$$

where N = number of (vertical) columns of the array that give rise to the (azimuth) angle θ_a and M = number of (horizontal) rows that generate the (elevation) angle θ_e. The above assumes the spacing between elements in the two directions is the same; but if they are not, the required modification is simple. Since array elements are not isotropic, the two-dimensional element factor should multiply this equation to obtain the antenna pattern.

Beam Steering and Array Feed Networks   The beam of a linear array can be steered in angle by changing the relative time delays between the elements. Consider, as in Fig. 9.15a, two elements of a many-element array spaced a distance d apart. The signal from a direction θ_0, relative to the normal to the two elements, arrives at element 2 before it arrives at element 1. If the signal is delayed at element 2 for a time $\Delta T = (d/c) \sin \theta$, it will be in time coincidence (congruent) with the signal at element 1. If they are added together, it is as though the "main beam" of this simple two-element array was pointed in the direction θ_0. Beam steering occurs by changing the time delay. Inserting variable true-time-delays at each element of a many-element phased array, however, can be quite complicated and is generally unattractive with available technology. Instead, it is much simpler to employ a (modulo 2π) phase shift equal to $\phi = 2\pi f_0 \, \Delta T = 2\pi(d/\lambda) \sin \theta_0$, where f_0 = frequency. The signals are then in phase rather than coincident in time. This is illustrated by Fig. 9.15b.

In a linear array, the phase shift that needs to be inserted at each of the elements in order to have all the signals with the same phase is $m\phi$, where m, an integer from 0 to

Figure 9.15 Two array elements spaced a distance d apart with a received signal arriving at an angle θ_0 measured with respect to the broadside direction. (a) Beam steering based on true time-delay; (b) beam steering using a phase shifter that is variable over the range from 0 to 2π radians.

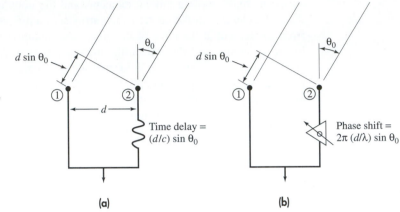

$N - 1$, is the number of the element relative to the reference element. This means that the phase difference between elements is ϕ. The normalized radiation pattern of a linear array of isotropic elements is

$$G(\theta) = \frac{\sin^2 [N\pi(d/\lambda)(\sin \theta - \sin \theta_0)]}{N^2 \sin^2 [\pi(d/\lambda)(\sin \theta - \sin \theta_0)]} \qquad [9.30]$$

The maximum of this pattern occurs when $\sin \theta = \sin \theta_0$; hence, θ_0 is the direction at which the main beam points. As before, the element pattern should multiply this equation to get the antenna radiation pattern. Thus the beam can be steered in an array by changing the phase shift at each element.

Feeding an Array   Variable phase shifters may be used at each element of a linear array to steer the beam as illustrated by the simple four-element array of Fig. 16a. This is called a *parallel-fed array*. The difference in phase between elements is $\phi = 2\pi(d/\lambda) \sin \theta_0$. When a series of power splitters, such as hybrid junctions, are used to create a tree-like structure as in the figure, it is sometimes called a *corporate feed,* since it vaguely resembles (when turned upside down) the organization chart of a corporation. Equal lengths of line between the elements and the transmitter/receiver are desired, but that is not always possible. The phase at each element should be the same (other than that introduced by the phase shifter). If the power is equally divided among all the elements and if the loss at each element is L_{ps}, then the entire loss in the parallel feed network is also given by L_{ps}.

A *series-fed* linear array is shown in Fig. 9.16b. Each phase shifter has the same phase, which means that only one steering command (the phase ϕ) need be generated, as compared to the $N - 1$ phase commands needed for the parallel-fed array. This is an advantage since it simplifies the computer that has to generate the phase commands. A serious disadvantage, however, of the series-fed array is its high loss. If the loss of each phase shifter is L_{ps}, then the loss through the array feed network is $(N - 1)L_{ps}$. Since it is not unusual for the loss of a phase shifter to be a significant fraction of a dB and since

Figure 9.16   Steering of a linear array with variable phase shifters: (a) parallel-fed; (b) series-fed from one end; (c) series-fed from the center.

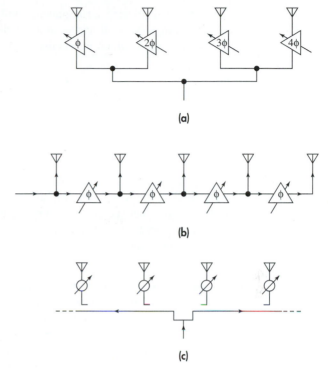

(a)

(b)

(c)

there might be many tens of elements in a linear array, the loss with a series feed is generally unacceptable.

There are two ways in which a series-fed array can be acceptable. One is when frequency scanning (Sec. 9.7) is used with low-loss waveguides connecting the elements. (The frequency scan array does not employ lossy phase shifters.) The other is when power amplifiers and low-noise receivers are placed between the phase shifter and the radiating element. There is still loss, but it is at low power level on transmit since it occurs before the power amplifier. On receive, the loss due to the phase shifter occurs after signal amplification so that it doesn't seriously affect the receiver overall noise figure.

The series-fed array, even when it is configured to have an acceptable low loss, is sensitive to changes in frequency. It has the properties of a frequency-scan array in that the direction of its beam will change with a change in frequency. Compensation for the shift of the beam position due to a change in frequency can be made in the computer by having it indicate the true beam-pointing direction for any given frequency. Inadvertent beam-steering with frequency can be avoided if the array is fed from the center as indicated in Fig. 9.16c. Although the beam is not shifted in angle, its beamshape will change with frequency.

The phase shifters in a two-dimensional parallel-fed planar array of $M \times N$ elements require $M + N - 2$ separate control signals. A two-dimensional series-fed array, however, requires but two control signals.

Grating Lobes Using an argument similar to the nonscanning array described previously, grating lobes will appear at an angle (or angles) θ_g whenever the denominator of Eq. (9.30) is zero, which means that

$$\pi \frac{d}{\lambda} (\sin \theta_g - \sin \theta_0) = \pm n\pi \qquad \text{[9.31]}$$

or

$$\left| \sin \theta_g - \sin \theta_0 \right| = n \frac{\lambda}{d}$$

From this equation it is found that the element spacing d should be no greater than half wavelength in order to avoid grating lobes. With $d = \lambda/2$, a grating lobe will only appear at $\theta_g = -90°$ when the main beam is steered to $\theta_0 = +90°$. Practical phased arrays, however, cannot scan $\pm 90°$. If the scan were limited to $\pm 60°$, Eq. (9.31) states that the element spacing should not be greater than 0.54λ.

Change of Beamwidth with Steering Angle As the beam of a phased array scans in angle θ_0 from broadside, its beamwidth increases as $1/(\cos \theta_0)$. This may be shown by assuming the sine in the denominator of Eq. (9.30) can be replaced by its argument, so that the radiation pattern is of the form $(\sin^2 u)/u^2$, where $u = N\pi(d/\lambda)(\sin \theta - \sin \theta_0)$. The $(\sin^2 u)/u^2$ antenna pattern is reduced to half its maximum value when $u = \pm 0.443\pi$. Denote by θ_+ the angle corresponding to the half-power point when $\theta > \theta_0$, and denote by θ_- the angle corresponding to the half-power point when $\theta < \theta_0$; that is, θ_+ corresponds to $u = +0.443\pi$ and θ_- to $u = -0.443\pi$. The $\sin \theta - \sin \theta_0$ term in the expression for u can be written[28]

$$\sin \theta - \sin \theta_0 = \sin (\theta - \theta_0) \cos \theta_0 - [1 - \cos (\theta - \theta_0)] \sin \theta_0 \qquad \text{[9.32]}$$

The second term on the right-hand side of this equation can be neglected when θ_0 is small (beam is near broadside), so that $\sin \theta - \sin \theta_0 \approx \sin (\theta - \theta_0) \cos \theta_0$. With this approximation, the two angles corresponding to the half-power (3 dB) point of the antenna pattern are

$$\theta_+ - \theta_0 = \sin^{-1} \frac{0.443\lambda}{Nd \cos \theta_0} \approx \frac{0.443\lambda}{Nd \cos \theta_0}$$

$$\theta_- - \theta_0 = \sin^{-1} \frac{-0.443\lambda}{Nd \cos \theta_0} \approx \frac{-0.443\lambda}{Nd \cos \theta_0}$$

The half-power beamwidth is

$$\theta_B = \theta_+ - \theta_- \approx \frac{0.886\lambda}{Nd \cos \theta_0} \qquad \text{[9.33]}$$

Thus when the beam is scanned an angle θ_0 from broadside, the beamwidth in the plane of scan increases as $(\cos \theta_0)^{-1}$. This expression, however, is not valid when θ_0 is large, and the array performance can be much worse. In addition to the approximation made in this derivation not being valid at large angles, mutual coupling effects can increase as the beam is scanned from broadside. At a scan angle of 60° from broadside, the beamwidth of a practical phased array antenna increases by more than the factor of 2 predicted from Eq. (9.33) and the sidelobe levels increase more than expected from simple theory.

Equation (9.33) applies for a uniform line-source distribution, which seldom is used in radar. With a cosine-on-a-pedestal aperture illumination of the form $a_0 + 2a_1 \cos(2\pi n/N)$ for a linear array of N elements with spacing d, the beamwidth is approximately[29]

$$\theta_B \approx \frac{0.886\lambda}{Nd \cos \theta_0}[1 + 0.636(2a_1/a_0)^2] \qquad [9.34]$$

where a_0 and a_1 are constants, and the parameter n in the aperture illumination represents the position of the element. Since the illumination is assumed to be symmetrical about the center element, n takes on values of $0, \pm1, \pm2, \ldots, \pm(N-1)/2$. The antenna aperture illuminations cover the span from uniform illumination to a tapered illumination that drops to zero at the ends of the array. (The effect of the array is assumed to extend a distance $d/2$ beyond each end element.) Although the above applies to a linear array, similar results are obtained for a planar aperture; that is, the beamwidth varies approximately inversely as $\cos \theta_0$.

A consequence of the beamwidth increasing with scan angle is that the antenna gain also decreases with scan angle as $\cos \theta_0$.

9.6 PHASE SHIFTERS

The shift in phase of a signal of wavelength λ transiting a line of length l at a velocity v is

$$\phi = 2\pi l/\lambda = 2\pi fl/v = 2\pi fl\sqrt{\mu\epsilon} \qquad [9.35]$$

where the frequency $f = v/\lambda$, μ = permeability and ϵ = permittivity. Usually the velocity of propagation v of electromagnetic waves is taken to be the velocity of light c; but with phase shifters it can be different. Here we have assumed for simplicity that the velocity of propagation corresponds to that in a TEM transmission line such as a coaxial cable, so that $v = 1/\sqrt{\mu\epsilon}$. The velocity of propagation of TE and TM waves propagating in waveguides is a bit more complicated than the above, but it is still proportional to $1/\sqrt{\mu\epsilon}$. Based on the far right-hand side of this equation, the various methods for obtaining a change in the phase shift may be summarized as follows:

- *Frequency f.* This is a relatively simple method for electronically scanning a beam. It was the first practical method for electronic beam steering and was at one time widely employed for many phased array radars. In spite of its simplicity, it is no longer popular since it restricts the use of bandwidth for other than beam-steering purposes and it is only practical for electronically steering the beam in one angular coordinate. Frequency scanning has been superseded by the development of other methods for phase shifting that do not have its limitations.

- *Line length l.* This may be accomplished by electronically switching in or out various lengths of transmission line to achieve the desired phase shift. Diodes are often used as the switches.

- *Permeability μ.* Ferrite, or ferrimagnetic, materials exhibit a change in permeability, and therefore a change in phase, when the applied magnetic field is changed. They have been popular for use at the higher microwave frequencies.

- *Permittivity ε.* The permittivity, or dielectric constant, of ferroelectric materials changes with a change in applied voltage. A change in the current of an electrical discharge also results in a change in the electron density which produces a change in permittivity.

- *Velocity v.* Changes in μ and ϵ cause the velocity of propagation to change; but a change in velocity can be had directly by changing the broad dimension of a rectangular waveguide, the so-called "*a*" dimension. By varying the "*a*" dimension of a rectangular waveguide, the proper phase change can be applied across an entire row of radiators of a linear array antenna to scan a beam in one angular coordinate. This form of rapid one-dimensional scanning was used for many years in *X*-band landing radars. It was called a *delta-a scanner* or an *eagle scanner*. The beam could be mechanically scanned over an angle of about 60° at a rate of 10 times per second.

All of the above have been employed or seriously considered as phase shifting devices for phased arrays. There are many other devices that can be used to obtain a phase shift for phased array radars, as has been mentioned in previous editions of this text; but the most popular are those that use ferrites or diodes.

Early electronic phase shifters were analog. Their phase shift could be continuously adjusted. Later they were replaced by digital phase shifting in which the values of phase took on discrete values, generally in binary steps. For example an *N*-bit phase shifter covers 360° of phase change in 2^N steps. Four-bit phase shifters with phase increments of 22.5° are commonly used, but digital phase shifters can have much finer quantization if needed. Although the analog phase shifter permits continuous variation of the phase shift, the relationship between its control current (or voltage) and phase is usually not linear, so that setting an analog phase shifter to a precise value of phase might not be as easily accomplished as obtaining similar or better accuracy with a digital device. Digital phase shifters have come to be the preferred method. Phase shifters have also been known as *phasors*.

Phase shifters for most phased array radar applications should be:

- Able to change phase rapidly (a few microseconds)

- Capable of handling high peak and high average power

- Require control signals that operate with little drive power (generally, one wouldn't want to use more power to drive the phase shifters than the total power that is radiated by the antenna)

- Low loss (a fraction of a dB if it is not used in an active aperture radar)

- Insensitive to changes in temperature

- Of small size (to fit within an element spacing of about a half-wavelength)

- Low weight (especially for airborne or mobile radars)

- Low cost (since the cost of a phase shifter is multiplied by the total number of phase shifters in the system).

There have been many types of phase shifters examined for radar application, and they possess these properties in varying degrees. No one type of phase shifter is sufficiently universal to meet the requirements of all applications.

Diode Phase Shifters[30-34]   The semiconductor diode works well as a switching device for radar phase shifters. They are capable of relatively high power and low loss, and they can be switched rapidly from one state to another (low impedance to high impedance, or vice versa). They are not significantly affected by normal changes in temperature; they can be switched with low control power; and they are compact in size. They lend themselves well to microwave integrated circuitry and are capable of being used over the entire range of frequencies of interest to radar, except their loss increases and their power handling decreases at the higher microwave frequencies.

There have been three methods by which diodes have been used: (1) digitally switched lines, (2) hybrid coupled, and (3) loaded-line. Each will be briefly discussed.

Digitally Switched Lines   A digital phase shifter can be obtained using a cascade of switched lines of length $\lambda/2$, $\lambda/4$, $\lambda/8$, and so forth. An N-bit phase shifter has N line lengths. Figure 9.17, for example, is a four-bit cascade of digitally switched phase shifters capable of switching in or out lengths of line equal to $\lambda/16$, $\lambda/8$, $\lambda/4$, and $\lambda/2$ to obtain a quantization level of $\lambda/16$, which corresponds to a phase increment of $360/16 = 22.5°$. Each phase bit consists of two lengths of line that provide the differential phase shift, and two single-pole, double-throw switches made up of four diodes. In this diagram, when the upper two switches are open, the lower two are closed, and vice versa. In the "zero" phase state, the phase shift is not zero, but is some residual amount ϕ_0, so that the two states are ϕ_0 and $\phi_0 + \Delta\phi_0$. The difference $\Delta\phi_0$ is the desired phase increment. The residual phase of the zero state has to be calibrated out in the radar system.

Hybrid Coupled   The hybrid-coupled phase bit, as shown in Fig. 9.18, uses a 3-dB hybrid junction with balanced reflecting terminations connected to the coupled arms. Two switches (diodes) control the phase change. The 3-dB hybrid junction has the property that a signal at port 1 is divided equally in power between ports 2 and 3, and no signal power appears at port 4. The diodes act to either pass or reflect the incident signals,

Figure 9.17   Digital phase shifter with four-bit diode-switched line lengths with $\lambda/16$ quantization. Particular arrangement shown gives 135° of phase shift (3/8 wavelength).

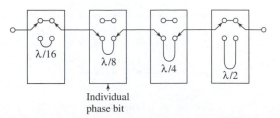

Individual phase bit

Figure 9.18 Hybrid-coupled phase bit.

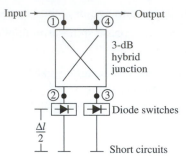

depending on the bias applied to the diode. When the diodes allow the signals to pass, they are reflected by short circuits located farther down the transmission lines. The reflected signals combine at port 4, but no reflected signal appears at port 1. If the diode impedances are such as to reflect rather than pass the signals, the total path length traveled is less. The difference Δl is the two-way path length with the diode switches open and closed, and is chosen to correspond to the desired increment of digitized phase shift. An *N*-bit phase shifter can be obtained by cascading *N* such hybrid junctions and diode switches, with different lengths of lines for each bit.

Loaded Line   This is a little different from the two diode phase shifters mentioned above. As shown in Fig. 9.19, it consists of a transmission line periodically loaded with spaced, switched impedances, or susceptances. Diodes are used to switch between the two states of susceptance. The spacing between diodes is one-quarter wavelength at the operating frequency. Adjacent quarter-wave-spaced loading-susceptances are equal and can take either of two values. If the magnitude of the susceptance is small compared to the characteristic impedance of the line, the quarter-wave spacing will result in cancellation of the reflections from any pair of symmetrical susceptances so that there will be matched transmission for either of the two susceptance conditions. Each pair of diodes spaced a quarter-wave apart produces an increment of the total phase required. Shunt capacitive elements increase the electrical length of the line and shunt inductive elements decrease its length. The number of pairs of shunt susceptances determines the total transmission phase shift. To obtain high power-handling capability, many such sections with small phase increments can be used so there are a large number of diodes available to share the power.

Figure 9.19 Periodically loaded-line phase shifter.

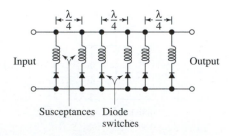

The advantage of the loaded line is its ability to handle larger power than other diode-based phase shifters. If the largest practical phase shift per diode pair is $\lambda/16$ (or 22.5°), 32 diodes would be needed to shift the phase 360°.

Comparison of Diode Phase Shifters The hybrid-coupled phase shifter generally has less loss than the other two, uses the least number of diodes, and can be made to operate over a wide band. The switched-lines phase shifter uses more diodes than the other types and has an undesirable phase-frequency response that can be corrected at the expense of a higher insertion loss. It is often used in solid-state TR modules where the phase shifting is done at low power ahead of the power amplifier on transmit and following the receiver front end on reception. For a four-bit phase shifter with a total phase change of 360°, the loaded line requires 32 diodes, the switched line 16, and the hybrid-coupled shifter needs only 8 diodes. The theoretical peak power capability of the switched-line device is twice that of the hybrid-coupled circuit since voltage doubling is produced by the reflection of the hybrid junction. The switched-line phase shifter has the greatest insertion loss, but its loss does not vary with the amount of phase shift as it does in the other two types of circuits.

Diode phase shifters have been built in practically all types of transmission-line media, including waveguide, coax, and stripline. Microstrip is useful for medium power devices because of its ease of manufacture and circuit reproducibility, as well as its reduced size, weight, and cost of production. Diode chips can be mounted directly on the substrate without the parasitic reactances of the diode package.

A multiple-bit diode phase shifter need not be constructed with just one type of phase shifting device. The loaded-line is often preferred for small phase increments because of its compact size. It is not as suitable for large phase increments because it is difficult to match in both states when large. For example, a four-bit phase shifter might use a loaded-line configuration for the 22.5 and 45° bits, and the hybrid-coupled reflection circuit for the 90 and 180° bits to obtain the minimum insertion loss with suitable bandwidth and power-handling capability.

PIN Diodes The PIN diode has been a popular choice for use in diode phase shifters since it can handle higher power than other diodes; it can be designed to have relatively constant parameters in either or its two states; and it can have switching times from a few microseconds for high-voltage diodes to tens of nanoseconds with low-voltage operation. (Typically, switching times of the order of one or a few microseconds are quite suitable for most radar applications.) As sketched in Fig. 9.20, the PIN diode consists of a thin slice of high-resistivity intrinsic semiconductor material sandwiched between heavily doped low-resistivity P^+ and N^+ regions. The intrinsic region acts as a slightly lossy dielectric at microwave frequencies, and the heavily doped regions are good conductors. When d-c biased in the reverse (nonconducting) state, it resembles a low-loss capacitor since it is essentially an insulator situated between two conductors. Its parallel-plate capacitance is determined by the dielectric of the intrinsic region and is independent of the reverse-bias voltage. The series resistance is determined by the resistivity and geometry of the metallic-like P and N regions. In the forward-bias (conducting) state, when appreciable current is passed, the injection of holes and electrons from the P and N regions,

Figure 9.20 PIN diodes and simplified equivalent circuit for forward and reverse states.

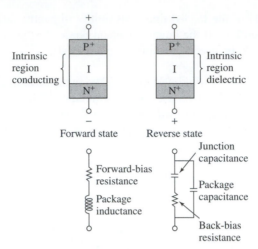

respectively, creates an electron-hole plasma in what was formerly the dielectric region. Thus the slightly lossy dielectric is changed to a fairly good conductor with the application of forward bias. The capacitive component of the circuit disappears and the equivalent circuit becomes a small resistance that decreases with increasing forward current. The resistance can vary from thousands of ohms at zero bias to a fraction of an ohm with tens of milliamperes bias current. With forward bias, the diode resembles a resistance of low value.

Varactor Phase Shifters   The varactor, or variable capacitance semiconductor, also can be used as the switch in a diode phase shifter. Its capacitance is varied by a change in voltage under reverse bias. It is capable of very rapid switching, of the order of a nanosecond, but its average-power handling is limited to a few tens of milliwatts as compared to about 100 W for a PIN diode.[35] The varactor peak-power rating is about 100 times less than that of the PIN. Instead of being used as a switch as is the PIN diode, the varactor can be employed as an analog (continuously variable) voltage-tuned phase shifter. This property has been used as an added module in a 6-bit digital diode phase shifter to provide a continuously variable phase change of from 0 to 11°.[36]

Monolithic Microwave Integrated Circuit (MMIC) Phase Shifters[37,38]   The diode phase shifters that have been discussed thus far have generally been implemented as hybrid microwave integrated circuits (MIC) in that the passive components are deposited on the surface of a low-loss dielectric substrate and the active semiconductor devices are either bonded or soldered to the passive circuit. Phase shifters can also be constructed using monolithic microwave integrated circuit (MMIC) technology in which the entire circuit of passive elements, active devices, and interconnections are incorporated into a single semiconductor substrate. A MMIC phase shifter can be of much smaller size and weight than similar devices in hybrid MIC. They provide better reliability, they are highly reproducible because of the absence of wire bonds, and they can be produced economically with high-volume production. Because of their small size they can be integrated on a single chip with other circuit functions such as power amplification, low-noise receiver front-end, and switching to form a compact T/R (transmit/receive) module for use in active array antennas. These advantages are accompanied, however, by the loss of flexibility in circuit tuning and troubleshooting that is available in hybrid MIC. This loss of flexibility to adjust (or tweak) the circuits means that more attention has to be paid to the use of computer-aided design to insure that the device, once manufactured, will do its job.

MMIC devices have usually utilized gallium arsenide (GaAs) metal semiconductor field-effect transistors (MESFET) as the switches for digital phase shifting. Although silicon technology has been extensively employed at the lower and middle microwave frequency regions, GaAs is preferred at the higher frequencies where MMIC techniques are employed. MESFETs are capable of rapid switching speeds of less than a fraction of a nanosecond, and they operate with relatively low d-c bias power.

The high-pass/low-pass phase-shifter configuration, Fig. 9.21, has been used with MESFETs as the switching elements for MMIC phase shifters. Its small size results from lumped elements being used, rather than distributed elements. A change in phase is obtained by switching between a high-pass filter and a low-pass filter. The insertion of a high-pass filter produces a phase advance and the insertion of a low-pass filter produces a phase delay.

Phase shifters using monolithic microwave integrated circuits at the higher microwave frequencies have been reported[38] to have wide bandwidth and constant phase shift over the band, insertion losses of 5 to 10 dB, and a maximum dimension of from a few millimeters to about a centimeter. The high loss has only a small effect since it occurs at a low power level before the power amplifier on transmit or after the low-noise amplifier on receive.

Ferrite Phase Shifters   A ferrite is a ceramic-like metal-oxide insulator material that possesses magnetic properties while maintaining good dielectric properties.[39–41] Its dielectric constant is in the range from 10 to 20. In contrast to ferromagnetic materials such as iron, ferrites are insulators rather than conductors and have a high resistivity which

Figure 9.21   High-pass/low-pass phase shifter using T-type networks, shown switched to the high-pass filter.[38]

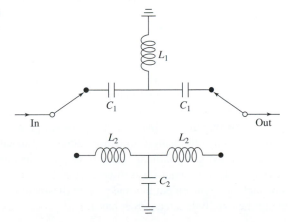

allows electromagnetic waves to propagate with low loss through the material. The term *ferrimagnetism* was introduced to describe the novel magnetic properties of these materials now known as ferrites.

Ferrite phase shifters are two-port devices that may be either analog or digital with either reciprocal or nonreciprocal characteristics. They are generally used at the higher microwave frequencies since their loss decreases with increasing frequency. Ferrites are generally preferred over diode phase shifters for radars above *S* band (except when the phase shifters are used before the power amplifier on transmit and after the low-noise amplifier on receive). At *S* band, either ferrites or diodes might be used. Below *S* band, the diode phase shifter usually is preferred.

The physics of propagation of electromagnetic energy in ferrite materials is not easy to describe, and will not be attempted here. The basic operation occurs by the interaction of electromagnetic waves with the spinning electrons of the ferrite material to produce a change in the microwave permeability of the ferrite, and therefore a change in phase. The magnetic permeability of a ferrite is anisotropic in that it must be represented by a complex tensor rather than a scalar. For this reason, the value of permeability and the resulting phase shift in a ferrite can depend on the direction of propagation. Some types of ferrite phase shifters, therefore, are *nonreciprocal* in that their phase change depends on the direction of propagation. This is different from the semiconductor phase shifters discussed earlier in this section, which were reciprocal devices. Nonreciprocal phase shifters have to be set differently for receiving than for transmitting.

There have been many different types of ferrite phase shifters developed, but those of most interest for radar include the latching, flux drive, and dual-mode phase shifters. One of the first successful ferrite phase shifters, the Reggia-Spencer shifter, will be described so as to illustrate some of the properties of ferrites, their limitations, and how these limitations were overcome in later types of ferrite devices. In spite of its shortcomings, the Reggia-Spencer device was used in an operational phased array radar at very high power (at the time, in the 1960s, when there was no better device available).

Reggia-Spencer Phase Shifter   This device consisted of a rod or bar of ferrimagnetic material suspended at the center of a section of rectangular waveguide. A solenoid was wound around the waveguide to provide a longitudinal magnetic field. A change in phase was obtained by changing the current flowing through the solenoid coil. It was a reciprocal, analog phase shifter that had a high *figure of merit* (defined as the *change of phase per dB of loss*) and was more compact than previous experimental ferrite phase shifters. It had two serious limitations, however. First, the location of the ferrite rod at the center of the waveguide meant it was out of contact with the metal waveguide walls. Thus it was difficult to conduct the dissipated heat away. Second, the time required to switch from one phase state to another was relatively long; hundreds of microseconds rather than the microsecond or two that is characteristic of diode phase shifters. Furthermore, this phase shifter was sensitive to changes in temperature so it usually had to be operated in a temperature-controlled environment. There were also hysteresis effects that had to be accommodated when the phase had to be changed.

The lack of a convenient thermal path to dissipate heat was overcome in one design[42] by having the axially located garnet bar directly cooled by a low-loss liquid dielectric that was allowed to flow along the surface of the garnet material. (A garnet is a ferrite with a different crystal structure than other ferrites.) The flow was confined by completely encapsulating the garnet in a teflon jacket so that the cooling liquid was in direct contact with the garnet bar. A *C*-band Reggia-Spencer phase shifter with this method of cooling operated over an 8 percent bandwidth at a peak power of 100 kW, average power of 600 W, insertion loss of 0.9 dB, and a VSWR of 1.25. The device was 2.4 by 2.1 by 8.2 inches and weighed 1.5 lb. It required, however, 125 μs to switch its phase, and at a switching rate of 300 Hz it used 16 W of switching power.

The long switching times for the Reggia-Spencer phase shifter were due to (1) the large inductance of the solenoid that provided the magnetic field and (2) the "shorted turn" effect caused by the metallic waveguide around which the solenoid was wrapped generating eddy currents in the metallic waveguide wall. There were things that could be done to reduce the switching time, but the Reggia-Spencer switching times were always much longer than those of other phase shifters.

Latching Ferrite Phase Shifter[40,43]   A latching ferrite phase shifter overcomes many of the limitations of the Reggia-Spencer device by taking advantage of the hysteresis loop of a magnetic material so as to latch, or lock, its permeability to one of the two remanent magnetization points on the ferrite material's *B-H* curve. It does not need a continuous holding current to maintain the phase shift; hence, its drive power might be an order of magnitude less than that of the Reggia-Spencer shifter. It is not as temperature sensitive, it has a much faster switching speed, and there is less of a problem caused by hysteresis in the ferrite. It also lends itself to implementation as a digital phase shifter. Figure 9.22a illustrates one bit of a latching ferrite phase shifter mounted in a waveguide. The ferrite is in the form of a rectangular toroid. The contact of the toroid with the walls of the waveguide allows the generated heat to be dissipated. The toroid, however, results in this device being nonreciprocal.

Figure 9.23 is a hysteresis loop, or *B-H* curve, for a magnetic material such as a ferrite. It is plot of the magnetization, or magnetic induction (units of flux density, or webers/m^2) as a function of the applied magnetic field (ampere-turns/m) for a toroidal-shaped section of ferrite. The applied magnetic field is proportional to the current in the drive wire, which can be considered a solenoid of one turn. When a sufficiently large pulse of current is passed through the drive wire threading the center of the toroid, the magnetization is driven to saturation. When the current is then reduced to zero, there exists a remanent magnetization B_r. Similarly, when a large current pulse of opposite polarity is passed through the drive wire, the ferrite becomes saturated with the opposite polarity, and when the current is reduced to zero the remanent magnetization of opposite sign is obtained. Thus a toroidal ferrite may take on two values of magnetization, $\pm B_r$, obtained by pulsing the drive wire with either a positive or a negative current pulse. The difference in the two states of remanent magnetization produces the differential phase shift. Only a short-duration current pulse is needed to set the phase of a latching phase shifter.

The amount of differential phase shift depends on the ferrite material and the length of the toroid. A digital latching phase shifter may be obtained by placing in cascade a number of separate toroids of the proper lengths. The lengths of each toroid are selected to provide a differential phase shift of 180°, 90°, 45°, 22.5°, and so on, depending on the

Figure 9.22 (a) Single bit of a latching ferrite phase shifter mounted in waveguide, showing the drive wire through the center of the toroid that establishes the magnetic field to latch the phase shift; (b) sketch of a five-bit latching ferrite phase-shifter.
| (From Wicker and Jones,[44] Courtesy IEEE.)

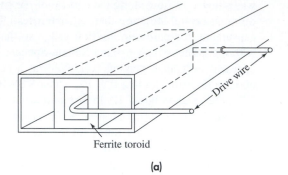

Ferrite toroid

(a)

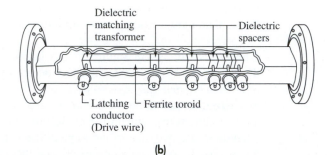

Dielectric matching transformer

Dielectric spacers

Latching conductor (Drive wire)

Ferrite toroid

(b)

Figure 9.23 Hysteresis loop, or *B-H* curve, of a ferrite toroid.

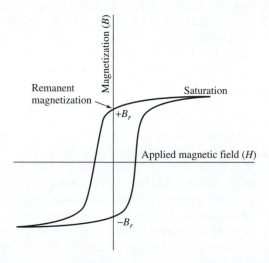

Magnetization (*B*)

Remanent magnetization

Saturation

$+B_r$

Applied magnetic field (*H*)

$-B_r$

number of bits required. The sketch in Fig. 9.22b illustrates a five bit latching ferrite phase shifter.[44] A separate drive wire is used for each bit. Impedance matching is provided at the input and output toroids. Filling the center slot of the toroids with a high dielectric-constant material produces a higher value of phase change per dB of loss (higher figure of merit) and lower switching power, but the lower will be the peak power that the device can handle before breakdown. The individual toroids usually are separated by thin dielectric spacers to avoid magnetic interaction. The drive wire is oriented for minimum RF coupling. It is also important that there be the proper mechanical contact between the toroid and the waveguide wall since an air gap can cause the generation of higher-order modes that result in relatively narrow-frequency-band, insertion-loss spikes. These unwanted air gaps have to be eliminated with care since excessive mechanical pressure on the material can cause magnetostriction that changes the magnetic properties of the material, especially if garnets are used. This type of latching phase shifter has also been called a *twin-slab toroidal phase shifter* since the major action is due to the two vertical branches of the toroids. The horizontal branches of the toroid do not contribute to the phase shift, but they are needed is to complete the magnetic circuit.

Introducing the applied magnetic field from within the waveguide via the single-turn drive wire eliminates the shorted-turn effect and avoids the long switching times that were characteristic of the original Reggia-Spencer phase shifter. Switching times of the order of microseconds become practical. Hysteresis was a nuisance to be tolerated in a Reggia-Spencer phase shifter, but the latching ferrite phase shifter takes advantage of the hysteresis loop to produce two discrete values of phase shift without the need for continuous holding power.

Different phase shifts must be used for transmit and receive with a nonreciprocal phase shifter. The nonreciprocal latching phase shifter, therefore, must be reset just after transmission is completed in order to receive the echo signals. The switching speeds of the latching phase shifter, which are of the order of microseconds, permit the rapid switching required. The phase shift for reception is obtained by simply reversing the polarity of the drive pulses that were used to set the phase for transmission. This reverses the direction of magnetization of the ferrite toroid, which is equivalent to reversing the direction of propagation. Although nonreciprocal phase shifters can be employed in many radar applications, they cannot be used in space-fed reflectarrays (Sec. 9.9) since the electromagnetic energy rapidly changes direction during both transmission and reception in such an antenna. Their use is also not practical in high-duty-cycle pulse doppler radars or in very short-range radars.

A nonreciprocal digital latching five-bit ferrite phase shifter was used in the S-band 3D radar known as the RAT 31/S built by Alenia of Rome, Italy.[45] It had the following characteristics: peak power = 7 kW, average power = 70 W, insertion loss <0.9 dB, VSWR < 1.3, bandwidth = 3 percent, switching time ≤2.5 μs, temperature tracking of the insertion phase = 0.6°/°C, and temperature range from 0 to 60°C. The rms value of the standard deviation of the insertion loss was 0.03 dB. The deviation of the insertion phase from its anticipated value was compensated in the path between phase shifter and antenna element. The rms value of the deviation of the phase of the smallest bit (nominally 12.25°, but actually an average value of 12.40°) was 1.13°; and that of the largest bit (nominally 180°, but actually an average of 199.8°) was 3.89°.

Twin-Toroid Latching Phase Shifter[46,47]   The latching phase shifter described above has been improved by the use of the twin toroid, the geometry of which is sketched in Fig. 9.24a. The two toroids are separated by dielectric which concentrates the RF energy in the center of the waveguide. The active ferrite regions (in which the nonreciprocal interaction with the RF field occurs) are the two vertical ferrite arms that are in contact with the dielectric in the center. The differential phase shift of the twin-toroid ferrite can be made independent of frequency, and it is capable of wide bandwidths. The twin-toroid phase shifter is said[48] to be easier to construct than the single-toroid device. Hord[46] gives the following characteristics for an X-band twin-toroid phase shifter: loss = 0.4 dB, switching time = 3 μs, switching energy = 100 μJ, and size = 0.27 by 0.18 by 2.3 inches. Bandwidth can be 10 percent or greater.

A variant of the twin-toroid employs what is called a grooved waveguide, as is illustrated by the cross section view of Fig. 9.24b.[49] Note that there are different gaps between the waveguide ridges. It has been said that this geometry increases the differential phase shift by 20 percent, decreases the insertion loss for 360° differential phase shift by about 10 to 30 percent (thus providing a better figure of merit), and allows better thermal conductivity and an increase in average-power capability.

Flux Drive[50]   The toroid ferrite phase shifter can be operated in an analog fashion to obtain digital phase-shift increments by varying the current of the drive pulse to provide different values of remanent magnetization. This is called *flux drive*. It has the further advantage of having reduced temperature sensitivity. A single long section of ferrite toroid is used that is capable of providing the total differential phase shift of 360°. The required digital phase increment is obtained by operating on a minor hysteresis loop, as indicated in Fig. 9.25. If $B_r(1)$, for example, were the remanent magnetization needed to produce a phase change of 180° (relative to the remanent magnetization $-B_r$), the amplitude and width of the driving pulse would be selected so as to rise to the point (1) on the hysteresis curve.

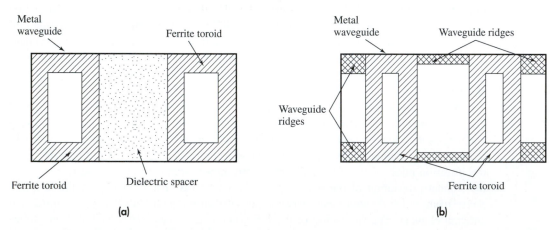

Figure 9.24   (a) Cross section of a twin-toroid ferrite phase shifter. (b) Cross section of a twin-toroid ferrite phase shifter with a grooved waveguide.

Figure 9.25   Hysteresis loop showing the operation of flux drive, where a single ferrite toroid is excited by discrete current pulses to produce digital phase-shift increments from what is basically an analog device.

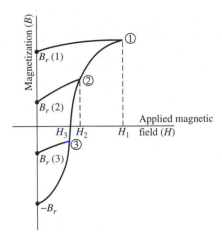

When the pulse current decays to zero, the magnetization falls back to the remanent value $B_r(1)$ along the indicated curve. The difference in phase between $B_r(1)$ and $-B_r$ determines the differential phase increment. With a different value of current pulse, a different value of remanent magnetization and a different phase shift are obtained. In this manner, the ferrite toroid is basically an analog device that can provide any phase increment. It acts as a digital phase shifter if the drive currents are digital. The length of the toroid can be made 15 to 20 percent greater than the normal value to allow for some shrinkage of the total available increment of magnetization due to temperature changes. When the drive output impedance is small, the effect of the temperature-caused variations in magnetization will be small.

Dual-Mode Ferrite Phase Shifters[46,51]   This is a latching phase shifter that is reciprocal, but without the limitations of the reciprocal Reggia-Spencer phase shifter. It is a variant of the Faraday rotation phase shifter[52] and the mechanical Fox phase shifter[53] (Faraday rotation is the rotation of the polarization, or electric field, when the wave propagates in a ferrite material in the presence of a magnetic field.) An outline of the configuration of a dual-mode ferrite phase shifter is sketched in Fig. 9.26. In the center portion is the ferrite bar that supports the propagation of circularly polarized waves. The bar is metallized to form a ferrite-filled waveguide that is accessible for dissipating the heat generated by the loss in the ferrite. A solenoid (not shown in the figure) is wound around the ferrite rod so as to apply an axial magnetic field that rotates the circular polarization to provide a phase change. A linearly polarized signal that enters the rectangular waveguide at the left-hand port is converted to circular polarization by a nonreciprocal circular polarizer (which is a ferrite quarter-wave plate, or quadrapole-field ferrite polarizer[54]). The applied axial magnetic field rotates the circular polarized wave in the ferrite bar, an action that imparts the desired phase shift. After propagating through the ferrite, the phase-shifted circular polarized wave is converted back to linear polarization by a second nonreciprocal polarizer. In a similar manner, a wave incident from the right is converted to circular polarization of the opposite sense by the nonreciprocal quarter-wave plate, and a

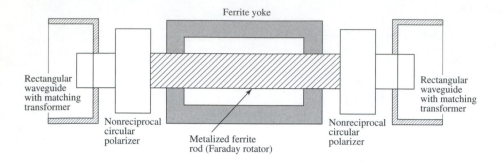

Figure 9.26   Outline of the configuration of a dual-mode ferrite phase shifter. The metalized ferrite rod in the center is based on the Faraday rotator.

phase shift occurs. Since both the sense of polarization and direction of propagation are reversed, the phase shift for a signal traveling from right to left is the same as that of a signal traveling from left to right. The magnetic circuit is completed externally by a temperature-stable ferrite yoke to permit latching of the magnetic field. Flux drive can be used to control the value of the remanent magnetization.

The dual-mode ferrite phase shifter is lightweight and capable of high average power. It also has a good figure of merit. Switching times are from 10 to 100 μs, which is longer than what is achieved with nonreciprocal ferrite phase shifters. Its longer switching times are due to the shorted-turn effect of the thin metallic film covering the ferrite rod. Hord[46] states that an X-band dual-mode phase shifter can have an insertion loss = 0.6 dB, switching speed = 100 μs, switching energy = 400 μJ, length = 1.6 in., and diameter = 0.48 in. Whicker and Young[55] indicate that dual-mode phase shifters are capable of 10 percent bandwidth, 1 kW peak and 100 W average power, and with latching (switching) speeds of 20 to 40 μs. Dual-mode phase shifters were used in the AN/TPN-19 X-band landing radar, where the phase shifter weight was 3.7 oz and had a phase error of 15°.[56]

Polarization-Insensitive Phase Shifters   It has been said[57] that the dual-mode phase shifter can be made to be insensitive to polarization; that is, have the same phase shift for differently polarized electromagnetic waves so that they can be used in phased array antennas that employ more than one polarization. Polarization insensitive phase shifters are of interest when the radar must use dual orthogonal polarizations to avoid the large loss of signal caused by Faraday rotation of the plane of polarization when VHF or UHF radar waves propagate through the ionosphere, when circular polarization is used for detecting targets in the rain, or in any other situation where a choice of more than one polarization is desired.[58]

Rotary-Field Phase Shifter[59]   This is similar to the dual-mode phase shifter mentioned above in that it also acts as a Faraday rotator to impart a phase shift. It is a reciprocal device, but is nonlatching. It is very accurate, being capable of phase errors of one degree or less, which is considerably better than many other types of phase shifters. Such accuracy is required for low sidelobe array antennas. Linear polarization in the input rectangular waveguide is converted to circular. The circular polarized wave propagates in the ferrite rod which completely fills a circular waveguide. A phase shift is obtained in the ferrite by a constant-magnitude magnetic bias that is rotated in space by the application of currents to two orthogonal windings on a ferromagnetic yoke fitted over the ferrite. This is accomplished with a pair of coils wound on a motor-like stator. The quadrapole field generated by the two coils can be smoothly rotated to any desired angle. The accuracy of the differential phase shift is determined by the ratio of the control currents in the two coils. The ferrite rod then acts as a half-wave plate whose orientation determines the amount of phase shift, which is similar to the function of the mechanical rotation of the half-wave plate (or 180° differential phase shift section) used in the original Fox phase shifter.[52] A rotation of the half-wave plate by an angle θ results in a 2θ-radian change in the time phase of the signal. After propagating through the ferrite rod and experiencing a phase change, the circular polarization is converted back to linear.

Boyd[60] states that at X band, a rotary-field phase shifter might have 0.5 dB loss, 10 percent bandwidth, rms phase error less than one degree, and switching time of 50 μs. The control power to the stator windings is less than 0.5 W. These devices are capable of moderate to high power, are less temperature sensitive than other ferrite phase shifters, have a phase shift that varies little with frequency over a wide band, are highly accurate, and their low weight makes them suitable for airborne application.

Other Phase Shifters   There have been many other phase shifters developed in the past, including other types of ferrite devices, electromechanical shifters, traveling wave tubes used as phase shifters, plasma devices, and ferroelectric phase shifters in which the dielectric constant of a ferroelectric material is a function of the applied electric field. The ferroelectric phase shifter has been said[61] to have high power capability, low drive power, voltage control of phase, and low production costs; but it has been difficult to obtain suitale ferroelectric materials to satisfy the important requirements of a phase shifter.

9.7 FREQUENCY-SCAN ARRAYS[62–64]

Because of its relative simplicity, the frequency-scan array was at one time the most popular form of phased array and was widely used. Its beam was steered by simply changing the radar frequency. It was especially popular for scanning a beam in one angular coordinate, such as with 3D air-surveillance radars.* A frequency-scan array has, however, significant limitations. The use of frequency for beam steering prevents the frequency domain from being used for other important purposes in radar, such as high range-resolution, electronic counter-countermeasures, and pulse-to-pulse frequency agility.

---

*A 3D air-surveillance radar is one which mechanically rotates in azimuth and scans one or more pencil beams in elevation or has multiple fixed beams in elevation for the purpose of measuring elevation angle. Other radars, of course, also can obtain three-dimensional data, but they are not usually "3D radars" in the sense of this definition.

Beam Steering by Change of Frequency   The frequency-scanned array is almost always series fed as depicted in Fig. 9.27. Although series-fed arrays using phase shifters have high loss (as was mentioned in Sec. 9.5), it is not the case here since only waveguide connects the elements. The loss in propagating through waveguide transmission line is low.

We next derive the relationship between the radar frequency and the beam steering angle. The difference in phase between two adjacent elements in the series-fed array of Fig. 9.27 is

$$\phi = 2\pi f l/v = 2\pi l/\lambda \qquad \qquad [9.36]$$

where f = frequency of the electromagnetic signal, l = length of line connecting adjacent elements (generally l is much greater than the distance between elements), v = velocity of propagation, and λ = radar wavelength. [Eq. (9.36) is basically the same as Eq. (9.35)]. For convenience in this simplified analysis, the velocity of propagation is taken to be c, the velocity of light. This applies for coaxial lines and other transmission lines which propagate a TEM mode. Waveguides, which are more often used than TEM lines for the end-fed transmission line, can have a velocity of propagation that varies with frequency (i.e., it is dispersive).

As described in Sec. 9.5, if the beam is to point in a direction θ_0, the phase difference ϕ between elements spaced a distance d apart must be equal to $2\pi(d/\lambda) \sin \theta_0$. In a frequency-scan array, it is advantageous for practical reasons to add an integral number m of 2π radians of relative phase change. Since phase is modulo 2π and m is an integer, $2\pi m = 0$; so that the inclusion of $2\pi m$ has no effect on the phase difference between elements. The addition of the $2\pi m$ radians phase is achieved with the length of line l that connects adjacent array elements. The reason for adding the fixed $2\pi m$ phase shift is that it allows a given scan angle to be obtained with a much smaller frequency change than if a line of length $d = \lambda/2$ were used. This will become evident from Eq. (9.38). Equating the phase difference $\phi + 2\pi m$ between adjacent elements that is required to scan the beam to an angle θ_0, to the phase shift [Eq. (9.36)] introduced by a transmission line of length l, results in

$$2\pi(d/\lambda) \sin \theta_0 + 2\pi m = 2\pi l/\lambda \qquad \qquad [9.37a]$$

Figure 9.27   Series-fed, frequency-scan linear array.

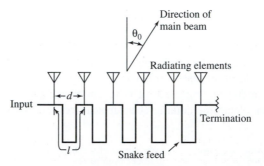

or

$$\sin \theta_0 = -\frac{m\lambda}{d} + \frac{l}{d} \qquad [9.37b]$$

When $\theta_0 = 0$, the beam points to broadside and the above equation yields $m = l/\lambda_0$, where λ_0 is the wavelength that points the beam to broadside. If the frequency corresponding to beam pointing at broadside is denoted f_0, the direction of beam pointing can be written

$$\sin \theta_0 = \frac{l}{d}\left(1 - \frac{\lambda}{\lambda_0}\right) = \frac{l}{d}\left(1 - \frac{f_0}{f}\right) \qquad [9.38]$$

From the above, the wavelength excursion $\Delta\lambda$ required to scan the beam over an angular region $\pm\theta_s$ is

$$\Delta\lambda = 2\,\lambda_0(d/l)\sin\theta_s \qquad [9.39]$$

This equation shows that the greater the ratio l/d, the smaller will be the wavelength excursion $\Delta\lambda$ required to cover a given angular region $\pm\theta_s$. The ratio l/d is usually called the *wrap-up factor*. (The beam position is symmetrical with wavelength, but it is not symmetrical as a function of frequency.) To scan the beam $\pm45°$ from the broadside direction requires a fractional wavelength change of 0.28 when the wrap-up factor is 5, and a 0.07 change when the wrap-up factor is 20 (fractional wavelength $= \Delta\lambda/\lambda_0$.)

Equations (9.38) and (9.39) apply for a TEM transmission line where the velocity of propagation is equal to the velocity of light. It is more usual, however, for waveguides to be used as transmission lines in this type of radar. Since the velocity of propagation in a waveguide depends on the frequency, a different and more complicated expression for the beam pointing angle results when waveguides are used instead of TEM lines. The velocity vs. frequency characteristic of waveguides can be used to good advantage to scan an angular region with less frequency change than indicated for a TEM line with the same wrap-up factor.

Grating lobes can occur in a frequency scan array, just as in other array antennas, when the electrical spacing between elements is too large. Equation (9.31) therefore applies. If we assume that a grating lobe can be tolerated at $\theta_g = -90°$ when the main beam is steered to the maximum scan angle $\theta_0 = +\theta_m$, then the following relationship applies

$$|1 + \sin\theta_m| < \lambda/d \qquad [9.40]$$

The onset of a grating lobe can limit the maximum angle the beam can be scanned.

Bandwidth Limitation   Equation (9.39) illustrates the need for large frequency tunability of the radar transmitter in order to employ frequency scanning, especially if l/d is small. A large bandwidth might cause a potential problem of interference with other electromagnetic systems. Interference among frequency-scan radars operating in the same band, however, might not be that serious a problem since such a radar dwells at any one frequency for only a short time. Of more significance, however, is the reduction in signal bandwidth that can be used with a frequency scan antenna as the l/d ratio increases. If a wideband signal is used, distortion of the main beam will result.

With a series-feed, such as was indicated by Fig. 9.27, the signal travels a total distance $(N - 1)l$ from one end of the array to the other, where N is the number of elements in the linear array and l is the length of the transmission line between adjacent elements. For example, consider an S-band frequency-scan radar ($\lambda_0 = 10$ cm) with a linear array antenna of 101 elements spaced one-half wavelength apart. The wrap-up factor is assumed to be 10. The width of the antenna is taken to be 5 m, so that the total length of the feed line is 10×5 m $= 50$ m. An impulse incident at the input of a TEM feed line would require 0.167 μs to travel down the 50 m feed line and reach the other end (assuming propagation at the velocity of light). This build-up time, or time to fill the array, has a similar effect on the radar's signal bandwidth as does the transient response time of the more familiar signal filter. Thus the time t_D for the signal to travel from one end of the antenna to the other will limit the bandwidth to $1/t_D$, or 6.7 MHz in this example. If the wrap-up factor were 20 instead of 10, the limitation on the bandwidth would be 3.3 MHz. Thus the greater the wrap-up factor, the less frequency excursion that need be used to provide a given angular coverage, but the more narrowband will be the radar. Another way to see the effect of too wide a bandwidth is to note that if the signal has a wide frequency spectrum, the beam will be smeared in angle. Frequency scanning, therefore, is generally not compatible with high-resolution radar that might require large bandwidths.

Various Forms of Frequency-Scanned Radars   There have been several methods by which frequency scanning has been employed, mostly as 3D air-surveillance radars that scan in one angular coordinate (elevation). Each of the following has been used in radar systems.

Single Scanning-Pencil-Beam   The original 3D frequency-scan radars used a single pencil-beam antenna to scan in elevation as they mechanically rotated 360° in azimuth, Fig. 9.28a. The antenna beam dwells at a particular frequency at one elevation beam position before moving on to the next position and a different frequency. If the pencil beam is narrow (order of 1 or 2°), and if long range and large elevation-angle coverage are required, the revisit times with a single scanning beam will likely be excessively long (perhaps more than a minute). For example, if a pencil-beam antenna had azimuth and elevation beamwidths of 1.5°, elevation coverage 30°, range coverage to 200 nmi, and 10 pulses integrated from each target, then the time to cover 360° in azimuth is 66.7 s. This is far too long. Thus a single narrow-beamwidth scanning-beam radar operating as described is not suitable as a long-range radar because of its long revisit time. This applies to any type of electronically steered single-beam phased array, not just frequency scan.

The scan time can be reduced by increasing the pulse repetition frequency (decreasing the time between pulses) as a function of elevation angle. A low prf would be used at low elevation angles to obtain the desired maximum unambiguous range. As the elevation angle increases, the maximum range decreases since aircraft do not fly above a certain altitude. The prf can be increased correspondingly at the higher elevation angles to decrease the total time it takes for the antenna to cover 360°.

The number of pulses per dwell can be reduced to decrease the total scan time. Fewer pulses mean a larger transmitter peak power and perhaps longer pulse widths to make up for the smaller energy available from the fewer number of echo pulses received from a target. The limit occurs when there is only one pulse per dwell, which has serious

Figure 9.28 Several beam configurations for frequency-scan phased arrays that scan in a single angle coordinate, usually elevation. (a) Single-beam scanning; (b) multiple-beam scanning; (c) multiple-frequency frequency-scanning (radiating on more than one frequency at each beam position); (d) within-pulse scanning, transmit; and (e) within-pulse scanning, receive.

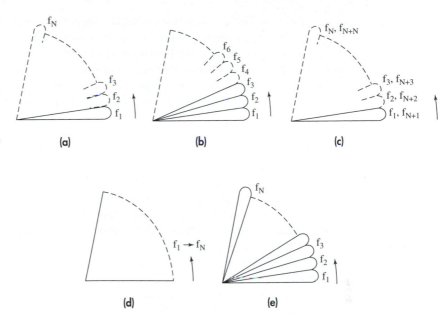

consequences for target detection. A single pulse per beam position can result in a large loss of two-way antenna gain if the target is near the half-power position of the antenna pattern when the pulse is transmitted. Another problem when using only one or a few pulses for target detection is that good doppler processing as required to eliminate clutter in an MTI radar cannot be achieved. Large MTI improvement factors (Sec. 3.7) require a large number of pulses to be processed (a longer time on target). Thus 3D long-range radars that employ one or a few pulses per beam position generally have poor, or no, clutter rejection.

Multiple-Beam Scanning   One method to reduce the time for the antenna to cover its surveillance volume is to simultaneously, or almost simultaneously, transmit more than one beam (at more than one frequency), Fig. 9.28b. In the example of this figure, three pulses might be radiated at frequencies f_1, f_2, and f_3, respectively, so as to cover three contiguous elevation beam positions. The next set of three pulses is transmitted at frequencies f_4, f_5, and f_6. In this example there are three pulses radiated nearly simultaneously, but the number of frequencies radiated as one burst typically might vary from 3 to 9. The time required to scan 360° in azimuth is reduced in proportion to the number of simultaneous beams (frequencies) radiated.

A disadvantage of this approach is that there must be a separate receiver and signal processor for each of the *n* beams radiated simultaneously. In a military radar that must use sidelobe cancelers to reduce the effects of jamming, a separate set of sidelobe cancelers is required for each beam. When there are a large number of simultaneous beams, the cost of the radar can increase significantly.

Even though multiple contiguous beams are formed, one might not be able to perform an accurate elevation-angle measurement by comparing the amplitudes received in adjacent beams. The reason is that amplitude-comparison angle measurement requires that the cross section of the target seen in the two adjacent beams to be the same. In a frequency-scan antenna, adjacent beams are at different frequencies. If the radar cross section of a target is a sensitive function of frequency, there can be a change in amplitude of the received echo signal due to changes in radar cross section that can result in an error when measuring angle by comparison of the amplitudes in adjacent beams.

Multiple-Frequency Frequency-Scan (or Multiple Mode)   In the frequency-scan radars described so far, each elevation beam corresponds to a fixed frequency. This is not good for a military radar since measurement of the frequency by a hostile intercept receiver can provide the elevation angle of the radar beam. Effective hostile jamming can be achieved in some types of frequency scan radars by concentrating the jammer's power over a narrow range of frequencies (angles) rather than force the jammer to cover the entire frequency range of the radar transmitter. Furthermore, the rigid relationship between frequency and elevation angle does not allow a target to be observed at two or more frequencies, something that is desired in order to decorrelate the target cross section for improved detection performance.

It is possible, however, for a frequency-scan array to radiate more than one frequency at the same elevation angle, Fig. 9.28c, assuming that the antenna and the rest of the radar are sufficiently broad band.[64] This can be seen from an examination of Eq. (9.37b). The factor m in this equation is an integer. As the frequency is increased, a beam will scan from the endfire direction (in the direction of the input) through broadside and to the endfire direction in the other direction (pointing to the termination of the array series feed). When the frequency is increased further, another beam will eventually form, corresponding to a higher value of m. The range of frequencies corresponding to one value of m has sometimes been called a *scan band*.

Only one beam at a time will be radiated if the grating lobe relation of Eq. (9.40) is satisfied. It can be shown that if an array radiates at a particular angle corresponding to a value m_1 in Eq. (9.37b) when the frequency is f_1, then for some other value of m, say m_2, a beam will be radiated at the same angle when the frequency is $f_2 = (m_2/m_1)f_1$. As an example, consider an array with spacing $d = 0.6\lambda_0$ and $l/d = 15$, which corresponds to $m_1 = l/\lambda_0 = 9$. From Eq. (9.37b), the array will scan over a region $\pm 30°$ as the frequency changes from $0.968f_0$ to $1.035f_0$, where f_0 is the frequency corresponding to the broadside position of the beam ($\theta_0 = 0$). As the frequency is increased further, the factor $m_2 = 10$ applies and the same angular region is scanned as the frequency varies from $1.075f_0$ to $1.149f_0$. For $m_3 = 11$, the corresponding frequency range is $1.183f_0$ to $1.264f_0$. The beams corresponding to different values of m are related to grating lobes discussed previously.

In addition to allowing better target detection and better electronic counter-countermeasures, the ability to operate similar frequency-scan radars in different parts of the frequency band can help in reducing mutual interference among nearby radars.

Within-Pulse Scanning   In the frequency-scan systems discussed above, the antenna beam dwells at each angular resolution cell (beamwidth) for one or more pulse-repetition intervals before moving to the next resolution cell. Another method that has been used for frequency scanning arrays is to radiate a single frequency modulated pulse that covers a frequency range wide enough to scan the beam over the entire elevation coverage, as in Fig. 9.28d. During each pulse, the antenna beam rapidly scans through all elevation angles. This is sometimes called *within-pulse scanning*. The transmitted waveform is similar to that of a linear FM pulse-compression radar, but it serves a different purpose. The frequency of an echo signal reflected from a target will be a function of its elevation angle. The receiver employs a bank of filters, each tuned to a different carrier frequency which in turn depends on the target's elevation angle, Fig. 9.28e. The number of filters depends on the antenna beamwidth and the total angular coverage. The bandwidth Δf_B of each filter is determined by the frequency change required to scan the antenna one beamwidth, which is then

$$\Delta f_B = \frac{df}{d\theta_0} \, \theta_B \approx \frac{df}{d\theta_0} \frac{\lambda}{D} \qquad [9.41]$$

where θ_B = beamwidth = λ/D, and D = aperture dimension. Rearranging Eq. (9.38), differentiating and substituting into the above gives

$$\Delta f_B = \frac{f}{f_0} \frac{\cos \theta_0}{(l/d)(D/c)} = \frac{f \cos \theta_0}{f_0 t_D} \qquad [9.42]$$

where t_D is the time for the signal to propagate through the transmission line of length $(l/d) \, D$. Thus in the vicinity of broadside ($\theta_0 = 0$), the received signal has a bandwidth $\Delta f_B \approx 1/t_D$, over which the signal is linearly frequency modulated. This bandwidth can be used for achieving a modest amount of pulse compression on receive. Normally, pulse compression cannot be used with frequency scan; but with within-pulse frequency scan, pulse compression processing can reduce the pulse to a width no smaller than the time t_D it takes for the signal to travel from one end of the array to the other. Another way to look at this is that the finite beamwidth of the scanning antenna causes the target to be illuminated with a changing frequency, which can be compressed on receive.

Back-to-Back 2D/3D Antennas   A 3D frequency-scan radar has the advantage of being able to obtain a measurement of the elevation angle (target height). A 2D radar does not obtain elevation angle, but since it has a longer time on target and more received pulses than does a 3D radar it has better MTI processing. In a military radar, a 2D radar can take advantage of the wide frequency range of a 3D frequency-scan transmitter to operate with greater flexibility in the choice of frequency so as to reduce the effectiveness of electronic jamming. The advantages of both 2D and 3D radars can be had by employing a 2D (fan beam) antenna back-to-back with a 3D (scanning pencil beam) antenna on the same rotating antenna mount. The agile-frequency transmitter can be switched between the two antennas to obtain the type of performance desired. A separate transmitter can be used for each antenna so as to achieve simultaneous, rather than time-shared, operation of the two.

Phase-Frequency Planar Array   Frequency scanning has been employed in the past for obtaining beam steering in one angle coordinate, with phase shifters to steer in the orthogonal angular coordinate. This is called a *phase-frequency* array in contrast to a *phase-phase* array which uses phase shifters to steer in both angle coordinates. In an $N \times N$ planar array a phase-frequency array needs only N phase shifters in addition to the fixed frequency-scan transmission lines, while a phase-phase array needs N^2 phase shifters. The phase-frequency array may be considered as a number of frequency-scan arrays placed side by side. The phase-frequency array was used in the early days of phased arrays when phase shifters might be said to have been primitive. As phase shifter technology improved, the all-phase-shifter array became more popular.

Frequency-Scanned Reflector Antennas[65,66]   A very different form of frequency-scanned antenna, compared to those described above, is one that employs a frequency-sensitive grating as the reflector surface. Those familiar with optics might recall the *diffraction grating* which has had an important history as an optical device. The optical diffraction grating is a planar or curved surface obtained by ruling many closely spaced parallel grooves on an optical surface (such as a polished metal mirror). It has the property that a beam incident on the grating will reflect at an angle that depends on the frequency. A similar property can be obtained at microwave frequencies with a periodic array of thin conducting elements etched on a dielectric substrate placed over a surface. When a beam is incident on such a surface, the angle at which it is reflected will depend on the frequency of the incident beam, Fig. 9.29. The dielectric substrate over which the periodic array is etched can be designed to convert most of the incident energy to the diffracted direction governed by its frequency rather than have it reflect in the specular direction given by Snell's law. The shape of the reflecting surface is not parabolic as is common for reflector antennas, but is determined by the properties of the grating and the need to suppress the direct wave whose reflection would be given by Snell's law. As seen in Fig. 9.29, the feed is offset from the normal to the reflector surface in order to achieve the frequency scanning properties. In one experimental demonstration, a 10 percent change in frequency resulted in a beam scan of 10°. The power reflected in the direction of the angle given by Snell's law (equal to the incident angle) was suppressed approximately 20 dB below the power in the frequency-scanned beam.

Frequency Scan in Two Coordinates   In principle, an array can be made to frequency scan a beam in two angular coordinates (a TV raster type of scan) by employing an array of slightly dispersive arrays fed from a single highly dispersive array. It has been said that a 90 by 20° sector can be scanned using a 30 percent frequency change.[67] Two-coordinate frequency scan is almost never used, however, since it requires a very large tunable frequency range of operation and it results in a very narrow signal bandwidth.

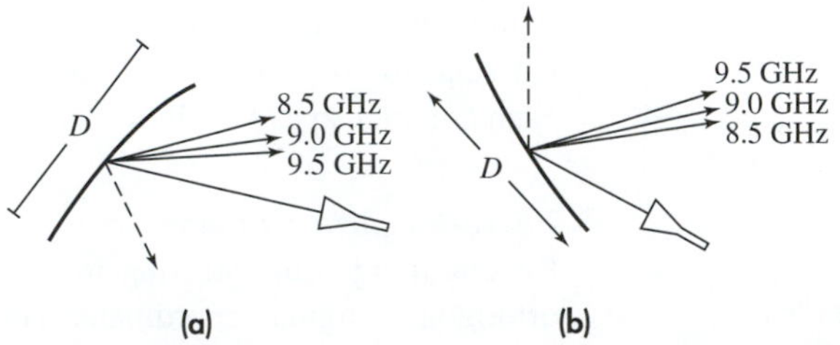

Figure 9.29   Geometries of a frequency-scanned reflector antenna employing a diffraction grating. In (a) the feed is positioned between the specular reflected beam and the diffracted beam; in (b) the specular and diffracted beams are reflected on the same side of the feed.
| (From Johansson et al.,[65] Copyright 1989 IEEE.)

Figure 9.30   Sketch of a frequency-scan array showing, on the left, a folded waveguide delay line (snake feed) feeding a set of waveguides with radiating slots cut in their narrow wall.

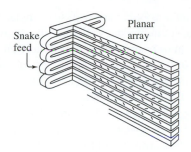

Transmission Lines for Frequency Scan   A popular form of transmission line for series-feeding a frequency-scan array has been the *snake feed,* shown at the left in Fig. 9.30. It has also been known as a *serpentine* or *sinuous feed.* Waveguide wrapped in the form of a helix has also been used, especially as the line feed for a parabolic cylinder reflector.

9.8   RADIATORS FOR PHASED ARRAYS

Types of Radiators   Many different types of radiating elements (antennas) have been used in phased array radars, but the most popular have been various types of dipoles, slots cut into a wall of a waveguide, notch radiators, patch radiators, and open-ended waveguides, Fig. 9.31.[68,69] The pattern of a radiating element differs when located in the midst of a phased array than when it is all by itself in free space.[70] The *radiation impedance* (impedance which accounts for the power radiated) also can change; for example, the radiation impedance of a dipole in free space is 73 ohms; but in an infinite array with half-wavelength element-spacing and a back screen of quarter-wave separation, it is 153 ohms when the beam is directed broadside.[71] These changes are due to the effects of mutual coupling among neighboring elements. Furthermore, the impedance and antenna pattern

Figure 9.31   Sketches of single radiating elements for phased array antennas. (a) Metal-strip dipole with transmission line; (b) outline of printed-circuit dipole (solid lines) showing the coupling structure (dashed lines); (c) slot cut in the narrow wall of a waveguide; the tilt determines the amount of energy coupled from the slot; (d) notch radiator in stripline, radiation is to the right in this figure; (e) rectangular patch radiator; and (f) open-end waveguide.

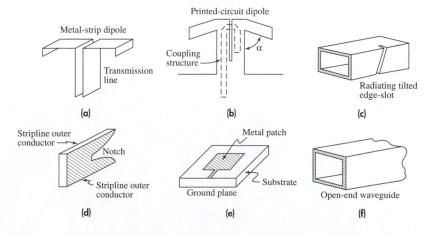

of a radiator in an array can vary with scan angle. Array antennas are of finite size so that the properties of an individual radiator will depend on where it is located within the array. An element at or near the edge experiences an environment different from an element near the center of the array. When trying to determine experimentally how a radiator will perform when located within a large array (prior to building the entire array), the element might be located at the center of an $n \times n$ array of identical elements. It has been said[72] that when using dipoles above a ground plane, an element in the middle of a 7×7 test array may be taken as typical of an element in a large array. With an array of open-ended waveguides the test array should be 9×9. Sometimes the test array might even be as large 11×11.

The well-known *dipole radiator* is a widely used antenna that might be described as "T" shaped. It consists of two collinear metallic rods, tubes, or strips arranged in-line, and is fed at its center from a two-wire transmission line. Figure 9.31a is a sketch of a dipole made of metal strip. Figure 9.31b illustrates a printed-circuit dipole.[73] Generally the dipole is a half-wavelength in dimension. Dipoles are more likely to be used at lower, rather than higher, radar frequencies. In addition to the conventional dipole, the dipole can have its arms bent back (like an arrowhead) so as to obtain wider angle coverage. The thick dipole can reduce mutual coupling between elements and provide a wider bandwidth. Crossed dipoles (two dipoles orthogonal to one another) can provide dual orthogonal linear polarizations or circular polarization. Printed-circuit dipoles are simple to fabricate, especially at the higher frequencies.[73] A dipole with a *director* (a passive rod placed in front of it) might be used to reduce mutual coupling. Dipoles are used with a reflecting ground screen, or its equivalent, to confine the radiation to the forward direction. The distance between dipole and reflecting screen might be in the vicinity of quarter-wavelength. The reflecting ground screen further modifies the properties of the dipole compared to its radiation in free space.

Slots cut into the walls of waveguides, Fig. 9.31c, are similar in many respects to dipoles. The waveguide in which the slots are cut serves as a low-loss line-feed for the array. At the higher microwave frequencies, slots are generally easier to construct than dipoles and can be accurately manufactured using numerically controlled milling machines. Edge-slots in the narrow wall of a waveguide are usually preferred over slots in the broad wall since the waveguide "sticks" (as they sometimes are called) may be stacked sufficiently close to avoid grating lobes in the orthogonal plane. The waveguide-slot array is more suited for scanning in one angular coordinate, which is why it has been popular for 3D radars. (The sketch in Fig. 9.30 was an example.) The power coupled out of the guide by a narrow-wall slot is a function of the angle at which the slot is cut. No power is coupled when the narrow-wall slot is perpendicular to the edges of the guide. The greater the tilt from the perpendicular, the greater will be the coupling. Thus in a phased array with a tapered aperture illumination (like a cosine on a pedestal), the tilt of the slotted radiators near the center will be greater than the tilt of the radiators near the outer ends of the array. When half-wave-spaced slots are fed in a series fashion with the energy propagating down the waveguide, the field inside the guide changes phase by 180° from element to element. The phases of every other slot must be reversed to cause the radiated energy to be in phase. This phase reversal can be accomplished by alternating the tilt of adjacent radiating elements. In a series-fed dipole array, the phase is reversed by rotating every other dipole 180°.

A *flared notch* antenna in stripline is indicated in Fig. 9.31d. These might be envisioned as starting with a dipole, tilting its two arms into a V shape, and then curving the arms of the V and smoothing the normally abrupt transition at the input. Such antennas might have a bandwidth of from 2 to 1 to 6 to 1.[74]

A *patch antenna*[75] consists of a thin metallic film bonded to a grounded dielectric substrate, Fig. 9.31e. Its shape is usually rectangular or circular, and it can be excited with microstrip. It has the advantage of being low profile, lightweight, and is easy and economical to manufacture. It can be mechanically robust and is readily employed with solid-state modules. Patch antennas, however, usually are not as broad band as other radiating elements.

Open-ended waveguides, Fig. 9.31f, are an extension of the waveguide sections in which the phase shifters are located. Their performance can be calculated or measured in a simple phased array waveguide simulator.[76] The waveguide may be loaded with dielectric to reduce its physical size in order to fit the element within the available space. If wide-angle scan is not required, the open-ended waveguides may be flared to form a horn radiator with greater directivity. An array of open-ended waveguides might be covered with a thin sheet of dielectric to better match the array to free space, as well as act as a radome to protect the array from the weather. A dielectric sheet across the face of the array, however, can result in coupling effects that modify the expected performance of the antenna.[77]

Other radiators that have been used with phased arrays include polyrods, Yagis, log-periodic antennas, spirals, and helices. Almost any type of radiator can be considered for application in a phased array; but the dipole, or its equivalent, probably has been the most popular.

Mutual Coupling   The analysis of the phased array used in this chapter, as well in many other books and publications on antennas, is based on a relatively simple model, as was used in Sec. 9.5. It simply combines in space the radiation from the individual antenna elements, taking account of their relative amplitude and phase. Maxwell's equations are not involved, which is why it should not be surprising that the simple theory is not adequate for predicting the performance of actual phased arrays. In particular, the simple theory does not account for interaction among the radiating elements. The current at a particular element depends on the amplitude and phase of the currents in many of its neighboring elements, as well as the original current applied by the antenna feed network. The effect of one element on the other is expressed by the term *mutual coupling.* When the antenna is scanned from broadside, mutual coupling can cause a change in antenna gain, shape of the antenna pattern, shape of the individual element patterns, sidelobe levels, and the radiation impedance.

A major effect of mutual coupling is the change in the impedance seen at the element due to the presence of nearby elements. This is important for properly matching the element as the beam is scanned. The purpose of matching is to avoid high voltage standing-wave ratios (VSWR) that can result at certain scan angles.

Much of the classical theory of mutual coupling[78-80] has been based on modeling the antenna as an infinite array so that all elements within the array see the same environment. This has been a widely applied model for predicting phased array antenna behavior even though an infinite array is not realistic, and it can led to questionable results at times. In most practical arrays a large fraction of the elements can be considered *edge elements,* or elements near the edges. For example, it was mentioned previously that an element must be placed in the center of a 9 by 9 or an 11 by 11 array to accurately determine the effect of the neighboring elements on its performance. With half-wave spacing this means that significant coupling will occur between a radiating element and all elements that are within $2\frac{1}{2}$ wavelengths (if the 11 by 11 array represents an infinite array). In a square array of 60 by 60 elements, approximately 30 percent are edge elements. With such a large fraction being edge elements, one has to use caution when applying the results of infinite array theory.

The energy applied to a radiating element can appear in the main beam, the sidelobes, or be returned to the transmitter. The infinite array theory generally *assumes* that as the antenna scans off broadside, the reduction in radiated power due to the reduction in gain of the radiating element (the element factor) causes energy to be returned to the transmitter rather than be radiated elsewhere in space. The energy that returns to the transmitter results in an increase in the VSWR seen at each element. When the beam is steered, therefore, to a null of the element factor, some theorists assume that all the power is returned to the transmitter, which can be catastrophic. The situation has been called *blindness*[81] or *lost beams.*[82] This is a well-accepted concept by some antenna theorists, but there remain unanswered questions about its applicability to real antennas. When the beam is steered to the null of the element factor, the main beam will be distorted or even disappear, but one can argue that most, if not all, of the energy of the main beam radiates into space in other directions (producing a weird radiation pattern) rather than return to the transmitter.

Experimentally, the VSWR has been observed to increase at some of the many elements of a phased array when the beam is scanned; but there seems to be little experimental evidence for all or a large part of the transmitter power being returned from the array when the main beam is scanned to the direction of an element factor null. One might be skeptical about theories based on the infinite array model that assert that *all* the power returns to the source at certain beam-steering angles. If the power is not returned to the transmitter it seems plausible to assert that it is radiated into space. One should not expect a neat main-beam pattern or low sidelobes when this happens. An example of an experimental measurement in which the power was found to radiate in space rather than be returned to the transmitter when the beam is scanned is described next.

Forward-Wave Interaction in an Array   Before leaving the subject of mutual coupling, an experimental measurement will be mentioned from the early 1960s which illustrates that further understanding of this subject is still required. Donald King and Harry Peters[83] reported measurements of small phased arrays using relatively high-gain closely spaced polyrod elements. Polyrods have seldom been used in phased arrays in the United States, but they illustrate the effects of mutual coupling that involve forward-traveling waves as well as diffraction effects and blockage. Measurements were made of a five-element array with polyrod radiators 6λ in length, spaced 3/4 wavelength apart. This is a rather close spacing for such high-gain endfire antenna elements, so it should not be expected that conventional array theory will apply. The E-plane pattern of a single 6λ polyrod in free space is shown in Fig. 9.32a. Its gain was 15.7 dB. The pattern of the center element of the five-element array is shown in Fig. 9.32b when the other four elements were terminated in a matched load. The pattern in the array is seen to be quite different from that in free space. It is broadened and becomes more rectangular in shape. The pattern of one

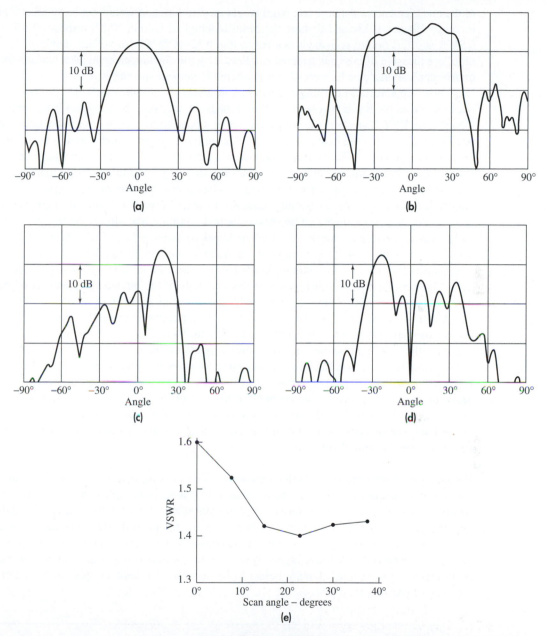

Fig, 9.32   Illustration of forward-wave mutual coupling in a five-element array with polyrod elements each six wavelengths long and spaced 3/4 wavelength apart. (a) measured *E*-plane radiation pattern of a single isolated polyrod element; (b) Measured *E*-plane radiation pattern of the center element with each of the other four elements terminated in a matched load; (c) pattern with the beam scanned to 15°; (d) pattern with the beam scanned to 37.5° (the main beam is not seen at its scan angle of 37.5°, but is "lost"); and (e) VSWR of the feed line of the center element as a function of the scan angle. Note there is no significant change in VSWR even for a scan angle of 37.5°.
I (From King and Peters,[83] Courtesy Microwave J.)

of the end elements was not given, but it would be different and would not be symmetrical in angle. Figures 9.32c and d show the beam scanned to 15 and 37.5°, respectively. The VSWR as a function of scan angle is given in Fig. 9.32e. (VSWR, or voltage standing-wave ratio, is a measure of how much signal is reflected by the antenna element back towards the transmitter.) Similar results were obtained when the element spacing was 1.5λ.

There are several interesting observations from these measurements. The VSWR did not increase as the beam was scanned from broadside, as would be expected from simple theory. The VSWR actually decreased, which is opposite to what is normally believed should happen when the beam of an array is scanned from broadside. There was no significant change in VSWR when the antenna was scanned to 37.5°. The antenna radiation pattern deteriorated considerably at this wide scanning angle and there was no main beam (the beam was "lost"), but there was energy radiated in other directions in space rather than reflected back to the transmitter, contrary to what is usually assumed in studies of mutual coupling. Thus there can be serious mutual coupling that affects the nature of the radiated pattern *without* energy being reflected back to increase the VSWR. Arrays of endfire elements such as polyrods, Yagis, log periodics, and axial-mode helices can experience similar mutual coupling due to forward-wave interaction, blockage, and diffraction rather than only the effects that appear at the terminals of the antenna elements. It might be expected that some of this behavior could also occur in an array of dipoles.

Closely spaced endfire elements have been used in radar antennas, but not often. The reason for mentioning this is that it gives evidence that what happens in the external near-region of the array and the interaction of the elements on the radiating aperture might be as important to understanding what is happening in an array as is the more familiar backward coupling that affects the circuit impedance, or VSWR, of the antenna. Further evidence of the importance of what takes place in front of the antenna is the effect of a dielectric slab or a periodic structure of baffles in front of the array, or even an array of open-ended waveguides, all of which can cause surface waves to travel across the array and contribute to mutual coupling effects.

Mutual Coupling and the Radar Systems Engineer   The existing theoretical analyses of mutual coupling for finite (realistic) phased arrays do not seem complete since they do not account sufficiently for the forward-wave coupling among the elements. In spite of this deficiency, there is sufficient theoretical basis to allow the phased array antenna engineer to approach the design of real systems. As is common with many antenna design problems, the experimental skill and ingenuity of antenna engineers in dealing with real-world design have allowed phased array radars to be successfully built in spite of the shortcomings of existing theory.

9.9 ARCHITECTURES FOR PHASED ARRAYS

The term "architecture" is often used in the discussion of phased array radars but it has not been officially defined and there does not seem to be universal agreement as to what it encompasses. Here we use the term to include the various ways that phased array radars can be configured or structured.

An extremely important part of any practical phased array system architecture is the means by which the power from the transmitter is efficiently divided and distributed to the radiating elements, and the reciprocal problem of combining the signals received at the elements and providing them to the receiver and signal processor. The structure that performs this function is called the *array feed*. Two major methods for this purpose are *constrained feeds* (a network of transmission lines, or power dividers) and *space feeds* (which resemble feeds for reflectors or lens antennas). The constrained feed is sometimes called a *corporate feed*. Analog multiple-beam-forming arrays, active-aperture arrays, and digital beam-forming also require special consideration of the array feed method. Array feeds should not introduce significant loss nor should they be of excessive weight and size. Since the array feed is not always one of the more "glamorous" or visible parts of an array, it is not always appreciated that any loss it introduces is equivalent to a loss in antenna power gain and has to be compensated by an increase in transmitter power or other, usually undesirable, means.

Constrained Feed   Simple illustrations of a parallel-fed and a series-fed constrained feed for one-dimensional linear arrays were shown in Fig. 9.16. The constrained feed is basically a $1 \times N$ power divider, where $N =$ total number of elements in the array. Figure 9.33 is an example of a constrained feed for an array in two dimensions. This particular example is a combination of a single parallel feed and a number of series feeds. In one dimension, the parallel feed is shown and in the other dimension series feeds are shown. Each element has its own phase shifter. The required phase shift to steer the beam in two dimensions can be determined for each element by the beam-steering computer and then distributed to each phase shifter. Alternatively, a small computation chip can be placed at each element to compute the phase required at that particular element, based on being told the azimuth and elevation angles to which the beam is to be steered. Feeds consisting of waveguide or coaxial transmission lines can handle high power with low loss and can be constructed with excellent precision. They can, on the other hand, be bulky and expensive. At the higher microwave frequencies (*L* band and above), strip lines which can be precisely fabricated with computer-aided manufacturing techniques are sometimes used.

The power distribution to the columns of the two-dimensional array of Fig. 9.33 is shown with a single parallel feed. The power in each column is distributed with a series

Figure 9.33   Planar array for scanning in two angular coordinates.

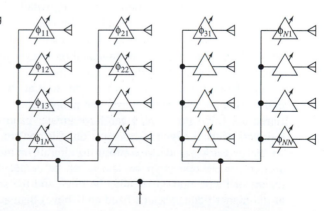

feed to the vertical elements. This is called a *parallel-series* feed. It the columns were fed with a parallel feed, it would be a *parallel-parallel* feed. There can also be a *series-series* feed. Series feeds may sometimes be easier to implement, especially in an active-aperture array where the transmitters and receivers are located between the feed and the radiators; but they have narrower bandwidth than the parallel feed.

Losses in constrained feeds are due to attenuation in the transmission lines as well as reflections from the junctions. Mailloux[84] states that the total length l of transmission line from an antenna element to the single array input is at least

$$l = (N^{1/2} - 1)d \qquad \textbf{[9.43a]}$$

This expression applies to a square array with total number of elements N and element spacing d. (The total length l can be longer than given by this equation, depending on the specific architecture.) The longer the length of line between antenna element and input, the greater will be the loss due to transmission line attenuation. Mailloux also states that the total number of power dividers (which also introduce loss) in series with each element is

$$\text{Total number of power dividers} = \log_2 N = 3.32 \log_{10} N \qquad \textbf{[9.43b]}$$

When the power splitters in a parallel-feed are four-port hybrid junctions, or the equivalent, the feed is said to be *matched*. Theoretically, there are no spurious signals generated by internal reflections in a matched feed. It is not always convenient or practical to use four-port hybrid junctions. Three-port tee junctions are sometimes used instead, for economic reasons, to provide the power splitting. Such a network cannot be perfectly matched in theory, and internal reflections can occur which can appear as spurious side-lobes in the radiation pattern.

Brick and Tile Assemblies   Brief mention should be made of two types of array architectures, one called *brick* and the other *tile*. They will not be described in sufficient detail to fully understand, but further information is available in the literature.[85,86] These terms relate to the manner in which the array is contructed in relatively larger sections to make assembly easier. They are used with a corporate feed. The array is often grouped into subarrays of rows, columns, or areas, with each subarray fed separately. In the brick construction, the array is assembled with circuits on boards that are mounted *perpendicular* to the array face. In tile construction, which has also been called *monolithic array construction,* the array is assembled with one or more layers *parallel* to the array face. The array face contains the radiating elements and semiconductor active subarrays. The phase shifter drivers might be mounted on a layer just below the radiating face, and another layer with the RF power dividers mounted below that. Brick construction utilizes greater depth than does tile construction, so it allows more room for circuits, better thermal dissipation, and more convenient maintenance. It is also compatible with dipole and flared-notch radiators which can have greater bandwidth than the flat microstrip patch-radiators of the tile type of construction. Tile construction has the advantage of being thin so that it can be made to conform to aircraft or missile surfaces. It can be folded and stowed for erection in space, and it can be compatible with robotic or other automatic means of fabrication. Generally, the brick and tile structures have been of more interest at the higher frequencies (X band and above) than at lower frequencies.

Monopulse Beams When multiple beams are needed for monopulse angle measurement with a constrained-feed phased array, the output of each receiving element can be split into three separate outputs that connect to three separate beam-forming networks. One output is used with a beam-forming system to provide the sum beam and the other two outputs are used with separate beam formers to generate the two angle beams. The two angle beams can have different aperture illuminations (weightings) than the sum beam so as to produce desirable difference patterns with low sidelobes and a good error-signal slope.

Space Feeds A space feed is similar to the feed of a reflector antenna. It enjoys the advantage of the relative simplicity that characterizes feeds for reflectors. There are two types of space feeds depending on whether the array is analogous to a lens or to a reflector.

Lens Arrays Although the lens array is considered first in this discussion, much of what is said about it is applicable to the reflectarray as well. The lens array, Fig. 9. 34a, is fed just as would a lens antenna. It is shown in Fig. 9.34a as a one-dimensional representation, but the space-fed array is almost always two-dimensional. The primary feed might be a single horn, a collection of horns, or a monopulse cluster of horns. (The space-fed array is described as a transmitting antenna, but an analogous description can be given as a receiving antenna.) An array of antenna elements collects the energy radiated by the feed and passes it through phase shifters which provide a phase correction to convert the incident spherical wave to a plane wave. The phase shifters also apply the phase shifts required to steer the plane wave to some angle off of broadside. The antenna elements on the opposite side of the lens array then radiate the beam into space. The feed illuminating the space-fed array provides a natural amplitude taper to produce lower sidelobes than would a uniform illumination. The feed may be placed off-axis to avoid reflections from the lens returning to the feed and producing a large VSWR.

Reflectarrays A space-fed reflectarray with an offset feed is diagrammed in Fig. 9.34b. The energy from the feed enters the antenna elements, passes through the phase shifters,

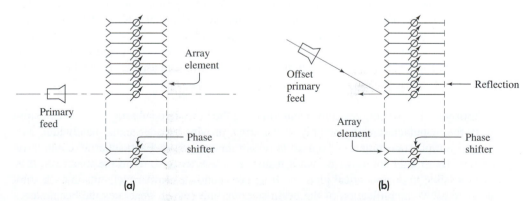

(a)　　　　　　　　　　　　　　　　　　**(b)**

Figure 9.34 Space-fed arrays. (a) Lens array; (b) reflectarray.

is reflected, and passes back through the phase shifters to be radiated as a plane wave in the desired direction. Because the energy passes through each phase shifter twice, a phase shifter need only be capable of half the phase shift needed for a lens array or a conventional array. The phase shifters, however, must be reciprocal. This can be a limitation since some ferrite phase shifters with excellent properties for use in phased arrays are nonreciprocal and therefore cannot be used in a reflectarray.

Comparison   Spillover radiation from the feed of a space-fed array can result in higher sidelobes at angles far from broadside than would be obtained with a constrained feed, unless some means are taken to minimize the spillover. Both sets of antenna elements at the front and back of the lens array require matching. This increases the matching problem of the array and can result in lower antenna efficiency. It is relatively straightforward, however, for the space-fed array to generate a cluster of beams, as for monopulse angle tracking, by use of multiple feed horns similar to those used for monopulse tracking with a reflector antenna. Compared to the constrained feed, space-fed arrays have the advantage of lower loss.

The lens array allows more freedom than the reflectarray in designing the feed assembly since there is no aperture blocking. The presence of a back surface in the reflectarray, however, not only allows better mechanical support and heat removal than in a lens array, but it also makes it easier to provide the needed control signals to the phase shifters. Space-fed arrays are generally cheaper than conventional arrays because of the omission of transmission-line feed networks. The space-fed array with a single transmitter and receiver is usually cheaper than an active-aperture array (to be described later) whose transmitters (and receivers) are distributed along the aperture.

A space-fed array may be cheaper than an array with a constrained feed or an active-aperture array, but it might not have the ability to control the aperture illumination sufficiently well to obtain ultralow sidelobes; and it is not capable of the very large powers possible with a conventional array where a transmitter can be placed at each of the many elements of the array.

Parallel-Plate Feeds   A folded pillbox antenna, a parallel-plate horn, or other similar microwave devices can be used to provide the power distribution to the antenna elements. These are called *reactive* feed systems. They are basically used with a linear array and can be considered as another form of space-fed array, but only for one dimension. They would have to be stacked to feed a planar array, and would thus be a heavy feed system.

Further information on the many methods for feeding a phased array can be found in Patton.[87]

Subarrays   It is sometimes convenient to divide the array into subarrays. These can simplify the manufacture and assembly of the array, provide broader signal-bandwidth, and allow multiple transmitters to be used to obtain greater power. Each subarray could have its own transmitter and receiver, but it is not necessary to do so to utilize subarrays. It is also possible to give identical phase steering commands to similar elements in each subarray to allow simplification of the beam-steering unit (which generates the beam steering commands). Because of the discrete nature of the subarrays, the phase distribution across the aperture has the character of a stair-step, with one step over each subaperture.[88] This can result in what are called *quantization lobes.* Such lobes can be reduced by either overlapping the subapertures or inserting a small amount of randomization[89] to the phases of the subapertures. The term subarray also has, at times in the past, been applied to the array feed networks of a constrained feed system that produce sum and difference patterns for monopulse angle measurement.[90]

Subarrays can achieve wide signal-bandwidth by employing a variable time-delay element at each subarray.[91] Although time-delay elements allow wideband operation, they have not been economically feasible to use at each radiating element of a large phased array. A compromise is to utilize them at each subarray.

The original U.S. Navy Aegis (AN/SPY-1) phased array radar system for ship air defense employed 32 transmitting and 68 receiving subarrays of different sizes.[92] Each of the 32 transmit subarrays had its own CFA (crossed-field amplifier) power amplifier.

Triangular Element Spacing[93,94]   For the most part, it has been assumed here that the radiating elements of a phased array are laid out in a square grid. A triangular, rather than a square, arrangement of elements, however, permits a savings in the total amount of elements needed in an array when the spacings are determined to avoid grating lobes (spurious beams comparable to the main beam that appear in the radiation pattern when the element spacings are too large). The reduction in the number of elements depends on the solid angle over which the main beam is scanned. For example, if the beam is to be scanned anywhere within a cone defined by a half-angle of 45°, the number of elements required with equilateral triangular spacing is 13.4 percent less than with square spacing. (The altitude of the equilateral triangle in this case is equal to the element spacing of a square grid.) The smaller the angular region to be covered, the less is the saving. Triangular spacings are more likely to produce higher sidelobes in some directions because of the phase quantization of digital phase shifters.[95] For most applications, however, these quantization lobes do not significantly limit system performance.

Active-Aperture Phased Arrays   There are several different ways to configure a phased array radar system. One traditional method is to use a high-power phase shifter at each antenna element, with a single high-power transmitter and a single receiver for the entire radar. This has sometimes been called a *passive aperture,* or passive array, in contrast to what is known as an *active aperture,* or active array. They are illustrated in Fig. 9.35a and b respectively. An active aperture has a transmitter (low or modest power) at each antenna element. There is also included at each element an individual receiver, phase shifter, duplexer, and control, as well as the RF power source. Thus an active-aperture phased array implies there is a miniature radar system at each of the array elements. The construction of the electronics at each element of an active-aperture array radar can be highly integrated as a module or with MMIC (monolithic microwave integrated circuitry) construction. The active-aperture module, Fig. 9.35c, is called the T/R (transmit/receive) module or transceiver module.

The passive aperture has had the advantage of usually being cheaper than an active aperture, but cost has not been the only criterion used in selecting a particular array architecture. There are factors in favor of the passive approach and factors in favor of the

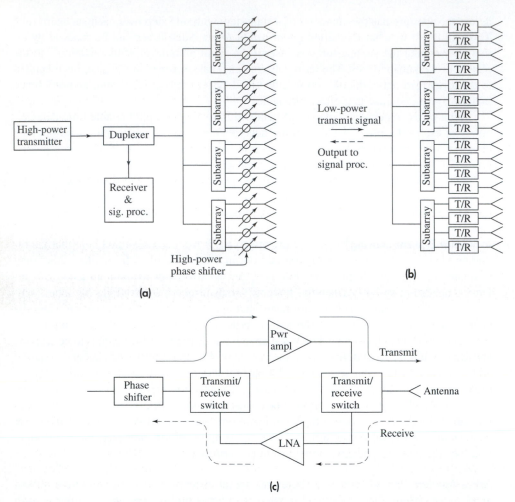

Figure 9.35 Comparison of the passive- and active-aperture array configurations, both shown with subarrays. (a) Passive-aperture phased array with a single high-power transmitter on the left and high-power phase shifters at each element. (b) Active-aperture array with T/R (transmit/receive) modules at each element. Sometimes there might be a booster amplifier on transmit at each subarray (that is bypassed on receive) and/or a time-delay element for increasing the signal bandwidth. (c) T/R module configuration, with a power amplifier on transmit and a low-noise amplifier (LNA) on receive.

active approach, and the choice as to which to use will depend on how the pros and cons balance for any particular application.

There are some generalities that can be cited regarding the relative costs of the two types of arrays. In the past it has been observed that it has often been cheaper to achieve a required total average power by employing a single high-power transmitter rather than obtain the same power by combining the outputs from a number of low-power sources. It has also been observed that high-power vacuum-tube transmitters usually have been more efficient than the solid-state transmitters employed in active-aperture arrays at the higher microwave frequencies. These advantages of a single transmitter might be balanced by the fact that in an active aperture the power from the individual, low-power sources is combined in space so there is no loss in distributing the power as there is in a passive aperture that uses a constrained feed. If the passive array is space-fed, the loss is less than would be experienced by an array with a constrained feed. In a passive aperture the phase shifters must be capable of handling higher power than the phase shifters in the active aperture. The phase shifter loss in a passive aperture is often some fraction of a dB, which is low. The phase shifter loss in an active aperture, however, can be much larger than the phase shifter loss in a passive array since it occurs at low power levels and can be made up by increasing the gain of the power amplifiers that are located between the phase shifter and the antenna element. Thus there is little effect of phase shifter loss on the performance of an active aperture. Similarly on receive, the loss introduced by the phase shifter in the passive aperture can degrade the receiver noise figure. Loss in the phase shifters of an active aperture on receive is less important since the phase shifters are preceded by a low-noise amplifier that determines the noise figure.

There are various corporate-fed beamformer architectures that can be used with active-aperture arrays, depending on (1) whether the amplitude taper is applied in the beamformer network or in the T/R modules, (2) the degree of reliability (mean time between failures) required, and (3) whether the array is narrow or wide band.[96]

It has also been claimed[97] that the distributed architecture of the active aperture can smooth the effect of pulse-to-pulse amplitude and phase variations introduced by the RF power source, and therefore increase the MTI improvement factor and obtain better detection of moving targets in clutter. Since the amplitude and phase variations tend to be random among the many modules of the active aperture, the fluctuations combine in a noise-like fashion to smooth the effect. This assumes that a single large power supply is not used to power the T/R modules.

The proponents of the active-aperture array architecture state that one of its chief attributes is that the total transmitter power will be less than that of a passive aperture (that employs a constrained feed) since it avoids the losses of the high-power phase shifters and feed system. This might make the cost of the active aperture radar less than a passive aperture, if the cost of the active aperture T/R modules is not too large. The proponents of the passive aperture, on the other hand, will argue that the high cost of T/R modules, especially at the higher microwave frequencies (such as X band), as well as the lower efficiency of solid-state transmitters, will offset the higher losses of the passive aperture to make the active aperture more expensive. They would also argue that the cost advantage is even more in favor of the passive aperture if a space-fed array is used.

Although one can debate whether the active aperture or the passive aperture is better, the choice—just like many other choices that have to be made in engineering—depends on the particular application and the particular constraints imposed. It is not always obvious without full analysis which approach results in a more cost-effective phased array radar.

Examples of Active-Aperture Phased Array Radars   The first "modern" phased array radar was the AN/FPS-85 satellite surveillance radar, which became operational at

Eglin Air-Force Base, Florida, in 1969.[98,99] In some respects, it can be said to have been the first active aperture phased array, in that it employed 5184 individual transmitter units, one at each of the radiating elements. It operated at UHF (centered at 442 MHz). Separate receiving and transmitting arrays were used since it was cheaper to employ two arrays rather than one array with duplexers. The receiving aperture was larger than the transmitting aperture and employed 19,500 receiving elements. Only 4660 of the elements in the receiving array were active (had receivers connected to them), the rest were inactive and were terminated. The receiving elements were arranged in a thinned, space-taper manner to reduce the number of receivers required while maintaining a suitable sidelobe level. The transmitters used a highly reliable tetrode as the final amplifier stage to produce a peak power output of 10 kW at each element. The total peak radiated power was 32 MW and the beamwidth was 1.4°. This radar was considered a success. It has been upgraded, but for a long time it continued to use a vacuum tube as the final stage at each transmitting element since it was cheaper to do so than convert to solid state. The radar has performed well its role in detecting, tracking, identifying, and cataloging earth-orbiting objects.

The first all solid-state active aperture phased array was the AN/FPS-115, more commonly known as Pave Paws.[100] It operated at UHF and was designed to detect subma-

Figure 9.36   (a) The U.S. Army's THAAD *X*-band Ground-Based Radar (GBR) active-aperture radar for tactical ballistic defense. The 25,000-element array antenna equipment is shown at the lower left. Just behind it is the electronics equipment unit that houses signal and data processing, uninterruptable power supplies, receiver/exciter, and waveform generator. To the right is the cooling unit. At the upper left-center is the operator control unit. Not shown is the 1.1-MW diesel generator prime power unit.
I (Courtesy, Raytheon, Inc.)

rine-launched ballistic missiles fired at the United States. It had a secondary mission to perform space surveillance. Pave Paws employed 1792 active elements arranged in a circular aperture 22.1 m (72.5 ft) in diameter, plus 885 dummy elements. The peak power per T/R module was 335 W, which produced a total peak power of 600 kW and an average power of 150 kW per face. A Pave Paws radar consisted of two faces to cover 240° in azimuth. There were 56 subarrays, each containing 32 modules feeding 32 radiating elements. Its range was said to be 3000 nmi for a 10 m^2 target. There were four operational Pave Paws radars in the United States. One of these, located in Georgia, was expected to be increased in capability by 10 dB (by employing more elements) and replace the AN/FPS-85. A larger version of Pave Paws has also replaced the parabolic torus reflector antennas in the Ballistic Missile Early Warning System (BMEWS).

The THAAD (Theater High Altitude Area Defense) radar, Fig. 9.36a, is an active aperture radar designed for ballistic missile defense.[101] It has also been known as the Ground based Radar, or GBR. Radars for ballistic missile defense have to perform target detection, acquisition, track, identification (recognition), discrimination (of reentry vehicles from decoys and chaff), and assessment of target kill as well as in-flight communication to the defensive missile. The THAAD GBR is an *X*-band radar with 25,344

(b) EL/M-2080 *L*-band active aperture radar for the Israeli Arrow Tactical Ballistic Missile Defense System.
I (Courtesy IAI/ELTA Electronics Industries, Ltd.)

radiating elements, each with its own gallium arsenide T/R module. In order to be able to operate with a wide-bandwidth signal, the aperture is divided into 72 subapertures, each containing 352 active elements. There is a time-delay steering element at each subarray to permit the use of wideband waveforms without distortion. The array aperture is 9.2 m^2 (almost 100 ft^2), which is quite a large aperture for an X-band phased array. Because there are so many of them, the T/R modules are a very important part of this radar (or any active-aperture array) and are the largest cost element of the array. It was said that "every \$100 saved in the T/R module cost corresponds to \$2.5M for the complete array." The entire array weighed over 46,000 pounds.

A different solid-state active-aperture phased array radar for theater ballistic missile defense is the Israeli L-band EL/M-2080 shown in Fig. 9.36b. It performs search, acquisition, and fire control as part of the stand-alone Arrow weapon system.[102] It is said to have detection ranges of hundreds of kilometers and can simultaneously track and engage many tens of missiles.

The active-aperture phased array has also been considered for airborne (fighter/attack) radar and for ship self-defense radar. In the airborne application the number of modules (and radiating elements) might be from 1000 to 2000, and for shipboard air defense there might be from 4000 to 8000 modules. A serious limitation of any fixed electronically steered phased array radar mounted in the nose of a fighter aircraft is its limited angle coverage. Although a phased array is usually said to be able to steer $\pm 60°$ coverage in angle, the main-beam gain decreases, the beam broadens, and the sidelobes rise significantly even before the beam approaches 60° from boresight. A fighter aircraft, however, requires its antenna beam to steer to even greater angles than 60°. This is practical to do with a mechanically steered antenna, but not with an electronically steered phased array. In the AN/APG-77 radar for the F-22 fighter aircraft, provision was made as a growth feature to allow installation of a "side array" on each side of the aircraft to allow coverage at angles beyond that available with a fixed electronically steered phased array antenna.[103] The addition of the side arrays, of course, increases the cost of an already expensive active-aperture radar.

Example of a Russian Phased Array Radar Architecture The Russians have generally employed a different approach from the United States to the design of their phased array airdefense radar systems. The U.S. Army's C-band Patriot and the U.S. Navy's S-band Aegis systems use multifunction phased array radars that perform the various radar functions required for air defense with a single system operating within a single frequency band. As discussed later in Sec. 9.14, this represents a compromise since the optimum frequencies for search and for track of aircraft are different. The Russians, on the other hand, use separate radars for each function, and the radars operate at frequencies more suitable for their particular function. Since Russia is a vast country requiring many air-defense systems, they emphasized a low-cost approach to radar design.

The Russian air-defense system S300V (NATO designation SA-12), used a 10,000-element X-band lens-array radar for multiple-target tracking and weapon guidance. The NATO designation of this X-band radar is *Grill Pan*. (The information in this subsection is taken from a paper by David Barton.[104]) The low cost and low RF loss experienced by this system was due, in part, to the separation of the surveillance and tracking functions rather than to their being combined in one multifunction array. The space-fed lens array utilized multimode monopulse horn-feeds, so it did not experience the larger loss that a constrained-feed system might have. Faraday rotation dual-mode ferrite phase shifters were used which operated with circular polarization. Rather than convert the normal linear polarization to circular for operation in the phase shifter, and back again to linear (which is normally done with the dual-mode ferrite phase shifter), the two polarization transformations characteristic of U.S. dual-mode ferrite phase shifters were omitted by having the radar transmit and receive circular polarization. The array received the orthogonal circular polarization; that is, it received left-hand circular if right-hand circular was transmitted. (The polarization of the echo from an aircraft when illuminated by one sense of circular polarization contains both right and left circular polarizations in roughly equal amounts.) The ferrite phase shifters were nonreciprocal, but the phase shifters did not have to be reset after transmission in order to receive since the same phase settings were applicable when the received signal was of a circular polarization orthogonal to that transmitted. Since the polarization on receive was different from that on transmit, the receiver was partially isolated from the transmitter. The isolation due to the orthogonal polarizations, plus the use of a rugged cyclotron-wave electrostatic amplifier as the receiver front-end, eliminated the need for a duplexer or solid-state receiver protector, further reducing the loss. (The electrostatic amplifier had a 3-dB noise figure and could withstand average leakage power of several hundred watts and much higher peak power.)

The Russian ferrite phase shifters had two sections in series to provide 720° of phase shift (instead of the more usual 360°). Each section had its own control coil. One coil was for setting the phase required to steer in elevation and the other for the phase to steer in azimuth. All the row coils were in series with each other to provide azimuth steering and all the column coils were in series to provide elevation steering. The 10,000 element array (100 × 100) required only 100 row drivers and 100 column drivers. There were no driver or logic circuits, data busses, or d-c power busses needed in the aperture for determining the combined phase shift for steering in two angles. The total two-way RF loss from transmitter to receiver, excluding propagation loss, was about 3 dB for the Grill Pan. This compares to the 7 to 12 dB losses found in comparable Western systems.

The lack of individual control of each phase shifter, however, can cause the phase errors at the elements to be correlated over entire rows or columns of elements. Loss of a driver can result in the loss of an entire row or column of elements. The simplicity and low cost of this method for setting phase shifters make it more difficult to achieve low sidelobes.[105]

In addition to lower system cost and more optimum frequency usage, another advantage in using multiple radars in an air-defense system rather than a single multifunction phased array is that the individual radars can employ long-dwell medium-prf and high-prf pulse doppler waveforms that are needed to detect moving aircraft and missile targets in heavy clutter.

The *X*-band Grill Pan described above was the target tracking and guidance radar for the SA-12 air-defense system. Air surveillance was performed with the *S*-band Bill Board radar, which is a scanning beam 3D radar using a phase-scanned planar array with slotted waveguide radiators. The array could be stowed for transport in the short time of one minute. The SA-12 also employed a separate sector-search radar for detection of tactical ballistic missiles. The Russian SA-10, a similar air-defense radar system, also deployed on a tower a horizon search radar called Clam Shell for detection of low-altitude targets.

According to Barton, this approach to air defense seems to reflect "the Russian military's insistence on high performance against targets of low cross section in environments containing rain, chaff, and other sources of clutter, an almost insoluble problem when the multifunction approach is adopted."

Simultaneous Multiple Beams (Analog)   As has been noted, the phased array can form a number of multiple simultaneous beams. This is important for monopulse angle tracking; but in this subsection multiple beams are considered to be more than what is normally needed for tracking. In principle, an N-element array can generate N independent beams. Multiple beams allow parallel operation and a higher data rate than with a single beam. The multiple beams may be fixed in space, steered independently, or steered as a group. They can be generated on transmit as well as on receive. When multiple beams are generated on receive, the transmit beam can have a wide radiation pattern that encompasses the coverage of the multiple receive beams. In the past, multiple beams were generated by analog components, but it is now advantageous to employ digital methods for beamforming. Digital beam-forming is not appropriate for transmit, but this is not necessarily a limitation since the transmitting antenna is relatively simple and employs a broad radiation-pattern (omnidirectional in some cases).

The simple linear array that generates a single beam can be converted to a multiple-beam array by attaching additional fixed phase shifts to the output of each element. Each beam to be formed requires one additional phase shift per element, as in Fig. 9.37. For

Figure 9.37   Simultaneous beam formation on receive (three beams shown). $\phi_0 =$ constant phase; $|\phi_1 - \phi_0| = |\Delta\phi| = |2\pi(d/\lambda) \sin\theta_0|$.

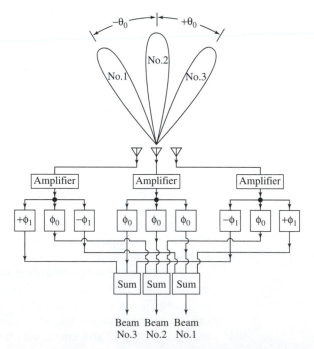

simplicity, the array in this figure is shown with but three elements, each with three sets of phase shifters. One set produces a beam broadside to the array ($\theta = 0$). Another set of three phase shifts generates a beam in the direction $\theta = +\theta_0$. The angle the beam points is determined by the relationship $\theta_0 = \sin^{-1}(\Delta\phi \, \lambda/2\pi d)$, where $\Delta\phi$ is the phase difference inserted between adjacent elements. Similarly, there is a set of phase shifters to produce a beam in the $-\theta_0$ direction. The receiving beam-forming networks may be at RF or at IF. At IF, tapped delay lines have been a convenient method to obtain the necessary phase shifts.

Butler Beam-Forming Array, or Butler Matrix[106]   This is an analog RF beam-forming network, an example of which is shown in Fig. 9.38. It consists of 3-dB directional couplers (or hybrid junctions) and fixed phase shifts to form N contiguous beams with an N-element linear array. The number N is an integer expressed as some power of 2, that is, $N = 2^p$. The 3-dB directional coupler is a four-port junction. A signal fed into one port will divide its power equally between two other ports and no power will appear at the fourth port. In the process, a 90° phase difference is introduced between the phases of the two equally divided signals. Likewise, a signal inserted at the fourth port will divide its power equally between the same two ports with a 90° relative phase difference, and no power will appear at the first port. The relative phase difference in this case is of opposite sign compared to the phase difference resulting from a signal introduced at the first port.

The example of a *Butler matrix* depicted in Fig. 9.38 is an eight-element array that produces eight independent beams. It utilizes 12 directional couplers and eight fixed phase

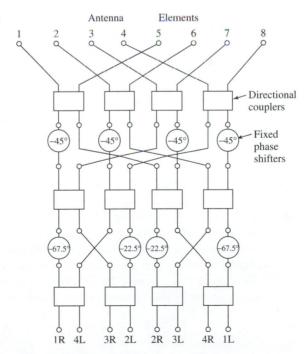

Figure 9.38   Eight-element Butler beam-forming matrix.

shifts. A Butler matrix has 2^p inputs and 2^p outputs. The number of directional couplers or hybrid junctions required for an N-element array is equal to $(N/2) \log_2 N$, and the number of fixed phase shifts is $(N/2) \log_2 (N - 1)$.

The Butler matrix is theoretically lossless in that no power is intentionally dissipated in terminations. There will always be, however, a finite insertion loss due to the inherent losses in the directional couplers, phase shifts, and transmission lines that make up the network. In a theoretically lossless, passive antenna radiating multiple beams, the radiation pattern and the crossover level of adjacent beams cannot be specified independently. With uniform illumination, as in the Butler matrix, the crossover level is 3.9 dB ($2/\pi$ in voltage) below the peak value of the beam. Crossover is independent of the beam position, element spacing, and frequency. The pattern is of the form $(\sin u)/u$, with -13.2 dB peak sidelobe.

The low crossover level of a Butler matrix is one of its disadvantages. If a lossless network could be achieved with a cosine illumination so as to reduce the peak sidelobe compared to that obtained with uniform illumination, the crossover level would be even worse (at a value of -9.5 dB); but the peak sidelobe would be -23 dB instead of -13.2 dB obtained with a uniform aperture illumination. By combining the output beams of the networks with additional circuitry, the Butler matrix can be modified to obtain an aperture illumination with lower sidelobes. The beamwidth will be widened, the gain lowered, and the network will no longer be theoretically lossless.

The flow diagram of the well-known fast Fourier transform (FFT) is similar to the circuit diagram of the Butler matrix.[107] The Butler matrix, however, was known before the FFT. Some of those familiar with the Butler matrix were surprised when they learned they were unknowingly using what would become an important procedure for the mathematical calculation of the Fourier transform. The FFT uses $(N/2) \log_2 N$ computations for an N-point transform, the same as the number of junctions needed for a Butler matrix.

The antenna equivalent of the conventional Fourier transform is called the *Blass beamforming network*. It requires N^2 couplers for N inputs and N outputs, just as the conventional Fourier transform requires N^2 computations for an N-point transform.

There have been other analog beam-forming methods considered for generating multiple beams with a phased array, as discussed in Sec. 8.7 of the second edition of this text. Most of them, as well as those mentioned above, are more suited for generating beams in one angular coordinate than in two coordinates.

Pattern Limitations in (Analog) Multiple-Beam Antennas   In the above discussion of the Butler matrix it was said that it was not possible to arbitrarily select the crossover level of adjacent beams. The crossover of 3.9 dB down from the peak of the beam is determined by the theoretical lossless nature of the Butler matrix beamformer. This crossover level, however, is a characteristic of any passive, lossless antenna that forms multiple independent beams when the aperture illumination is uniform, as was indicated by Warren White.[108] He showed that a passive lossless beam-forming antenna requires that the individual beam patterns be orthogonal in space, which means that

$$\int_0^{2\pi} d\theta \int_{-\pi/2}^{+\pi/2} E_j(\theta,\phi)\, E_k^*(\theta,\phi)\, \cos\phi\, d\phi = 0 \qquad \text{[9.44]}$$

where θ = the longitude angle on the unit sphere, ϕ = the latitude angle, $E_j(\theta,\phi)$ = the radiation pattern associated with the jth input terminal, and $E_k^*(\theta,\phi)$ = the complex conjugate of $E_k(\theta,\phi)$. Independent orthogonal beams mean that when two or more beam input ports are simultaneously excited, the resulting radiation is a linear superposition of the radiations that would be obtained when the ports are excited separately. In addition, when a signal is applied to one port it should have no output at the other ports. An antenna which is lossless and passive means that the radiated power is the same as the input power.

As has been mentioned, the $(\sin u)/u$ pattern produced by a uniform illumination is an example of a pattern with orthogonal properties and produces a crossover of 3.9 dB between adjacent beams. With a cosine-squared illumination, however, White stated that the crossover with a lossless passive multibeam antenna is 15.4 dB down, and for a Hamming distribution it is 18.4 dB down. These are unacceptably high values. They apply to one-dimensional antennas. For two-dimensional antennas with pencil beams, the crossover in dB is double.

White suggested that these poor values of beam crossover can be avoided by use of active elements (amplifiers) inserted between the antenna radiating elements and the resistive beam-forming network. The sensitivity of the receiving antenna is established by the active circuits (low-noise amplifiers) at each element. Any losses that follow the amplifier do not affect the overall receiving system sensitivity. This method for generating efficient multiple-beam array antennas is similar to what is done in a digital beam-forming array, as will be described later in this section.

Systems Considerations for Multiple Beam-Forming Arrays A surveillance radar with a broad-beamwidth transmitting antenna (beamwidth θ_T) and with N fixed, narrow receiving beams (each of beamwidth θ_R) that cover the same angular region (so that $N\theta_R = \theta_T$) can have performance equivalent to a conventional radar that uses a single rotating transmit-receive beam of width θ_R. Equivalent performance means, in this case, that the two different types of radars will be able to detect the same size targets at the same range and have the same revisit time, which, in the case of a mechanically rotating antenna, is the antenna scan time, or the time to rotate 360° in azimuth. This requires that the signal integration time for the radar with fixed multiple beams also be the same as the time for the rotating antenna to make one revolution. The transmitting antenna gain of the multiple-beam radar is $1/N$th that of the scanning single-beam system. However, the observation time available to each of the fixed beams of the multiple-beam system is increased by $\theta_T/\theta_R = N$. If the signal is integrated without loss over this time, the reduction in transmitting antenna gain in the multiple-beam system is just compensated by the increased signal energy obtained because of its much longer integration time. The assumption of no integration loss is correct for perfect pre-detection (coherent) integration. When post-detection integration is used, however, there will be a theoretical integration loss so that the two radar systems will not be of exactly equivalent performance.

In a radar with fixed multiple receiving beams and a single broad transmitting beam, the two-way sidelobes are not as low as with a single-beam scanning radar. In some applications where low two-way sidelobes are important, it may be desirable to decrease the antenna sidelobes of the multiple receiving beams more than normal.

Multiple beam-forming array radars that use analog beam-forming are usually more expensive and require more equipment than a conventional scanning surveillance radar. They also do not seem to offer significant advantages in performance or capabilities over a conventional scanning radar. For these reasons they have seldom been considered for operational radar system applications. The production of multiple beams using digital beam-forming, to be discussed next, can also be expensive, but the use of digital technology offers advantages not readily available otherwise. Digital beam-forming, therefore, is of much more interest for multiple-beam radar applications than are analog beam-forming methods.

Digital Beam-Forming (DBF)[109–113]   Almost all modern phased array radars utilize some sort of digital phase shifting to form a beam; but the term *digital beam-forming* (DBF) usually means something different—the formation of multiple receive beams by digitizing the outputs of the receiving array elements and forming beams by means of a digital processor. Although a single beam can be formed in this manner, digital beam-forming has more often been considered when multiple, simultaneous, directive beams are wanted. When digital beam-forming is employed to generate multiple receiving beams, the accompanying transmitting antenna has a broad beamwidth covering the same total angular region as the multiple receiving beams. The outputs from the multiple beams can be processed in parallel by the radar.

The basic configuration of a digital beam-forming receiving array might be as shown in Fig. 9.39. At each antenna element there is a receiver whose analog signal is digitized by the A/D converter. The A/D converter is a critical component since it can place a limit

Figure 9.39   Basic configuration of a digital beam-forming receiving array.

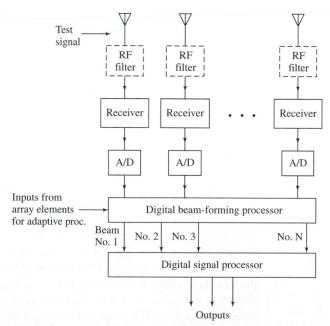

on the system bandwidth and dynamic range. The lower the frequency at which the A/D converter operates, the greater can be its dynamic range and bandwidth. For this reason the input to the receiver is usually heterodyned (downconverted) to a lower frequency, filtered, and amplified to a power level suitable for the A/D converter. The receivers must be closely matched to one another so that the relative amplitude and phase of the signals at each element are preserved. If other than a uniform aperture illumination is desired (for example, to reduce the antenna sidelobes), amplitude weights can be applied to the digital output of each element and the weighted signals combined in the processor to produce the antenna beam. If a linear phase weight is applied to the digitized signal at each element, the antenna beam can be made to appear as if steered to a different angular direction. In this manner the digital processor can produce a multitude of receiving beams, each pointing in a different direction. An antenna beam, of course, is not actually formed in space by the processor. The pattern is evident as the variation of the output response of the digital signal processor as a function of angle.

Since the signal-to-noise ratio is established at the digital output of each receiver, there is no loss in signal-to-noise ratio when manipulating the digital outputs to form multiple beams. There can be any number of closely spaced antenna beams formed without degradation in SNR. Digital beam-forming, therefore, does not suffer the limitations inherent in analog beam forming, which are the low crossover of adjacent beams and the loss that occurs when trying to obtain lower peak sidelobes than the -13.2 dB of a uniform illumination.

The operation of the beam former is to take the outputs from each element, apply a complex weight (an amplitude and/or phase) to each, and then sum them to provide the output signal. The form of this output is similar to a digital Fourier transform (or inverse digital Fourier transform). When the phase weights are expressed in terms of the beam pointing angle θ_0, the relative response of the weighted outputs of the N-array elements as a function of the angle θ is represented as

$$g(\theta,\theta_0) = \sum_{n=1}^{N} a_n s_n \exp\left[j2\pi n(d/\lambda)(\sin\theta - \sin\theta_0)\right] \qquad \text{[9.45]}$$

where a_n = weight (amplitude for sidelobe control), s_n = output of the nth element, d = element spacing, λ = wavelength, and θ_0 = direction of the maximum beam response. In this manner, $M \leq N$ multiple beams can be formed. For each of the M beams there is a temporal sequence of digital numbers (data) that represent the temporal signals received at each beam position.

The computation of the Fourier transform in real time by the signal processor when there are many elements in the array can require a large number of operations per second. The computation requirements can be significantly decreased if the fast Fourier transform (FFT) is used. The FFT requires that the number of elements be some integer power of 2 (i.e., $N = 2^p$ and p = integer). There is less control of the beam patterns, however, with the FFT than if the conventional digital Fourier transform is used. The patterns of each FFT beam (in $\sin\theta$ space) are the same, the peak sidelobes will be high, and the crossover of adjacent beams is predetermined (by the orthogonality constraint) and may be lower than desired. Thus with the FFT one trades control of the radiation pattern for ease of computation.

Two-Dimensional Beam-Forming   The above description of digital beam-forming assumed a linear array forming beams in one angular dimension. With a two-dimensional rectangular array of $M \times N$ elements, beam-forming in two angular dimensions can be accomplished with the two-dimensional FFT. An N-point FFT is taken on each of the M rows to obtain $N \times M$ outputs. This is followed by an M-point FFT on each of the N columns to provide $M \times N$ beam outputs.[112]

Baseband versus IF Digitizing   In any radar the digitizing of the received signal can be done either in the IF stage of the receiver or at baseband (zero IF) with I (in-phase) and Q (quadrature) channels. The I and Q channels at baseband are obtained in a similar manner to that described in Chapter 3 for an MTI radar (Fig. 3.29). It will be recalled that the received signal is divided into two baseband channels. The two channels are identical except that the phase of the reference oscillator in one differs from the reference phase in the other by 90°. The bandwidth of the signals in each of the two baseband channels is half that of the signal bandwidth; but two A/D converters are necessary, one in each channel. The smaller bandwidth of each A/D can sometimes be of advantage as opposed to requiring digitizing with the full signal bandwidth in the IF stage. To obtain good performance, the phase difference between the I and the Q channels should not deviate significantly from 90°. This requires careful design or some form of feedback control to maintain the precise phase relationship between the two channels. The signals in the two channels must also be well matched in amplitude.

Digitizing of the received signal may also be done directly in the IF of the receiver. There is only one channel and one A/D converter, so that the problem of balancing the phases and amplitudes of the I and Q channels is not present. The A/D converter has to be capable of greater bandwidth (sampling rate) than with a baseband I,Q arrangement. P. Barton[109] states that the minimum sampling rate of an A/D converter in the IF has to be 5.4 times the signal (half-power) bandwidth. (The sampling rate has to be greater than the theoretical Nyquist rate of $2 \times$ bandwidth because of the need to avoid distortion of the signal spectrum caused by the folding of the spectrum around zero frequency.) This compares with a minimum sampling rate of 1.4 times the signal bandwidth required for each of the two A/D converters of the equivalent baseband implementation.

A/D Converters   The sampling rate and the dynamic range of an A/D converter can set limits on what might be achieved in a digital processor. In practical A/D converters the greater the sampling rate (bandwidth), the smaller will be the number of bits into which the A/D converter can digitize a signal. In addition, as the bandwidth or number of bits increase, the size and cost of the converter can also increase. The limitations of A/D converters are practical ones, but their capabilities have been continually improved over the years. Section 3.5 gives values of the performance of A/D converters, but it is always risky to state in a text such as this what the state of the art of such devices might be. The prudent engineer, of course, should always consult current manufacturers catalogs for up-to-date information.

Other Characteristics of Digital Beam-Forming   In addition to the attractive aspects of digital beam-forming (DBF) array radars mentioned previously in this subsection, the following are further favorable attributes of such radars.

Self-calibration and error correction—Errors in the phase and amplitude that occur in the analog portion of the receivers of a DBF can be compensated with relative ease in the digital portion. This may be obtained by injecting a precise RF signal at the front-ends of each receiver, or by placing an external RF source at a known position in the near or far field of the antenna, or by use of the echo signal from a well-defined scatterer of the transmitter signal.[110] This precise RF signal is used as the standard to adjust the phase and amplitude at each element.

Low antenna sidelobes—The ability to digitally self-calibrate the array allows the potential for achieving low or even ultralow receiving antenna sidelobes after digital processing. The effect of mutual coupling can also be compensated. It has been said:[114] "Mathematically, the compensation consists of a matrix multiplication performed on the received signal vector. This, in effect, restores the signals as received by the isolated elements in the absence of coupling. An attractive feature is that this matrix is fixed and thus is valid for all desired pattern shapes and scan directions. Although it may be difficult to realize in analog form, it can be readily implemented in a digital beam-forming antenna system."

Adaptive nulling—The flexibility of a DBF array allows nulls to be placed in the antenna radiation pattern in the direction of noise sources (jammers) so as to reduce the noise that enters the receiver. The placement of nulls in a receiving antenna pattern to cancel unwanted noise sources that enter via the antenna sidelobes is a well known technique, and can be done with much simpler systems than a DBF array.[115] Sidelobe cancellation, for example, can be readily achieved with a mechanically scanned reflector antenna and a relatively small number of auxiliary elements. But when one has a DBF array radar, it can be implemented to cancel noise sources that appear in the antenna sidelobes in a manner different from the conventional sidelobe canceler or a fully adaptive array, and with advantages not found in radars that do not have multiple beams. Nulls are achieved in "beamspace" by using one or more formed beams properly attenuated to adaptively cancel noise sources rather than using one or more omnidirectional elements.[116,117,110] If there are J noise sources to be nulled, the angular location of each source is found in a conventional manner and J directive beams are then formed to adaptively cancel each source. Use of beamspace allows cancellation without disturbance to the main beam or the main-beam sidelobes, except in the immediate vicinity of the noise sources. The use of beamspace for sidelobe cancellation has not normally been used in the past with conventional radar because of the complexity in forming multiple independent beams by other methods. In a DBF array, however, this is not a consideration since the forming of multiple independent beams is what such a system normally does anyway.

Adaptive nulling of clutter as a function of range—Nulls can be formed adaptively in those directions where there are large clutter echoes as well as in those directions in which there are noise sources. Unlike noise, however, clutter might be limited in range (for example, a large mountain or a patch of chaff). In such cases, range-dependent antenna pattern nulls can be formed only around those individual range cells containing localized clutter or chaff.

Correction for failed elements[111]—The complete failure of a sufficient number of antenna elements can seriously degrade the performance of a low sidelobe array antenna. It has been said[118] that it is possible to compensate, however, for the loss of antenna elements in a digital beam-forming receive array by using simple linear operations with the outputs of a small group of good elements within the array. By properly utilizing the signals received at P elements of the array when P signals are received from different incident directions, it is possible to reconstruct the signal that would have appeared at the failed elements, if all of the array antenna elements have the same radiation pattern and some other restrictions apply. The technique has been said to work when the incident signal directions are not precisely known or even when they are only known to be within a broad angular sector.

Flexible data rates—As has been mentioned, the digital beam-forming array which looks everywhere can have a data rate that varies with the operational situation. This is unlike a conventional rotating radar whose data rate is fixed by the antenna rotation rate. At long ranges, a high data rate is not as important for an air-defense surveillance radar as it is at short ranges where weapons are engaged. Thus a lower data rata can be employed at long range which means that the integration times can be increased and improved target detection can be obtained without an increase in transmitter power. The flexibility of a DBF array to provide unrestricted data rates is important to a military weapon-control radar that should operate with high data rates during an engagement or when the target is seen to maneuver.

Simultaneous multiple functions—In an ubiquitous radar (one which looks everywhere all the time with fixed multiple receive beams) that uses digital beam-forming, it is possible to perform multiple functions simultaneously rather than sequentially. A conventional phased array such as Patriot or Aegis, on the other hand, has to time-share its various functions. Sometimes such radars run out of time to perform all the various functions required of them, so that some functions with lower priority have to be neglected in favor of more important ones. In an air-defense system, for example, priority will go to targets actually being engaged by missiles or to the searching for low-altitude pop-up targets, rather than to long-range surveillance. The ubiquitous radar with DBF, however, can perform its various functions simultaneously so long as it doesn't run out of computer capability.

Improved noncooperative target recognition—The ability to see everywhere all the time with whatever data rate is required is of benefit for those methods of noncooperative target recognition that depend on an observation time longer than normally needed for detection. The imaging of a ship or an aircraft by use of inverse synthetic aperture radar, for example, requires that the aspect of the target change sufficiently so that recognizable doppler-frequency shifts from different parts of the target can be isolated (resolved). Time is needed to allow for the target aspect to change, and time is something available with a DBF array. Recognition of helicopter targets based on the transient "flash" of the rotating blades when they are briefly oriented perpendicular to the radar line of site requires that the radar dwell long enough on the target to detect this phenomena.[119] The fixed beams of DBF provide this flexibility in target observation time.

Lower probability of intercept—It is relatively easy for an intercept receiver to detect the radiated signals of conventional radars at long ranges. To reduce a radar's detectability to a hostile intercept receiver, its peak power should be made as low as possible. The radiated energy should be spread over a wide angular region, over a long time interval, and over a wide frequency band. The ubiquitous DBF array offers the ability to spread the radiated energy over a wide angular region, something not possible with a scanning directive transmitting beam.

A beam-forming array that produces many multiple beams has capabilities not readily obtained with a conventional radar that employs a mechanically scanning antenna. Digital beam-forming offers advantages over analog beam-forming in generating multiple beams in that the same digital outputs from each array element are reused to readily generate multiple beams as well as perform other types of spatial and temporal processing. Digital processing has increased the feasibility and capability of the ubiquitous radar.

Examples of Digital Beam-Forming Arrays   Digital beam-forming has been employed in both HF over-the-horizon radar and in 3D air-surveillance radar. Neither of these applications employ the full benefits of digital beam-forming arrays as described above, so they are not indicative of what might be done with this radar concept.

The U.S. Navy's Relocatable Over-the-Horizon Radar (ROTHR), AN/TPS-71, is designed to detect aircraft and ships at ranges from 500 to 2000 nmi.[120] It employs a receiving linear-array antenna 2.7 km in length with a total of 372 monopole antenna elements. At each antenna element there is a receiver that converts the signal to an IF frequency. The digital outputs of these 372 antenna elements are used to form 16 contiguous receiving beams that can be placed anywhere within the radar's angular coverage. Both spatial processing (beam forming) and temporal processing (doppler filtering and matched filtering) are performed with the same digital data from each receiving antenna element.

The *SMART 3D radar*, developed by Signaal of the Netherlands, is a 3D air-surveillance radar with an azimuth-rotating antenna which generates a number of simultaneous multiple-beams in elevation to provide a measurement of elevation angle using digital beam-forming. The original version of this radar was at S band, a later version was at L band. The L band radar formed 16 beams and the S-band version 14 beams.

9.10 MECHANICALLY STEERED PLANAR ARRAY ANTENNAS

Mechanically steered planar arrays offer important advantages not available with conventional reflector antennas in some radar applications. They have been employed at the lower radar frequencies (VHF) for air-surveillance applications, at microwave frequencies for 3D radars which need to obtain a measurement of target elevation angle, in airborne radars that operate from the nose of the aircraft, in missiles, civil marine radars, and in low-sidelobe antennas such as used for AWACS. The large number of elements in an array aperture allows better control of the aperture illumination and therefore, better control of the antenna radiation pattern.

Mechanically Rotating Arrays for Air Surveillance   The first radar antennas developed by the United States in the 1930s, such as the Army's 100-MHz SCR-270 and the Navy's 200 MHz CXAM were mechanically steered phased arrays. At these low radar frequencies, it was natural to use planar arrays of dipoles since they were consistent with the communications antenna technology of that time. Also, they were well suited for the VHF frequencies at which the early radars operated. As radar frequencies increased, the parabolic reflector was introduced since it was simpler than a planar array when there had to be a large number of dipoles. The German World War II Wurtzberg radar that operated at 550 MHz employed a parabola. The first U.S. microwave air-defense radar, the S-band SCR-584, also used a parabolic reflector. The parabolic reflector was well-known in optics, and it was not difficult to translate its technology and theory from optics to microwave frequencies.

The early air-surveillance radars were 2D in that they measured azimuth angle and range, but not elevation angle (or height). When pencil-beam 3D radars (such as the AN/SPS-39) first appeared in the 1950s to obtain elevation angle, they originally used a parabolic cylinder antenna with a line feed. These were followed by planar arrays (as in the AN/SPS-48) consisting of rows of slotted waveguides to obtain multiple beams in elevation. These original 3D systems utilized frequency-scanned phased arrays to electronically steer one or more beams in one angular coordinate (elevation). The other angle coordinate was obtained by mechanically rotating the entire antenna 360° in azimuth.

Almost all modern 3D air-surveillance radars employ a planar array antenna of some type that is mechanically steered in azimuth and which uses some form of electronic steering or beam-forming in elevation. Mechanical rotation is satisfactory for the air-surveillance application since it is not necessary to have rapid beam switching among many targets as it is in the weapons control application. The mechanically rotated planar array not only provides a convenient means for obtaining electronic steering in elevation, but the availability of many elements in the antenna also provides more flexibility in achieving the desired radiation pattern.

Mechanically Steered Slotted Planar Arrays[121–123]   This type of planar array antenna is widely used for radars mounted in the nose of an aircraft. An example is shown in Fig. 9.40. Such antennas are common for military radars used in fighter and attack aircraft, airborne weather radar, and for missiles. An important advantage of this type of antenna for application in the nose radome of an aircraft is that it can be made relatively thin. This allows a larger diameter antenna to be mechanically scanned within an aircraft's nose radome than is possible with a relatively thick parabolic antenna and a feed that projects a distance from the reflector surface.

Microwave planar array antennas generally employ radiators that are slots cut into waveguide, a simple example of which was shown in Fig. 9.31c. They may be edge slots cut into the narrow wall or slots cut into the broad wall. The use of slots cut into the narrow wall allows the rows of an array to be spaced closer than if they were cut into the broad wall. The slot is the so-called *Babinet equivalent* of the dipole. Its pattern resembles that of a dipole, except that a vertical slot radiates horizontal polarization and a vertical dipole radiates vertical polarization. The waveguide feed structure lies directly behind the radiating slots of the array antenna.

There are two basic methods for structuring a series-fed mechanically scanned slotted waveguide radiator. One is called the standing wave, or resonant, configuration. The other is the traveling wave, or nonresonant, configuration. The spacings between elements in a resonant array is half a guide wavelength. (The guide wavelength is the wavelength measured in the waveguide rather than in free space.) The beam is broadside to the aperture, but the half-wavelength spacing between elements means that impedance mismatches

Figure 9.40   Mechanically steered planar array antenna for the AN/APG-68 airborne radar found on the F-16 aircraft.
| (Courtesy Northrop Grumman Corp.)

at the elements can accumulate and cause a high VSWR (large mismatch). This results in restricting the number of elements in a linear array to less than 20,[124] which might be too low for most radar applications. Resonant arrays also have narrow bandwidths. The nonresonant array usually has an element spacing greater than half a guide wavelength and so does not experience the high VSWR or the narrow bandwidth of the resonant array. The waveguide of the nonresonant array, however, has to be terminated with a matched load to absorb the fraction of the input power that is not coupled to the other elements. The amount of power lost because of the matched load is about 5 to 10 percent of the power at midband.[121] Thus the array efficiency will be less than unity. (If the array is not well matched, a portion of the signal can be reflected from the termination and radiate as a high sidelobe in some spurious direction.) The direction of the peak of the beam radiated by a series-fed nonresonant array varies with frequency; similar to a frequency-scan array, but much less dramatic. With a wavelength λ, the beam is directed at an angle θ given by

$$\sin \theta = \lambda/\lambda_g - \lambda/2d \qquad [9.46]$$

where λ_g = guide wavelength, and d = element spacing. Yee and Voges[123] state that in most nonresonant slot arrays, $2d$ is selected to be greater than λ_g so that the angle of the beam will move toward the load-end of the array as the frequency is increased. The frequency dependence of the beam direction might need to be taken into account in some applications.

The mechanically scanned planar array can be designed to provide monopulse beams in two angular coordinates.

Endfire Arrays   In the above we have considered the array that radiates its main beam perpendicular (broadside) to the aperture. It is also possible to radiate a beam parallel to the aperture. These are known as *endfire array antennas.* If we consider the dipole as the radiating element, it must be oriented so that its element pattern allows radiation in the endfire direction, and the spacing between elements and the phase shift at each element allow propagation in the endfire direction. For example, the spacing between elements in an equal-amplitude linear endfire array fed from one end might be a quarter-wavelength, with phase shifts of $\pi/2$ radians between elements, to give an antenna pattern with most of its energy oriented in one endfire direction. In this example, the phase is retarded progressively by the same amount as that experienced by the traveling wave from one element to the next. (A phase retardation of 0.6π radians actually produces a higher directivity for this endfire antenna, according to the Hansen-Woodyard criterion, as mentioned by Kraus.[125])

The Yagi-Uda antenna, which originated in Japan, is a simple and inexpensive example of an endfire array. It consists of a single driven dipole plus several spaced parallel rods that form an endfire array, Fig. 9.41. Each rod may be thought of as a short-circuited dipole. The rod located behind (to the left of) the driven dipole acts to reflect the energy to the forward direction. One or more spaced rods in front of the driven dipole direct the energy forward. The rods are known as either *reflectors* or *directors*. The reflector might have a length of about a half-wavelength and be spaced a quarter-wavelength behind the driven dipole. The directors are slightly smaller (by about 10 percent) with spacings about a third of a wavelength.

Endfire antennas can be arrayed and have been used in radars, especially at VHF and UHF. An example is the antenna for the E2C Airborne Early Warning radar (that was shown in Fig. 3.45a). An advantage of the endfire antenna for this application is that a narrower beamwidth can be obtained in the vertical dimension without the need for a large vertical aperture. The beamwidth of an endfire antenna in the dimension orthogonal to its

Figure 9.41   Sketch of a Yagi-Uda endfire antenna. (The mechanical mount is insulated from the dipoles.) For radar application they have been used in a linear array configuration; one example is that of the E2C AEW radar shown in Fig. 3.45a.

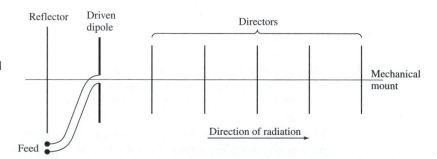

longitudinal axis is proportional to the square root of the antenna length, as compared to a conventional broadside antenna where the beamwidth is directly proportional to the antenna size.

Rotating Electronically Steered Phased Arrays   At times it has been suggested[126,127] that a single face of an electronically steered phased array be mechanically rotated in azimuth but electronically steered in azimuth and elevation. Whether the single rotating phased array is an attractive alternative to a four-faced fixed phased array depends on the particular application and the assumptions that are made. Caution should be exercised, however, when considering this approach since in some cases it is conceivable that a rotating single phased array face might provide the worst attributes of both the phased array and the mechanically rotated antenna rather than the best of both. The mechanically trainable phased array, described next, is different and has some important operational advantages.

Trainable Phased Arrays   A trainable phased array is one which is electronically steered in both azimuth and elevation over a wide angular sector, but which is mounted so as to be mechanically positioned to cover a desired sector. (Once in position it remains fixed rather than continuously rotated.) It is a convenient method for using an array for missile-range instrumentation. An example is the transportable *C*-band AN/MPS-39 MOTR (multiple object tracking instrumentation radar) developed by Lockheed-Martin at Moorestown, N. J. (shown in Fig. 4.1b). The 12-ft-diameter space-fed lens array with 8359 elements is mounted on a precision elevation-over-azimuth tracking pedestal so as to achieve coverage of a 60° cone anywhere within the full hemisphere. Its beamwidth is one degree and its measured antenna gain is just under 46 dB. In missile range applications, many targets might have to be tracked simultaneously; including the firing aircraft, the target missile or drone, surface-to-air missiles, air-to-air missiles, and other aircraft that might be on the range (range safety). The MOTR can simultaneously track up to 10 targets with an absolute angle accuracy of 0.2 mils rms and a range accuracy of 1 yd rms. Prior to such radars, a separate air-surveillance radar and more than one mechanical tracking radars had to be used.

The trainable phased array is also of advantage for ship air-defense when the agility of a phased array is needed. Traditionally, when a phased array is used on board a naval ship (for example, the *S*-band Aegis system or the U.S. Navy's original *S*-band AN/SPS-33 phased array radar system), there are four planar array faces mounted around the ship to provide 360 degrees of all-around coverage. Although the designer might want to configure the phased array so that all four apertures can operate simultaneously, the need to reduce the cost of such radars has sometimes resulted in having only one or two transmitters and receivers time-shared among the four array apertures. This might result in acceptable performance, but it is less than what could be obtained if the four faces were able to simultaneously operate all the time.

An alternative method for configuring an air-defense phased array radar system when four complete phased array faces are too costly (or even when they are not), is to employ two trainable arrays instead. An example used by the Russian navy on Cruisers is the trainable array radar shown in Fig. 9.42 which is known by its NATO nomenclature as the Top Dome SAN-6, or by its Russian/Soviet name of RIF. In a Russian Navy Cruiser,

Figure 9.42 Soviet/Russian electronically scanned trainable phased array pulse doppler radar whose Russian name is RIF. Its NATO designation is Top Dome and is part of the SAN-6 shipboard surface-to-air missile system. It is said to employ row-and-column beam steering. The antenna assembly includes a hemispherical radome. In addition to the main phased array there is a wide-angle phased array for acquisition of the surface-to-air missiles just after launch.

one Top Dome radar is located fore and a second one is located aft. (A truism is that if something is important to have on a naval ship—such as a gun, missile system, or radar—there ought to be at least two of them to insure that one is available when needed.) The two trainable array radars can cover attacks simultaneously within any two 90° sectors. An important advantage with this configuration is that one or more attacks from within a single 90° sector can be engaged by both radars simultaneously. A four-face phased array, on the other hand, can bring only one array face to bear to defend against such an attack.

Two trainable phased arrays cannot engage a multiple simultaneous attack from more than two 90° sectors. The likelihood of this being a serious concern is small. The ability to mount a simultaneous attack from three different directions over 360° in azimuth is certainly possible, but having all targets appear simultaneously at the target ship from different directions is very difficult to do. If they are not simultaneous, the ship's air-surveillance radar with 360° of azimuth coverage can be expected to detect and recognize such attacks and the two trainable arrays can be scheduled to engage them without over-lap. Also, it is seldom that a major naval ship that carries an expensive phased array radar for air defense operates by itself; so if there is a large multiple attack, one ship does not have to handle the total attack all by itself. Thus employing trainable phased arrays offers advantages not found with the traditional four-faced phased array system.

9.11 RADIATION-PATTERN SYNTHESIS[128]

A radar antenna radiation-pattern is required to have a specified beamwidth along with acceptably low sidelobe radiation. In some cases, the antenna radiation pattern must provide a desired contour, or shape, over a specified angular region. An example is the cosecant-squared elevation pattern of an air-surveillance radar. The aperture illumination for the squinted beams of an amplitude-comparison monopulse tracking radar, on the other hand, must have a suitable sum pattern with low sidelobes, as well as a suitable difference pattern with low sidelobes and a large slope at beam crossover. This section reviews some of the methods available to the radar antenna designer to achieve the radiation patterns necessary to produce the desired radar performance.

As was mentioned previously in Sec. 9.3, the radiation pattern is determined by the distribution of current across the aperture. We have called the distribution of current the aperture illumination. Equation (9.10) gave the (electric field strength) radiation pattern $E(\phi)$ of a linear one-dimensional antenna in one angle-coordinate ϕ as the inverse Fourier transform of the aperture illumination $A(z)$. Similarly the radiation pattern of a two-dimensional planar antenna is given as a two-dimensional inverse Fourier transform of its aperture illumination. We shall first consider the problem of obtaining a desired main-beam pattern with acceptable low sidelobes and then the problem of obtaining shaped radiation patterns.

Obtaining a desired antenna pattern is slightly different for a continuous aperture (such as a reflector) than a phased array that consists of many individual elements. One can sometimes approximate the continuous (reflector) aperture illumination with a discrete (array) aperture illumination, and vice versa. The discussion of antenna patterns in this section is done chiefly for a linear one-dimensional aperture or for rectangular apertures where the illumination is separable; that is, $A(x,z) = A(x)A(z)$. When the illumination for an array is considered, the array is assumed to have uniformly spaced isotropic elements with element spacing generally taken to be half-wavelength. The radiating elements of a real array are not isotropic but have some element pattern $E_e(\theta)$. If the desired array antenna pattern is $E_d(\theta)$, the pattern to be found is $E_d(\theta)/E_e(\theta)$ when using a technique based on the assumption that the elements are isotropic.

Patterns with a Desired Beamwidth and Low Sidelobes   Obtaining a pencil-beam or a fan-beam radiation pattern is not usually a synthesis problem. Instead, the patterns obtained from various aperture illuminations are calculated and a suitable one is selected. Table 9.1 in Sec. 9.3 lists a number of antenna patterns for several types of aperture illuminations that can be expressed in analytical form. These were considered in the past since their analytical form permitted the corresponding radiation patterns to be readily calculated. With modern computers, however, using aperture illuminations just because they are readily integrated is no longer necessary. Thus these aperture illuminations are not now generally used. The table is useful in that it illustrates how the maximum gain, beamwidth, and maximum sidelobe level vary with change in shape of the aperture illumination. The more tapered the aperture illumination (that is, the more rapidly its amplitude falls off as a function of the distance from the center of the aperture) the lower will be the sidelobe level, but the lower will be the antenna gain and the wider will be the beamwidth. In practice, other aperture illuminations, such as Taylor illuminations, are usually used rather than those given in Table 9.1.

Taylor Aperture Illumination   For a specified maximum sidelobe level, the antenna pattern which has all of its sidelobes equal produces the narrowest beamwidth (where the beamwidth is measured by the angular distance between the first nulls that define the main beam). This is known as a Dolph-Chebyshev pattern since it was first shown by C. L. Dolph, a mathematician working at the U.S. Naval Research Laboratory during World

War II, who based it on equating the Chebyshev polynomial to the polynomial describing the pattern of an array antenna.[129] In spite of its desirable properties, the Dolph-Chebyshev pattern is seldom used for radar antennas since it is unrealizable with arrays containing other than a small number of elements. As the antenna size increases, the currents required at the ends of the aperture become nonmonotonic and large compared with the currents along the rest of the aperture. More of a restriction is the fact that these large currents are required to occupy a very narrow spatial region at the ends of the aperture, too narrow to be obtained with an actual antenna. This inability to achieve the theoretical aperture illumination sets an upper limit to the size of an antenna that can have a Dolph-Chebyshev pattern and therefore sets a lower limit to the width of the main beam that can be achieved.

A realizable approximation to the Dolph-Chebyshev aperture illumination was devised by T. T. Taylor.[130] The Taylor aperture illumination, as it is known, has been widely used for radar antennas. It produces a pattern with equal-amplitude sidelobes of a specified value, but only in the near vicinity of the main beam. Unlike the equal-sidelobe level of the Dolph-Chebyshev pattern, the sidelobes of the theoretical Taylor pattern are of uniform amplitude only within an angular region ϕ defined by

$$|(D/\lambda) \sin \phi| < \bar{n}$$

where $\bar{n}$ = integer, D = antenna dimension, and λ = wavelength. With a linear aperture, the sidelobes decrease as $1/\sin \theta$ with increasing angle θ outside this region (similar to the fall-off of a pattern with a uniform illumination). Hence $\bar{n}$ divides the sidelobes into a uniform region, which straddles the main beam, and a decreasing sidelobe region. The number of equal sidelobes on each side of the main beam is equal to $\bar{n} - 1$. The integer $\bar{n}$ is usually a small number. (Sometimes it is difficult to observe either on a calculated or an actual antenna pattern a region of equal sidelobes in the vicinity of the main beam, yet they are a part of the theoretical Taylor pattern.) The beamwidth of a Taylor pattern will be broader than that of a Dolph-Chebyshev. If the Taylor design sidelobe level is -25 dB, a value of $\bar{n} = 5$ gives a beamwidth almost 8 percent greater than the theoretical Dolph-Chebyshev. With $\bar{n} = 8$, the beamwidth is 5.5 percent greater.

The Taylor pattern is specified by (1) the peak design sidelobe level and (2) the value of $\bar{n}$. The integer $\bar{n}$ can take on only a small range of values for a given design sidelobe level. Taylor states that $\bar{n}$ must be at least 3 for a design sidelobe of -25 dB and at least 6 for a design sidelobe of -40 dB. The larger the value of $\bar{n}$ the sharper will be the beam. On the other hand, it cannot be too large since the same realizability difficulties will arise as with the Dolph-Chebyshev. A suitable criterion for obtaining a realizable Taylor pattern is to choose an illumination that decreases monotonically from the center out to the ends of the aperture and has a zero derivative at the ends of the aperture. The illumination need not be zero amplitude at the ends but can have a finite value (a pedestal). Taylor showed that a distribution with a pedestal, or nonzero value at the edges, is more effective in producing low sidelobes.

A rough guide to the selection of the parameter $\bar{n}$ has been given by Hansen.[131] For example, when the aperture illumination is monotonic, he states that the value of $\bar{n}$ equals 5 for a peak sidelobe of -25 dB, 9 for a peak sidelobe of -35 dB, and 11 for a peak sidelobe of -40 dB. The aperture efficiencies for these three examples are, respectively, 0.91, 0.82, and 0.77. (Hansen later added that the maximum value of $\bar{n}$ for a monotonic illumination is 17 for a peak sidelobe of -50 dB and 23 for a peak sidelobe of -60 dB.[132]) Thus care must be exercised in the selection of the sidelobe level and the value of $\bar{n}$ for a Taylor pattern. Although the Taylor pattern was developed as a realizable approximation to the Dolph-Chebyshev, in practice it seldom resembles the theoretical equal-sidelobe pattern.

Taylor aperture illuminations can also be obtained for a circular aperture.[133] Figure 9.43 illustrates the nature of the Taylor circular aperture illumination and their corresponding radiation patterns. (This is from an Institute for Defense Analyses report[134] that was not widely circulated, but some of the information appears in the *IEEE AP-S Transactions.*[135]) Shown in Fig. 9.43a are four circular-aperture antenna patterns with the same half-power beamwidth, $70\lambda/D$, but with different values of the Taylor $\bar{n}$ parameter varying from 3 to 15. As $\bar{n}$ increases, the peak sidelobe decreases. From these patterns it would appear that one would want to select a large value of $\bar{n}$. There is a problem with this, however, as can be seen from their corresponding aperture illuminations in Fig. 9.43b. One might be able to realize the required aperture illumination for $\bar{n} = 3$, and might even be able to roughly approximate the illumination for $\bar{n} = 7$. But at the higher values of $\bar{n} = 10$ and 15, it is not likely that one can achieve the necessary aperture illuminations at the edge of the aperture.

There have been several other methods for selecting the aperture illumination for a conventional antenna pattern, as can be found in many of the texts[74,136] on antennas; but the Taylor seems to have been the most popular for radar applications.

Bayliss Illumination[137,138]   Difference patterns are used in monopulse tracking radars along with the sum pattern, as was mentioned in Chap. 4. It has been said before in this text that the sum pattern which produces maximum gain is of the form $(\sin u)/u$, and is obtained with a uniform aperture illumination. The symbol $u = \pi(D/\lambda) \sin\theta$, where D = aperture dimension and λ = wavelength. If one forms a difference pattern by starting with a uniform aperture illumination and subtracting one half of the aperture from the other half (that is, $A(x) = -1$ for $-D/2 < x < 0$, and $+1$ for $0 < x < +D/2$), then the difference pattern is proportional to $(1 - \cos u)/u$. The peak sidelobe is 10.6 dB below the peak of the beam, which is relatively high. The optimum illumination for a difference pattern (one that produces maximum slope, or minimum error) is linear-odd over the aperture. That is, the aperture illumination is a straight line that passes through zero at the center of the aperture and has a maximum (say, for example, $+1$) at one edge and a minimum (-1) at the other edge. Its peak sidelobe of -8.3 dB is even worse than that of the uniform-illumination difference pattern. When a single aperture illumination is used to obtain both the sum and difference patterns, an aperture illumination has to be found which represents a suitable compromise for the gain of the sum pattern, the peak sidelobes of both patterns, and the slope of the difference pattern. A better approach, when permitted, is to use an antenna which can support independent sum and difference patterns.

When the difference pattern of a monopulse antenna can be selected independently of the sum pattern, as in a phased array, the criterion is to obtain the maximum angle accuracy commensurate with a desired sidelobe level. The Bayliss illumination has been popular for this purpose. It is based on the same principles as the Taylor illuminations. As with the Taylor, the Bayliss illumination[137] depends on the peak sidelobe level and the

Figure 9.43
(a) Taylor circular aperture radiation patterns each having a beamwidth in degrees of $70\lambda/D$, but with different values of $\bar{n}$. (b) The corresponding aperture illuminations for the patterns of (a).
| (From W. White[134] Courtesy Institute for Defense Analyses.)

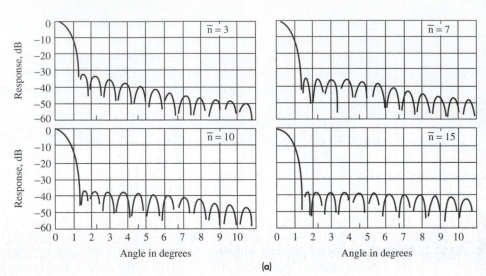

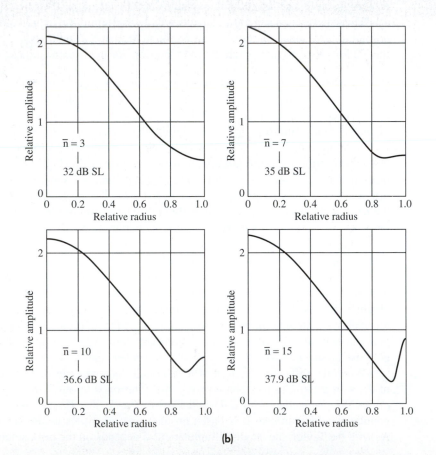

number $(\bar{n} - 1)$ of equal sidelobes. Similar restrictions apply on the selection of these two parameters as with the Taylor illuminations.

The Bayliss difference pattern has also been applied to circular apertures, and can be found in any one of several antenna books, including those authored by Mailloux,[74] Hansen,[69] and by Elliott.[136]

Shaped Antenna Patterns   In the above, we have considered aperture illuminations for obtaining suitable radiation patterns when a pencil beam or a simple fan beam was required. Sometimes it is necessary to form shaped antenna patterns where the patterns are wide compared to the minimum beamwidth (approximately λ/D radians) that can be obtained with an aperture of dimension D. An important example is the cosecant-squared beam discussed in Sec. 2.11. Other examples are a "square-top" pattern and an elevation pattern with a sharp cutoff at the horizon to minimize surface reflections.[139] In this subsection, we briefly describe two methods for synthesizing such patterns.[128] The methods for finding the aperture illumination to achieve some desired pattern are similar to the methods for finding the filter frequency response function to produce a desired time waveform.

Fourier Synthesis   Since the antenna pattern is, in theory, given by the inverse Fourier transform of the aperture illumination [as in Eq. (9.10)], the aperture illumination required to achieve a desired antenna pattern can be found by taking the Fourier transform of the desired pattern, which is

$$A(z) = \frac{1}{\lambda} \int_{-\infty}^{+\infty} E(\phi) \exp\left(-j2\pi \frac{z}{\lambda} \sin \phi\right) d(\sin \phi) \qquad [9.47]$$

where the symbols are the same as described for Eq. (9.10). The limits of the integration actually are finite since $|\sin \phi| \leq 1$. The Fourier transform is such that when the angular region is finite the Fourier transform of $A(z)$ is infinite. Since we have only a finite aperture, the actual radiation pattern will only be approximate, and can be shown to be[140]

$$E_a(\phi) = \frac{D}{\lambda} \int_{-\infty}^{+\infty} E(\xi) \frac{\sin\left[\pi(D/\lambda)(\sin \phi - \sin \xi)\right]}{\pi(D/\lambda)(\sin \phi - \sin \xi)} d(\sin \xi) \qquad [9.48]$$

where $E_a(\phi)$ is the Fourier-integral pattern which approximates the desired pattern $E(\phi)$ when $A(z)$ is restricted to a finite aperture of dimension D. The angle ξ is the variable of integration. The approximation to the antenna pattern obtained on the basis of the Fourier integral for continuous apertures (or the Fourier-series method for discrete array antennas) has the property that the mean-square deviation between the desired pattern $E(\phi)$ and the approximate pattern $E_a(\phi)$ is a minimum. The larger the aperture, the better will be the approximation.

The Fourier series may be used to approximate the pattern of a discrete array, just as the Fourier integral may be used to approximate the pattern of a continuous aperture. The Fourier series method is restricted in practice to arrays with element spacing in the vicinity of a half-wavelength. Spacings larger than a wavelength produce undesired grating lobes. Spacings much smaller than a half-wavelength result in so-called "supergain" radiation patterns (beamwidths much smaller than λ/D radians) that are not realizable since they are a consequence of an overly simplified model of radiation.

Woodward-Levinson Method   This is the spatial-domain analogy to the well-known sampling theorem for temporal signals. The classical sampling theorem for time waveforms states: A band-limited signal $s(t)$ with no frequency components greater than B Hz is determined by its amplitude at a series of points spaced $1/2B$ apart in time. The analogous sampling process applied to an antenna is that the radiation pattern $E(\phi)$ from an antenna with a finite aperture D is determined by a series of amplitudes spaced in angle λ/D apart. Figure 9.44a shows a pattern $E(\phi)$ and the sampled points λ/D radians apart. The sampled values $E_s(n\lambda/D)$, which determine the antenna pattern are shown in *b*. An antenna pattern $E_a(\phi)$ can be constructed from the sample values $E_s(n\lambda/D)$ using a pattern of the form $(\sin u)/u$ about each of the sample values, where $u = \pi(D/\lambda) \sin \phi$. The $(\sin u)/u$ pattern is called the composing function. The antenna pattern is given by

$$E_a(\phi) = \sum_{n=-(N-1)/2}^{(N-1)/2} E_s(n\lambda/D) \frac{\sin \left[\pi(D/\lambda)(\sin \phi - n\lambda/D)\right]}{\pi(D/\lambda)(\sin \phi - n\lambda/D)} \qquad [9.49]$$

Figure 9.44   (a) Radiation pattern $E(\phi)$ with sampled values λ/D radians apart in angle, where $\lambda =$ wavelength and $D =$ aperture dimension; (b) sampled values $E_s(n\lambda/D)$, which specify the antenna pattern of (a); (c) reconstructed pattern $E_a(\phi)$ using $(\sin u)/u$ composing function to approximate the desired radiation pattern $E(\phi)$.

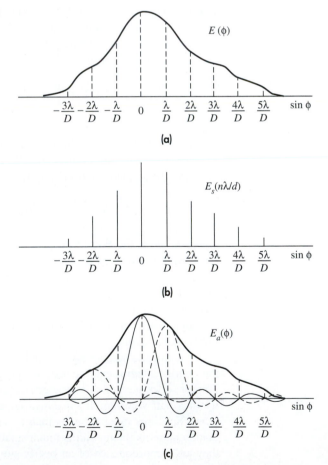

where N is the total number of samples, assumed to be odd. Thus the antenna pattern is constructed from a sum of individual $(\sin u)/u$ patterns spaced λ/D radians apart, each weighted in amplitude according to the sample values $E_s(n\lambda/D)$, as illustrated in Fig. 9.44c. The aperture illumination corresponding to the radiation pattern of Eq. (9.49) is

$$A(z) = \frac{1}{D} \sum_{n=-(N-1)/2}^{(N-2)/2} E_a(n\lambda/D) \exp\left(-j2\pi nz/D\right) \qquad [9.50]$$

The difference between the Woodward-Levinson method and the Fourier-integral synthesis method is that the former gives an antenna pattern which exactly fits the desired pattern at a finite number of points and the latter gives a radiation pattern whose mean-square deviation from the desired pattern is a minimum.

Cosecant-Squared Antenna Pattern   This is an antenna with its beam shape proportional to the cosecant-squared of the angle in one plane (usually in elevation) and with a conventional narrow beam in the orthogonal plane. The original reason for using an antenna pattern whose elevation pattern was proportional to $\csc^2 \theta_e$ was to obtain an echo signal that did not vary with range so long as the target flew at a constant altitude and the earth could be assumed to be flat (Sec. 2.11). Many air-surveillance radars have been designed with such an elevation pattern. An even more important reason for using a cosecant-squared elevation pattern is that it allows the radar to provide coverage of targets at high altitudes and shorter ranges much more efficiently than would a conventional fan-beam antenna pattern. In most modern air-surveillance radars, the coverage is modified even further to allow for the curvature of the earth and to provide more radiation at high elevation angles and short range than would be obtained with a cosecant-squared pattern.[141] This additional coverage at high angles is necessary to compensate for the reduction in gain at short ranges caused by the use of sensitivity time control (STC). With STC the receiver gain is reduced at short ranges to attenuate the large clutter echoes that might appear. The cosecant-squared pattern, or a slight variation, has also been important for airborne surface-surveillance radars that must provide relatively uniform coverage of the surface.

The methods available for designing shaped beams, such as the Woodward-Levinson or the Fourier integral, can be applied to obtain the aperture illumination required for a cosecant-squared pattern or its variations. The cosecant-squared pattern may be approximated with a reflector antenna by modifying the shape of the reflecting surface. For example, the upper half of a parabolic reflector can be a parabola that reflects energy from the feed at its focus in a direction parallel to the axis of the parabola axis—as in any other parabolic reflector. The lower half, however, is distorted (tilted slightly forward) to direct a portion of the radiated energy in the upward direction. This is a very simple method, which can be adequate for some applications.

A cosecant-squared antenna pattern may also be obtained by (1) feeding a parabolic surface with two or more horns, (2) a linear array, (3) a parabolic cylinder fed by a linear array or a line source, or (4) a point-source feeding a reflector surface with double curvature. An example is the antenna for the U.S. Air Force's AN/TPN-19 S-band Airport Surveillance Radar which employed an offset paraboloid reflector fed from 12 feedhorns in a vertical line-feed with the uppermost feedhorn located at the focal point of the paraboloid.

Loss in Gain   The gain of a cosecant-squared antenna will be less than the gain of a conventional antenna from which it was derived. A very approximate estimate of the reduction in gain for such an antenna is given as

$$\frac{G_{\text{csc}}}{G} = \frac{\theta_0}{\theta_0 + \sin^2 \theta_0 (\cot \theta_0 - \cot \theta_m)} \approx \frac{1}{2 - \theta_0 \cot \theta_m} \qquad [9.51]$$

where G = the gain of a rectangular antenna pattern of width θ_0 that radiates uniformly from $\theta = 0$ to $\theta = \theta_0$, and G_{csc} = gain of the cosecant-squared antenna. The radiation decreases proportional to $\csc^2 \theta$ from θ_0 to θ_m, the maximum angle at which the cosecant-squared pattern is applied. (There is no radiation beyond θ_m.) The approximate expression on the right of Eq. (9.51) applies for small θ_0 and large θ_m. (In addition, the assumption that the beamwidth θ_0 is that of a rectangular-shaped beam can also affect the accuracy of this expression.) For example, if $\theta_0 = 6°$ and $\theta_m = 20°$, the gain of the cosecant-squared antenna is decreased by 2.2 dB compared with a rectangular beam 6° in width. A modified cosecant-squared pattern to allow for the coverage of the earth and to account for the action of the STC will have even lower gain.

Theoretical and Actual Antenna Patterns   The discussion of theoretical antenna patterns in this section was based on the Fourier transform relationship between the antenna pattern and the aperture illumination. It has been widely accepted and widely used, but it has limitations. The Fourier transform relationship applies fairly well to the main beam of the antenna and to the region near the main beam. It is less accurate the farther one goes in angle from broadside, and it does not faithfully account for energy radiated near or beyond ±90°.

In a reflector antenna the spillover radiation from the feed, blockage by the feed or feed supports, the radiation diffracted by the reflector, and any leakage through a mesh reflector surface are not accounted for by the Fourier transform. The far-field pattern of any antenna also is affected by nearby structures or other obstructions that can block or diffract the energy radiated by the antenna. The blockage of an antenna by a mast on a ship, for example, might result in a −40 dB peak sidelobe being increased to as high as −15 to −20 dB or greater. Errors in the phase and amplitude across an antenna aperture due to either mechanical or electrical inaccuracies will increase the far-out sidelobe levels over what is predicted from the classical Fourier transform relationship. This is a subject that will be described next.

9.12   EFFECT OF ERRORS ON RADIATION PATTERNS

Experimentally measured patterns of actual antennas often deviate from the theoretically calculated pattern, especially in the region of the far-out sidelobes. Generally, the fault does not lie with the theory, but in the fact that it is not possible to reproduce precisely the necessary aperture illumination specified by synthesis theory. There are small, but ever-present, errors in the fabrication of the antenna and in how it is fed. These contribute unavoidable perturbations in the aperture illumination and result in a pattern different in detail from the one expected.

Errors in the aperture illumination may be either *systematic* or *random*. The former are predictable (usually), but the latter are not and can only be described in statistical terms. Examples of systematic errors include (1) aperture blocking by the feed and its supports in a reflector antenna, (2) mutual coupling among the elements of an array, (3) quantization lobes due to the discrete value of the phase in a digital phase shifter, (4) spurious lobes due to mismatch in a constrained feed for an array, and (5) periodicities introduced in an antenna during the manufacturing process. Random errors include (1) errors in the machining or manufacture of the antenna due to the finite precision of construction techniques, (2) the precision with which a phase shifter can be set to its required phase, (3) errors incurred in adjusting an array, (4) random distortions of the antenna surface, and (5) mechanical and electrical (phase) variations caused by temperature gradients or wind (and in some cases gravity) across the antenna. Although random errors throughout the antenna may be relatively small, their effect on the sidelobes (which are also small) may be relatively large compared to the small levels of the sidelobes. Systematic errors do not differ much from one antenna to another in any particular design constructed by similar methods. Random errors, on the other hand, can differ from one antenna to the next even though they may be of the same design and are constructed similarly. The effect of random errors, therefore, must be discussed in terms of statistical averages found over many similar antennas.

When no specific guidance is available, the antenna designer often assumes that the antenna should radiate a wavefront that differs in phase from the desired wavefront by no more than $\pm\lambda/16$, where λ = wavelength. Because of the two-way propagation from a reflector surface, the mechanical accuracy of a reflector antenna surface must be held within $\pm\lambda/32$. As we shall see, it is possible to obtain more precise criteria for antenna errors, especially when low or ultralow sidelobes are desired. Most of the discussion of errors in this section will concern random errors rather than systematic errors. (If the systematic errors are known, their effect on the antenna pattern can be ascertained by taking the Fourier transform of the actual aperture illumination, including the effects of systematic errors.) The discussion of the effects of random errors in reflectors is separated from the effects of errors in phased arrays.

Random Errors in Reflector Antennas   The classical work on the effects of random errors on antenna patterns was due to the pioneering efforts of John Ruze.[142] For small phase errors he showed that the gain G of a circular aperture with mean-square phase-error $\overline{\delta^2}$ is approximately

$$G = G_0 \exp\left[-\overline{\delta^2}\right] = \rho_a(\pi D/\lambda)^2 \exp\left[-(4\pi\epsilon/\lambda)^2\right] \qquad \text{[9.52]}$$

where G_0 is the gain of the antenna without errors; the phase error δ, in radians, is with respect to the mean phase plane; ρ_a is the aperture efficiency; D is the diameter of the circular antenna; and ϵ is the rms error of the reflector surface in the same units as the wavelength λ. In the above, the expression $4\pi A_e/\lambda^2$ (discussed early in this chapter) was substituted for the gain G_0. For a given reflector size D, the gain increases as the square of the frequency when the errors are small, until the exponential term becomes significant. Differentiating Eq. (9.52), setting it equal to zero, and solving for wavelength gives the wavelength at which the maximum gain is obtained for an rms error ϵ, which is

$$\lambda_m = 4\pi\epsilon \qquad \text{[9.53]}$$

At this wavelength, the gain will be 4.3 dB below what it would have been in the absence of errors. The maximum gain of an antenna due to phase errors is then

$$G_{max} = \frac{\rho_a}{43}\left(\frac{D}{\epsilon}\right)^2$$

[9.54]

For wavelengths shorter than λ_m, the gain drops off rapidly with decreasing wavelength (increasing frequency).

The gain of a reflector antenna is limited by the mechanical tolerance to which its surface can be constructed and maintained when in operation. The most precise antennas, under benign, controlled conditions, seem to be limited in practice to a precision of about one part in 20,000. From Eq. (9.54) the diameter of such an antenna is about 1600 wavelengths for maximum gain. Its beamwidth would be 0.04° and it would have a gain of about 68 dB. Special purpose nonradar antennas have been constructed with slightly better tolerances, but these generally have some means for measuring the antenna surface and correcting the surface automatically with feedback control while the antenna is operating. Such antennas are operated in controlled environments that may not be suitable for operational radar applications. In practice, therefore, radar antennas are seldom larger in dimension than approximately 300 wavelengths, which corresponds to a beamwidth of about 0.2°.

The construction tolerance of a reflector antenna is often described by its "peak" error, rather than its rms error. The ratio of the peak to the rms error is found in practice to be about 3:1. This truncation of errors occurs since large errors are usually corrected during manufacture.

With small phase errors the exponential factor in the gain expression of Eq. (9.52) can be approximated by

$$G \approx G_0 \left(1 - \overline{\delta^2}\right)$$

[9.55]

If the loss in antenna gain is to be less than 1 dB, this simple expression says that the rms phase variation about the mean phase surface should be less than 0.45 radian, or 26°. This is equivalent to an rms distance error of $\lambda/14$. For shallow reflector antennas, however, the two-way propagation path requires that the rms deviation of the surface from its true value be one half this value, or $\lambda/28$, for a 1 dB reduction of gain.

Ruze showed that under certain conditions the radiation pattern of a reflector antenna which is distorted by a large number of gaussian-shaped "bumps" can be expressed as

$$G(\theta,\phi) = G_0(\theta,\phi)e^{-\overline{\delta^2}} + (2\pi C/\lambda)^2 e^{-\overline{\delta^2}} \sum_{n=1}^{\infty} \frac{\left(\overline{\delta^2}\right)^n}{n!n} e^{-(\pi Cu/\lambda)^2/n}$$

[9.56a]

where C is the correlation distance of the error (the size of the region on the aperture where the errors cannot be considered independent) and u in this case is $\sin\theta$. The coordinate system for this equation is the classical coordinates of antenna theory, as was shown in Fig. 9.4, with the antenna lying in the x,y plane. The error current in one region of the antenna (the correlation distance) is assumed independent of error currents in other regions. The size of the correlation distance affects both the magnitude and the direction of the spurious radiation that results from the presence of errors. For small error, when only the first term of the series ($n = 1$) need be considered, Eq. (9.56a) becomes

$$G(\theta,\phi) = G_0(\theta,\phi)e^{-\overline{\delta^2}} + (2\pi C/\lambda)^2\overline{\delta^2}e^{-(\pi Cu/\lambda)^2} \qquad \text{[9.56b]}$$

The first term of the above equation [as well as Eq. (9.56a)] represents the no-error pattern reduced by a factor dependent on the mean-square phase error. The second term represents the average value of the sidelobes that are generated by the phase errors (not the average of the peaks, but the *average*). Near the main beam the sidelobes are determined mainly by the inverse Fourier transform of the aperture illumination [Eq. (9.10)], but eventually these drop below the error sidelobes and at angles far from the main beam the errors determine the sidelobe level. When the error sidelobes are dominant, the average sidelobe level is independent of angle.

Other observations about errors in reflector antennas are:

1. Ruze's original analysis[143] showed that the error sidelobes are proportional to the mean-square error and to the square of the correlation distance measured in wavelengths.

2. If errors are unavoidable, they should be kept small in extent; that is, for the same mechanical tolerance, the antenna with the smaller correlation distance will give lower sidelobes than an antenna with a larger correlation distance. An error stretching most of the length of the antenna is likely to have a worse effect than a localized bump or dent of much greater amplitude. Thus small disturbances such as screws and rivets on the reflector surface will have relatively little effect on the antenna radiation pattern.

3. An increase in frequency increases both the phase errors and the correlation distance (measured in wavelengths). Therefore the gain of a constant-area antenna does not increase as rapidly as the square of the frequency when errors are a factor.

Since the radiation pattern in the far-out sidelobe region is more likely to depend on the accuracy with which the antenna is constructed rather than the particular aperture illumination selected, the mechanical engineer, the skilled machinist, and technician are very important in realizing in practice a satisfactory antenna pattern.

Errors in Arrays   In the above analysis of reflector-antenna errors, only the effect of the phase errors were considered. In an array antenna, however, other factors may enter to cause degradation of the radiation pattern. These include errors in the amplitude and phase of the current at each element of the array, missing or inoperative elements, rotation or translation of the element from its correct position, errors in the phase provided by the phase shifter, effects of a quantized phase shift, and variations in the individual element patterns because of mutual coupling. These errors can result in a decrease in antenna gain, increase in sidelobe level, generation of spurious sidelobes, and a shift in the location of the main beam.

It is not possible to predict the pattern of an antenna unless the actual errors experienced by that particular antenna are known. The average, or expected, value of a radiation pattern of an *ensemble* of antennas of the same type can be computed based on the rms values of the random errors. The statistical description of the radiation pattern cannot be applied to any particular antenna of the ensemble, but applies to the entire collection of similar antennas whose errors are described by the same statistics. Usually the average pattern is computed, but other statistical descriptions can be obtained if desired.

Average Radiation Pattern Due to Errors   The ensemble-average radiation intensity pattern of a uniform array of M by N isotropic elements arranged on a rectangular grid with equal spacing between elements is given by[144]

$$\overline{|f(\theta,\phi)|^2} = P_e^2 e^{-\overline{\delta^2}} |f_0(\theta,\phi)|^2 + \left[\left(1 + \overline{\Delta^2}\right) P_e - P_e^2 e^{-\overline{\delta^2}} \right] \sum_{m=1}^{M} \sum_{n=1}^{N} i_{mn}^2 \qquad [9.57]$$

where

$$P_e = \text{probability of an element being operative (or the fraction of elements that remain operating)}$$

$$\delta = \text{phase error, radians (described by a gaussian probability density function)}$$

$$|f_0(\theta,\phi)|^2 = \text{no-error pattern}$$

$$\Delta = \text{relative amplitude error (as a fraction of } i_{mn})$$

$$i_{mn} = \text{no-error current at the } mn\text{th element}$$

Similar to the error pattern of a reflector antenna discussed earlier [Eq. (9.56)], the first term is the no-error pattern reduced by a factor which depends on the phase errors and the fraction of operative elements. The second term represents a statistical average side-lobe level due to the phase and amplitude errors and the fraction of elements that are operating. It also depends on the aperture illumination as given by the currents i_{mn}. This second term is independent of angle, and can be thought of as a statistical omnidirectional pattern which we shall call the error sidelobes. This second term causes the far-out side-lobes of the radiation pattern to be higher in the presence of errors as compared to the no-error pattern; but the shape of the main beam and the near-in sidelobes are not significantly affected by these errors, other then by the exponential term which is usually small. [Sometimes the rms error in amplitude is expressed in dB. When in dB it is not the mean square value which is in dB, but the value of $[1 - (\overline{\Delta^2})^{1/2}]^2$. For example, an rms amplitude error of 0.1 is equivalent to an error of 0.9 dB.]

If $P_e = 1$ (no missing elements) and if the errors are small, the normalized radiation intensity obtained by dividing Eq. (9.57) by the maximum radiation intensity at the center of the main beam, $|f_0(0,0)|^2$, is

$$\overline{|f(\theta,\phi)|^2} \approx |f_{0n}(\theta,\phi)|^2 + \left(\overline{\Delta^2} + \overline{\delta^2} \right) \frac{\displaystyle\sum_m \sum_n i_{mn}^2}{\left(\displaystyle\sum_m \sum_n i_{mn} \right)^2} \qquad [9.58]$$

The second term indicates that the larger the number of elements, the smaller will be the error-sidelobe level. The main-beam intensity, being coherent, increases as the square of the number of elements, while the error sidelobes, being noncoherent, increase only directly with the number of elements. The gain of a broadside array of isotropic elements is

$$G_0 = \frac{\left(\displaystyle\sum_m \sum_n i_{mn} \right)^2}{\displaystyle\sum_m \sum_n i_{mn}^2} \qquad [9.59]$$

When i_{mn} = constant, $G_0 = MN$, which states that the gain of an array of isotropic elements with uniform illumination is equal to the total number of elements. The normalized pattern of Eq. (9.58) then can be expressed as

$$\overline{|f_n(\theta,\phi)|^2} \approx |f_{0n}(\theta,\phi)|^2 + \frac{\overline{\Delta^2} + \overline{\delta^2}}{G_0} \qquad [9.60]$$

The average error-sidelobes is given by the second term. The greater the antenna gain, the less will be the effect of errors on the sidelobes. [Sometimes the denominator of the second term of the above equation is given as πG_0, where π is the gain of the so-called "ideal" element factor.]

The component parts of the normalized ensemble-average pattern as given by Eq. (9.60) are sketched in Fig. 9.45. The ordinate is in dB. The horizontal line shown G_0 (in dB) below the peak of the main beam is the radiation that would be produced by an isotropic antenna with the same power output as the directive antenna. The horizontal line shown $G_0/(\overline{\Delta^2} + \overline{\delta^2})$ (in dB) below the peak is the average value of the error sidelobes, which is independent of angle. The error sidelobes have little effect near the main beam, but they are the dominant factor affecting the far-out portion of the radiation pattern. The measured sidelobes of any actual array antenna, of course, would not be constant as shown in this figure. Instead they would have the usual shape expected of sidelobe radiation, but their ensemble-average value would be constant with angle.

Gain Reduction   By substituting the radiation intensity of Eq. (9.57) into the definition of gain (or directivity) of Eq. (9.3), it can be shown that

$$G/G_0 = \frac{P_e}{(1 + \overline{\Delta^2}) \exp(\overline{\delta^2})} \approx \frac{P_e}{1 + \overline{\Delta^2} + \overline{\delta^2}} \qquad [9.61]$$

This states that the relative reduction in gain is independent of the number of elements and depends only on the fraction of elements that are operative and the mean-square value

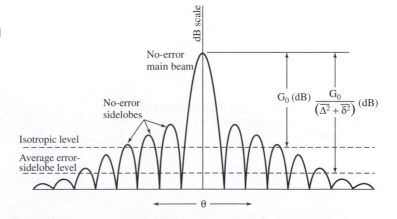

Figure 9.45   Qualitative sketch of the no-error radiation pattern and the average error-sidelobe level, as indicated by Eq. (9.60).

of the relative-amplitude and phase errors. When $P_e = 1$ and $\Delta = 0$, this expression, which applies for small phase errors, is similar to that of Eq. (9.55) for the reflector antenna.

Pointing Error Random phase and amplitude errors in the aperture illumination can give rise to an error in the pointing of the main beam.[145,146] If the aperture illumination is uniform across an M by M square array, the statistical rms beam pointing error in radians is

$$\delta\theta_0 = \frac{\sqrt{3}\sigma}{2\pi(d/\lambda)M^2} \qquad [9.62]$$

where σ = rms value of the normalized error current assuming Rayleigh distributed errors, d = element spacing, λ = wavelength, and M = number of elements along one dimension of a square array. According to this expression, the effect of errors on the beam-pointing accuracy generally is small.

Error-Sidelobe Statistics In the above we have considered the average value of the error sidelobes. The radar system engineer, however, is often more concerned about the peak sidelobe level rather than the average. The actual peak value cannot be predicted, but it can be described on a statistical basis. Usually one wants to determine (or specify) the probability that a sidelobe will not exceed a particular desired value. The approach outlined here is taken from James K. Hsiao.[147,148] He described three methods for finding the effect of errors on the peak sidelobe. These assume that the statistics of the radiation pattern with errors can be described by the Rice probability density function, something we have used in Sec. 2.5 when discussing the statistical detection of a signal in noise. Although all three methods are similar, we shall consider here only the second that he describes in his paper.

From the Rice cumulative probability of sidelobe level, Hsiao obtains a set of curves, shown in Fig. 9.46, for various values of the cumulative probability P that range from 0.90 to 0.98. This figure relates the peak sidelobe level (abscissa) and the amplitude and phase errors of the array (ordinate). The abscissa is actually the ratio of the design sidelobe level to the desired sidelobe level in dB. (The design sidelobe level is always less

Figure 9.46 Curves for determining the rms amplitude error σ_δ and the rms phase error σ_ϕ as a function of the ratio of the *design* sidelobe level to the *desired* sidelobe level (abscissa) for various values of the cumulative probability P that the sidelobes will be less than the design sidelobe level.

The parameter $D = \sum i_{nm}^2/$(desired sidelobe level), where i_{mn} is the amplitude of the current at the *mnth* element.

| (Due to James Hsiao, taken from refs. 147 and 148.)

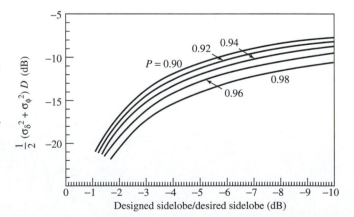

than the desired sidelobe level.) The ordinate in Fig. 9.46 includes a factor D, as well as the phase and amplitude rms errors. Hsiao defines D as

$$D = \sum_n \sum_m i_{nm}^2/(\text{desired sidelobe level}) \qquad [9.63a]$$

where i_{nm} is the amplitude of the current at the nmth element, as determined by the desired aperture illumination. It is assumed that the illumination function i_{nm} is normalized such that $\sum_n \sum_m i_{nm} = 1$. Following Hsiao, we consider the example of a 100-element linear array designed with a Chebyshev aperture illumination to provide equal sidelobes -40 dB below the maximum value of the main beam. (As said before in this chapter, the Chebyshev pattern with such low sidelobes is likely to be unrealizable. It is used here to illustrate the procedure.) The design sidelobe level is then -40 dB. In this example we wish to keep the peak sidelobe of the array to a value no greater than -37 dB (the desired sidelobe) with a probability of 0.90. The ratio of the design sidelobe level to the desired sidelobe level (the abscissa of Fig. 9.46) is -3 dB. With an abscissa of -3 dB and a probability $P = 0.90$, we find the ordinate of Fig. 9. 46 to be -14 dB. The aperture illumination for the 40 dB sidelobe Chebyshev pattern results in $\sum_n i_n^2 = -19$ dB. The value

of D is then $-19 + 37$ dB $= 18$ dB and the sum of the mean square amplitude and phase errors is -29 dB. If the phase and relative amplitude errors are made equal, then the rms phase tolerance is $1.44°$ and the rms (relative) amplitude tolerance is 0.025, or 0.22 dB. Such tolerances are quite demanding and not easy to achieve. A plot is shown in Fig. 9.47 of the no-error -40 dB Chebyshev pattern along with a pattern where there are errors that give sidelobes with a cumulative probability of 90 percent that they will not exceed -37 dB.

If the design sidelobe level were -45 dB instead of -40 dB, and the desired sidelobe level were still -37 dB, one would find that the tolerance would be $2.6°$ for phase and 0.045 for amplitude. These are also quite demanding, but not as much as with the higher value of design sidelobe.

Figure 9.47 The design no-error Chebyshev pattern for a 100-element linear array with a -40 dB design sidelobe is shown along with the resulting desired pattern due to errors when $P = 0.9$, desired sidelobe level $= -37$ dB, rms amplitude error $\sigma_\delta = 0.025$, and rms phase error $\sigma_\phi = 1.44°$.
ı (Due to James Hsiao, taken from refs. 147 and 148.)

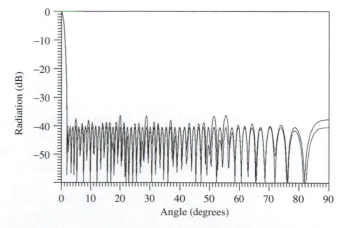

These are relatively tight tolerances. They result from not only requiring a low side-lobe level, but also because there are only a relatively few antenna elements (100 in this example of a linear array). The tolerances for a large planar array would be easier to achieve. As an illustration, Hsiao shows that a planar array with a directive gain of 40 dB, desired sidelobe of −40 dB, design sidelobe of −43 dB, and a probability of 0.9 that the sidelobes will not exceed the desired value results in an rms phase tolerance of 6.6° and an rms amplitude tolerance of 0.12. If the design sidelobe were reduced to −48 dB, the rms phase tolerance becomes 12.3° and the rms amplitude becomes 0.21.

When using a lower design sidelobe to ease the error tolerances, a price has to be paid. The lower design sidelobes result in an increase in the antenna half-power beamwidth and an even larger increase in the width between the first nulls of the antenna pattern. Alternatively, a larger aperture is required if the beamwidth is to remain constant.

A slightly different approach to determining the error tolerances required for keeping the sidelobes below a specified value was given by Cheston and Frank.[149]

Effect of Digital Phase Shifter Quantization[150,151]   Phase shifters, whether analog or digital, will usually have a phase shift that is not exactly what one thinks has been set. There will always be some error, and its effects on the antenna pattern can be determined by the equations given in the above subsection. The deliberate quantization of phase that results with the use of digital phase shifters, however, introduces a different type of "error" in the desired aperture illumination, and produces pattern degradation similar to that produced by a random error.

Reduction in Gain   The gain of an array antenna due to phase errors can be obtained from Eq. (9.61) by setting $P_e = 1$, $\overline{\Delta^2} = 0$, and assuming $\overline{\delta^2}$ is small, which results in

$$G = G_0\left(1 - \overline{\delta^2}\right) = G_0\left(1 - \frac{\pi^2}{3 \times 2^{2B}}\right) \qquad \text{[9.63b]}$$

The right-hand portion is obtained by assuming that the phase error of a digital phase shifter of B bits is described by a uniform probability density function (Sec. 2.4) that extends over an interval $\pm \pi/2^B$. From Eq. (9.63b) a three-bit phase shifter causes a reduction in gain of 0.23 dB and a four-bit phase shifter has a gain reduction of 0.06 dB. Thus, on the basis of the loss in antenna gain caused by digital phase quantization, a three- or four-bit phase shifter should be satisfactory for most purposes.

Increase in Sidelobes   In addition to a reduction in main-beam gain, the quantization of the phase in digital phase shifters can result in an increase in the rms sidelobe level. With the assumptions that (1) the energy lost by the reduction in main-beam gain shows up as an increase in the rms sidelobe level, (2) the element gain is the same for the main beam and the sidelobes (within the region of space scanned by the array, (3) an allowance of one dB for the reduction in gain due to the aperture illumination, and (4) one dB for scanning degradation; then the sidelobe level due to phase quantization is

$$\text{rms sidelobe level} \approx \frac{5}{2^{2B}N} \qquad \text{[9.64]}$$

where N = total number of elements in the array. If an array has 4000 elements, a three-bit phase shifter would give rms sidelobes of 47 dB below the main beam, and a four-bit phase shifter gives -53 dB sidelobes. Thus three or four bits should be sufficient for most large arrays, except when very low sidelobes are desired.

Peak Quantization Sidelobe   The above assumed a random distribution of phase error across the aperture for the purpose of computing the rms sidelobe level. The actual phase distribution with digital phase shifters, however, is likely to be periodic which gives rise to spurious quantization lobes, similar to grating lobes but with smaller amplitude. Peak sidelobes sometimes are of more concern to the radar system engineer than are the rms sidelobes. The peak quantization lobe relative to the main beam when the phase error has a triangular repetitive distribution is

$$\text{peak quantization lobe} = 1/2^{2B} \qquad [9.65]$$

This applies when the main beam points close to broadside and there are many radiator elements within the period of quantized phase error. The position θ_q of the quantization lobe in this case is

$$\sin \theta_q \approx (1 - 2^B) \, \theta_0 \qquad [9.66]$$

where θ_0 is the angle to which the main beam is steered.

Equation (9.65) is an optimistic estimate for the peak lobe. The greatest phase quantization lobe is said to occur when the element spacing is exactly one half the phase quantization period or an exact multiple thereof. With an element spacing of one-half wavelength, the quantization lobe will appear at $\sin \theta_q \approx \sin \theta_0 - 1$, and will have a value of

$$\text{peak quantization lobe} \approx \frac{\pi^2}{4} \frac{1}{2^{2B}} \frac{\cos \theta_q}{\cos \theta_0} \qquad [9.67]$$

The peak sidelobes due to the phase quantization of the digital phase shifter can be significant. Attempts should be made to reduce them if their presence is objectionable. One method for reducing the peak sidelobe is to randomize the phase quantization. A constant phase shift can be inserted in the path to each element, with a value that differs from element to element by amounts that are unrelated to the bit size. The added phase shift is then subtracted in the phase command sent to the phase shifter. (With a space-fed array, such as the lens array or the reflectarray, decorrelation is inherent in the array geometry.)

Beam-Pointing Error   The maximum pointing error $\Delta \theta_0$ due to quantization, according to C. J. Miller[151] is

$$\Delta \theta_0 = \theta_B \frac{\pi}{4} \frac{1}{2^B} \qquad [9.68]$$

where θ_B is the beamwidth. A four-bit phase shifter, for example, allows an angle error of $\Delta \theta_0 / \theta_B = 0.05$. Small steering increments are possible with quantized phase shifters. A linear array of 100 elements, for instance, can be steered in increments of about 0.01 beamwidth with three-bit phase shifters.[149]

9.13 LOW-SIDELOBE ANTENNAS

The highest sidelobe of an antenna pattern is usually, but not always, the first sidelobe adjacent to the main beam. A conventional reflector antenna might have a maximum sidelobe of about 23 to 28 dB below the peak of the main beam. Sidelobes much lower than −35 to −40 dB with conventional reflector antennas are difficult to obtain by normal methods. There are some radar applications, however, that require much lower sidelobes. An example is the airborne high-prf pulse doppler radar, such as AWACS that was discussed in Sec. 3.9. A high-prf radar sees many multiple-time-around clutter echoes that enter the radar receiver through the antenna sidelobes. Such sidelobe clutter can be large enough to limit the performance of an airborne doppler radar. It was the need for a low-sidelobe antenna for AWACS that led Westinghouse (now Northrop Grumman) radar antenna engineers to be the first to successfully demonstrate in the mid-1960s sidelobes almost three orders of magnitude lower than was the practice with conventional antennas. A low-sidelobe antenna is also helpful in combating hostile noise jamming that enters the receiver via the sidelobes. It also aids in combating antiradiation missiles (ARM) that home on the radar's radiation, and in making more difficult the task of a hostile intercept receiver. Low sidelobes, however, come with a price. Such antennas need to be more precise, are more complex, their beamwidth is widened, and they have to operate in a clear environment.

Table 9.2 lists typical performance of low-sidelobe antennas as given by Evans and Schrank.[152] (In their paper, the authors state that "it is probably possible to do 5 dB better with tuning of phase and amplitude during the test." I have taken the liberty to add this 5 dB to the numbers that originally appeared in their paper.) The phased arrays in this table are not scanned in angle.

There has been no generally accepted definition of what are low sidelobes. Schrank[153] proposed that low sidelobes be defined as from −30 to −40 dB and that ultralow sidelobes be below −40 dB. These values probably should be lower. One might consider antennas with sidelobes of the order of −40 dB to be low, and antennas with −50 dB or lower to be ultralow sidelobe antennas.

The peak sidelobe is not always a good measure of the difficulty involved in obtaining low sidelobe levels. The larger the antenna gain, the easier it is (relatively) to obtain low sidelobes. A better measure of the difficulty for an antenna engineer to achieve low sidelobes is how far the peak sidelobe level is below the isotropic value, which we take

Table 9.2 Performance of Low-Sidelobe Antennas

Type of Antenna	Peak Sidelobe	RMS Sidelobe	Bandwidth
Slotted waveguide	−50 dB	−60 dB	10%
Corporate-fed array	−45 dB	−55 dB	60%
Reflector	−45 dB	−55 dB	60%

I Adapted from Evans and Schrank, Ref. No. 152, *Microwave Journal*.

here to be the level which is $1/G_0$ down from the peak, where G_0 = antenna gain. Low or ultralow sidelobe antennas might have peak sidelobes from 10 to 20 dB below isotropic. Thus if an antenna has a gain of 30 dB, its sidelobes might be made as low as -40 to -50 dB. From the second term of Eq. (9.60) it can be seen that the mean square error $\overline{\Delta^2} + \overline{\delta^2}$ must be small compared to the desired peak sidelobe level times the antenna gain G_0. Thus the larger the gain the larger can be the errors for a given sidelobe level relative to the main beam.

To achieve low sidelobes, an antenna must (1) employ an aperture illumination that will theoretically provide the required sidelobe level and be practical to implement, (2) be constructed and maintained with high precision, and (3) have no blocking of the aperture by nearby structures. Both array antennas and reflector antennas can be made to have low sidelobes, but it is easier to do with an array than with a reflector.

Low-Sidelobe Aperture Illuminations The Taylor aperture illumination, discussed in Sec. 9.11, has often been used for obtaining low sidelobe antenna patterns. It will be recalled that the Taylor illumination is characterized by a parameter $\bar{n}$, such that the first $\bar{n} - 1$ sidelobes closest to the main beam are equal. Beyond the $\bar{n}$th sidelobe, the sidelobe level decreases. For a circular aperture, the Taylor far-out sidelobes fall off as $(\sin \theta)^{-3/2}$. As was mentioned in Sec. 9.11, the parameter $\bar{n}$ must be chosen so that the aperture illumination is realizable.

Figure 9.48, from a report by Warren White,[154] is an example of how the normalized beamwidth β of a Taylor pattern varies with the peak sidelobe level and the parameter $\bar{n}$

Figure 9.48 Normalized beamwidth β versus the sidelobe level and the value of $\bar{n}$, where the half-power beamwidth $\theta_B = \beta\lambda/D$. (From White.[154] Courtesy Institute for Defense Analyses.)

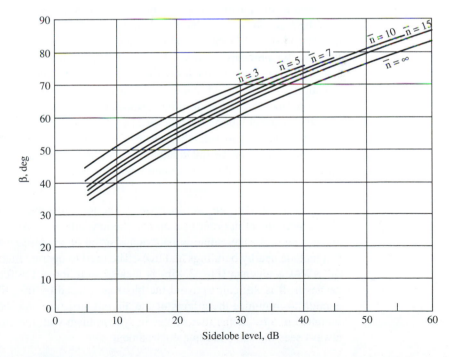

for a continuous circular aperture. The half-power beamwidth is $\theta_B = \beta\,(\lambda/D)$. The curve for $\bar{n} = \infty$ applies for the Chebyshev illumination, which is not realizable. Comparing a pattern with $\bar{n} = 5$ and a -25 dB sidelobe level with a pattern having $\bar{n} = 15$ and a -60 dB sidelobe level, the half-power beamwidth increases by a factor of almost 1.4 when lowering the sidelobes level from -25 dB to the level of -60 dB. The null width of the Taylor distribution increases even faster; over the same range of sidelobe levels, the null width increases by a factor of about 1.9. Thus one of the costs of low sidelobes is that the beamwidth increases as the sidelobe level is decreased. The gain also decreases with decreasing sidelobes. If the same beamwidth is to be maintained as the sidelobes are lowered, the antenna aperture must be increased in size.

Previously, Fig. 9.43 in Sec. 9.11 illustrated that the Taylor aperture illuminations are not always realizable because of the difficulty in obtaining the required shape of the currents at the edges of the aperture. The value of $\bar{n}$ must be chosen appropriately in order to achieve a pattern that is practical. A suitable criterion for selecting a realizable Taylor illumination is that it have a monotonically decreasing illumination, the slope at the edge of the aperture should be zero, and it should not turn positive. Ludwig[155] states that for a circular aperture the value of $\bar{n}$ should be no greater than 7 for -40-dB sidelobes, 11 for -50-dB sidelobes, and 16 for -60-dB sidelobes. The value of $\bar{n}$ should also not be too small. It should be at least 3 for -30-dB sidelobes and 4 for -40-dB sidelobes.[156] [Note these values of $\bar{n}$ for a circular Taylor aperture-illumination differ slightly from those given in Sec. 9.11 for a line-source Taylor illumination.]

Achieving the Low-Sidelobe Pattern   In the above we indicated it is necessary to obtain an aperture illumination that can be implemented. In addition, the desired illumination must be maintained. The phase and amplitude tolerance on the aperture illumination must be determined, as in the previous section. As was indicated with the discussion of Fig. 9.47 in the previous section, the required error tolerances can be quite demanding. Schrank[153] points out that for a low-sidelobe array antenna, one must control the systematic errors and the mutual coupling in addition to the random errors. Mutual coupling can be compensated by appropriate computer-aided design. Systematic errors must be controlled by careful fabrication and serious attention to tolerances. Systematic errors generally affect the near-in sidelobes and can generate spurious lobes. Random errors, as we have seen, affect the far-out sidelobes. According to Schrank, it is the random errors that ultimately limit the ability to obtain low sidelobes.

Blocking or Masking of the Aperture   The low-sidelobe antenna must be located in a clear environment in order to maintain the low sidelobes. Any obstruction in front of the antenna can alter the radiation pattern and result in an increase in sidelobes. Obstructions can include nearby buildings and trees. The need to operate in a clear environment is why the AWACS antenna (Fig. 3.45b) is located well above the fuselage of the aircraft that carries it. It is difficult to avoid the blockage caused by the tail, but the antenna is high enough to minimize the effects of the aircraft structure and the engines. The antenna is mounted in a rotodome that rotates in synchronism with the antenna, so that the antenna always sees the same radome environment.

A classic example of the effect of aperture blocking on antenna sidelobes is that caused by the masts and other superstructure on a ship.[157] No matter how low the side-lobes of an antenna might be in free space, when a mast is situated so as to block a portion of its radiation, the sidelobes can increase considerably. Blockage of a shipboard radar antenna can be avoided by mounting the antenna at the top of the mast. This is something that can be accommodated during the design of a new ship, but it is difficult to mount a heavy antenna at the top of a mast on an old ship which very likely has used up all of its margin for topside weight and moment.

There are two factors that cause the sidelobes of an antenna to increase when blockage occurs. One is that a part of the beam has been masked, which is equivalent to having a part of the aperture illumination excised. In other words the effective aperture illumination has been modified. The other effect is that the obstruction can scatter the radiated energy in new directions, so that target echoes or clutter echoes might appear as false targets in an erroneous direction. The radar thinks the scattered echo is in the direction at which the main beam points at the time, when the energy is from some other direction because it was scattered by the mast. Shaping of the mast or covering it with absorbing material can reduce the scattering that produces false echoes, but it does not reduce the distortion of the pattern caused by excising part of the radiated energy.

It is usually difficult to avoid degradation to the sidelobes and the main beam by blockage of a mast. Masts are often of steel, but masts made of dielectric will also cause similar blockage effects. The mast is often modeled as a cylinder for purposes of calculating its scattering effects, but actual shipboard masts are much more complex than simple cylinders. One method to avoid aperture blocking is to employ a four-face phased array antenna distributed around the ship so that none of its faces look into a mast. Even this is not perfect, however, since the ship's superstructure or deck might intercept part of the beam when the ship encounters large pitch or roll angles because of sea conditions.[158,159]

Effect of a Radome on Sidelobes   The phase and amplitude of a signal can change when propagating through a radome, and can limit how low the antenna sidelobes can be. The periodicity of a metal space-frame geodesic dome precludes its use with very low side-lobe antennas. A air-inflated "bag" of thin dielectric material will allow lower sidelobes than can the space-frame geodesic dome. An airborne radar antenna located in the nose of an aircraft behind an aerodynamically shaped radome also can inhibit achieving low sidelobes. The radar designer does not have much control over this type of airborne radome since its first priority is to maintain structural integrity and conform to aerodynamical requirements. In addition, the radome must be able to withstand bird strikes, rain erosion, and lightning strikes. For lighting protection the antenna might be enclosed in a cage of metal rods. A radome can cause phase and amplitude changes in the radiated wave that result in loss and distortion. Energy that is scattered by the inner surface of the radome can result in a sidelobe that is known as the *radome flash*. In some cases[160] the peak value of the radome flash was found to be of the order of 40 dB below the main beam. Thus it can be more difficult to achieve low sidelobes with an antenna inside a radome. The effect the radome can have on an antenna pattern must be determined and corrected if low sidelobes are to be achieved. It should be kept in mind that the effects of a radome mounted above the fuselage can be successfully handled, as it was with the impressive low-sidelobe antenna and radome employed on AWACS.

No-Sidelobe Array In the quest for a low-sidelobe antenna, it might be asked what would an antenna be like if there were no sidelobes whatsoever. Such an antenna pattern can be produced by a linear array with half-wave spacing and with element currents given by the coefficients of the binomial expansion. This no-sidelobe pattern is attributed to J. S. Stone who obtained patents for it in the late 1920s. Leon Ricardi[161] gives its beamwidth as $0.975/(N - 1)^{1/2}$ rad and its directive gain as $1.77N^{1/2}$, where N is the number of elements. Although this antenna has no sidelobes it has a very fat main beam (the radiated energy has to go somewhere), the ratio of the current at the center of the array aperture to the current at the edge element is quite large for values of N that may be of interest, and its gain increases only as the square root of the aperture size. The currents at the elements of a nine-element binominal array, for example, are proportional to 1, 8, 28, 56, 70, 56, 28, 8, 1. The ratio of the current at the center to that at the edge is 70 to 1. For many reasons, the antenna with absolutely no sidelobes produces poor patterns and is not something that a radar engineer should aspire to use as a design goal.

It has been said that the gaussian antenna radiation pattern can be used to approximate the pattern of a no-sidelobe binomial illumination when the number of elements in the binomial array is greater than five and if the standard deviation of the gaussian pattern is made equal to $(L\lambda/8)^{1/2}$, where L is the antenna dimension.[162]

Examples of Low-Sidelobe Antennas The first low-sidelobe antenna was for the AN/APY-1 S-band radar that was the basis for the AWACS (E3A) airborne warning and control system.[152] As was mentioned, an antenna with exceptionally low sidelobes was needed for such an application in order to limit the large clutter echoes that might enter the receiver via the antenna sidelobes. The antenna, Fig. 9.49, consisted of 30 slotted waveguides, called *sticks*. These were fed from one end (series fed). The antenna was 24 ft (7.3 m) in width and 5 ft (1.5 m) in height, and was enclosed in a rotodome that was mechanically rotated in azimuth at 6 rpm. It is a rugged antenna, as is needed to maintain the mechanical tolerances.

Slotted Array The series-fed slotted array, of which the AWACS end-fed array was an example, is well suited as a low-sidelobe antenna since the slots provide a convenient means for achieving the necessary aperture illumination. The slots can be milled with precision computer-controlled machines, the structure can be made mechanically sturdy, and there is no blockage of the aperture as would be the case with a conventional reflector antenna. To achieve the desired low sidelobes, the mutual coupling between the slots in each stick and the coupling between sticks must be properly taken into account. The slots may be in the broadwall or the sidewall of the waveguide slots. (The AWACS antenna used sidewall slots.) Sidewall slots must be tilted at an angle in order to achieve coupling of energy from the waveguide. The tilt of the slots causes cross-polarized radiation, which must be surpressed in some system applications. Periodic errors due to the introduction of the necessary phase reversals in adjacent slot radiators (in order to provide the necessary phase in a traveling-wave waveguide array) must also be surpressed.

Figure 9.49   Slotted-array low-sidelobe antenna for the AWACS, AN/APY-1 radar.
(Courtesy Northrop Grumman Corp.)

The end-fed, or any series-fed, slotted array is usually not of wide bandwidth since the direction of the radiated beam will squint (change angle) when the frequency is changed. With a narrowband signal this is not a serious problem since the direction the beam points relative to the antenna is known and can be compensated accordingly when extracting the angle measurement. When a wideband signal, however, is radiated by such an antenna, the beam will broaden, or smear. If a wide signal bandwidth is required in a low-sidelobe antenna, the corporate-fed array can be used.

Corporate-Fed Planar Array   Operation over a wide bandwidth can be obtained with a corporate-fed array if equal path lengths are used connecting the antenna input to each element of the array. With equal path-lengths, changes in frequency do not cause changes in phase between the radiating elements, as they do in a series-fed array. The feed network of the corporate-fed array is not as simple as that of a series-fed array. There are couplers and branch lines in each path to the radiating elements, and they must be exceptionally precise in order to achieve the precision aperture illuminations required for low sidelobes. The use of dielectrics in such a feed system should be kept to a minimum since they can produce a phase change with a change in frequency. Strip transmission lines, with the strips supported by a minimum of dielectric, can be employed if their power handling is satisfactory. The precise tolerances and complexity of the corporate-fed array can result in greater cost and complexity than the series-fed slotted array, but this is the price that has to be paid to have low sidelobes with wide bandwidth. Because of its greater complexity, Table 9.2 indicates that the corporate-fed array is more likely to have higher sidelobes than the slotted array.

Electronically Steered Low-Sidelobe Arrays   It is more difficult to achieve low side-lobes when the array must be electronically steered. (The wide vertical beamwidth of the AWACS antenna can be steered electronically over a limited range of elevation angles, but this is different from an antenna that scans over wide angles in both azimuth and elevation.) Phase shifters introduce error and if they are digital, they must contain a large number of bits in order to suppress spurious sidelobes. The phase shifters might have to have six to eight bits rather than the three to four bits acceptable for antennas with conventional sidelobe levels. The effects of mutual coupling are more difficult to correct since mutual coupling will change when the antenna beam is electronically scanned. Arrays made up of subarrays also can produce periodic errors that can result in high sidelobes. Subarrays, especially large ones, can be a problem when low sidelobes are desired.

The additional cost of low-sidelobe phased arrays has been examined by W. Patton.[163] Based on analyzing three different arrays, each with 40 dB directive gain, but with sidelobes differing by steps of 6 dB, he concluded that the cost of building a phased array increased about 2.3 percent for each dB the sidelobe level is reduced when the size of the array is held constant and about 3.2 percent for each dB of sidelobe reduction when the antenna beamwidth is held constant.

FASR, an Electronically Steered, Low-Sidelobe Array Radar[164]   This was one of the first low-sidelobe electronically steerable phased arrays. FASR (Fixed Array Surveillance Radar) was an experimental UHF radar developed by the Naval Research Laboratory to demonstrate how low-sidelobe antennas can be obtained in a shipboard environment by placing four fixed phased-array faces around a ship so as to avoid blockage by masts or superstructure. A single array antenna was 32 by 12.5 ft with 297 dipole radiators arranged in 27 columns and 11 rows. Although not small in physical size, the array contained a small number of elements and was not of high gain. Thus it was more difficult to achieve low sidelobes (tighter tolerances were required) than in an antenna with higher gain (and thus more elements). FASR was designed for −40 dB sidelobes and used six-bit digital diode phase-shifters. The 5 by 12° beam scanned 120° in azimuth and 90° in elevation. Monopulse sum and difference beams were generated. The desired sidelobes of −40 dB were achieved at broadside, but they increased slightly when the beam was scanned. The difference pattern also achieved −40 dB sidelobes with respect to the sum pattern.

Parabolic Reflector[165]   There are many advantages of a parabolic reflector for radar applications; however, it is more difficult to achieve low sidelobes with a reflector. Among the things that need to be done to obtain low sidelobes are:

1. A solid rather than a mesh reflector surface should be used to avoid leakage in the back direction.

2. The feed system should illuminate the edges of the reflector with low energy, not only to obtain a highly tapered illumination, but also to minimize the sidelobe energy caused by spillover.

3. Spillover radiation from the edges of the reflector can be attenuated by the appropriate placement of absorbing materials or shields.

4. The feed system might have to consist of more than one horn or radiator in order to properly control the aperture illumination.

5. There can be no aperture blockage by the feed system, which leads to the use of an offset reflector.

6. The mechanical tolerances of the reflector surface have to be better than that of an array by a factor of two because of the two-way path on reflection from the surface.

Scudder[166] described the design of an *S*-band 3D air-surveillance radar using an offset reflector 20 ft wide by 12 ft high with a bandwidth of 600 MHz. The azimuth beamwidth was 1.4°. Seven corrugated feed horns were positioned to provide seven overlapping beams in elevation from 0° to 20°. More than one feed horn was used to form each elevation beam. Absorbers were placed at both the top and bottom edges of the reflector to suppress spillover radiation. Based on measurements of a one-tenth scale model operating at K_a band, the peak near-in sidelobes were −40 dB and decreased rapidly to less than −50 dB, with the wide-angle sidelobes below −60 dB.

Another example is that described by Williams et al.[167] Their elliptical shaped reflector had a major axis of 45 wavelengths and a minor axis of 15 wavelengths producing an azimuth beamwidth of 1.7° and an elevation beamwidth of 5°. It was feed from an offset array of four conical horns. The aperture illumination was designed to produce a peak sidelobe of −50 dB. The price paid for this low sidelobe design was that the beamwidth was broadened by a factor of 1.47 compared to the beamwidth that would have been produced by a uniformly illuminated circular aperture. An *X*-band model produced a peak sidelobe of −43 dB, which was attributed by the authors to spillover radiation from the feed support struts.

An offset-fed parabolic cylinder has some advantage over other reflector antennas for producing low sidelobes because the line source feed (which may be a linear array or a pill box) allows much better control of the aperture illumination than when a conventional parabolic reflector is used with one or several horn feeds.[168] If a corporate-fed linear array is used with equal lengths of lines to the radiators, the antenna can have much broader bandwidth than a series-fed array.

Measurement of Low-Sidelobe Radiation Patterns A good pattern range is needed to accurately measure the radiation pattern of a low-sidelobe antenna, especially if the depth of the nulls are of interest.[169] The pattern range must be sufficiently large so that the curvature of the wavefront due to the finite distance does not affect measurement accuracy. The rule of thumb used by antenna engineers for conventional antennas is that the distance between the antenna and the pattern-measuring source should be at least $2D^2/\lambda$, where D is the antenna dimension and λ is the wavelength. This is adequate for sidelobes down to about -30 dB; but not for lower sidelobe levels. Hacker and Schrank[170] state, however, that the distance of $2D^2/\lambda$ can be satisfactory for the accurate measurement of the wide-angle (far-out) sidelobes, but it does not accurately measure the first one or two near-in sidelobes. If, on the other hand, the entire antenna pattern is to be determined to an accuracy of better than 0.5 dB, the pattern range must be $8D^2/\lambda$ when the pattern is a modified Taylor with first sidelobe of -50 dB. Using the Taylor $\bar{n}$ linear aperture illuminations, Hansen added to this by determining that that if an error of 1 dB or less is required when measuring a -40 dB sidelobe, the measurement distance should be $6D^2/\lambda$.[131] For a -60 dB sidelobe design, the distance should be $12D^2/\lambda$.

System Implications of Low Sidelobes   Low antenna sidelobes are important for achieving the desired performance of pulse doppler radars in the face of heavy clutter. They can also be useful for reducing the effects of sidelobe jamming and making the job of hostile intercept receivers and antiradiation missiles more difficult. Low sidelobes, however, do not come without cost—both monetary and performance.

Compared to the conventional parabolic reflector, low-sidelobe antennas are more expensive, less rugged, more likely to be heavier, require better mechanical and electrical tolerances, and are harder to maintain. As has been mentioned, the lower the sidelobes the less the antenna gain and the wider will be the main beam. If a larger antenna cannot be used to maintain the beamwidth, the wider beam means poorer angular resolution and accuracy, greater susceptibility to main-beam jamming, and larger clutter echoes received in the main beam. In an airborne doppler radar that is concerned with detecting low velocity ground-moving vehicles with doppler frequencies near that of the main-beam clutter, a narrow antenna beamwidth with conventional sidelobe levels may be more important than a low sidelobe antenna with an increased beamwidth.

To avoid negating the benefits of low sidelobes, the antenna must be operated in a clear environment without obstructions or nearby objects that block or scatter the radiated energy. When a low-sidelobe antenna must be operated within a radome, special considerations need to be given to the design of the radome in order that the sidelobes are not degraded. Usually the radome will rotate along with the antenna (a rotodome) in order to maintain the same radome environment seen by the antenna as it rotates.

As with most things, low-sidelobe antennas have both good and bad effects. They are not universally applicable and should be used only when their desirable features outweigh their disadvantages.

9.14 COST OF PHASED ARRAY RADARS

A phased array radar generally costs more than a conventional radar that employs a mechanically scanned antenna but it also can provide unique capabilities not available with other antennas. It has been used chiefly in military applications where the unique features of a phased array may sometimes compensate for its higher cost. It has seldom been employed for civilian radar applications since there are few applications where its greater expense can be justified when competitive market forces make price an important consideration. There has been much interest in reducing the costs of phased arrays to make them more competitive for both military and civilian applications.

Factors Affecting the Cost of a Phased Array Radar   Tang and Brown[171] state that the cost problem of a phased array is attributable to the following:

- The large number of discrete components in a conventional phased array that have to be individually fabricated, assembled, tested, and installed.

- The production yield of components, especially solid-state amplifiers that use monolithic high-power chips.

- The labor involved.

- The limited production quantity of any particular radar system which does not justify large capital investments for a dedicated production line with special production tooling that could reduce manufacturing costs.

These four categories are only a part of the problem. Cost depends on the particular radar architecture, the degree to which multifunction operation is used, computer software for operating the radar, and the radar frequency. Each of these will be briefly reviewed.

Effect of Radar Architecture on Cost   Section 9.9 discussed the various architectures for phased array radars, and some mention was made of the effect of the architecture on radar cost. It is difficult to be quantitatively accurate regarding predictions of radar cost, but some generalizations can be made. For example, a space-fed array using a single high-power vacuum tube transmitter has been less expensive than a corporate-fed array or an active aperture solid-state phased array. Also, the row-column control of phase shifters is usually cheaper than when each phase shifter is controlled individually.

There was a brief description in Sec. 9.9 of how Russian (Soviet) air-defense phased array systems achieved lower cost. Cory[105] summarized these as:

- Minimizing the total number of phase shifter modules.

- Simplifying the radar architecture (such as by using a space-fed array).

- Designing simple and inexpensive components.

- Minimizing the size and complexity of the control system.

- Simplifying the feed design.

In addition the Russians used several low-cost radars, each performing a single air-defense function rather than the more complex multifunction phased array that has been the more usual practice in the U.S.

Multifunction Radar and Cost   Because of its flexibility and rapid beam steering, an electronically steered phased array antenna can be used to perform multiple radar functions including search, track, weapon control, missile guidance, target recognition, and perhaps others. The ability to perform multiple functions with one phased array radar has been a major selling point of those who market radar systems that employ phased arrays. One should be cautious, however, since a single multifunction phased array radar might not be the best approach for all radar applications. This is especially true for air defense.

The basic problem with a multifunction electronically steered phased array radar for air defense is that compromises must be accepted when a single radar is used for both surveillance and weapon control. It is well known among radar system engineers that the lower microwave frequencies are more suited than the higher frequencies for long-range air-surveillance radars. The frequency of choice is usually L band (1.215–1.4 GHz). On the other hand, the higher microwave frequencies are more desirable for weapon control, with X band the usual choice (8.5–10 GHz). When a single phased array radar is required to perform both surveillance and weapon control, a compromise choice of a single frequency has to be made, generally somewhere between L and X bands. The U.S. Navy's Aegis air-defense system is at S band and the U.S. Army's Patriot air-defense system is at C band, yet they basically have the same mission (except one is on ships and the other is land mobile). When the beamwidths must be the same no matter what the frequency (which implies that the antenna gain is independent of frequency) the antenna aperture will be smaller at the higher frequencies. Thus performing the surveillance function at S band results in less range performance (for a given average transmitter power) than if it were at L band. Air surveillance at higher frequencies (such as C band) result in even less range performance. Also, the higher the frequency the less will be the available doppler space (to detect moving targets in clutter) because of blind speeds. When weapon control radars operate at C band or S band, the antenna beamwidths will be wider than they would be at X band, resulting in poorer angle accuracy and less ability to deal with multipath effects from surface reflections. Thus, even if cost were of no concern whatsoever, one usually has to accept lesser performance in both surveillance and weapon control when a multifunction radar is employed for air defense at a single frequency band.

It is not always necessarily true that an air-defense radar system with multiple radars at different frequencies has to be more expensive than a single-frequency multifunction system which has the same performance. Both Barton[104] and Cory[105] have written about the benefits of the Russian design approach to air-defense radar systems that employ several simple cost-effective phased array radars to perform the various functions of surveillance, weapon control, and low-altitude detection.

In some radar system applications the optimum frequencies for surveillance and for track might be the same, so that multifunction radars do not have the same limitations that are experienced with their use for air defense. Space surveillance for the detection and tracking of satellites is one example where both the surveillance and the tracking functions can be performed at the same frequency. (UHF is a good choice, as in the AN/FPS-85 and Pave Paws.) Thus no significant compromises in performance need be made when using a single multifunction radar for the space surveillance functions of detection and tracking.

Multifunction radars that use mechanically scanned planar array antennas have been well suited for military combat aircraft and, in the past, have been the norm. A modern radar for a fighter/attack aircraft has to perform a number of functions, maybe from 6 to 9, for air-to-air purposes and a similar number of different functions for air-to-ground purposes. There have been no significant limitations (other than having sufficient time) in using the same airborne radar to perform all the many functions at X band. The multifunction electronically steered phased array can also be used for airborne fighter/attack radar application, but a single phased array face in the nose of a fighter/attack aircraft (usually limited to less than $\pm 60°$ in angle) cannot provide as large an angular coverage as can a mechanically steered planar array antenna. (Large angular coverage is especially important for the dog-fight role.) Two or more phased arrays might be used for increased coverage, but they result in increased system size and weight. Mechanical antennas thus can be competitive, and in some ways superior, to electronically steered phased arrays for military airborne applications since they can do what is required of an airborne radar antenna at less cost and less weight than an electronically steered phased array.

Offensive bomber aircraft, such as the B-1B, have also employed the multifunction electronically steered phased array to perform the many radar functions unique to the bomber. Such radars have been more expensive than conventional mechanically scanned radars for the same purpose.

Computer Software and Cost   The computer can be a significant part of the cost of a versatile multifunction phased array radar. It has not been the computer hardware but the software that can be a sizable fraction of the total radar development cost. In early long-range multifunction phased array radar systems, software cost was about 30 to 40 percent of the total. Over time, computer software has become better and easier to obtain, but it is still a significant factor in achieving a successful phased array radar system. The design of the computer software for phased arrays must enter the radar system development process at an early stage, with sufficient time and funds allowed for it to be successfully completed. Without sophisticated computer control, a phased array can do very little.

Effect of Frequency on Phased Array Radar Cost   In general, the lower the frequency of an air-surveillance radar that uses a phased array antenna, the lower the cost. The rationalization is as follows. A passive phased array is assumed, one with a single receiver and a single high-power transmitter, that requires a specified power-aperture product. (That is, the average transmitter power times the antenna area is a constant, Sec. 2.13.) The cost of an antenna element and a phase shifter is more or less independent of frequency, but the cost of an array is proportional to the number of elements. In this comparison it is further assumed that the antenna must have the same gain (same elevation and azimuth beamwidths) at whatever frequency is selected for its operation. Thus a 9000-MHz (X-band) phased array radar would have the same number of elements as a 450-MHz (UHF) radar. With the above assumptions, the array antennas at the two different frequencies would cost approximately the same. The aperture of the 450 MHz radar, however, is 20 times larger in linear dimension and 400 times larger in area than that of the X-band radar. Since the power-aperture product in this example is assumed to be the same at both frequencies for equal performance, the transmitter power for the X-band radar must be 400 times that of the UHF radar, with the result that the cost of the X-band radar will be many times greater than that of the radar at UHF. The effect of a higher frequency on cost is probably even greater for active-aperture phased array radars that employ a T/R module at each element containing its own solid-state transmitter, receiver, phase shifter, and duplexer. The cost of a T/R module is likely to be greater at the higher frequencies rather than be relatively independent of frequency as in the passive array. The above argument assumed a surveillance radar where $P_{av}A$ is constant. In a tracking radar, generally the product $P_{av}A^2$ is a constant with frequency. This makes it even more likely that an equivalent radar at a lower frequency will cost significantly less than one at a higher frequency.

The above has been a very simplistic argument with some very gross assumptions. There may be other requirements the radar must meet which require that a phased array radar operate at higher frequencies in spite of higher cost (such as if it has to fit into the nose of an aircraft). Nevertheless, it is often true that the lower the frequency the more affordable will be the phased array radar.

Reducing Phased Array Cost   This subsection summarizes some of the guidelines for lowering the cost of phased array radars.[172]

1. *Operating at as low a frequency as the application will permit.*

2. *Emphasizing low-loss design.* This might seem obvious or trivial, but it has not always been given sufficient attention.

3. *Time sharing a four-faced phased array with a single transmitter.* This is done as a cost-saving measure, but it can affect the overall performance of the system.

4. *Employing the active aperture architecture only when it is appropriate.* The losses in the active aperture array are less than the losses in a passive array with a constrained feed. It is not always obvious, however, that the total system cost of an active aperture radar will be lower or its performance better than other phased-array architectures.

5. *Single transmitter.* It has usually been true that the greater the RF average power from a single device, the lower will be its cost per watt.

6. *Space-fed arrays.* Lower loss and less complexity of the space-fed array can result in lower cost.

7. *Attention to computer issues.* The cost of the computer software for a phased array and the time required to generate it can be significant.

8. *Trainable arrays.* If an application requires 360° of coverage, one might not want to have the expense of four identical phased array radar systems to provide the total coverage. As mentioned in no. 3 above, one or two transmitters might be time-shared among the four apertures of an array radar system. The combination of two trainable arrays and a conventional 2D air-surveillance radar could be less expensive that a full four-face phased array.

9. *Avoiding a multifunction array.* There can be, in some applications, other less costly approaches that can perform more effectively the same mission.

It was mentioned that phased array radars have the advantage of being more readily hardened to withstand nuclear blast effects than a mechanically steered radar. Hardening only adds to the already large costs of phased arrays. In the U.S. development of intercontinental ballistic missile defense systems in the 1960s and early 1970s, the high cost of a fully hardened system was a factor in leading to the ABM (antiballistic missile) treaty between the U.S. and the Soviet Union.

It might be mentioned that the life-cycle costs of military systems, which were not considered in the above, include the cost of development, procurement, installation, training, operating, and maintenance. A further consideration is that the development cost of a radar usually is a small fraction of the total life-cycle costs of the system.

9.15   OTHER TOPICS CONCERNING PHASED ARRAYS

Bandwidth of a Phased Array Antenna   Two different kinds of bandwidths need to be considered for phased arrays. One is the instantaneous, or signal, bandwidth, which is an indication of the maximum bandwidth of a signal that the array can handle without distortion. Usually, it is difficult to obtain an array signal bandwidth of more than a few percent. The other is the operating, or tunable, bandwidth over which a narrowband signal can be received (or transmitted) without distortion.

Signal Bandwidth   Figure 9.15a showed a two element array receiving a signal that arrives at an angle θ_0 relative to broadside. The signal appears at element 2 before it appears at element 1. If a delay line of the proper length is inserted at element 2, the two signals will coincide and add without loss. There is no theoretical limitation to the signal bandwidth in this case when delay lines are used to bring signals from the various elements of the array into time coincidence. As has been said in the original discussion of Fig. 9.15, time delays inserted at each of the many elements of a large array have not been not practical. Instead, the delay line is replaced with a phase shifter that is limited in phase to the range 0 to 2π radians. Signals can be phase coherent so long as they overlap in time. But the signals do not overlap with 0 to 2π phase shifters during the transient build-up time. Thus phase shifters produce coherent addition only for narrowband (long time duration) signals.

The limitation on bandwidth when phase shifters are used in an array is dependent on the rise time, or build-up time, of the signal as it transits across the array. The transient response of the incident signal as it builds up across the array has the same effect on signal bandwidth as the transient response, or build-up time, of a conventional filter. The build-up time of a linear array of dimension D when a signal is incident on the array at an angle θ_0, is $(D \sin \theta_0)/c$, where c is the velocity of propagation. If, for example, the angle of arrival $\theta_0 = 45°$, and $D = 30$ ft, the transient build-up time is about 22 ns. The signal bandwidth is thus limited to the reciprocal of the transient build-up time, or about 45 MHz. There is zero build-up time for a signal that arrives from the broadside direction ($\theta_0 = 0$); hence, there is no theoretical bandwidth limitation. (This assumes that the signals from all the antenna elements are summed with equal-length transmission lines.)

Another aspect of signal bandwidth has to do with the change of phase with a change in frequency. The phase shift ϕ required to steer a beam to a given direction is assumed to be independent of frequency. If the value of ϕ is chosen so as to point the beam to a direction θ_0 when the frequency is f_1, the beam will point to a new direction when the frequency is f_2. If it is assumed that the signal's frequency spectral width must not cause the beam to scan more than $\pm$ one-fourth beamwidth, Cheston and Frank[173] show that

signal relative-bandwidth in percent = broadside beamwidth in degrees   **[9.69]** where the signal relative-bandwidth in percent is 100 times the absolute bandwidth divided by the RF carrier frequency, or $(B/f_0) \times 100$. This expression is based on the array having an equal-path-length feed and the beam is scanned to an angle of 60°.

Although it has been impractical to employ delay lines at each element of an array, they have sometimes been used at the subarrays to increase bandwidth. Phase shifters are used at each element in addition to the delay lines at each subaperture. This reduces the complexity of an array compared to one with delay lines at every element, but it also has less bandwidth than a true time-delay array. Subarrays with delay lines increase the bandwidth of a linear array in proportion to the number of subarrays, compared to the bandwidth of an array that does not have subarrays.[174] The sidelobes will increase, however, with the use of subarrays, which may not be desirable for some applications.

Operating, or Tunable, Bandwidth   Although the build-up time of an array limits the signal bandwidth that it can handle without distortion, it is possible to operate an array over a very wide band of frequencies by retuning; that is, by re-setting the phase shifters to new values when the frequency (of a narrowband signal) is changed. Such an array can radiate different narrowband signals one at a time at different frequencies by readjusting the phase shifters with each new frequency. The operating bandwidth of the array can be quite large and might be limited only by the onset of grating lobes. This assumes that the antenna elements and other components of the array are wideband.

It has been reported[175] that an array containing 4096 open-ended waveguide radiators with a triangular arrangement of elements was capable of operating over a 30 percent frequency band and over a scan volume of more than 120° in both azimuth and elevation. A wide-angle impedance matching dielectric sheet was placed in front of the array. (The frequency at which this array operated was not given.)

Computer Control of an Array   Although a radar with a conventional mechanically scanned antenna can operate without computer control, the multifunction electronically scanned phased array must be controlled by a computer if it is to achieve its full potential. An important task of a computer is to generate the phase shifter commands for each element of the array to steer the beam in the desired direction. But this is only a small part of what an array computer must accomplish. A much more demanding task is to effectively manage the various radar functions required of the array. The computer hardware for the multifunction operation of an array usually is not as much a concern as is the computer software needed to generate and schedule the various waveforms, data rates, and processing without degradation of performance.

The demands on the computer that controls the phased array radar vary with the application. Generally, the various radar tasks that have to be performed by a phased array under computer control must be done sequentially rather than simultaneously. In a phased array radar for air defense, for example, it is the job of the computer to allow the radar to track a large number of targets as well as search a large volume of space within a specified time for the detection of targets. The tracking data rate depends on whether or not the target is considered hostile and is being engaged. The revisit time during search, for example, might need to be one or two seconds when looking for pop-up, low-altitude targets that first appear over the horizon at short ranges (perhaps 8 to 20 nmi). With long-range targets (150 to 200 nmi), the revisit time can be greater (perhaps 10 s) since there is more time for an air-defense system to react to a long-range threat than to a short-range threat. (A sea-skimming missile flying at Mach 3 at very low altitude might first appear above the radar horizon at a range of 10 nmi from the radar. If the detection decision is made almost instantaneously, then there is then less than 20 s available to destroy the missile before it reaches its objective.) If the radar is to detect tactical ballistic missiles at long ranges, the search patterns and data rates will be different from those used for aircraft targets. Thus the computer must program different search procedures depending on the range and type of target expected.

Once a target is detected, the data is used to update an existing track or initiate a new track. When the radar is performing its search function, the scanning must be interrupted periodically in order to radiate one or more pulses in the direction of known targets already held in track. Since the target is in track, the radar will know the approximate time the echo is supposed to arrive back at the radar. At that time, the phased array beam can be pointed to the direction of the target so as to receive the echo signal. Search and track are therefore accomplished in an interleaved manner. The computer has to be programmed to be able to accomplish this efficiently; that is, to search the required volume with the required revisit times and to track a large number of targets without serious degradation.

In spite of best efforts, there will likely be radar functions that become overloaded when the number of targets is large. This occurs because the various functions of a conventional phased array are performed sequentially in time. Everything cannot be done at once, so priorities must be assigned to the various functions to be performed. Those with less importance are performed at a lower data rate—or maybe, not at all. Among the highest priority functions are those involved with the engagement of threatening hostile attacks. This includes tracking of the attacker and providing guidance information to the intercepting missile. Time critical functions, such as short-range horizon-search for low altitude threats are also of high priority. Next in priority might be the tracking of confirmed hostile threats. Of lower priority is the tracking of known friendly targets. Low priority is also given to above-horizon search for targets at long range. These are only a few of the many tasks or functions that a phased array radar has to perform as part of an air-defense system.[176,177]

As the number of targets increases or as the demands of the air-defense increase, the computer can become increasingly overloaded and it will be more difficult for the phased array radar to perform all of its tasks with equal effectiveness. The problem has sometimes been described as there not being enough microseconds in a second to do all that is required of the phased array radar. One author[178] described the problem as not being able "to get round the sky quickly enough." Among the many factors that can cause the phased array to be overloaded and result in compromised performance are (1) a relatively large number of angular resolution cells that the radar must examine, (2) the need for the radar to detect small moving targets in large clutter, which requires that it dwell in each direction long enough to obtain good doppler filtering to suppress the clutter, and (3) a relatively long time to acquire targets and to initiate tracks.

The overloading of the functions performed by a conventional phased array radar is less of a problem with a radar that uses digital beam forming to look everywhere all the time since the various functions can be performed simultaneously (in parallel) rather than sequentially (in series), as was discussed in Sec. 9.9.

Radant Phased Array[179,180]   Radant, originally developed by Thomson-CSF of France, is a different method for employing diodes or microelectromechanical switches (MEMS) to produce agile beam steering. Instead of using diode phase shifters in the conventional manner, Radant employs strips of metalized diodes arranged in columns to produce phase shift in one angular dimension by changing the voltage applied to each strip of diodes. The principle of the Radant diode lens for scanning in one angular coordinate is illustrated in Fig. 9.50. The vertical strips of diodes that produce the phase shifts are illuminated from the back by a plane wave as might be generated by a simple planar array. A parabolic reflector or a lens also might be used to illuminate the Radant, as might space feeding (but not offset space feed).

Radant can be thought of as a lens whose index of refraction (or dielectric constant) can be varied by appropriately biasing each string of diodes.[181] Biasing the PIN diodes can provide the desired change in susceptance to change the "index of refraction" of the lens consisting of many diodes. A number of planes of diodes are used to obtain the total phase shift. In a Radant array antenna with the geometry of Fig. 9.50, there might be 20 planes of diodes having a total thickness of 4 inches at *X* band.

As described, beam steering occurs in only one plane. Figure 9.51 illustrates two-dimensional steering. In this arrangement the first lens steers the beam in the horizontal direction. The plane of polarization is then rotated 90°, and a second lens oriented 90° to lens no. 1 steers the beam in the vertical. Control of two-dimensional beam steering is done with row and column commands so that with a $N \times M$ element array there need be only $N + M$ commands rather than $N \times M$. An advantage claimed for the Radant antenna is that it can be of lower cost than other array configurations.[182]

A Radant lens electronically steered phased array radar, called the RBE2, was developed by Thomson-CSF for the French Rafale multirole combat aircraft built by Dassult Aviation.[183]

Ferroelectric Phased Arrays   A change in phase can be had by a change in the dielectric constant (permittivity) of the material in which the electromagnetic signal propagates. Materials whose dielectric constant varies with the d-c voltage applied across it are known

Figure 9.50   Principle of the Radant antenna for scanning in one angle coordinate, horizontal in this case.
(Courtesy of Jaganmohan Rao, Naval Research Laboratory Radar Division.)

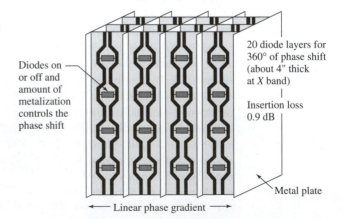

Diodes on or off and amount of metalization controls the phase shift

20 diode layers for 360° of phase shift (about 4" thick at *X* band)

Insertion loss 0.9 dB

Metal plate

Linear phase gradient

Figure 9.51   Method of obtaining a two-coordinate beam steering Radant antenna.
(Courtesy of Jaganmohan Rao, Naval Research Laboatory Radar Division.)

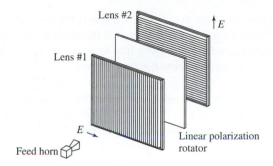

as *ferroelectrics*. Ferroelectric phase shifters have been investigated in the past, but they were difficult to match because of their large dielectric constant and they have higher losses than might be desired. With time, ferroelectric materials have improved and they have shown more promise.

An interesting approach to a ferroelectric phased array based on a lens array with bulk phase shifting is that of Jay Rao and colleagues.[184] Figure 9.52 is a very simplified sketch of the operation of a ferroelectric lens array that steers in one plane. It is made up of ferroelectric slabs sandwiched between conducting plates that apply the d-c voltage that determines the dielectric constant and the phase shift. Changing the voltage on the conducting plates changes the direction of the beam. Beam steering in two orthogonal planes is obtained by rotating the plane of polarization 90° and using a second lens array oriented 90° to the first lens, as was illustrated in Fig. 9.51 for the Radant lens. Similarly the ferroelectric array can be illuminated with a conventional non-steerable planar array or by a horn feed, just as for Radant. The material used was a bulk oxide-ceramic composite of barium strontium titanate oxide (BSTO) with a typical dielectric constant between 90 and 120.

Theory and experimental measurements indicate that a ferroelectric array at X band can operate over the frequency range from 8 to 12 GHz with a voltage standing-wave

Figure 9.52   Basic configuration of a ferroelectric lens array for steering in the horizontal plane with horizontal polarization.[184]
(Provided by J. Rao and D. Patel of the Naval Research Laboratory Radar Division.)

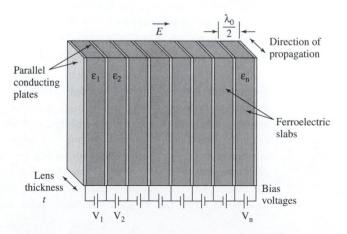

ratio less than two, and with a loss between one and two dB.[184] The advantage of this form of phased array compared to other phased arrays is its potential for low cost. It uses row and column steering that requires far fewer phase-shift control signals than an array which requires control signals for each element of the array. It is said to have smaller lens thickness, higher power capability, simpler beam steering controls, and use less power to control the phase shift than a Radant lens array. The use of row-column bulk phase steering instead of individual phase shifters at each element can make it more difficult, however, to achieve low sidelobe levels.

Conformal Array Antennas   A long-sought, but difficult to achieve, desire of radar system engineers is to be able to place array elements anywhere on a surface of a relatively arbitrary shape and obtain a directive beam with good sidelobes, good efficiency, and which can be conveniently scanned electronically. An array on a nonplanar surface is called a *conformal array*. Such an array, if practical, might be configured to conform to the nose, wing, or fuselage of an aircraft or missile. Most of the work on conformal arrays has been for relatively simple shapes such as the cylinder, hemisphere, cone, or truncated cone.[185] It has been difficult to achieve a practical conformal array, except in simple geometries. There are good reasons why almost all operational phased arrays are planar.

In some conformal geometries, equal element-to-element spacing is not practical. Because of the nonplanar surface the element polarization can vary from one point on the array to another and with the beam pointing direction. The aperture illumination cannot be separated into two orthogonal patterns, as it can in a rectangular array. The boresight of a difference pattern might vary with angle. The calculation of mutual coupling is more difficult with a nonplanar surface. The issue is not whether a conformal array can be built, but whether some other solution is better.

If one wanted to have a cylindrical antenna, one way to achieve it would be to use four planar arrays arranged in a square and cover the structure with a cylindrical radome—and not let anyone look inside. If the change in beamwidth and gain with scan angle experienced by planar arrays were of concern, adjacent faces of a four-face phased array might be used cooperatively to maintain the beam relatively unchanged with scan. As the beam from one planar array is scanned off broadside, some of the transmitter power can be applied to an adjacent face and diverted to scan in the same direction as the original face.[186] Thus each face might work in conjunction with its two adjacent faces to maintain the beam shape almost independent of scan angle.

Thinned, or Unequally Spaced, Arrays[187,188]   A normal phased array antenna has its elements spaced about a half-wavelength apart for good performance. When the elements are, on average, spaced much greater than half-wavelength the array is said to be thinned, or unequally spaced. (This assumes that a wide angular coverage is required; otherwise it would be known as a limited-scan array.) The beamwidth is determined by the electrical size (in wavelengths) of the array but its gain and sidelobe levels are determined by the number of elements that remain. A thinned array will have about the same beamwidth as a filled array, but its gain will be reduced in proportion to the number of elements removed. Its peak and average sidelobes will increase. [The simple expression for antenna gain, $G = \pi^2/\theta_a\theta_b$ of Eq. (9.5b), doesn't apply to thinned arrays.] Many methods have been tried to determine element spacings that will result in acceptable antenna patterns. Two of the more successful will be described briefly.

Dynamic Programming   One method is to have a computer calculate, for a fixed number of elements, the antenna patterns for all possible locations of the elements within the array and determine which is best. This method of total enumeration, is impractical because of the large number of combinations that are possible. A more practical approach is *dynamic programming,* an optimization method that can produce an equivalent result to total enumeration under certain conditions. Its advantage is that it is much less computer intensive. Dynamic programming determines the optimum solution to a multistage problem by optimizing each stage of the problem on the basis of the input to that stage. It is a good method for finding the spacings of a thinned array when the number of elements is not too large.[189]

Density Taper   The other method, which is applicable to large linear or large planar arrays, is to employ a *density taper.*[190] Consider a uniform grid of possible element locations with equal spacing of one-half wavelength. The desired amplitude illumination for a conventional filled array is used as the model for determining the *density* of equal-amplitude elements. That is, the density of equal-amplitude elements is made to approximate the desired aperture illumination. The choice of whether or not to include an element in a possible location may be made statistically or deterministically. In one design, a one-degree beamwidth circular array which would have 7800 elements if completely filled, had 3773 elements when a density taper was employed based on a 30-dB Taylor amplitude illumination.[187] This represents a thinning of 52 percent, where the degree of thinning is defined as the ratio of the number of elements removed from a filled array divided by the original number of elements. Density taper was used in designing the receiving aperture of the AN/FPS-85 and the Cobra Dane (AN/FPS-108) space-surveillance phased array radars.

System Degradation   Thinning of a phased array will produce serious undesirable characteristics for many radar applications when the degree of thinning is too high. The gain is significantly reduced and there will be high peak and average sidelobes.

A conventional filled phased array will have almost all of its radiated energy within its main beam. Only a few percent of the radiated energy will appear in the sidelobes. With a highly thinned array, however, the reverse is true. Too much energy is wasted in the sidelobes. An array with 90 percent thinning might have about 90 percent of its energy in the sidelobes. If clutter is a problem, as it is in high-prf pulse doppler radars, the high sidelobes throughout space can result in high levels of clutter entering via the sidelobes.

Thinning may look attractive at first glance because it seems to allow a narrow antenna beamwidth with a reduced number of elements, but one does not usually get something for nothing. It should be attempted only with one's eyes wide open to the consequences.

9.16 SYSTEMS ASPECTS OF PHASED ARRAY RADARS

Attractive Attributes of Phased Array Radar The electronically steered phased array antenna is of interest since it can provide capabilities not readily available with other types of antennas. Its advantages are summarized below.

Interialess, rapid beam-steering. The beam from an array can be scanned, or switched from one position to another, in a time determined by the switching speed of the phase shifters. A diode phase shifter, for example, allows the beam to be switched in several microseconds or less. Ferrite phase shifters provide switching speeds that are slightly longer.

Multiple, independent beams. A single array aperture can generate independent simultaneous beams on receive, as described for the digital beam-forming phased array discussed in Sec. 9.9. On the other hand, multiple simultaneous beams on transmit are difficult to obtain, which is why a broad transmitting beam sometimes is used in conjunction with a number of contiguous narrow receive beams. For almost simultaneous tracking of many targets, a simpler method is to rapidly switch a single transmitting beam through a sequence of positions by means of a time-sequenced burst of pulses, with each pulse steered to a different direction. Rapidly acting phase shifters are needed, as well as an application that does not require a short minimum range. Since the targets are in track, their directions are known. On reception the receive beam is switched at the proper time to the direction from which the echo is expected so as to update a target already in track.

Potential for large peak and/or average power. Each element of an array can have its own individual transmitter with the outputs combined in "space" to obtain a large total power. (The power per element will be limited by the need for each individual transmitter or T/R module to fit within the space available between adjacent elements.) The active-aperture radar, Sec. 9.9, is an example. In addition to being able to achieve a large radiated power, an array with a transmitter at each element avoids the loss that can occur when the power from a single high-power transmitter has to be divided and distributed to each radiating element.

Control of the aperture illumination. Since the aperture illumination is determined by the currents at a large number of individual radiating elements across the array, a particular antenna radiation pattern is much easier to obtain with an array than with other antennas. This is important when shaped beams or very low sidelobes are desired. Separate monopulse sum and difference patterns, each with its own optimum characteristics, can also be obtained with arrays.

Adaptive processing. Adaptive arrays are designed to automatically adjust the aperture illumination to place nulls in the antenna pattern in the direction of external noise sources and/or clutter echoes. Full array adaptivity, however, in which each element of a large array is part of the adaptivity process, has been too expensive in the past to implement. The sidelobe canceler is an example of a practical adaptive system that requires the use of only a few auxiliary low-gain antennas. When sidelobe canceling is employed in a phased array, several of the array's elements can be used as the auxiliary antennas.

Lower radar cross section. Because of its flat surface, the phased array can have a lower radar cross section than conventional reflector antennas, when illuminated by a radar with a frequency lower than the radar frequency. The flat face of a phased array can produce a large specular reflection, but the array can be tilted so that its specular scattering is directed to an angle where it is less likely to be detected. The effect of the tilt on beam steering can be compensated in the commands to the individual phase shifters. On the other hand, the radar cross section of a planar array might not be low when illuminated by a radar at a frequency higher than that for which it was designed.

Flush aperture shape. The flat surface of an array permits it to be flush mounted and to be hardened to resist the effects of blast.

Multiple functions. The agile beam-steering offered by a phased array allows a single array radar to be time-shared (sequenced) among more than one radar function.

Electronic beam stabilization. The ability to steer the beam electronically allows stabilization of the beam-pointing direction when the radar is on a ship or aircraft that is subject to roll, pitch, and/or yaw. This avoids the need for heavy mechanical stabilization machinery, but it also requires that the array be able to steer the beam over wider angles than when it is mechanically stabilized.

One other advantage sometimes claimed for a phased array is that it degrades gracefully when failures occur in the system. Since there can be many elements in the array (several thousand to several tens of thousands, or more), the effect of the failure of a few individual elements is small. It has been said[191] that for an active-aperture phased array "typically, 5 percent or more [T/R] module failures can be tolerated while maintaining acceptable performance as a multimode radar." There are several reservations that need to be mentioned, since graceful degradation is not guaranteed. First, there can be failure modes in a phased array radar that can affect a large number or all of the elements, or even be catastrophic. Second, no matter how "graceful" the failure of an array might be, sooner or later there will come a time when failures finally have to be replaced. Third, if the buyer of the radar (especially if the radar is for the government) is told by the radar company's marketing department that the radar can operate satisfactorily when a significant fraction of its elements fail, it is likely that at some time during the development of the radar—when serious overruns in money or time occur—the margin that allows for graceful degradation might be quietly removed. In spite of difficulties, graceful degradation should be designed into a phased array radar and the radar systems engineer should try to make sure it is not removed.

Limitations of Phased Arrays   As with most things in life, the desirable benefits of the phased array do not come without their price—and the price is sometimes measured in more than just dollars.

Complexity   The phased array radar is much more complex than a radar with a reflector antenna. In addition to being made up of many thousands of individual elements, there must be means for insuring that the phase and amplitude at each element are what they were designed to be; and, if not, there needs to be means to readjust them to their correct value. If the phased array is to perform the functions of multiple radars, there must be more equipment behind the aperture than if only a single radar function were being performed. The antenna aperture of a phased array radar is much like the "tip of an ice berg." There is a lot more than what is normally visible. The complexity of the array also increases the problems associated with its *maintainability,* achieving high *reliability,* and assuring that its *availability* is such that it will be able to operate when needed.

Software Intensive   Everything that the phased array does is commanded by a computer. The cost of the computer software to control the beam steering of the array and the various radar functions can be a significant portion of the total radar system cost, especially if the radar must perform the multiple functions of search, track, and weapon control.

Cost   The cost of phased arrays was discussed in Sec. 9.14, and nothing more need be said here other than the high cost of phased array radars has limited their use to applications where the customer has been willing to pay the higher costs in order to obtain the special attributes of an array radar. For this reason, the phased array that scans in two angular coordinates has seldom been used for other than military applications.

System Limitations   An advantage of a phased array is that it can be time shared to perform multiple functions. Some of the functions performed by an array, however, might take more time than is available. Three radar tasks that usually require more time than usual to accomplish properly are doppler processing, target acquisition, and long-range surveillance.

Doppler processing (as in MTI or pulse doppler radar) is used to detect moving targets in heavy clutter. As has been seen in Chap. 3, the longer the dwell time, the better can moving targets be separated from clutter. When a target is detected by the surveillance waveform of the array and is then designated for tracking, the tracking beam must be accurately directed to the direction of the target so that the beam can rapidly acquire the target without having to search for it. (A weapon control radar using mechanical trackers, on the other hand, usually can take the time—a few seconds—to search a limited angular region to find and acquire the target to be tracked. The phased array that has to perform multiple functions generally doesn't have this luxury.) Long-range air-surveillance requires the radar beam to remain in a fixed direction until the potential echoes from all ranges can return to the radar. These and other radar tasks, such as some forms of non-cooperative target recognition and burnthrough against ECM jamming, can require long dwell times that could overload the scheduling of the array and cause it to omit or delay tasks of lower priority.

More than Just Firepower   One of the reasons the phased array has been used for military air-defense applications is that it has been said to provide increased *firepower.* Firepower has not always been a well-defined term, but it has been defined in Webster's Ninth New Collegiate Dictionary as "the capacity to deliver effective fire on a target." In addition to firepower, a balanced air-defense system must avoid *leakage* (the penetration of the defense by an attacker when the kill probability is too low), *saturation* (which means that the attack is so large and occurs within such a short time that the defense becomes overwhelmed and cannot engage all targets), and *exhaustion* (when the defense runs out of missiles before it runs out of attackers). The phased array radar addresses only one of these three, which is saturation.

9.17 OTHER ANTENNA TOPICS

This section considers some miscellaneous antenna topics that did not seem to fit in other sections of the chapter. They are discussed in no particular order of importance.

Mirror-Scan Antenna, or Inverse Cassegrain[192] The radiated beam of the antenna configuration shown in Fig. 9.53 can be rapidly scanned over a wide angle by mechanical movement of a light weight planar mirror called a twist reflector. This has been known by many names, including mirror-scan antenna, mirror-track antenna, polarization-twist Cassegrain, flat-plate Cassegrain, parabolic reflector with planar auxiliary mirror, and inverse Cassegrain. The parabolic reflector shown on the left in Fig. 9.53 is made up of parallel wires spaced less than a half-wavelength apart. (The wires are usually supported by a low-loss dielectric material.) For purposes of discussion, assume they are oriented vertically. The thin parallel vertical wires of the parabolic reflector make it sensitive to vertical incident polarization (when the E field is vertical). The parabola will completely reflect linear vertical polarization and be transparent to linear horizontal polarization. If the energy radiated by the feed in the center of the figure is vertically polarized (E field parallel to the vertical wires of the parabolic reflector), it will be completely reflected and directed towards a planar reflector (a mirror) called the *twist reflector*. The twist reflector has the property that it imparts a 90° rotation of the plane of polarization to the energy reflected from it. The polarization of the energy reflected from the twist reflector will then be horizontal and will pass through the parabolic reflector with negligible attenuation. The

Figure 9.53   Geometry of the polarization-twist mirror-scan antenna, using a polarization sensitive parabolic reflector and a planar polarization-rotating twist-reflector. Rapid scanning of the beam in azimuth and elevation is accomplished by mechanical movement of the lightweight planar twist-reflector.

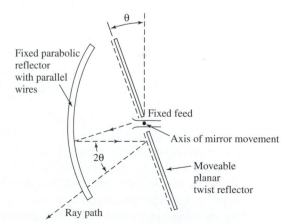

radiated beam is steered in angle by mechanically rotating the low inertia twist reflector. When the twist reflector is rotated by an angle θ, the radiated beam is rotated through an angle 2θ. The beam can be rapidly scanned over an angle of $\pm 90°$ without the need for microwave rotary joints.

One method of making a twist reflector is to orient a grating of thin wires 45° to the incident polarization and placed a quarter-wavelength in front of the planar reflecting surface. This type of construction is limited to a bandwidth of about 10 percent. Much broader bandwidth is possible with other constructions. A meanderline polarizer backed by a reflector surface, for example, can achieve an octave bandwidth.[193] A mirror-scan antenna with a twist reflector using a number of log-periodic layered structures demonstrated operation over a frequency range from 2 to 12 GHz.[194]

The mirror-scan antenna has been widely used by many countries, especially the former Soviet Union, for land-based, shipborne, and airborne radar applications.

Beam Steering of a Reflector Antenna by Movement of the Feed   The beam of a parabolic reflector antenna can be scanned by laterally displacing the feed from the focus of the antenna. Generally the beam can only be scanned a few beamwidths off axis before the antenna pattern degrades significantly.[195,196] The antenna gain decreases and the sidelobes increase; and in radar applications it is often the increase in sidelobe level that determines how far the beam can be scanned rather than the decrease in gain. The larger the f/D ratio (f = focal length and D = antenna diameter) the greater in angle the beam can be scanned, but it is still quite limited.

A *spherical reflector* can produce a slightly larger scan angle when positioning the feed off the focal point, but spherical reflectors have aberrations that cause high sidelobes. The Arecibo antenna in Puerto Rico used for radar and radio astronomy employs a large spherical reflector 1000 ft in diameter. It has a specially designed feed that corrects for aberrations so that the UHF beam can be steered about 20° off axis.

A *parabolic torus* is a reflector antenna that is generated by rotating the parabolic section of Fig. 9.7 over an arc of a circle whose center is on the axis of the parabola.[197,198] Thus the vertical profile of the parabolic torus is a parabola and its horizontal profile (the plane in which the beam is scanned) is a circle of radius r. The radius r of the circular contour of the torus is made large enough so that the portion of the antenna surface illuminated by the feed does not appreciably differ from a true parabola. In other words, even though the horizontal contour of the torus is circular, the portion of the aperture illuminated by the feed approximates a parabolic shape, which is why the antenna generates satisfactory radiation patterns. Only a portion of the reflector surface is illuminated at any beam position. The main use of a parabolic torus is to rapidly scan a beam in a single plane by either mechanically moving a single feed along a circle whose radius is $r/2$, or by switching among many fixed feeds located on the circle of radius $r/2$. The latter method of scanning was used with the original antenna for the Ballistic Missile Early Warning System (BMEWS) radar, a high-power UHF radar for the detection of intercontinental ballistic missiles at ranges over 2000 nmi. Its beam was mechanically scanned 120° in azimuth in two seconds. The BMEWS antenna was 165 ft high by 400 ft wide, and although it is a large antenna, its cost was relatively low since it was a fixed structure and only the feeds were switched by an organ-pipe scanner.[199] The BMEWS parabolic torus antennas have been replaced by a phased array based on the technology of the Pave Paws radar.

The parabolic torus also was used in the past for naval height-finder radars, such as the AN/SPS-30, which rapidly scanned a horizontal fan beam in the vertical plane to extract an aircraft target's elevation angle.

The torus reflector antenna also has been used to generate two spaced antenna beams separated in azimuth, in addition to the main antenna beam, for the AN/SPQ-9B shipboard radar which uses a single mechanically rotating reflector, Fig. 9.54. The purpose of the two additional azimuth beams is to allow confirmation looks on a target after detection by the main beam so as to more rapidly establish a track on a target. After a single pass by the target, this allows the air-defense system to begin to establish a track on the target.

It has been reported[200] that a torus-like reflector with an elliptical contour rather than a circular one (but with a parabolic contour in the orthogonal plane) can have wide-angle scanning with good performance and yet be much smaller than the conventional parabolic torus discussed in the above. With a modified contour in the plane of scan, this approach has also been demonstrated to scan the beam using an offset-fed reflector to avoid aperture blockage.[201]

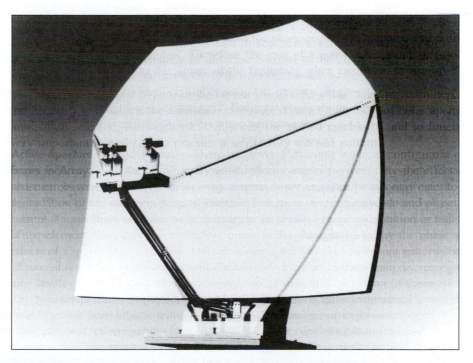

Figure 9.54   Parabolic torus reflector used with the AN/SPQ-9B shipboard radar to obtain three beams separated in azimuth.
| (Provided by L. Leibowitz and B. Cantrell of the Naval Research Laboratory Radar Division.)

Lens Antennas[202]   Microwave lenses may be constructed from dielectric materials, artificial dielectrics, or metal plates (waveguide media) to cause a focusing action similar to that of an optical lens. Dielectric lenses (the microwave analogy of the optical lens) are generally heavy and difficult to obtain with uniform properties. Lenses constructed from artificial dielectrics[203] can be made lighter but are usually poor conductors of heat so that it might be difficult to dissipate the heat generated in such materials when they are used for high-power radar transmitting antennas.

The metal-plate waveguide lens is constructed from side-by-side parallel-plate waveguides.[204] The phase velocity in a parallel-plate waveguide is greater than that in free space so that its index of refraction (and its dielectric constant) is less than unity, which is why they can be used to make lenses. They were used in early monopulse tracking radars to avoid aperture blockage caused by the large feed systems needed with reflector antennas.

The Luneburg lens[205] differs from optical lenses and other microwave lenses in that it is spherical and its index of refraction η is not uniform, but varies with distance from the center of the sphere as

$$\eta = \left[2 - (r/r_0)^2\right]^{1/2} \qquad [9.70]$$

where r is the radial distance and r_0 is the radius of the lens. It has the property that a plane wave incident on the sphere is brought to a focus on the surface at the diametrically opposite side. Likewise, a transmitting point source on the surface of the sphere emerges as a plane wave on passing through the lens. The beam of a Luneburg lens can be steered by moving the feed along the surface of the lens. With multiple feeds, it can form multiple simultaneous beams which was why it was seriously considered as the antenna for the Nike Zeus ballistic missile defense system, an early intercontinental ballistic missile defense system concept conceived in the late 1950s by Bell Telephone Laboratories for the U.S. Army (but never reached deployment).

The Fresnel zone-plate lens is an interesting form of lens that is simple, of small thickness, light weight, low loss, and low cost, especially for use at millimeter waves.[206] It has not had significant microwave radar application, however.

The lens antenna generally is less efficient than a reflector since unwanted reflections occur from both the front and rear surfaces of the lens. It is not as easy to support mechanically as is a reflector. Dielectric lenses have problems in dissipating heat when radiating high power. The lens antenna has interesting attributes, but it is not often used for radar application.

Radomes[207–209]   Mechanical engineers can design a ground-based or shipborne antenna to be structurally strong enough to operate in high winds, icing, and other adverse weather conditions. It is often much cheaper and better, however, to enclose the antenna in an electromagnetically transparent protective shield called a *radome*. An antenna enclosed by a radome can be lighter and have a smaller drive motor than an antenna exposed to the elements.

Radomes for ground-based radars are often in the shape of a sphere (for example, a three-quarters sphere). The sphere is a good mechanical structure and offers aerodynamic advantages in high winds. Precipitation particles blow around a sphere rather than impinge upon its surface, so that snow or other frozen particles are not readily deposited. Antennas mounted on aircraft must be housed within a radome that does not interfere with the aerodynamics of flight, be strong enough to be a part of the aircraft's structure, as well as not distort the antenna pattern.

A radome with good electromagnetic properties should be of low loss, have an adequate bandwidth, not raise the sidelobe level significantly, provide a low VSWR, a low antenna noise temperature, and not cause the boresight (pointing direction) to shift. In highly accurate tracking radars, the radome must also not increase the rate of change of the shift in the boresight.

There are two types of radomes that have been used for ground-based and shipboard radars: the rigid self-supporting radome and the air-supported radome.

Rigid Radomes   An example of a rigid radome is the *space frame,* an example of which is shown in Fig. 9.55. This type of radome consists of a three-dimensional lattice of primary load-bearing structural members enclosed with thin dielectric panels (0.02–0.04 inch thick). The panels can be made of teflon-coated fiberglass and can be very thin, even for large radomes, since they do not carry the main loads or stresses. This type of construction whereby a spherical structure is constructed from flat panels of simple geometric shapes is sometimes called a *geodesic dome.* The supporting framework can be of steel, aluminum, or plastic. Metal structures are superior in electrical performance compared to plastic or fiberglass since they can be made smaller because of their increased strength. Smaller thickness means less aperture blocking. Metal space frames are generally cheaper

Figure 9.55   Rigid space-frame radome for the Argos-10 air-surveillance radar. (Courtesy Alfonso Farina and Alenia Marconi Systems.)

and easier to fabricate, transport, and assemble, and can be used for larger diameter configurations.

Aluminum structural members are typically used for the space frame. They are larger than steel of equivalent strength, but they are lightweight, noncorrosive, and require no maintenance. It is important that the plastic panels be able to repel water (hydrophobic) rather than absorb water. A hydrophobic material will cause the water on the radome surface to form into beads rather than coat the surface in a sheet or a film. Water in the form of beads usually has dimensions small compared to the radar wavelength and do not cause as much of an effect as a water film.[210] In some radomes, the exterior surface is coated with a white radar-transparent paint, such as Hypalon, to reduce the interior temperature rise caused by solar radiation. A material known as Tedlar has also been used as a surfacing material.

A metal space-frame radome might be made up of individual triangular panels with a relatively uniform pattern. Instead of panels of uniform shape, there can be a quasi-random selection of different panel sizes and shapes to minimize the periodicity of the structure and avoid the generation of spurious sidelobes obtained with a periodic structure. The randomization of the space frame also makes it less sensitive to polarization. A metal space-frame radome might typically have a transmission loss of 0.5 dB and cause the antenna sidelobes to increase an average of 1 dB at the −25 dB level. The boresite might be shifted less than 0.1 mrad and the antenna noise temperature might increase less than 5 K. Space-frame radomes can be larger than 150 ft in diameter, and they can be used at any microwave radar frequency. Typically they can be designed for wind speeds of 150 mph, but they have also been designed to withstand winds as high as 300 mph.

Rigid radomes also have been made of fiberglass-reinforced solid plastic laminates and as a rigid shell sandwich, but these are usually of smaller size than the space frame. Solid laminate radomes can be up to 35 ft in diameter and operate at frequencies up to 3 GHz. Panels are sometimes arranged in an "orange peel" geometry. Large-diameter sandwich radomes, which can be as large as 80 ft, are usually constructed with an A-type sandwich (described later), but are often narrowband.[207] Some L-band sandwich radomes for long-range FAA enroute ARSR radars have been said to have transmission efficiencies of 98 percent (less than 0.1 dB).[211]

Weather Effects on Rigid Radomes   An important advantage of the rigid radome is its ability to withstand the rigors of severe climate. Rime ice, the prevalent type of icing in the Arctic region, has little or no effect on most radomes. Although it tends to collect on many types of structures and can obtain large thickness, both theory and experiment show a lack of rime-ice formation on a spherical radome.[212] The trajectories of water droplets in the air stream flowing around a large spherical radome do not impinge upon the surface. Droplets of freezing rain, on the other hand, are large and almost 100 percent of freezing rain can collect on the radome's surface. Dry snow does not stick to cold surfaces and is generally not a problem. Wet snow can stick to the radome and affect its transmission properties. Removal of snow by thermal means is expensive; but on the smaller radomes snow can be removed mechanically by tying a rope at the top of the radome and having someone walk the rope around the radome to knock off the snow.

Liquid water can collect on a radome as a film due to condensation or rainfall if the surface is non-hydrophobic. The films can be very thin and still cause attenuations of several dB or more depending on the frequency and the amount of water. Effenberger et al.[213] state that at X band a water film can result in 8 dB of added transmission loss. If the water is in the form of droplets instead of a film, the added transmission loss is reduced to 1 dB. These losses compare to a loss of a dry air-supported radome of 0.1 dB and a dry metal space frame of 0.7 dB. Thus it is important that materials used for a radome not absorb water and not allow water to form a film on its surface. Radomes should be checked periodically to insure that their hydrophobic properties do not weaken with time.

Air-Supported Radomes   The first large radomes for ground-based radar antennas that appeared shortly after World War II were constructed of a strong, flexible rubberized airtight material supported by air pressure from within. Since the radome material is relatively thin and uniform, it approximates the electrically thin shell that provides very low loss (less than 0.1 dB) and very small boresight error. The air-supported radome is constructed of gore-shaped fabric sections with the gores or seams in the vertical direction. Reliable operation depends on the use of uninterruptable power supplies and redundant blower systems. Teflon fiberglass is a commonly used material.[210] Air-supported radomes can be folded into a small package which makes them suitable for transportable radars requiring mobility and quick assembly and disassembly times. At a prepared site, a 50-ft air-supported radome can be installed in about one or two hours.[214] They are also of interest when wideband operation is important.

The life of an air-supported radome is limited by exposure to ultraviolet light, surface erosion, and the constant flexing of the material in the wind. The life can be increased by the use of better materials such as neoprene-coated nylon and teflon-coated fiberglass. In high winds the material can be damaged by flying debris. The rotation of the antenna might have to be stopped to prevent the fabric from being blown against the antenna and torn. Nevertheless, these structures are designed to withstand winds of 100 mph, and in special applications it has been extended to 200 mph.[215] Maintaining the internal pressure in high winds can sometimes be difficult. Frequent and costly maintenance is another problem. The air-supported radome has superior electrical properties to the rigid spaceframe radome, but the latter is much more rugged.

Aircraft Radomes[208,216]   The shape of the radome used in the nose of a military fighter/attack aircraft is determined primarily by aerodynamic requirements rather than by electromagnetics. It is often an ogive or some other similar conical shape. Because of the aerodynamic shape of the nose radome, the angle of incidence of the beam on the surface of the radome depends on the scan angle and might be from 0 to 80°, or more. Also, the incident polarization can vary with angle of scan. This is unlike the situation with spherical shaped ground-based radomes where the incidence angle and polarization are relatively independent of the scan angle. Thus there can be distortions in the antenna pattern, spurious sidelobes, and angle errors that depend on the angle of scan. Airborne radomes, especially those used at supersonic speeds, can be subjected to mechanical stress and aerodynamic heating so severe that the electromagnetic requirements of a radome made of dielectric materials might have to be sacrificed in order to obtain sufficient mechanical strength to survive. Rain impinging on an unprotected radome in flight can cause structural damage within minutes due to erosion at high subsonic speeds. Rain erosion is reduced by coating the radome surface with a shock-absorbing rubber-like material such as neoprene or polyurethane elastomer. A lightning strike can puncture a hole in the radome wall and severely damage the radar equipment inside. The damaging effects of lightning are reduced by placing conductors on the external surface of the radome to divert the lightning to these conductors rather than puncture the radome wall. If a large number of conductors is necessary to insure that lightning is diverted, the performance of the radar might be affected. Other factors in radome design are avoidance of static charge build-up and damage due to bird strike or the impact of hail. The radome must, in addition, withstand high temperatures and not be too heavy. It has been said[216] "Few, if any, other components [of an airborne radar system] have such a variety of conflicting requirements as does a radome."

Radome Wall Construction   The following are some wall constructions that have been used in radomes:

> *Thin wall.* The wall is electrically thin compared to the radar wavelength. If the wall physical thickness is *d,* with a dielectric constant ϵ, and a wavelength λ, then a thin wall radome is characterized by $d < 0.05\lambda/\sqrt{\epsilon}$. A thin wall radome has good electrical properties, but it can be weak structurally.

> *Half wavelength.* This is a solid dielectric surface whose electrical thickness is approximately a half-wavelength. Theoretically, the halfwave-thick surface is nonreflecting and has no loss other than ohmic losses of the material. It is of limited bandwidth, as well as limited in the range of incidence angles over which the electromagnetic energy can be transmitted with minimal reflection. There can be multiple half-wave surfaces.

> *A-sandwich.* A three-layer wall consisting of two thin relatively high-dielectric-constant skins separated by a low-dielectric-constant core whose thickness is approximately one-quarter wavelength. The skins are glass reinforced plastic laminates and are thin compared to a wavelength. The core might be a honeycomb or a foam.

> *B-sandwich.* The inverse of the A-sandwich with quarter-wavelength skins having a dielectric constant lower than that of the core.

> *C-sandwich.* Two back-to-back A-sandwiches. It is used when the ordinary A-sandwich does not provide sufficient strength.

> *Multilayer.* A general term for more layers than that of the C-sandwich.

Metallic Radomes   A thin metal sheet with periodically spaced openings (such as slots) has a bandpass characteristic and can be used for radomes. The metallic structure overcomes the mechanical limitations of radomes made of dielectric materials, yet has good electrical properties.[217] It also is better able to distribute frictionally induced heating and better able to withstand the stresses caused by rain, hail, dust, and lightning. This type of radome is a *frequency selective surface* that reduces interference because of its bandpass characteristic. It also reduces the nose-on radar cross section of the aircraft when viewed by radar systems not within its own passband.[218] The penalty for its use, however, is narrow bandwidth. A frequency selective surface can be more sophisticated and have better properties than a thin metallic sheet with simple periodic slots.[219]

Rotodomes   In some cases the radome is made to rotate in synchronism with the antenna. It is then called a *rotodome*. An example is the AWACS AN/APY-1 antenna, Fig. 3.45b, where the antenna and radome are designed together so that the antenna can maintain the very low sidelobes required of a high-prf pulse doppler radar. A rotodome is also used with the ASDE-3, see Fig. 1.9, the Airport Surface Detection Equipment found at the top of the control tower of major airports for monitoring taxiing aircraft and ground traffic at airports.

Adaptive Antennas[220–224]   An adaptive antenna, usually an array, is one that senses the received signals incident across its aperture and adjusts the phase and amplitude of the aperture illumination to maximize the signal-to-external-noise ratio or signal-to-clutter ratio. Adaptive arrays usually require some prior knowledge of the desired signal and the nature of the noise or clutter to be rejected. When one speaks of a *fully adaptive array* it means one where each element of the array is part of the adaptive process. There have been many investigations of the theory and algorithms to perform adaptive array processing, but the difficulty and cost of implementing fully adaptive processing at each element of a large array antenna has been prohibitive. Until large fully adaptive arrays become more practical, they will be mainly of academic interest. There are, however, two important radar applications where adaptive array technology has proven to be practical and important. These occur when only a relatively few adaptive elements are needed. One such application is the *sidelobe canceler;* the other is the AEW radar that employs *space-time adaptive processing.*

Sidelobe Cancellation[225]   This adaptively places nulls in the antenna radiation pattern in the direction of a limited number of noise jammers so as to reject the noise before it enters the receiver. A small number of omnidirectional (or wide beamwidth) auxiliary antenna elements are placed on or in the near vicinity of the main radar antenna. Typically there might be from 3 to 6 auxiliary elements used for the sidelobe canceler. The main radar antenna can be a reflector or a phased array. In theory, one auxiliary element (one degree of freedom) can create one adaptive null; but, in practice, especially when multipath propagation occurs, two auxiliary elements might be needed in some instances to place a suitable null in the direction of one noise source. The sidelobe canceler has been a successful application of adaptive antennas and has been applied operationally to a number of radar systems.

Space-Time Adaptive Processing (STAP)   Military airborne MTI radars, or Airborne Early Warning (AEW) radars, must be able to cancel ground clutter echoes which have a non-zero doppler velocity with respect to the moving radar, along with hostile jamming that enters the radar via the antenna sidelobes. According to J. Ward[226] "Space-time adaptive processing (STAP) refers to multidimensional adaptive filtering algorithms that simultaneously combine the signals from the elements of an array antenna and the multiple pulses of a coherent radar waveform, to suppress interference and provide target detection." Fred Staudaher,[227] of the Naval Research Laboratory, described the differences between spatial processing, temporal processing, and space-time processing as follows: "Spatial adaptive array processing combines an array of signals received at the same instant of time that are sampled at the different spatial locations corresponding to the antenna elements. Temporal adaptive array processing combines an array of signals received at the same spatial location (e.g., the output of a reflector antenna) that are sampled at different instances of time, such as several periods for an adaptive MTI. Space-time adaptive array processing combines a two-dimensional array of signals sampled at different instances of time and at different spatial locations."

STAP and other forms of antenna adaptive processing are more subjects in circuit design and algorithm development than antennas.

The Quest for Superresolution   The ability of a radar to resolve two targets in angle depends on their relative radar cross section, signal-to-noise ratio, antenna beamwidth, the phase difference between the two signals, and the criterion used to establish resolution. It has been generally accepted that two equal targets can be resolved in angle when they are separated by eight-tenths of a beamwidth, provided the signal-to-noise ratio is large enough for good detection. Resolution can be better than this with high signal-to-noise ratios and a criterion which acknowledges that there can be a phase difference between the two signals. Every now and then, however, there have been different proposals for obtaining better angular resolution with radar systems—all without true success thus far!

A technique that has had impressive claims made for it is known as *spectral estimation* or *superresolution*.[228] It is also known by some of the many algorithms that are used, such as the maximum entropy method, autoregression, Burg algorithm, and others. These angular resolution methods apply for noncoherent sources, such as independent noise radiators (jammers or radio stars). Superresolution methods are basically the same as adaptive antennas which place sharp nulls in the direction of noise sources. Superresolution and adaptive antennas use the same algorithms and the same hardware, and the plots of their outputs are the same except that one is plotted upside down with respect to the other. That is, the nulls of the adaptive antenna become the narrow spikes of superresolution when inverted.

Superresolution may resolve closely spaced noise-like sources, but it does not reliably resolve the echoes from multiple targets illuminated by the same radar. The echoes from targets illuminated by the same radar have a phase relationship among each other and are thus coherent. Superresolution, or spectral estimation, algorithms employ nonlinear mathematical operations. When multiple echo signals from the same radar are subject to nonlinear processing, they can produce spurious signals that do not allow good resolution capabilities. Thus superresolution does not provide improved angular resolution with radar echo signals. This was first stated in the radar literature by A. W. Rihaczek.[229]

Other "Superresolution" Concepts   In the traditional antenna literature, one can often find discussed the concept of a "superdirective" array antenna (formerly called supergain).[230] This is defined as an array antenna with a directivity higher than that obtained when the same antenna has a uniform aperture illumination. (Its antenna illumination efficiency is said to significantly exceed 100 percent.[231]) Many reasons have been offered why such an antenna is not practical (narrow bandwidth, high Q, large aperture currents, high loss, and extremely precise tolerances), but supergain antennas require aperture illuminations that must change amplitude (spatially) across the aperture faster than can be expected of a signal operating at the given RF frequency. Superdirectivity appears to result from the simple algebraic models of an antenna pattern rather than as a solution to Maxwell's equations applied to a real radar antenna.

In the 1960s the *multiplicative array* was the "superresolution" technique that caused excitement, for a while. In a multiplicative array the outputs of the individual radiating elements were combined in a nonlinear manner rather than linearly. The nonlinear manipulation of the aperture illumination results in an apparently narrower antenna pattern. For example, the pattern of an *N*-element array can be expressed by a polynomial of degree *N*. If the output of one half of the array is multiplied by the output from the other half, the resulting expression is a polynomial of order $(N^2/4)$, which when plotted appears as a much narrower pattern of an array with $N^2/4$ elements rather than *N*. This narrower pattern looks exciting as a means for obtaining improved resolution by simple multiplicative processing of the array output. When two closely spaced targets are examined, however, with such multiplicative processing the resolution is not that of a larger array and, to make matters worse, there will be spurious signals generated because of the nonlinear mathematical operation. (The nonlinear operation of squaring or cubing the antenna pattern also will make it narrower and appear to provide better resolution than a conventional antenna, but when multiple targets are present, the resulting pattern is not the superposition of the individual patterns, but much worse.)

There have also been attempts to achieve improved resolution by what was called "data restoration," which smoothed the received aperture illumination and extrapolated it beyond the physical confines of the antenna. This also did not produce the significant improvement in resolution that was desired.

Thus one should not expect to obtain significantly better angular resolution with an antenna by some nonlinear form of antenna processing. There seems to be no magic radar resolution algorithm. The only method that has worked in the past for obtaining improved resolution when the electrical size of the aperture cannot be increased is to increase the signal-to-noise ratio and recognize in the resolution procedure that the echoes signals can be of different phase.

Microelectromechanical Switches in Phased Arrays[232,233]   A microelectromechanical switch, or MEMS, is a small, low-inertia fast-acting switch activated by an electrostatic field. The switching mechanism may be in the form of a cantilever, rotary, or membrane configuration. In one example, the upper contact is a 0.3-μm aluminum membrane suspended across polymer posts. This suspended membrane is 4 μm above a substrate surface with the bottom contact of 0.7-μm gold or aluminum metal layer. On top of this metal layer is a thin dielectric layer, typically 0.1 μm (1000 Å) of silicon nitride. It is not a metallic contact switch, but switches by providing a change in capacitive impedance. The dielectric on the bottom part of the switch makes contact with the metallic portion of the suspended membrane, which eliminates the problem of striction that would occur if two metallic layers came into contact.

MEMS can have very wide bandwidth and can be made to operate with signals from a few MHz to 40 GHz. Measured insertion loss is less than 0.2 dB per switch. The switch can activate in 2 to 5 μs and it can handle RF power up to 10 watts. The pull-in voltage is from 10 to 30 volts. Since a MEMS is activated by a d-c electrostatic field, no d-c current is required and power consumption is small. The energy required to activate a switch is on the order of 10 nJ.

The MEMS can be used in the same way that diode switches are used in the description of the digital phase shifter, shown in Fig. 9.17, to switch in and out fixed length of lines to obtain various phase shifts. In this type of X-band phase shifter the loss in a four-bit phase shifter is from 1.2 to 2 dB. Its size is approximately 6 mm by 9 mm. These phase shifters can be fabricated on silicon wafers. Hundreds of phase shifters can be built on a single 8-in. wafer, which makes them inherently low cost.

It has been said[234] that the MEMS for electronically scanned phased array radars has the potential to reduce the cost, weight, and power consumption for such systems when the array size exceeds 10,000 elements.

REFERENCES

1. Kraus, J. D. *Antennas,* 2nd ed. New York: McGraw-Hill, 1988, Sec. 10–12.

2. Probert-Jones, J. R. "The Radar Equation in Meteorology." *Quart. J. Roy. Meteor. Soc.* 88 (1962), pp. 485–495.

3. Stutzman, W. L. "Estimating Directivity and Gain of Antennas." *IEEE Antennas and Propagation Magazine* 40 (August 1998), pp. 7–11.

4. Evans, G. E. *Antenna Measurement Techniques,* Artech House, Boston, MA, 1990, p. 115.

5. Cutler, C. C., A. P. King, and W. E. Kock. "Microwave Antenna Measurements." *Proc. IRE* 35 (December 1947), pp. 1462–1471.

6. Sherman, J. W. "Aperture-Antenna Analysis." *Radar Handbook,* 1st ed., M. Skolnik (Ed.). New York: McGraw-Hill, 1970, Chap. 9.

7. Mints, M. Ya., Ye. D. Prilepskiy, and V. M. Zaslonko. "Optimization of the Radiation Power Concentration Factor of an Antenna with a Circular Aperture and a Maximally Flat Radiation Pattern." *Soviet J. of Communications Technology and Electronics* 34 (May 1989), pp. 33–39.

8. Silver, S. *Microwave Antenna Theory and Design,* vol. 12 of the M. I. T. Radiation Laboratory Series. New York: McGraw-Hill, 1949, Chap. 6.

9. Silver, S. Ref. 8, Sec. 6.5.

10. Sherman, J. W. Ref. 6, Sec. 9.2

11. Bodnar, D. G. "Materials and Design Data." *Antenna Engineering Handbook,* 3rd ed., R. C. Johnson (Ed.). New York: McGraw-Hill, 1993, Chap. 46, Sec. 46–5.

12. Jasik, H. "Fundamentals of Antennas." *Antenna Engineering Handbook,* 3rd ed., R. C. Johnson (Ed.), New York: McGraw-Hill, 1993, Chap. 2, Sec. 2.7.

13. Johnson, R. C. *Designer Notes for Microwave Antennas.* Boston: Artech House, 1991, Sec. A.12.

14. Cutler, C. C. "Parabolic Antenna Design for Microwaves." *Proc. IRE* 35 (November 1947), pp. 1284–1294.

15. Sciambi, A. F. "The Effect of the Aperture Illumination on the Circular Aperture Antenna Pattern Characateristics." *Microwave J.* 8 (August 1965), pp. 79–84.

16. Olver, A. D., P. J. B. Clarricoats, A. A. Kishk, and L. Shafai. *Microwave Horns and Feeds.* New York: IEEE Press, 1994.

17. Ruze, J. "Feed Support Blockage Loss in Parabolic Antennas." *Microwave J.* 11 (December 1968), pp. 76–80.

18. Kildal, P-S, E. Olson, and J. A. Aas. "Losses, Sidelobes, and Cross Polarization Caused by Feed-Support Struts in Reflector Antennas: Design Curves." *IEEE Trans.* AP-36 (February 1988), pp. 182–190.

19. Rudge, A. W., and N. A. Adatia. "Offset-Parabolic-Reflector Antennas: A Review." *Proc. IEEE* 66 (December 1978), pp. 1592–1618.

20. Cook, J. H., Jr. "Earth Station Antennas." *Antenna Engineering Handbook,* 3rd ed., R. C. Johnson (Ed.), New York: McGraw-Hill, 1993, Chap. 36, pp. 36-8 to 36-10.

21. Terada, M. A., and W. L. Stutzman. "Design of Offset-Parabolic-Reflector Antennas for Low Cross-Pol and Low Sidelobes." *IEEE Antennas and Propagation Magazine* 35, no. 6 (December 1993), pp. 436–449.

22. Terada, M. A., and W. L. Stutzman. "Computer-aided Design of Reflector Antennas." *Microwave J.* 38 (August 1995), pp. 64–73.

23. Weiss, H. G. "The Haystack Microwave Research Facility." *IEEE Spectrum* 2 (February 1965), pp. 50–59.

24. Hannan, P. W. "Microwave Antennas Derived from the Cassegain Telescope." *IRE Trans.* AP-9 (March 1961), pp. 140–153.

25. Josefsson, L. G. "A Broad-Band Twist Reflector." *IEEE Trans.* AP-19 (July 1971), pp. 552–554.

26. Lewis, B. L., and J. P. Shelton. "Mirror Scan Antenna Technology." *Record of the IEEE 1980 International Radar Conf.,* Arlington, VA, pp. 279–283. IEEE Publication 80CH1493-6 AES.

27. Dwight, H. B. *Tables of Integrals and Other Mathematical Data.* New York: Macmillan, 1947, Equation No. 420.3.

28. Bickmore, R. W. "A Note on the Effective Aperture of Electronically Scanned Arrays." *IRE Trans.* AP-6 (April 1958), pp. 194–196.

29. Elliott, R.S. "The Theory of Antenna Arrays." *Microwave Scanning Antennas,* R. C. Hansen, (Ed.). New York: Academic, 1966, vol. II, Chap. 1.

30. Koul, S. K., and B. Bhat. *Microwave and Millimeter Wave Phase Shifters, Vol. II, Semiconductor and Delay Line Phase Shifters.* Boston: Artech House, 1991.

31. Garver, R. V. *Microwave Diode Control Devices.* Boston: Artech House, 1976.

32. Tang, R., and R. W. Burns. "Phased Arrays." *Antenna Engineering Handbook,* 3rd ed., R. C. Johnson (Ed.). New York: McGraw-Hill, 1993, Chap. 20, pp. 20-36 to 20-44.

33. White, J. F. *Microwave Semiconductor Engineering.* New Jersey: Van Nostrand, 1982.

34. Temme, D. H. "Diode and Ferrite Phaser Technology." *Phased Array Antennas,* A. A. Oliner and G. H. Knittel (Eds.). Boston: Artech House, 1972, pp. 212–218.

35. Koul, S. K., and B. Bhat. Ref. 30, Chap. 11.

36. Andricos, C., I. J. Bahl, and E. L. Griffin. "C-Band 6-Bit Monolithic Phase Shifter." *IEEE Trans.* MTT-33 (December 1985), pp. 1591–1596.

37. Koul, S. K., and B. Bhat. Ref. 30, Chap. 12.

38. Shenoy, R. P. "Phased Array Antennas." *Advanced Radar Techniques and Systems,* G. Galati (Ed.). Peter Peregrinus, 1993, Chap. 10, Sec. 10.27.

39. Stark, L. "Microwave Theory of Phased Array Antennas—A Review." *Proc. IEEE* 62 (December 1974), pp. 1661–1701.

40. Koul, S. K., and B. Bhat. *Microwave and Millimeter Wave Phase Shifters, Vol. I, Dielectric and Ferrite Phase Shifters.* Boston: Artech House, 1991.

41. Rodrique, G. P. "A Generation of Microwave Ferrite Devices." *Proc. IEEE* 76 (February 1988), pp.121–137.

42. Stark, L., R. W. Burns, and W. P. Clark. "Phase Shifters for Phased Arrays," *Radar Handbook,* 1st ed., M. Skolnik (Ed.), New York: McGraw-Hill, 1970, Chap. 12.

43. Whicker, L. R. (Ed.). *Ferrite Control Devices, Vol. 2, Ferrite Phasers and Ferrite MIC Components.* Boston: Artech House, 1974.

44. Wicker, L. R., and R. R. Jones. "A Digital Current Controlled Latching Ferrite Phase Shifter." *IEEE 1965 International Convention Record,* pt. V, pp. 217–223.

45. Cattgasrin, G., et al. "A Digital Ferrite Phase-Shifter for High Power S-Band Operation." *Rivista Tecnica, Selenia* 8, no. 2 (1982), pp. 29–34.

46. Hord, W. E. "Microwave and Millimeter-Wave Ferrite Phase Shifters." *Microwave J.* 1989 State of the Art Reference, pp. 81–93.

47. Ince, W. J., and E. Stern. "Nonreciprocal Remanence Phase Shifters in Rectangular Waveguide." *IEEE Trans.* MTT-15 (February 1967), pp. 87–95.

48. Koul, S. K., and B. Bhat. Ref. 40, Sec. 4.8.

49. Junding, W., et al. "Analysis of Twin Ferrite Toroidal Phase Shifter in Grooved Waveguide." *IEEE Trans.* MTT-42 (April 1994), pp. 616–621.

50. DiBartolo, J., W. J. Ince, and D. H. Temme. "A Solid State 'Flux Drive' Control Circuit for Latching-Ferrite-Phaser Applications." *Microwave J.* 15 (September 1972), pp. 59–64.

51. Koul, S. K., and B. Bhat. Ref. 40, Sec. 5.6.

52. Fox, G. A., S. E. Miller, and M. T. Weiss. "Behavior and Application of Ferrites in the Microwave Region." *Bell System Tech. J.* 34 (January 1955), pp. 5–103.

53. Fox, A. G. "An Adjustable Wave-Guide Phase Changer." *Proc. IRE* 35 (September 1947), pp. 1489–1498.

54. Yansheng, X., and J. Zhengchang. "Dual-Mode Latching Ferrite Devices, Part I." *Microwave J.* 29 (May 1986), pp. 277–280.

55. Whicker, L. R., and C. W. Young, Jr. "The Evolution of Ferrite Control Components." *Microwave J.* 21 (November 1978), pp. 33–37.

56. Ince, W. J. "Recent Advances in Diode and Ferrite Phaser Technology for Phased-Array Radars, Part II." *Microwave J.* 15 (October 1972), pp. 31–36.

57. Yansheng, X., and J. Zhengchang. "Dual-Mode Latching Ferrite Devices, Part II." *Microwave J.* 29 (May 1986), pp. 282–286.

58. Monaghan, S. R., and M. C. Mohr. "Polarization Insensitive Phase Shifter for Use in Phased-Array Antennas." *Microwave J.* 12 (December 1969), pp. 75–80.

59. Hord, W. E. "Design Considerations for Rotary-Field Ferrite Phase Shifters." *Microwave J.* 31 (November 1988), pp. 105–115.

60. Boyd, C. R., Jr. "Progress in Ferrite Phase Shifters," Microwave Applications Group brochure, Santa Maria, CA (no date).

61. Varadan, V. K. "A Novel Microwave Planar Phase Shifter." *Microwave J.* 38 (April 1995), pp. 244–254.

62. Ajioka, J. S. "Frequency-Scan Antennas." *Antenna Engineering Handbook,* R. C. Johnson (Ed.), New York: McGraw-Hill, 1993, Chap. 19.

63. Hammer, I. W. "Frequency-Scanned Arrays." *Radar Handbook,* 1st ed., M. Skolnik (Ed.). New York: McGraw-Hill, 1970, Chap. 13.

64. Begovich, N. A. "Frequency Scanning." *Microwave Scanning Antennas, Vol. III,* R. C. Hansen (Ed.). New York: Academic, 1966, Chap. 2.

65. Johansson, F. S., L. G. Josefsson, and T. Lorentzon." A Novel Frequency-Scanned Reflector Antenna." *IEEE Trans.* AP-37 (August 1989), pp. 984–989.

66. Johansson, F. S. "Frequency-Scanned Gratings Consisting of Photo-Etched Arrays." *IEEE Trans.* AP-37 (August 1989), p. 996–1002.

67. Croney, J. "Doubly Dispersive Frequency Scanning Antenna." *Microwave J.* 6 (July 1963), pp. 76–80.

68. Tang, R. "Practical Aspects of Phased Array Design." *Antenna Handbook,* Y. T. Lo and S. W. Lee (Ed.). New York: Van Nostrand Reinhold, 1988, Chap. 18, pp. 18-6 to 18-11.

69. Hansen, R. C. *Phased Array Antennas.* New York: John Wiley, 1998, Chap. 5.

70. Hansen, R. C. Ref. 69, Sec. 8.2.

71. Stark, L. "Comparison of Array Element Types." *Phased Array Antennas, Proc. 1970 Phased Array Antenna Symp.* Dedham, MA: Artech House, 1972, pp. 51–67.

72. Tang, R., and R. W. Burns. "Phased Arrays." *Antenna Engineering Handbook,* 3rd ed., R. C. Johnson (Ed.). New York: McGraw-Hill, 1993, Chap. 20.

73. Edward, B., and D. Rees. "A Broadband Printed Dipole with Integrated Balun." *Microwave J.* 35 (May 1987), pp. 339ff.

74. Mailloux, R. J. *Phased Array Antenna Handbook.* Boston: Artech House, 1994.

75. Richard, W. F. "Microstrip Antennas." *Antenna Handbook,* Y. T. Lo and S. W. Lee (Eds.). New York: Van Nostrand Reinhold, 1988, Chap. 10.

76. Lewis, L. R. "Phased Array Elements—Part 2." In *Practical Phased Array Antenna Systems,* E. Brookner (Ed.). Boston: Artech House, 1991, Lecture 5.

77. Cheston, T. C., and J. Frank. "Phased Array Radar Antennas." *Radar Handbook,* 2nd ed., M. Skolnik (Ed.). New York: McGraw-Hill, 1990, Chap. 7, pp. 7.31–7.32.

78. Mailloux, R. J. Ref. 74, Chap. 6.

79. Hansen, R. C. Ref. 69, Chap. 7.

80. Hannan, P. W. "The Element-gain Paradox for a Phased-Array Antenna." *IEEE Trans.* AP-12 (July 1964), pp. 423–433.

81. Pozar, D. M., and D. H. Schaubert. "Scan Blindness in Infinite Arrays of Printed Dipoles." *IEEE Trans.* AP-32 (June 1984), pp. 602–610.

82. Byron, E. V., and J. Frank. "'Lost Beams' from a Dielectric Covered Phased-Array Aperture." *IEEE Trans.* AP-16 (July 1968), pp. 494–499.

83. King, D. D., and H. J. Peters. "Element Interaction in Steerable Arrays." *Microwave J.* 6 (February 1963), pp. 73–77.

84. Mailloux, R. J. Ref. 78, p. 314.

85. Kinsel, J., B. J. Edward, and D. E. Rees. "V-Band Space-Based Phased Arrays." *Microwave J.* 30 (January 1987), pp. 89–102.

86. Mailloux, R. J. "Phased Array Architecture." *Proc. IEEE* 80 (Jaunary 1992), pp. 163–172. See also Mailloux, Ref. 74, Sec. 5.3.1.

87. Patton, W. T. "Array Feeds." In *Practical Phased-Array Antenna Systems,* E. Brookner (Ed.). Boston: Artech House, 1991, Lecture 6.

88. Hansen. R. J. Ref. 69, Secs. 2.3.4 and 2.3.5.

89. Smith, M. S., and Y. C. Guo. "A Comparison of Methods for Randomizing Phase Quantization Errors in Phased Arrays." *IEEE Trans.* AP-31 (November 1983), pp. 821–828.

90. Cheston, T. C., and J. Frank. Ref. 77, Sec. 7.8.

91. Cheston, T. C., and J. Frank. Ref. 77, pp. 7.53–7.55.

92. Patton, W. T. "Compact, Constrained Feed Phased Array for the AN/SPY-1." In *Practical Phased-Array Antenna Systems,* E. Brookner (Ed.). Boston: Artech House, 1991, Lecture 8.

93. Sharp, E. D. "A Triangular Arrangement of Planar-Array Elements that Reduces the Number Needed." *IRE Trans.* AP-9 (March 1961), pp. 126–129.

94. Cheng, D. H. S. "Characteristics of Triangular Lattice Arrays." *Proc. IEEE* 56 (November 1968), pp. 1811–1817.

95. Nelson, E. A. "Quantization Sidelobes of a Phased Array with a Triangular Element Arrangement." *IEEE Trans.* AP-17 (May 1969), pp. 363–365.

96. Agrawal, A. K., and E. L. Holzman. "Beamformer Architectures for Active Phased-Array Radar Antennas." *IEEE Trans.* AP-47 (March 1999), pp. 432–442.

97. Holzman, E. L., A. K. Agrawal, and J. G. Ferrante. "Active Phased Array Design-for High Clutter Improvement Factor." *1996 IEEE International Symposium on Phased Array Systems and Technology,* October 15–18, 1996, IEEE Catalog Number 96TH8175, pp. 44–47.

98. Reed, J. E. "The AN/FPS-85 Radar System." *Proc. IEEE* 57 (March 1969), pp. 324–335.

99. Grimes, M. D., J. M. Major, and T. J. Warnagiris. "Peak Power Tailoring and Phase Nulling of the AN/FPS-85 Radar." *SPIE* 2154, *Intense Microwave Pulses II,* pp. 241–246, 1994.

100. Brookner, E. *Aspects of Modern Radar.* Boston: Artech House, 1988, Sec. 2.2.1.2, p. 198, and pp. 279–281.

101. Sarcione, M., et al. "The Design, Development and Testing of the THAAD (Theater High Altitide Area Defense) Solid State Phased Array (formerly Ground Based Radar)." *1996 IEEE International Symp. on Phased Array Systems and Technology,* October 15–18, 1996, IEEE Catalog Number 96TH8175, pp. 260–265.

102. Dryer, et al. "EL/M 2080 ATBM Early Warning and Fire Control Radar System." *1996 IEEE International Symp. on Phased Array Systems and Technology,* October 15–18, 1996, IEEE Catalog Number 96TH8175, pp. 11–16.

103. Malas, J. A. "F-22 Radar Development." *Proc. IEEE 1997 NAECON* 2, pp. 831–839, IEEE Catalog no. CH36015-97.

104. Barton, D. K. "The 1993 Moscow Air Show." *Microwave J.* 37 (May 1994), pp. 24ff.

105. Corey, L. E. "A Survey of Russian Low Cost Phased-Array Technology." *1996 IEEE International Symp. on Phased Array Systems and Technology,* October 15–18, 1996, IEEE Catalog Number 96TH8175, pp. 255–259.

106. Ajioka, J. S., and J. L. McFarland. "Beam-Forming Feeds." Chap. 19, *Antenna Handbook,* Y. T. Lo and S. W. Lee (Ed.). New York: Van Nostrand Reinhold, 1988. See also references 86 to 100, Chap. 8. In *Introduction to Radar Systems,* 2nd ed. M. Skolnik.

107. Shelton, J. W. "Fast Fourier Transform and Butler Matrices." *Proc. IEEE* 56 (March 1968), p. 350.

108. White, W. D. "Pattern Limitations in Multiple-Beam Antennas." *IRE Trans.* AP-10 (July 1962), pp. 430–436.

109. Barton, P. "Digital Beam Forming of Radar." *IEE Proc.* 127, Pt. F., No. 4 (August 1980).

110. Steyskal, H., and J. F. Rose. "Digital Beamforming for Radar Systems." *Microwave J.* 32 (January 1989), pp. 121ff.

111. Steyskal, H. "Digital Beamforming at Rome Laboratory." *Microwave J.* 39 (February 1996), pp. 100ff.

112. Farina, A. *Antenna-Based Signal Processing Techniques for Radar Systems.* Boston: Artech House, 1992, Sec. 2.7.

113. Skolnik, M. "Improvements for Air-Surveillance Radar." *Proc. 1999 IEEE Radar Conf.* April 20–22, 1999, IEEE Catalog Number 99CH36249, pp. 18–21.

114. Steyskal, H., and J. S. Herd. "Mutual Coupling Compensation in Small Array Antennas." *IEEE Trans.* AP-38 (December 1990), pp. 1971–1975.

115. Lewis, B. L., F. F. Kretschmer, and W. W. Shelton. *Aspects of Radar Signal Processing.* Norwood, MA: Artech House, 1986, Chap. 3.

116. Brookner, E., and J. M. Howell, "Adaptive-Adpative Array Processing." *Proc. IEEE* 74 (April 1986), pp. 602–604.

117. Gabriel, W. F. "Using Spectral Estimation Techniques in Adaptive Processing Antenna Systems." *IEEE Trans.* AP-34 (April 1986), pp. 291–300.

118. Mailloux, R. J. "Array Failure Correction with a Digitally Beamformed Array." *IEEE Trans.* AP-44 (December 1996), pp. 1543–1550.

119. Wirth, W. D. "Long Term Integration for a Floodlight Radar." *1995 IEEE International Radar Conf.* Arlington, VA, pp. 698–703.

120. Headrick, J. M. "HF Over-the-Horizon Radar." *Radar Handbook,* M. Skolnik (Ed.). New York: McGraw-Hill, 1990, Chap. 24.

121. Sparks, R. A. "Systems Applications of Mechanically Scanned Array Antennas." *Microwave J.* 31 (June 1988), pp. 26–48.

122. Richardson, P. N., and H. Y. Yee. "Design and Analysis of Slotted Waveguide Antenna Arrays." *Microwave J.* 31 (June 1988), pp. 109ff.

123. Yee, H. Y, and R. C. Voges. "Slot-Antenna Arrays." *Antenna Engineering Handbook.* 3rd ed., R. C. Johnson (Ed.), New York: McGraw-Hill, 1993, Chap. 9.

124. Watson, C. K., and K. Ringer. "Feed Network Design for Airborne Monopulse Slot-Array Antennas." *Microwave J.* 31 (June 1988), pp. 129ff.

125. Kraus, J. D. *Antennas,* 2nd ed., New York: McGraw-Hill, 1988, Sec. 4.6.

126. Butler, J. M., A. R. Moore, and H. D. Griffiths. "Resource Management for a Rotating Multi-Function Radar." *Radar-97, 14-16 October 1997,* IEE Publication No. 449, pp. 568–572.

127. Billam, E. R. "Rotating vs Fixed Active Arrays for Multifunction Radar." *Radar-97, 14-16 October 1997,* IEE Publication No. 449, pp. 573–575.

128. Mailloux, R. J. Ref. 74, Chap. 3, "Pattern Synthesis for Linear and Planar Arrays."

129. Dolph, C. L. "A Current Distribution for Broadside Arrays Which Optimizes the Relationship between Beamwidth and Side Lobe Level." *Proc. IRE* 34 (June 1946), pp. 335–348; also discussion by H. J. Riblet, vol. 35, pp. 489–492.

130. Taylor, T. T. "Design of Line-Source Antennas for Narrow Beamwidth and Low Side Lobes." *IRE Trans.* AP-3 (January 1955), pp. 16–28.

131. Hansen, R. C. "Linear Arrays." *Handbook of Antenna Design,* vol. 2, A. W. Rudge, K. Milne, A. D. Oliver, and P. Knight (Eds.). London: Peter Peregrinus, 1983, Chap. 9.

132. Hansen, R. C. "Measurement Distance Effects on Low Sidelobe Patterns." *IEEE Trans.* AP-32 (June 1984), pp. 591–594.

133. Taylor, T. T. "Design of Circular Apertures for Narrow Beamwidths and Low Sidelobes." *IRE Trans.* AP-8 (January 1960), pp. 17–22.

134. White, W. D. "Desirable Illuminations for Circular Aperture Arrays." Institute for Defense Analyses, Arlington, VA, Research paper P-351, IDA Log No. HQ 67–6476, December 1967.

135. White, W. D. "Circular Aperture Distribution Functions." *IEEE Trans.* AP-25 (September 1977), pp. 714–716.

136. Elliott, R. S. *Antenna Theory and Design.* Englewood Cliffs, NJ: Prentice-Hall, 1981.

137. Bayliss, E. T. "Design of Monopulse Antenna Difference Patterns with Low Sidelobes." *Bell System Tech. J.* 47 (May–June 1968), pp. 623–650.

138. Hansen, R. C. *Phased Array Antennas.* New York: John Wiley, 1998, Sec. 3.7.

139. Lopez, A. R. "Sharp Cutoff Radiation Patterns." *IEEE Trans.* AP-27 (November 1979), pp. 820–824.

140. Ruze, J. "Physical Limitations on Antennas." MIT Research Lab. Electronics Tech, Rept. 248, Oct. 20, 1952; or see p. 255 of the 2nd ed. of this text.

141. Shrader, W. W., and V. Gregers-Hansen. "MTI Radar." *Radar Handbook,* 2nd ed., M. Skolnik (Ed.). New York: McGraw-Hill, 1990, Chap. 15, Fig. 15.64.

142. Ruze, J. "Antenna Tolerance Theory—A Review." *Proc. IEEE* 54 (April 1966), pp. 633–640.

143. Ruze, J. "Physical Limitations on Antennas." MIT Research Lab. Electronics Tech. Rept. 248, Oct. 30, 1952.

144. Skolnik. M. I. "Nonuniform Arrays." *Antenna Theory,* Pt I, R. E. Collin and F. J. Zucker (Eds.). New York: McGraw-Hill, 1969, Chap. 6, Sec. 6.6.

145. Rondinelli, L. A. "Effects of Random Errors on the Performance of Antenna Arrays of Many Elements." *IRE Natl. Conv. Record* 7, pt. 1 (1959), pp. 174–187.

146. Lichter, M. "Beam-Pointing Errors of Long Line Sources." *IRE Trans.* AP-8 (May 1960), pp. 268–275.

147. Hsiao, J. K. "Design of Error Tolerance of a Phased Array." *Electronic Letters* 21, no. 19 (September 12, 1985), pp. 834–836.

148. Hsiao, J. K. "Array Sidelobes, Error Tolerance, Gain, and Beamwidth." Naval Research Laboratory, Washington, D.C., Report 8841, September 28, 1984.

149. Cheston, T. C., and J. Frank. "Phased Array Radar Antennas." Chap. 7, *Radar Handbook,* 2nd ed., M. I. Skolnik (Ed.), New York: McGraw-Hill, 1990, p. 7.41.

150. Cheston, T. C., and J. Frank. Ref. 149, Sec. 7.6.

151. Miller, C. J. "Minimizing the Effects of Phase Quantization in an Electronically Scanned Array." *Proc. of Symp. on Electrically Scanned Array Techniques and Applications,* Rome Air Development Center Technical Documentary Report RADC-TDR-64–225, vol. 1, pp. 17–38, July, 1964.

152. Evans, G. E., and H. E. Schrank. "Low-Sidelobe Radar Antennas." *Microwave J.* 26 (July 1983), pp. 109–117.

153. Schrank, H. E. "Low Sidelobe Phased Array Antennas." *IEEE APS Newsletter* (April 1983), pp. 5–9.

154. White, W. D. "Desirable Illuminations for Circular Aperture Antennas, Research Paper P-351." Institute for Defense Analyses, Arlington, VA, December 1967.

155. Ludwig, A. C. "Low Sidelobe Aperture Distribution for Blocked and Unblocked Circular Apertures." *IEEE Trans.* AP-30 (September 1982), pp. 933–946.

156. Taylor, T. T. "Design of Circular Apertures for Narrow Beamwidth and Low Sidelobes." *IRE Trans.* AP-8 (January 1960), pp. 17–22.

157. Green, T. J. "The Influence of Masts on Ship-Borne Radar Performance." *Radar-77, IEE (London) Conference Publication* no. 155 (October 1977), pp. 405–408.

158. Mangulis, V. "Effective Sidelobe Levels Due to Scatterers." *IEEE Trans.* AES-15 (May 1979), pp. 325–333.

159. Mangulis, V. "Antenna Sidelobes in the Presence of Flat Reflectors." *IEEE Trans.* AES-30 (October 1994), pp. 1122–1125.

160. Scorer, M. "The Calculaton of Radome Induced Sidelobes." *Radar-77,* IEE (London) Conference Publication No. 155 (October 1977), pp. 414–418.

161. Ricardi, L. J. "Radiation Properties of the Binomial Array." *Microwave J.* 15 (December 1972), p. 20.

162. Krall, A. D., D. G. Jablonski, and J. Coughlin. "Radiation Properties of a 'Gaussian' Antenna." *Microwave J.* 27 (May 1984), pp. 283 & 288.

163. Patton, W. T. "Low-Sidelobe Antennas for Tactical Phased-Array Radars." *RCA Engineer* 27 (Sept./Oct 1982), pp. 31–36.

164. Maine, E. E., Jr., and J. M. Willey. The Fixed Array Surveillance Radar, private communication.

165. Schrank, H. E. "Low Sidelobe Reflector Antennas." *IEEE APS Newsletter,* (April 1985), pp. 5–16.

166. Scudder, R. M. "Advanced Antenna Design Reduces Electronic Countermeasures Threat." *RCA Engineer* 23 (Feb./Mar. 1978), pp. 61–65.

167. Williams, N., P. Varnish, and D. J. Browning. "Reduced Cost Low Sidelobe Reflector Antenna Systems." *IEE (London) International Conf. Radar-82* (October 1982), pp. 351–354.

168. Fante, R. L., P. R. Franchi, N. R. Kerweis, and L. F. Dennett. "A Parabolic Cylinder Antenna with Very Low Sidelobes." *IEEE Trans.* AP-28 (January 1980).

169. Evans, G. E. *Antenna Measurement Techniques.* Boston: Artech House, 1990.

170. Hacker, P. S., and H. E. Schrank. "Range Distance Requirements for Measuring Low and Ultralow Sidelobe Antenna Patterns." *IEEE Trans.* AP-30 (September 1982), pp. 956–966.

171. Tang, R., and R. Brown. "Cost Reduction Techniques for Phased Arrays." *Microwave J.* 30 (January 1987), pp. 139–146.

172. Skolnik, M. "The Radar Antenna—Circa 1995." *J. Franklin Inst.* 332B, no. 5 (May 1995), pp. 503–519.

173. Cheston, T. C., and J. Frank. "Phased Array Radar Antennas." Chap. 7, *Radar Handbook,* M. Skolnik (ed.) New York: McGraw-Hill, 1990, Chap. 7, Sec. 7.7.

174. Mailloux, R. J. Ref. 74, Sec. 8.3.

175. Smolders, A. B. "Design and Construction of a Broadband Wide-Scan-Angle Phased-Array Antenna with 4096 Radiating Elements." *1996 IEEE International Symp. on Phased Array Systems and Technology,* Boston, MA, October 15–18, 1996, pp. 87–92.

176. Baugh, R. A. *Computer Control of Modern Radars.* published by RCA Corp., (now Lockheed Martin) Aegis Department, Moorestown, NJ, 1973.

177. Huizing, A. G., and A. A. F. Bloemen. "An Efficient Scheduling Algorithm for a Multifunction Radar." *1996 IEEE International Symp. on Phased Array Systems and Technology,* Boston, MA, October 15–18, 1996, pp. 359–364.

178. Billam, E. R. "The Problem of Time in Phased Array Radar." *Radar-97,* October 14–16, 1997, IEE Publication No. 449, pp. 563–567.

179. Bony, Gilbert. "Electrically Controlled Dielectric Panel Lens," United States Patent No. 3,708,796, Jan. 2, 1973.

180. Skolnik, M. I. "The Radar Antenna—Circa 1995." *J. Franklin Institute,* 332B, No. 5 (1995), pp. 503–519.

181. Chekroun, C., et al. "RADANT. New Method of Electronic Scanning." *Microwave J.* 24 (February 1981), pp. 45–53.

182. Rao, J. B. L., G. V. Trunk, and D. P. Patel. "Two Low-Cost Phased Arrays." *1996 IEEE International Symp. on Phased Array Systems and Technology,* Boston, MA, October 15–18, 1996, pp. 119–124.

183. Colin, Jean-Marie. "Phased Array Radars in France. Present & Future." *1996 IEEE International Symp. on Phased Array Systems and Technology,* Boston, MA, October 15–18, 1996, pp. 458–462.

184. Rao, J. B. L., D. P. Patel, and V. Krichevsky. "Voltage-Controlled Ferrolelectric Lens Phased Arrays." *IEEE Trans.* AP-47 (March 1999), pp. 458–468.

185. Mailloux, R. J. "Conformal and Low-Profile Arrays." *Antenna Engineering Handbook,* 3rd ed., R. C. Johnson (Ed.). New York: McGraw-Hill, Chap. 21, 1993.

186. Hsiao, J. K. "Approximation of a Conformal Array with Multiple, Simultaneously Excited Planar Arrays," Naval Research Laboratory, Washington, D. C., Report 7442, July 28, 1972.

187. Skolnik, M. I. "Nonuniform Arrays." *Antenna Theory,* pt. I, R. E. Collin and F. J. Zucker (Eds.). New York: McGraw-Hill, 1969, Chap. 6.

188. Mailloux, R. J. *Phased Array Antenna Handbook.* Boston: Artech House, 1994, Sec. 2.4.

189. Skolnik, M. I., G. Nemhauser, and J. W. Sherman. "Dynamic Programming Applied to Unequally Spaced Arrays." *IEEE Trans.* AP-12 (January 1964), pp. 35–43.

190. Skolnik, M. I., J. W. Sherman, and F. C. Ogg. "Statistically Designed Density-Tapered Arrays." *IEEE Trans.* AP-12 (July 1964), pp. 408–417.

191. Hall, W. P., Jr., and R. D. Nordmeyer. "Active-Element, Phased Array Radar: Affordable Performance for the 1990s." *IEEE National Telesystems Conf. Proc.,* Atlanta, GA, pp. 193–197, March 26 and 27, 1991.

192. Cross, D. C., D. D. Howard, and J. W. Titus. "Mirror-Antenna Radar Concept." *Microwave J.* 29 (May 1986), pp. 323–335.

193. Orleansky, E., C. Samson, and M. Havkin. "A Broadband Meanderline Twistreflector for the Inverse Cassegrain Antenna." *Microwave J.* 30 (October 1987), pp. 185–192.

194. Lewis, B. L., and J. P. Shelton. "Mirror Scan Antenna Technology." *1980 IEEE International Radar Conf.,* Washington, D.C., pp. 279–283, April 1980.

195. Ruze, J. "Lateral Feed Displacement of a Paraboloid." *IEEE Trans.* AP-13 (September 1965), pp. 660–665.

196. Imbriale, W. A., P. G. Ingerson, and W. C. Wong. "Large Lateral Feed Displacement in a Parabolic Reflector." *IEEE Trans.* AP-22 (November 1974), pp. 742–745.

197. Kelleher, K. S., and H. H. Hibbs. "A New Microwave Reflector." Naval Research Laboratory, Washington, D.C., Report 4141, 1953.

198. Kelleher, K. S., and G. Hyde. "Reflector Antennas." *Antenna Engineering Handbook,* 3rd ed., R. C. Johnson (Ed.). New York: McGraw-Hill, 1993, Chap. 17, pp. 17-46 to 17.52.

199. Kelleher, K. S. "Electromechanical Scanning Antennas." *Antenna Engineering Handbook.* 3rd ed., R. C. Johnson (Ed.). New York: McGraw-Hill, 1993, Chap. 18, pp. 18-23 to 18-24.

200. Rappaport, C. M., and W. P. Craig. "High Aperture Efficiency Symmetric Reflector Antennas with up to 60° Field of View." *IEEE Trans.* AP-39 (March 1991), pp. 336–344.

201. Craig, W. P., C. M. Rappaport, and J. S. Mason. "A High Aperture Efficiency, Wide-Angle Scanning Offset Reflector Antenna." *IEEE Trans.* AP-41 (November 1993), pp. 1481–1490.

202. Bodnar, D. G. "Lens Antennas." *Antenna Engineering Handbook,* 3rd ed., R. C. Johnson (Ed.). New York: McGraw-Hill, 1993, Chap. 16.

203. Harvey, A. F. "Optical Techniques at Microwave Frequencies." *IEE Proc.* 106, pt. B (March 1959), pp. 141–157. Contains an extensive bibliography.

204. Kock, W. E. "Metal Lens Antennas." *Proc. IRE* 34 (November 1946), pp. 828–836.

205. Luneburg, R. K. *Mathematical Theory of Optics.* Berkeley: University of California, 1964. (Originally mimeographed lecture notes, Brown University Graduate School, Providence, R.I., 1944.)

206. Wiltse, J. C., and J. E. Garrett. "The Fresnel Zone Plate Antenna." *Microwave J.* 34 (January 1991), pp. 101–114.

207. Huddleston, G. K., and H. L. Bassett. "Radomes." *Antenna Engineering Handbook,* 3rd ed., R. C. Johnson (Ed.), New York: McGraw-Hill, 1993, Chap. 44.

208. Walton, J. D., Jr. *Radome Engineering Handbook.* New York: Marcel Dekker, 1970.

209. Schrank, H. E., G. E. Evans, and D. Davis. "Reflector Antennas." *Radar Handbook,* 2nd ed., M. Skolnik (Ed.). New York: McGraw-Hill, 1990, Chap. 6. Sec. 6.9.

210. Electronic Space Systems Corporation (Essco) Web site, 1998.

211. Hughes, D. "New FAA Radomes to Have 98% Transmission Efficiency." *Aviation Week & Space Technology* (January 10, 1994), pp. 66–67.

212. Vitale, J. A. "Large Radomes." *Microwave Scanning Antennas,* Vol. 1." R. C. Hansen (Ed.). New York: Academic, 1964, Chap. 5.

213. Effenberger, J. A., R. R. Strickland, and E. B. Joy. "The Effect of Rain on a Radome's Performance." *Microwave J.* 29 (May 1986), pp. 261–274.

214. Punnett, M. S. "Developments in Ground Mounted Air Supported Radomes." *IEEE 1977 Mechanical Engineering in Radar Symposium,* Nov. 8–10, 1977, Arlington, VA, pp. 40–45, IEEE Publication 77CH1250-0 AES.

215. Advertising brochure "Engineered Fabric Structures," from Birdair Structures Division, North Bennington, Vermont (no date).

216. Conti, D. A. "Special Problems Associated with Aircraft radomes." *IEE Proc.* 128, Pt. F, no. 7 (December 1981), pp. 412–418.

217. Pelton, E. L., and B. A. Munk. "A Streamlined Metallic Radome." *IEEE Trans.* AP-22 (November 1974), pp. 799–803.

218. Knott, E. F., J. F. Shaeffer, and M. T. Tuley. *Radar Cross Section,* 2nd ed. Boston: Artech House, 1993, Sec. 10.4.

219. Mittra, R., C. H. Chan, and T. Cwik. "Techniques for Analyzing Frequency Selective Surfaces—A Review." *Proc. IEEE* 76 (December 1988), pp. 1593–1615.

220. Farina, A. *Antenna-Based Signal Processing Techniques for Radar Systems.* Boston: Artech House, 1992.

221. Nitzberg, R. *Adaptive Signal Processing for Radar.* Boston: Artech House, 1992.

222. Widrow, B., and S. D. Stearns. *Adaptive Signal Processing.* Englewood Cliffs, NJ: Prentice-Hall, 1985.

223. Gabriel, W. F. "Adaptive Arrays—An Introduction." *Proc. IEEE* 64 (February 1976), pp. 239–272.

224. Gabrial, W.F. "Adaptive Processing Array Systems." *Proc. IEEE* 80 (January 1992) pp. 1521–162.

225. Howells, P. W. "Explorations in Fixed and Adaptive Resolution at GE and SURC." *IEEE Trans.* AP-24 (September 1976), pp. 575–584.

226. Ward, J. "Space-Time Processing for Airborne Radar." MIT Lincoln Laboratory Technical Report 1015, December 13, 1994.

227. Staudaher, F. M. "Airborne MTI." *Radar Handbook,* M. Skolnik (Ed.). New York: McGraw-Hill, 1990, Chap. 16, Sec. 16.8.

228. Gabriel, W. F. "Spectral Analysis and Adaptive Array Superresolution Techniques." *Proc. IEEE* 68 (June 1980), pp. 654–666.

229. Rihaczek, A. W. "The Maximum Entropy of Radar Resolution." *IEEE Trans.* AES-17 (January 1981), p. 144.

230. Hansen, R. C. *Phased Array Antennas.* New York: John Wiley, 1998, Chap. 9.

231. *IEEE Standard Dictionary of Electrical and Electronics Terms.* New York: IEEE, 1988.

232. Pugh, M. L., et al. "Electromechanically Scanned Arrays Using Micro Electro Mechanical Swith (MEMS) Technology." *Proc. 5th International Conf. on Radar Systems,* Brest, France, May, 1999, Subassemblies Section.

233. Smith, J. K., F. W. Hopwood, and K. A. Leahy, "MEM Switch Technology in Radar." *Record of the IEEE 2000 International Radar Conference*, Alexandria, VA, pp. 193–198.

234. Norvell, B. R., et al. "Micro Electro Mechanical Switch (MEMS) Technology Applied to Electrically Scanned Arrays (ESA)." *Proc. International Radar Symp. (IRS 98),* Munich Germany, September 15–17, 1998, vol. II, German Institute of Navigation.

PROBLEMS

9.1 (a) Derive the expression for the field-intensity pattern for a uniformly illuminated line-source aperture of dimension D. (b) Make a rough sketch of its radiation (power) pattern (with the ordinate in dB). (c) If the antenna dimension is 60 wavelengths, what is the width between the first nulls that define the main beam? (d) What is its half-power beamwidth?

9.2 The pattern of problem 9.1 also is the pattern from either of the two principal planes of a uniformly illuminated square aperture. Derive the field-intensity pattern in the diagonal plane of the square aperture, where the aperture illumination is triangular. (The triangular illumination is also known as a *gabled illumination.* It can be expressed as $A(z) = 1 - \dfrac{2}{D_g}|z|$, where $|z| \le D_g/2$, and D_g is the diagonal of the square aperture.)

9.3 Determine and roughly sketch the field-intensity pattern for an antenna with the following aperture illumination pattern:

$$A(z) = -1 \text{ for } -D/2 < z < 0$$

$$A(z) = 0 \text{ for } z = 0$$

$$A(z) = +1 \text{ for } 0 < z < +D/2$$

where D = aperture dimension. What is the peak sidelobe level? (This is the difference pattern that would be obtained from an antenna whose sum pattern is of the form $(\sin x)/x$ obtained with a uniform aperture as in problem 9.1.)

9.4 Calculate and sketch the field-intensity pattern produced by an aperture illumination of a line source of dimension D given by $A(z) = \cos(\pi z/D)$. What is its null width (width between the two nulls defining the main beam), and what is the level of the first sidelobe?

9.5 In what manner is the field-intensity pattern $E(\phi)$ of an antenna, Eq. (9.10), and its aperture illumination $A(z)$ related to the time waveform $s(t)$ and the spectrum $S(f)$? Identify the analogous pairs of parameters between these two relationships?

9.6 Efficiency is generally defined as the ratio of the output power to the input power. The "aperture efficiency" of a reflector antenna, Eq. (9.9), is not defined in this manner and the term aperture "efficiency" can therefore be misleading. Discuss the effect of the aperture efficiency on radar performance and why it should not be interpreted as an indicator of a power loss.

9.7 Why does a parabolic surface make a good reflector antenna?

9.8 When might each of the following parabolic reflector antennas be used: (a) paraboloid, (b) section of a paraboloid, (c) parabolic cylinder, (d) parabolic torus, (e) offset-fed reflector, (f) Cassegrain, and (g) mirror-scan antenna? When might (h) a spherical reflector or (i) a lens antenna be used?

9.9 *Background.* In a Cassegrain antenna, Fig. 9.11, the feed is located at (or near) the apex of the primary paraboloid reflector. The radiation from the feed is reflected by a secondary reflector in front of the primary reflector. The radiation is returned to the primary reflector where it is reradiated in the forward direction. Blockage of the radiated energy occurs because of the obstruction by the secondary reflector and by the interception of energy by the feed. If the secondary reflector is made smaller so as to reduce its blockage, the feed has to be made larger so as to illuminate without spillover the smaller secondary reflector. Similarly, if the feed is made smaller to avoid blockage, the secondary reflector has to be made larger in order to intercept the wide-angle energy from the feed. If the secondary reflector and the feed are both circular with a diameter of S and d, respectively, the total blockage is due to their combined area $(\pi/4)(S^2 + d^2)$. *Problem:* What should the relationship be between the diameter S of the feed and the diameter d of the secondary reflector in order to minimize the total blockage in a Cassegrain antenna? (You may assume that the beamwidth of the feed is λ/d radians.)

9.10 List the five basic methods available for obtaining a phase shift, and give an example of a phase shifter based on each.

9.11 When might ferrite phase-shifters be used in an electronically steered phased array antenna, and when might diode phase-shifters be used?

9.12 (a) If the minimum range of radar is to be no greater than 1.2 nmi, what should be the maximum switching time of a phase shifter so that the array is ready to receive after it has transmitted? (Assume that the pulse duration is small and can be neglected.) (b) Select a type of ferrite phase shifter that can probably meet this requirement. (c) What limitation might occur with your selection? (d) If the minimum allowable range were increased to

12 nmi, how might your answer for (b) and (c) change? (e) What type of ferrite phase shifter can be used when the radar operates with dual orthogonal linear polarizations? (f) What type of ferrite phase shifter might be used in a reflectarray?

9.13 (a) Derive the array factor for a uniformly spaced linear array of N isotropic elements. [Suggest you start with Eq. (9.22); but instead of a summation of sinewaves, it will be easier to use the exponential relation such as $1 + e^{j\phi} + e^{j2\phi} + e^{j3\phi} \cdots$ where $\phi = 2\pi(d/\lambda) \sin \theta$, as in Eq. (9.22). You also will have to recall or rederive the expression for the sum of a geometric series that you learned in high school.] (b) At what angles will grating lobes appear (over a range of $\pm90°$) when the element spacing is four wavelengths? (c) Compare, in words, the pattern produced by the linear array of equal amplitude elements to the pattern produced by a continuous line source with uniform illumination (as in problem 9.1) and the same aperture size.

9.14 When the beam of a phased array antenna is electronically steered to an angle θ_0 from broadside, show that its beamwidth varies inversely as $\cos \theta_0$.

9.15 (a) Show that grating lobes will not appear in a steered phased array if the element spacing is less than one-half wavelength. (b) What should the element spacing be when $\theta_0 = 0$ if there can be grating lobes at $\pm30°$, but not at smaller angles?

9.16 (a) What wrap-up factor is required in a frequency-scan array to scan the beam over an angle of $\pm50°$ using no more than a total tunable (relative) bandwidth of 5 percent? (Assume a TEM transmission line where the velocity of propagation is the velocity of light.) (b) If there are 80 elements in the frequency scan array with an element spacing of 10 cm and a wrap-up factor the same as computed for part (a), what is the time required for a signal to fill the array aperture? (c) What bandwidth does this correspond to?

9.17 Consider the series-fed linear array of Fig. 9.16b. How far will the beam deviate from the broadside direction when the frequency is changed by 20 percent from the design frequency? (Assume the transmission feed-line propagates in the TEM mode so that the velocity in the line is c, the velocity of light.)

9.18 A frequency-scan array has an element spacing $d = 5$ cm, aperture dimension $D = 3$ m, and a feed system with a wrap-up factor $= 16$. As the beam is frequency scanned past the target, the echo will be frequency modulated with a bandwidth Δf_B [Eq. (9.42)]. (a) If a frequency $f = 1.05 f_0$ points the beam to 30°, what is the spectral width of the echo signal due to the linear FM modulation induced on the echo? (b) If pulse compression processing is used on receive to take advantage of the frequency modulation of the echo signal, what will be the compressed pulse width?

9.19 Why do you think problems might occur when attempting to apply the theory of the infinite array to a finite size array?

9.20 Compare the corporate-fed (passive) phased array, the active-aperture phased array, and the space-fed phased array with respect to loss, transmitter efficiency (Sec. 10.3), relative prime power required for the transmitter, and any other factors of concern.

9.21 The Dolph-Chebyshev antenna illumination produces an antenna pattern with a narrow beamwidth and all sidelobes equal. It would appear to be a good antenna pattern except that it is not practical. What is it about the Dolph-Chebyshev illumination that makes it unrealizable, especially with large gain?

9.22 What approximate reduction in gain [Eq. (9.51)] results when an antenna pattern is rectangular from 0 to 3° and is shaped to have a cosecant-squared variation of gain over the angular region from $\theta_0 = 3°$ to $\theta_m = 25°$?

9.23 For a fixed-size reflector antenna with an rms surface error ϵ: (a) Show that the wavelength that results in maximum antenna directivity is $\lambda_m = 4\pi\epsilon$. [This is Eq. (9.53).] (b) How does this relate to the usual "rule of thumb" for reflector antenna tolerances which states that the rms error ϵ should be less than $\lambda/32$, where λ = wavelength? (c) Determine the loss of gain at the wavelength λ_m compared to the gain of a perfect (no error) antenna. (d) What is the maximum gain that is achievable for $D/\epsilon = 1000$ and $D/\epsilon = 10,000$? (Assume the aperture efficiency is 1.) (e) In your opinion, how high an antenna gain might be achieved with a practical radar antenna (and explain your answer)?

9.24 If a one-dB reduction in antenna gain is allowed due to errors in a 100-element linear array, what is the phase error (in degrees), the amplitude error (in dB), and the fraction of missing elements that can be tolerated when each one of these three factors is the only one contributing to the gain reduction? (That is, determine the phase error with the amplitude error zero and no missing elements, and so on.)

9.25 *Background.* A periodic error in an antenna will produce multiple equally spaced beams (or sidelobes) in $\sin \theta$ space similar to the formation of grating lobes in an array antenna, except that these periodic error sidelobes are much smaller in gain. *Problem.* An antenna is suspected of having a periodic error since there are prominent sidelobes in its radiation pattern at angles of ±37.3 and ±53.9°. If there is a periodic error, there will be other closer-in sidelobes; but in this problem we assume that these closer-in lobes are masked by the normal antenna sidelobe radiation. Based on the observed sidelobes whose directions are given above, what is the period (spacing) of this periodic error?

9.26 (a) With a uniformly illuminated planar array of 1000 isotropic elements, what should be the rms value of the phase error (in degrees) and the rms value of the amplitude error (in dB) to make the average error-sidelobes equal to −50 dB? (Assume that the contribution from the phase error equals that from the amplitude error.) (b) Repeat for an array antenna with 10,000 elements.

9.27 What determines the number of bits to be used in a digitally switched phased shifter?

9.28 According to G. Evans [Ref. 169, p. 115] the accuracy with which the antenna gain can be measured is commonly ± 0.5 dB. What is the maximum rms phase error in a reflector antenna that results in a loss of antenna gain of 0.5 dB?

9.29 (a) In a phased array with a square aperture of 100 by 100 elements, with half-wave spacing between elements, what is the rms angle error (in degrees) when the rms value of the normalized error current is $\sigma = 0.4$? (b) What fraction of a beamwidth would this be if the aperture illumination were uniform?

9.30 (a) What are the characteristics of an aperture illumination that can achieve ultralow sidelobes? (b) When might ultralow sidelobes be needed? (c) What antenna characteristics have to be sacrificed for ultralow sidelobes? (As a start, see Table 9.1) (d) What factors ultimately limit the sidelobe levels that can be achieved in practice?

9.31 When might a radar systems engineer decide not to use an ultralow sidelobe antenna?

9.32 Why is a phased array antenna more suitable than a reflector antenna as an ultralow side-lobe antenna?

9.33 Compare the advantages and disadvantages for shipboard air defense of a traditional four-faced phased array radar and a system consisting of two trainable phased array radars and one 2D rotating air-surveillance radar with 360° azimuth coverage.

9.34 (a) What are the advantages and limitations of operating a ground-based air-surveillance radar under a radome? (b) Compare the air-inflated radome and the rigid geodesic radome for this application.

9.35 What types of antennas might be used for the detection and tracking of hostile ballistic missiles? (This is a question not just about antennas, but also about how radar systems might be applied for ballistic missile defense. There is no simple unique answer to this question, and there has not been general agreement as to the best approach.)

9.36 Why are phased array radars for air surveillance generally cheaper at the lower radar frequencies than at the higher frequencies?

9.37 For the application of air defense, compare the advantages and disadvantages of a single multifunction phased array that operates in one frequency band with a system having a separate air-surveillance radar and a separate phased array weapon control (tracking) radar, each operating in different bands.

9.38 A phased array has a beamwidth of 2° when pointed to broadside. If it is required to scan to an angle of 60° from broadside, what is the maximum signal bandwidth in MHz that a radar can have that operates at a center frequency of 3.3 GHz?

9.39 What does one have to do to obtain an electronically steered phased array with a large instantaneous signal bandwidth?

9.40 *Background.* It has sometimes been suggested (usually by non-radar administrators) that increasing the frequency of an air-defense phased array radar will result in a smaller size antenna for the same beamwidth (which is true), and thus provide a smaller radar system (which might not be true). If a smaller system were to result it can be made more mobile if a ground based system or, if a ship-based system, it can be employed on smaller ships. A smaller radar, however, does not result when the frequency is increased and all other requirements remain the same. *Problem.* What is wrong with reasoning that concludes that a smaller radar system will result if the array antenna is reduced in size by operating at a higher frequency? [To illustrate your answer, you can, if you wish, assume an S-band multifunction phased array radar (3.5 GHz) having a 12 by 12 foot aperture and 10 kW of average power. If the radar is increased to 35 GHz, the aperture is reduced in size to 1.2 by 1.2 foot.]

9.41 The typical *L*-band 2D air-surveillance radar usually has its maximum elevation coverage extending to about 20 to 40°, depending on the particular radar. If it is required to extend coverage of the radar to higher elevation angles, there are reasons why it might be better to employ a separate antenna at a different frequency to fill the surveillance hole above the radar. (a) What are some reasons why a separate radar might be used rather than attempt to increase the elevation coverage of the 2D antenna? (b) If the elevation hole extends from 30° elevation angle to the zenith at 90°, what type of scanning patterns might be used? (There can be more than one choice of scanning pattern.) (c) What frequency band might be used for this hole-filler (and explain the reason for your selection)? (d) What type of antenna might be used for the hole-filler radar?

chapter
10

Radar Transmitters

10.1   INTRODUCTION

The radar systems engineer would like a transmitter to provide sufficient energy to detect a target, be easily modulated to faithfully produce the desired waveforms, generate a stable signal so that doppler signal processing can be performed without transmitter noise masking the doppler-shifted received signal, provide the needed signal bandwidth and tunable bandwidth, be of high efficiency, be of high reliability, be easy to maintain, have long life, be able to operate with a minimum of personnel, be of a size and weight suitable for the intended application, and be of affordable price. All of the above can be obtained, but seldom all together in one transmitter. Compromise is necessary.

Some radar transmitters have to generate large peak power as well as large average power; but it is the average power (which relates to energy) that is the measure of radar performance rather than peak power. It was seen in Chap. 2 that the range of a radar is proportional to the fourth root of the radar transmitter's average power. To increase the range of a radar by one order of magnitude (a factor of ten) the transmitter power has to be increased by ten thousand. Although there have been radars with average powers greater than a million watts, power cannot be increased without limit since high-power transmitters are heavy, take up space, and can consume much prime power (the power taken from the local power company) or fuel for motor-driven electrical generators.

An indicator of the performance of a radar is the product of the antenna area times the transmitter's average power. Tom Weil[1] described quite well the problem of choosing between high transmitter power and a large antenna as follows:

It obviously would not make sense for a radar to have a huge, costly antenna and a tiny, inexpensive transmitter, or vice versa, because doubling the tiny part would allow cutting the huge part in half, which would clearly reduce total system cost. Thus, minimizing total system cost requires a reasonable *balance* between the costs of these two subsystems. The result, for any nontrivial radar task, is that significant transmitter power is always demanded by the system designers.

Radar transmitters have been based on either a power amplifier, such as a klystron, or a power oscillator, such as a magnetron. In the early days of microwave radar in the 1940s and 1950s, the magnetron power oscillator was used almost exclusively since it was the only high-power microwave tube available at the time. It did an outstanding job in making microwave radar a reality in World War II, but it had many serious limitations. Magnetrons are noisy devices that limit the MTI improvement factor that can be obtained. Although they can produce high peak power (megawatts), they are not capable of large average power, and their signal output cannot be readily modulated to produce pulse-compression waveforms. All of these disadvantages are overcome with amplifiers such as the klystron, traveling wave tube, and the transistor. Modern high-performance radars almost always employ some sort of power amplifier as the transmitter. The magnetron appears to be limited to those applications where its relatively small size and lower cost are more important than its limitations.

Most of the discussion in this chapter is about the RF power source. A transmitter, however, is more than just the active RF power source. It includes the exciter and driver amplifiers that provide the signal to be amplified if the power source is an amplifier. If the transmitter generates a pulse waveform, a pulse modulator of some type is needed (except for RF power sources that are self-pulsed by the input waveform, as are transistors). There must be a d-c power supply for generating the necessary voltages and currents to operate the RF power device; means to remove the heat dissipated, including a heat exchanger when liquid cooling is used; protection devices for dissipating high-voltage arc discharges; safety interlocks; monitoring devices; isolators; high-voltage cable; insulating-oil tanks (to immerse high-voltage cathode bushings to prevent corona and high-voltage breakdown); and lead shielding of X-rays when high voltages are used. Not all high-power radar transmitters need all of the above, but an RF power source is useless without the ancillary devices required to make it function.

The efficiencies of RF power sources typically might range from about 10 percent to about 60 percent. This is the *RF conversion efficiency,* defined as the ratio of RF power output available from the device to the d-c power input to the electron stream. It is the efficiency of interest to the tube or RF power source designer, but it is not the best measure for the radar systems engineer. A better measure is the *transmitter system efficiency,* which is the ratio of the RF power available from the transmitter to the total power needed to operate the transmitter. The total power includes the power to generate the electrons at the cathode, the power to generate any electromagnetic fields required for containing the electron beam, the power to cool the device, and any other power needed for the proper operation of the transmitter. If, for example, the RF conversion efficiency were 40 to 50 percent, the transmitter system efficiency might be 20 to 25 percent, or less. Thus one usually doesn't want to start with a power source whose RF conversion efficiency is only 10 to 15 percent, unless the power is so low that efficiency is not an important consideration.

For maximum efficiency, most high-power RF power sources operate saturated, meaning they are either completely on or completely off, with no intermediate power levels. This is all right for a radar that generates a rectangular-like pulse waveform. There are times, however, when it might be desired to have an amplitude-tapered, or shaped, pulse (for example, to reduce the time sidelobes in pulse compression waveforms or to minimize the effects of RF interference to other users of the electromagnetic spectrum). Highly shaped transmission waveforms are seldom found in high-power microwave radar systems because of their lower efficiency. Transistor amplifiers can be operated in what is called class-A operation so that there is a linear relationship between the output and input signal amplitudes. The efficiency of a class-A amplifier, however, is much less than that which would be achieved with the same device operated class-C. (Class-C amplifiers are nonlinear and are self-pulsing in that they generate pulses when the RF drive is turned on and off.) Thus, a microwave radar transmitter is almost always operated in saturation and not as a linear device.

High reliability and long life are important for a transmitter. The life of most RF power sources can be many thousands or many tens of thousands of hours, as will be discussed when describing the individual devices later in this section. If a transmitter's mean time between failures (MTBF) is not as long as expected, factors other than the RF power source are often at fault. Likely candidates are fans and blowers, the wrong type of coolant, RF connectors and coolant fittings damaged, coolant lines clogged, and leads that are broken, mishandled, or abused. Conservative mechanical and electrical design and procurement practices that guarantee reliability from suppliers are needed for a trouble-free transmitter (or anything else). The user of RF power sources also needs to help in avoiding less than the achievable reliability. For example, A. S. Gilmour[2] states "If the truth were known, it could well be that over 50 percent of all failures are the fault of the users, rather than the tube manufacturers."

Summary of Radar RF Power Sources   The different RF power sources available for high-power radar application include the klystron, traveling wave tube, solid-state transistor amplifier, Twystron, magnetron, crossed-field amplifier, amplitron, grid-controlled vacuum tube, extended interaction amplifier, gyrotron, and others. None provide all the desirable features that might be wanted. Some are no longer as popular as they once were. The choice depends in large part on the application and its constraints. Each of these devices will be briefly summarized below and described in more detail later in the chapter. All except the magnetron are power amplifiers. The gyrotron can be either an amplifier or an oscillator.

Klystron   This is an excellent radar tube when it can be employed. It has high gain and good efficiency, and is capable of higher average and peak power than most other RF power sources. It can have wide bandwidth (in the vicinity of 8 to 10 percent relative bandwidth*) when its power is large, long life (tens of thousands of hours), low interpulse noise, and good stability for doppler processing. When large peak powers are generated it requires very high voltage and X-ray shielding.

---

*Relative, or fractional, bandwidth in percent = $(\Delta f/f_0) \times 100$, where Δf = absolute bandwidth and f_0 = center frequency.

Traveling Wave Tube (TWT)   TWTs have slightly less power, slightly less gain, and slightly less efficiency than a klystron; but they are capable of wide bandwidth, especially at modest power levels. At high power levels the TWT bandwidth is lower than what can be achieved at lower powers, but it is still relatively large.

Hybrid Klystrons   These are similar to klystrons, but with one or more of the resonant cavities replaced by multiple cavities similar to what are used in a TWT. There are at least three versions: the *Twystron,* the *extended interaction klystron,* and the *clustered cavity klystron.* Bandwidths can be 15 to 20 percent, or more. Extended interaction klystrons have also been used for modest power millimeter-wave transmitters.

Solid-State Transistor Amplifiers   These are capable of wider bandwidth than most other RF power sources. They operate with low voltages, ease of maintenance, and have promise of long life. They are inherently of low power so that a large number of individual devices must be combined to generate sufficient power for most radar applications. For adequate efficiency, they should be operated at high duty cycles, which require that they generate long pulses and employ pulse compression.

Magnetron   The magnetron is generally smaller in size and utilizes lower voltage than the klystron; but its average power is limited and it has poor noise and stability characteristics, which restrict its ability to cancel clutter when used in an MTI radar.

Crossed-Field Amplifiers   These are capable of high power, good efficiency, and wide bandwidth, but are of relatively low gain (about 10 dB). Lower voltage, just as with the magnetron, means that X-rays are not usually a problem. Crossed-field amplifiers are generally noisier and less stable than most other RF power sources.

Grid-Control Tubes   These are modern versions of the classical triode and tetrode vacuum tubes originally introduced early in the twentieth century. They are a good power source for UHF radars, but they have been largely replaced by solid-state devices.

Microwave Power Module   This is a combination of a modest power TWT driven by a solid-state device that competes favorably with high-power solid-state modules in some applications.

Gyrotrons   These can produce very high power in the millimeter wave region, but they require large magnetic fields. They have had only modest application in operational radar systems.

RF power sources may be grouped into four general classes: (1) linear-beam tubes, (2) solid state, (3) crossed-field tubes, and (4) others not included in the first three. The klystron, traveling wave tube, magnetron, and crossed-field amplifier are *slow-wave devices* in that the phase velocity of the electromagnetic wave in the RF structure is slowed so as to be approximately equal to the velocity of the electron stream in order for (d-c) energy from the electron beam to transfer to electromagnetic energy in the (RF) signal.[3] The gyrotron, on the other hand, is a *fast-wave device* in that the phase velocity of the electromagnetic wave exceeds the speed of light in the interaction region.

10.2 LINEAR-BEAM POWER TUBES

In the linear-beam tube the electrons emitted from the cathode are formed into a long cylindrical beam that receives the full potential energy of the electric field before the beam enters the RF interaction region. The klystron, traveling wave tube, Twystron, and extended interaction amplifier are examples of linear-beam tubes. The last two are basically hybrid devices that combine the technology of the klystron with the RF structure of the TWT. An axial magnetic field is used in linear-beam tubes to confine the electron beam and keep electrons from hitting the RF structure. Transit-time effects, which can limit conventional vacuum tube performance at high frequencies, are used to good advantage in linear-beam tubes to *density modulate* the uniform d-c electron beam to create bunches of electrons from which RF energy can be extracted.

Linear-beam tubes can produce much higher power than other power sources. Klystrons are capable of more than a megawatt of average power. High power is a result, in part, of their larger size and high voltages. Thomson-CSF in France produced a UHF klystron that delivered more than one MW CW power. It was 5 m (16.4 ft) long and weighed 1400 kg (3000 pounds).[4] On the other hand, *X*-band klystron and TWT transmitters producing many kilowatts of average power can be made light enough and small enough to fly in the nose of military fighter/attack aircraft.

Klystron   A sketch of the principal parts of a klystron is shown in Fig. 10.1. At the left is the cathode which emits a stream of electrons that is formed into a narrow cylindrical beam by the *electron gun*. The electron gun consists of the cathode that is the source of electrons, a modulating anode or other beam-control electrode to provide a means for turning the beam on and off to generate pulses, and the anode. The electron emission density at the surface of the cathode is less than that required for the electron beam, so a large-area cathode surface is used and the emitted electrons are caused to converge to a narrow beam of high electron density. The multiple RF cavities, which correspond to the *LC* resonant circuits of conventional lower-frequency amplifiers, are at anode potential. Electrons are not intentionally collected by the anode, as in some other tubes; instead they are

Figure 10.1   Representation of the principal parts of a three-cavity klystron amplifier.

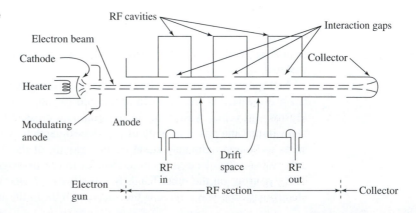

removed by the collector electrode (shown on the right) after the beam has given up its RF energy to the output RF cavity.

The RF input signal is applied across the *interaction gap* of the first cavity. Those electrons which arrive at the gap when the input signal voltage is a maximum (peak of the sinewave) experience a voltage greater than those electrons which arrive at the gap when the input is at a minimum (trough of the sinewave). Thus the electrons that see the peak of the sinewave are speeded up and those that see the trough are slowed down. The process whereby some electrons are speeded up and others slowed down is called *velocity modulation* of the electron beam. In the *drift space,* electrons that are speeded up during the peak of one cycle catch up with those slowed down during the previous cycle. The result is that the electrons of the velocity-modulated beam become "bunched," or density modulated, after traveling through the drift space. A klystron usually has one or more appropriately placed intermediate cavities to enhance the bunching of the electron beam, which increases the gain. If the interaction gap of the output cavity is placed at the point of maximum bunching, power can be extracted from the density-modulated beam. The gain of a klystron might be 15 to 20 dB per stage when synchronously tuned (all cavities tuned to the same frequency), so that a four-cavity (three stage) klystron might provide over 50 dB gain.

After the bunched electron beam delivers its RF power to the output cavity, the energy of the electron beam that remains is dissipated when the spent electrons are removed by the collector. The energy dissipated by the collector is energy lost and reduces the efficiency of the tube. If the collector is insulated from the body of the tube and a negative voltage is applied to the collector, the electrons in the spent beam will have lower kinetic energy so that less heat is produced when they impact upon the collector.[5] This results in an increase in the efficiency of the tube. There is a spread, however, in the velocities of the electrons in the beam; so if the potential is too negative, some of the slower velocity electrons will be returned to the walls of the RF section of the tube and be collected as body current, with a decrease in efficiency. This problem is overcome by employing a collector with several segments insulated from one another and with different negative potentials so that electrons with different velocities can be separated and collected at their optimum potential. Figure 10.1 shows a single-stage collector, but both the klystron and TWT usually employ multiple-stage depressed collectors for greater efficiency. The multiple stages (three or four might be typical in a radar tube) are at intermediate voltages, which allow catching the spent electrons at a voltage near optimum.

According to Weil,[1] a klystron with a peak power of 1 MW requires a voltage of about 90 kV. Gains might vary from 30 to 70 dB, bandwidths from 1 to 8 percent, and efficiencies from 40 to 60 percent (with depressed collectors).

A long solenoid (not shown in Fig. 10.1) with iron shielding around its outside diameter surrounds the high-power klystron to provide an axial magnetic field that confines the electrons to a relatively long, thin beam and prevents the beam from dispersing. Cooling might have to be provided for the electromagnets. In a high-power klystron, from 2 to 5 percent of the beam power might normally be intercepted by the interaction structure, or body of the tube. If the beam were not properly confined by the external magnetic field, the stray electrons impinging on the structure of the tube could cause it to overheat and possibly be destroyed. Protective circuitry is normally employed to remove the electron beam voltage in the event the magnetic field fails to keep the beam properly focused.

The electron beams of klystrons and TWTs also can be confined with permanent magnets. They do not require power or cooling, and the various protective circuits needed with solenoids are eliminated. Permanent magnets have been used with high power tubes, but they are quite heavy.[6,7] A significant reduction in weight, however, can be obtained with a periodic permanent-magnet (PPM) focusing system that consists of a series of magnetic lenses. These lenses employ washer-shaped disk magnets separated by iron pole pieces. The PPM replaces the uniform field of a solenoid with a periodic, essentially sinusoidal, field having the same rms value as the uniform field. Samarium cobalt is an example of a magnet material that has been widely used for tubes requiring permanent magnets. In the past, PPM focusing was usually not suited for large average-power tubes, but it has been successfully applied to high peak-power klystrons, as described next.

High-power klystrons have been used ever since the 1950s for linear accelerators to generate energetic beams of charged particles for research on the physics of high-energy particles. Many advances in klystron capability have been obtained from the development of klystrons for this purpose. These advances have, of course, been of benefit to radar as well. Although the invention of the klystron amplifier was reported in 1939, before the invention of the magnetron, it was not used or further developed significantly during World War II. It did not find its way into radar application until the development of a 20-MW peak power klystron, used in one of the first linear accelerators, was reported in 1953 by Stanford University. Thus the high-power klystron, which is a very important power source for radar, was a by-product of basic research in science. Work continued at the Stanford Linear Accelerator Center (SLAC) to develop high peak-power klystrons for electron-positron colliders. The klystron for the Next Linear Collider (NLC)[8] is at a frequency of 11.424 GHz, with a peak power of 60 to 75 MW using periodic permanent magnets made of neodymium-iron boron. The PPM with 10 pounds of permanent magnets replaces a 1/2-ton, 10-kW focusing solenoid. The NLC uses 6528 klystrons, which means that a total of 65 MW of solenoid power is avoided. The use of PPM, as well as having a tube with 60 percent efficiency, reduces "the NLC electric power bill by tens of millions of dollars per year." The tube requires an electron beam voltage of 490 kV. The average life of the S-band klystrons used for the previous Stanford Linear Collider (SLC) was 50,000 hours, and it is expected that a similar life will be obtained with the X-band NLC klystron. The manufacturing cost was said to be $30,000 per tube. If the tube delivers 75 MW with a pulse width of 1.5 μs and a prf of 120 Hz, its average power is 13.5 kW.

Bandwidth of a Klystron   The frequency of a klystron is determined by its resonant cavities. When all the cavities are tuned to the same frequency, the gain of the tube is high but the bandwidth is narrow, usually a fraction of one percent for a tube of modest power output. This is called *synchronous tuning*. To maximize the klystron's efficiency the next to last (penultimate) cavity is tuned upward in frequency and is outside the passband. Although the gain is reduced by about 10 dB, the improved electron bunching results in greater efficiency and in 15 to 25 percent more output power.[9] Broadbanding of a multicavity klystron may be accomplished by *stagger tuning* the cavities, similar to the method for broadbanding a conventional multistage IF amplifier. Stagger tuning a klystron is not precisely analogous to stagger tuning an IF amplifier because of interactions among the cavities that can cause the tuning of one cavity to affect the tuning of the others. The VA-87 four-cavity S-band synchronously tuned klystron amplifier with a 20 MHz bandwidth and a 61 dB gain can have a 27 MHz bandwidth and a 57.6 dB gain when tuned for maximum power.[10] When stagger tuned it has a 77-MHz bandwidth (2.8 percent) and a gain of 44 dB.

Theory shows that the bandwidth of a klystron can be significantly increased by increasing its power and its beam perveance (which is defined as the beam current I divided by beam voltage V to the 3/2 power, or $p = I/V^{3/2}$). A 10 MW peak power klystron, for example, can have an 8 percent bandwidth, as compared to a 200 kW tube which might have a 2 percent bandwidth, and a 1 kW tube having only a 0.5 percent bandwidth.[10] High-power multicavity klystrons can be designed with bandwidths as large as 10 to 12 percent.

Frequency Changing, or Tuning[6,9]   Conventional narrowband klystrons may have their frequency changed mechanically over a relatively wide frequency range. The individual cavities of a klystron can be changed in frequency (or tuned) by having a flexible wall in the resonant cavity (tuning range of about 2 or 3 percent), by a movable capacity element in the cavity (10 to 20 percent tuning range) or by a sliding contact movable cavity wall.

It can be tedious to tune a multiple-cavity klystron because of the interactions among the several cavities. Conventional gang-tuning is complicated since the resonant cavities do not have the same tuning rates. The *channel tuner mechanism*[6] avoids the problem of frequency tracking of the resonant cavities by pretuning the cavities, generally at the factory. The tuning information is stored mechanically within the tuner mechanism. When a particular frequency is selected, the tuner mechanism provides the correct tuner position for each cavity to furnish the desired klystron frequency response. The channel tuner mechanism is in a box attached to the klystron with gears to simultaneously set the tuning plungers at each cavity to their predetermined positions for a given frequency. The tuning plungers can be actuated manually or remotely by push buttons and a servomotor. The frequency can be changed in seconds.

Power   Some of the highest power radar transmitters have used klystrons. The ability of a klystron to produce higher power than other microwave power sources is, in part, due to its geometry. The regions of beam formation, RF interaction, and beam collection are separate. Each can be designed to best perform its own particular function independently of the others. The cathode, for example, is outside the RF field and need not be restricted to sizes that are small compared to a wavelength. Large cathode area and large interelectrode spacings may be used to keep the emission current densities and voltage gradients to reasonable values. The only function of the collector is to dissipate heat. It can be a shape and size best suited for satisfying the average or peak power requirements without regard for conducting RF currents, since none are present.

Efficiency   It is said in Ref. 8 that Robert Symons found that the best values of efficiency reported for klystrons worldwide followed the relation

$$\text{RF efficiency (percent)} = 90 - 20 \times \text{microperveance} \qquad [\textbf{10.1}]$$

where the microperveance is the perveance $I/V^{3/2}$ times 10^6, where I = beam current and V = beam voltage. Thus the lower the microperveance (or perveance) the higher will be the klystron efficiency. The perveance affects other properties of the klystron, including its bandwidth and power. Higher efficiency often requires, therefore, a reduction in band-width and lower power.

Reliability and Life   High-power transmitters employing power vacuum tubes have sometimes had the unwarranted reputation for poor reliability and short life. There is much evidence to the contrary for the klystron tube. Gilmour[11] reports the mean time between failures (MTBF) of eleven different applications of klystrons in radar systems (not iden-tified by type or power). The MTBF for these examples varied from 75,000 hours to 5000 hours, with an average value of about 37,000 hours for all eleven applications. (There are 8760 hours in a year.) The VA-842 high-power klystron tube used by the U.S. Air Force in the original Ballistic Missile Early Warning System (BMEWS) had a demonstrated life in excess of 50,000 hours. Symons[12] reports that one of the BMEWS tubes he designed in 1958 was still operating after 240,000 hours when the radar in Greenland was replaced by the solid-state Pave Paws radar.

An S-Band Klystron   The venerable VA-87 klystron built by Varian (now Communica-tions & Power Industries) was widely used in the FAA's S-band Air Surveillance Radars commonly found at major airports. It was a six-cavity tube tunable from 2.7 to 2.9 GHz, the frequency band reserved for air-traffic control radars. It had a peak power of from 0.5 to 2 MW, average power of from 0.5 to 3.5 kW, 50 dB gain, 45 percent efficiency, and a one-dB bandwidth of 39 MHz. It demonstrated a mean-time-between-failure (MTBF) rate of 72,000 hours. A similar tube was used in the Nexrad Doppler Weather Radar, but with bandwidth from 2.7 to 3.0 GHz.

Traveling Wave Tubes (TWT)   Like the klystron, the traveling wave tube is also a linear-beam tube with the cathode, RF circuit, and collector separated from one another. The klystron and the TWT were invented at different times in different parts of the world, but they are similar to one another. There is continuous interaction of the electron beam and the RF field over the entire length of the propagating structure of the traveling wave tube. In the klystron, on the other hand, the interaction occurs only at the gaps of a relatively few resonant cavities. The chief characteristic of a TWT is that it has wide bandwidth. Low power TWTs with a helix slow-wave RF structure are capable of octave bandwidths. With the high peak powers required of most radar applications, the bandwidths available with high-power TWTs are, however, much less than an octave.

   The major parts of a TWT are indicated in Fig. 10.2. A helix is shown for the slow-wave RF structure even though the helix is seldom used in TWTs found in radar appli-cations. The electron beam is similar to that of the klystron. Both the TWT and the kly-stron employ the principle of velocity modulation to cause the electron-beam current to be periodically bunched (density modulation). The electron beam passes thorough the RF interaction circuit known as the *slow-wave structure,* or periodic delay line. The velocity of propagation of the RF signal is slowed down by the periodic delay line so that it is nearly equal to the velocity of the electron beam. This is the reason that the helix and

Figure 10.2   Representation of the principal parts of a traveling-wave tube.

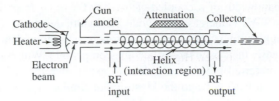

other microwave circuits used in TWTs are called slow-wave structures. The synchronism between the electromagnetic wave propagating along the slow-wave structure and the d-c electron beam propagating inside the helix results in a cumulative interaction which transfers d-c energy from the electron beam to increase the energy of the RF wave, causing the wave to be amplified. Just as in the klystron, an axial magnetic field keeps the electron beam from dispersing as it travels down the tube.

After delivering their d-c energy to the RF field on the slow-wave structure, the electrons are removed by the collector, which is usually a multistage depressed collector, as was described for the klystron. It is easier to design a depressed collector for a TWT than for a klystron since the spent electron beam of a TWT might have a 20 percent spread in velocity, but the klystron might have a velocity spread of almost 100 percent.[5] Because the efficiency of a conventional TWT is usually lower than that of a klystron, the increase in efficiency in the TWT provided by the depressed collector has a greater relative effect than with a klystron.

Although a helix is shown in Fig. 10.2 as the slow-wave structure, it is seldom found in TWTs used for radar. The helix TWT is limited to voltages of about 10 kV and a peak power output of a few kilowatts,[1] which is generally too low for most radar applications. Other types of RF slow-wave structures have to be employed instead, and these do not have as wide a bandwidth as the helix. A modification of the helix known as the ring-bar circuit can be used if the peak power is less than about 100 to 200 kW. One example is the Raytheon QKW-1671A, a tube suitable for air-surveillance radars. It has a peak power of 160 kW, duty cycle of 0.036, 70-μs pulse width, 45-dB gain, and a 100-MHz bandwidth. The Air Force Cobra Dane phased array radar uses 96 QKW-1723 ring-bar TWTs, each with a peak power of 175 kW and average power of 10.5 kW. The Cobra Dane operates from 1175 to 1375 MHz.

Powers greater than 200 kW are obtained with the coupled-cavity circuit, of which the so-called "cloverleaf" is an example. The bandwidth, however, is less than that of lower power TWTs. The individual unit cells of the coupled-cavity circuit resemble klystron resonant cavities. Several tens of these klystron-like cavities are used for the slow-wave structure of a high-power TWT.[13] There is no direct coupling between the cavities of a klystron, but in the traveling wave tube, coupling is provided by a long slot in the wall of each cavity. There are two slots in each cavity (input and output) that are 180° apart in rotational position so they act similar to a folded waveguide.

An example of a TWT using a cloverleaf coupled-cavity slow-wave structure is the S-band VA-125A. It is liquid cooled and is capable of 3 MW peak power over a 300-MHz bandwidth, 0.002 duty cycle, 2-μs pulse width, and a gain of 33 dB. It was originally designed to be used interchangeably with the VA-87 klystron, except that the VA-125 TWT has a wider bandwidth and requires a larger power input signal because of its lower gain.

The bandwidth of a coupled-cavity TWT can be from 10 to 15 percent. When the power of a TWT is increased, its wide bandwidth decreases. On the other hand, when the power of a klystron is increased, its narrow bandwidth widens. As the power of these two tubes increase, their bandwidths become comparable. With high power, the klystron tends to be the preferred tube since it doesn't experience the stability problems of the TWT.

Although the TWT and the klystron are similar in many respects, a major difference is that there is feedback along the slow-wave structure of the TWT, but the back coupling of the RF energy in the klystron is negligible. If there is a mismatch at the input of the TWT and if sufficient energy is fed back to the input, undesired oscillations can result. To reduce the amount of feedback energy to an insignificant level, attenuation has to be inserted in the slow-wave structure. The attenuation may be distributed or lumped, but it is usually found within the middle third of the tube. The loss introduced to attenuate the feedback also reduces the power of the forward-traveling wave, and is therefore undesirable. The loss in the forward wave can be avoided by the use of *severs,* which are short internal terminations designed to dissipate the reverse-directed power without seriously affecting the power in the forward direction. The number of severs depends on the gain of the tube; one sever is used for each 15 to 30 dB of tube gain.[1]

The efficiency of TWTs is less than that of the klystron because of the necessity for including attenuation or severs. Efficiency is also reduced by the presence of relatively high RF power over an appreciable fraction of the entire structure.

In some traveling-wave tubes with coupled-cavity circuits, oscillations appear for an instant during the turn-on and turn-off portions of the pulse.[1] These are called *rabbit-ear oscillations* because of their characteristic appearance when the RF envelope of the pulse waveform is displayed visually on a CRT. They are undesirable in some military applications since they might provide a distinctive feature for recognizing a particular radar. Weil[1] describes some of the ways rabbit-ear oscillations can be avoided.

TWT MTBF   The mean time between failures (MTBF) is given by Gilmour[11] for nine different types of coupled-cavity TWTs. (The type of TWT, frequency, and power are not mentioned.) The MTBFs of these nine classes vary from a high of 17,800 hours to 2200 hours, with an average of 7000 hours for all nine classes of tubes. TWTs for space applications, which are of lower power than radar tubes, are said by Gilmour to have MTBFs of the order of one million hours.

Hybrid Variants of the Klystron   By judiciously combining the best features of the klystron and the traveling wave tube, an RF power source can be obtained which has bandwidth, efficiency, and gain flatness better than either the conventional klystron or TWT. This is achieved by replacing one or more of the klystron resonant cavities with broader bandwidth cavities that are more like the coupled-cavity circuits used in traveling wave tubes. There have been three variants of the klystron in which this is done: the *Twystron,* the *extended interaction klystron,* and the *clustered-cavity klystron.* Such combinations of klystrons and TWTs are sometimes called *hybrid tubes.*[14]

Twystron   The bandwidth of a klystron is limited by the output resonant cavity. It cannot be made broadband without a decrease in efficiency. Since coupled-cavity slow-wave circuits have broader bandwidth than klystron resonant cavities, replacing the output cavity of a klystron with a TWT coupled-cavity circuit can significantly increase the bandwidth as well as achieve a slight increase in efficiency. Although the output is a TWT slow-wave circuit, the driver portion of the tube (the intermediate cavities and the input cavity) consists of resonant cavities that are stagger tuned. Such a tube is called a Twystron, a trademark name of Varian (now Communications & Power Industries, or CPI). The VA-145 Twystron has demonstrated a 14 percent 3-dB bandwidth (12 percent 1-dB bandwidth), 48 percent efficiency, and 41 dB gain at midband.[1]

Extended Interaction Klystron (EIK) In this device, the single-gap resonant cavity of a klystron is replaced by a resonated slow-wave TWT-like circuit. The use of slow-wave coupled resonators can be applied to the prior cavities as well as the output cavity. The extended interaction amplifier (EIA) klystron can have a high average power; for example, 1 MW CW at *X* band using a five-cavity resonator.[15] It has broader bandwidth than a klystron, but less than that of a TWT. EIKs also have been used for low-power millimeter wave tubes. A 150-W average power 95-GHz klystron, advertised by CPI Canada, is claimed to have a 1.5-kW peak power, 500-MHz bandwidth, 25 percent efficiency, 45-dB gain, and to weigh 4.5 kg. There is also an extended interaction oscillator, or EIO, which has been used at millimeter wavelengths.

Clustered-Cavity Klystron The technique of grouping cavities was extended in what has been called a *clustered-cavity klystron* (CCK). In this tube, Fig. 10.3, the individual intermediate cavities of a multicavity klystron are replaced by pairs or triplets of artificially loaded low-*Q* cavities with *Q*s of one half or one third that of the single cavity they replace.[16] Similar groups of cavities are used in both the EIA and the CCK, but there is

Figure 10.3 Comparison of a conventional staggered-tuned klystron (top) and a clustered-cavity klystron (bottom). The shaded circular regions represent bunching of the electrons.
(From Symons and Vaughan,[16] Copyright 1994 IEEE.)

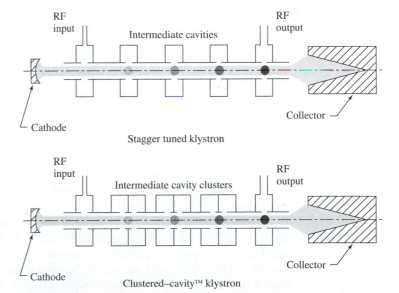

no inductive or other coupling between cavities in the CCK as there is in the EIA. Theory indicates that bandwidths of 30 percent should be obtained in megawatt klystrons using fifteen intermediate cavities in triplets. In practice, 20 percent bandwidths have been observed. This form of structure also provides the greatest bandwidth in the shortest length. Clustered-cavity klystrons might be more complex and costly than a klystron, but they are less complex and of less cost than a comparable TWT or Twystron.[1] Symons,[12] the inventor of the clustered-cavity klystron, states that two of these tubes can be used instead of the two narrower-band klystrons in the AWACS radars. Redundant operation is provided, without a large weight penalty, since either of the clustered-cavity klystrons provide full operational capability similar to the redundancy commonly employed in FAA radars.

10.3 SOLID-STATE RF POWER SOURCES

Use of Solid-State Amplifiers for Radar Transmitters   The solid-state RF power generation device usually of interest for radar application has been the transistor amplifier, both silicon bipolar and gallium arsenide FET. An individual transistor amplifier device is inherently of low power and low gain, but it operates with low voltages and has high reliability.

A single microwave transistor might have an average power capability from a few watts to over a hundred watts, depending on the frequency and the duty cycle. The lower the frequency the greater can be the power. To increase the power, transistors may be operated in parallel, and with more than one stage to increase the gain. A single power *module* might, for example, consist of eight transistors, with four operating in parallel as the final stage, two in parallel as the next to last stage, and two in series as the driver stages. To achieve the high powers required for most radar applications, the outputs of many solid-state devices have to be combined in some manner. Combining of many devices can be achieved with microwave circuitry or by combining in space (radiating from many individual antenna elements of an array antenna).

Solid-state power devices of a given average power cannot be operated at high peak powers as can vacuum tubes. According to Borkowski,[17] "a microwave transistor capable of perhaps 50-W average power cannot handle much more than 100 to 200 W of peak power without overheating during the pulse." For this reason solid-state amplifiers for radar generally operate in the vicinity of 0.1 duty factor, instead of the 0.001 to 0.01 duty factors common with high-power vacuum tube RF power sources. Thus when solid-state devices are employed in radar transmitters they have long pulse widths and require pulse compression to obtain useful range resolution. Long pulses are not always desired by the radar engineer, but they have been accepted as one of the prices to be paid for the use of solid state.

There are at least four ways that solid-state devices can be employed in radar: (1) as a transmitter for a low-power application, (2) as a high-power transmitter where a large number of individual transistors are combined with microwave circuitry, (3) with many modules distributed on a mechanically steered planar array (such as a 3D radar), and (4) with a module at each of the many elements of an electronically scanned phased array (also called an active aperture). In the last two, the power from the many solid-state transmitter modules is "combined in space."

Low-Power Transmitter   The solid-state device is used as a direct replacement for a vacuum tube when the radar waveform is of low power and of high duty cycle or CW. Examples are the FM-CW radar altimeter, doppler (police) speedmeter, and the airborne doppler navigator. The solid-state transmitter has been highly successful in such applications. It has been difficult, however, for solid-state to replace the small magnetron in the civil marine radars found on many ships and pleasure boats because this radar market is highly competitive and low price is important for success. The same appears to be true for the absence of solid-state transmitters for the nonradar application of microwave ovens for the household market.

High-Power Transmitter   The solid-state transmitter has replaced the high-power vacuum tube in some air-surveillance radars. A large number of transistors are combined to produce a single output that feeds a conventional antenna. (This was at one time called a "solid-state bottle," but such transmitters are housed in cabinets which do not resemble "bottles.") Two such transmitters will be described.

One of the first solid-state radars to have its tube transmitter replaced by solid state was the AN/SPS-40, a modest UHF 2D shipboard air-surveillance radar used mainly by United States ASW (antisubmarine warfare) destroyers to provide conventional air-surveillance for keeping track of ASW aircraft.[18] It was developed by Westinghouse, Baltimore (now known as Northrop Grumman). This was a good example of a direct replacement since the tube transmitter operated with a long pulse (60 μs), pulse compression, and a moderate duty cycle (0.018), so that the solid-state transmitter could utilize the same waveform as did the original radar system. The basic transistor building block operated from 400 to 450 MHz, with 400- to 500-W peak power, 8-dB gain, and 55 percent efficiency. A module consisted of two stages with a total of 10 silicon bipolar transistors that produced 2500-W peak power out when the input was 120 W. There were 112 of these modules combined in two groups of 56 each to produce 250-kW peak power and 4 kW of average power. Each of the two groups of 56 modules was housed in its own cabinet. There was a third cabinet with the driver, power supplies, and some other devices. The transmitter was designed so that no damage occurred when a full short circuit was applied across the load. Both liquid and forced air cooling were used; and if the liquid cooling was lost, the transmitter could operate with 80 percent power (200-kW peak) with only air cooling. The loss of one module reduced the transmitter power output by 0.08 dB. The transmitter had good reliability due in part to its built-in spare modules. The solid-state transmitter for this radar cost more and was larger than the vacuum tube transmitter it replaced; but it was considered a success.

The Ramp (Radar Modernization Project) radar system was an L-band (1250 to 1350 MHz) air-traffic control Primary Surveillance Radar (PSR) located at major airports across Canada.[19] It was developed by Raytheon Canada and had a range of 80 nmi and an altitude coverage of 23,000 ft against a 2 m^2 aircraft target with 80 percent probability of detection. It used a solid-state transmitter with a peak power of 28 kW and an average power of 1.2 kW, which corresponded to a duty cycle of 0.068.

There were a total of 14 modules used in the RAMP PSR. Each module consisted of 42 identical class-C 100-W peak-power silicon bipolar transistors arranged in a 2-8-32 configuration to produce 2350 W of peak power. As described by Merrill,[20] "The transistors were arranged in a one-driving-four 'unit amplifier' format, with eight unit amplifiers in parallel so that 10 transistors were drivers while 32 were output devices." The 50-pound air-cooled module had a measured efficiency greater than 25 percent and a power gain greater than 16 dB.[17]

The 14 modules were combined as pairs to form seven transmitting channels. Only six of the seven channels were needed to meet the system requirement for a minimum peak power of 21 kW. The extra seventh channel permitted maintenance and repair to be performed on a failed channel while the remaining six channels were continuously available. The extra channel, therefore, allowed the radar to maintain a high availability. [The theoretical power output when N out of 7 channels are operating is $P_{out} = P_7 (N/7)^2$, where P_7 = the power delivered by all seven channels.] The antenna for this radar was a 33 ft wide by 22 ft high reflector with 33.5-dB gain. It rotated at 12 rpm.

Modules Arranged on a Mechanically Scanned Planar Array, AN/TPS-59   Individual transmitter modules can be arranged on a mechanically scanning array antenna by placing one module at each element, but it has been more usual in such a radar to employ one module at each row of the antenna. This is the arrangement in the AN/TPS-59,[17] a transportable *L*-band 3D air-surveillance radar developed by GE, Syracuse, N.Y. (now Lockheed Martin) for the U.S. Marine Corps for air defense and ground-control of intercept (GCI). It was designed to detect a 1-m^2 fluctuating target within a 200-nmi range with 90 percent probability of detection, but it was also required to cover a volume out to 300 nmi in range and an altitude up to 100,000 ft. The rotating planar array antenna is 30 ft high by 15 ft wide and consisted of 54 rows, each with 24 dipole elements. The radar can operate within 1200 to 1400 MHz with 54-kW peak power and 9.7-kW average power, which results in the relatively high duty cycle of 0.18. A single pencil beam is electronically scanned over an elevation angle of 20° as the antenna rotates 360° in azimuth.

At each row is a transceiver, which is a miniature radar containing transmitter, receiver preamplifier, duplexer, phase shifter for steering in elevation, logic control, cooling, and power supply. Each transmitter module has ten 100-W amplifier units consisting of two 55-W silicon bipolar transistors (with 7-dB gain) driven by a smaller 25-W device. There are a total of 540 modules on the antenna.

Fixed-site variants of this radar are known as the GE-592 and the AN/FPS-117. The latter uses a 24 ft by 24 ft antenna with 44 rows. It is a minimally attended radar whose antenna operates within a radome for use in northern regions. The tactical mobile version of this radar is the TPS-117, which was shown in Fig. 1.8.

Active Aperture, Electronically Steered Phased Array   An example of the composition of a T/R module as might be used for an active-aperture phased array antenna is illustrated in Fig. 10.4. The T/R switches select between the transmitter and the receiving paths. The circulator, which might be the largest and heaviest component on the module, performs the function of the duplexer. The receiver protector (which is a diode limiter) provides further protection of the low-noise amplifier (LNA). The same phase shifter is

Figure 10.4   Example of the composition of a T/R module that might be used for an active-aperture phased array radar.

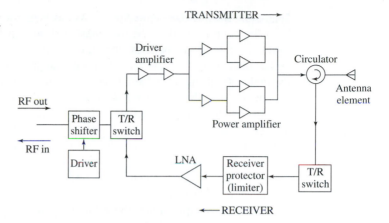

used for both transmitting and receiving. Many other parts of a T/R module are not shown in this illustration, such as the module controller and the power conditioner.[21] The module controller obtains a beamsteering command from a central computer and calculates the correct settings for the phase shifter. To minimize power consumption, the power amplifiers and the LNA might be gated off when the controller is on. The module controller might also perform self-testing and reporting of the status of the module. The power conditioner is important in keeping the efficiency of the module as high as practical.

The Pave Paws electronically steered array radar, also known as the AN/FPS-115, is a UHF radar developed by Raytheon that was the first all solid-state active aperture electronically steered phased array radar. Its function was to detect and warn of sea-launched ballistic missiles fired at the United States. This radar was discussed in Sec. 9.9. It operated from 420 to 450 MHz with a peak power of 600 kW and an average power of 150 kW, which corresponds to a duty cycle of 0.25. Its diameter was 72.5 ft, with room to grow to 102 ft. An individual module delivered 340 W of peak power with 39 percent efficiency.[17,22] The pulse width was 16 ms. There were two phased array faces per site to cover approximately 240° in azimuth. Each antenna face had the capability of operating with 5354 elements, but only 1792 active transceiver modules were used for transmission. Extra elements and a narrow beam were used on receive. (The remaining elements were for future growth.) A transmitting module was made up of seven Class-C silicon bipolar transistors in a 1-2-4 amplifier configuration. It has been said[22] that "system performance is maintained with as many as 200 modules per face inoperative." An enlarged version of the Pave Paws radar replaced the parabolic torus antennas used in the original Ballistic Missile early Warning System (BMEWS). This version had 850-kW peak power with a 0.30 duty cycle.

THAAD Ground Based Radar   This is an example of an active-aperture phased array radar with 25,344 elements, a very large number. It was discussed in Sec. 9.9 and its picture shown in Fig. 9.36. With such a large number of elements, the cost of an individual T/R module must be kept small. If, for example, the cost of an individual T/R module were $1000 each, the total cost of just the modules for this radar would be twenty-five million dollars.

Solid-State Devices Used in Radar[17]   As has been mentioned, the transistor amplifier has been the device usually used for radars with high-power solid-state transmitters. At the lower microwave frequencies, the silicon bipolar transistor is usually used; and at the higher microwave frequencies, it has been the gallium arsenide (GaAs) FET transistor. At the higher frequencies, the solid-state power device can also be incorporated as part of a microwave monolithic integrated circuit (MMIC).

The silicon bipolar transistor has been used at microwave frequencies below about 3 GHz (S band). According to Olson,[23] a typical internally matched silicon bipolar transistor operating as a class-C amplifier over the frequency range from 2.7 to 2.9 GHz with a pulse width of 50 μs, 10 percent duty cycle and a supply voltage of 40 V, can have a minimum power output greater than 100 W, a minimum gain of 6.5 dB, and a minimum efficiency of 35 percent.

The power output of the silicon bipolar transistor decreases with increasing frequency. At the higher microwave frequencies, the gallium arsenide FET, often in the form of a MESFET (metal semiconductor field-effect transistor) is capable of greater power than the silicon bipolar transistor. Transistors should be operated at a high duty cycle since the peak power output for pulsed operation cannot be significantly increased over that of CW operation. At *X* band, the power output of such devices might be 10 W. Other devices that have been considered for solid-state power sources include[23] GaAs HEMT (high electron mobility transistor); GaAs-based pseudomorphic HEMT, or PHEMT; GaAs heterojunction bipolar transistor (HBT); and devices employing unconventional materials such as silicon carbide and semiconductor diamond for high temperature, high power operation.[24]

At the higher microwave frequencies where compactness in size is desired, the microwave monolithic integrated circuit (MMIC) has been of interest for T/R modules. Active and passive circuit elements are formed on a semi-insulating semiconductor substrate, usually GaAs to create system architectures that are difficult to realize with less integrated technologies. The benefits of MMIC are due in large part to the batch processing of both the active and passive components on the same substrate. Borkowski[17] lists the advantages of MMIC for radar as low cost, increased reliability, increased reproducibility, and small size and weight. The nonrecurring costs of engineering design of MMICs, however, can be high and design might require a relatively long time. Since MMICs are not well suited for tweaking of the circuits once manufactured, the designs must be tolerant to variations in the processing. Olson[23] states that the power available from MMICs is about 10 W in the frequency range from 3 to 10 GHz and then decreases with increasing frequency at a rate of 6 dB per octave.

Advantages of Solid State   The solid-state transistor amplifier has been of interest for radar transmitter applications because of the following:[17]

*   Individual solid-state devices have long MTBF (mean time between failures).
*   Maintenance is relatively easy with the modular construction of solid state. (A defective module is pulled out and replaced by another.)
*   Very wide bandwidths can be obtained (up to 50 percent or more).
*   No cathode heater is required (no warm-up time and no heater power to reduce the overall transmitter efficiency).

- Solid-state devices operate at much lower voltages (order of tens of volts) than RF power tubes (order of tens of kilovolts).

- No pulse modulator is required. (When operated as a class-C amplifier, the transistor is self-pulsing in that it automatically turns on when the RF drive signal is applied and it automatically turns off when the drive signal is turned off.)

- Solid-state transistor amplifiers have low noise and good stability (important for detection of small targets in the presence of large clutter echoes).

Another attribute often claimed for a solid-state radar is that many individual devices can fail without significant effect on the overall transmitter power (graceful degradation). The power output in dB varies as 20 log r, where r = ratio of the number of operating devices to the total number of devices.[17] This is correct in principle, but in practice there can be catastrophic failure modes for a solid-state transmitter and, eventually, the modules that fail must be replaced, even if they fail "gracefully." Except for its long pulses and high duty cycle the solid-state transistor is well suited for use in an active-aperture phased array where each element contains its own transmitter/receiver module.

When solid-state transmitters were first proposed for radar, it was said they would be lighter in weight and lower in cost than vacuum tube transmitters. It is not obvious that this has occurred. In some cases in the past when a solid-state transmitter replaced a high-power vacuum-tube transmitter in an existing radar, the solid-state transmitter was heavier and cost more.

Systems Implications of Solid-State Devices   Pulse radars typically have been characterized in the past by low duty-cycle waveforms with typical values of duty cycle ranging from approximately 0.001 to 0.01. Power vacuum tubes are well suited for low duty cycles. For a given average power their peak power can be increased by a factor of 1000 or more with little penalty other than the practical problem of making the insulators able to stand off the higher voltages. Semiconductor power devices, on the other hand, cannot be efficiently operated at low duty cycles. For a given average power, the peak power might be less than ten times the average power. Thus replacing a vacuum-tube transmitter with a solid-state device usually means the radar must use high duty-cycle waveforms. High duty cycles mean long pulse widths which have the disadvantage of long minimum ranges. When a short minimum range is important, more than one pulse width might have to be transmitted. The long pulses of solid-state transmitters require pulse compression to achieve good range resolution. The technology of pulse compression has been widely used in radar, but it does have some limitations that short-pulse waveforms do not have. Seldom, however, is the cost of pulse compression and the cost of multiple waveforms considered as a solid-state transmitter cost, even though they increase the total cost of the radar system and are not needed with many vacuum-tube transmitters.

There are at least two reasons why the cost of radars with solid-state transmitters is often higher than those with comparable vacuum-tube transmitters.[12] One is that the efficiency of solid-state transmitters is generally less than that of a high-power vacuum-tube transmitter. The other is that the cost of obtaining power is greater when the total power is obtained with more than one power-source unit. An advantage of low-power solid-state devices for computers and low-power transmitters is that they can be made very compact.

A lot can be placed on a small chip. When the solid-state device has to handle high power, however, as it would for a radar transmitter, there can be a problem in dissipating the heat generated by the power sources. Solid-state devices that have to handle high power have to be spread out over a greater extent of circuit board area to avoid exceeding the heat transfer limits. The size and weight of the solid-state amplifier are therefore determined by the power densities that the amplifier can handle rather than by the size of the individual components. Thus the advantages of photolithographic fabrication of high-density low-power solid-state integrated circuits are not available when the power is high. Spreading the solid-state devices over a wide area in order to provide for heat dissipation can result in lower efficiency due to higher power-combining losses in transmission lines and in combiners. Dissipation of the higher heat levels requires heavier heat sinks and results in a heavier transmitter.[12]

It has been known for a long time that the cost of a single high-power vacuum tube varies as the square root of the output average power. Thus, the lowest cost tube transmitter for a given total power output is the transmitter that uses a single high-power tube rather than multiple tubes. Symons[12] indicates that the cost of a transmitter made up of multiple devices, such as is necessary when using high-power solid-state devices, increases almost linearly with the number of devices. At high powers, therefore, it should be expected that the single vacuum-tube transmitter will be of lower cost than a solid-state transmitter made up of many modules generating the same total power.

Individual solid-state devices can have a much longer life and lower failure rate than an individual vacuum-tube power source. The life of a solid-state transmitter, however, is determined not by the life of a single transistor or a single module but by the life of all the modules and the many other components that make up a transmitter. Vacuum-tube transmitters, when well designed and properly operated, have been known to achieve very long lifetimes, mentioned previously in this chapter. Maintenance of a solid-state transmitter should be easier than maintenance of a high-power vacuum tube transmitter, but the life of a properly designed and operated vacuum tube transmitter should be quite long and not be a serious system problem.

Compared to the high-power vacuum tube, the solid-state transmitter has advantages, but it also has some serious limitations. It is not obvious that current solid-state transmitter technology will cause the high power RF vacuum tube to disappear. Solid-state will be used when its particular advantages are more important than its limitations and higher cost. As happens often in engineering, the radar systems engineer has a number of choices when it comes to selecting the type of RF power source to use for a particular radar system application. The solid-state transmitter is just one of many possibilities that have to be considered for any particular application—unless the customer insists otherwise.

10.4 MAGNETRON

The magnetron has been the only high-power RF power source used for radar that is a power oscillator rather than a power amplifier. It is a crossed-field device in that its electric field and its magnetic field are perpendicular to one another. The compact size and efficient operation of the magnetron at microwave frequencies allowed radars to be small enough to fly in military aircraft, be mobile for ground warfare, and even be used on submarines.

Coaxial Magnetron   A major improvement in the power, efficiency, stability, and life of the original magnetron architecture came about with the coaxial magnetron introduced in the 1960s. The key difference was the incorporation of a built-in stabilizing cavity surrounding the conventional magnetron. Figure 10.5 is a sketch of the cross section of the circular geometry of a coaxial magnetron. At the center is the "fat" oxide-coated cathode. Surrounding the cathode are a number of RF resonant cavities defined by the radial vanes. Between the cathode and the resonant cavities is the interaction space where the electrons interact with the d-c electric field and the static magnetic field in such a manner that the electrons give up their d-c energy to the RF field. The crossed electric and magnetic fields cause the electrons to be "bunched" almost as soon as they are emitted from the cathode. After bunching, the electrons move along in a traveling-wave field that is almost the same speed as that of the electrons.

The frequency of a coaxial magnetron can be changed by mechanically moving one of the end plates, called a tuning piston, of the stabilizing cavity. (The end plate is located in the plane of the paper of Fig. 10.5 and is not shown.) The tuning piston can be positioned mechanically from outside the vacuum by means of a vacuum bellows.

There is also an inverted form of the coaxial magnetron (an inside-out) version with the anode and resonant cavities in the center and the cathode around the outer perimeter of the tube. It is supposed to provide better performance at higher frequencies when the cavity becomes small and the regular type of coaxial magnetron would result in a small cathode.

Figure 10.5   Cross-sectional sketch of a coaxial cavity magnetron.

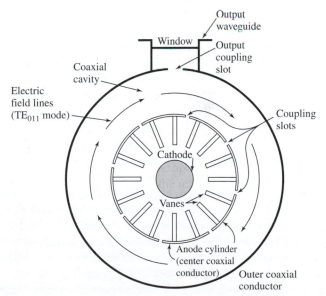

π Mode   It is not easy to describe the theory of operation of a magnetron in a simple manner, so no attempt is made to do so here. A magnetron can oscillate at a number of different, closely spaced frequencies due to various possible configurations of the RF field that can exist between the cathode and the resonant cavities. These different RF field configurations, along with coupling among the many cavity resonators of the magnetron, result in different modes of oscillation. The magnetron can shift, almost unpredictably, from one mode to another (which means from one frequency to another) as the voltage changes or as the input impedance that the magnetron sees changes. The shift from one mode to another (often called *moding*) is especially bad since it can occur when the radar antenna scans and views different environments. It is important to avoid moding.

The preferred magnetron mode of operation is the so-called *π mode* that occurs when the RF field configuration is such that the RF phase alternates 180° (π radians) between adjacent cavities. The advantage of the π mode is that its frequency can be more readily separated from the frequencies of the other possible modes. (An *N*-cavity magnetron has $N/2$ possible modes of oscillation. The π mode oscillates at only a single frequency, but the other modes can oscillate at two different frequencies, so that the magnetron can oscillate at a total of $N-1$ different frequencies.)

In the coaxial magnetron, the output of every other resonant cavity is coupled to a stabilizing cavity that surrounds the anode structure, as indicated in Fig. 10.5. The output power is then coupled from the stabilizing cavity. The cavity operates in the TE_{011} mode with the electric lines closed on themselves and concentric with the circular cavity. The RF current at every point on the circumference of the cavity has the same phase, so that the alternate slots which couple to the stabilizing cavity are of the same phase as required for π-mode operation.

Coaxial Magnetron Life   The power that can be produced by a magnetron depends on its size. A larger size means more resonators, which then makes it more difficult to separate the various modes of oscillation in a conventional magnetron. The coaxial magnetron, however, with stabilization controlled by the TE_{011} outer cavity permits stable operation with a larger number of resonant cavities, and thus with greater power. The anode and cathode structures of a coaxial magnetron can also be bigger than those of a conventional magnetron, which further allows operation at larger powers. The larger structures permit more conservative design, with the result that the coaxial magnetron exhibits longer life and better reliability than conventional magnetrons, in addition to having a more stable operation. The operating life of a coaxial magnetron tube can be between 5000 and 10,000 hours, which represents a five- to twenty-fold increase compared to conventional magnetrons.[25] It has been said[26] that a VMS-1143 S-band coaxial magnetron operating at 3 MW of peak power in an AN/FPS-6 height-finder radar exceeded a life of 50,000 hours. This tube was likely one-of-a-kind, but it indicates that a magnetron does not necessarily have to have a short life, as once was the case.

Just after World War II and during the 1950s, the life of the early magnetrons was as low as 200 hours mean time between failures, which probably explains why some have had the impression that power vacuum tubes were inherently unreliable and of short life. The demonstrated long lifetimes of the coaxial magnetron and even longer lifetimes of the modern klystron and TWT linear-beam tubes offer ample evidence that the use of power vacuum tubes need not result in unreliable radars. (The reader might try to recall how often, when watching TV or listening to the radio, the transmission shut down and went off the air because of a failure in the tube transmitter.)

Systems Aspects of the Magnetron   The magnetron transmitter was used in a large number of different types of radar transmitters. At one time it was the most popular radar transmitter. Its use, however, has diminished greatly because of the more demanding requirements of modern radars that it cannot readily meet, but which can be satisfied by other RF power sources. Weil[27] describes the problems encountered with the use of the magnetron in his chapter "Transmitters" in the *Radar Handbook.* The major limitations of the magnetron are its limited average power and poor ability to see moving targets in heavy clutter.

Although the magnetron can produce a peak power of several megawatts, its average power is limited to about one or two kilowatts. This may be sufficient for some medium-range radars and for civilian air-traffic control radars that use large antennas, but it is not large enough for many military radar applications. The magnetron is usually smaller in size than many other types of RF power sources, but this is due in part to its low average power.

During World War II, air-surveillance radars did not have an MTI (moving target indication) capability, except for the S-band AN/CPS-1 MEW (Microwave Early Warning) radar which only became available in small numbers near the end of the war. In order for radar to detect aircraft in World War II, targets had to be in the clear outside of the clutter. The bomber aircraft of that time unknowingly cooperated since they were not designed to fly at low altitude below the radar coverage. The early analog MTI radars could use only a single or a double delay-line canceler, so they were limited in MTI improvement factor (or clutter attenuation) to about 20 dB. The magnetron itself limits the improvement factor to perhaps 30 or 40 dB. Therefore, the magnetron was not the limiting factor in early MTI radar performance. This changed when digital signal-processing of MTI signals became available that allowed much better values of MTI improvement factor. Now power amplifiers such as the klystron, TWT, or the transistor have to be used—and not the magnetron—in order to detect small moving targets in heavy clutter, consistent with the full capabilities of digital doppler-signal-processing.

There are several other factors that are not favorable with the magnetron. Its pulse widths are limited from just under 0.1 μs to about 100 μs, but this is usually not a problem. However, modulating the pulse with frequency or phase to achieve pulse compression is quite difficult with a magnetron and has not been done operationally. The signal is not coherent from pulse to pulse so that in MTI radar the phase of the coho (reference oscillator) in the receiver has to be reset every time a new pulse is transmitted. The magnetron frequency drifts with time, which requires that the frequency of the local oscillator in the receiver be continually tuned to the transmitter's frequency (whatever it might be). Magnetrons are noisy and can produce electromagnetic interference at frequencies outside their design frequency range.

Marine Radar Magnetrons   The magnetron has proven to be a tube well suited for civil marine radars. These are small devices that generate peak powers between 3 and 75 kW with low average powers of a few watts to a few tens of watts. The marine radar customer demands reliability. When a commercial vessel goes to sea, its captain wants the radar to still be operating when the ship returns to its home port. An example of a magnetron for a civil marine radar is the third generation MG5241 manufactured by EEV of Chelmsford, England. It is an 18-cavity *X*-band magnetron, which produces a peak power of 12.5 kW with an efficiency of 43 percent. The manufacturer claims an expected typical life of over 10,000 h and guarantees a minimum life of 3000 h. Its weight is 625 g (1.4 lb) and it has a volume of 315 cm^3. It operates at a fixed frequency within the band 9.38 to 9.44 GHz.

10.5 CROSSED-FIELD AMPLIFIERS

The crossed-field amplifier (CFA) resembles the magnetron in that it employs a magnetic and electric field that are perpendicular to one another.[1,28] It is also similar in appearance to the magnetron, except that the RF circuit is interrupted to provide the input and output connections, Fig. 10.6. CFAs have high efficiency (40 to 60 percent), use lower voltage than linear-beam tubes, and are lighter in weight and smaller in size. They are of wide band (10 to 20 percent), with high peak and average power, and have good phase stability; but their gain is relatively low. They are sometimes used with a high-gain, but lower power, TWT that serves as the driver for the CFA.

CFA Operation There are several different types of crossed-field amplifiers, and they all employ a slow-wave circuit, cathode, and input and output ports. For radar, CFAs usually have the form diagrammed in Fig. 10.6, which is reentrant with distributed emission. Distributed emission means that, like the magnetron, the cathode is adjacent to the full length of the RF structure. Electrons are emitted from the cylindrical cathode, which is coaxial to the RF slow-wave circuit that acts as the anode. The electrons, under the action of the crossed electric and magnetic fields, form into rotating electron (space-charge) bunches, or spokes. These bunches drift along the slow-wave circuit in phase with the RF signal

Figure 10.6 Simple representation of a crossed-field reentrant CFA showing the drift region and the control electrode.

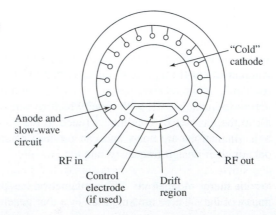

and transfer their d-c energy to the RF wave to produce amplification. The spent electrons that remain after their energy is extracted are collected by the slow-wave anode structure. The electrons that are not collected after their energy is extracted at the output are permitted to reenter the RF interaction area at the input, which is the reason such a tube is called reentrant. Some of the reentering electrons contain modulation (bunched electrons) that will be amplified in the next pass around the RF circuit. To prevent this, a drift space is included between the output and input ports. In the drift space, space charge forces cause the electron bunches to disperse, removing any modulation that accompanies the reentering electrons.

Cold-Cathode Emission   In high-power CFAs, the electrons can be generated by cold-cathode field emission rather than thermionic emission with a heated cathode. Cold-cathode emission requires the presence of both the d-c voltage between cathode and anode as well as the RF drive signal applied to the tube. The initiation of electron-emission build-up results from the relatively small number of free electrons present near the cathode surface. Emission is sustained by those electrons not collected by the anode and which are returned to the cathode by the action of the RF field and the crossed electric and magnetic fields. When these electrons strike the cathode they produce secondary electrons that maintain the electron emission process. There is little pulse-to-pulse time jitter in the starting process, and the buildup time is quite rapid (<10 ns).[28]

Insertion Loss   A CFA has low insertion loss, perhaps less than 0.5 dB. Sometimes this can be an advantage in a multistage transmitter. By omitting the application of d-c voltage to the final stage of a multistage transmitter, the lower level RF drive can be fed directly through the final stage with little attenuation. This allows a radar with such a transmitter to have two power levels, which might be of interest for some system applications.

The low insertion loss of a CFA means that the RF drive power will appear at the output tube with little attenuation. In a low-gain amplifier, such as the CFA, the input power that appears at the output can be a sizable fraction of the total, perhaps one-tenth or more. The *conversion efficiency* of a CFA is defined as

$$\text{Efficiency} = \frac{\text{RF power output} - \text{RF drive power}}{\text{d-c power input}} \qquad \text{[10.2]}$$

When the RF drive power is included in the output power rather than omitted as it is in Eq. (10.2), it is sometimes called the *power-added efficiency*. That is, power-added efficiency is the total RF power out divided by the d-c power in. Tube engineers like to quote the power-added efficiency instead of Eq. (10.2), since it results in a higher value.

Forward- and Backward-Wave CFAs   The interaction of the electron bunches with the RF signal may be with either a forward traveling wave or a backward traveling wave. The type of interaction is determined by the slow-wave circuit. A forward-wave interaction takes place when the phase velocity and the group velocity of the propagating signal along the slow-wave circuit are in the same direction. (The *group velocity* is the velocity with which energy is propagated along the slow-wave circuit, and the *phase velocity* is the velocity of the RF signal on the slow-wave circuit as it appears to the electrons.) To achieve amplification, the phase velocity must be near the velocity of the electron stream. A backward-wave interaction, as in the CFA device known as the *amplitron,* takes place when the phase velocity and the group velocity are in opposite directions. The forward-wave CFA can operate over a broad range of frequency with a constant anode voltage, and with only a small variation in the output power. On the other hand, the power output of a backward-wave CFA, with a constant anode voltage, varies with frequency. It is like a voltage-tuned amplifier. The power output can vary 100 percent for a 10 percent change in frequency.[29] It is possible, however, with conventional modulator techniques to operate a backward-wave CFA over a wide band with little change in output power. The line-type modulator can be compensated for the power variation with frequency and can hold the variation of output power within acceptable levels. When the anode voltage is properly adjusted, the bandwidth of a backward-wave CFA might be 10 percent. Operation of a forward-wave CFA is more like that of a TWT, and it can obtain bandwidths up to about 20 percent.[28]

High-Gain CFA[1,30]   The gain of conventional pulsed crossed-field amplifiers is typically 8 to 15 dB. By designing the cold cathode as a slow-wave circuit, and introducing the RF drive at the cathode emitting surface, it is possible to achieve 15 to 30 dB of gain in a high-power pulse CFA with power, bandwidth, and efficiency comparable to conventional designs. The RF output is taken from the anode slow-wave circuit. This is known as a *high-gain CFA* or a *cathode-driven CFA.* Its higher gain means that less drive power is required. The cathode-driven CFA can also be made to have lower noise than a conventional CFA by 10 to 20 dB if a slightly different configuration is used. It has been said[31] that signal-to-noise ratios greater than 70 dB/MHz were obtained, which are claimed to be comparable to linear beam tubes and 20 to 30 dB better than conventional CFAs. However, both high gain and low noise cannot be obtained simultaneously with the same configuration.

A 1.2-MW peak power *S*-band cathode-driven CFA operating from 3.1 to 3.5 GHz (12 percent bandwidth) achieved 23-dB gain at saturation and 30 dB at reduced peak power, with an efficiency of 53 percent when employing a drive power of 6 kW.[29,32]

Modulating a CFA   The CFA can be pulsed-modulated by turning on and off the high voltage between the anode (at ground potential) and the cathode (at a large negative potential). This is called *cathode pulsing.* A forward-wave CFA with cold-cathode electron emission, however, can be pulsed without the need for a high-power modulator as required for cathode pulsing. The d-c operating voltage is applied continuously between cathode and anode. The tube remains inactive until the application of the RF input pulse starts the electron emission process, and amplification then takes place. At the end of the RF drive-pulse the electrons remaining in the tube must be cleared from the interaction area to avoid causing feedback that generates oscillation or noise. In reentrant CFAs, the electron stream can be collected after removal of the RF drive-pulse by mounting an electrode in the cathode, but insulated from it, in the region of the drift space between the RF input and output ports. This is called a *cutoff,* or *control electrode,* and was indicated in Fig. 10.6. A short positive pulse is applied to the cutoff electrode at the termination of the pulse to quench the remaining electron current. This method of modulation is called *d-c* *operation.* Weil[1] has said that d-c operation has been seldom used because it requires a large capacitor bank to limit droop of the pulse and because an arc in a d-c operated tube requires a crowbar protective device to quench the arc, which interrupts the operation for a few seconds instead of for only a single pulse.

It is also possible to turn the CFA on and off with just the RF drive pulse, without the need for a positive pulse applied to the cutoff electrode at the end of the drive pulse. The is called *RF keying,* and is a simple method for modulating a CFA. It has not been widely used, however, since there are factors other than modulator size that determine the best method of modulation.[1]

System Implications   CFAs in the past have been employed for high-power air-surveillance radars, for power applied at the subarrays of a high-power phased array radar, and as a power booster following a magnetron oscillator. The low gain of CFA, however, requires that there be multiple stages. When used in an amplifier chain, the CFA is generally found in only one or two of the highest power stages. It can be preceded by a medium-power, high-gain traveling-wave tube, a combination that takes advantage of the best qualities of both tube types. The TWT provides high gain and the CFA allows high power to be obtained with high efficiency and lower voltage.

The electrons in the rotating space-charge bunches do not have identical velocities so that random currents are induced in the slow-wave structure, which generates broadband noise. The noise levels in a CFA, therefore, will be higher than those of a linear-beam tube by about 20 to 30 dB.[32]

The possibility of pulsing the CFA in a simpler manner than with a cathode pulser has been an attractive feature; but d-c operation and RF keying have limitations that have made cathode-pulsing preferred even though it is heavier. Since high voltage remains on the tube between pulses with d-c operated CFAs, serious levels of noise may be generated even though there is only a small amount of beam current flowing through the tube. With cathode pulsing, on the other hand, high voltage is removed between pulses so that noise is not normally encountered. The increased interpulse noise of the CFA without a cathode pulser, as well as its high level of in-band spurious noise can prevent attaining good MTI performance (large MTI improvement factors) and low time-sidelobes in pulse-compression systems.

The CFA does not seem to generate as much interest as it once did, and it appears to have been overtaken by linear-beam tubes when high performance is required.

10.6   OTHER RF POWER SOURCES

In addition to the RF power sources already mentioned, there have been several other RF devices that been used or proposed for radar application.

Microwave Power Modules (MPM)[33]   The microwave power module, outlined in Fig. 10.7, combines in a single unit a solid-state MMIC (monolithic microwave integrated circuit) amplifier driving a moderate-power helix traveling-wave tube, along with an integrated

Figure 10.7 Microwave power module.

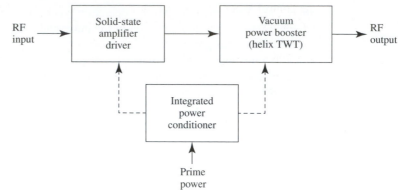

power conditioner in a compact lightweight package. It provides an RF power source with high efficiency, wide instantaneous bandwidth, low noise, and average power levels from several tens to several hundreds of watts. It is smaller and lighter than comparable TWT and solid-state RF power sources and is also capable of operating at high ambient temperatures. The gain of an MPM (nominally 50 dB) is divided between the solid-state driver and the TWT power booster in the ratios from 20/30 to 30/20. The MPM is claimed to be competitive to a TWT or a solid-state power amplifier for radar applications as well as for electronic warfare. It leverages the advantages of both solid-state and vacuum-tube technologies while minimizing their disadvantages.

The MPM seems best suited for the higher microwave frequencies, from S band to K_u band, approximately 2 GHz to 40 GHz. It is expected that an MPM operating over a frequency range from 6 to 18 GHz would produce a peak power of 100 W (up to a 100 percent duty cycle) and be able to be packaged in a space 5/16 in × 4 in × 6 in. The MPM is claimed to have an efficiency three times better than that of a comparable solid-state power amplifier and an improvement in noise 30 dB better than a TWT alone. It has been suggested that the MPM can be employed in *C*- and *X*-band phased arrays, synthetic aperture radars for unmanned airborne vehicles (UAV), missile seekers, and airborne turbulence-warning radars.

Although the MPM has been considered mainly for wideband (nonradar) applications operating over one or more octaves, it can also be used for radar with more usual bandwidths. An example is a C-band module described by Smith, Armstrong, and Duthie[34] which can operate over the band from 4 to 6 GHz with an efficiency greater than 35 percent and CW power greater than 125 W. If its operation is restricted to frequencies from 4.6 to 4.9 GHz, the output CW power is 200 W and its efficiency is 50 percent. It is of low weight and small size, producing more than 70 W per pound and 4 W per cubic inch. A limitation for radar application is that it requires a high duty factor (preferably unity or perhaps 50 percent).

Grid-Control Tubes   This is the microwave version of the classical triode or tetrode vacuum tube. These tubes employ a cathode to generate electrons, an anode to collect them, and one (if a triode) or two (if a tetrode) control grids in between. A voltage applied to the control grid acts as a gate, or valve, to control the number of electrons traveling from the cathode to the anode. By varying the voltage on the control grid, the number of electrons that reach the plate also varies. The process by which the electron density of the electron stream is modulated by the signal on the control grid to produce amplification is called density modulation. The grid-controlled tube is capable of high power, wide bandwidth, good efficiency, and inherent long life; but it is of low or moderate gain. It can be used only at the lower radar frequencies.

The performance of density modulated grid-control vacuum tubes is limited by the time it takes for the electrons to transit from the cathode to the anode. The transit time should be small compared to the period of the RF signal to be amplified. For this reason, grid-control tubes have been limited to frequencies below 1 GHz. To minimize transit-time effects, the complete RF input and output circuits and the electrical interaction system are within the vacuum envelope.

Grid-control tubes have been used with success in HF over-the-horizon radar, and in VHF and UHF radars, including the U.S. Navy's E2C airborne early warning radar,[35] and the U.S. Air Force's AN/FPS-85 satellite surveillance radar.[36] Outside of radar, the grid-control vacuum tube has been widely used in commercial VHF and UHF TV transmitters. It is not likely, however, that the grid-control tube will be used much for radar in the future. Solid-state transmitters, even though they may cost more than vacuum tubes, seem to be preferred by those who buy radars at the frequencies where grid-control vacuum tubes have been used previously.

Inductive Output Tube (IOT), or Klystrode[3]   This device dates back to work in 1939 by Andrew Haeff,[37] who tried to extend density modulated vacuum tubes to microwave frequencies. Haeff called his device an *Inductive Output Tube* (IOT). He described an IOT that produced 100-W CW power at 450 MHz with 35 percent efficiency and 10-dB gain. This was quite good for its time. Nothing further happened since interest then was mainly on velocity modulated linear-beam tubes.

The IOT was independently reinvented about 30 years later by D. H. Preist and M. B. Shrader under the name *Klystrode*. The name was chosen to signify that the device resembled the klystron in the region between the anode and the collector and it resembled a tetrode in the region between cathode and anode. According to Priest and Shrader,[38] Haeff realized that the conventional triodes and tetrodes were limited by their use of intercepting grids, so in his IOT he replaced the wire grids with an aperture that did not intercept the electrons. A coaxial magnetic field confined the electron stream, as in a klystron or TWT. The action of the grid was to density modulate the electrons, as in a conventional triode, to form bunches of electrons. RF energy was removed from the bunched electron beam by passing it through a resonant cavity that extracted the kinetic energy of the high-velocity electrons. The spent electrons were not collected by the resonators, but by a separate collector. The IOT was like the klystron, except that the density modulation of the electron beam was performed by a grid rather than by an input resonant cavity and drift space that induced velocity modulation on the electrons as in the klystron.

The Klystrode was developed mainly for UHF TV. It can produce many tens of kilowatts average power at high efficiencies (50 to 60 percent) with power gains of 18 to

25 dB. Although it has not been used in radar, it has the potential to provide better performance for UHF radars than the tetrodes used previously.

Constant-Efficiency Amplifier (CEA)   One reason for mentioning the IOT and the Klystrode in the above is that they can be modified to produce an amplifier whose efficiency is approximately independent of the power output. Such a device would be of interest for radar when it is desired to shape the radar pulses in order to reduce the time sidelobes of pulse-compression waveforms or to reduce the out-of-band interference generated by a rectangular waveform. Conventional radar RF power sources, such as discussed in this chapter, cannot operate with shaped pulses without a loss in efficiency. A so-called *Constant Efficiency Amplifier* (CEA), however, can be obtained by combining the Inductive Output Tube (IOT) with a multistage depressed collector similar to that used in klystrons and TWTs.[12] The CEA was developed for the television industry. It is claimed that with a CEA the prime power requirements of a TV transmitter can be reduced to one-half compared to a conventional tube transmitter[39] and to one third the prime power of a silicon-carbide solid-state transmitter.[40] The CEA, however, does not operate at frequencies higher than UHF.

Gyrotrons[41–43]   The power available from the microwave-radar power sources discussed thus far in this chapter decreases as the frequency is increased. This is due to the resonant structures of the devices becoming smaller with increasing frequency (decreasing wavelength) and the difficulty in removing heat dissipated in small structures. Consequently the power output of a particular type of generator varies approximately inversely as the square of the frequency. The *gyrotron,* on the other hand, does not have this type of frequency dependence since it does not employ a resonant slow-wave structure. Instead, it is based on a fast-wave structure such as a smooth circular tube (one where the phase velocity of the electromagnetic wave is greater than the speed of light). The diameter of the gyrotron circuit can be several wavelengths and the electron beam need not be placed close to the RF structure. Since the size limitations of conventional microwave power sources with resonant circuits are not present in gyrotrons, their power handling capability can be considerably greater. The gyrotron is of interest as a potential source of high peak and average power at millimeter wavelengths. It has also been considered for operation at microwave frequencies, but it has not been able to compete with more conventional microwave power sources.

The gyrotron, also known as a *cyclotron resonance maser,* employs a strong externally applied axial magnetic field to cause electrons within the circular fast-wave structure to rotate at the electron cyclotron frequency, which is $\omega_c = eB_0/m\gamma$, where e = electron charge, m = electron rest mass, B_0 = axial magnetic field, γ is the relativistic factor which is $[1 + (e/mc^2)V_0]$, c = velocity of light, and V_0 = beam voltage. The beam voltage and the corresponding electron velocity are high enough to cause relativistic effects. The electrons follow helical paths around the lines of the magnetic field in the presence of an electromagnetic wave with a transverse component of electric field. The electrons become phase-bunched in their cyclotron orbits as a result of the relativistic mass change of the electrons. Electrons that lose energy to the electromagnetic wave become lighter and accumulate phase lead and catch up with the electrons that gain energy and become heavier and accumulate phase lag.

The frequency of the gyrotron is determined by the magnetic field and not by the characteristic size of the interaction region as it is in microwave power tubes. The magnetic field must be quite large in order to generate cyclotron oscillations at the higher frequencies. For this reason, the magnets used for millimeter wavelength gyrotrons are usually superconducting, which can be a burden for some applications, especially if the device has to be operated in a cryostat at liquid helium temperatures. However, the development of magnets based on high-temperature superconducting materials and efficient closed-cycle refrigerators, or cryocoolers, offers the possibility of using supercooled gyrotrons in mobile platforms such as ships and aircraft.

Since the diameter of the gyrotron RF circuit is normally large compared to the wavelength of the electromagnetic wave it generates, it is possible to have a higher order mode or multiple modes of the electromagnetic field. Operation in more than one mode can result in operation at more than one frequency. The design of a gyrotron requires that care be given to insure stable, single-frequency operation in only one mode.

The gyrotron oscillator at 94 GHz can produce CW power greater than 100 kW and peak pulse power of 1 MW, with efficiencies of about 30 percent. A quasi-optical gyrotron was tunable from 80 to 130 GHz (a half octave) by varying the magnetic field.[44] The power was relatively constant over this frequency range, averaging about 60 kW.

The gyrotron can be operated as an amplifier as well as an oscillator. Generally, more power can be obtained from a gryrotron oscillator than the gyrotron amplifier, but the amplifier might have some advantages in radar when doppler processing is important—just as at microwave frequencies. The electron beam of a *gyroklystron* passes through two or more resonant cavities with standing-wave fields separated by drift spaces. A *gyroTWT* is one which operates with traveling wave fields, similar to a microwave TWT. The gyroklystron has a smaller relative bandwidth than the gyroTWT, but it is capable of higher gain, efficiency, and output power. A *gyrotwystron* operates similarly to a microwave twystron in that it uses standing-wave fields to bunch the electrons and a traveling-wave field to extract the energy. The relative bandwidths of millimeter-wave gyrotron amplifiers are generally less than the relative bandwidths that can be obtained with microwave power amplifiers.

The specifications for a particular experimental gyroklystron designed for radar operation at a center frequency of 94 GHz required that it have an average power of 10 kW, peak power of 80 kW, efficiency of 20 percent, and bandwidth of 600 MHz.[45]

Multiple-Beam Klystrons[46] In a conventional klystron with a single electron beam, the power can be increased by increasing the already high beam voltage. Instead of increasing the beam voltage to obtain greater power, it is possible to employ many electron beams that pass through individual channels located in a single multichannel drift tube. The total power is the sum of the power extracted from each of the lower-current electron beams. The number of beams has been from 6 to 61. Such a power generator is known as a *multiple-beam klystron* (MBK).

The significant reduction in beam voltage results in reduced size and weight compared to a conventional klystron of comparable power. Its magnet and power supply are smaller and lighter. The geometry of the multiple beams of the MBK allows an increase in bandwidth because of an increase in perveance. The lower voltage also can eliminate the need for lead shielding to screen against X-ray radiation.

The high-power multiple-beam klystron was first seriously examined in the U.S. in the 1950s, but interest was not sustained because the needs for high power were satisfied by other more conventional types of klystrons. Tube engineers in the Soviet Union began seriously investigating the MBK in the 1970s, and were successful in producing RF power sources that have been widely used in Russian radar systems. An example of an MBK produced and marketed by the Russian Company ISTOK is their IKS-9007, a six-cavity, 36-beam klystron. It operates at 3.3 GHz with a 200-MHz bandwidth (6 percent), peak power from 500 to 800 kW, duty cycle of 0.02, gain of 40 dB, and an efficiency of 40 to 50 percent. The beam voltage is 28 to 32 kW. The klystron tube weight is 25 kg and the solenoid magnet weight is 95 kg, which is said to be 2 to 3 times less than single-beam klystrons of similar performance.

ISTOK has also applied the multiple-beam technology to the traveling wave tube and the inductive output tube.

10.7   OTHER ASPECTS OF RADAR TRANSMITTERS

Pulse Modulators[47,48]   The function of the modulator is to turn the transmitter on and off to generate the desired waveform. When the waveform is a pulse, the modulator is some-times called a *pulser*. Each RF power source has its own particular characteristics that de-termine the type of modulator to be used. The magnetron modulator, for instance, has to handle the full pulse power. The transistor amplifier, on the other hand, requires no mod-ulator at all since the transistor is turned on and off by the presence or absence of the in-put pulse. The full power of the klystron and the traveling-wave tube can be switched by a modulator handling only a small fraction of the total electron-beam power if the tubes are designed with a modulating anode or a shadow grid. The crossed-field amplifier (CFA) is often cathode-pulsed, requiring a full-power modulator. Some CFAs, however, can be d-c operated, which means they can be turned on by the start of the RF input pulse and turned off by a short, low-energy pulse applied to a cut-off electrode. Similar to the tran-sistor amplifier, some lower power CFAs require no modulator since they can be turned on and off by the start and stop of the input RF pulse.

The basic elements of one type of radar modulator is shown in Fig. 10.8. Energy from an external power source is accumulated in the energy-storage element at a slow rate dur-ing the interpulse time. The charging impedance limits the rate at which energy is deliv-ered to the storage element. When the pulse is ready to be formed, the switch is closed and the stored energy is quickly discharged through the load to form the d-c pulse that is applied to the RF power device. During the discharge part of the cycle, the charging im-pedance prevents energy from the storage element from being returned to the energy source.

Line-Type Modulator   In this device a delay line, or pulse forming network (PFN), is used as the storage element. The switch can be a hydrogen thyratron, mercury ignitron, a silicon controlled rectifier (SCR), saturable reactor, or other device that can initiate the discharge of the PFN to form a rectangular pulse. The shape and duration of the pulse are

Figure 10.8 Basic elements of one type of radar pulse modulator.

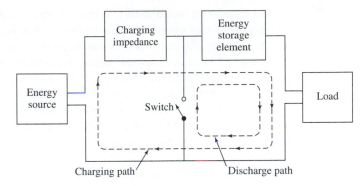

determined by the passive elements of the PFN. The switch has no control over the pulse shape other than to initiate it. The pulse ends when the PFN has discharged sufficiently to allow the switch to close and recover its voltage hold-off capability. The trailing edge of the pulse is usually not sharp since it depends on the discharge characteristics of the PFN. The line-type modulator is simple, compact in size, and can tolerate abnormal load conditions. It has been widely used in the past for magnetron pulsing.

Active-Switch Modulator   The switching element in this type of modulator has to be able to be turned off as well as on. Originally the switch was a vacuum tube and the modulator was called a hard-tube modulator to distinguish it from the gas-tube switch often used in a line-type modulator. In addition to the vacuum tube, the switch can be a semiconductor device such as a silicon controlled rectifier.

There are three types of active-switch modulators: (1) cathode pulsers that control the full beam power of the RF tube, (2) mod-anode pulsers that are required to switch at the full beam voltage of the RF tube but with little current, and (3) grid pulsers that operate at a far smaller voltage than that of the beam.

Unlike the line-type modulator, the switch in the active-switch modulator controls both the beginning and the end of the pulse. The energy storage element is a capacitor. To prevent droop in the pulse shape due to the exponential nature of a capacitor discharge, only a small fraction of the stored energy is extracted from the capacitor for the pulse to be delivered to the tube. In high-power transmitters with long pulses the capacitance must be very large. A large capacitance might be obtained with a collection of capacitors known as a *capacitor bank.*

The active-switch modulator permits more flexibility and precision than the line-type modulator. It can provide excellent pulse shape, varying pulse durations and prfs, including mixed pulse lengths and bursts of pulses with close spacings. It is, however, of greater complexity and weight than a line-type modulator.

Crowbar Protective Device   Power tubes can develop internal arc discharges with little warning. When an arc occurs in an unprotected device, the capacitor bank discharges large currents through the arc and the tube can be damaged. The tube can be protected with a device called an *electronic crowbar,* which places a short circuit across the capacitor bank to transfer its stored energy. When a sudden surge of current due to a fault in a protected power tube is sensed, the crowbar switch is activated within a few microseconds. The current surge also causes the circuit breaker to open and deenergize the primary source of power. The name "crowbar" is derived from the analogous action of placing a heavy conductor, like a crowbar, directly across the capacitor bank. A crowbar is required for a high-power active-switch modulator because of the large amount of energy that is stored in its capacitor bank. Crowbars are not usually needed with line-type modulators which store less energy in their pulse-forming network. Line-type modulators are designed to discharge safely all the stored energy each time a pulse is generated.

Transmitter Noise and Spectrum[49]   An RF power source can produce spurious, unwanted outputs as harmonics of the fundamental frequency, adjacent-band (out-of-band) noise, and in-band noise. Harmonics and adjacent-band noise can be reduced 30 to 60 dB by using high-power filters. Shaping of the pulse to make it more rounded and less rectangular reduces out-of-band signal energy. In-band spurious signals and noise cannot be readily filtered since these unwanted signals are within the frequency range of the desired signal spectrum. The in-band noise is greater in some RF power sources than in others. For example, Weil states[49] that the noise level in a 1-MHz bandwidth of a conventional CFA is typically 50 to 60 dB down, but is down 90 dB or better in linear-beam tubes.

Section 3.7 discussed the effects of equipment instabilities on the amount of clutter cancellation, or improvement factor, that can be achieved in MTI radars. The typical noise level in conventional CFAs can set a limit on the achievable MIT improvement factor to perhaps the vicinity of 45 dB or so. Linear beam tubes, on the other hand, are capable of very high MTI improvement factors except for limitations introduced by their modulators and power supplies. The ripple on the modulator voltage and the variation of the high-voltage power supply (HVPS) must be sufficiently small to obtain the large improvement factors needed in high-performance radars.

In a staggered prf MTI system (Sec. 3.3) the variation of the interpulse period causes a variation in the HVPS voltage, which can be a significant source of transmitter instability. The consequent reduction in improvement factor that would result needs to be corrected, as indicated by Weil.[49]

The transmitter and its modulator can also distort pulse compression waveforms and introduce spurious time-sidelobes. Active-switch modulators are more likely to allow low pulse-compression time-sidelobe levels as compared with the time sidelobes produced by line-type modulators.

REFERENCES

1. Weil, T. A. "Transmitters." In *Radar Handbook,* 2nd ed. M. Skolnik (Ed.). New York: McGraw-Hill, 1990, Chap 4.

2. Gilmour, A. S., Jr. *Microwave Tubes.* Norwood, MA: Artech House, 1986, Chap. 16.

3. Granatstein, V. L., R. K. Parker, and C. M. Armstrong. "Vacuum Electronics at the Dawn of the Twenty-First Century." *Proc. IEEE* 87 (May 1999), pp. 702–718.

4. News article from *Microwaves & RF,* (November 1984), p. 31.

5. Smith, M. J., and G. Phillips. *Power Klystrons Today.* New York: John Wiley, 1995, Sec. 7.2.3.

6. Staprans, A. "Linear Beam Tubes." In *Radar Technology,* E. Brookner (Ed.). Boston: Artech House, 1977, Chap. 22.

7. Gilmour, A. S., Jr. Ref. 2, Chap. 4.

8. Phillips, R. M., and D. W. Sprehn. "High-Power Kystrons for the Next Linear Collider." *Proc. IEEE* 87 (May 1999), pp. 738–751.

9. Staprans, A., E. W. McCune, and J. A. Ruetz. "High-Power Linear-Beam Tubes." *Proc. IEEE* 61 (March 1973), pp. 299–330.

10. Dodds, W. J., T. Moreno, and W. J. McBride, Jr. "Methods for Increasing the Bandwidth of High Power Microwave Amplifiers." *IRE WESCON Conv. Rec.* 1, pt. 3 (1957), pp. 101–110.

11. Gilmour, A. S., Jr. Principles of Traveling Wave Tubes. Boston: Artech House, 1994, Sec. 18.4.

12. Symons. R. S. "Tubes: Still Vital After All These Years." *IEEE Spectrum* 35 (April 1998), pp. 52–63.

13. Gilmour, A. S., Jr. Ref. 11, Chap. 13.

14. Gilmour, A. S., Jr. Ref. 2, Chap. 11.

15. Luebke, W. and G. Caryotakis. "Development of a One Megawatt CW Klystron." *Microwave J.* 9, no. 8 (August 1966), pp. 43–47.

16. Symons, R. S., and J. R. M. Vaughan. "The Linear Theory of the Clustered-Cavity™ Klystron." *IEEE Trans.* PS-22 (October 1994), pp. 713–718.

17. Borkowski, M. T. "Solid-State Transmitters." *Radar Handbook,* 2nd ed., M. Skolnik (Ed.). New York: McGraw-Hill, 1990, Chap. 5.

18. Lee, K., C. Corson, and G. Mols. "A 250 kW Solid State AN/SPS-40 Radar Transmitter." *Microwave J.* 26 (July 1983), pp. 93–105.

19. Dyck, J. D., and H. R. Ward. "RAMP's New Primary Surveillance Radar." *Microwave J.* 27 (December 1984), pp. 105–113.

20. Merrill, P. R. "A 20 kW Solid-State L-Band Transmitter for the RAMP PSR Radar." *Microwave J.* 31 (March 1988), pp. 165–173.

21. Chilton, R. H. "MMIC T/R Modules and Applications." *Microwave J.* 30 (September 1987), pp. 131–146.

22. Hoft, D. J. "Solid State Transmit/Receive Module for the Pave Paws Phased Array Radar." *Microwave J.* 21 (October 1978), pp. 33–35.

23. Olson, F. A. "Microwave Solid-State Power Amplifier Performance: Present and Future." *Microwave J.* 38 (February 1995), pp. 24–46.

24. Trew, R. J., J-B Yan, and P. M. Mock. "The Potential of Diamond and SiC Electronic Devices for Microwave and Millimeter-Wave Power Applications." *Proc. IEEE* 79 (May 1991), pp. 598–620.

25. Butler, N. "The Microwave Tube Reliability Problem." *Microwave J.* 16 (March 1973), pp. 41–42.

26. Advertisement of the Electron Device Group, Varian Beverly Division, Beverly, MA. (Varian is now known as CPI, Inc.)

27. Weil, T. A. Ref. 1, Sec. 4.2.

28. Gilmour, A. S., Jr. Ref. 2, Sec. 13.3.

29. Weil, T. A. "Comparison of CFA's for Pulsed-Radar Transmitters." *Microwave J.* 16 (June 1973), pp. 51–54, 72.

30. Kaisel, S. F. "Microwave Tube Technology Review." *Microwave J.* 20 (July 1977), pp. 23–42.

31. Anonymous. "Cathode-Driven Crossed-Field Amplifer." *Microwave J.* 31 (February 1988), pp. 208–209.

32. Sivan, L. *Microwave Tube Transmitters.* London: Chapman & Hall, 1994, Sec. 7.4.

33. Abrams, R. H., Jr. "The Microwave Power Module: A 'Supercomponent' for Radar Transmitters." *Record of the 1994 IEEE National Radar Conf.,* Atlanta, GA, pp. 1–6, IEEE No. 94CH3359–7.

34. Smith, C. R., C. M. Armstrong, and J. Duthie. "The Microwave Power Module: A Versatile RF Building Block for High-Power Transmitters." *Proc. IEEE* 87 (May 1999), pp. 717–737.

35. Yingst, T. E., et al. "High-Power Gridded Tubes—1972." *Proc. IEEE* 61 (March 1973), pp. 357–381.

36. Reed, J. E. "The AN/FPS-85 Radar System." *Proc. IEEE* 57 (March 1969), pp. 324–335.

37. Haeff, A. V. "An Ultra-High-Frequency Power Amplifier of Novel Design." *Electronics* 10 (February 1939), pp. 30–32.

38. Preist, D. H., and M. B. Shrader. "The Klystrode—An Unusual Transmitting Tube with Potential for UHF-TV." *Proc. IEEE* 70 (November 1982), pp. 1318–1325.

39. Symons, R. S. "The Constant Efficiency Amplifier." *NAB Broadcast Engr. Conf. Proc.* (1997), pp. 523–530.

40. Symons, R. S., et al. "The Constant Efficiency Amplifier—A Progress Report." *NAB Broadcast Engr. Conf. Proc.,* 1998.

41. Granatstein, V. L., and I. Alexoff. *High-Power Microwave Sources.* Boston: Artech House, 1987.

42. Gilmour, A. S., Jr. Ref. 2, Chap. 14.

43. Felch, K. L., et al. "Characteristics and Applications of Fast-Wave Gyrodevices." *Proc. IEEE* 87 (May 1999), pp. 752–781.

44. Manheimer, W. M. "On the Possibility of High Power Gyrotrons for Super Range Resolution Radar and Atmospheric Sensing." *Int. J. Electronics* 72, nos. 5 and 6 (1992), pp. 1165–1189.

45. Blank, M., B. G. Danly, and B. Levush. "Circuit Design of a Wideband W-Band Gyroklystron Amplifier for Radar Applications." *IEEE Trans.* PS-26 (June 1998), pp. 426–432.

46. Gelvich, E. A., et al. "The New Generation of High-Power Multiple-Beam Klystrons." *IEEE Trans.* MTT-41 (January 1993), pp. 15–19. See also the ISTOK Web Site at www.istok.com.

47. Weil, T. A. Ref. 1, Sec. 4.8.

48. Ewell, G. W. *Radar Transmitters.* New York: McGraw-Hill, 1981.

49. Weil, T. A. Ref. 1, Secs. 4.6 and 4.7.

PROBLEMS

10.1   One way to define the efficiency of a transmitter is RF power out, P_{out}, divided by the prime power in, P_{in}. (a) Plot the power dissipated, $P_{dis} = P_{in} - P_{out}$, as a function of the transmitter efficiency, ϵ, for a fixed power out. [Make the ordinate the ratio (power dissipated)/(power out).] (b) If the output power has to be 30 kW, what power will be dissipated with a transmitter having a 15 percent efficiency? (c) If the transmitter efficiency can be increased to 50 percent, what is the amount of power to be dissipated? (d) What disadvantages occur with low efficiency?

10.2   (a) In a solid-state transmitter (a solid-state "bottle") with 300 modules, what would be the reduction in output power if 20 modules were to fail? (b) What fractional reduction in radar range would this cause?

10.3   (a) If 10 percent of the modules in an active-aperture phased array radar fail, what would be the reduction in transmitter power? (b) What would be the reduction in the maximum radiation power density? (c) What would be the reduction in radar range?

10.4   For an air-traffic control radar application, compare the advantages and disadvantages of a solid-state transmitter, klystron transmitter, and magnetron transmitter.

10.5   (a) If one wanted a radar transmitter with a 10 percent bandwidth, what options are available to the radar system designer and which RF power source appears the most desirable? (You may have to make some assumptions about the application.) (b) If one wanted a radar transmitter with 40 percent bandwidth, what options are available and which of these options might you choose? (Include the reason for your choice.) (c) Are there any undesirable consequences with your choice for (b) that you might have to live with?

10.6   What are the advantages and disadvantages of the gyroklystron (amplifier) for radar application at 94 GHz (millimeter waves)?

10.7   How can a tube designer achieve a large bandwidth with a klystron type of power tube?

10.8   When might the systems engineer choose to use a traveling wave tube over a klystron for a high-power radar application?

10.9 For a high-power UHF radar transmitter application, compare the advantages and disadvantages of solid state, the grid-control vacuum tube, the constant-efficiency amplifier, and the klystron.

10.10 When might a magnetron be desirable for use in radar applications?

10.11 What factors might be involved when a radar systems engineer tries to choose among the crossed-field amplifier, TWT, and Twystron for some radar application?

10.12 If the R&D genie were to grant a radar systems engineer many wishes, what improvements might the radar systems engineer want to have for a radar transmitter?

11

Radar Receiver

11.1 THE RADAR RECEIVER

The function of the receiver in early radar systems was to extract the weak echo signals that appeared at the antenna terminals and amplify them to a level where they could be displayed to a radar operator who then made the decision as to whether or not a target echo signal was present. The modern radar receiver still has to extract the weak echo signals and amplify them, but it does much more. It employs a matched filter (Sec. 5.2) whose purpose is to maximize the peak-signal-to-mean-noise-ratio and discriminate against unwanted signals whose waveforms are different from those transmitted by the radar. When the clutter echoes are large enough to mask desired target echoes, the receiver also has to incorporate means for separating the moving targets from stationary clutter echoes by recognizing the doppler frequency shift of the moving targets (Chap. 3).

In modern radars the decision whether a target is present or absent is seldom made by an operator viewing on a display the unprocessed output of a receiver. Instead, the detection decision is made automatically based on threshold detection (Sec. 5.5). Information about a target's location in range and angle can also be extracted automatically instead of manually by an operator. In an operational air-surveillance radar, tracking of targets is no longer performed by an operator marking with a grease pencil on a radar display the location of blips (targets) from scan to scan and calculating the target speed and estimating its direction. Targets are acquired and tracked automatically (Sec. 4.9) and only processed tracks are displayed to the operator or sent to some automatic device, such as an air-traffic control system or weapons control computer, for further use. When the radar cannot remove all the clutter echoes, constant false alarm rate (CFAR) circuitry is employed to prevent the tracking computer from becoming overloaded when trying to establish tracks using clutter echoes. The receiver is also the place where external interference and hostile electronic-countermeasures are kept from interfering with the detection of targets.

Thus, in addition to detection and amplification of signals, a radar receiver performs many other functions either directly as part of the receiver or in conjunction with it. These other functions include signal processing, information extraction, data processing, electromagnetic compatibility, and electronic counter-countermeasures. (The modern receiver might be thought of as the *receiver/processor*.) Sometimes the display is considered part of the receiver system. In this and other radar books, these other functions are often considered separately from the discussion of the receiver. The interested reader will find a more thorough review of the radar receiver by John W. Taylor, Jr. in the *Radar Handbook*.[1]

The radar receiver is almost always a *superheterodyne,* or superhet. It was shown in the block diagram of Fig. 1.4 and briefly described in Sec. 1.3. The essential characteristic of a superheterodyne is that it converts the RF input signal to an intermediate frequency (IF) where it is easier than at RF to achieve the necessary filter shape, bandwidth, gain, and stability. An advantage of the superheterodyne receiver is that its frequency can be readily changed by changing the frequency of the local oscillator (LO). The first stage, or *front-end,* of a radar superheterodyne receiver can be an RF low-noise amplifier (LNA) such as a transistor.

Before the availability of low-noise transistors, the receiver front-end was the mixer stage without an RF amplifier preceding it. In some applications the mixer stage might still be desired as the receiver front-end instead of a low-noise amplifier. A receiver with a mixer as the first stage has a greater dynamic range than one with a low-noise amplifier, which might be important when large MTI improvement factors are needed to remove clutter echoes. The extra dynamic range available with a mixer as the front-end can also be of value in reducing the likelihood of receiver saturation when large signals or jamming are present. The larger receiver noise figure of a mixer might be compensated with greater transmitter power and/or a bigger antenna, both of which are beneficial when a military radar is faced with hostile noise jamming. Although the mixer front-end might have some advantage over a low-noise transistor amplifier front-end, a low-noise amplifier as the first stage of a superheterodyne receiver generally seems to be preferred by those who buy radars.

A high-performance air-surveillance radar sometimes employs more than one type of receiver. Each would share the front-end, mixer, and IF stages. One receiver might be a linear amplifier and envelope detector for detection of targets in the clear (no competing clutter). A second receiver might be for doppler processing to remove clutter, as in an MTI radar. It would use *I* and *Q* channels and digital signal processing to filter the moving targets. A third receiver might be a log-FTC (Sec. 7.8), or something similar, to aid in detecting targets located within moving weather clutter beyond the range of surface clutter.

It was said in Chap. 2 that if the radar designer wishes to increase the detection range of a radar, the chief factors available are the average power of the transmitter and the area of the antenna. The classical radar equation also indicates that the range can be increased by reducing the receiver noise figure; but, in practice, the noise figures of radar receivers are already quite low and any further decrease can produce marginal results and, sometimes, unwanted effects. Further lowering of the noise figure might not be justified if it significantly increases receiver cost, lowers the dynamic range, subjects the device to a greater risk of burn-out, and results in less reliability. A very sensitive receiver also allows more interference to enter the receiver. Sometimes an increase in interference is a price that may be worth the benefits of improved sensitivity, but there can be limits.

A radar receiver has to have sufficient gain to increase the level of the weak echo signal to where it is large enough to be processed or displayed. In the superheterodyne the total receiver gain is divided between the IF and the video amplifiers. The receiver should have adequate dynamic range (where the receiver is linear) so that large clutter echoes do not cause the receiver to saturate and reduce the MTI improvement factor. It must not introduce unwanted phase or amplitude changes that could distort the echo signals. It must be protected from overload, saturation, and damage (burnout) by strong unwanted signals. Timing and reference signals are needed to properly extract target information and take advantage of the doppler frequency shift of echo signals.

A limitation with early radar receivers when vacuum tubes were the only available technology was that they were relatively large in size. The size of radar receivers is no longer a problem with modern technology. The trend is to make the receiver as digital as is practical, with analog components confined to the RF and perhaps the IF.

There can be many demands on the radar receiver, but the receiver designer has responded well to the challenges and there exists a highly developed state of receiver technology. Radar receiver design and implementation may not always be an easy task, but receiver designers have usually been able to provide the radar systems engineer the means to accomplish the desired objectives.

11.2   RECEIVER NOISE FIGURE[2,3]

Definition   The receiver noise figure was described in Sec. 2.3 as a measure of the noise produced by a practical receiver compared to the noise of an ideal receiver. The noise figure of a linear network may be defined as either

$$F_n = \frac{N_{\text{out}}}{kT_0 B_n G} \quad \text{or} \quad \frac{S_{\text{in}}/N_{\text{in}}}{S_{\text{out}}/N_{\text{out}}} \qquad [11.1]$$

where N_{out} = available output noise power; $kT_0 B_n = N_{\text{in}}$ = available input noise power; k = Boltzmann's constant = 1.38×10^{-23} J/deg; T_0 = standard temperature of 290 K (approximately room temperature); B_n = noise bandwidth defined by Eq. (2.3); $G = S_{\text{out}}/S_{\text{in}}$ = available gain; S_{out} = available output signal power; and S_{in} = available input signal power. The term "available power" refers to the power that would be delivered to a matched load. (The term "available" will be understood in the following discussion of noise figure and is not mentioned further.) The product $kT_0 = 4 \times 10^{-21}$ W/Hz.

The reason for a standard temperature T_0 in the definition of noise figure is to refer measurements made under differing temperature conditions to a common basis of comparison.

Equation (11.1) permits two different, but equivalent, interpretations of the noise figure. It may be considered (right-hand side) as the degradation of the signal-to-noise ratio as the signal passes through the network, or (left-hand side) it may be interpreted as the ratio of the noise-power out of the actual network to the noise-power out of an ideal network that amplifies the input thermal noise and introduces no additional noise of its own. The noise figure of Eq. (11.1) can be expanded as

$$F_n = \frac{kT_0B_nG + \Delta N}{kT_0B_nG} = 1 + \frac{\Delta N}{kT_0B_nG} \qquad [11.2]$$

where ΔN is the additional noise introduced by the practical (nonideal) network.

The noise figure is commonly expressed in decibels; that is, $10 \log F_n$. The term *noise factor* has also been used at times instead of noise figure. The definition of noise figure assumes that the input and output of the network are matched. In some devices, less noise is obtained under mismatched, rather than matched, conditions. In spite of definitions, such networks would be operated so as to achieve the maximum output signal-to-noise ratio.

Noise Figure of Networks in Cascade   Consider two networks in cascade, each with the same noise bandwidth B_n but with different noise figures and gain, Fig. 11.1. Let F_1, G_1 be the noise figure and gain, respectively, of the first network and F_2, G_2 be similar parameters for the second network. The problem is to find F_0, the overall noise-figure of the two networks in cascade. From the definition of noise figure given by Eqs. (11.1) and (11.2), the output noise N_{out} of the two networks in cascade is

N_{out} = noise from network 1 at output of network 2 + noise ΔN_2 introduced by network 2

$$= F_0kT_0B_nG_1G_2 = F_1kT_0B_nG_1G_2 + \Delta N_2 = F_1kT_0B_nG_1G_2 + (F_2 - 1)\, kT_0B_nG_2$$

which results in

$$F_0 = F_1 + \frac{F_2 - 1}{G_1} \qquad [11.3]$$

It is not sufficient that the first stage of a low-noise receiver have a low noise figure. The second stage must also have a low noise figure or, if not, the gain of the first stage needs to be large. Too large a first-stage gain, however, is not always desirable since the dynamic range of the receiver is reduced by the gain G_1 of the low-noise amplifier. If the first network is not an amplifier, but is a diode mixer, the gain G_1 should be interpreted as a number less than unity (a loss).

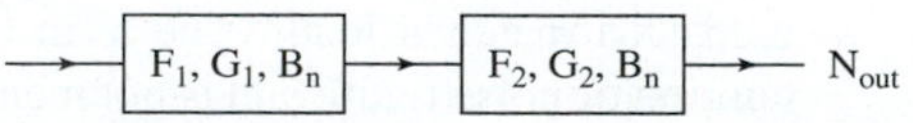

Figure 11.1   Two networks in cascade with different noise figures and gains, but the same noise bandwidths.

The noise figure of N networks in cascade may be shown to be

$$F_0 = F_1 + \frac{F_2 - 1}{G_1} + \frac{F_3 - 1}{G_1 G_2} + \cdots + \frac{F_N - 1}{G_1 G_2 \cdots G_{N-1}} \qquad \text{[11.4]}$$

Similar expressions may be derived when the bandwidth and/or temperature of the individual networks are not the same.[4]

Noise Figure Due to Loss in the Transmission Line   Any losses in the RF portion ahead of the receiver front-end result in an increase in the apparent overall noise figure. Such losses can be due to the transmission line between antenna and receiver, the duplexer, receiver protector, rotary joint, preselector filter, STC if applied at RF, monitoring devices, and the radome. The noise figure due to these RF losses, obtained from the second part of the definition of Eq. (11.1), is equal to the RF loss L_{RF}. (This can be seen since L_{RF} is the loss in signal-to-noise ratio as the signal travels from the antenna to the receiver.) It can also be obtained from the first part of Eq. (11.1) since the noise out of a lossy transmission line is $kT_0 B_n$, and its gain $G = 1/L_{RF}$.

If the loss in the transmission line and its associated devices is incorporated in the receiver noise figure, it should not also be a part of the system losses. Most radar analyses treat these losses as system losses rather than as part of the receiver noise figure. There are some RF devices that are likely to be closely associated with the low-noise receiver. These include the circulator that provides isolation, the receiver protector, waveguide to coax transition, and receiver performance monitoring (which might produce a loss). When a noise figure for a receiver is quoted in a publication or given in a manufacturer's catalog, it might not always be obvious what is included. It might be the receiver noise figure without the associated receiver losses or it might include the loss of those devices that are a close part of the receiver, as mentioned above. With a mixer front-end receiver the noise figure quoted has sometimes been that of the mixer alone and not of the entire receiver. There seems to be no standard method for reporting receiver noise figure.

Noise Temperature   The noise introduced by a network may also be expressed as the *effective noise temperature*, T_e, defined as the (fictional) temperature at the input of the network, that accounts for the additional noise ΔN at the output. Therefore $\Delta N = kT_e B_n G$ and from Eq. (11.2) we have

$$F_n = 1 + \frac{T_e}{T_0} \qquad \text{[11.5]}$$

$$T_e = (F_n - 1)T_0 \qquad \text{[11.6]}$$

The *system noise temperature* T_s is defined as the effective noise temperature of the receiver including the effects of antenna temperature T_a. If the receiver effective noise temperature is T_e, then

$$T_s = T_a + T_e = (F_s - 1)T_0 \qquad \text{[11.7]}$$

where F_s is the *system noise figure*. This equation also defines the system noise figure when it includes the effects of the antenna temperature T_a and receiver effective noise temperature T_e.

The effective noise temperature of a receiver consisting of a number of networks in cascade is

$$T_e = T_1 + \frac{T_2}{G_1} + \frac{T_3}{G_1 G_2} + \cdots \qquad [11.8]$$

where T_i and G_i are the effective noise temperature and gain of the ith network.

The effective noise temperature and the noise figure both describe the same characteristic of a network. The effective noise temperature generally is used to describe the noise performance of very low-noise receivers, lower than might be of interest for radar. It is also preferred by some radar engineers and many receiver designers as being more useful than noise figure for analysis purposes. The noise figure, however, seems to be the more widely used term to describe radar receiver performance, and is used in this text for that purpose.

11.3 SUPERHETERODYNE RECEIVER

The discussion of the superheterodyne receiver in this section does not include all aspects of the receiver, but only with those component parts that have an effect on the radar system design. This includes the low-noise RF amplifier, the mixer, receiver dynamic range, the $1/f$ noise at IF, oscillator noise, and the detector.

Low-Noise Front-End   The first stage of a superheterodyne receiver for radar application can be a transistor amplifier. At the lower radar frequencies the silicon bipolar transistor has been used. Gallium-arsenide field-effect transistors (FET) are found at the higher frequencies. Other types of transistors also can be used, depending on the trade-off between the desired noise figure and the ability of the transistor to withstand burnout. An X-band transistor can provide a noise figure of about one dB and can withstand a leakage peak power of 0.2 W.[5] With a diode limiter ahead of the transistor, the peak power can be as great as 50 W before burnout. The diode limiter increases the noise figure about 0.5 dB at X band and 0.2 dB at C band. The lower the frequency the lower can be the transistor noise figure. At C band the noise figure might be around 0.6 dB. These values are more than adequate for radar. (Early microwave radars had noise figures of 12 to 15 dB and radars in the 1960s had noise figures of 7 to 8 dB.) It is not necessary for the radar systems engineer to have extremely low noise figures in most radar applications, especially when the unavoidable losses in the transmission line between receiver and antenna are considered. If improved radar system performance is of concern, it is probably more fruitful to try to reduce some of the many system losses that occur elsewhere in a radar rather than try to reduce further the noise figure of the low-noise amplifier (LNA). It is usually good enough.

Prior to the low-noise transistor amplifier, the parametric amplifier and the maser were available as low-noise receiver front-ends. Although their noise figures were low (lower than those of transistors, which came later), they were seldom used operationally for radar. They were expensive, of large size, and often did not have sufficient dynamic range. Until low noise transistor amplifiers were developed, the radar receiver seldom employed an RF amplifier stage except perhaps at UHF or lower frequencies. Before the low-noise transistor, the mixer was the receiver front-end. As already mentioned, a mixer as the front-end without a low-noise amplifier preceding it is a valid option for some radar applications in spite of its higher receiver noise figure.

Achieving low receiver noise is no longer the problem it once was, and designers of high-performance radar receivers usually are more concerned with obtaining large dynamic range and low oscillator noise.

Mixers[6]   Whether or not it is used as the front-end, the mixer is a key element in a superheterodyne receiver since it is the means by which the incoming RF signal is converted to IF (intermediate frequency). When the down conversion from RF to IF is performed in one step, it is called single conversion. Sometimes the down conversion is done in two steps with two mixers and IF amplifiers. This is known as dual conversion. Dual conversion superheterodyne receivers are used to avoid some forms of interference and (spoofing) electronic countermeasures. The mixer should have a low conversion loss, introduce little additional noise of its own, minimize spurious responses, and not be susceptible to burnout, especially when it is used as the front-end without a low noise amplifier ahead of it. An integral part of the mixer is the local oscillator.

Noise Figure of a Mixer Used as a Front End   The noise figure of a mixer is determined by its conversion loss and noise-temperature ratio. The conversion loss of a mixer is defined as

$$L_c = \frac{\text{available RF power}}{\text{available IF power}} \qquad [11.9]$$

It is a measure of the efficiency of the mixer in converting RF signal power into IF. The conversion loss of typical microwave diodes in a conventional single-ended mixer configuration varies from about 5 to 6.5 dB. Schottky diodes in an image recovery mixer have been reported to have a minimum conversion loss of 3.5 dB over a narrow band and a 4-dB conversion loss over a 10 percent bandwidth at S band.[7] The noise-temperature ratio of a mixer (not to be confused with the effective noise temperature) is defined as

$$t_r = \frac{\text{actual available IF noise power}}{\text{available noise power from an equivalent resistance}} \qquad [11.10]$$

or

$$t_r = \frac{F_m k T_0 B G_c}{k T_0 B} = F_m G_c = \frac{F_m}{L_c}$$

where F_m = mixer noise figure and L_c = conversion loss = $1/G_c$. The noise temperature ratio of a mixer varies inversely with the IF frequency from about 100 kHz down to a small fraction of a hertz. At a frequency of 30 MHz, the noise temperature ratio might range from 1.2 to 2.0. Generally, the lower the conversion loss, the larger is the noise temperature ratio.

The noise figure of a mixer based on Eq. (11.10) is $F_m = L_c t_r$. It is, however, not a complete measure of the sensitivity of a receiver with a mixer front-end. The overall noise figure depends not only on the mixer stage, but also on the noise figure of the IF amplifier. The latter becomes a significant factor in the overall noise figure since the mixer has a loss rather than a gain. Using Eq. (11.3), the noise figure of the first network (the mixer) is $F_1 = L_c t_r$ and its gain is $G_1 = 1/L_c$. The noise figure of the second network is that of the IF amplifier, so that $F_2 = F_{IF}$. The receiver noise figure with a mixer front-end is then

$$F_R = F_1 + \frac{F_2 - 1}{G_1} = L_c t_r + (F_{IF} - 1)L_c = L_c(t_r + F_{IF} - 1) \qquad [11.11]$$

(This does not include any losses in the RF transmission line.) If, for example, the conversion loss of the mixer were 5.5 dB, the IF noise figure 0.5 dB, and the noise-temperature ratio 1.2, the receiver noise figure would be 6.7 dB. For low noise-temperature diodes, the receiver noise figure is approximately equal to the conversion loss times the IF noise figure.

Some manufacturers have used Eq. (11.11) to determine the noise figures of the mixers listed in their catalogs. Other manufacturers have used the expression $L_c t_r$, which is lower than that of Eq. (11.11). When using mixer noise figures quoted in advertising brochures or in the literature, one should be aware of how it was determined.

Types of Mixers[8,9]   An ideal mixer is one whose output is proportional to the product of the RF echo signal and the local oscillator (LO) signal. The mixer provides two output frequencies that are the sum and difference of the two input frequencies, or $f_{RF} \pm f_{LO}$, assuming $f_{RF} > f_{LO}$. The difference frequency $f_{RF} - f_{LO}$ is the desired IF frequency. The sum frequency $f_{RF} + f_{LO}$ is rejected by filtering. There are however, two possible difference frequency signals at the IF when a signal appears at the RF. One is $f_{IF} = f_{RF} - f_{LO}$, assuming the input RF signal is of greater frequency than the LO frequency. The other possible difference frequency occurs when the RF signal is at a lower frequency than the LO frequency such that $f_{IF} = f_{LO} - f_{RF}$. If one of these is at the desired signal frequency, the other is the *image frequency*. Signals and receiver noise that appear at the image frequency need to be rejected using either an RF filter or an image-reject mixer described later in this subsection.

A relatively simple mixer is the *single-ended mixer*, which uses a single diode, as in Fig. 11.2a. The diode terminates a transmission line and the LO is inserted via a directional coupler. A low-pass filter, not shown, following the diode allows the IF to pass while rejecting the RF and LO signals. In a single-ended mixer the image frequency is short-circuited or open-circuited so as to avoid having the noise from the image frequency affect the mixer output.

The diode of a mixer is a nonlinear device and, in theory, can produce intermodulation products at other frequencies, called *spurious responses*. These occur for any RF signal that satisfies the relation[10]

$$mf_{RF} + nf_{LO} = f_{IF} \qquad [11.12]$$

where m and n are integers such that $m,n = \ldots, -2, -1, 0, 1, 2, \ldots$. These are unwanted since they appear within the radar receiver bandwidth. Spurious responses that are

Figure 11.2 Types of mixers: (a) single-ended mixer, (b) balanced mixer, (c) image-rejection mixer.

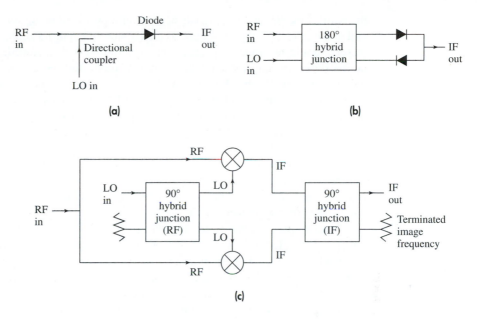

(a)

(b)

(c)

IF outputs due to the action of the mixer should not be confused with spurious signals, or spurs, that are due to the LO or the receiver power supply and can occur even in the absence of an RF signal. Taylor[7] describes the so-called *mixer chart*, which allows one to determine the combinations of the RF and LO frequencies that are free of strong spurious components. Such a chart indicates the bandwidth available for the mixer as a function of the ratio of the RF and LO frequencies. Taylor points out that the nature of the spurious responses are such that single-conversion receivers generally provide better suppression of spurious responses than double-conversion receivers. The third-order intermodulation product generally affects the dynamic range of the receiver, and is mentioned later under the discussion of dynamic range. There also can be other spurious, or intermodulation, responses from a mixer when two or more RF signals are present at the mixer's input and produce responses within the IF bandwidth.

Noise that accompanies the local oscillator (LO) signal in a single-ended mixer can appear at the IF frequency because of the nonlinear action of the mixer. This noise can be eliminated by inserting a narrowband RF filter between the LO and the mixer. It also has to be a tunable filter if the LO frequency is also tunable. A method to eliminate LO noise that doesn't have these disadvantages is a *balanced mixer*. The balanced mixer also can remove much of the mixer intermodulation products.

A diagram of a balanced mixer is shown in Fig. 11.2b. It can be thought of as two single-ended mixers in parallel and 180° out of phase. At the left of the figure is a four-port junction such as a magic-T, hybrid junction, 3-dB coupler, or equivalent. (Either a 90° or a 180° hybrid can be employed; here it is 180°.) In Fig. 11.2b the LO is applied to one port and the RF is applied to a second port. The signals inserted at these two ports appear in the third port as their sum and in the four port as their difference. A diode mixer is at the output of each of the other two ports. The hybrid junction has the property that the sum of the RF and LO signals appears at a port containing one of the diode mixers, and at the other port the difference of the RF and LO appears at the diode. The two diode mixers should have identical characteristics and be well matched. The IF signal is obtained by subtracting the outputs of the two diode mixers. In Fig. 11.2b, the balanced diodes are shown reversed so that the IF outputs can be added to obtain the required difference between the two channels. Local-oscillator AM noise at the two diode mixers will be in phase and will be canceled at the output. This mixer configuration also suppresses the even harmonics of both the RF and LO signals.

A *double-balanced mixer* (not shown) utilizes four diodes in a ring, or bridge, network to reduce the LO reflections and noise at the RF and IF ports, achieve better isolation between the RF and LO ports, reject spurious responses and certain intermodulation products, provide good suppression of the even harmonics of both the RF and LO signals, and permit wide bandwidth.[11]

In an *image-rejection mixer*, Fig. 11.2c, the RF signal is split and fed to the two mixers. The LO is fed into one port of a 90° hybrid junction that produces a 90° phase difference between the LO inputs to the two mixers. On the right is an IF hybrid junction that imparts another 90° phase difference in such a manner that the signal frequency and the image frequency are separated. The port with the image signal can be terminated in a matched load. According to Maas,[12] to reduce the image frequency by 20 dB requires that the phase error of the image-rejection mixer be less than 10° and the gain imbalance to be less than 1 dB. Dixon states[13] that the image-rejection mixer provides only about 30 dB of image rejection, which might not be sufficient for some applications. The image-rejection mixer is capable of wide bandwidth, and is restricted only by the frequency sensitivity of the structure of the microwave circuit. It is attractive because of its high dynamic range, good VSWR, low intermodulation products, and less susceptibility to burnout. The noise figure of the image-rejection mixer as well as the balanced mixer will be higher than that of a single-ended mixer because of the loss associated with the hybrid junctions.

The *image-recovery mixer* is an image-rejection mixer designed to reduce the mixer conversion loss by properly terminating the diode in a reactance at the image frequency. Sometimes the lower conversion loss is offset by an increase in noise temperature, a mismatch at the IF, and higher intermodulation products. The improvement using image enhancement is about 1 or 2 dB; hence, the mixer needs to be of low loss so as not to negate the benefit.[14]

Dynamic Range   There seems to be no unique definition for the *dynamic range* of a radar receiver. It can generally be described as the ratio of the maximum input signal power to the minimum input signal power the receiver can handle without degradation in performance. "Degradation in performance," however, is not easy to define since it depends on the application. The minimum signal is sometimes taken to be the receiver rms noise level, which depends on the receiver bandwidth. The *minimum detectable signal* S_{min} might be selected as the minimum in the definition of dynamic range; but it also depends on bandwidth and apparently it is seldom used for this purpose.

The maximum signal might be the signal that causes the receiver to saturate (the output no longer increases with an increase in input). Saturation, however, is gradual as the signal increases in power, so the signal level that results in saturation is not precise. The maximum signal is more usually defined by the acceptable amount of gain compression, which is the deviation of the gain curve (output vs. input) from a straight line. The signal that causes a gain compression of one dB is commonly used for defining the maximum signal. Another criterion for the maximum signal power is based on the onset of *intermodulation distortion*. Intermodulation distortion generally occurs with large signals in the later stages of the receiver. The mixer generates a unique form of intermodulation product called *spurious responses*. It occurs when a harmonic of the local oscillator frequency mixes with a harmonic of the RF signal frequency to create difference frequencies that appear within the IF bandwidth. *Third-order intermodulation* occurs when two equal-amplitude signals within the receiver pass-band at two different frequencies f_1 and f_2 are input to the receiver and produce at the output the frequencies $2f_1 - f_2$ and $2f_2 - f_1$. The maximum signal power is then specified by the third-order intermodulation that can be tolerated when two signals are present within the pass band. The *third-order intermodulation* product is difficult to eliminate by filtering when the two frequencies are close to one another. Another possible indication of the maximum signal is when the echo signal at the mixer approaches the power level of the local oscillator. (The local oscillator power should be at least 7 dB greater than the largest received signal.[15]) No matter what definition is used, the dynamic range is almost always expressed in dB.

A large dynamic range is important if receiver saturation is to be avoided. Once the receiver saturates it can take a finite time to recover before targets again can be detected. Furthermore, when clutter is large enough to saturate the receiver, the MTI improvement factor will be reduced (Sec. 3.7). Saturation of the receiver by clutter echoes causes weak target echoes to be suppressed and not detected even though there might have been adequate improvement factor otherwise. In high-performance radars that must detect small moving targets in the presence of large clutter echoes by doppler processing, the receiver dynamic range must be at least equal to the required improvement factor.

As an example of the variation of the target echo signal power that might be experienced by a radar receiver, assume that an air-surveillance radar has to detect aircraft at ranges from 4 to 200 nmi. This corresponds to a variation in signal power of $(200/4)^4$, which is 68 dB. The average cross section of aircraft might vary from 2 to 100 m^2 (a variation of 17 dB), and the fluctuations in cross section might range over 30 dB. Adding all three factors, the variation of the total target echo signal might be 115 dB, more or less. This might be an extreme value, but radars that have to detect low cross-section targets could require even greater dynamic range.

Clutter echoes might vary over a range from 60 to 70 dB, or more. The use of STC (sensitivity time control), where the receiver gain is made to vary with time (Sec. 7.8), can reduce the variation in the target echo signal as well as the clutter echo signal. Not all radars, however, can employ STC. Pulse doppler radars, for example, cannot.

The mixer stage is often the limiting factor in dynamic range. A radar receiver that uses a doppler filter bank will have a higher dynamic range because of the narrower bandwidth of each filter. Pulse compression can also increase the dynamic range in proportion to the pulse compression ratio, if the clutter seen by the time sidelobes is not too large. The wider the bandwidth of the receiver (the IF stage) the less will be the dynamic range because of the greater likelihood that mixer intermodulation products (spurious responses) will be within the frequency band to limit the maximum signal that can be received. A wide bandwidth, as mentioned, also increases the noise level, which reduces dynamic range.

Large dynamic range may be obtained in some radar applications by inserting variable attenuation into the receiver as needed to keep the receiver from being overloaded, but this solution is limited to situations where rapid changes in the input signal are not expected.

Flicker Noise, or 1/f Noise   There exists in semiconductors a noise mechanism whose spectral density is inversely proportional to the frequency. It is called *flicker noise* or *1/f noise*,[16] and can be of importance at the lower frequencies. It is quite different from thermal or shot noise, which are independent of frequency. Flicker noise occurs in semiconductor devices such as diodes or transistors, and also in vacuum tubes with oxide-coated cathodes. The frequency relationship of flicker noise is more like $1/f^{\alpha}$, where α varies between 0.8 and 1.3, but it is more common to characterize it as $\alpha \approx 1$.[17] This relationship holds for very low frequencies, lower than might be of practical interest for radar.

The $1/f$ noise is not important for radar receivers whose IF frequencies are greater than a few hundred kilohertz. This is the case for most radar IF frequencies. It can be a factor limiting sensitivity, however, in radars that employ a *homodyne receiver*, also known as a superheterodyne receiver with zero IF. Homodyne receivers are sometimes used in CW radars because of their simplicity. The decrease in sensitivity due to $1/f$ noise at low frequencies might be tolerated for very short-range systems; but when maximum performance is necessary, the effect of the $1/f$ noise can be avoided by use of a superheterodyne receiver with an IF frequency where $1/f$ noise is low.

Oscillator Stability   In conventional pulse radars that do not perform doppler processing, stability of the local oscillator, or LO, cannot be ignored but it is usually not a major concern. However, when doppler processing is used to detect moving targets in clutter, as in the MTI radar, the LO has to be quite stable in order to reliably detect the doppler shift. This is why the LO in an MTI radar is called a *stalo*, or stable local oscillator. The MTI improvement factor that can be achieved with a magnetron oscillator transmitter is limited to modest values, so that the demands on oscillator stability can readily be met when a magnetron is used. Power amplifiers such as the klystron, TWT, and the transistor, however, allow much larger improvement factors than a magnetron. Thus greater demands are placed on the stability of the stalos used in such radars. Some high-prf pulse doppler radars that have to detect small targets in the midst of large clutter might encounter clutter echoes that could be 100 dB, or greater, than the target echoes, and thus require highly stable RF sources.

MTI and pulse doppler radars that employ a power amplifier use the sum of the receiver stalo and the coho as the input signal to the power amplifier. (This was indicated in Fig. 3.7 for the MTI radar.) Since the stalo is at a much higher frequency than the coho, it is the stalo that usually sets the limits on what can be achieved. The stalo can have a greater effect on performance than the power amplifier of the transmitter.[18] Thus we only consider the stalo here.

Phase Noise   Instability or phase noise in a stalo can be caused by power supply ripple; mechanical and acoustic vibrations from fans, motors, and cooling systems; or by vibrations of the platform (such as an aircraft or ship); as well as spurious responses and noise from the stalo itself. Phase noise is usually considered in the frequency domain, but in the time domain it can be thought of as being due to the deviation of the oscillator signal from a perfect sinewave. There is also amplitude-modulation noise associated with oscillators, but AM noise is usually small compared to phase noise. If not, it can be reduced by balanced mixers or other means.

In Sec. 3.7 the effect of equipment instabilities on MTI performance was mentioned. There it was shown that a pulse-to-pulse change in phase, $\Delta\phi$, limits the improvement factor of a two-pulse MTI to $I_f = (\Delta\phi)^{-2}$. When an MTI or pulse doppler radar uses many pulses to perform doppler filtering, this simple expression for improvement factor no longer applies. A different model has to be considered.

The reader is referred to Fig. 3.36 for an example of the spectrum of an oscillator as might be used in a mixer. (There is also further discussion of oscillator stability for doppler radars where this figure appears in Sec. 3.7.) In addition to the narrow spike at d-c due to the carrier (not shown in the figure), there is a noise spectrum that decreases monotonically with increasing frequency. There are also spikes, or *spurs*, that are usually caused by the power supply or vibrations. At the higher frequencies the phase noise levels off and is characterized by a uniform noise floor. The ordinate in Fig. 3.36 is the noise power within a one-hertz bandwidth relative to that of the carrier. It should be multiplied by the receiver bandwidth to obtain the actual power at the receiver.

Although Fig. 3.36 might be the noise from a stalo, it also represents the noise radiated by the transmitter since the stalo is a major part of the signal that excites the power amplifier transmitter. This noise may seem far down from the peak of the carrier, but the spectrum of the echo from stationary clutter is the same as the spectrum of the transmitter. (Internal motion of the clutter can further increase the spectrum of the received echo signal.) As mentioned, clutter can be quite large compared to the weak moving target echo. The MTI or pulse doppler radar may be able to attenuate the main clutter line at d-c, but the clutter spectrum often has components at frequencies where doppler-shifted echoes from moving targets are expected. These components can mask the desired target echoes. Good performance of an MTI or a pulse doppler radar requires that the transmitter spectrum, and the clutter-echo spectrum it produces, be low enough to detect the slowly moving weak targets that are of interest. It would not be unusual to find that oscillator noise can be the limiting factor in some high-performance radars that must detect low-speed, low cross-section moving targets in heavy clutter. Good oscillator design is therefore important for achieving good radar performance.

The effect of phase noise can be determined by measuring the phase-modulation spectrum of the stalo and using it to obtain the MTI improvement factor. The procedure will not be given here. It is outlined by Taylor[18] and given in more detail with examples by Goldman.[19] Since the stalo is part of the transmitter as well as part of the receiver, the effect of phase noise on MTI performance will be range dependent. At the shorter ranges, or time delays, greater stalo noise can be tolerated at frequencies closer to the carrier (lower target doppler frequency shifts) than at longer ranges. For this reason, the effect of stalo stability needs to be computed for several ranges.

Oscillator phase noise can be a serious limitation to the performance of a modern high-performance MTI or a pulse doppler radar. Its effect has to be found with measurements and analysis more complicated than was indicated in Sec. 3.7.

Types of Stable Oscillators[20,21]   Almost all of the oscillators used for stable sources can be thought of as consisting of an amplifier, a resonant circuit that determines the frequency and the phase noise, and feedback to generate oscillation. The amplifier is often a transistor. The following is a brief listing of the various oscillators that have been considered for use as stable sources.

Crystal Oscillator   The mechanically vibrating piezoelectric quartz crystal has been an important device for producing stable oscillators ever since the early days of commercial radio.[22] A piezoelectric material is one which mechanically deforms along one crystal axis when an electric potential is applied along another axis. Conversely an electrical potential is obtained when a mechanical deformation occurs. The piezoelectric crystal is used as the resonator in the feedback circuit of a transistor oscillator. It is often mounted in a small-size temperature-controlled oven and isolated from vibrations. It is a very stable source at low frequencies (10 to 180 MHz), but its output can be multiplied in frequency to provide stable signals in the microwave region.

Frequency Multiplier   A low-frequency stable oscillator can be multiplied to a higher frequency by applying its signal to a nonlinear device such as a diode or varactor to generate harmonics of the fundamental frequency. A filter is used to select the desired harmonic. The phase noise power, however, increases as the square of the frequency-multiplication ratio. For example, when a 10-MHz stable source is multiplied to 10 GHz, its noise is increased by 60 dB. In addition, there can also be additive phase noise produced in frequency multipliers. In spite of the increase in noise with multiplication in frequency, multiplication is a good method for taking advantage of the excellent stability of a low-frequency sources to obtain a stable oscillator at radar frequencies.

Dielectric Resonator Oscillator (DRO)   The resonant circuit in this type of oscillator is a dielectric material such as a sapphire crystal, ceramic,[23] or titanate in a regular geometric form that acts as a microwave resonant cavity. The high dielectric constant of the resonator allows it to be much smaller in size compared to a metallic cavity resonator. It is among the most stable of room-temperature oscillators. Because of its small size it has a relatively high Q and may be quite rigid so as to reduce its sensitivity to shock and vibration. When the dielectric resonator is made larger to obtain even higher values of Q and enhanced frequency stability, it might be more sensitive to temperature changes and vibration. The DRO has been a popular device for application as a low-noise, stable oscillator at microwave frequencies.

SAW Oscillator   The surface acoustic wave (SAW) device can also be used as the resonator in a feedback oscillator. SAW oscillators can be quite small and can be obtained from about 100 MHz to 3 GHz. Ewell[20] states that the phase noise of a SAW oscillator can be worse than that of a frequency-multiplied crystal oscillator at low offset frequencies (1 kHz for example) from the carrier, but it can be better at high frequency offsets (greater than 10 kHz).

YIG Oscillator   A small sphere of yttrium iron garnet (YIG) suspended within a resonant cavity with an applied magnetic field can act as the resonant device of an oscillator. The resonant frequency of a spherical YIG crystal depends only on the applied magnetic field and not on its dimensions. It has a relatively high level of phase noise, but it has the advantage of being tunable by changing the applied magnetic field.

Klystron Oscillator and Gunn Oscillator   The reflex klystron oscillator (originally used as the local oscillator of many a World War II radar receiver) and the Gunn diode oscillator are two very different type of devices. Both have relatively high phase noise, but when coupled to a high-Q external resonant cavity they can be of high stability. The use of superconducting high-Q cavities can produce "extremely low phase noise levels."[20]

High-Temperature Superconducting Oscillators   As was mentioned in Sec. 3.7, the phase noise of oscillators can be improved by employing very low loss superconductive resonators, especially those that are superconductive at the temperature of liquid nitrogen, 77 degrees.[24]

Direct Digital Synthesis[25]   A *frequency synthesizer* produces one or more frequencies over a wide spectrum by translating the stable frequency of a precision frequency source, such as a crystal-controlled oscillator. In *direct synthesis* a single precision oscillator is multiplied and/or divided to obtain a desired frequency. When this process is performed digitally, it is called *direct digital synthesis* (DDS).[26] A DDS can generate the multiple frequencies needed for the stalo, coho, a second LO if dual conversion is used, and timing frequencies. It can also provide linear or nonlinear FM for pulse compression systems. A DDS generally uses a phase accumulator (to establish the time sequence), sine lookup table (to establish the amplitude of the signal waveform), a digital-to-analog converter, low pass filter, and frequency multiplier or heterodyne to translate to a higher frequency.[27] It has the advantage of extremely fast frequency switching, small size steps in frequency, excellent phase noise, reasonably good spurious performance, transient-free (phase continuous) changes in frequency, flexibility in applying modulation, and it achieves its good performance in a small volume.

A/D Converters   The A/D converter, which changes analog signals to digital signals, is an important component of digital processing. There are many different ways it has been implemented.[28] Its performance for radar is judged by the number of bits into which it can quantize a signal and the sampling rate at which it can operate. As was mentioned in Sec. 3.5 (where the effect of the A/D converter on MTI performance was discussed and some examples of performance were given), the number of bits into which the A/D converter can quantize a signal decreases as the sampling rate, or bandwidth, increases. Thus the larger the bandwidth of the signal the more difficult it is to maintain good performance. The A/D converter sometimes can be a limitation in wideband radar or when large clutter attenuation is required.

Bandpass Sampling at IF   Digital signal processing that is conducted at baseband (video) requires two baseband A/D converters and an in-phase and a quadrature channel. Although baseband digital processing has been widely used, there are limitations. The two baseband converters have to be well balanced over a wide dynamic range and there cannot be significant phase errors between the two channels (the phase difference between the two channels cannot differ significantly from 90°). Waters and Jarrett[29] indicate that these problems do not appear if the A/D conversion is performed in the bandpass portion of the receiver at IF. The in-phase and quadrature components are obtained by a single A/D converter from the samples taken directly from the original IF signal. The phase errors between the two channels are considerably smaller with IF sampling than with baseband sampling. Although only one channel is needed in bandpass sampling, its sampling rate has to be greater than that of the A/D converters used in baseband sampling. Further discussion can be found in Sec. 3.5.

Digital Radar Receiver   There does not seem to be a unique, well-accepted definition of a digital receiver. A digital radar receiver, ideally, could be thought of as one that is completely digital with a wide dynamic range A/D converter that operates directly on the signal received at the antenna terminals. This would be followed by a highly capable computer to perform the functions found in a radar receiver. It is difficult, however, to achieve such a receiver with the bandwidth and large dynamic range required for high-performance microwave MTI and pulse doppler radars. More realistically, a digital radar receiver might be one that uses an analog RF amplifier and mixer, and even analog IF circuitry, followed by an IF A/D converter and digital video processing.

A different and more practical definition of a digital radar receiver was proposed by Wu and Li,[30] who stipulate that such a receiver have two significant differences compared to analog radar receivers. It should utilize (1) a direct digital synthesizer (DDS) as the local oscillator and (2) direct bandpass sampling at IF before detection, with all subsequent processing being done digitally.

In addition to a high-speed A/D converter with many bits of quantization, a digital receiver requires the digital processing to have sufficient speed to operate in real time and to have a large enough information storage memory. There is little doubt that the major advances in radar and its increased applicability since the 1970s have been due to the phenomenal advances in digital processing technology. It is likely that "digits" will continue to be a major driver of future advances in radar performance.

Russian Cyclotron Wave Electrostatic Amplifier[31–33]   Solid-state amplifiers have been a popular choice for the front-end of a radar receiver, but they are not the only choice. A Russian receiver development, called the cyclotron wave electrostatic amplifier (CWESA), has been a popular receiver for certain types of radars because of characteristics not available with other devices. It is also more usually known as an *electrostatic amplifier* (ESA). The ESA is said to have a low noise figure, bandwidths of 5 to 10 percent, linear phase variation with frequency, and other attributes suitable for a receiver front end; but its uniqueness is that it can sustain a high level of input power without additional protection and it can recover quickly from overload. Duplexers or receiver protectors are not needed.

In the ESA an electrostatic cyclotron wave is launched on a thin electron beam in an input structure; it is amplified in an intermediate structure, and then coupled to an output structure. The thin electron beam at the cathode might have dimensions of 0.03 by 0.7 mm and a current of 250 to 280 μA. The theory[31] of this device will not be summarized here except to say that a longitudinal magnetic field is required so that the input signal, when coupled to the electron beam, results in cyclotron motion of the electrons. Permanent magnets are used to reduce weight. At S or C bands, these units are said to weigh approximately 2 kg, have an approximate volume of one liter, and a power consumption of 1 to 1.5 W. A 1.0 dB noise figure was achieved at frequencies up to 3 GHz and a 2.4 dB noise figure at 10 GHz.

When a large signal appears at the input to the ESA, it causes a large reflection coefficient (a large VSWR) so that the signal is entirely reflected and is not absorbed, which is unlike diode receiver protectors that absorb the input energy. Thus these receivers have been used in radars without additional duplexers or diode receiver protectors. When the overload is removed the device quickly returns to service, typically in about 20 ns at frequencies above S band.[31,32] Longer values of recovery time are experienced at lower frequencies. It has been claimed that in radar applications the ESA can withstand peak powers of 10 kW and average powers of 300 W at frequencies above S band, and higher powers at lower frequencies.

Sometimes a transistor amplifier is added as a second stage to obtain higher gain. Such a combined ESA and transistor, operating from 7 to 7.4 GHz with a single high-voltage supply of 400 V in addition to small filament and transistor amplifier supply voltages, produced a noise figure of 3.4 dB and a gain of approximately 23 dB. It could withstand in excess of 5 kW of peak and 150 W of average power at the input and recover in less than 50 ns. When a transistor second stage is used, such a device sometimes is called an *electrostatic combined amplifier* (ESCA).

A tunable version of the ESA was said to demonstrate very rapid tuning over a 50 percent bandwidth, with an instantaneous bandwidth of 1 percent.

The rapid recovery time of this amplifier makes it attractive for use with pulse doppler radars which require a high prf. Pulse doppler radars operate with high duty cycles so there is little range-space available. Long recovery times reduce the available range-space. If the duty cycle were 10 percent and the pulse width were 1 μs, a diode protector recovery time of 1 μs would significantly increase the receiver dead time and increase the minimum range. The 20 ns recovery time of the ESA would hardly be noticed. The duty cycles of high-prf pulse doppler radars can be as high as 0.3 to 0.5, which makes receiver recovery time even more important.

The testing of a 200-MHz bandwidth X-band ESA combined with a transistor second stage as an ESCA receiver for a high-prf pulse doppler radar was described by Ewell.[31] Pulse repetition rates were from 1 kHz to several hundred kHz and pulse durations less than 1 μs. For this application, the ESCA was considered superior to conventional gas TR tubes whose recovery times were too long and too unpredictable. They were also superior to multipactor discharges which had good recovery time but had high spike leakage that required varactor diode receiver protectors. They also were costly and required additional components such as an oxygen generator, ion pump, and cooling system. Ewell's measurements appear to confirm the consistency of this device to meet pulse doppler radar system requirements. It provided protection from overload, fast recovery time, linearity, and electronic control of dynamic range.

Because of its size, the ESA is not suitable for most applications of active-aperture phased array radar; but there are many important radar applications where an active aperture is not necessary. The ESA is attractive in those radar systems where a single or only a few receiver channels are used. At the time of publication of the cited references for this subsection, there were roughly ten thousand of these devices manufactured and in use in a number of systems around the world, mainly in Russia and China. An example is its use in the Russian S300 PMU air defense and anti-tactical ballistic missile (ATBM) system (NATO designation SA-10). This employs an X-band space-fed phased array radar with pulse doppler waveform designed to operate in high clutter and electromagnetic countermeasure environments.[31] The ESA is also found in the Russian S300V (NATO designation SA-12) air defense system. Barton points out that the electrostatic amplifier tube helps make the total RF loss in these Russian receiving systems significantly lower than the loss found in comparable Western systems.[34]

Phase Detector, Phase-Sensitive Detector   In Fig. 3.7 of Sec. 3.1, the *phase detector* was introduced as the device in an MTI receiver that extracted the doppler frequency shift of an echo signal. It compared the echo signal to a reference signal (the coho) which was coherent with the MTI transmitter signal. In the MTI phase detector, it is the rate of change of phase of the echo signal with time that is of interest since it determines the doppler frequency shift of the echo from a moving target. In Fig. 4.4 of Sec. 4.2, the *phase-sensitive detector* was shown in the amplitude-comparison monopulse tracking radar to allow the extraction of the sign of the angle error along with its magnitude. The input to this detector was the angle-error signal and the signal from the sum channel which acted as the reference. In both the MTI radar and the monopulse tracker two sinusoidal voltage inputs were available to a nonlinear device. The two were coherent with respect to one another in that they could be thought of as being from the same source. In both these detectors, one of the two voltages is the reference and the other is the received echo signal.

Taylor[35] points out that the distinction between a phase detector and a phase-sensitive detector is not always clear because of the similarity of the analog circuits that perform these two functions. He states that it is generally agreed that a phase detector is one in which only phase information is present in the output; a phase-sensitive detector is one in which both phase and amplitude information are in the output; and a mixer when phase, amplitude, and frequency information are present in the output. He also points out that "doppler frequency shifts are excepted in this convention."

Krishnam[36] indicates that the difference between these two detectors is in the actual operating conditions and not the hardware. He states that it had been usual to assume that the reference and the signal are of the same amplitude for the phase detector. For the phase-sensitive detector it was common to assume that the reference is much larger than the signal. He then shows that other assumptions can be made. He denotes $V_1 = E_1 \sin \omega t$ as the reference and $V_2 = E_2 \sin (\omega t + \phi)$ as the signal. For his particular detector model, he then shows that when E_1 is exactly equal to E_2, the output is $E_0/E_1 = 2(|\cos \phi/2| - |\sin \phi/2|)$, which is approximately linear with respect to ϕ over the range $0 < \phi < \pi/2$. Under these conditions, the device can be considered as a phase detector. When $E_2 >> E_1$ (signal is large compared to the reference), the output is $E_0 = 2E_1 \cos \phi$, which also is a phase detector. When $E_2 < E_1$ so that the reference is larger than the largest signal E_2, the device is shown to be a "perfectly linear" phase-sensitive detector with an output $E_0 = +2E_2$ when $\phi = 0$, and $E_0 = -2E_2$ when $\phi = \pi$.

Example of a Receiver   One seldom finds in the radar literature a paper on the design of a radar receiver. For some reason, receiver designers do not prepare such papers, or the journal editors and referees do not accept them. There is, however, at least one paper of which I am aware that describes the receiver for the original Aegis AN/SPY-1A shipboard air-defense system.[37] The receiver is in two parts. One part is the on-array portion which contains the low-noise amplifiers and related components. It is mounted to the rear of each of the four antenna faces of Aegis to minimize pre-RF amplifier losses. The other part is located both in the fore and aft deckhouses and contains components with minimal impact on the noise figure. There are eleven receiver channels: three monopulse tracking channels, one sidelobe interference blanker channel, six auxiliary ECCM sidelobe canceler channels, and one auxiliary channel that acts as a spare. Each channel has two inputs so they can be time-shared between two antenna arrays to reduce cost.

There is too much in the paper to adequately summarize here, but it is recommended as being one of the few examples available that provides an overview of radar receiver engineering not usually found in radar texts.

11.4   DUPLEXERS AND RECEIVER PROTECTORS

A pulse radar can time share a single antenna between the transmitter and receiver by employing a fast-acting switching device called a *duplexer*. On transmission the duplexer must protect the receiver from damage or burnout, and on reception it must channel the echo signal to the receiver and not to the transmitter. Furthermore it must accomplish the switching rapidly, in microseconds or nanoseconds, and it should be of low loss. For high-power applications, the duplexer is a gas-discharge device called a TR (transmit-receive) switch. The high-power pulse from the transmitter causes the gas-discharge device to break down and short circuit the receiver to protect it from damage. On receive, the RF circuitry of the "cold" duplexer directs the echo signal to the receiver rather than the transmitter. Solid-state devices have also been used in duplexers. In a typical duplexer application, the transmitter peak power might be a megawatt or more, and the maximum safe power that can be tolerated by the receiver might be less than a watt. The duplexer, therefore, must provide more than 60 to 70 dB of isolation between the transmitter and recovery with negligible loss on transmit and receive.

The duplexer cannot always do the entire job of protecting the receiver. In addition to the gaseous TR switch, a receiver might require diode or ferrite limiters to limit the amount of leakage that gets by the TR switch. These limiters, which have been called *receiver protectors*, also provide protection from the high-power radiation of other radars that might enter the radar antenna with less power than necessary to activate the duplexer, but with greater power than can be safely handled by the receiver. There might also be a mechanically actuated shutter to short-circuit and protect the receiver whenever the radar is not operating. Sometime the entire package of devices has been known as a *receiver protector.*[38] The term is ambiguous, since receiver protector is also the name for the diode limiter or similar device that follows the duplexer for the purpose of reducing the leakage power passed by the duplexer. In this text the term receiver protector is used to denote a limiter that follows the duplexer. The duplexer, receiver protector, and other devices for preventing receiver damage are better known as the *duplexer system*, so as to prevent confusion by the same term (receiver protector) being used to describe the entire receiver protection system as well as one part of it.

Balanced Duplexer   The balanced duplexer, shown in Fig. 11.3, is based on the short-slot hybrid junction which consists of two sections of waveguides joined along one of their narrow walls with a slot cut in the common wall to provide coupling between the two.[39] (The short-slot hybrid junction may be thought of as a broadband directional coupler with a coupling ratio of 3 dB.) Two TR tubes are used, one in each section of waveguide. In the transmit condition, Fig. 11.3a, power is divided equally into each waveguide by the first hybrid junction (on the left). Both gas-discharge TR tubes break down and reflect the incident power out the antenna arm as shown. The short-slot hybrid junction has the property that each time power passes through the slot in either direction, its phase is advanced by 90°. The power travels as indicated by the solid lines. Any power that leaks through the TR tubes (shown by the dashed lines) is directed to the arm with the matched dummy load and not to the receiver. In addition to the attenuation provided by the TR tubes, the hybrid junctions provide an additional 20 to 30 dB of isolation.

On reception the TR tubes do not fire and the echo signals pass through the duplexer and into the receiver as shown in Fig. 11.3b. The power splits equally at the first junction and because of the 90° phase advance on passing through the slot, the signal recombines in the receiving arm and not in the arm with the dummy load.

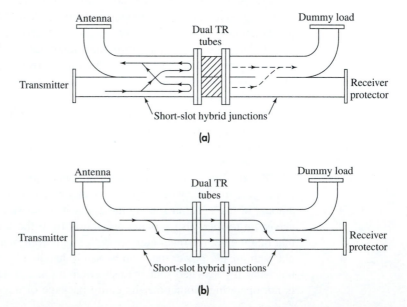

Figure 11.3   Balanced duplexer using dual TR tubes and two short-slot hybrid junctions. (a) Transmit condition and (b) receive condition.

The balanced duplexer is a popular form of duplexer with good power handling capability and wide bandwidth.

TR Tube   The TR tube is a gas-discharge device designed to break down and ionize quickly at the onset of high RF power, and to deionize quickly once the power is removed. One construction of a TR consists of a section of waveguide containing one or more resonant filters and two glass-to-metal windows to seal in the gas at low pressure. A noble gas like argon in the TR tube has a low breakdown voltage, and offers good receiver protection and relatively long life. TR tubes filled only with pure argon, however, have relatively long deionization times (long recovery times) and are not suitable for short-range applications. Adding water vapor or a halogen gas to the tube speeds up the deionization time, but such tubes have shorter lifetimes than tubes filled only with a noble gas. Thus a compromise must usually be accepted between fast recovery time and long life.

To insure reliable and rapid breakdown of the TR tube on application of high power, an auxiliary source of electrons is supplied to the tube to help initiate the discharge. This may be accomplished with a "keep-alive," which produces a weak d-c discharge that generates electrons that diffuse into the TR where they assist in triggering the breakdown once RF power is applied by the transmitter. An alternative is to include a small source of radioactivity, such as tritium (a radioactive isotope of hydrogen), which produces low-energy-level beta rays to generate a supply of electrons.[40] The tritium is in compounded form as a tritide film. The radioactive source, sometimes called a *tritiated ignitor*, has the advantage of not increasing the wideband noise level as does a keep-alive discharge (by about 50 K) and has longer life (by an order of magnitude), but it passes more leakage energy so that it requires one or more cascaded PIN diode limiter stages to further attenuate the leakage.[41] The tritium ignitor needs no active voltages, so it allows the receiver protector to function with the radar off without the need for a mechanical shutter to protect the radar from nearby transmissions. Being a radioactive device, however, does cause concern about its handling and disposal. The combination of the tritium-activated TR followed by a diode limiter has been called a *passive TR-limiter*.

The TR is not a perfect switch; some transmitter power always leaks through to the receiver. The envelope of the RF leakage might be similar to that shown in Fig. 11.4. The short-duration, large-amplitude *spike* at the leading edge of the leakage pulse is the result of the finite time required for the TR to ionize and break down. Typically, this time is of the order of 10 nanoseconds. After the gas in the TR tube is ionized, the power leaking through the tube is considerably reduced from the peak value of the spike. This portion

Figure 11.4   Leakage pulse through a TR tube.

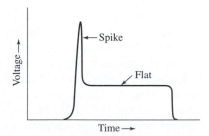

of the leakage pulse is called the *flat*. Damage to the receiver front-end may result when either the energy contained within the spike or the power in the flat portion of the pulse is too large. The spike leakage of TR tubes varies with frequency and power and whether or not the tube is primed with electrons, but might be "typically" about one erg. The attenuation of the incident transmitter power might be of the order of 70 to 90 dB.

A fraction of the transmitter power incident on the TR tube is absorbed by the discharge. This is called *arc loss*. It might be 0.5 to 1 dB in tubes with water vapor and 0.1 dB or less with argon filling. On reception, the TR tube introduces an insertion loss of about 0.5 to 1 dB. The life of a TR tube is determined more by the amount of leakage power it allows to pass or when its recovery time becomes excessive rather than by its physical destruction or wear.

Solid-State Receiver Protectors, Diode Limiters   Improvements in receiver sensitivity sometimes are obtained with front-ends and mixers that are more sensitive to damage from RF leakage. Such sensitive devices require better protection from the RF leakage of conventional duplexers. A PIN diode limiter placed in front of the receiver helps reduce the leakage and act as a *receiver protector*. A diode limiter passes low power with negligible attenuation, but above some threshold it attenuates the signal so as to maintain the output power constant. This property can be used for the protection of radar receivers in two different implementations depending whether the diodes are operated unbiased (self-actuated) or with a d-c forward-bias current. Unbiased operation without the use of an external current supply is also known as *passive*. It has the advantage of almost unlimited operating life, fast recovery time, no radioactive priming, and versatility to perform multiple roles.[38] Its chief limitation is its low power handling. A passive solid-state limiter for *X*-band WR90 waveguide with 7 percent bandwidth and a 1-μs pulse width had a peak power capability of 10 kW and a CW power capability of 10 W.[42] Its insertion loss was 0.6 dB, leakage power was 10 mW, and a 1.0-μs recovery time. When used with a 40-μs pulse width, this limiter could withstand 2 kW of peak power and 300 W CW, with a loss of 0.8 dB.

Biasing of the diodes during the high-power pulse, also known as *active*, is capable of handling a great deal more power than when operated passively. The diode is biased into its low impedance mode prior to the onset of the transmitter pulse. Although the active diode-limiter offers many advantages for use with duplexers, it does not protect the receiver when the bias is off. It thus offers poor protection against nearby asynchronous transmissions that arrive at the radar during the interpulse period or when the radar is shut down.

Incorporation of Sensitivity Time Control[43,44]   In Sec. 7.8 the use of sensitivity time control (STC) was described as a method to reduce the effects of nearby large clutter echoes without seriously degrading the detection of desired targets at short range. STC is the programmed change of receiver gain with time, or range. At short ranges the receiver gain is lowered to reduce large nearby clutter echoes. As the pulse travels out in range the gain is increased until there are no more clutter echoes.

There are advantages for having the STC in the RF portion of the radar just ahead of the receiver. STC can be applied by biasing the diodes of a receiver protector to provide a time-varying attenuation without adding to the receiver noise figure. There is no increase in insertion loss to obtain the STC action, above that inherent in the design of the receiver protector. The PIN diode stages provide the self-limiting action during transmit and the STC function during receive. The nonlinear nature of the diode requires a linearizing circuit to achieve the desired variation of attenuation with time. The STC variation depends on the nature of the terrain seen by the radar. A digitally controlled STC drive with random access memory allows the radar designer to employ different STC response profiles according to the various types of terrain that might be seen by the radar.

Varactor Receiver Protectors   With fast-rise-time, high-power RF sources, the receiver protector may be required to self-limit in less than one nanosecond. This can be achieved with fast-acting PN (varactor) diodes. A number of diode stages, preceded by plasma limiters, might be employed. In one design, an *X*-band passive receiver protector was capable of limiting 1-ns rise time, multikilowatt RF pulses to 1-W spike levels.[45]

Ferrite Limiters   The ferrite limiter has very fast recovery time (can be as low as several tens of nanoseconds), and if the power rating is not exceeded, it should have long life. The spike and flat leakage are low and it has been able to support a peak power of 100 kW;[46] but the insertion loss is usually higher (1.5 dB) and the package is generally longer, heavier, and more expensive than other receiver protectors. Except for the initial spike, the ferrite limiter is an absorptive device rather than a reflective device (as is a gas-tube TR) so that the average power capability of these devices can be a problem. Air or liquid cooling might be required. A diode limiter usually follows the ferrite limiter to reduce the leakage at high peak power.

Pre-TR Limiter[38,47]   A pre-TR is a gaseous tube placed in front of a solid-state limiter. The function of the pre-TR is to reduce the power that has to be handled by the diode limiter. (It is similar to what was called a passive TR-limiter earlier in this section.) The pre-TR gas tube has high power handling capability, can operate with long pulses, has very fast recovery time, contains a radioactive priming source, but has limited operating life. Very high average power levels may require liquid cooling of the pre-TR mount. End of life for a pre-TR tube usually is caused by the increased recovery times that result from the cleanup of the gas within the tube.

The pre-TR tube can be a quartz cylinder filled with chlorine or a mixture of chlorine and an inert gas. Chlorine, a halogen gas, has a very rapid recovery time; typically a fraction of a microsecond for pulse widths up to 10 μs. The tube is mounted in a waveguide iris. In some cases, the quartz pre-TR tube can be designed to be field replaceable once it reaches end of life.

Mutipactor[46,48]   The recovery times of high-power duplexers discussed thus far are from a fraction of a microsecond to several tens of microseconds. By employing the principle of multipacting, a recovery time as short as 5 or 10 ns is possible. Fast recovery time is important for high prf and high duty cycle radars. The multipactor is a vacuum tube and does not have the long recovery characteristics of a gas-filled tube. It contains surfaces capable of large secondary electron emission upon impact by electrons. The secondary emission surfaces are biased with a d-c potential. The presence of RF energy causes electrons to make multiple impacts that generates by secondary emission a large electron cloud. The electron cloud moves in phase with the oscillations of the applied RF electric field to absorb energy from the RF field. RF power is dissipated thermally at the secondary emission surfaces, and the device requires liquid cooling to remove the absorbed power. Since it is a vacuum device, the recovery time of the multipactor is extremely fast. The flat-leakage power passed by the multipactor is often high enough to require a passive diode limiter to follow it. The multipactor offers no protection when the power is turned off. It has the disadvantage of being complex in that it requires liquid cooling, an ignitor electrode to ensure that multipacting starts quickly, an oxygen source to maintain the magnesium oxide surface that provides the secondary emission electrons, and a pump to maintain a good vacuum.

Solid-State Duplexers   There has always been a desire to replace the gas-discharge duplexer with an all-solid-state duplexer because of the potential for long life, fast recovery time, no radioactive priming, and versatility. Although passive operation is desired, it is limited in power. The lowest loss and highest power handling are obtained with active circuits in which the PIN diodes are switched in synchronism with the transmitter pulses. Generally, diodes that can handle high power will have longer recovery times and tend to have higher leakage power—so that they might require additional stages of lower level limiters with increased loss and increased cost. A failure of the active drive circuit, however, could cause destruction of the diode switches as well as the receiver.

   Several examples of all solid-state duplexers have been described in the literature. An *L*-band self-switching duplexer design used four PIN diodes that were biased by four fast-acting decoupled varactor detector diodes.[49] These detector diodes bias the PIN diodes into conduction in a time considerably shorter than the rise time of the RF power pulse. The device could handle 100-kW peak power with 100-W average power and a 3-μs pulse width. Its insertion loss was 0.5 dB. The duplexer was followed by a low-power multiple stage varactor limiter that reduced the spike and flat leakage of 2.8 kW and 32 W peak respectively to levels low enough that low-noise amplifiers were adequately protected. The recovery time was about 15 μs. A UHF solid-state duplexer also using four diodes was reported to have 300-kW peak power, 5-kW average power, 60-μs pulse width, and an insertion loss of 0.75 dB. A *C*-band solid-state duplexer with 16 PIN diodes was capable of 1-MW peak power with a 14-μs pulse width, 0.01 duty cycle, and insertion loss of less than 1 dB.[50] This device was followed by an additional low-power diode switch with an insertion loss of about 0.6 dB. It provided an isolation of 60 dB, making the total isolation of the duplexer system over 100 dB.

Circulators as Duplexers   The ferrite circulator is a three- or four-port device that can, in principle, offer isolation of the transmitter and receiver. In the three-port circulator, the transmitter may be connected to port 1. It radiates out of the antenna connected to port 2. The received echo signal from the antenna is directed to port 3 which connects to the receiver. The isolation between the various ports might be from 20 to 30 dB, but the limitation in isolation is determined by the reflection (due to impedance mismatch) of the transmitter signal from the antenna that is then returned directly to the receiver. For

Table 11.1 Comparison of various types of duplexing devices

Device	Recovery Time	Average Power	Peak Power
TR tube	<1 μs to 100 μs		1 MW
Pre-TR	50 ns to 1 μs	50 kW	5 MW
Diode limiter	50 ns to 10 μs	1 kW	100 kW
Ferrite limiter	20 ns to 120 ns	10 W	100 kW
Multipactor	1 ns to 20 ns	500 W	80 kW
Electrostatic amplifier	20 ns	300 W, or higher	10 kW, perhaps as high as 500 kW

example, if the VSWR (voltage standing wave ratio) of the antenna were 1.5 (a pretty good value), about 4 percent of the transmitter power will be reflected by the antenna and return to the receiver. This corresponds to an isolation of 14 dB. If the VSWR were 2.0, the effective isolation is only 10 dB. To limit damage, a good receiver protector needs to be included. Circulators can be made to withstand high peak and average power; but large power capability generally comes with large size and weight. For example, an *S*-band differential phase-shift waveguide circulator that weighs 80 pounds has essentially the same insertion loss, isolation, and bandwidth of an *S*-band miniature coaxial Y-junction circulator that weighs 1.5 oz.[51] The larger circulator, however, can handle 50 kW of average power while the smaller circulator is rated at 50 W. (The ratio of powers exceeds the ratio of weights.)

Small-size circulators, usually in conjunction with a receiver protector, often are used as the duplexer in solid-state TR modules for active aperture phased arrays. (Note that, unfortunately, the term "TR" has been used for both T/R modules and TR duplexer gas-discharge tubes.)

Summary of Performance   Table 11.1 summarizes the performance of various duplexer devices in term of recovery time and power handling. This table is adapted from the paper by Bilotta,[38] but modified from information in other references cited previously. The values in this table depend on frequency and other factors so they should not be considered as absolute limits, but only as an approximate guide.

11.5 RADAR DISPLAYS

Originally the radar display had the important purpose of visually presenting the output of the radar receiver in a form such that an operator could readily and accurately detect the presence of a target and extract information about its location. The display had to be designed so as not to degrade the radar information and to make it easy for the operator to perform with effectiveness the detection and information extraction function. It was not uncommon for an operator to employ a grease pencil to mark on the face of a cathode-ray-tube display the location of a target from scan to scan and manually extract the target speed and direction. As digital signal processing and digital data processing improved, more and more of the detection and information extraction process was performed automatically by electronic means so that the role of the operator was less. Processed detection and target information now are displayed to the operator who has little responsibility for making the actual detection decision. Instead of displaying only detections, many surveillance radars display target track vectors along with auxiliary alphanumeric information to an operator.

When the display is connected directly to the output of the radar receiver without further processing, the output is called *raw video*. When the receiver output is first processed by an automatic detector or an automatic detector and tracker before display, it is called *synthetic video* or *processed video*. The requirements for the display differ somewhat depending whether raw or processed video is displayed. Some radar operators prefer to see on a display the raw video lightly superimposed on the processed video.

In many cases the operator does not see the unprocessed output of the radar. An example is the Nexrad doppler weather radar in which the radar measures three parameters in each resolution cell: the amplitude of the echo signal (proportional to its radar cross section), the mean radial velocity (from the doppler frequency shift) of the meteorological scatterers, and the variance of the radial velocity (a measure of the motion of the individual scatterers within the resolution cell). These three meteorological echo parameters in each resolution cell are passed to a computer that generates the many different types of weather products such as maps of precipitation, wind shear at various horizontal and vertical planes, mesocyclones, tornadoes, prediction of flooding, and many others.

The radar display is now more like the familiar television monitor or computer display that shows the entire scene continuously rather than just indicate the echoes from the region currently illuminated by the narrow antenna beam. Thus the role of the display has changed as the need for operator interpretation has decreased.

Types of Display Presentations   The IEEE Standard on Radar Definitions includes 19 different types of display formats.[52] Most date to World War II and many are now seldom used. The standardized definitions do not cover all possible display formats. Given below are some of the more popular formats that have been employed. The IEEE uses the term "display" in its definitions but here we use either "scope" or "display" depending on what is perceived to be the more common usage. The following definitions are not precisely identical with the IEEE definitions, but they are consistent with them.

A-scope. *A deflection-modulated rectangular display in which the vertical deflection is proportional to the amplitude of the receiver output and the horizontal coordinate is proportional to range (or time delay).* This display is well suited to a staring or manually tracking radar, but it is not appropriate for a continually scanning surveillance radar since the ever-changing background scene makes it difficult to detect targets and interpret what the display is seeing.

B-scope. *An intensity-modulated rectangular display with azimuth angle indicated by one coordinate (usually horizontal) and range by the orthogonal coordinate (usually vertical).* It has been used in airborne military radar where the range and angle to the target are more important than concern about distortion in the angle dimension.

C-scope. *A two-angle intensity-modulated rectangular display with azimuth angle indicated by the horizontal coordinate and elevation angle by the vertical coordinate.* One application is for airborne intercept radar since the display is similar to what a pilot might see when looking through the windshield. It is sometimes projected on the windshield as a heads-up display. The range coordinate is collapsed on this display so a collapsing loss might occur, depending how the radar information is processed.

E-scope. *An intensity-modulated rectangular display with range indicated by the horizontal coordinate and elevation angle by the vertical coordinate.* The E-scope provides a vertical profile of the radar coverage at a particular azimuth. It is of interest with 3D radars and in military airborne terrain-following radar systems in which the radar antenna is scanned in elevation to obtain vertical profiles of the terrain ahead of the aircraft. The E-scope is related to the RHI display.

PPI-display, or plan-position indicator. *An intensity-modulated circular display in which echo signals from reflecting objects are shown in plan view with range and azimuth angle displayed in polar (rho-theta) coordinates to form a map-like display.* Usually the center of the display is the location of the radar. A *sector-scan PPI* might be used with a forward-looking airborne radar to provide surveillance or ground mapping over a limited azimuth sector. An *offset PPI* is one where the origin (or location of the radar) is at a location other than the center of the display. This provides a larger display area for a selected portion of the coverage. The location of the radar with an offset PPI may be outside the face of the display.

RHI-display, or range-height indicator. *An intensity-modulated rectangular display with height (target altitude) as the vertical axis and range as the horizontal axis.* The scale of the height coordinate is usually expanded relative to the range coordinate. It has been used with meteorological radars to observe the vertical profile of weather echoes.

In addition, imaging radars such as synthetic aperture radar (SAR) and side-looking airborne radar (SLAR) generally display their output as a strip map with range as one coordinate and cross-range as the other coordinate. With the expanding graphics technology available from the computer industry, there is much more flexibility available in displaying radar information than previously.

Cathode Ray Tube Display   The cathode ray tube (CRT), the origin of which dates back to the end of the nineteenth century, has been widely used as a radar display. There are two basic types of CRT displays. One is the *deflection-modulated CRT*, such as the A-scope, in which a target is indicated by the deflection of the electron beam. The A-scope displays the receiver output amplitude as a function of range, or time. An example was shown in Fig. 7.21, which illustrated the effect of frequency agility on clutter and target echoes. The other type is the *intensity-modulated CRT*, in which an echo is indicated by intensifying the electron beam and presenting a luminous spot on the face of the CRT. An example is the PPI shown in Fig. 1.5. The TV CRT is also an example of an intensity-modulated display.

In general, a deflection-modulated display has the advantage of simpler circuits, and a target may be more readily discerned in the presence of noise, clutter, or interference. On the other hand, an intensity-modulated display, such as the PPI, has the advantage of presenting data in a more convenient and easily interpreted form. The deflection of the beam or the appearance of an intensity-modulated spot of a radar display caused by the presence of a target is commonly referred to as a *blip*.

Even though the CRT display has been widely used in radar, as well as in TV and in computers, it is by no means ideal. It employs a relatively large vacuum tube, which is a disadvantage compared to other types of displays. The entire display with its necessary circuits and controls can be even larger. The amount of information that can be presented is limited by the spot size of the electron beam. The number of resolution cells (pixels) per diameter might be one or two thousand, or more. In some high-resolution radars the number of resolvable range cells from the radar might be greater than the number of resolution cells available on the PPI. The result can be a collapsing loss. Increasing the CRT diameter does not necessarily increase the number of resolvable pixels since the spot diameter varies linearly with screen diameter. Another limitation of the intensity-modulated CRT is that its inherent dynamic range, or contrast ratio, might be of the order of 10 dB. This can cause blooming of the display by large targets so as to mask the blips from nearby smaller targets.

The decay characteristics of CRT phosphors are important when an operator views the screen to detect targets and extract information. The decay time of the visual information displayed should be long enough to allow the operator to detect the echo, yet short enough that the information painted on one scan does not interfere with the information entered from the succeeding scan. When processed information rather than raw video is displayed, the display characters might be dots, vectors to indicate direction and velocity, alphanumerics, or other appropriate symbols.

The improvements in electronic circuitry that have allowed major advances in signal and data processing have also benefited the CRT display. Character generators can be on a chip rather than occupy a bulky box. A complete deflection system can be placed on a chip. High-voltage power supplies that used to occupy a cubic foot and weigh 50 pounds[53] have been decreased considerably. Digital memories are small enough to replace the bulky analog scan converter. The required decay times for a PPI display need not depend on the decay characteristics of the phosphor, but an artificial persistence can be achieved with electronic circuitry that controls the refresh rate of the display. In the past, CRTs often had to be viewed in a darkened room or with a hood, but the brightness of CRTs has been increased so that they can be used in ambient light or in sunlit aircraft cockpits. There have been significant advances in other types of displays due to the demands of TV and computers, but the CRT has been able to make significant advances as well. In spite of limitations, the CRT has been a competitive display because of its ruggedness, cost, color capability, ability to operate over wide temperature ranges, its wide viewing angle, and ability to conveniently display the type of information obtained by a radar.

In a conventional PPI display when raw, unprocessed video is displayed to an operator, some background noise should be present since it improves the operator's ability to make a detection decision. A completely "black scope" has reduced sensitivity compared to one with some background noise. This applies to a radar with raw video and not to a display presenting processed data where the detection decision is made by automatic circuitry without the intervention of an operator.

Stroke and Raster Displays   The conventional stroke PPI display is generated in synchronism with the rotating antenna rather than all at once as it might appear in a photo. The photo of a PPI display is usually made by opening the shutter of a camera and holding it open for one or more scans, as was done in Fig. 1.5. An operator viewing a normal PPI, on the other hand, sees a rotating radial line, or strobe, that rotates in synchronism with the scanning antenna. The trace of the rotating strobe with a raw-video display with no echoes present is normally dim, but is brightened to indicate the location of an echo signal when detected by the radar. The brightened blip fades with time depending on the persistence characteristics of the phosphor or the refresh characteristics of the electronic circuitry. The operator focuses his or her attention on the rotating radial strobe line to detect targets. This type of display is called a *stroke* display. In a stroke display the operator concentrates on that portion of the display in the vicinity of the strobe line since that is usually all that will be strong enough to be seen.

A TV-like display based on a raster scan* to provide a continuous picture of the radar output has some advantages over the stroke display. It can be made brighter than a stroke display. Information from other sensors such as other radars, the Air Traffic Control Radar Beacon System (ATCRBS), military identification friend or foe (IFF), low light level TV, forward-looking IR (FLIR), collision avoidance systems, or information from civil or military data links, can all be combined on one display. Maps of the region viewed by the radar as well as alphanumeric information and graphics can also be superimposed on a raster display, in addition to the processed radar video and raw video. A scan converter is required to change the format of the stroke display to that of a raster TV-like display.

Scan Converter[54]   A scan converter changes the r,θ coordinates of a PPI into the x,y coordinates of a raster (TV-like) display. The r,θ coordinates are natural for the radar, but the x,y coordinates are more natural for viewing the output of the radar on a display. Early analog scan-converters were bulky, had low resolution, and poor presentation of gray scales. They were of limited utility compared to modern digitally generated scan converters in which the r,θ polar coordinates (range and azimuth) of the radar information are converted to rectangular x,y coordinates and stored in digital memory to generate a raster TV-like display. The raster display can be presented continuously to the operator since it can be refreshed at a rapid rate. If desired, an artificial decay can be inserted to imitate the decay characteristics of natural phosphors. Alternatively, there need be no decay and the image can be frozen for the time equal to the radar revisit time, and then updated. Displays with 2560×2048 pixels can be accommodated. The outputs of multiple radars can be shown together with appropriate symbology even though they may be air-surveillance radars with considerably different antenna rotation rates (revisit times) and weapon control radars with pie-shaped angular coverage of a limited sector. The use of a

| *A raster is a scan pattern in which an area is scanned from side to side in lines from top to bottom.

scan converter usually does not seriously extend the latency of the display; that is, the time between echo detection and its display is held to a minimum. The format of a raster display can be that of a TV display or that of a computer monitor. The advantage of a TV-display format is that it can be recorded on tape with an inexpensive consumer video cassette recorder (VCR), viewed with a conventional TV monitor, and easily remoted using standard video cabling.

Flat Panel Displays (FPD)[55]   Interest in the flat panel display for radar evolved from its successful development for commercial computer and TV applications. There have been several different types of FPDs produced or explored, but not all are suitable for radar.

The *liquid crystal display* (LCD) has been widely used for nonradar applications where low weight, volume, and power consumption are important as in laptop computers, watches, instruments, and calculators. The LCD does not emit light of its own, but operates by controlling the light that either passes through it or reflects off it. Usually the light is directed from behind, and the display is said to be *backlit*. There are two types of liquid crystal displays: the passive-matrix LCD and the active-matrix LCD. In the latter, a thin-film transistor is associated with each pixel of the display. The passive-matrix has seen much wider nonradar application than the active matrix because of its lower cost, but the active-matrix LCD (AMLCD) has much higher resolution, better image quality, it can display in color, and has faster response (greater video bandwidths). Thus the active-matrix LCD has more potential for radar applications than does the passive-matrix.

Other types of FPDs are the *plasma display* which can produce large flat full-color displays, electroluminescent displays, light-emitting diodes, and field emitter displays.

Flat panel displays such as the AMLCD and the plasma display have several important advantages over conventional CRTs. They are smaller, lighter, occupy less volume (reduced depth), and require less power than CRTs. In addition they are expected to have better reliability and reduced life-cycle cost. For most radar applications, however, they have to be more rugged than for commercial applications in that they usually have to withstand greater shock and vibration, as well as extremes in temperature.

The FPD is especially well suited for cockpit displays in military airborne applications and is replacing the CRT in many airborne systems.[56] In addition to presenting radar information, military cockpit displays must also handle data provided by electronic warfare sensors, command and control information for situation awareness, navigation information, alphanumeric data, graphics, and others.

Color in Radar Displays   The availability of color in a radar display allows another "dimension" for the presentation of information. It can aid in providing a clear, easily understood picture of the situation as observed by radar. It is also an "attention getter" to alert the operator to something special or dangerous. Different colors can be used to indicate such things as the outputs of different radars presented on the same display; the outputs from multiple beams of a 3D radar; the areas of adverse weather with color coding of the weather by rain intensity; range rings; target tracks along with single detections; identification information from civil ATCRBS and/or military IFF; superimposed video maps of the area being observed by radar; and superimposed raw video. It can also be used to indicate the altitude or cross section of individual radar

Table 11.2   An Example of the Display Colors Used for an Airborne Weather Avoidance Radar[57]

Storm Intensity	Rainfall Rate (mm/h)	Rainfall Rate (dBZ)	Display Color
Drizzle	0.25 mm/h	13 dB	Black
Light	1.0 mm/h	23 dB	Green
Moderate	4.0 mm/h	33 dB	Yellow
Industry standard pilot alert	11.5 mm/h	40 dB	Red

The parameter dBZ was explained in Sec. 7.6. The "industry standard pilot alert" is the rainfall rate above which there might be hail that could damage an aircraft or cause sufficient turbulence to disturb passengers. Pilots are advised to stay clear of such areas.

echoes by color coding the target blips or by use of alphanumeric color symbols inserted on the display.

An example of the use of color is in the airborne weather-avoidance radar display, where the intensity of precipitation is designated by a distinctive color. A listing of the colors used by one radar manufacturer to indicate storm intensities is shown in Table 11.2.[57]

The original tricolor shadow-mask cathode-ray tube used for color TV did not have the resolution capability of monochrome displays or the penetration color tube which used a multilayer screen. This has changed with the increasing demands for high resolution computer color graphics as well as high-definition TV. Although a monochrome display with various shades of gray can be made to exhibit much of the same information that a color display can, color is capable of providing a greater number of distinguishable shades than can a monochrome display, is more pleasing, and has been widely accepted.

REFERENCES

1. Taylor, J. W., Jr. "Receivers." *Radar Handbook*, M. Skolnik (Ed). New York: McGraw-Hill, 1990, Chap. 3.

2. Mumford, W. W., and E. H. Scheibe. *Noise Performance Factors in Communication Systems*. Dedham, MA: Horizon House—Microwave, Inc., 1968.

3. Pettai, R. *Noise in Receiving Systems*. New York: John Wiley, 1984.

4. Goldberg, H. "Some Notes on Noise Figures." *Proc. IRE 36* (October 1948), pp. 1205–1214.

5. Heil, T., B. Roehrich, and J. Hakoupian. "Advances in Receiver Front-End and Processing Components." *Microwave J.* 40 (January 1997), pp. 174–180.

6. Maas, S. A. *Microwave Mixers*, 2nd ed. Boston: Artech House, 1993.

7. Neuf, D. "Extended Dynamic Range Mixers." *Applied Microwave & Wireless* (Winter 1996), pp. 24–39.

8. Taylor, J. W., Jr. Ref. 1, Sec. 3.4.

9. Eaves, J. L., and E. K. Reedy. *Principles of Modern Radar*. New York: Van Nostrand Reinhold, 1987, Chap. 7.

10. Maas, S. A. "Microwave Mixers in the 90s." *Microwave J. 1990 State of the Art Reference*, pp. 61–72.

11. Maas, S. A. Ref. 6, Sec. 7.3.

12. Maas, S. A. Ref. 6, Sec. 7.3.5.

13. Dixon, R. C. *Radio Receiver Design*. New York: Marcel Dekker, 1998, Sec. 6.4.

14. Maas, S. A. Ref. 6, Sec. 4.6.1.

15. Dixon, R. C. Ref. 13, Sec. 6.5.

16. van der Ziel, A. "Unified Presentation of $1/f$ Noise in Electronic Devices: Fundamental $1/f$ Noise Sources." *Proc. IEEE* 76 pp. (March 1988), pp. 233–258.

17. Halford, D. "A General Model for f^{α} Spectral Density Random Noise with Special Reference to Flicker Noise $1/f$." *Proc. IEEE*. 56 (March 1968), pp. 251–257.

18. Taylor, J. W., Jr. Ref. 1, Sec. 3.5.

19. Goldman, S. J. *Phase Noise Analysis in Radar Systems Using Personal Computers*. New York: John Wiley, 1989.

20. Ewell, G. W. "Stability and Stable Sources." In *Coherent Radar Performance Estimation*, Scheer, J. A., and J. L. Kurtz (Eds.). Boston: Artech House, 1993, Chap. 2.

21. Losee, Ferril. *RF Systems, Components, and Circuits Handbook*. Boston: Artech House, 1997, Chap. 16.

22. Terman, F. *Radio Engineering*. New York: McGraw-Hill, 1937, Sec. 70.

23. Elmi, N., and M. Radmanesh. "Design of Low-Noise, Highly Stable Dielectric Resonator Oscillators." *Microwave J.* 39 (November 1996), pp. 104–112.

24. Khanna, A. P. S., M. Schmidt, and R. B. Hammond. "A Superconducting Resonator Stablized Low Phase Noise Oscillator." *Microwave J.* 34 (February 1991), pp. 127–130.

25. Galani, Z., and R. A. Campbell. "An Overview of Frequency Synthesizers for Radars." *IEEE Trans.* MTT-39 (May 1991), pp. 782–790.

26. Kroupa, V. F. (Ed.). *Direct Digital Frequency Synthesizers*. Piscataway, NJ.: IEEE Press, 1999.

27. Crawford, J. A. *Frequency Synthesizer Design Handbook*. Boston: Artech House, 1994, Secs. 7.3 and 7.5.

28. Hoeschele, D. F. *Analog-to-Digital and Digital-to-Analog Conversion Techniques*, 2nd ed., New York: John Wiley, 1994.

29. Waters, W. M., and B. R. Jarrett. "Bandpass Signal Sampling and Coherent Detection." *IEEE Trans.* AES-18 (November 1982), pp. 731–736.

30. Wu, Y., and J. Li. "The Design of Digital Radar Receivers." *Proc. 1997 IEEE National Radar Conf.* pp 207–210; also reprinted in the *IEEE AES Systems Magazine* 13 (January 1998) pp. 35–41.

31. Manheimer, W. M., and G. Ewell, "Cyclotron Wave Electrostatic and Parametric Amplifiers." Naval Research Laboratory, Washington, D.C., Memorandum Rep. MR/6707-97-7910, February 28, 1997.

32. Budzinsky, Yu. A., and S. P. Kantyuk. "A New Class of Self-Protecting Low-Noise Microwave Amplifiers." *1993 IEEE International Microwave Symp. Digest*, Atlanta GA., vol. 2, p. 1123, June, 1993. See also information on the ISTOK Web Site http://www.istok.com.

33. Manheimer, W. M., and G. W. Ewell. "Electrostatic and Parametric Cyclotron Wave Amplifiers." *IEEE Trans.* PS-26 (August 1998), pp. 1282–1296.

34. Barton, D. K. "The 1993 Moscow Air Show." *Microwave J.* 37 (May 1994), pp. 24–39.

35. Taylor, J.W., Jr. Ref. 1, Sec. 3.10.

36. Krishnam, S. "Diode Phase Detectors." *Electronic & Radio Engr.* 36 (February 1959), pp. 45–50.

37. Socci, R. J. "The Aegis Radar Receiver." *Microwave J.* 21 (October 1978), pp. 38–47.

38. Bilotta, R. F. "Receiver Protectors: A Technology Update." *Microwave J.* 40 (August 1997), pp. 90–96.

39. Riblet, H. J. "The Short-Slot Hybrid Junction." *Proc. IRE* 40 (February 1952), pp. 180–184.

40. Golde, H. "Radioactive (Tritium) Ignitors for Plasma Limiters." *IEEE Trans.* ED-19 (August 1972), pp. 917–928.

41. Golde, H. "What's New with Receiver Protectors?" *Microwaves* 15 (January 1976), pp. 44–52.

42. Roberts, N. "A Review of Solid-State Radar Receiver Protection Devices." *Microwave J.* 34 (February 1991), pp. 121–125.

43. Ratliff, P. C., W. Cherry, M. J. Gawronski, and H. Goldie. "L-Band Receiver Protection Using Sensitivity Time Control." *Microwave J.* 19 (January 1976), pp. 57–60.

44. Goldie, H. "Combined Receiver Protector, AGC Attenuator and Sensitivity Time Control Device." *United States Patent* 4,194,200, March 18, 1980.

45. Nelson, T. M., and H. Goldie. "Fast Acting X-band Receiver Protector Using Varactors." *IEEE MTT Symp. Digest* (1974), pp. 176–177.

46. Brown, N. J. "Modern Receiver Protection Capabilities with TR-Limiters." *Microwave J.* 17 (February 1974), pp. 61–64.

47. "Product and Engineering Data, Receiver Protectors," Communication and Power Industries, Beverly Microwave Division, Beverly, MA, Brochure no. EDB-2417/273 (no date, but circa 1995)

48. Ferguson, P., and R. D. Dokkem. "For High-Power Protection . . . Try Multipacting." *Microwaves* 13 (July 1974), pp. 52–53.

49. Patel, S. D., and H. Goldie "A 100 kW Solid-State Coaxial Limiter for L-band, Part I." *Microwave J.* 25 (December 1981), pp. 61–65; Part II, vol. 26, (January 1982), pp. 93–97.

50. Hamilton, C. H. "A 1 MW C-band PIN Diode Duplexer." *1978 Conf. Proc. Military Microwaves*, pp.103–107, Microwave Exhibitions and Publishers, Ltd., Sevenoaks, Kent, England.

51. Rodrigue, G. P. "Circulators from 1 to 100 GHz." *Microwave J. 1989 State of the Art Reference*, vol. 32, pp. 115–132.

52. IEEE Standard Radar Definitions, *IEEE Std 686–1997,* Piscataway, NJ.

53. Wurtz, J. E. "CRT Update." *IEEE EASCON-77*, paper 12-2, 1977.

54. Some of the information in this subsection was obtained from the advertising literature of Folsom Research, Rancho Cordovia, CA, and from Robert W. Cribs, the CEO of Folsom Research.

55. Werner, K. "U. S. Display Industry on the Edge." *IEEE Spectrum* 32 (May 1995), pp. 62–69.

56. Hopper, D. G. (Ed.). *Cockpit Displays V: Displays for Defense Applications.* Proc. SPIE (International Society for Optical Engineering) 3363, 1998, Bellingham, WA.

57. Aires, R. H., and G. A. Lucchi. "Color Displays for Airborne Weather Radar." *RCA Engineer* 23 (February/March 1978), pp. 54–60.

PROBLEMS

11.1 (a) Find the overall noise figure of a superheterodyne receiver consisting of a low-noise RF amplifier with noise figure of 1.4 dB and gain of 15 dB, a mixer with 6.0-dB conversion loss and noise-temperature ratio of 1.2, and an IF amplifier with a noise figure of 1.0 dB. (b) What would be the noise figure of the receiver in (a) if the RF low-noise amplifier had a gain of 30 dB instead of 15 dB? (c) What would be the overall receiver noise figure if the IF amplifier noise figure in part (a) were 3.0 dB instead of 1.0 dB, and do you think this change is significant?

11.2 (a) Derive the overall noise figure of a receiver with noise figure F_r that is preceded by an RF device with a loss L_{RF}. (b) What is the overall noise figure of a transmission line and duplexer, which have a loss of 1.2 dB, connected to a receiver whose noise figure is 2.3 dB?

11.3 The greater the gain of the RF low-noise amplifier, the lower will be the overall noise figure. What adverse effect, however, occurs with an increase in the gain of the RF low-noise amplifier?

11.4 Show that the noise figure of a mixer is approximately the product of its conversion loss and the IF amplifier noise figure, when the diode mixer has a low noise-temperature ratio.

11.5   (a) Show that when a receiver of noise figure F_{RF} is attached to an antenna with antenna temperature T_a, the system noise figure F_s [Eq. 11.7] is

$$f_s = \frac{T_a}{T_0} + F_r$$

where T_0 is the standard temperature 290 K. (b) What is the system noise figure if the antenna temperature is 300 K, transmission line loss is 1.5 dB, and the receiver noise figure is 2.6 dB?

11.6   Consider a radar system with a receiver noise figure of 1.0 dB preceded by a transmission line with a loss of 0.5 dB. If the antenna temperature is 300 K, how important is it, in general, to attempt to reduce the receiver noise figure from 1.0 dB to 0.5 dB?

11.7   (a) Show that a radome with a loss L at a physical temperature T_{rd}, used with an antenna whose noise temperature in the absence of the radome is T_a', has an antenna noise temperature given by

$$T_a = \frac{T_a'}{L} + T_{rd}\frac{(L-1)}{L}$$

(b) Starting with the above, derive the *change* in antenna noise temperature $\Delta T_a = T_a - T_a'$ due to the presence of the radome. (c) Using the result of (a), what is the system noise figure? (d) What is the system noise figure when the receiver has a noise figure of 2.6 dB, a transmission line loss of 1.5 dB, an antenna with a noise temperature of 110 K, radome at a physical temperature of 310 K, and a loss of 0.6 dB through the radome?

11.8   (a) Find the noise bandwidth [Eq. (2.3)] of a network whose frequency response function $H(f) = (1 + jf/B)^{-1}$, where B is the half-power bandwidth. (b) The above network is a single-stage low-pass RC filter. What is the expression for the noise bandwidth of a single-stage RLC bandpass network? (Do by inspection.) (c) Find the noise bandwidth for a band-pass filter with a gaussian shaped response $\exp[-a^2(f - f_0)^2]$, with $f > 0$.

11.9   Why might a double-conversion superheterodyne receiver be used instead of a single-conversion receiver? What limitation might there be in using a double-conversion receiver?

11.10   What effect does the local oscillator have on the receiver's dynamic range?

11.11   A receiver with a mixer front end has a noise figure of 6.6 dB. A low-noise amplifier (LNA) with a noise figure of 1.2 dB and gain of 10 dB is inserted ahead of the mixer to reduce the overall receiver noise figure. (a) How much of the new receiver noise figure is due to the mixer noise, and by how much has the dynamic range of the receiver been reduced? (b) If the gain of the LNA were increased to 20 dB, what would be the receiver noise figure and the decrease in dynamic range?

11.12   Why is a diode-limiter following the duplexer sometimes used as a receiver protector?

11.13   What duplexer options are available for a pulse doppler radar with a 10 percent duty cycle and a 0.1-μs pulse duration?

11.14   What limitations might there be in using an all-solid-state duplexer?

11.15 What is the usual cause, or criterion, for the end-of-life of a gas-discharge TR tube?

11.16 Consider a high prf pulse doppler radar with a 1-μs pulse width and a 10 percent duty cycle. (a) If a receiver protector is used that has a 1.5-μs recovery time, what fraction of the range coverage is blanked out? (b) If an electrostatic amplifier with a 30-ns recovery time is used, what fraction of the range coverage would be blanked out?

INDEX